I dedicate the third edition of this practical guide to Robyn Shimizu. Robyn and I have worked together for many years, performing bench work; training students, post-doctoral fellows, and medical residents; presenting workshops and seminars; handling consult; performing studies; and preparing manuscripts for publication. It's been a wonderful collaboration and sharing of information throughout our careers. A very special thanks to Robyn for sharing this educational adventure; hopefully our many students have found these contributions helpful.

PRACTICAL GUIDE TO

Diagnostic Parasitology

THIRD EDITION

PRACTICAL GUIDE TO Diagnostic Parasitology

THIRD EDITION

Lynne S. Garcia, MS, MT(ASCP), CLS(NCA), F(AAM)
LSG & Associates, Santa Monica, California

Washington, DC

WILEY

Editorial Correspondence: ASM Press, 1752 N Street, NW, Washington, DC 20036-2904, USA
Registered Offices: John Wiley & Sons, Inc., 111 River Street, Hoboken, NJ 07030, USA
For details of our global editorial offices, customer services, and more information about Wiley products, visit us at www .wiley.com.
Wiley also publishes its books in a variety of electronic formats and by print-on-demand. Some content that appears in standard print versions of this book may not be available in other formats.

Library of Congress Cataloging-in-Publication Data

Names: Garcia, Lynne Shore, author. | American Society for Microbiology.
Title: Practical guide to diagnostic parasitology / Lynne S. Garcia.
Description: 3rd edition. | Hoboken, NJ : Wiley, 2021. | Includes bibliographical references and index.
Identifiers: LCCN 2021001642 (print) | LCCN 2021001643 (ebook) | ISBN 9781683670391 (hardback) | ISBN 9781683670407 (adobe pdf) | ISBN 9781683673620 (epub)
Subjects: MESH: Parasitic Diseases–diagnosis | Clinical Laboratory Techniques
Classification: LCC QR255 (print) | LCC QR255 (ebook) | NLM WC 695 | DDC 616.9/6075–dc23
LC record available at https://lccn.loc.gov/2021001642
LC ebook record available at https://lccn.loc.gov/2021001643

Cover image: A group of red blood cells (magenta) with the central cell infected by *Plasmodium falciparum*, the parasite that causes the deadliest form of malaria. Courtesy of Dr. Michał Pasternak, WEHI (Walter and Eliza Hall Institute), https://imaging.wehi.edu.au/our-work/new-views-malaria-parasites.

Cover and interior design: Susan Brown Schmidler

Set in 9.5/12.5pt MinionPro by SPi Global, Chennai, India

Printed in Singapore
M114055_260321

Contents

SECTION 7

Parasite Identification 265

Preface

As we move forward into the 21st century, the field of diagnostic medical parasitology continues to see some dramatic changes, including newly recognized pathogens, changing endemicity of familiar pathogens, disease control challenges, geographic and climate changes that support the spread of parasitic disease, new methodologies, expansion of diagnostic testing, and an ongoing review of the approach to and clinical relevance of this type of diagnostic testing on patient care within the managed care environment, as well as the world as a whole.

The third edition of the *Practical Guide to Diagnostic Parasitology* is organized to provide maximum help to the user, particularly from the bench use perspective. New aspects of the field have been addressed throughout, and many new figures and plates have been added, including extensive color images. All of the changes for this edition are based on the need for readers to update their information related to diagnostic medical parasitology and specifically issues involving laboratory techniques. With continued emphasis on regulatory requirements related to chemical disposal and the use of mercury compounds, laboratories are being required to develop skills using substitute fixatives that are prepared without the use of mercury-based compounds. In most cases, organism identification is comparable; an example of a rare exception is one in which the number of organisms present is quite low. This is a prime example of a change where "different" has been acceptable and relevant, not necessarily "good" or "bad." It is also important to remember the large number of variables relevant to diagnostic parasitology testing.

Section 1 of this new edition, on the philosophy and approach to diagnostic parasitology, has been expanded to include discussions on neglected tropical diseases, the impact of global climate change, population movements, potential outbreak testing, the development of laboratory test menus, and the risk management issues related to "stat" testing. The Section 2 discussion of organism classification and relevant tables has been expanded and updated to provide the user with current information related to changes in nomenclature and the overall importance of the various parasite categories to human infection.

In Section 3, expanded information on stool specimen fixatives and testing options has been provided. This information is valuable for any laboratory that is reviewing collection and testing options related to fixative compatibility with the routine ova and parasite examination, as well as fecal immunoassays and the newer molecular parasite panels. It is always important to check the literature from the relevant company to confirm the FDA status of any new product. The discussion on blood collection, including the pros and cons of current changes from finger-stick blood to venipuncture, has been greatly enhanced, particularly related to potential problems with blood parasite morphology and lag time issues. Additional tables serve to summarize much of this new information.

New tables and information have been added to Sections 4 and 5, including a number of new algorithms. Section 6 is one of the most important sections in the book, with extensive revisions related to the most commonly asked questions regarding diagnostic parasitology methods. Additional techniques have been included (molecular test panels), as well as new information related to reporting results and the importance of report comments. This section's "FAQ" format makes it easy for the reader to use the expanded information.

Section 7 has been greatly expanded, including the addition of extensive color figures. Figures and life cycle diagrams have also been expanded and updated. Section 8 presents information on potential problems with organism differentiation from one another and from possible artifacts. The section has been significantly expanded with new tables and image plates to illustrate these differences. Section 9 contains numerous tables and a set of identification keys that summarize identification aids and organism characteristics. As with earlier sections, the information is presented to assist the bench microbiologist with routine diagnostic testing methods.

Many laboratories are reviewing all microbiological services, and specific questions are being asked related to diagnostic parasitology options. Some of these questions include the following: what laboratories should be performing this type of testing, when should testing be performed, what tests should be performed, and what factors should be considered when developing test menus. There are also ongoing discussions related to the development of automated parasite panels and how the use of these panels could impact many of the routine procedures now in use.

Laboratories are also reviewing specimen collection options, particularly as they relate to their geographic area and types of patients serviced. This kind of analysis is beneficial to all concerned, not only in helping laboratories to understand the specimen collection options, but how they relate to test orders, diagnostic testing, and results impacting patient care.

With changes in collection, testing, reporting, and interpretation options, it is critical to remember that this information needs to be shared with the laboratory's client base, particularly if the test orders and results are to be used for the best-quality patient care. Although there are many ways to approach diagnostic parasitology testing, it is mandatory that the laboratory and user both understand the pros and cons of the methods selected and the importance of test menu names, particularly in terms of procedure limitations and test name relevance for billing functions. The use of different approaches to parasitology diagnostic testing is acceptable; however, the benefits and drawbacks must be thoroughly understood by all participants. There may be legitimate reasons why different approaches are used by different laboratories; however, cost containment must not be the sole factor in selecting methods.

Another consideration is the fact that not all clinical laboratories will continue to perform diagnostic parasitology testing. This may be due to financial considerations, lack of skilled personnel, etc. With increased emphasis on cross-trained individuals, the technical expertise required to identify these parasites by using routine microscopy may be lacking. Even with the use of molecular diagnostics, these tests are not capable of covering the entire spectrum of organisms that may be present as pathogens. However, as more of these newer automated molecular panels become commercially available, the use of nonmicroscopic methods will increase. Many laboratories now include both the ova and parasite examination and various fecal immunoassays on their routine test menus; on the basis of patient histories and symptoms, appropriate orders may focus on one or the other of these options. An important consideration in deciding whether to send out parasitology testing, or to maintain the testing in-house, relates to stat testing (collection, processing, testing, reporting of thick and thin blood films, and the examination of cerebrospinal fluid and other specimens for the presence of free-living amebae). These tests must be handled as stat; the time required from collection to reporting must be considered prior to moving these procedures off-site or sending specimens to a reference laboratory.

Based on the many changes in clinical laboratories within the past few years and many years' experience with teaching and diagnostic bench work, it is my hope that the information contained in this third edition will provide valuable information for the user. This guide is not designed to serve as a diagnostic parasitology text or to contain all possible diagnostic test options, but to help the user make some sense of a field for which training has become almost nonexistent. I have included a section on commonly asked questions about diagnostic medical parasitology and hope that this discussion will be of practical value to the user; the answers to some of these questions are often difficult to find, even in a more comprehensive book. Again, let me emphasize that different approaches to laboratory work are not always "good" or "bad." The key to success is making sure that both the laboratory and clients understand the pros and cons of each collection, testing, and reporting option and that educational information is provided for all clients on an ongoing basis. Two of the most important functions of the clinical laboratory in the future will be educational and consultative, particularly when laboratory services are within the microbiology area. The importance of well-trained bench microbiologists cannot be underestimated.

A final point is that infectious diseases, particularly parasitic infections, play a huge role in the world's overall health and economy. As travel increases, we anticipate seeing many more people who will be infected with parasites that may not be endemic to the specific area where they live. Continued vector and disease control efforts will remain on the high-priority list, especially when seen within the context of global health. It is hoped that parasitologists and microbiologists, including those who have diagnostic skills, will be available to support these global initiatives.

ACKNOWLEDGMENTS

I thank the hundreds of colleagues over the past years who have shared their thoughts and suggestions regarding this fascinating field of diagnostic parasitology. There are too many of you to name—you all know who you are, and we all recognize the pitfalls, as well as the fun, in providing diagnostic services in this field of microbiology.

I also thank the many bench techs and microbiologists who have tackled this field over the last 40+ years, including those who attended workshops and seminars; your contributions to the growth and expansion of diagnostic parasitology have been significant. Discussions of questions asked, problems for resolution, and reviews of testing options have been invaluable in shaping our approach to diagnostics in this field. This type of interaction helped all of us keep an open mind when reviewing options and possible new ways to approach the work.

Over the years, our association with many companies has also been extremely valuable in helping to understand test development, test trials, and relevance of results to the ultimate user within the diagnostic laboratory. Again, these interactions have helped maintain some balance and perspective on new options and their relevance to improved patient care.

A challenge to all of us who are still actively working in this area of diagnostic microbiology: Serve as a mentor to some of the young people entering the field of microbiology. The number of personnel trained in this field continues to decline. Until parasite morphology is no longer required for differentiation and diagnosis, skilled microscopists will remain valuable members of the microbiology team and mandatory for the practice of diagnostic medical parasitology.

I would also like to thank Christine Charlip, the Director of ASM Press (now partnered with John Wiley & Sons to copublish ASM Press titles), and members of the editorial staff, including the developmental editor, Ellie Tupper, the copyeditor, Jennifer Schaffer, and Editorial Rights Coordinator Lindsay Williams; they are outstanding professionals. Their many contributions always help the author "look good," and I appreciate their collaboration. Perhaps retirement will just have to wait a while longer.

Above all, my very special thanks go to my husband, John, for his love and support for this and other projects over the years. I could never have taken on these challenges without his help, understanding, and wonderful sense of humor.

Note: Images in this book credited to CDC PHIL and CDC DpDx were obtained by courtesy of the Centers for Disease Control and Prevention Public Health Image Laboratory (https://phil.cdc.gov/) and CDC DpDx – Laboratory Identification of Parasites of Public Health Concern (https://www.cdc.gov/dpdx/index.html).

About the Author

Lynne Shore Garcia is the director of LSG & Associates, a firm providing training, teaching, and consultation services for diagnostic medical parasitology and health care administration. A former manager of the UCLA Clinical Microbiology Laboratory, she is a sought-after speaker (nationally and internationally) and author of hundreds of articles, book chapters, and books including two ASM Press books, *Clinical Laboratory Management, Second Edition* and *Diagnostic Medical Parasitology, Sixth Edition*. She served as Editor-in-Chief of the ASM Press *Clinical Microbiology Procedures Handbook, Third Edition*, and was a senior editor for ASM's *Clinical Microbiology Reviews* journal. She serves as a reviewer for a number of journals, provides consulting to the CAP Microbiology Resource Committee, was chair of the CLSI Parasitology Subcommittee, and is a Fellow of The American Academy of Microbiology. Lynne is the 2009 recipient of the ASM bioMérieux Sonnenwirth Award for Leadership in Clinical Microbiology.

SECTION 1

Philosophy and Approach to Diagnostic Parasitology

Practical Guide to Diagnostic Parasitology, Third Edition. Lynne S. Garcia
 DOI: 10.1128/9781683673637.ch01

Neglected Tropical Diseases

The term "neglected tropical diseases" (NTDs) was first used in the early 2000s, primarily reflecting the lack of research funds and limited interest of the health care and pharmaceutical industries in investing in affordable drugs for these diseases. One of the earliest conferences on NTDs was organized by Médicins sans Frontières in early 2002 (1). From 2003 to 2007, key steps were taken to develop a framework for tackling NTDs in a coordinated and integrated way (2). In 2005 and 2006, two important articles (3, 4) provided a list of 15 NTDs, 13 of which were deemed of particular importance in terms of annual mortality rates and global burden. This list of 15 diseases formed the initial scope of *PLoS Neglected Tropical Diseases*. Included were nine helminth infections (cysticercosis/taeniasis, drancunculiasis [guinea worm], echinococcosis [added by WHO], foodborne trematodiasis, lymphatic filariasis, onchocerciasis, schistosomiasis, the three main soil-transmitted helminthiases [ascariasis, hookworm infection, and trichuriasis]), three protozoal infections (Chagas' disease, human African trypanosomiasis, and leishmaniasis), scabies and other ectoparasite infections (added by WHO), and three bacterial infections (Buruli ulcer, leprosy, and trachoma) (Table 1.1) (2).

A subsequent review by Hotez and colleagues titled "Control of Neglected Tropical Diseases," published in the *New England Journal of Medicine* in 2007, clearly demonstrated that the term "neglected tropical diseases" had become mainstream (5). In October 2007, the Public Library of Sciences published the inaugural issue of a new open-access journal, *PLoS Neglected Tropical Diseases* (6). As of mid-November 2020, more than 8,200 original research papers, editorials, expert opinions, viewpoints, and other magazine-type articles on NTDs have been published.

In a publication from 2017, Dr. Hotez discussed the tremendous progress towards neglected tropical disease control or even elimination. However, there are important gaps, nine of which are discussed below; parasitic infections have been emphasized for this list (7).

The first group of problems is linked to the geopolitics of the NTDs.

1. **Regional significance**. There are several NTDs that are very important in the areas where they occur; however, they are generally ignored by the global community. Examples include loiasis in Central Africa and mucocutaneous leishmaniasis in the New World.
2. **Political unrest in the Old World.** Second only to poverty, conflict may have the largest social impact on NTDs. Both cutaneous and visceral leishmaniasis outbreaks are now arising in Syria, Iraq, Afghanistan, Sudan, and South Sudan. Cutaneous leishmaniasis has now reached hyperendemic proportions in current and former ISIS occupation zones, and through forced human emigrations, this NTD may spill over into Lebanon, Turkey, and Jordan.
3. **Political destabilization in the New World.** As Venezuela's health system continues to decline, we have seen the resurgence or reemergence of malaria and NTDs such as Chagas' disease and schistosomiasis.
4. **Climate change and its impact on vector-borne and zoonotic NTDs.** Climate change, along with poverty, war, and population movements, produces detrimental effects which include the increase and spread of NTDs.
5. **NTDs in "wealthy" nations.** The poor living in the Group of 20 (G20) nations—and also Nigeria (richer than the bottom three or four G20 nations)—account for a majority of the world's disease burden for poverty-related neglected diseases and NTDs. These numbers include millions of Americans living in the United States with an NTD and significant but often unrecognized levels of poverty and disease in Europe and Australia.

The second group is related to coverage gaps and providing universal access to treatment.

6. **Female genital schistosomiasis.** Female genital schistosomiasis is one of the most common gynecologic conditions of women who live in poverty in Africa and is one of Africa's most important cofactors in its AIDS epidemic.
7. **Patient access to essential drugs for Chagas' disease**. Today, most cases of Chagas' disease occur in Latin America's three large economies: Argentina, Brazil, and Mexico. However, more than 90% of people with *Trypanosoma cruzi* infection do not have access to treatment.
8. **Mass drug administration (preventive chemotherapy).** This includes drugs against scabies, lymphatic filariasis, onchocerciasis, and schistosomiasis.
9. **Research and development (R&D)** for a single approach rather than multiple new approaches for control and disease elimination. For malaria, we will need to pursue several R&D approaches, including new drugs, diagnostics, vaccines, and vector control approaches.

A subsequent article by Dr. Hotez and colleagues in 2018 anticipated a number of challenges related to the emergence and reemergence of these diseases (8). These challenges include stress from climate change and catastrophic weather events, regional conflicts over shifting and limited resources, such as water, and the development and spread of urban helminth infections (schistosomiasis and toxocariasis), foodborne trematode infections, cysticercosis, protozoan infections, and zoonotic toxoplasmosis.

Why Perform Diagnostic Parasitology Testing?

Travel

With the increase in world travel and access to varied populations and geographic areas, we continue to see more "tropical" diseases and infections outside areas of endemicity due to the rapidity with which people and organisms can be transmitted from one place to another. Travel has also become accessible and more affordable for many people throughout the world, including those whose overall health status is in some way compromised. The increased transportation of infectious agents and potential human carriers, particularly via air travel, has been clearly demonstrated during the last few years. It has also been well documented that vectors carrying parasitic organisms can be transported via air travel in baggage and in the unpressurized parts of the plane itself; once released, these infected vectors can then transmit these parasites to humans, even in areas where the infections are not endemic.

Population Movements

In many parts of the world, particularly where conflict is ongoing, there continue to be large population movements. Such movements include refugee migrations to and from areas of endemic parasitic diseases (9, 10). Often, in refugee situations, living conditions are very poor and medical limitations may lead to high levels of parasitic disease and severe illness. Also, migrants may move into countries and geographic areas where serious parasitic infections are generally nonendemic, including Europe and parts of North America. Even if these individuals are uninfected when entering these areas, travel home to visit relatives may result in infections that can be imported when they return.

Control Issues

Control of parasites that cause disease is linked to a number of factors, including geographic location, public health infrastructure, political stability, available funding, social and behavioral customs and beliefs, trained laboratory personnel, health care support teams, environmental constraints, degree of understanding of organism life cycles, and opportunities for control and overall commitment. Often control efforts do not cross political or geographic boundaries; unfortunately, vectors and other carriers of infectious agents do not "play by the rules," and as a result, these boundaries are meaningless in the context of disease control.

Climate Change

With the continued increase in the global temperature, worldwide climate changes are leading to an overall increase in infectious diseases, vector populations, and ranges of endemicity of both parasites and vectors. Global warming enhances the potential spread of tropical parasitic infections, specifically those due to parasites such as *Plasmodium* spp., *Leishmania* spp., and *Trypanosoma* spp. (11). Examples of vectors whose range is increasing include *Anopheles, Aedes,* and *Culex* mosquitoes, hard ticks, and triatomid bugs. Another example is the vectors of schistosomiasis (12).

Epidemiologic Considerations

When newer infectious agents and/or diseases are recognized, there is often very little information available regarding the organism life cycle, potential reservoir hosts, and environmental requirements for survival. Priorities may change, and epidemiologic considerations may have been moved lower on the priority list in areas of the world where they were considered important in the past; unfortunately, funding often plays a role in decisions that impact disease control measures.

Compromised Patients; Potential Sex Bias Regarding Infection Susceptibility; Aging

With the tremendous increase in the number of patients whose immune systems are compromised by underlying illness, chemotherapy, transplantation, AIDS, or age, we are much more likely to see increasing numbers of opportunistic infections, including those caused by parasites. Also, we continue to discover and document organisms that were thought to be nonpathogenic but can cause serious disease in the compromised host. When the possible cause of illness in this patient population is being assessed, the possibility of parasitic infections must be considered as part of the differential diagnosis.

Various studies have revealed a bias toward males regarding susceptibility to and severity of parasitic diseases. Although a number of external factors influence the exposure to infection sources among males and females, one recurrent factor suggests that hormonal influence impacts the simultaneous increase in disease occurrence and hormonal activity during the aging process. However, to date, very few controlled studies have been performed. Hormones are suspected to play a role in parasitic disease processes such as amebiasis, malaria, leishmaniasis, toxoplasmosis, and schistosomiasis (13).

There are various examples of the relationship of parasite pathogenesis and the aging population. Parasite biomass, endothelial activation, and microvascular dysfunction are associated with severe disease in *Plasmodium knowlesi* malaria and likely contribute to pathogenesis. The association of each of these processes with aging may account for the greater severity of malaria observed in older adults in regions of low endemicity (14).

Approach to Therapy

As new etiologic agents are discovered and the need for new therapeutics increases, more sensitive and specific diagnostic methods to assess the efficacy of newer drugs and alternative therapies will become mandatory. Skilled laboratorians, physicians, public health personnel, and other health care team members will be required to think globally in terms of infectious diseases caused by bacterial, fungal, parasitic, and viral agents, particularly when certain parasitic infections require very specific therapeutic regimens.

Who Should Perform Diagnostic Parasitology Testing?

Laboratory Personnel

Diagnostic procedures in the field of medical parasitology require a great deal of judgmental and interpretative experience and are, with very few exceptions, classified by the Clinical Laboratory Improvement Act of 1988 (CLIA '88) as high-complexity procedures. Currently, very few procedures can be automated for routine laboratory use, and organism identification relies on morphologic characteristics that can be difficult to differentiate. Although morphology can be "learned" at the microscope, knowledge about the life cycle, epidemiology, infectivity, geographic range, clinical symptoms, range of illness, disease presentation depending on immune status, and recommended therapy is critical to the operation of any laboratory providing diagnostic services in medical parasitology. As laboratories continue to downsize and reduce staff, cross-training will become more common and critical to financial success. Maintaining expertise in fields such as diagnostic parasitology has become more difficult, particularly when standard manual methods are used. Also, the lower the positive rate for parasitic infections, the more likely it is that the laboratory will generate both false-positive and false-negative laboratory reports. It is important for members of the health care team to thoroughly recognize areas of the clinical laboratory that require experienced personnel and why various procedures are recommended above others.

Nonlaboratory Personnel

Health care delivery settings where physicians provide parasitology diagnostic testing occasionally provide "simple" test results (CLIA '88 waived tests) based on wet-mount examinations. However, in spite of the CLIA classification of these diagnostic methods, wet-mount examinations are often very difficult to perform, and results are often incomplete or incorrect. Currently, there are no specific "over-the-counter" testing methods for parasitic infections; however, the future may see some newer diagnostic developments in this area. The key to performance of diagnostic medical parasitology procedures is formal training and experience. As the laboratory setting continues to change, it is important to recognize that these changes will require a thorough understanding of the skills required to perform diagnostic parasitology procedures and the pros and cons of available diagnostic methods. **Laboratories will have a number of diagnostic options; whatever approach is selected by an individual laboratory, the clinical relevance of the approach must be thoroughly understood and conveyed to the client user of the laboratory services.**

Where Should Diagnostic Parasitology Testing Be Performed?

Inpatient Setting

Most diagnostic parasitology procedures can be performed either within the hospital setting or at an offsite location. There are very few procedures within this discipline that must be performed and reported on a stat basis. **Two procedures fall into the stat category: request for examination of blood films for the diagnosis of malaria or other blood parasites and examination of cerebrospinal fluid (CSF) for the presence of free-living amebae, primarily *Naegleria fowleri*. Any laboratory providing diagnostic parasitology procedures must be prepared to examine these specimens on a stat basis 7 days a week, 24 h a day (orders, specimen collection, processing, examination, and reporting).** Unfortunately, these two procedures can be very difficult to perform and interpret; cross-trained individuals with little microbiology training or experience will find this work difficult and subject to error, and this will cause severe risk management issues for the laboratory. It has been well documented that automated hematology instrumentation lacks the sensitivity to diagnose malaria infections, particularly since most patients seen in an emergency room have a very low parasitemia. **However, even a low parasitemia can be life-threatening in infections with *Plasmodium falciparum* and *Plasmodium knowlesi*.**

Outpatient or Referral Setting

Diagnostic laboratories outside the hospital setting are very appropriate settings for this type of diagnostic testing; the test requests, for the most part, are routine and are batch tested rather than being tested singly. With very few exceptions, stat requests are not relevant and are not sent to such laboratory locations; therefore, immediate testing and reporting are not required.

Decentralized Testing

Point-of-care testing within the hospital (ward laboratories, intensive care units, emergency rooms, and bedside) is usually not considered appropriate for diagnostic parasitology testing; one exception might be the emergency room, where patients with malaria may first present with fever and general malaise. Alternative sites (outpatient clinics, shopping malls, senior citizen groups, and others) are generally not considered appropriate settings for diagnostic parasitology testing, although relevance might be dictated by geographic location and the development of newer, less subjective methods.

Physician Office Laboratories

As mentioned above, the majority of physician office laboratories are not involved in diagnostic parasitology testing; however, as more molecular biology-based (nonmicroscopic) methods are developed, they may become more widely used in this setting. One example is fecal immunoassay methods, specifically designed to detect antigens of *Cryptosporidium* spp., *Giardia lamblia*, the *Entamoeba histolytica*/*Entamoeba dispar* group, and *Entamoeba histolytica*. Rapid tests for the detection of *Plasmodium* spp., particularly *Plasmodium falciparum*, are also currently available (one is FDA cleared in the United States). However, it is critical that the pros and cons of these tests be thoroughly understood prior to patient testing and reporting.

Over-the-Counter (Home Care) Testing

Currently, no diagnostic tests for medical parasitology are available for this potential market. However, outside the United States, some of these options are more likely to be available (none are currently FDA cleared).

Field Sites

Field sites are very relevant for diagnostic parasitology testing, particularly in many areas of the world where instrumentation and automation are not routinely found within clinical laboratories. As the methodology becomes less expensive and easier to use and interpret, testing sites outside the routine laboratory may become more relevant, particularly when associated with epidemiologic studies.

What Factors Should Precipitate Testing?

Travel and Residence History

Although travel history is generally considered in terms of weeks or months, a number of parasitic infections involve potential exposure many years earlier. The patient may become symptomatic years after having left the area of endemic infection. Therefore, it is important to consider long-range history, as well as the previous few weeks or months. This is particularly true when one is considering various places where the patient may have lived prior to becoming symptomatic. The more information the laboratory has regarding past organism exposure, the more likely it is that the causative agent will be identified and the infection confirmed. **It is often imperative that the laboratory follow up with specific questions for the physician; routine information received with the test request may be minimal, at best.**

Immune Status of the Patient

Certain parasites can cause severe illness in debilitated patients and should be considered when these patients present with relevant symptoms. Infections that may cause few or no symptoms in an immunocompetent host may cause prolonged illness or death in an immunocompromised patient. Unfortunately, information regarding the patient's immune status is not always readily available. Client education can help to increase awareness of possible infections with human parasites.

Clinical Symptoms

When infectious diseases are suspected, there are a number of possible etiologic agents. When a patient presents with gastrointestinal symptoms, it is difficult to tell whether the cause is infectious and, if so, which microbe might be responsible. These symptoms often have many noninfectious causes, so laboratory findings can be extremely valuable in confirming a suspected infectious organism. The same diagnostic procedures can also be used to rule out specific etiologic agents.

Documented Previous Infection

Many parasitic infections are difficult to cure or may not cause symptoms on a continual basis; information regarding past exposure or prior documented disease will be valuable for the laboratory. Knowing that the patient may be experiencing a relapse can guide the laboratory in detecting the suspected organism.

Contact with Infected Individuals

In situations where multiple reports of symptomatic patients are confirmed, contacts of these infected patients should be tested, particularly during a potential outbreak. An example of this type of situation is outbreaks of diarrhea in the nursery school setting. When *Cryptosporidium* is identified as the causative agent, all nursery school attendees, employees of the school, and family members are often tested for infection. Another example involves a group of individuals who experience the same symptoms at similar times after attending a function where food was served; the causative agent might be confirmed as *Cyclospora cayetanensis*. These situations have public health significance within the community.

Potential Outbreak Testing

In potential outbreak situations, laboratories that perform certain tests on request only may revise their protocol and begin to test all specimens for a particular suspected parasite. A potential *Cryptosporidium* outbreak might require a change from testing on request only to screening all fecal specimens submitted for parasite testing for this particular organism. Often this occurs after consultation among various groups, such as health care providers, public health personnel, water company personnel, and pharmacy purchasing agents (who may report an increase in the purchase of antidiarrheal medication).

Occupational Testing

The most common example of occupational testing involves food handlers and routine testing for intestinal parasites. This practice is less common than in the past, probably due to financial constraints. Each city, county, and state has specific regulations and/or recommendations.

Therapeutic Failure

With few exceptions, patients are generally retested after therapy to confirm therapeutic efficacy. If testing reveals that the infection has not been eradicated, there may be several reasons. In some cases, the patient may not have taken the medication correctly or may have failed to take the recommended number of doses; these reasons are more likely than the presence of a drug-resistant organism. However, there are certainly examples where drug resistance is possible and is the more likely reason, often depending on the geographic area involved. Another reason might involve the timing between therapy and posttreatment checks for cure; if the time lag is extended, the patient's infection may also represent reinfection. This is particularly true if the patient's living conditions, site, potential parasite and/or vector exposure, and other epidemiological considerations are not modified.

What Testing Should Be Performed?

Routine Tests

"Routine" can imply a widely used, well-understood laboratory test; it can also imply a low- or moderate-complexity method, rather than a high-complexity procedure. Routine diagnostic parasitology procedures could include the ova and parasite examination (O&P exam), preparation and examination of blood films and pinworm tapes or paddles, occult-blood tests, and examination of specimens from other body sites (urine, sputum, duodenal aspirates, urogenital specimens, etc.).

The selection and use of routine test procedures often depend on a number of factors, including geographic area, population served, overall positivity rate, client preference, number of test orders, staffing, personnel experience, turnaround time requirements, epidemiology considerations, clinical relevance of test results, and cost. Routine tests generally have a wide range of both sensitivities and specificities. As an example, the O&P exam (which involves direct wet mount, concentration, and permanent staining of the smear) could be considered a routine test method for the detection of a number of different intestinal protozoa and helminth infections; this procedure is moderately sensitive but relatively nonspecific. Monoclonal antibody-based test methods tend to be very specific (generally for a single organism, such as *Giardia lamblia*) and more sensitive than the routine O&P exam for specific intestinal protozoa. However, the test results are limited in scope; either the organism is present or it is not, and none of the other possible etiologic agents have been ruled in or out.

Diagnostic laboratories generally offer tests on request; an example is testing for the presence of *Cryptosporidium* spp. However, if a potential waterborne outbreak was suspected, this laboratory might change its approach and begin testing all stool specimens submitted for an O&P exam rather than testing only specimens accompanied by a specific test request for *Cryptosporidium* spp. These decisions require close communication with other entities, as described above for potential outbreak testing.

Special Testing and Reference Laboratories

Special procedures, such as parasite culture, are usually performed in limited numbers of reference laboratories. These procedures require the maintenance of positive control cultures used for quality control checks on all patient specimens; they also require special expertise and time. Many clinical laboratories do not meet these requirements. Although some standardized reagents are now commercially available, many clinical laboratories choose to send their requests for serologic testing for parasitic diseases to other laboratories. Often, the Centers for Disease Control and Prevention (CDC) performs serologic testing on specimens submitted to a given state's department of public health. Generally, specimens for parasitic serologic testing are not submitted directly to CDC but instead are submitted through state public health laboratories. In an emergency situation, consultation with the county or state public health laboratory may allow shipment of a specimen directly to CDC.

Specialized Referral Test Options—DPDx and Other Sites

DPDx (https://www.cdc.gov/dpdx/index.html; accessed 8 June 2019) is a website developed and maintained by CDC's Division of Parasitic Diseases and Malaria. This site provides an interactive and rapid exchange of information with two primary functions. The first function is a reference and training function encompassing parasite reviews, collection, and shipping of clinical specimens. The second function is diagnostic assistance.

Many consultants now provide identification assistance via transmitted microscopic images. There is a wide group of colleagues that share these inquiries and provide input to the requestor. Many of these individuals respond to routine questions with no charge. We all learn from this experience. Many sites also include extensive case histories (Lynne Garcia) (www.med-chem.com [accessed 8 June 2019]; also see the Medical Chemical Corporation site option, Para-Site Online). Another option is the Parasite Wonders blog by Dr. Bobbi Pritt (https://parasitewonders.blogspot.com; accessed 4 July 2020). Other sites can be located by searching "parasite case studies."

Other (Nonmicrobiological) Testing

Test results from other procedures performed in a clinical laboratory can be very helpful when one is trying to diagnose a parasitic infection. Specific examples are routine urinalysis, hematology procedures including a complete blood count, and various chemistry profiles. These results often provide supporting data consistent with a suspected parasitic infection.

What Factors Should Be Considered in Development of Test Menus?

Physical Plant

Provided that equipment requirements are met, most clinical laboratory space designed for microbiology procedures can be used for diagnostic parasitology testing. In smaller facilities, this work can be incorporated into a routine microbiology laboratory. Another consideration is the physical location of the laboratory with respect to the source of clinical specimens. If the distance from the collection site to the laboratory is a consideration, the use of appropriate specimen preservatives must be incorporated into patient specimen collection protocols to ensure that accurate laboratory results will be obtained.

Client Base

Recognition and identification of groups of clients served may dictate the methods and range of diagnostic testing available. Requests for testing for parasites may be minimal in a hospital setting where many procedures are related to elective surgery. In contrast, test requests originating from a large medical center with extensive outpatient clinics may require a broader range of testing and expertise.

Customer Requirements and Perceived Levels of Service

Depending on the client base, patient complexity, history of test requests, and physician interests, the laboratory may be required to provide minimal testing that includes the most commonly performed parasitology procedures. This type of laboratory would generally not be considered a consultative resource; it would need to identify a consultative laboratory to assist with more unusual tests and/or test interpretations. The range and complexity of available tests would also depend on the laboratory's definition of its role in the local, regional, national, or international health care arena.

Personnel Availability and Level of Expertise

Most procedures performed in the diagnostic parasitology laboratory require extensive microscopy training and experience. They are categorized as high-complexity tests by CLIA '88 and are frequently performed by licensed technologists. Based on microscopy examinations, these procedures require a great deal of interpretation. Although cross-training provides some help with certain procedures, including specimen processing, the necessary interpretive skills are not learned in a week or two and can be easily lost without practice. For this reason, it is important to have a minimum of one person who is not only skilled at performing the procedures but also capable of interpreting the findings and providing client training and consultation.

Equipment

The level of equipment required for diagnostic parasitology work is minimal; however, the one expense that should not be limited is that of one or more microscopes with good optics.

Each microscope should be equipped with high-quality (flat-field) objectives (10×, 40×, 50×, 60×, and 100× oil immersion objectives). The oculars should be a minimum of 10×.

Depending on the range of immunoassay testing available, a fluorescence microscope or enzyme immunoassay reader might be desirable. The availability of this equipment varies tremendously from one laboratory to another, and the equipment may be shared with other groups within the laboratory.

Another option is a fume hood, in which the staining could be performed; this is not required, but it is recommended, particularly if the laboratory is still using xylene for dehydration of permanent stained fecal smears.

The rest of the equipment is quite common and can be shared with other areas within the laboratory. Such equipment includes refrigerators, freezers, and pipette systems.

Budget

In general, approximately 70% of a microbiology laboratory budget is related to personnel costs. Although diagnostic procedures in the parasitology area are labor-intensive and may require a microscope with good optics, in general budget costs are minimal. Costs tend to increase when newer immunoassay procedure kits are brought into the laboratory; however, these increased supply costs may be balanced out by diminished labor costs. Each laboratory will have to decide which procedures to offer, which tests can be performed in a batch mode, how many procedures will be ordered per month, what length of turnaround time is required, whether stat testing is possible, and what options exist for referral laboratories, as well as taking educational initiatives and client preferences into consideration.

Although diagnostic parasitology can be an important part of the microbiology laboratory, it is just one section within the total laboratory context and should be analyzed as such for cost containment and clinical relevance. As more automated molecular methods become available, multiple-organism panel testing costs and clinical relevance will need to be considered.

Risk Management Issues Associated with Stat Testing

There are two circumstances in diagnostic medical parasitology that represent true stat testing situations (encompassing orders, specimen collection, processing, examination, and reporting). One is a suspected case of primary amebic meningoencephalitis (PAM) caused by *Naegleria fowleri* or granulomatous amebic encephalitis (GAE) caused by *Acanthamoeba* spp., *Balamuthia mandrillaris*, or *Sappinia diploidea*, and the other situation is any case where thick and thin blood films are requested for testing for blood parasites, possibly those that cause malaria. Extensive discussions of these organisms can be found in reference 15. These tests need to be available on a 24-h/day, 7-days/week basis.

Primary Amebic Meningoencephalitis

Amebic meningoencephalitis caused by *N. fowleri* is an acute, suppurative infection of the brain and meninges. With extremely rare exceptions, the disease is rapidly fatal in humans. The period between contact with the organism and onset of clinical symptoms, such as fever, headache, and rhinitis, may vary from 2 to 3 days to as long as 7 to 15 days.

The amebae may enter the nasal cavity by inhalation or aspiration of water, dust, or aerosols containing the trophozoites or cysts. The organisms then penetrate the nasal mucosa, probably through phagocytosis of the olfactory epithelium cells, and migrate via the olfactory nerves to the brain. Data suggest that *N. fowleri* directly ingests brain tissue by producing food cups, or amebostomes, in addition to carrying out contact-dependent cytolysis, which is mediated by a heat-stable hemolytic protein, heat-labile cytolysis, and/or phospholipase enzymes. Cysts of *N. fowleri* are generally not seen in brain tissue.

Early symptoms include vague upper respiratory distress, headache, lethargy, and occasionally olfactory problems. The acute phase includes sore throat, stuffy, blocked, or discharging nose, and severe headache. Progressive symptoms include pyrexia, vomiting, and stiffness of the neck. Mental confusion and coma usually occur approximately 3 to 5 days prior to death. The cause of death is usually cardiorespiratory arrest and pulmonary edema.

PAM can resemble acute purulent bacterial meningitis, and these conditions may be difficult to differentiate, particularly in the early stages. The CSF may have a predominantly polymorphonuclear leukocytosis, increased protein concentration, and decreased glucose concentration like that seen with bacterial meningitis. Unfortunately, if the CSF Gram stain is interpreted incorrectly (identification of bacteria as a false positive), the resulting antibacterial therapy has no impact on the amebae, and the patient usually dies within several days. In recent years, fatal cases of PAM have been associated with the use of neti pot sinus irrigation using nonsterile water (16).

Extensive tissue damage occurs along the path of amebic invasion; the nasopharyngeal mucosa shows ulceration, and the olfactory nerves are inflamed and necrotic. Hemorrhagic necrosis is concentrated in the region of the olfactory bulbs and the base of the brain. Organisms can be found in the meninges, perivascular spaces, and sanguinopurulent exudates.

Clinical and laboratory data usually cannot be used to differentiate pyogenic meningitis from PAM, so the diagnosis may have to be reached by a process of elimination. A high index of suspicion is often mandatory for early diagnosis. **All aspects of diagnostic testing (ordering, specimen collection, processing, examination, and reporting) should be carried out stat.** Although most cases are associated with exposure to contaminated water through swimming or bathing, this is not always the case. The rapidly fatal course of 3 to 6 days after the beginning of symptoms (with an incubation period of 1 day to 2 weeks) requires early diagnosis and immediate chemotherapy if the patient is to survive.

Analysis of the CSF shows decreased glucose and increased protein concentrations. Leukocyte counts may range from several hundred to >20,000 cells per mm^3. Gram stains and bacterial cultures of CSF are negative; however, the Gram stain background can incorrectly be identified as bacteria, thus leading to incorrect therapy for the patient.

A definitive diagnosis could be made by demonstration of the amebae in the CSF or in biopsy specimens. Either CSF or sedimented CSF should be placed on a slide under a coverslip and observed for motile trophozoites; smears can also be stained with Wright's or Giemsa stain. CSF, exudate, or tissue fragments can be examined by light microscopy or phase-contrast microscopy. Care must be taken not to mistake leukocytes for actual organisms or vice versa. It is very easy to confuse leukocytes and amebae, particularly when one is examining CSF by using a counting chamber, hence the recommendation to use a regular slide and coverslip. Motility may vary, so the main differential characteristic is the spherical nucleus with a large karyosome.

Specimens should never be refrigerated prior to examination. When the CSF is centrifuged, low speeds ($250 \times g$) should be used so that the fresh, unpreserved trophozoites are not damaged. Although bright-field microscopy with reduced light is acceptable, phase microscopy, if available, is recommended. Use of smears stained with Giemsa or Wright's stain or a Giemsa-Wright's stain combination can also be helpful. If *N. fowleri* is the causative agent, only trophozoites are normally seen. If the infecting organism is *Acanthamoeba* spp., cysts may also be seen in specimens from individuals with central nervous system (CNS) infection. Unfortunately, most cases are diagnosed at autopsy; confirmation of these tissue findings must include culture and/or special staining with monoclonal reagents in indirect fluorescent-antibody procedures. Organisms can also be cultured on nonnutrient agar plated with *Escherichia coli.*

In cases of presumptive pyogenic meningitis in which no bacteria are identified in the CSF, the computed tomography appearance of basal arachnoiditis (obliteration of basal cisterns in the precontrast scan with marked enhancement after the administration of intravenous contrast medium) should alert the staff to the possibility of acute PAM.

The amebae can be identified in histologic preparations by indirect immunofluorescence and immunoperoxidase techniques. The organism in tissue sections looks very much like an *Iodamoeba bütschlii* trophozoite, with a very large karyosome and no peripheral nuclear chromatin; the organisms can also be seen with routine histologic stains.

Organisms can be cultured on nonnutrient agar plated with *E. coli* (15). The trophozoites begin feeding on bacteria and grow to cover the agar surface in 1 to 2 days at 37°C. The presence of the protozoa can be confirmed by examining the agar surface using an inverted microscope or with a conventional microscope by inverting the plate on the stage and focusing through the agar with a 10× objective (Figure 1.1). *N. fowleri*, the causal agent of PAM, undergoes transformation to a pear-shaped flagellate, usually with two flagella but occasionally with three or four flagella; the flagellate stage is a temporary nonfeeding stage and usually reverts to the trophozoite stage (Figure 1.2). *N. fowleri* trophozoites are typically ameba-like and move in a sinuous way. They are characterized by a nucleus with a centrally located, large nucleolus. The trophozoites are also characterized by the presence of a contractile vacuole that appears once every 45 to 50 s and discharges its contents. The contractile vacuole looks like a hole or a dark depression inside the trophozoite and can be easily seen upon examination of the plate under the 10× or 40× objective. When the food supply is exhausted, *N. fowleri* trophozoites differentiate into spherical, smooth-walled cysts (15).

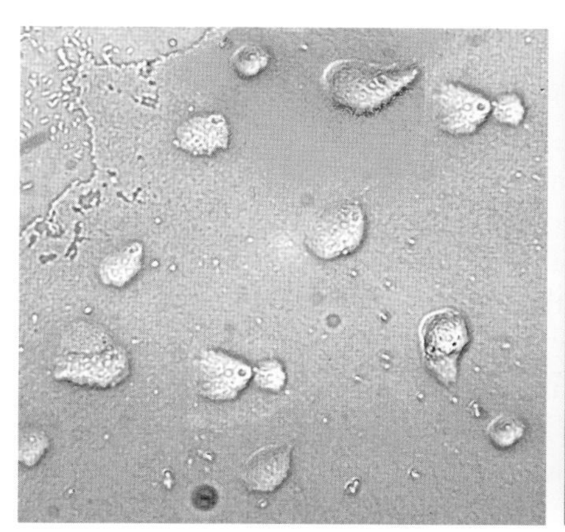

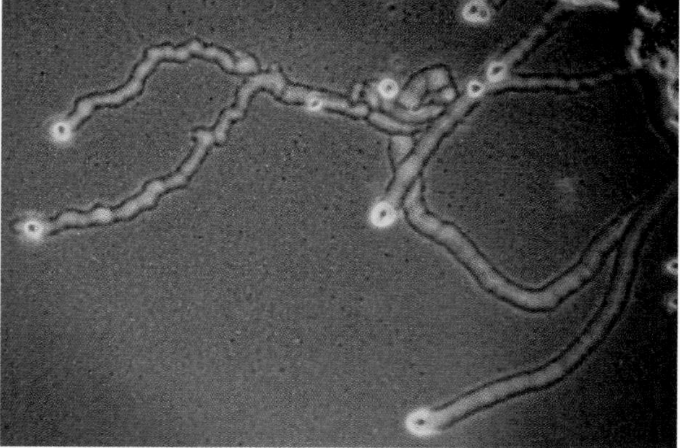

Figure 1.1 Free-living amebae on nonnutrient agar seeded with *E. coli*. Left, trophozoites; right, motility tracks on an agar plate. (Courtesy of Lillian Fritz-Laylin, UMass-Amherst, Amherst, MA.)

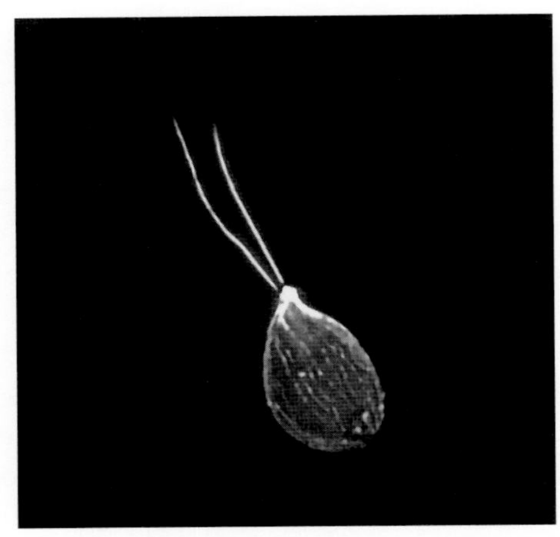

Figure 1.2 When placed in distilled water (enflagellation test), *N. fowleri*, the causal agent of PAM, undergoes transformation to a pear-shaped flagellate, usually with two flagella but occasionally with three or four flagella; the flagellate stage is a temporary nonfeeding stage and usually reverts to the trophozoite stage. Reprinted from Fritz-Laylin LK, Ginger ML, Walsh C, et al, Res Microbiol **162**(6):607–618, 2011, with permission from Elsevier, Ltd.

Granulomatous Amebic Encephalitis and Amebic Keratitis

The most characteristic feature of *Acanthamoeba* spp. is the presence of spine-like pseudopods called acanthapodia. Several species of *Acanthamoeba* (*A. culbertsoni*, *A. castellanii*, *A. polyphaga*, *A. astronyxis*, *A. healyi*, and *A. divionensis*) cause GAE, primarily in immunosuppressed, chronically ill, or otherwise debilitated persons. These patients tend to have no relevant history involving exposure to recreational freshwater. *Acanthamoeba* spp. also cause amebic keratitis, and it is estimated that to date approximately 1,000 cases of *Acanthamoeba* keratitis have been seen in the United States.

GAE caused by freshwater amebae is less well defined and may occur as a subacute or chronic disease with focal granulomatous lesions in the brain. The route of CNS invasion is thought to be hematogenous, with the primary site being the skin or lungs. In this infection, both trophozoites and cysts can be found in the CNS lesions. An acute-onset case of fever, headache, and pain in the neck preceded by 2 days of lethargy has also been documented. The causative organisms are probably *Acanthamoeba* spp. in most cases, but it is possible that others are involved, such as *Balamuthia mandrillaris* and *Sappinia diploidea*.

Cases of GAE have been found in chronically ill or immunologically impaired hosts; however, some patients apparently have no definite predisposing factor or immunodeficiency. Conditions associated with GAE include malignancies, systemic lupus erythematosus, human immunodeficiency virus (HIV) infection, Hodgkin's disease, skin ulcers, liver disease, pneumonitis, diabetes mellitus, renal failure, rhinitis, pharyngitis, and tuberculosis. Predisposing factors include alcoholism, drug abuse, steroid treatment, pregnancy, hematologic disorders, AIDS, cancer chemotherapy, radiation therapy, and organ transplantation. This infection has become more widely recognized in AIDS patients, particularly those with a low $CD4^{+}$ cell count.

Laboratory examinations similar to that for *N. fowleri* can be used to recover and identify these organisms; the one exception is recovery by culture, which has not proven to be as effective with GAE patients infected with *B. mandrillaris*.

Request for Blood Films

Malaria is one of the few parasitic infections considered immediately life-threatening, and a diagnosis of *P. falciparum* or *P. knowlesi* malaria should be considered to indicate a medical emergency, because the disease can be rapidly fatal. Any laboratory providing the expertise to identify malarial parasites should do so on a stat basis (24 h/day, 7 days/week).

Patients with malaria can present for diagnostic blood work when they are least expected. Laboratory personnel and clinicians should be aware of the stat nature of such requests and the importance of obtaining some specific patient history information. On microscopic examination of the blood films, the typical textbook presentation of various *Plasmodium* morphologies may not be seen by the technologist. The smears should be examined at length and under oil immersion. The most important thing to remember is that even though a low parasitemia may be present on the blood smears (in patients with no prior exposure to malaria and in the presence of residual antibody), the patient may still be faced with a serious, life-threatening disease.

It is important for both physicians and laboratorians in areas where malaria is not endemic to be aware of the problems associated with malarial diagnosis and to remember that symptoms are often nonspecific and may mimic other medical conditions. Physicians must recognize that travelers are susceptible to malarial infection when they visit a country where malaria is endemic and that they should receive prophylactic medication.

With the tremendous increase in the number of people traveling from the tropics to malaria-free areas, the number of imported malaria cases is also on the rise. There have been reports of imported infected mosquitoes transmitting the infection among people who live or work near international airports. It is also possible that mosquitoes can reach areas far removed from the airports. The resulting illness has been termed "airport malaria," i.e., malaria that is acquired through the bite of an infected anopheline mosquito by persons with apparently no risk factors for the disease. Unfortunately, unless a careful history is obtained, the diagnosis of malaria can be missed or delayed. Tests to exclude malaria should be considered for patients who work or live near an international airport and who present with an acute febrile illness. The potential danger of disseminating the mosquito vectors of malaria via aircraft is well recognized; however, modern disinfection procedures have not yet eliminated the risk of vector transportation. Not only can insects survive nonpressurized air travel, but also, they may be transported further by car or other means after arrival at the airport.

We usually associate malaria with patients having a history of travel within an area where malaria is endemic. However, other situations that may result in infection involve the receipt of blood transfusions, use of hypodermic needles contaminated by prior use (for example, by intravenous-drug users), possibly congenital infection, and transmission within the United States by indigenous mosquitoes that acquired the parasites from imported infections. Also, for a number of different reasons, organism recovery and subsequent identification are frequently more difficult than the textbooks imply. It is very important that this fact be recognized, particularly when one is dealing with a possibly fatal infection with *P. falciparum*. It is important to ensure that clinicians are familiar with the following issues.

Automated Instrumentation

Potential diagnostic problems with the use of automated hematology differential instruments have been reported. Some cases of malaria, as well as *Babesia* infection, have been completely missed by these methods. The number of fields scanned by a technologist on instrument-read smears is quite small; thus, failure to detect a low parasitemia is almost guaranteed. In cases of malaria and *Babesia* infection, after diagnosis had been made on the basis of smears submitted to the parasitology division of the laboratory, all previous smears examined by the automated system were reviewed and found to be positive for parasites. Failure to make the diagnosis resulted in delayed therapy. These instruments are not designed to detect intracellular blood parasites, and the inability of the automated systems to discriminate between uninfected erythrocytes and those infected with parasites may pose serious diagnostic problems in situations where the parasitemia is ≤0.5%.

Patient Information

When requests for malarial smears are received in the laboratory, some patient history information should be made available to the laboratorian. This history can be obtained by asking the ordering physician important questions such as the following:

1. Where has the patient been, and what was the date of return to the United States? (Where do you live and where do you work? [relevant for detecting airport malaria])
2. Has malaria ever been diagnosed in the patient before? If so, which species was identified?
3. What medication (prophylaxis or otherwise) has the patient received, and how often? When was the last dose taken?
4. Has the patient ever received a blood transfusion? Is there a possibility of other transmission via needle (drug user)?
5. When was the blood specimen drawn, and was the patient symptomatic at the time?
6. Is there any evidence of a fever periodicity?

Answers to such questions may help eliminate the possibility of infection with *P. falciparum*, *P. knowlesi*, or *Plasmodium vivax* (which can cause serious sequelae), usually the only three species that can cause severe disease and, in the case of *P. falciparum* or *P. knowlesi*, can rapidly lead to death.

Conventional Microscopy

Often, when the diagnosis of malaria is considered, only a single blood specimen is submitted to the laboratory for examination; however, **single films or specimens cannot be relied upon to exclude the diagnosis,** especially when partial prophylactic medication or therapy is used. Partial use of antimalarial agents may be responsible for reducing the numbers of organisms in the peripheral blood and lead to a blood smear that contains few organisms and a conclusion that reflects a low parasitemia when in fact serious disease is present. Patients with a relapsing case or an early primary case can also have few organisms in the blood smear. It is recommended that both thick and thin blood films be prepared immediately, and at least 300 oil immersion fields should be examined on both films before a negative report is issued. Since one set of negative smears does not rule out malaria, additional blood specimens should be examined over a 36-h period. **Although Giemsa stain has been recommended for all parasitic blood work, the organisms can also be seen if other blood stains, such as Wright's stain or any of the rapid blood stains, are used.** Blood collected with the use of EDTA anticoagulant is preferred over heparin; however, if the blood remains in the tube for approximately an hour or more, true stippling might not be visible within the infected erythrocytes (e.g., those infected with

P. vivax). When EDTA is being used, if blood is held for more than 2 h prior to blood film preparation, several artifacts may be seen; after 4 to 6 h, some of the parasites will be lost. During the time when the parasites are in the tube of blood, they continue to grow and change according to the life cycle for that species. Also, when anticoagulants are used, it is important to remember that the proper ratio of blood to anticoagulant is necessary for good organism morphology—the tube should be filled with blood. Both thick and thin blood films should be prepared immediately after receipt of the blood. If the specimen is sent to a reference laboratory, both the thick and thin blood films, as well as the tube of blood (room temperature), should be sent. Since this test is always considered a stat request, it is also important to know what turnaround times are available from the reference laboratory.

All requests for malaria diagnosis are considered stat requests, and specimens should be collected, processed, examined, and reported accordingly. Although other diagnostic tests can be ordered, any request for examination of blood films should include a possible diagnosis of malaria; thus, these requests are always considered stat. Not only should the blood collection be considered stat, but also, the processing and examination of both thick and thin blood films should be performed immediately on receipt of the blood. Often, immunologically naive individuals with no prior exposure to malaria can present to the emergency room or clinic with symptoms such as fever and malaise and a relevant travel history to an area of the world where malaria is endemic. These patients can have very vague symptoms, but they have the potential to become very ill with malaria, even with a low parasitemia (0.0005% to 0.1%).

Remember, when reporting results (genus and species), provide the stages of malaria present (rings, developing trophozoites, schizonts, and/or gametocytes); this information is very relevant for those providing therapy for the patient.

References

1. **Nelson K.** 2002. Stimulating research in the most neglected diseases. *Lancet* **359:**1042.
2. **WHO.** 2010. *Working to overcome the global impact of neglected tropical diseases. First WHO report on neglected tropical diseases.* World Health Organization, Geneva, Switzerland.
3. **Molyneux DH, Hotez PJ, Fenwick A.** 2005. "Rapid-impact interventions": how a policy of integrated control for Africa's neglected tropical diseases could benefit the poor. *PLoS Med* **2:**e336.
4. **Hotez PJ, Molyneux DH, Fenwick A, Ottesen E, Ehrlich Sachs S, Sachs JD.** 2006. Incorporating a rapid-impact package for neglected tropical diseases with programs for HIV/AIDS, tuberculosis, and malaria. *PLoS Med* **3:**e102.
5. **Hotez PJ, Molyneux DH, Fenwick A, Kumaresan J, Sachs SE, Sachs JD, Savioli L.** 2007. Control of neglected tropical diseases. *N Engl J Med* **357:**1018–1027.
6. **Hotez P.** 2007. A new voice for the poor. *PLoS Negl Trop Dis* **1:**e77.
7. **Hotez PJ.** 2017. Ten failings in global neglected tropical diseases control. *PLoS Negl Trop Dis* **11:**e0005896.
8. **Hotez PJ, Fenwick A, Ray SE, Hay SI, Molyneux DH.** 2018. "Rapid impact" 10 years after: the first "decade" (2006-2016) of integrated neglected tropical disease control. *PLoS Negl Trop Dis* **12:**e0006137.
9. **Smith DJ, Conn DB.** 2015. Importation and transmission of parasitic and other infectious diseases associated with international adoptees and refugees immigrating into the United States of America. *Biomed Res Int* **2015:**763715.
10. **Saroufim M, Charafeddine K, Issa G, Khalifeh H, Habib RH, Berry A, Ghosn N, Rady A, Khalifeh I.** 2014. Ongoing epidemic of cutaneous leishmaniasis among Syrian refugees, Lebanon. *Emerg Infect Dis* **20:**1712–1715.
11. **Rossati A, Bargiacchi O, Kroumova V, Zaramella M, Caputo A, Garavelli PL.** 2016. Climate, environment and transmission of malaria. *Infez Med* **24:**93–104.

12. **Stensgaard AS, Vounatsou P, Sengupta ME, Utzinger J.** 2019. Schistosomes, snails and climate change: current trends and future expectations. *Acta Trop* **190:**257–268.

13. **Bernin H, Lotter H.** 2014. Sex bias in the outcome of human tropical infectious diseases: influence of steroid hormones. *J Infect Dis* **209**(Suppl 3):S107–S113.

14. **Barber BE, Grigg MJ, William T, Piera KA, Boyle MJ, Yeo TW, Anstey NM.** 2017. Effects of aging on parasite biomass, inflammation, endothelial activation, microvascular dysfunction and disease severity in *Plasmodium knowlesi* and *Plasmodium falciparum* malaria. *J Infect Dis* **215:**1908–1917.

15. **Garcia LS.** 2016. *Diagnostic Medical Parasitology*, 6th ed. ASM Press, Washington, DC.

16. **Yoder JS, Straif-Bourgeois S, Roy SL, Moore TA, Visvesvara GS, Ratard RC, Hill VR, Wilson JD, Linscott AJ, Crager R, Kozak NA, Sriram R, Narayanan J, Mull B, Kahler AM, Schneeberger C, da Silva AJ, Poudel M, Baumgarten KL, Xiao L, Beach MJ.** 2012. Primary amebic meningoencephalitis deaths associated with sinus irrigation using contaminated tap water. *Clin Infect Dis* 55:e79–e85.

Table 1.1 Common features of the neglected tropical diseases[a]
Ancient afflictions that have burdened humanity for centuries
Poverty-promoting conditions
Associated with stigma
Rural areas of low-income countries and fragile states
No commercial markets for products that target these diseases
Interventions, when applied, have a history of success

[a] See reference 2.

SECTION 2

Parasite Classification and Relevant Body Sites

Practical Guide to Diagnostic Parasitology, Third Edition. Lynne S. Garcia
 DOI: 10.1128/9781683673637.ch02

Although common names are often used to describe parasites and parasitic infections, these names may refer to different parasites in different parts of the world. To eliminate these problems, a binomial system of nomenclature is used in which the scientific name consists of the genus and species designations.

It is appropriate to use the most commonly accepted name (*Giardia lamblia*); the proficiency testing organism lists can be followed as well. When a replacement name begins to appear on the proficiency test list, then it is appropriate to notify clients and institute the name change. Remember, it is very important to notify all clients prior to making any name changes.

Based on life cycles and organism morphology, classification systems have been developed to indicate the relationship among the various parasite species. Closely related species are placed in the same genus, related genera are placed in the same family, related families are placed in the same order, related orders are placed in the same class, and related classes are placed in the same phylum, one of the major categories in the animal kingdom. As one moves up the classification schema, each category is more expansive than the previous one; however, each category still has characteristics in common.

Parasites of humans are classified in several major divisions (see Table 2.1). These include the Protozoa (amebae, flagellates, ciliates, apicomplexans, and microsporidia [now classified with the fungi]), the Nematoda (roundworms), the Platyhelminthes (flatworms: cestodes, trematodes), the Pentastomida (tongue worms), the Acanthocephala (thorny-headed worms), and the Arthropoda (insects, spiders, mites, ticks, and so on). Although these categories appear to be clearly defined, there may be confusion in attempting to classify parasites, often due to the lack of confirmed specimens. If organisms recovered from humans are very rare, it is difficult to determine their correct taxonomic positions. Type specimens must be deposited for study before a legitimate species name can be given. Also, even when certain parasites are numerous, they may represent strains or races of the same species with slightly different characteristics.

Reproductive mechanisms have been used as a basis for determining species definitions, but there are many exceptions within parasite groups. Another difficulty in species recognition is the ability and tendency of an organism to alter its morphologic forms according to age, host, or nutrition, which often results in several names being assigned to the same organism. An additional problem involves alternation of parasitic and free-living phases in the life cycle; these phases may be very different and difficult to recognize as belonging to the same species. However, newer molecular methods of grouping organisms have often confirmed taxonomic conclusions reached hundreds of years earlier by experienced taxonomists. As studies in parasitic genetics, immunology, and biochemistry continue, species designations will be defined more clearly by highly sophisticated molecular techniques.

No attempt has been made to include every possible organism here; only those that are considered clinically relevant in the context of human parasitology are addressed. Some human infections are represented by very few cases; however, they are well documented and are included here. Further information is provided in the tables at the end of this section.

Protozoa (Intestinal)

Amebae, Stramenopiles

Amebae are single-celled organisms characterized by having pseudopods (motility) and trophozoite and cyst stages in the life cycle. The cell's organelles and cytoplasm are enclosed by a cell membrane, such that the cell obtains its food through phagocytosis. However, there are some exceptions in which a cyst form has not been identified. In environments which are potentially lethal to the cell, an ameba may become dormant by surrounding itself with a protective membrane to become a cyst. The cell remains in this form until it encounters more favorable conditions, at which time the organism excysts to release trophozoites. While in the cyst form, amebae do not replicate and may die if they are unable to excyst for a lengthy period.

Amebae are usually acquired by humans via fecal-oral transmission or mouth-to-mouth contact (*Entamoeba gingivalis*). In most species, after several nuclear divisions occur, comparable division of the cytoplasm follows excystation. *Entamoeba histolytica* is the most significant organism in this group (Figure 2.1).

Representative amebae include *Entamoeba histolytica, Entamoeba dispar, Entamoeba moshkovskii, Entamoeba bangladeshi, Entamoeba coli, Entamoeba hartmanni, Entamoeba gingivalis, Endolimax nana,* and *Iodamoeba bütschlii.*

Although *Blastocystis* spp. are enteric protozoan parasites that are commonly found worldwide, the classification is undergoing review. Currently, *Blastocystis* is grouped with the Stramenopiles (brown algae, slime nets, diatoms, and water molds) (Figure 2.2).

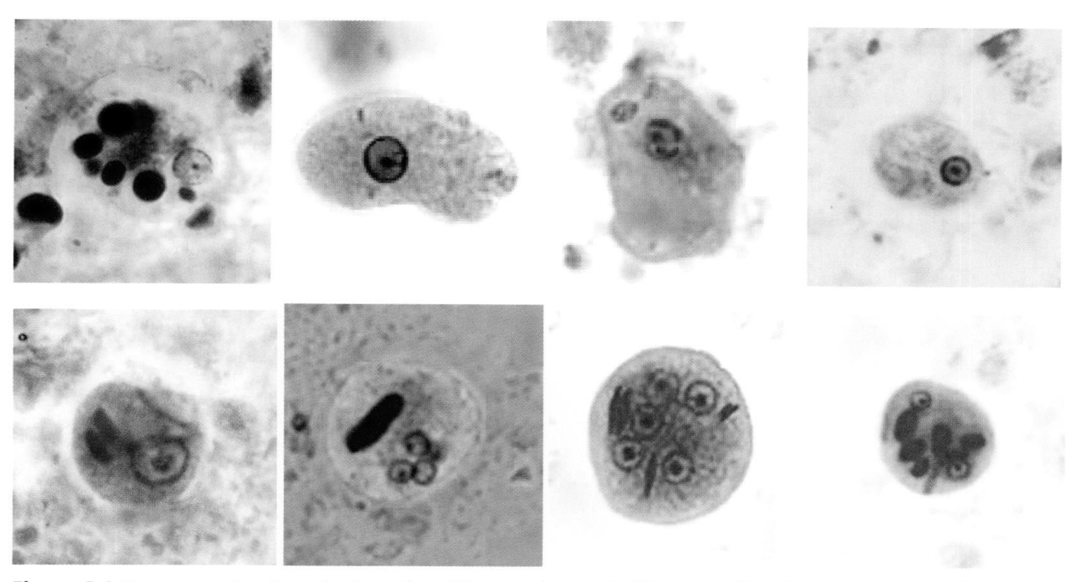

Figure 2.1 Representative intestinal amebae (*Entamoeba* spp.). **(Top row)** Trophozoites: *Entamoeba histolytica, Entamoeba histolytica/E. dispar, Entamoeba coli, Entamoeba hartmanni.* **(Bottom row)** Cysts: *Entamoeba histolytica/E. dispar* precyst, *Entamoeba histolytica/E. dispar, Entamoeba coli, Entamoeba hartmanni.*

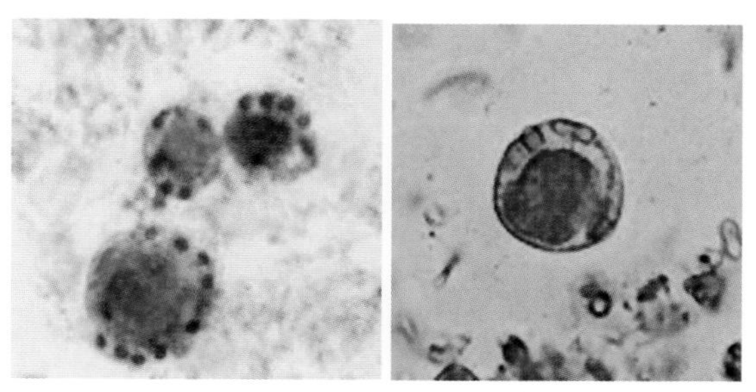

Figure 2.2 Intestinal Stramenopiles (*Blastocystis* spp.). Central-body forms are shown; note the nuclei around the margins of the central-body area (green in left organisms, orange/iodine in right organism).

Blastocystis has extensive genetic diversity and infects humans and many other animals. Statistically, it may be the most common intestinal parasite recovered. Some of the nine subtypes that are found in humans are considered pathogens, while others are probably nonpathogenic, a situation that leads to different opinions regarding pathogenicity. Unfortunately, these subtypes cannot be differentiated on the basis of microscopic morphology, and they may eventually be divided into strains, species, and/or subspecies. Since there may be a relationship between numbers present and symptoms, this is one of the few parasites whose numbers should be specified in the report (rare, few, moderate, many, or packed). It is recommended that quantitation be determined from the permanent stained smear.

Flagellates

Flagellates move by means of flagella and are acquired through fecal-oral transmission. With the exception of organisms in the genus *Pentatrichomonas*, flagellates have both trophozoite and cyst stages in the life cycle (Figure 2.3). Reproduction is by longitudinal binary fission. *Giardia lamblia* (also known as *G. duodenalis* and *G. intestinalis*) is the most common pathogen in this group and is one of the most commonly found intestinal parasites. However, when permanent stained smears are routinely performed, *D. fragilis* is also found to be as common as or more common than *Giardia*. Representative organisms include *G. lamblia*, *D. fragilis*, *Pentatrichomonas hominis*, *Chilomastix mesnili*, *Enteromonas hominis*, and *Retortamonas intestinalis*.

Ciliates

Ciliates are single-celled protozoa that move by means of cilia and are acquired through fecal-oral transmission. *Balantidium coli* (also known as *Neobalantidium coli*) is the only human pathogen in the group (Figure 2.4). Hosts include pigs, wild boars, rats, primates

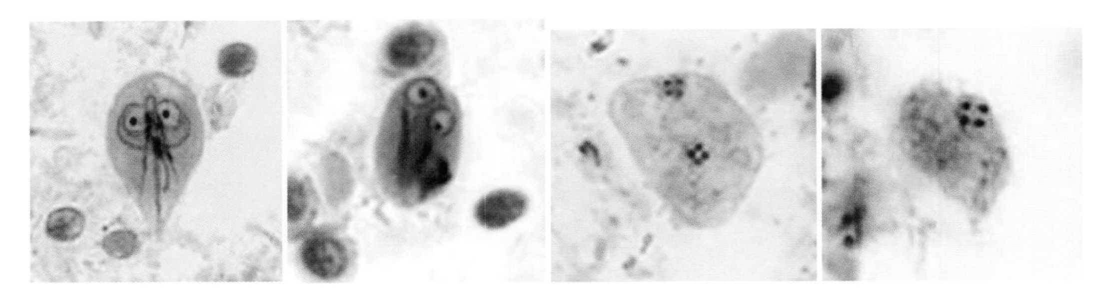

Figure 2.3 Representative intestinal flagellates. *Giardia lamblia* (also called *G. duodenalis* or *G. intestinalis*) trophozoite; *Giardia* cyst; *Dientamoeba fragilis* trophozoite; *D. fragilis* trophozoite (note: trophozoites can have either a single nucleus or two nuclei; nuclear chromatin is fragmented into several granules). The recently confirmed *D. fragilis* cyst can be seen in Section 7.

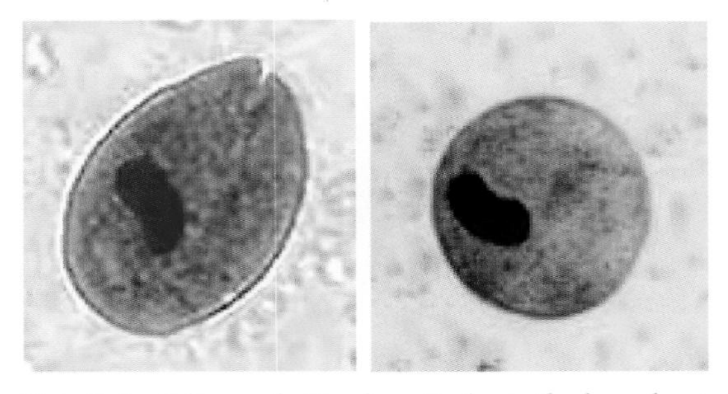

Figure 2.4 The intestinal ciliate *Balantidium coli*. Trophozoite (note the large bean-shaped macronucleus and cytostome at upper right); cyst with single bean-shaped macronucleus.

(including humans), horses, cattle, and guinea pigs. Infection is transmitted within or between these species by fecal-oral transmission of the infective cysts. Pigs are the most significant reservoir hosts, although they show few if any symptoms. Following ingestion, excystation occurs in the small intestine, and the trophozoites colonize the large intestine. The cilia beat in a coordinated rhythmic pattern, and the trophozoite moves in a spiral path. They have both trophozoite and cyst stages in the life cycle, and both stages contain a large macronucleus and a smaller micronucleus. These protozoa have a distinct cell mouth (cytostome) and cytopharynx and a less conspicuous cytopyge (anal pore). These organisms are considerably larger than the majority of the intestinal protozoa and can be mistaken for debris or junk when seen in a permanent stained smear. The concentration wet preparation examination is recommended for microscopic examination.

Apicomplexa (Including Coccidia)

Apicomplexa are microscopic, spore-forming, single-celled, obligately intracellular parasites, which means that they must live and reproduce within an animal cell (*Toxoplasma gondii* and *Cryptosporidium* spp.) (Figure 2.5). These protozoa are acquired by ingestion via contaminated food and/or water. In some cases, these organisms disseminate to other body sites, particularly in severely compromised patients. These protozoa have both asexual and sexual cycles, and the infective stages include oocysts, all of which can be acquired through fecal-oral transmission. The **coccidia** are classified in the Apicomplexa and include *Cyclospora cayetanensis*, *Cystoisospora belli* (formerly *Isospora belli*), and *Sarcocystis* spp.

Microsporidia (Now Classified with the Fungi)

Currently, the most difficult intestinal parasites to diagnose are the microsporidia (spore size range, 1 to 2.5 μm); the development of molecular biology-based methods should

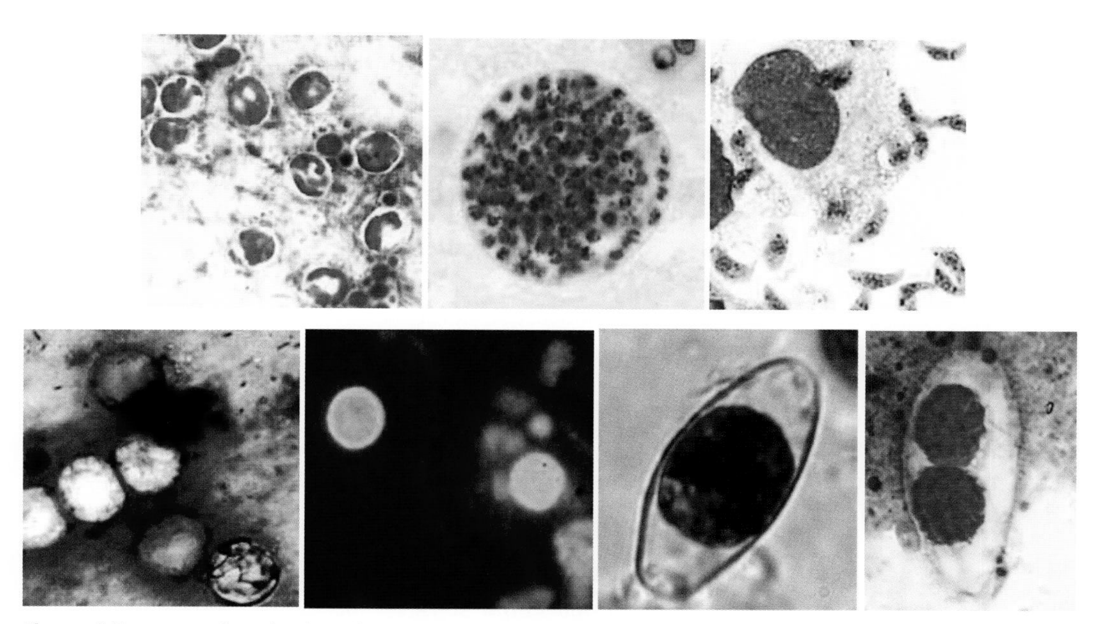

Figure 2.5 Apicomplexa (top) and coccidia (bottom). **(Top row)** *Cryptosporidium* oocysts; *Toxoplasma gondii* bradyzoites; *T. gondii* tachyzoites (note the typical crescent shape). **(Bottom row)** *Cyclospora cayetanensis* oocysts; *C. cayetanensis* oocyst autofluorescence; *Cystoisospora belli* immature oocyst; *C. belli* mature oocyst.

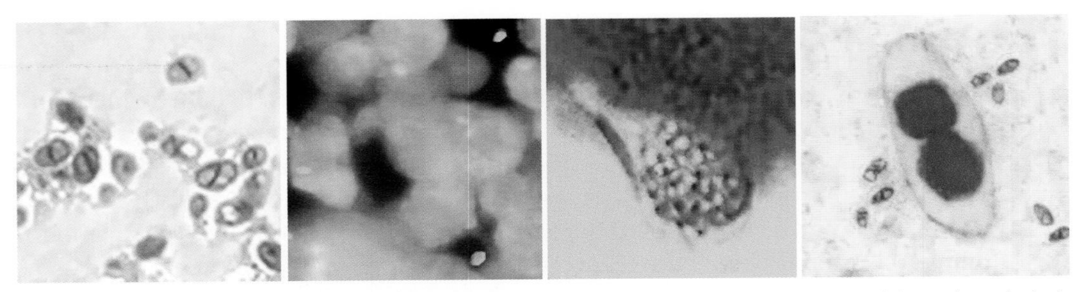

Figure 2.6 Microsporidia (fungi). Microsporidian spores (note the cross/diagonal line of the polar tubules); spores in urine sediment, visualized by calcofluor white staining (can be both intracellular or outside the cells); spores in a specimen from an eye (other body sites); combination infection with microsporidian spores and a mature *Cystoisospora belli* oocyst.

provide more specific and sensitive methods (Figure 2.6). These organisms have also been documented to disseminate from the intestinal tract to other body sites, including the kidneys and lungs. Routine parasitology stains are not useful; modified trichrome stains have been developed specifically for these organisms. Compared with the more common Wheatley's trichrome for routine stool staining, the modified trichrome stain contains a 10-fold-higher concentration of the main dye, chromotrope 2R. The infective form is called the spore; each spore contains a polar tubule that is used to penetrate new host cells, thus initiating or continuing the life cycle. Infections are acquired through ingestion, inhalation, or direct inoculation of spores from the environment. Currently, members of at least two genera have been documented to cause human infection in the intestinal tract (*Encephalitozoon intestinalis* and *Enterocytozoon bieneusi*).

Protozoa (Other Body Sites)

Amebae

With the exception of *Entamoeba gingivalis* (found in the mouth), nonintestinal amebae are pathogenic, free-living organisms that may be associated with warm, freshwater environments (Figure 2.7). They have been found in the central nervous system, the eyes, and other body sites. Amebae that invade the central nervous system (*Naegleria fowleri*) can cause severe, life-threatening infection that often ends in death within a few days. Other amebae in this group can cause more chronic central nervous system disease (*Acanthamoeba* spp., *Balamuthia mandrillaris*, and *Sappinia diploidea*, particularly in immunocompromised patients). *Acanthamoeba* can also cause keratitis; untreated cases can result in blindness.

Flagellates

Trichomonas vaginalis is found in the genitourinary system and is usually acquired by sexual transmission (Figure 2.8). *Trichomonas tenax* can be found in the mouth and is considered nonpathogenic.

Apicomplexa (Including Coccidia)

Parasites within the Apicomplexa (*Cryptosporidium* spp.) are particularly important in the compromised patient and can cause life-threatening disease. These organisms can disseminate from the intestinal tract to other body sites. They may also infect many individuals who have

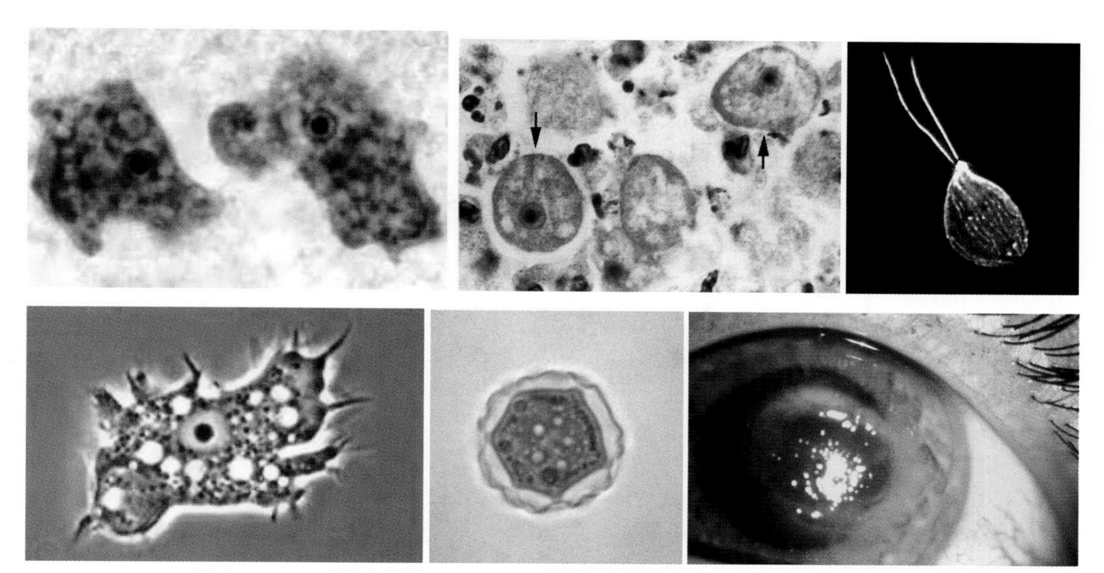

Figure 2.7 Protozoa (*Naegleria fowleri*, top) and amebae (*Acanthamoeba* spp., bottom) from other body sites. **(Top row)** Trophozoites (note the smooth, rounded pseudopods); trophozoites in brain tissue (arrows); nucleated stage from culture protocol. **(Bottom row)** Trophozoite (note the spiky pseudopods); cyst (note the double wall); example of *Acanthamoeba* keratitis.

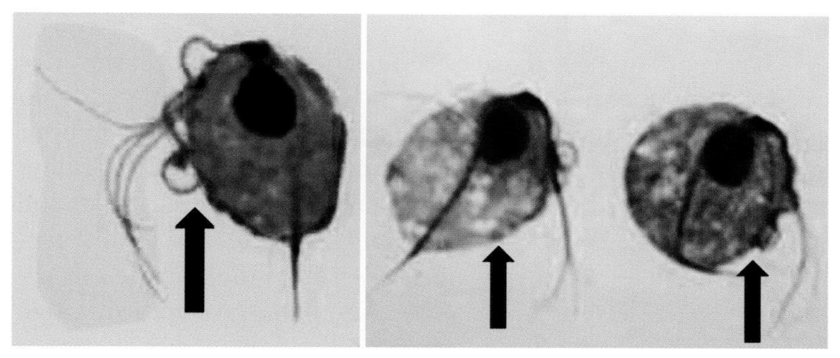

Figure 2.8 Flagellates from other body sites. *Trichomonas vaginalis* (urinary genital system) trophozoites are shown. Note that the undulating membrane (arrows) goes approximately halfway down the organism, not to the bottom as in *Pentatrichomonas hominis* (gastrointestinal tract).

relatively few symptoms. In immunocompetent patients, symptoms may be minimal or absent; however, in compromised patients, sequelae may be very serious and even life-threatening.

Microsporidia (Now Classified with the Fungi)

As mentioned above, microsporidia include the most difficult organisms to diagnose (size range, 1 to 2.5 µm). Dissemination from the intestine to other body sites has been well documented. Modified trichrome stains have been developed specifically for detection of these organisms, since routine parasitology stains for fecal specimens are not very effective for microsporidial spores. Optical brightening agents such as calcofluor are also recommended; although they are very sensitive, they are nonspecific. Currently, at least 18 species have been identified as human parasites (see example of eye infection in Figure 2.6). Genera include *Anncaliia*, *Brachiola*, *Encephalitozoon*, *Enterocytozoon*, *Microsporidium*, *Nosema*,

Pleistophora, *Trachipleistophora*, *Vittaforma*, and *Tubulinosema*. *Encephalitozoon* and *Enterocytozoon* tend to be in the gastrointestinal tract but can disseminate to other tissues.

Protozoa (Blood and Tissue)

Apicomplexa (Including Sporozoa)

All sporozoa are arthropod borne and are in the Apicomplexa. The genus *Plasmodium* includes parasites that undergo exoerythrocytic and pigment-producing erythrocytic schizogony in vertebrates and a sexual stage followed by sporogony in mosquitoes (Figure 2.9). *Babesia* spp. are tick borne and can cause severe disease in patients who have been splenectomized or otherwise immunologically compromised (Figure 2.10). Diagnosis may be somewhat more difficult than for the intestinal protozoa, particularly if automated blood differential systems are used; the microscopic examination of both thick and thin blood films is recommended (see Figure 2.10). **Note: examination of blood for these parasites is considered a stat test request; coverage needs to be available 24/7.** Representatives in the genus *Plasmodium* include *P. vivax*, *P. ovale wallikeri*, *P. ovale curtisi*, *P. malariae*, *P. falciparum*, and *P. knowlesi*. Organisms in the genus *Babesia* include *B. microti*, *B. divergens*, additional organisms from the West Coast of the United States, including *B. duncani*, and others that are not yet classified to the species level.

Flagellates

Leishmaniae

The leishmaniae have undergone extensive classification revisions. However, from a clinical perspective, recovery and identification of the organisms are still related to body sites such as the macrophages of the skin (cutaneous), the skin and mucous membranes (mucocutaneous), the skin (American tegumentary leishmaniasis; primarily cutaneous lesions; southern United States to northern Argentina), and the whole reticuloendothelial system (visceral—bone marrow, spleen, liver) (Figure 2.11). Recovery of leishmanial amastigotes is limited to the site of the lesion in infections other than those caused by the *Leishmania donovani* complex (visceral leishmaniasis). These protozoa have both amastigote (mammalian host) and promastigote (sand fly) stages in the life cycle. Reproduction in both forms occurs by binary longitudinal division. Their primary hosts are vertebrates; *Leishmania* commonly infects hyraxes, canids, rodents, and humans. Representative organisms include *Leishmania tropica*, *L. major*, *L. mexicana*, *L. braziliensis*, and *L. donovani*.

Trypanosomes

The trypanosomes are normally identified to the species level based on geographic exposure history and clinical symptoms. These protozoa are characterized by having, at some time in the life cycle, the trypomastigote form with the typical undulating membrane and free flagellum at the anterior end (Figure 2.12). Unfortunately, the longer the duration of the infection, the more difficult it may be to confirm the diagnosis. The organisms that cause African sleeping sickness (*Trypanosoma brucei gambiense* and *T. b. rhodesiense*) generally cause different disease entities; one tends to be chronic and more typically the patient appears to have sleeping sickness (*T. b. gambiense*), and the other causes a more fulminant disease, often leading to death before typical sleeping sickness symptoms can develop (*T. b. rhodesiense*).

The etiologic agent of American trypanosomiasis (formerly South American trypanosomiasis) is *T. cruzi*, which contains amastigote and trypomastigote stages (in the mammalian host) and the epimastigote form (in the arthropod host). Human American

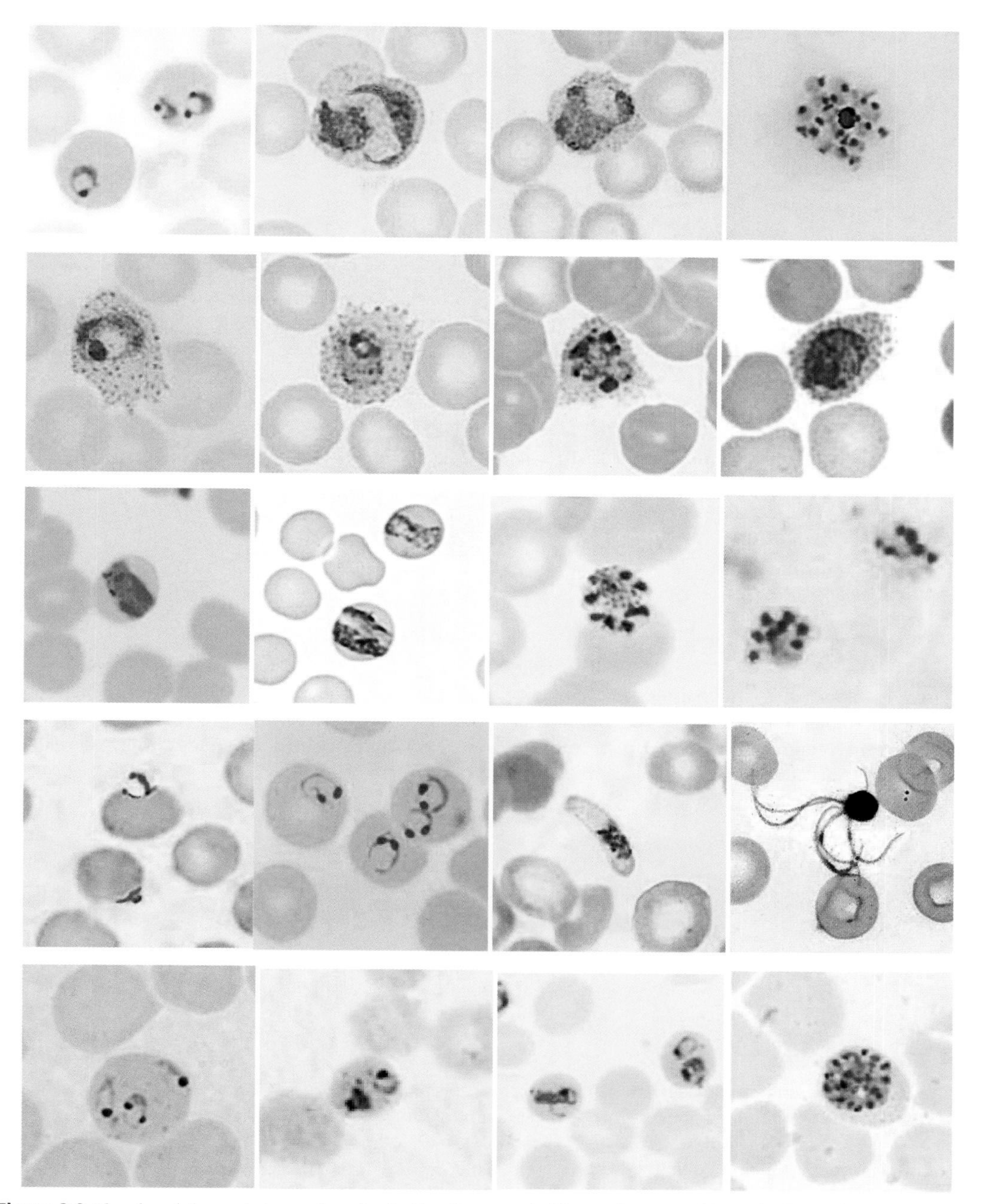

Figure 2.9 Blood and tissue Apicomplexa (including Sporozoa): *Plasmodium* spp. **(Row 1)** *Plasmodium vivax* rings (note that double rings can occur in this species); developing *P. vivax* trophozoite; developing trophozoite; mature schizont. **(Row 2)** *Plasmodium ovale* developing trophozoite (note Schüffner's dots, i.e., true stippling); developing trophozoite (note fimbriated edges); developing schizont; gametocyte. **(Row 3)** *Plasmodium malariae* typical developing trophozoites (note band form configuration); band form; mature schizont (daisy head configuration [merozoites arranged in a circle around the malarial pigment]); two mature schizonts seen in a thick blood film. **(Row 4)** *Plasmodium falciparum* trophozoites (note that the ring at the top appears to be poking through the red blood cell (RBC); multiple rings in "headphone" configuration within RBCs; crescent-shaped gametocyte (note that although the parasite is within the RBC, the RBC outline is not visible); exflagellation of the male gametocyte (can occur in any of the five species if male gametocytes are present in the blood sample). **(Row 5)** *Plasmodium knowlesi* rings (which resemble *P. falciparum*); band form; band form; mature schizont (note that band forms and schizont "mimic" *P. malariae*; often these infections are mistaken for a mixed infection of *P. falciparum* and *P. malariae*).

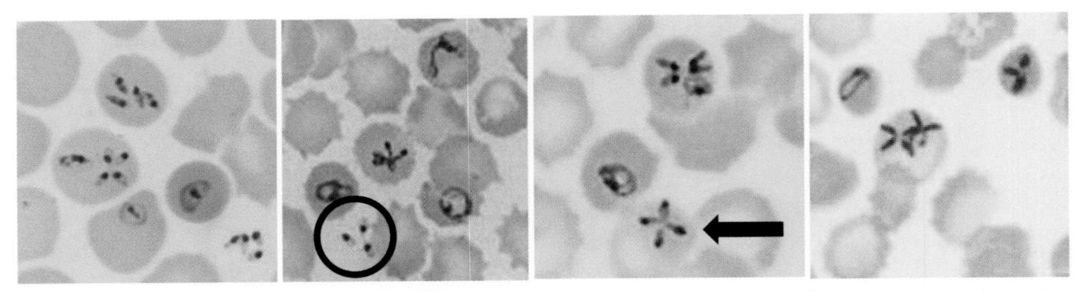

Figure 2.10 Blood and tissue Apicomplexa (including Sporozoa): *Babesia* spp. *Babesia* rings (note that a few of the rings are outside the RBCs); rings (note that there are a few rings outside the RBC [circle]; this is more common and is rarely seen in malaria unless the parasitemia is very high); rings with a Maltese cross configuration (arrow); rings.

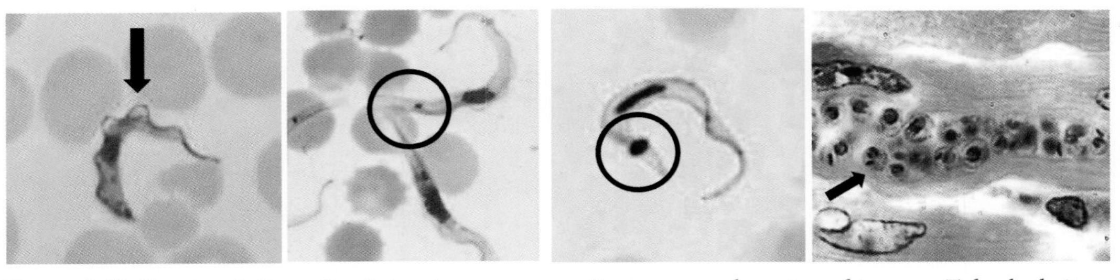

Figure 2.11 Blood and tissue flagellates: leishmaniae. Examples of cutaneous *Leishmania* lesions; amastigotes within a macrophage; bone marrow containing amastigotes (note the nucleus [red arrow] and the kinetoplast [black arrow]) (courtesy of CDC DpDx, https://www.cdc.gov/dpdx/leishmaniasis/index.html); developing organisms within a culture system.

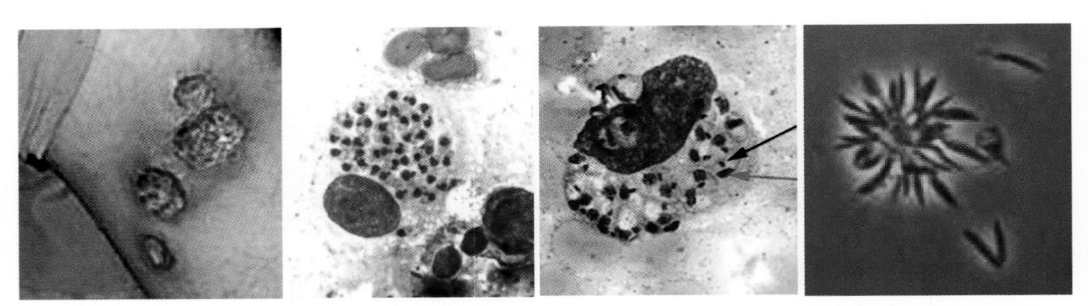

Figure 2.12 Blood and tissue flagellates: trypanosomes. *Trypanosoma brucei gambiense* or *T. b. rhodesiense* (African trypanosomiasis) trypomastigote (note the undulating membrane [arrow]); two trypomastigotes (note the very small kinetoplast [circle]); *Trypanosoma cruzi* (American trypanosomiasis) trypomastigote (note the very large kinetoplast [circle]); *T. cruzi* amastigotes in cardiac muscle (arrow; the large dot is the nucleus, and the small bar-shaped structure is the kinetoplast).

trypanosomiasis, or Chagas' disease, is a potentially fatal disease of humans. It has two forms, a trypomastigote found in human blood and an amastigote found in tissues. The acute form usually goes unnoticed, and infection may present as a localized swelling at the site of entry of the parasites in the skin. The chronic form may develop 10 to 20 years after infection. This form affects internal organs (e.g., the heart, esophagus, colon, and peripheral nervous system), and patients may die from heart failure. In 2007, the Red Cross began screening donor blood units for antibody to this parasite. Also, the geographic range of *T. cruzi* in the United States continues to expand and now includes all of Texas. Infected triatomid bug vectors are also present in other states, such as California.

Nematodes (Intestinal)

The largest number of helminthic parasites of humans belongs to the roundworm group (Figure 2.13). Nematodes are elongate-cylindrical and bilaterally symmetrical, with a triradiate symmetry at the anterior end. They have an outer cuticle layer, no circular muscles, and a pseudocoelom containing all systems (digestive, excretory, nervous, and reproductive). These organisms are normally acquired by ingestion of their eggs or skin penetration by larval forms from the soil. Nematodes commonly parasitic in humans include *Trichuris trichiura*, *Necator americanus*, *Ancylostoma duodenale*, *Enterobius vermicularis*, *Ascaris lumbricoides*, *Strongyloides stercoralis*, and *Trichostrongylus* spp.

Nematodes (Tissue)

Some tissue nematodes are rarely seen within the United States; however, some are more important and are found worldwide (Figure 2.14). Diagnosis may be difficult if the only specimens available are obtained through biopsy and/or autopsy, and interpretation must be based on examination of histologic preparations. Examples include infections caused by *Trichinella* spp., *Toxocara* spp., *Baylisascaris procyonis* (raccoon roundworm), *Ancylostoma* spp., *Angiostrongylus* spp., and *Gnathostoma* spp.

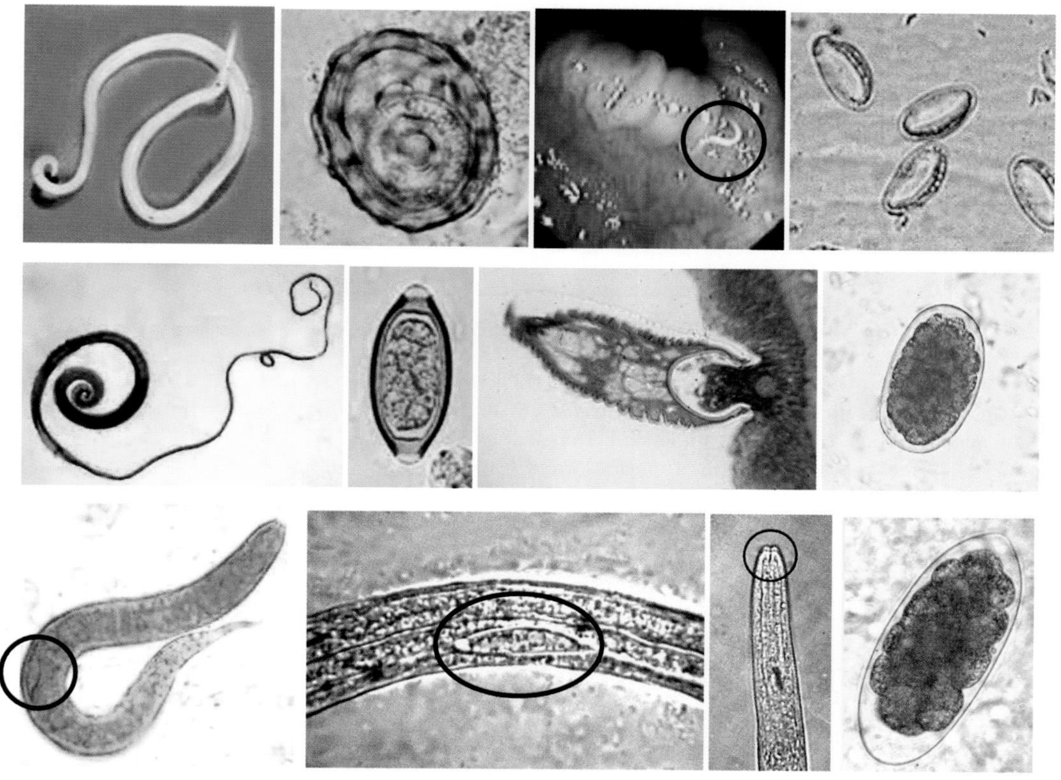

Figure 2.13 Nematodes (intestinal). **(Top row)** Adult *Ascaris lumbricoides* male; *Ascaris* fertilized egg; *Enterobius vermicularis* (pinworm) on perianal skin (circle); pinworm eggs. **(Middle row)** Adult *Trichuris trichiura*; *T. trichiura* egg (note the polar plugs at each end); *Necator* or *Ancylostoma* hookworm (note that the anterior end is embedded in the mucosa); hookworm egg. **(Bottom row)** *Strongyloides stercoralis* rhabditiform larva (a genital packet of cells can be seen [circle]); rhabditiform larva (a genital packet of cells can be seen [oval]); rhabditiform larva with a short buccal capsule/mouth opening (circle); *Trichostrongylus* sp. egg.

Figure 2.14 Nematodes (tissue). **(Row 1)** *Trichinella* sp. encysted larva in muscle; *Trichinella* sp. encysted in muscle; three images showing *Toxocara* sp. eggs. **(Row 2)** Creeping eruption from dog/cat hookworm larvae infecting a human (oval); creeping eruption on toes; *Angiostrongylus* infective larva. **(Row 3)** Three images of *Gnathostoma* spp. (courtesy of CDC DpDx, https://www.cdc.gov/parasites/gnathostoma/); linear larva migrans lesion on skin of breast (adapted from Frean J, *Trop Med Infect Dis* 5[1]:39, 2020, CC-BY 4.0, https://creativecommons.org/licenses/by/4.0/). **(Row 4)** *Baylisascaris procyonis* egg (note resemblance to other nematode eggs, such as *Ascaris* and *Toxocara*); bolus of adult worms from a raccoon; autopsy of raccoon showing *Baylisascaris* adult worms; cross section of *B. procyonis* larva in brain tissue section (courtesy of the CDC Public Health Image Library).

Nematodes (Blood and Tissue)

Blood and tissue nematodes (filarial worms) are arthropod borne (Figure 2.15). The adult worms tend to live in the tissues or lymphatics of the vertebrate host. Diagnosis is made on the basis of the recovery and identification of the larval worms (microfilariae) in the blood, other body fluids, or skin. While circulating in peripheral blood or cutaneous tissues, the microfilariae can be ingested by blood-sucking insects. After the larvae mature, they can escape into the vertebrate host's skin when the arthropod takes its next blood meal. The severity of disease due to these nematodes varies; however, elephantiasis may

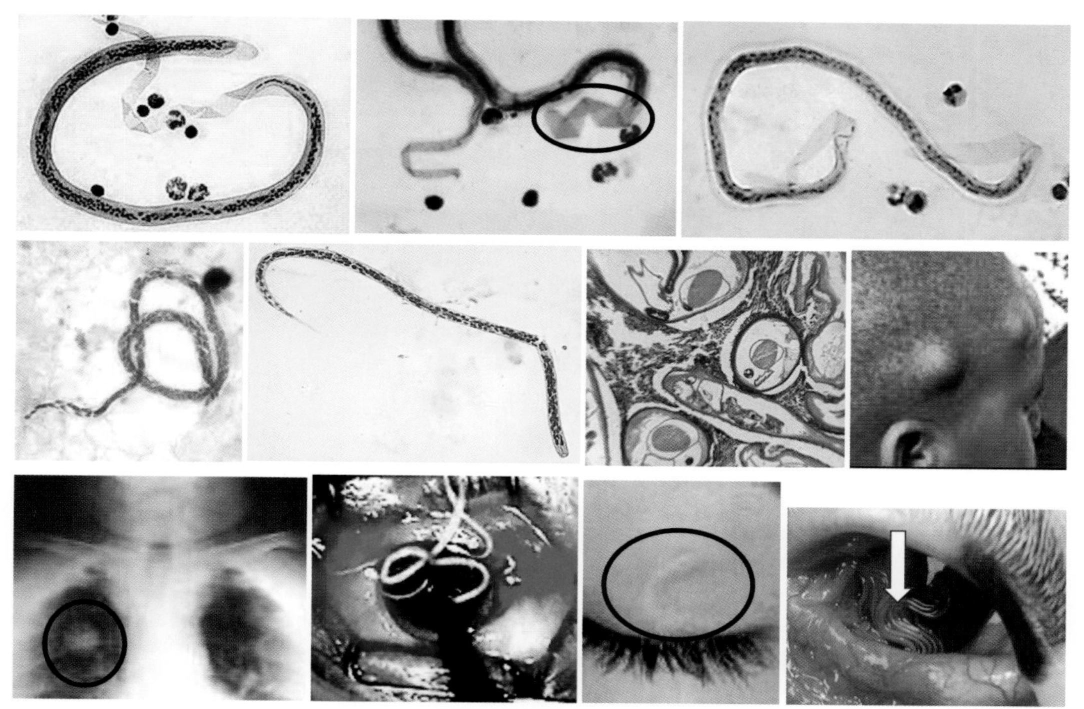

Figure 2.15 Nematodes (blood and tissue): filarial worms. **(Top row)** *Wuchereria bancrofti* microfilaria; *Brugia malayi* microfilaria (note that the sheath stains pink with Giemsa stain [oval]); *Loa loa* microfilaria. **(Middle row)** *Mansonella perstans* microfilaria, patient from Cameroon (courtesy of CDC DpDx, https://www.cdc.gov/dpdx/mansonellosis/index.html); *Onchocerca volvulus* microfilaria; cross section of *Onchocerca* nodule; *Onchocerca* nodules on the head (from the collection of Herman Zaiman, "A Presentation of Pictorial Parasites"). **(Bottom row)** *Dirofilaria* sp. in human lung ("coin lesion" [circle]); removal of *Dirofilaria* worm from eye; *Dirofilaria* worm in eyelid; *Thelazia* sp., multiple worms in dog eye (arrow) (Otranto D, Dutto M, *Emerg Infect Dis* **14**[4]:647–649, 2008, https://www.ncbi.nlm.nih.gov/pmc/articles/PMC2570937/).

be associated with some of the filarial worms. Specific organisms include *Wuchereria bancrofti, Brugia* spp., *Loa loa, Mansonella* spp., *Onchocerca volvulus, Dirofilaria* spp., and *Thelazia* spp.

Cestodes (Intestinal)

The adult form of the tapeworm is acquired through ingestion of the larval forms contained in poorly cooked or raw meats or freshwater fish (*Taenia* spp. and *Diphyllobothrium* spp.) (Figure 2.16). *Dipylidium caninum* infection is acquired by the accidental ingestion of dog fleas infected with the larval tapeworms. Both *Hymenolepis nana* and *H. diminuta* are transmitted via ingestion of certain infected arthropods (fleas and beetles). Also, *H. nana* can be transmitted through egg ingestion (its life cycle can bypass the intermediate beetle host). The adult tapeworm consists of a chain of egg-producing units called proglottids, which develop from the neck region of the attachment organ, the scolex. Food is absorbed through the worm's integument. The intermediate host contains the larval forms, which are acquired through ingestion of the adult tapeworm eggs. Humans can serve as both the intermediate and definitive hosts in *H. nana* and *Taenia solium* infections.

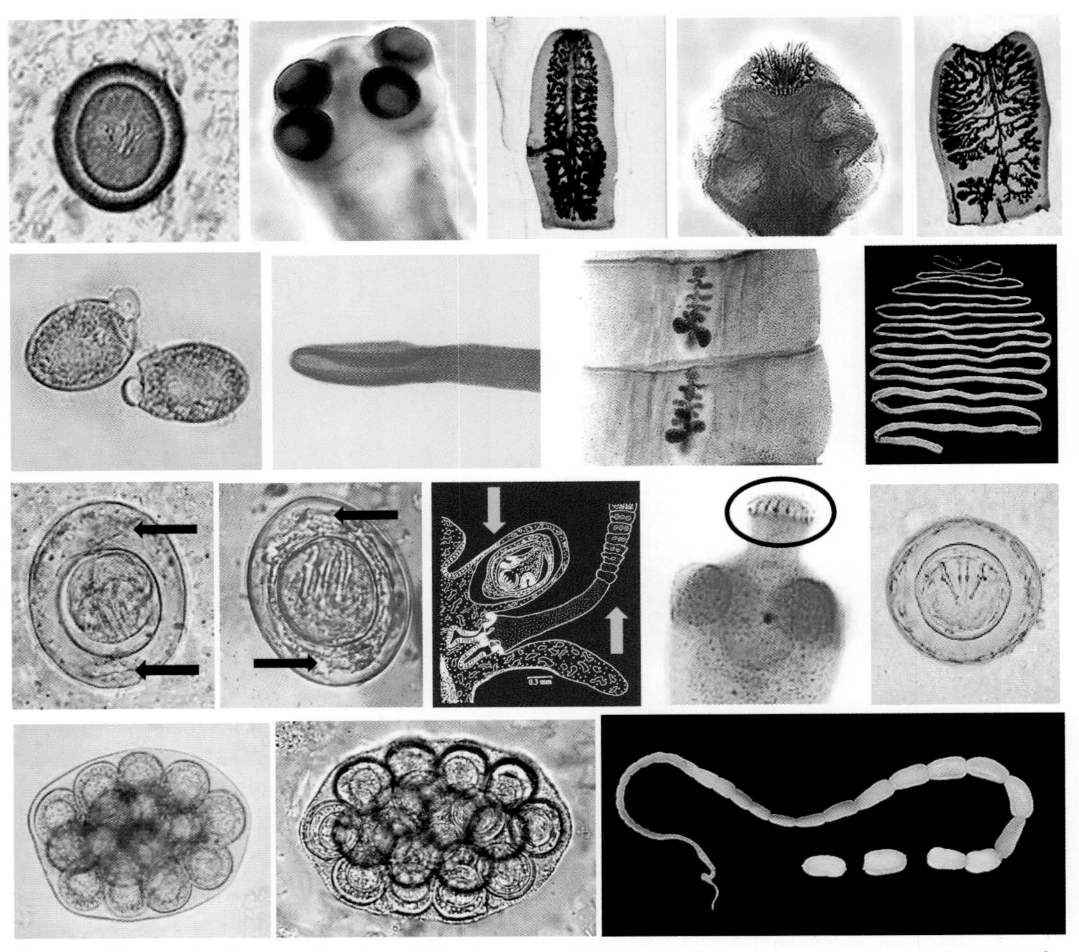

Figure 2.16 Cestodes (intestinal). **(Row 1)** *Taenia* sp. egg (without special staining, *Taenia saginata* cannot be distinguished from *Taenia solium*); *T. saginata* scolex (note four suckers and no hooks); *T. saginata* gravid proglottid (note the large number of lateral uterine branches; the count for one side is >12); *T. solium* scolex (note suckers and hooks); *T. solium* gravid proglottid (note the small number of lateral uterine branches; the count on one side is <12). **(Row 2)** *Diphyllobothrium latum* operculated eggs; *Diphyllobothrium* scolex (note the grooves rather than suckers); *Diphyllobothrium* gravid proglottids (note that they are wider than long, unlike the *Taenia* sp. gravid proglottids); adult *Diphyllobothrium* worm (worm length can reach 30 ft). **(Row 3)** *Hymenolepis nana* (dwarf tapeworm) egg (note the polar filaments that lie between the oncosphere/larva and the egg shell [arrows]); *H. nana* life cycle can occur within the human after egg ingestion—no intermediate host required (note the developing larval form [upper arrow] and adult tapeworm [lower arrow]; *H. nana* scolex showing the retractile rostellum with a circle of hooks (oval) (the adult tapeworm is rarely seen); *Hymenolepis diminuta* (rat tapeworm) egg (note the absence of polar filaments between the developing oncosphere and the egg shell). **(Row 4)** *Dipylidium caninum* (dog tapeworm) egg packets (individual eggs resemble *Taenia* sp. eggs; however, they are passed in a packet of ~10+ eggs); adult *D. caninum* tapeworm (when individual proglottids are fresh, they resemble cucumber seeds; when dry, they resemble rice grains).

Cestodes (Tissue)

The ingestion of certain tapeworm eggs or accidental contact with certain larval forms can lead to tissue infection with *Taenia solium*, other *Taenia* spp., *Echinococcus* spp., *Diphyllobothrium* spp., and *Spirometra mansonoides* (Figure 2.17).

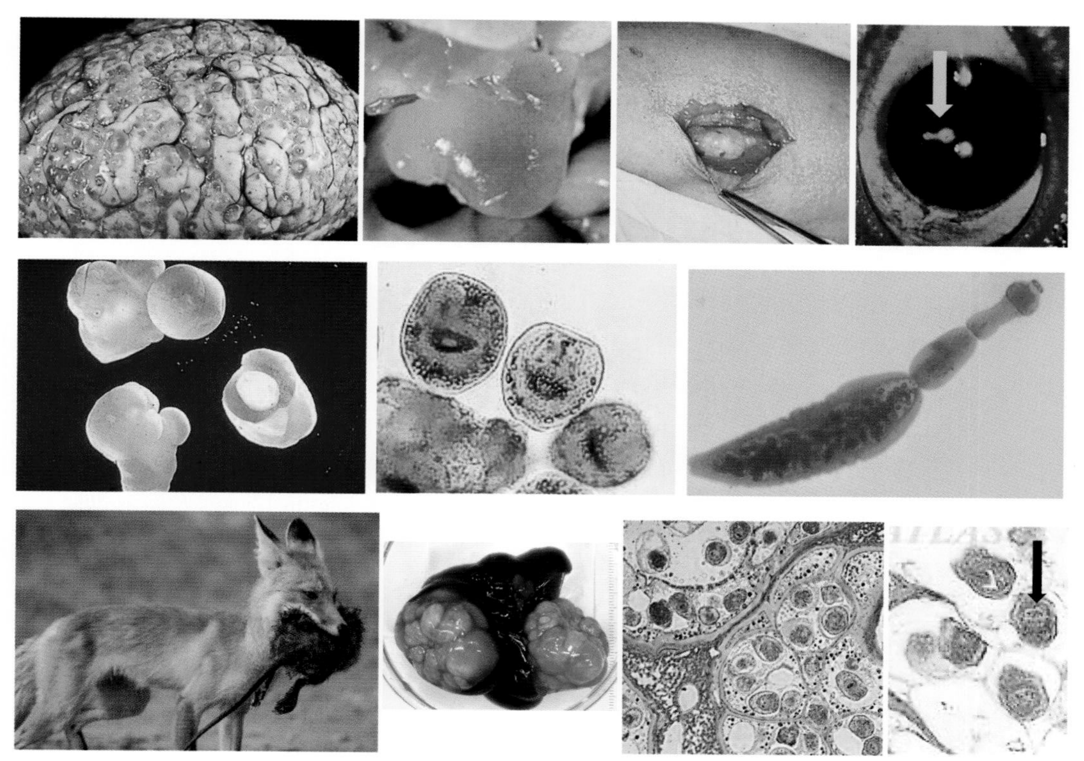

Figure 2.17 Cestodes (tissue). **(Top row)** Brain with multiple cysticerci of *Taenia solium*; individual cysticercus at the base of the brain; *T. solium* cysticercus being removed from the arm; cysticercus of *T. solium* in the pupil of the eye (arrow) (courtesy of the CDC Public Health Image Library). **(Middle row)** *Echinococcus granulosus* daughter hydatid cysts; immature protoscolices (dark lines are the hooklets); adult tapeworm found in the carnivore (often dog). **(Bottom row)** Wolves and foxes carry adult worms of *Echinococcus multilocularis* (courtesy of the Ontario *Echinococcus multilocularis* website, http://blog.healthywildlife.ca/echinococcus-multilocularis-ontario-website/); rodents carry intermediate forms (note multiple cysts that are found in liver (https://www.ncbi.nlm.nih.gov/pmc/articles/PMC4804384/); *E. multilocularis* in human tissue section (note that there is no limiting cyst wall; morphology is more like a metastatic cancer); enlarged image showing the individual protoscolices (note hooklets that resemble a bar across the organism [arrow]).

Trematodes (Intestinal)

Trematodes are flatworms and are exclusively parasitic (Figure 2.18). With the exception of the schistosomes (blood flukes), flukes are hermaphroditic. They may be flattened, and most have oral and ventral suckers. All of the intestinal trematodes require a freshwater snail to serve as an intermediate host; these infections are foodborne (acquired by ingestion of freshwater fish, mollusks, or plants) and are emerging as a major public health problem (more than 40 million people are infected with intestinal and liver or lung trematodes). Specific examples include *Fasciolopsis buski*, *Heterophyes heterophyes*, and *Metagonimus yokogawai*.

Trematodes (Liver and Lungs)

The hermaphroditic liver and lung trematodes also require a freshwater snail to serve as an intermediate host; these infections are foodborne (acquired by ingestion of freshwater fish, crayfish, crabs, or plants) (Figure 2.19). Public health concerns include cholangiocarcinoma associated with *Clonorchis* and *Opisthorchis* infections, severe liver disease

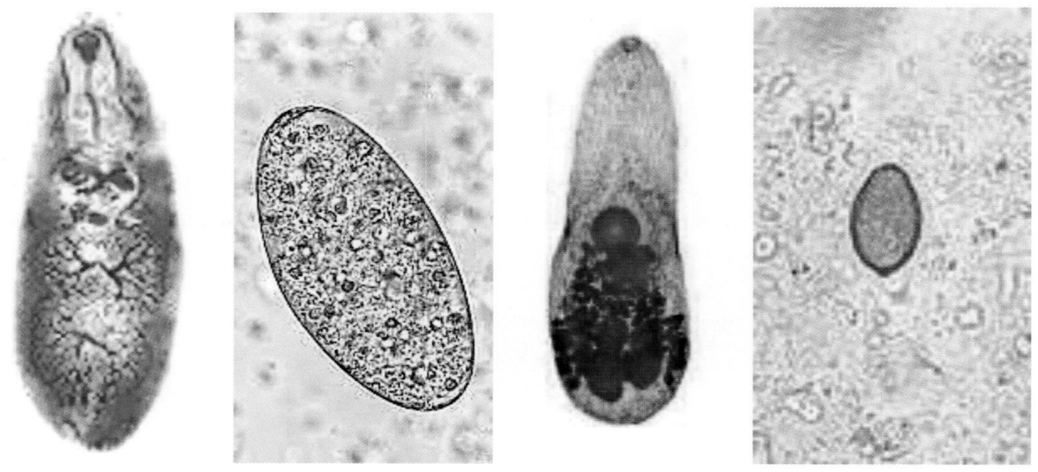

Figure 2.18 Trematodes (intestinal). *Fasciolopsis buski* adult fluke (giant intestinal fluke); *F. buski* fluke egg, which is quite large (130 to 140 by 80 to 85 μm); *Heterophyes heterophyes* adult fluke; *Heterophyes* fluke egg, which is quite small (27 to 30 by 15 to 17 μm).

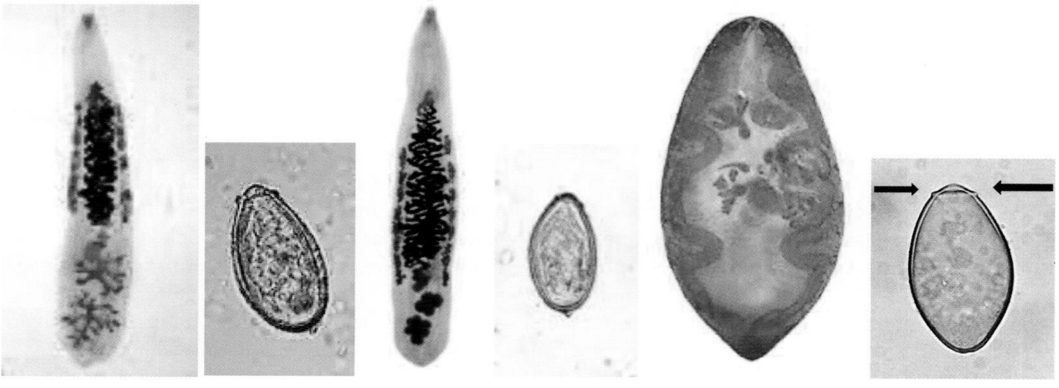

Figure 2.19 Trematodes (liver and lungs). *Clonorchis sinensis* adult fluke (which lives in bile ducts and is the cause of cholangiocarcinoma); *C. sinensis* egg (28 to 35 by 12 to 19 μm); *Opisthorchis viverrini* adult fluke (which also causes cholangiocarcinoma); *O. viverrini* egg (average, 27 by 15 μm); *Paragonimus* sp. adult fluke (found in lungs); *Paragonimus* egg (80 to 120 by 45 to 65 μm; eggs can be found in respiratory specimens) (note the opercular shoulders on the egg [arrows]; the operculum fits into the top like a teapot lid).

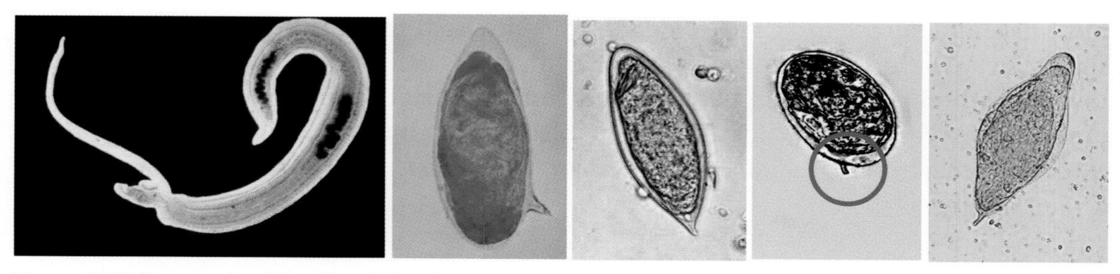

Figure 2.20 Trematodes (blood). Adult schistosomes (note that the slender female worm is lying in the gynecophoral canal of the thicker male; they mate for life); *Schistosome mansoni* egg (note the large lateral spine); *Schistosoma haematobium* egg (note the prominent terminal pointed spine); *Schistosoma japonicum* egg (note very small lateral spine [circle]); *Schistosoma intercalatum* egg (although this egg also has a terminal spine, note the bulge in the center of the egg). The eggs of *Schistosoma mekongi* are quite similar to those of *S. japonicum*, but smaller (50 to 80 by 40 to 65 μm) (no image shown).

associated with *Fasciola* infections, and the misdiagnosis of tuberculosis in those infected with *Paragonimus* spp.

Trematodes (Blood)

The sexes of blood trematodes (schistosomes) are separate, and infection is acquired by skin penetration by the cercarial forms that are released from freshwater snails (Figure 2.20). The males are characterized by having an infolded body that forms the gynecophoral canal in which the female worm is held during copulation and oviposition. The adult worms reside in the blood vessels over the small intestine, large intestine, or bladder. Although these parasites are not endemic in the United States, there are patients in the United States who may have acquired schistosomiasis elsewhere. Schistosomiasis is a chronic disease, and many infections are subclinically symptomatic, with mild anemia and malnutrition being common in areas where the infection is endemic. Acute schistosomiasis (Katayama's fever) may occur weeks after the initial infection, especially infection by *S. mansoni* and *S. japonicum*. Signs may include abdominal pain, cough, diarrhea, high eosinophilia, fever, fatigue, and hepatosplenomegaly. Representative species include *S. mansoni*, *S. japonicum*, *S. mekongi*, and *S. haematobium*.

Pentastomids

Pentastomids are found in a separate phylum, Pentastomida, and are called tongue worms (Figure 2.21). Human infections have been reported from Africa, Europe, Asia, and the Americas. When humans serve as the intermediate hosts, the infective larvae die *in situ*. However, when mature larvae (often encysted) are ingested, they may migrate from the stomach, attach themselves to nasopharyngeal tissues, develop into adult pentastomids, and cause symptoms of halzoun syndrome. Symptoms include throat discomfort, paroxysmal coughing, sneezing, and occasionally dysphagia and vomiting. Pentastomids isolated from humans include *Armillifer* spp., *Linguatula serrata*, and *Sebekia* spp.

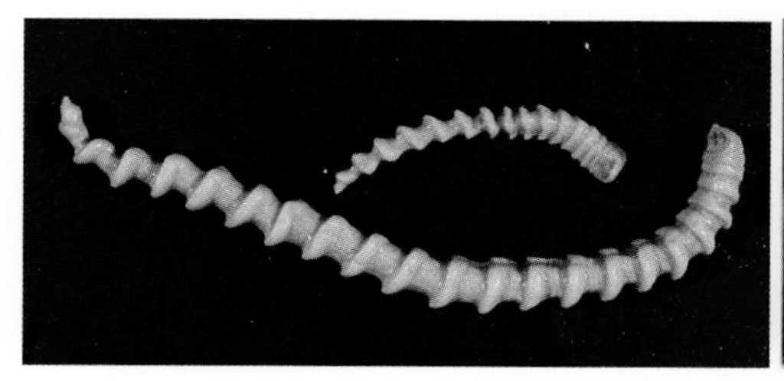

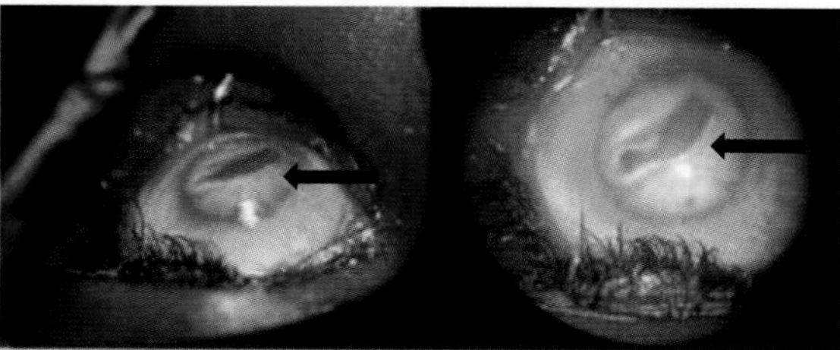

Figure 2.21 Pentastomids (tongue worms). (Left) Male (small) and female (large) *Armillifer* sp. Human parasitism by pentastomes is called pentastomiasis and may occur in the viscera, where immature forms develop in the liver, spleen, lungs, eyes, and other organs or in the nasopharyngeal area. *Linguatula serrata* occurs in canids, and in some parts of the Middle East, more than 50% of stray dogs are infected. These animals harbor adults and larvae in their upper respiratory tracts and shed ova in their nasal secretions, saliva, and feces onto wet vegetation or into water. Ova are ingested by herbivorous intermediate hosts and by humans, who can be infected directly by contact with dogs' nasal secretions or contaminated food, developing visceral pentastomiasis and becoming intermediate hosts. Humans can also be infected by ingesting infected offal from ruminants, in which case they develop nasopharyngeal pentastomiasis (also called halzoun), becoming temporary definitive hosts. (Right) Annulated foreign body in the anterior chamber of the eye (arrows). On removal, the pentastomid nymph was identified as *Armillifer grandis*. (Reprinted from Sulyok M, Rózsa L, Bodó I, et al, *PLoS Negl Trop Dis* **8**(7):e3041, 2014, https://doi.org/10.1371/journal.pntd.0003041. © 2014 Sulyok et al., CC-BY 4.0, https://creativecommons.org/licenses/by/4.0/.)

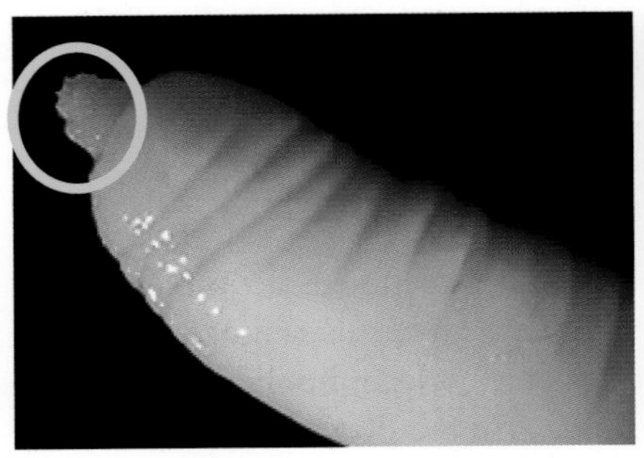
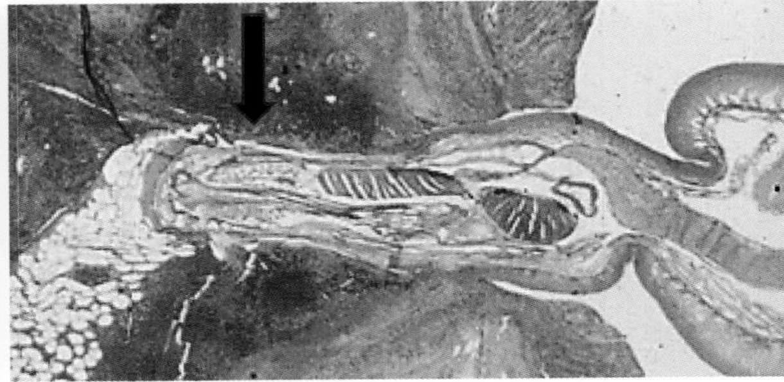

Figure 2.22 Acanthocephala (thorny-headed worms). (Left) *Macracanthorhynchus hirudinaceus* (note the cylindrical, retractive proboscis (yellow circle) that has rows of hooked spines (spines are difficult to see unless the proboscis is in a more protruded position and the magnification is increased). (Right) Cross-section of the intestine of a pig, stained with H&E, showing the anterior end of an adult *M. hirudinaceus* (arrow) (both images courtesy of the CDC Public Health Image Library).

Rare eye infections have been reported; inflammation is minimal, and the infection is probably the result of direct eye contact with water containing pentastomid eggs. Even more rare is human infection caused by a pentastomid larva belonging to the genus *Sebekia*. The adult worms are parasites found in the respiratory tract of reptiles, and the larval forms are found in the viscera, in the muscles, and along the spinal cord of freshwater fish. There may be serpiginous burrows surrounded by an intense erythematous zone and a 30% eosinophilia. The infection is possibly acquired by the ingestion of water containing eggs of the parasites or the ingestion of raw or improperly cooked fish. Diagnosis is made by identifying the pentastomid in a biopsy specimen or at autopsy.

Acanthocephala

The Acanthocephala (thorny-headed worms) are closely related to the tapeworms (Figure 2.22). These worms are diecious and tend to have a retractable proboscis, which is usually armed with spines. The larvae require an arthropod intermediate host, and the adult worms are always parasites in the intestines of vertebrates. Two of these organisms are parasitic in humans: *Macracanthorhynchus hirudinaceus* and *Moniliformis moniliformis*. Human infection is acquired from the ingestion of infected insects (various beetles and cockroaches). Very few cases have been reported in the literature; however, symptoms include abdominal pain and tenderness, anorexia, and nausea. In some cases, adult worms have been passed in the stool.

Suggested Reading

Beaver CB, Jung RC, Cupp EW. 1984. *Clinical Parasitology*. Lea & Febiger, Philadelphia, PA.

Carroll KC, Pfaller MA, Landry ML, McAdam AJ, Patel R, Richter SS, Warnock DW (ed). 2019. *Manual of Clinical Microbiology*, 12th ed. ASM Press, Washington, DC.

Garcia LS. 2016. *Diagnostic Medical Parasitology*, 6th ed. ASM Press, Washington, DC.

Gibson DI. 1998. Nature and classification of parasitic helminths, p 453–479. *In* Collier L, Balows A, Susman M (ed), *Topley & Wilson's Microbiology and Microbial Infections*, 9th ed. Oxford University Press, New York, NY.

Goddard J. 2007. *Arthropods of Medical Importance*, 5th ed. CRC Press, New York, NY.

Table 2.1 Classification of human parasites

- **I. Protozoa**
  - **A. Intestinal amebae**
    - *Entamoeba histolytica*
    - *Entamoeba dispar*[a]
    - *Entamoeba moshkovskii*
    - *Entamoeba bangladeshi*
    - *Entamoeba hartmanni*
    - *Entamoeba coli*
    - *Entamoeba polecki*
    - *Endolimax nana*
    - *Iodamoeba bütschlii*
    - *Blastocystis* spp. (Stramenopile)
  - **B. Intestinal flagellates**
    - *Giardia lamblia*[b]
    - *Chilomastix mesnili*
    - *Dientamoeba fragilis*
    - *Pentatrichomonas hominis*
    - *Enteromonas hominis*
    - *Retortamonas intestinalis*
  - **C. Intestinal ciliates**
    - *Balantidium coli* (*Neobalantidium coli*)
  - **D. Intestinal Apicomplexa, coccidia, and microsporidia**
    - **1. Apicomplexa**
      - *Cryptosporidium parvum*
      - *Cryptosporidium hominis*
      - *Cryptosporidium* spp.
    - **2. Coccidia**
      - *Cyclospora cayetanensis*
      - *Cystoisospora belli*
      - *Sarcocystis hominis*
      - *Sarcocystis suihominis*
    - **3. Microsporidia**
      - *Enterocytozoon bieneusi*
      - *Encephalitozoon* (*Septata*) *intestinalis*
  - **E. Sporozoa and flagellates from blood and tissue**
    - **1. Sporozoa (causing malaria and babesiosis)**
      - *Plasmodium vivax*
      - *Plasmodium ovale*
      - *Plasmodium malariae*
      - *Plasmodium falciparum*
      - *Plasmodium knowlesi*
      - *Babesia* spp.
    - **2. Flagellates (leishmaniae and trypanosomes)**
      - **a. Old World *Leishmania* species**
        - *Leishmania* (*Leishmania*) *tropica*
        - *Leishmania* (*Leishmania*) *major*
        - *Leishmania* (*Leishmania*) *aethiopica*
        - *Leishmania* (*Leishmania*) *donovani*
        - *Leishmania* (*Leishmania*) *archibaldi*
        - *Leishmania* (*Leishmania*) *infantum*
      - **b. New World *Leishmania* species**
        - *Leishmania* (*Leishmania*) *mexicana*
        - *Leishmania* (*Leishmania*) *amazonensis*
        - *Leishmania* (*Leishmania*) *pifanoi*
        - *Leishmania* (*Leishmania*) *garnhami*
        - *Leishmania* (*Leishmania*) *venezuelensis*
        - *Leishmania* (*Leishmania*) *chagasi*
        - *Leishmania* (*Viannia*) *braziliensis*
        - *Leishmania* (*Viannia*) *colombiensis*
        - *Leishmania* (*Viannia*) *guyanensis*
        - *Leishmania* (*Viannia*) *lainsoni*
        - *Leishmania* (*Viannia*) *naiffi*
        - *Leishmania* (*Viannia*) *panamensis*
        - *Leishmania* (*Viannia*) *peruviana*
        - *Leishmania* (*Viannia*) *shawi*
      - **c. Old World *Trypanosoma* species**
        - *Trypanosoma brucei gambiense*
        - *Trypanosoma brucei rhodesiense*
      - **d. New World *Trypanosoma* species**
        - *Trypanosoma cruzi*
        - *Trypanosoma rangeli*
  - **F. Amebae and flagellates from other body sites**
    - **1. Amebae**
      - *Naegleria fowleri*
      - *Sappinia diploidea*
      - *Acanthamoeba* spp.
      - *Hartmannella* spp.
      - *Balamuthia mandrillaris*
      - *Entamoeba gingivalis*
    - **2. Flagellates**
      - *Trichomonas vaginalis*
      - *Trichomonas tenax*
  - **G. Apicomplexa and microsporidia from other body sites**
    - **1. Apicomplexa**
      - *Toxoplasma gondii*
    - **2. Microsporidia**
      - *Nosema ocularum*
      - *Pleistophora* spp.
      - *Pleistophora ronneafiei*
      - *Trachipleistophora hominis*
      - *Trachipleistophora anthropophthera*
      - *Anncaliia vesicularum*
      - *Anncaliia* (*Brachiola*) *algerae*
      - *Anncaliia* (*Brachiola*) *connori*
      - *Encephalitozoon cuniculi*
      - *Encephalitozoon* (*Septata*) *intestinalis*
      - *Encephalitozoon hellem*
      - *Enterocytozoon bieneusi*
      - *Vittaforma corneae* (*Nosema corneum*)
      - *Microsporidium*[c]
- **II. Nematodes (roundworms)**
  - **A. Intestinal**
    - *Ascaris lumbricoides*
    - *Enterobius vermicularis*
    - *Ancylostoma duodenale*
    - *Necator americanus*
    - *Strongyloides stercoralis*
    - *Strongyloides fuelleborni*
    - *Trichostrongylus colubriformis*
    - *Trichostrongylus orientalis*
    - *Trichostrongylus* spp.
    - *Trichuris trichiura*
    - *Capillaria philippinensis*

Table continues on next page

Table 2.1 Classification of human parasites (*continued*)

B. Tissue
- *Trichinella spiralis*
- *Trichinella* spp. (*T. britovi, T. murrelli, T. nativa, T. nelsoni, T. papuae, T. pseudospiralis, T. zimbabwensis*)
- *Toxocara canis* and *Toxocara cati* (visceral and ocular larva migrans)
- *Ancylostoma braziliense* and *Ancylostoma caninum* (cutaneous larva migrans)
- *Dracunculus medinensis*
- *Angiostrongylus cantonensis*
- *Angiostrongylus costaricensis*
- *Gnathostoma spinigerum*
- *Anisakis* spp. (larvae from saltwater fish)
- *Phocanema* spp. (larvae from saltwater fish)
- *Contracaecum* spp. (larvae from saltwater fish)
- *Hysterothylacium*
- *Porrocaecum* spp.
- *Capillaria hepatica*
- *Thelazia* spp.
- *Ternidens diminutus*

C. Blood and tissues (filarial worms)
- *Wuchereria bancrofti*
- *Brugia malayi*
- *Brugia timori*
- *Loa loa*
- *Onchocerca volvulus*
- *Mansonella ozzardi*
- *Mansonella streptocerca*
- *Mansonella perstans*
- *Dirofilaria immitis* (usually found in lung lesions, eyes; in dogs, heartworm)

III. Cestodes (tapeworms)

A. Intestinal
- *Diphyllobothrium latum*
- *Diplogonoporus* spp.
- *Dipylidium caninum*
- *Hymenolepis nana*
- *Hymenolepis diminuta*
- *Taenia solium*
- *Taenia saginata*
- *Taenia asiatica*

B. Tissue (larval forms)
- *Taenia solium*
- *Echinococcus granulosus*
- *Echinococcus multilocularis*
- *Echinococcus vogeli*
- *Echinococcus oligarthrus*
- *Multiceps multiceps*
- *Spirometra mansonoides*
- *Diphyllobothrium* spp.

IV. Trematodes (flukes)

A. Intestinal
- *Fasciolopsis buski*
- *Echinostoma ilocanum*
- *Echinochasmus perfoliatus*
- *Heterophyes heterophyes*
- *Metagonimus yokogawai*
- *Gastrodiscoides hominis*
- *Phaneropsolus bonnei*
- *Prosthodendrium molenkempi*
- *Spelotrema brevicaeca*
- *Plagiorchis* spp.
- *Neodiplostomum seoulense*

B. Liver and lung
- *Clonorchis (Opisthorchis) sinensis*
- *Opisthorchis viverrini*
- *Opisthorchis felineus*
- *Dicrocoelium dendriticum*
- *Fasciola hepatica*
- *Fasciola gigantica*
- *Metorchis conjunctus*
- *Paragonimus westermani*
- *Paragonimus kellicotti*
- *Paragonimus africanus*
- *Paragonimus uterobilateralis*
- *Paragonimus miyazakii*
- *Paragonimus mexicanus*
- *Paragonimus caliensis*

C. Blood
- *Schistosoma mansoni*
- *Schistosoma haematobium*
- *Schistosoma japonicum*
- *Schistosoma intercalatum*
- *Schistosoma mekongi*
- *Schistosoma malayi*
- *Schistosoma matteei*

V. Pentastomids (tongue worms); see Arthropods below

A. Tissue (larval forms)
- *Armillifer* spp.
- *Linguatula serrata*
- *Sebekia* spp.

B. Nasopharyngeal
- *Armillifer* spp.
- *Linguatula serrata*

VI. Acanthocephalans (thorny-headed worms)

A. Intestine
- *Macracanthorhynchus hirudinaceus*
- *Moniliformis moniliformis*

VII. Arthropods

A. Arachnida
- Scorpions
- Spiders (black widow, brown recluse)
- Ticks (*Dermacentor, Ixodes, Argas,* and *Ornithodoros* spp.)
- Mites (*Sarcoptes* spp.)

B. Crustacea
- Copepods (*Cyclops* spp.)
- Crayfish, lobsters, and crabs

C. Pentastomida
- Tongue worms

D. Diplopoda
- Millipedes

E. Chilopoda
- Centipedes

F. Insecta
- Phthiraptera (sucking lice [*Pediculus* and *Phthirus* spp.])
- Blattaria (cockroaches)
- Hemiptera (true bugs [*Triatoma* spp.])
- Coleoptera (beetles)
- Hymenoptera (bees, wasps, etc.)
- Lepidoptera (butterflies, caterpillars, moths, etc.)
- Diptera (flies, mosquitoes, gnats, and midges [*Phlebotomus, Aedes, Anopheles, Glossina, Simulium* spp., etc.])
- Siphonaptera (fleas [*Pulex* and *Xenopsylla* spp., etc.])

Table 2.1 (continued)

[a] The name *Entamoeba histolytica* is used to designate the true pathogen, while the name *E. dispar* is now being used to designate the nonpathogen. However, unless trophozoites containing ingested red blood cells (*E. histolytica*) are seen, the two organisms cannot be differentiated on the basis of morphology seen in permanent stained smears of fecal specimens. Fecal immunoassays are available for detecting the *Entamoeba histolytica/E. dispar* group or for differentiating between the two species. *E. moshkovskii* and *E. bangladeshi* cannot be differentiated from *E. histolytica* with no ingested red blood cells or *E. dispar*. The *Entamoeba histolytica/E. dispar* complex or group includes *Entamoeba histolytica, E. dispar, E. moshkovskii*, and *E. bangladeshi*; however, *E. moshkovskii* and *E. bangladeshi* are not routinely reported.

[b] Although some individuals have changed the species designation for the genus *Giardia* to *G. duodenalis* or *G. intestinalis*, there is no general agreement. Therefore, for this listing, the name *Giardia lamblia* is retained.

[c] This designation was not considered a true genus but a catch-all for organisms that had not been assigned to the genus and/or species levels. However, *Microsporidium ceylonensis* and *Microsporidium africanum* are now accepted as members of the genus infecting humans.

Table 2.2 Cosmopolitan distribution of common parasitic infections[a]

Protozoa

- **Intestinal**
  - *Blastocystis* spp. (Stramenopile)
  - *Cryptosporidium parvum*
  - *Cryptosporidium hominis*
  - *Cyclospora cayetanensis*
  - *Dientamoeba fragilis*
  - *Entamoeba histolytica*
  - *Entamoeba dispar*[b]
  - *Giardia lamblia*
  - *Cystoisospora belli*
  - Microsporidia
- **Tissue**
  - *Toxoplasma gondii*
  - Microsporidia
- **Other**
  - *Acanthamoeba* spp.
  - Microsporidia
  - *Naegleria fowleri*
  - *Trichomonas vaginalis*
- **Cestodes**
  - *Hymenolepis nana*
  - *Taenia saginata*

Nematodes

- **Intestinal**
  - *Ascaris lumbricoides*
  - *Enterobius vermicularis*
  - Hookworm
  - *Strongyloides stercoralis*
  - *Trichuris trichiura*
- **Tissue**
  - *Trichinella* spp.

[a] These parasites are distributed across the world, being found in North America, Mexico, Central America, South America, Europe, Africa, Asia, and Oceania.

[b] *Entamoeba histolytica* is used to designate the true pathogenic species, while *E. dispar* is now being used to designate the nonpathogenic species. However, unless trophozoites containing ingested red blood cells (*E. histolytica*) are seen, organisms in the *Entamoeba histolytica/E. dispar* complex cannot be differentiated on the basis of morphology in the permanent stained smear. The *Entamoeba histolytica/E. dispar* complex (or group) includes *Entamoeba histolytica, E. dispar, E. moshkovskii*, and *E. bangladeshi*; however, *E. moshkovskii* and *E. bangladeshi* are not routinely reported. Fecal immunoassays are available for detecting the *Entamoeba histolytica/E. dispar* group or for differentiating the two species.

Table 2.3 Body sites and possible parasites recovered[a]

Site	Parasite(s)
Blood	
Red cells	*Plasmodium* spp.
	Babesia spp.
White cells	*Leishmania* spp.
	Toxoplasma gondii
Buffy coat	All blood parasites
Whole blood or plasma	*Trypanosoma* spp.
	Microfilariae
Bone marrow	*Leishmania* spp.
	Trypanosoma cruzi
	Plasmodium spp.
	Toxoplasma gondii
Central nervous system	*Taenia solium* (cysticerci)
	Echinococcus spp.
	Naegleria fowleri
	Acanthamoeba and *Hartmannella* spp.
	Balamuthia mandrillaris
	Sappinia diploidea
	Toxoplasma gondii
	Microsporidia
	Trypanosoma spp.
	Baylisascaris procyonis
	Angiostrongylus cantonensis
	Toxocara spp. (visceral larva migrans)
Cutaneous ulcers	*Leishmania* spp.
	Acanthamoeba spp.
	Entamoeba histolytica
Eyes	*Acanthamoeba* spp.
	Toxoplasma gondii
	Loa loa
	Microsporidia
	Dirofilaria spp.
	Thelazia spp.
	Toxocara spp. (ocular larva migrans)
Intestinal tract	*Entamoeba histolytica*
	Entamoeba dispar, *E. moshkovskii*, *E. bangladeshi*
	Entamoeba coli
	Entamoeba hartmanni
	Endolimax nana
	Iodamoeba bütschlii
	Blastocystis spp. (Stramenopile)
	Giardia lamblia
	Chilomastix mesnili
	Dientamoeba fragilis

Table 2.3 (continued)	
Site	**Parasite(s)**
Intestinal tract (continued)	*Pentatrichomonas hominis*
	Balantidium coli
	Cryptosporidium parvum
	Cryptosporidium hominis
	Cryptosporidium spp.
	Cyclospora cayetanensis
	Cystoisospora belli
	Enterocytozoon bieneusi
	Encephalitozoon intestinalis
	Ascaris lumbricoides
	Enterobius vermicularis
	Hookworm
	Strongyloides stercoralis
	Trichuris trichiura
	Hymenolepis nana
	Hymenolepis diminuta
	Taenia saginata
	Taenia solium
	Taenia asiatica
	Diphyllobothrium latum
	Clonorchis (*Opisthorchis*) *sinensis*
	Paragonimus spp.
	Schistosoma spp.
	Fasciolopsis buski
	Fasciola hepatica
	Metagonimus yokogawai
	Heterophyes heterophyes
	Angiostrongylus costaricensis
Liver and spleen	*Echinococcus* spp.
	Entamoeba histolytica
	Leishmania spp.
	Microsporidia
	Capillaria hepatica
	Clonorchis (*Opisthorchis*) *sinensis*
Lungs	*Cryptosporidium* spp.[b]
	Echinococcus spp.
	Paragonimus spp.
	Toxoplasma gondii
	Helminth larvae
Muscles	*Taenia solium*, *T. asiatica* (cysticerci)
	Trichinella spp.
	Onchocerca volvulus (nodules)
	Trypanosoma cruzi
	Microsporidia[c]

Table continues on next page

Table 2.3 Body sites and possible parasites recovered[a] (*continued*)

Site	Parasite(s)
Skin	*Leishmania* spp.
	Onchocerca volvulus
	Microfilariae
	Ancylostoma spp. (cutaneous larva migrans)
Urogenital system	*Trichomonas vaginalis*
	Schistosoma spp.
	Microsporidia
	Microfilariae

[a] This table does not include every possible parasite that could be found in a body site. However, the most likely organisms have been listed (trophozoites, cysts, oocysts, spores, adults, larvae, eggs, amastigotes, and trypomastigotes).

[b] Disseminated in severely immunosuppressed individuals.

[c] The genera *Pleistophora* and *Trachipleistophora* have been documented to occur in muscles.

SECTION 3

Collection Options

Practical Guide to Diagnostic Parasitology, Third Edition. Lynne S. Garcia
 DOI: 10.1128/9781683673637.ch03

Various collection methods are available for specimens suspected of containing parasites or parasitic elements (see Table 3.1). The decision about which method to use should be based on a thorough understanding of the value and limitations of each. The final laboratory results are based on parasite recovery and identification and will depend on the initial handling of the organisms. Unless the appropriate specimens are properly collected and processed, these infections may not be detected. Therefore, specimen rejection criteria have become much more important for all diagnostic microbiology procedures. Diagnostic laboratory results based on improperly collected specimens may require inappropriate expenditures of time and supplies and may also mislead the physician. **As a part of any continuous quality improvement program for the laboratory, the generation of test results must begin with stringent criteria for specimen acceptance or rejection.**

Clinically relevant testing also depends on the receipt of appropriate test orders from the physician (see Table 3.2). The laboratory is not authorized to order tests; this function is the responsibility of the physician. Depending on the patient's history, very specific diagnostic tests are recommended. It is very important that physician clients become familiar with the test order options available from the laboratory testing menu. They must also have an understanding of the pros and cons of each test when considered within the context of the patient's symptoms and clinical history. Without the appropriate test orders and collection procedures, test results may be misleading or even incorrect. **It is important for the laboratory to provide appropriate and complete information to all clients in order to ensure quality patient care.**

Safety

All fresh specimens should be handled carefully, since each specimen represents a potential source of infectious material. Safety precautions should include the following: proper labeling of fixatives; specific areas designated for specimen handling (biological safety cabinets may be necessary under certain circumstances); proper containers for centrifugation; acceptable discard policies; appropriate policies regarding eating, drinking, or smoking within the working areas; and, if applicable, correct techniques for organism culture and/or animal inoculation. In general, standard precautions as outlined by the Occupational Safety and Health Act must be followed when applicable, particularly when one is handling blood and other body fluids.

Collection of Fresh Stool Specimens

Stool specimens should always be collected before barium is used for radiological examination. Specimens containing barium are unacceptable for processing and examination, and intestinal protozoa may be undetectable for 5 to 10 days after barium is given to the patient. Certain substances and medications also interfere with the detection of intestinal protozoa; these include mineral oil, bismuth, antibiotics, antimalarial agents, and nonabsorbable antidiarrheal preparations. After administration of any of these compounds, parasites may not be recovered for one to several weeks. After the administration of barium or antibiotics, specimen collection should be delayed for 5 to 10 days or for at least 2 weeks, respectively.

Collection Method

Fecal specimens should be collected in clean, wide-mouth containers; often a waxed cardboard or plastic container with a tight-fitting lid is selected for this purpose. The specimens should not be contaminated with water or urine, because water may contain free-living organisms (including arthropod larvae and free-living nematodes) that can be mistaken for human parasites, and urine may destroy motile organisms. Stool specimen containers should be placed in plastic bags for transport to the laboratory for testing. The specimens should be labeled with the following information: the patient's name and identification number, the physician's name, and the date and time the specimen was collected (if the laboratory is computerized, the date and time may reflect arrival in the laboratory, not the actual collection time). The specimen must also be accompanied by a request form indicating which laboratory procedures are to be performed; in some cases, this information will be computerized and will be entered into the system in the nursing unit or in the clinic. Although it is helpful to have information concerning the presumptive diagnosis or relevant travel history, this information rarely accompanies the specimen. If such information is relevant and necessary to maximize diagnostic testing, the physician may have to be contacted for additional patient history.

Number of Specimens To Be Collected

Standard Approach

In the past, it was recommended that a normal examination for stool parasites before therapy include three specimens: two specimens collected from normal movements and one collected after the use of a cathartic such as magnesium sulfate or Fleet's Phospho-Soda. The purpose of the laxative is to stimulate some "flushing" action within the gastrointestinal tract. A cathartic with an oil base should not be used, and a stool softener (taken either orally or as a suppository) is usually inadequate for obtaining a purged specimen. The use of a laxative prior to collection of the third specimen is much less common than in years past. Also, if the patient already has diarrhea or dysentery, the use of any laxative would be contraindicated and uncommon.

When a patient is suspected of having intestinal amebiasis, collection of six specimens may be recommended. The examination of six specimens ensures detection of approximately 90% of amebic infections. However, considering cost containment measures, the examination of six specimens is rarely requested.

Other Approaches

During the past few years, recommendations regarding the collection, processing, and testing of stool specimens for diagnostic parasitology have been under review. New suggestions and options have arisen as a result of cost containment measures, limited reimbursement, and the elimination of mercury-based compounds for stool preservatives. The number of nonmercury preservative choices, collection systems, concentration devices, and immunoassays has increased dramatically.

It is important to realize that different laboratories will select different approaches. These differences should not be categorized as "right or wrong" or "acceptable or unacceptable"; they are merely different! To assume that there is only one correct approach for the examination of stool specimens is neither appropriate nor realistic. There are many factors to consider before selecting the approach for your own laboratory. Some of the considerations include (in no particular order) client base, physician ordering patterns, number of specimens received per month, cost, presence or absence of appropriate equipment, current and possible methods (including the new immunoassays such as enzyme immunoassay [EIA], fluorescent-antibody assay [FA], and rapid tests [membrane flow cartridge devices]), availability of expert microscopists, collection options, selection of preservative-stain combinations, reimbursement issues, client education, area of the world where the laboratory is located, and emphasis on the most common infections (helminth or protozoa or both) seen in that geographic location.

When laboratory test menus are being developed, the pros and cons of the approaches selected need to be thoroughly understood, and diagnostic tests, potential results, and reporting formats should be carefully explained to all clients. As an example, if the results of a stool examination are based on a concentration sediment examination only, this information must be conveyed to the physician. Many intestinal protozoa are missed when this diagnostic test approach is used, and it is important for the physician to recognize the limitations of such testing. Most physicians receive very little, if any, exposure to medical parasitology in medical school, and many newer physicians trained as generalists or family practitioners also have limited parasitology training or experience (see Table 3.1).

It is usually realistic to assume that patients are symptomatic if they are submitting stool specimens for diagnostic parasitology testing. In an excellent article by Hiatt et al., the premise tested was that a single stool specimen from a symptomatic patient would be sufficient to diagnose infections with intestinal protozoa. However, with additional stool examinations, the yield of intestinal protozoa from symptomatic patients increased dramatically (*Entamoeba histolytica*, 22.7% increase; *Giardia lamblia*, 11.3% increase; and *Dientamoeba fragilis*, 31.1% increase). This publication demonstrates the problems associated with performing only a single stool examination (using the ova and parasite examination [O&P exam]). If the patient becomes asymptomatic after examination of the first stool specimen, it may be acceptable to discontinue the series of stool examinations (this should be a clinician decision).

Available options are compared to the gold standard, which includes a series of three stool examinations (direct exam, concentration, and permanent stained smear for a fresh specimen; concentration and permanent stained smear for a preserved specimen). The single-specimen pros and cons are discussed above (symptomatic versus asymptomatic patients). A suggestion has been made to pool three specimens and to perform a single concentration and a single permanent stained smear on the pooled specimen sample. Depending on the number of positives and negatives, this may be a viable option, but some organisms may be missed. Another testing option for three pooled specimens is to perform a single concentration and three separate permanent stained smears. This approach would probably increase the yield of intestinal protozoa over the previous option. Another suggestion involves placing a sample of each of three stools in a single vial of preservative. This collection approach would require only a single vial, but it is very likely that the vial would be overfilled and that mixing and the ratio of fixative to stool would be inaccurate.

Fecal immunoassays could also be used for *G. lamblia*, the *E. histolytica*/*E. dispar* group, *E. histolytica*, or *Cryptosporidium* spp. However, information about patient history is rarely received with the clinical specimen; fecal testing based on the risk group or recent activity of the patient is impossible without sufficient information. Testing by immunoassay procedures should be performed on request; however, client education is critical for successful implementation of this approach. Collection and testing options and their pros and cons can be seen in Tables 3.1 and 3.2.

The Clinical Laboratory Standards Institute document *Procedures for the Recovery and Identification of Parasites from the Intestinal Tract* (approved guideline M28-2A) was updated in 2005. Various stool collection, processing, and testing options are also included in that publication.

In summary, laboratories performing diagnostic parasitology testing must decide on the test methods that are relevant for their own operations based on a number of variables mentioned above. It is unrealistic to assume or state that one approach is applicable for every laboratory; however, it is important to thoroughly understand the options within your test menu and to convey this information to your clients once your approach has been selected for implementation. Prior discussion with clients, written educational memos, meetings, and examples of revised report formats are highly recommended prior to implementation. **Based on available testing options, it is recommended that laboratories include both the O&P exam and various fecal immunoassays in their test menus.**

Collection Times

A series of three specimens should be collected on separate days. If possible, the specimens should be collected every other day; otherwise, the series of three specimens should be collected within no more than 10 days. If a series of six specimens is requested, these specimens should also be collected on separate days and within no more than 14 days. Many organisms, particularly the intestinal protozoa, are shed sporadically and do not appear in the stool in consistent numbers on a daily basis; thus, a series of two (minimum) or three specimens is recommended for an adequate examination. Multiple specimens should not be collected from the same patient on the same day. One possible exception would be a patient who has severe, watery diarrhea such that any organisms present might be missed because of a tremendous dilution factor related to fluid loss. These specimens should be accepted only after consultation with the physician. It is also not recommended for the three specimens to be submitted one each day for three consecutive days; however, use of this collection time frame would not be sufficient reason to reject the specimens.

Although three stool specimens are recommended, some laboratories are more willing to accept two specimens, primarily because of cost savings and the assumption that if the patient is symptomatic, the presence of any organisms is likely to be confirmed by testing two specimens. However, it is important that physicians and laboratorians understand the pros and cons of the two approaches. Both collection approaches are being used by diagnostic laboratories; however, statistically, three specimens remain more accurate than two.

Posttherapy Collection

Patients who have received treatment for a protozoan infection should be checked 3 to 4 weeks after therapy, and those treated for *Taenia* infections should be checked 5 to 6 weeks after therapy; these recommendations have been used for many years. If the patient remains asymptomatic, the posttherapy specimens may not be collected, often as a cost containment measure. If the patient becomes symptomatic again, additional specimens can be submitted. If fecal immunoassays are ordered and parasites are present, the patient should be tested 7 to 10 days posttherapy. It usually takes approximately a week for antigen to be eliminated from the stool. **It is important to remember that if the posttherapy specimens are collected too long after therapy, the presence of parasites or parasite antigen may represent a reinfection.**

Specimen Type, Stability, and Need for Preservation

Fresh specimens are required for the recovery of motile trophozoites (amebae, flagellates, or ciliates). The protozoan trophozoite stage is found in patients with diarrhea; the gastrointestinal tract contents move through the system too rapidly for cyst formation to occur. Once the stool specimen is passed from the body, trophozoites do not encyst but may disintegrate if not examined or preserved within a short time after passage. However, most helminth eggs and larvae, coccidian oocysts, and microsporidian spores can survive for extended periods. Since it is impossible to know which organisms might be present, it is recommended that the most conservative time limits be used for parasite preservation and recovery. Liquid specimens should be examined within 30 min of passage, not 30 min from the time they reach the laboratory. If this is not possible, the specimen should be placed in one of the available fixatives. Soft (semiformed) specimens may contain a mixture of protozoan trophozoites and cysts and should be examined within 1 h of passage; again, if this time frame is not possible, preservatives should be used. Immediate examination of formed specimens is not as critical; in fact, if the specimen is examined any time within 24 h after passage, the protozoan cysts should still be intact (see Table 3.3).

Currently, fresh or frozen fecal specimens are required for fecal immunoassays for the *E. histolytica*/*E. dispar* group and *E. histolytica* (either as a single-organism test or when combined with other organisms such as *G. lamblia* or *Cryptosporidium* spp.).

Summary: Collection of Fresh Stool Specimens	
Advantages	**Disadvantages**
There are certain procedures that can be performed on fresh stool specimens only; however, they tend to be specialized procedures and not usually required.	Occupational Safety and Health Act regulations (standard precautions) should be used for handling all specimens; pathogens may be present without the use of fecal fixatives.
Trophozoite motility may be visible.	Motile organisms are not always seen; direct wet mounts are much less sensitive than permanent stained smears.
Interfering substances (e.g., barium, mineral oil, or antibiotics) should be avoided when fresh stool specimens are collected.	Unfortunately, the laboratory may not have this information; test results will generally be poor.
No need to handle fecal preservatives	Contamination of fresh stool with urine or water should be avoided; the use of fecal preservatives may actually be easier than collecting/handling fresh stool.
Recommendation for collection: two (minimum) or three specimens collected, one every other day or within a 10-day time frame; see Table 3.1 for options and pros/cons.	Again, the use of fecal fixatives may be recommended over collection of fresh stools.
Liquid stool should be examined or preserved within 30 min of passage (trophozoites). Soft stool should be examined or preserved within 1 h of passage (trophozoites and cysts[a]). This approach maximizes the ability to see motile trophozoites. Formed stool should be examined or preserved within 24 h of passage.	These time limits are rarely met; thus, the use of fecal preservatives is recommended.
Fresh or frozen fecal specimens are required for *E. histolytica*/*E. dispar* group and *E. histolytica* fecal immunoassays (either as a single-organism test or combined with other organisms, such as *G. lamblia* or *Cryptosporidium* spp.). Fresh, frozen, or preserved specimens can be used for *G. lamblia* and *Cryptosporidium* spp.; specimens submitted in Cary-Blair transport medium are also acceptable.	Due to the freeze-thaw cycle, frozen specimens cannot be used for the FA procedures for *G. lamblia* and/or *Cryptosporidium* spp. because of destruction of the actual cysts and/or oocysts that are visual proof of a positive specimen.

[a] *Dientamoeba fragilis* trophozoites can be found in formed stool specimens.

Preservation of Stool Specimens

Overview of Preservatives

If there are likely to be delays from the time of specimen passage until examination in the laboratory, the use of preservatives is recommended. To preserve protozoan morphology and to prevent the continued development of some helminth eggs and larvae, the stool specimens can be placed in preservative either immediately after passage (by the patient using a collection kit) or once the specimen is received by the laboratory. Several fixatives are available: formalin, sodium acetate-acetic acid-formalin (SAF), Schaudinn's fluid, and single-vial systems (see Table 3.3). Many of the fixative options contain polyvinyl alcohol (PVA), which is a plastic powder that serves as an adhesive to "glue" the stool onto the glass slide; PVA is inert and has no preservative qualities. **Regardless of the fixative selected, use of the appropriate ratio of fixative to stool (3 parts fixative to 1 part stool) and adequate mixing of the specimen and preservative are mandatory.** Although many products are commercially available, the most commonly used preservatives are discussed below. They are all available from various scientific supply houses.

When selecting an appropriate fixative, keep in mind that a permanent stained smear is required for a complete examination for parasites. If the physician orders a fecal immunoassay such as FA, EIA, or the rapid-flow method, you will need to confirm that the fixative you are using is compatible with the immunoassay you have selected. It is also important to remember that disposal regulations for compounds containing mercury are becoming more strict; each laboratory must check applicable state and federal regulations to help determine fixative options.

Formalin

Formalin is an all-purpose fixative that is appropriate for helminth eggs and larvae and for protozoan cysts. Two concentrations are commonly used: 5%, which is recommended for preservation of protozoan cysts, and 10%, which is recommended for helminth eggs and larvae. Although 5% is often recommended for all-purpose use, most commercial manufacturers provide 10%, which is more likely to kill all helminth eggs. To help maintain organism morphology, the formalin can be buffered with sodium phosphate buffers, i.e., neutral formalin. Selection of specific formalin formulations is at the user's discretion. Aqueous formalin permits examination of the specimen as a wet mount only, a technique much less accurate than a stained smear for the identification of intestinal protozoa. The most common preparation is 10% formalin, prepared as follows.

Formaldehyde (USP) .... 100 ml (or 50 ml for 5%)
Saline solution, 0.85% NaCl .... 900 ml (or 950 ml for 5%)

Dilute 100 ml of formaldehyde with 900 ml of 0.85% NaCl solution (distilled water may be used instead of saline solution).

Note: Formaldehyde is normally purchased as a 37 to 40% HCHO solution; however, for dilution, it should be considered to be 100%.

If you want to use buffered formalin, the recommended approach is to mix thoroughly 6.10 g of Na_2HPO_4 and 0.15 g of NaH_2PO_4 and store the dry mixture in a tightly closed bottle. For 1 liter of either 10 or 5% formalin, 0.8 g of the buffer salt mixture should be added.

Protozoan cysts (not trophozoites), coccidian oocysts, microsporidian spores, and helminth eggs and larvae are well preserved for long periods in 10% aqueous formalin. Hot (60°C) formalin can be used for specimens containing helminth eggs, since in cold formalin some thick-shelled eggs may continue to develop, become infective, and remain viable for long periods; however, this approach is not practical for routine clinical laboratories. Several grams of fecal material should be thoroughly mixed in 5 or 10% formalin.

Summary: Formalin	
Advantages	**Disadvantages**
Good overall fixative for stool concentration	Does not preserve trophozoites well; must be used with a second vial for the permanent stain or a single-vial system can be substituted for formalin vial (single-vial systems can be used for both concentration and permanent stained smear)
Easy to prepare, long shelf life	Does not adequately preserve organism morphology for an acceptable permanent stained smear
Formalinized stool can be used with some of the immunoassay detection kits (*Giardia lamblia* and *Cryptosporidium* spp.)	Currently cannot be used with fecal immunoassay methods for the *Entamoeba histolytica*/*E. dispar* group or *Entamoeba histolytica* (true pathogen)
Once formalin use has been monitored for formalin vapor, levels do not have to be rechecked unless the number of specimens processed dramatically increases or formalin use within the microbiology laboratory is modified. **(Note: Formalin use within microbiology laboratories rarely comes close to the allowable limits.)**	Not all laboratories are aware of the regulations for monitoring formalin exposure.

Sodium Acetate-Acetic Acid-Formalin (SAF)

Both the concentration and the permanent stained smear can be performed from specimens preserved in SAF, and the formula has the advantage of not containing mercuric chloride, as is found in Schaudinn's fluid and mercuric chloride-based fixatives containing PVA. It is a liquid fixative, much like the 10% formalin described above. The sediment is used to prepare the permanent smear, and it is frequently recommended that the stool material be placed on an albumin-coated slide to improve adherence to the glass.

SAF is considered a "softer" fixative than Schaudinn's fluid or mercuric chloride-based PVA-containing fixatives. The organism morphology is not quite as sharp after staining as that of organisms originally fixed in solutions containing mercuric chloride. **The pairing of SAF-fixed material with iron hematoxylin staining provides better organism morphology than does staining of SAF-fixed material with trichrome unless a specialized trichrome method specifically for use with SAF is available.** Although SAF has a long shelf life and is easy to prepare, the smear preparation technique may be a bit more difficult for less experienced personnel who are not familiar with fecal specimen techniques. Laboratories that have considered using only a single preservative have selected this option. Helminth eggs and larvae, protozoan trophozoites and cysts, and coccidian oocysts and microsporidian spores are preserved using this method.

▶ **SAF fixative**

Sodium acetate . . . . 1.5 g
Acetic acid, glacial . . . . 2.0 ml
Formaldehyde, 37 to 40% solution . . . . . 4.0 ml
Distilled water . . . . 92.0 ml

To make Mayer's albumin, mix equal parts of egg white and glycerin. Place 1 drop on a microscope slide and add 1 drop of SAF-preserved fecal sediment (from the concentration procedure). After mixing, allow the smear to dry at room temperature for 30 min prior to staining. Albumin can also be obtained commercially.

Summary: SAF	
Advantages	**Disadvantages**
Can be used for concentration and permanent stained smears	Poor adhesive properties; albumin-coated slides recommended; there may be some adhesion problems unless smears are thoroughly dry prior to staining
Contains no mercury compounds	Protozoan morphology is better if iron hematoxylin stains are used for permanent stained smears (trichrome fair); a trichrome modified for use with SAF is also available but not commonly used.
Easy to prepare; long shelf life	May be a bit more difficult to use; however, this does not seem to be a limiting factor
SAF-preserved stool can be used with the new immunoassay methods (*Giardia* and *Cryptosporidium*)	Contains formalin; some laboratories are trying to remove formalin from the microbiology section; formalin may interfere with some molecular testing options, including the multiplex panels (bacterial, viral, and parasitic organisms)
One-vial system, less expensive	

Schaudinn's Fluid

Schaudinn's fluid (which contains mercuric chloride) was one of the original stool fixatives and is used with fresh stool specimens or samples from the intestinal mucosal surface. Many laboratories that receive specimens from in-house patients (which have fewer problems with delivery times) often select this approach. Permanent stained smears are then prepared from fixed material. A concentration technique using Schaudinn's fluid-preserved material is also available but is not widely used. Due to the difficulties (sources and cost) related to mercury disposal, this fixative is being phased out by most laboratories. Although mercury substitutes are available, the overall protozoan morphology may not be as precise as that seen when mercury-based fixatives are used. However, the key question is, "Can the organisms present be identified?," not "How much better is the morphology compared to that seen with mercury-based fixatives?"

▶ **Mercuric chloride, saturated aqueous solution**

Mercuric chloride ($HgCl_2$) .... 110 g
Distilled water .... 1,000 ml

Use a beaker as a water bath; boil (use a hood if available) until the mercuric chloride is dissolved; let stand several hours until crystals form.

▶ **Schaudinn's fixative (stock solution)**

Mercuric chloride, saturated aqueous solution ... 600 ml
Ethyl alcohol, 95% .... 300 ml

Immediately before use, add glacial acetic acid, 5 ml/100 ml of stock solution.

Summary: Schaudinn's Fluid	
Advantages	**Disadvantages**
Fixative for smears prepared from fresh fecal specimens or samples from the intestinal mucosal surfaces	Not generally recommended for use in concentration procedures
Provides excellent preservation of protozoan trophozoites and cysts	Contains mercuric chloride, creating a disposal problem; being phased out for most laboratories
For many years considered the gold standard	Poor adhesive qualities with liquid or mucoid specimens

Schaudinn's Fluid Containing PVA (Mercury Base)

PVA is a plastic resin that can be incorporated into Schaudinn's fixative (or other liquid fixatives). **The PVA powder is not a fixative but serves as an adhesive for the stool material; i.e., when the stool-PVA mixture is spread onto the glass slide, it adheres because of the PVA component.** Fixation is still accomplished by the Schaudinn's fluid itself. Perhaps the greatest advantage in the use of PVA is the fact that a permanent stained smear can be prepared. Fixatives containing PVA are highly recommended as a means of preserving cysts and trophozoites for later examination. The use of fixative-PVA also permits specimens to be shipped (by regular mail service) from any location in the world to a laboratory for subsequent examination. PVA is particularly useful for liquid specimens and should be used at a ratio of 3 parts PVA to 1 part fecal specimen. The formula is as follows.

PVA	10.0 g
Ethyl alcohol, 95%	62.5 ml
Mercuric chloride, saturated aqueous	125.0 ml
Acetic acid, glacial	10.0 ml
Glycerin	3.0 ml

Mix the liquid ingredients in a 500-ml beaker. Add the PVA powder (stirring is not recommended). Cover the beaker with a large petri dish, heavy wax paper, or foil and allow the PVA to soak overnight. Heat the solution slowly to 75°C. When this temperature is reached, remove the beaker and swirl the mixture for 30 s until a homogeneous, slightly milky solution is obtained.

Summary: Schaudinn's Fluid Containing PVA (Mercury Base)

Advantages	Disadvantages
Can be used to prepare permanent stained smears and perform concentration techniques (see also Disadvantages). Throughout the discussion on fixatives, PVA refers to LV-PVA (low-viscosity PVA).	*Trichuris trichiura* eggs and *Giardia lamblia* (*G. duodenalis, G. intestinalis*) cysts are not concentrated as easily as from formalin-based fixatives; *Strongyloides stercoralis* larval morphology is poor (better to use formalin-based preservation); *Cystoisospora* (*Isospora*) *belli* oocysts may not be visible from preservatives containing PVA plastic powder (adhesive); usually combined with a formalin collection vial
Provides excellent preservation of protozoan trophozoites and cysts	Contains mercury compounds (Schaudinn's fluid)
Long shelf life (months to years) in tightly sealed containers at room temperature	PVA may turn white and gelatinous when it begins to dehydrate or when refrigerated.
Allows specimens to be shipped to any laboratory for subsequent examination	Difficult to prepare in the laboratory
Specimens preserved in fixatives containing PVA cannot be used with fecal immunoassay kits; also, they are not compatible with the multiplex organism panels.	For many reasons, this fixative is being phased out in most laboratories.

Schaudinn's Fluid Containing PVA (Copper Base, Zinc Base)

There has been a great deal of interest in developing preservatives without the use of mercury compounds, and substitute compounds now provide the quality of preservation necessary for comparable protozoan morphology on the permanent stained smear. Copper sulfate has been tried but does not provide results equal to those seen with mercuric chloride. However, zinc sulfate has proven to be a good mercury substitute and is used with trichrome stain or iron hematoxylin. Although zinc substitutes have become widely available, each manufacturer has a proprietary formula for the fixative.

Note: The important question is not, "How beautiful are the organisms?" but, "Can you tell which organisms are present?" With some training, microscopists can identify the organisms, although the morphology is not as clear as that seen using mercury compounds. Unfortunately, parasitology microscopy is not a perfect science; we probably miss rare organisms even when using mercury-based fixatives.

Summary: Schaudinn's Fluid Containing PVA (Copper Base, Zinc Base)	
Advantages	**Disadvantages**
Can prepare permanent stained smears and perform concentration techniques	Overall protozoan morphology of trophozoites and cysts is poor when they are preserved in the copper sulfate-based fixative, particularly compared with that of organisms preserved with mercuric chloride-based fixatives.
Many workers prefer the zinc substitutes over those prepared with copper.	Zinc-based fixatives are a better alternative than copper.
Does not contain mercury compounds. Cost per vial and selected stain vary considerably.	Staining characteristics of protozoa is not consistent (some good, some poor); organism identification may be more difficult, particularly with small protozoan cysts present in low numbers.

Single-Vial Collection Systems (Other than SAF)

Several manufacturers now have available single-vial stool collection systems, similar to the SAF or modified fixatives containing PVA. Some of these formulations are advertised as "ecologically friendly" and do not contain either mercury or formalin. From the single vial, both the concentration and permanent stained smear can be prepared. It is also possible to perform fecal immunoassays (EIA, FA, or the rapid-flow assay) with samples from some of these vials (those not containing PVA). Be sure to ask the manufacturer about all three assays (concentrate, permanent stained smear, and immunoassay procedures) and for specific information indicating that there are no formula components that would interfere with any of the three methods. Like the zinc substitutes, these formulas are proprietary.

Summary: Single-Vial Collection Systems (Other than SAF) (with or without PVA)	
Advantages	**Disadvantages**
Can be used to prepare permanent stained smears and perform concentration techniques	Overall protozoan morphology of trophozoites and cysts is not as good as that of organisms preserved with mercuric chloride-based fixatives; morphology similar to that seen with modified PVA options (zinc base, NOT copper base).
Can be used in fecal immunoassay procedures (with some exceptions)	Staining characteristics of protozoa are not consistent (some good, some poor); identification of *Endolimax nana* cysts may be difficult (fixatives containing PVA are not compatible with fecal immunoassays).
Do not contain mercury compounds (Ecofix, Unifix, STF, and others that may be available)	**Some of the commercially available single-vial options may not perform as advertised; it is important to see examples of intestinal protozoa preserved and stained from these fixatives; ask for other user contacts prior to changing.**

Universal Fixative (Total-Fix)

The Total-Fix stool collection kit is a single-vial system that provides a standardized method for untrained personnel to properly collect and preserve stool specimens for the detection of helminth larvae and eggs, protozoan trophozoites and cysts, coccidian oocysts, and microsporidian spores. Concentrations, permanent stains, most fecal immunoassays, and some molecular methods (including some of the multiplex panels) can be performed from a Total-Fix-preserved specimen. Total-Fix is a mercury-, formalin-, and PVA-free fixative that preserves parasite morphology and helps with disposal and monitoring problems encountered by laboratories (www.med-chem.com [accessed 10 June 2019]). Total-Fix is similar to Unifix and Z-PVA (zinc PVA), common fixatives that have been commercially available and used in many laboratories since 1992. This fixative does not require PVA or albumin adhesives to stick the fecal material onto the glass slide.

Some of the automated multiplex panel platforms have been testing parasitology specimens preserved in Total-Fix; this fixative will be included in some of the new package inserts for the parasitology panels.

Summary: Universal Fixative (Total-Fix)	
Advantages	**Disadvantages**
Can be used to prepare permanent stained smears and perform concentration technique. **The description of "Universal Fixative" indicates a comprehensive range of applications for parasitology stool methods.**	Organism morphology will not always equal that seen with the mercuric chloride gold standard; however, for the most part, organism morphology is comparable to that of organisms preserved using mercury.
Overall protozoan morphology of trophozoites and cysts is quite good and in some cases equals results obtained with mercuric chloride-based fixatives.	Specimen processing for the routine O&P exam is a bit different, but not complicated.
Can be used for fecal immunoassays (EIA, FA, rapid cartridges) and special stains for coccidia and microsporidia (modified acid-fast, modified trichrome)	Be sure to read the package inserts; some validation may be required if switching from another fixative.
Does not contain mercury compounds or formalin; does not require PVA or albumin for adhesive	Fecal smears must be absolutely dry prior to staining (to be effective without using any adhesive [PVA or albumin]).
Can be used for some molecular testing; preliminary results (including with the automated multiplex panel platforms) have been excellent; some new panels will include Total-Fix in the package inserts.	Be sure to check with the manufacturer; some validation may be required.
Single-vial system; less costly than the two-vial approach	

Quality Control for Preservatives

Preservatives for fecal specimens are checked for quality control by the manufacturer prior to sale, generally with living protozoa. If you prepare your own fixatives, you can use the following approach for quality control. The specimen used for quality control presented below is designed to be used with fixatives from which permanent stained smears will be prepared (Schaudinn's fluid, fixatives containing PVA or SAF, or the single-vial systems). However, this quality control specimen can also be used in a concentration; the white blood cells (WBCs) can be seen in the concentrate sediment (sedimentation concentration) or in the surface film (flotation concentration).

1. Obtain a fresh anticoagulated blood specimen, centrifuge, and obtain a buffy coat sample (try to find a specimen with a high WBC count).
2. Mix approximately 2 g of soft, fresh fecal specimen (normal stool, containing no parasites) with several drops of the buffy coat cells.
3. Prepare several fecal smears and fix immediately in Schaudinn's fluid to be quality controlled.
4. Mix the remaining mixture of feces and buffy coat in 10 ml of preservative to be quality controlled.
5. Allow 30 min for fixation, and then prepare several fecal smears. Allow to dry thoroughly (60 min at room temperature or 30 to 60 min in an incubator [approximately 35°C]).
6. Stain slides by normal staining procedure.
7. After staining, if the WBCs appear well fixed and display typical morphology and color (cell structural details, stain intensity), one can assume that any intestinal protozoa placed in preservative from the same lot would also be well fixed, provided that the fecal sample was fresh and fixed within the recommended time limits.
8. Record all quality control results. If the WBC morphology does not confirm good fixation, then describe the results and indicate what corrective action procedures were used (e.g., repetition of the test, preparation of new fixative).

Procedure Notes for Use of Preservatives (Stool Fixative Collection Vials)

1. Most of the commercially available kits have a "fill to" line on the vial label to indicate how much fecal material to add to ensure adequate preservation of the fecal material. However, patients often overfill the vials; remember to turn the vials away from your face when you open them. There may be excess gas in the vials that may create aerosols once the vial lids are opened.
2. Although the two-vial system (one vial of 5 or 10% buffered formalin [concentration] and one vial of mercuric chloride-based fixative containing PVA [permanent stained smear]) has always been the gold standard, most laboratories are now using other options, such as single-vial collection systems. Changes in the selection of fixatives are based on the following considerations.
   a. Problems with disposal of mercury-based fixatives and the lack of multilaboratory contracts for disposal of such products
   b. The cost of a two-vial system compared with the cost of a single collection vial
   c. Selection of specific stains (trichrome or iron hematoxylin) to use with specific fixatives
   d. Whether the fecal immunoassay kits (EIA, FA, or rapid cartridges [*Giardia*, *Cryptosporidium*, and *E. histolytica*]) and/or multiplex panel systems including parasites can be used with stool specimens preserved with that particular fixative

Procedure Limitations for Use of Preservatives (Stool Fixative Collection Vials)

1. Adequate fixation still depends on the following factors:
   a. Meeting recommended time limits for lag time between passage of the specimen and fixation

b. Use of the correct ratio of fixative to specimen (3:1)

c. Thorough mixing of the fixative and specimen (**once the specimen is received in the laboratory, any additional mixing will not counteract the lack of fixative-specimen mixing and contact prior to that time**)

2. Unless the appropriate stain is used with each fixative, the final permanent stained smear may be difficult to examine (organisms may be hard to see and/or identify). Examples of appropriate combinations include the following:

a. Schaudinn's fluid or fixatives including PVA with trichrome or iron hematoxylin stain

b. SAF fixative with iron hematoxylin stain

c. Single-vial mercuric chloride substitute systems with trichrome, iron hematoxylin, or company-developed proprietary stains matched to their specific fixatives

Collection of Blood

Collection and Processing

Depending on the parasite life cycle, a number of parasites may be recovered in a blood specimen, either whole blood, buffy coat preparations, or various types of concentrations. These parasites include *Plasmodium*, *Babesia*, *Trypanosoma*, *Leishmania donovani*, and microfilariae. Although some organisms may be motile in fresh whole blood, species are normally identified from the examination of permanent stained blood films, both thick and thin films (see Table 3.4). Blood films can be prepared from fresh whole blood collected with no anticoagulants, anticoagulated blood, buffy coat cells, or sediment from the various concentration procedures (see Table 3.5).

Blood can be collected by either finger stick or venipuncture. Venous blood should be collected in a tube containing EDTA. Multiple thick and thin blood films from the blood or buffy coat should be prepared and examined immediately after receipt of the blood by the laboratory, and multiple blood examinations should be performed before blood-borne parasite infection is ruled out. Unless you are positive that you will receive well-prepared thick and thin blood films from finger stick blood, request a tube of anticoagulated blood (EDTA anticoagulant [lavender top] is preferred). The tube should be filled with blood to provide the proper blood/anticoagulant ratio. For detection of stippling, the smears should be prepared within 1 h after the specimen is drawn. After that time, stippling may not be visible on stained films; however, the overall organism morphology is still acceptable. Most laboratories routinely use commercially available blood collection tubes; preparation of EDTA collection tubes in-house is neither necessary nor cost-effective. However, if the need should arise, EDTA (Sequestrene) can be prepared and tubed as follows: dissolve 5 g of EDTA in 100 ml of distilled water, aliquot 0.4 ml into tubes, and evaporate the water. This amount of anticoagulant is sufficient for 10 ml of blood. One can also use 20 mg of EDTA (dry) per tube (20 mg/10 ml of blood).

The time when the specimen was drawn should be clearly indicated on the tube of blood and also on the result report. The physician will then be able to correlate the results with any fever pattern or other symptoms that the patient may have. There should also be a comment on the test result report that is sent back to the physician that one negative specimen does not rule out the possibility of a parasitic infection.

Stat Test Requests and Risk Management Issues

All requests for malaria diagnosis are considered stat requests, and specimens should be ordered, collected, processed, examined, and reported accordingly. All requests for examination of blood films should include a possible diagnosis of malaria; thus, these requests are always considered stat (ordering, collection, processing, examination, and reporting). Not only should the blood collection be considered stat, but also the processing and examination of both thick and thin blood films should be performed immediately on receipt of the blood (see Table 3.6). Immunologically naive individuals with no prior exposure to malaria often present to the emergency room or clinic with symptoms such as fever and malaise and a relevant travel history to an area of the world where malaria is endemic. **These patients often have vague symptoms, but they have the potential to become very ill with malaria, even with a low parasitemia (0.0005% to 0.1%).**

Collection of Specimens from Other Body Sites

Although clinical specimens for examination can be obtained from many other body sites, these specimens and the appropriate diagnostic methods are not used as commonly as routine stool specimens. The majority of specimens from other body sites (see Table 3.7) would be submitted as fresh specimens for further testing. More information is given in the discussion of specific procedures in Section 5.

Suggested Reading

Beaver PC, Jung RC, Cupp EW. 1984. *Clinical Parasitology*, 9th ed. Lea & Febiger, Philadelphia, PA.

Brooke MM, Goldman M. 1949. Polyvinyl alcohol-fixative as a preservative and adhesive for protozoa in dysenteric stools and other liquid material. *J Lab Clin Med* **34:**1554–1560.

Cartwright CP. 1999. Utility of multiple-stool-specimen ova and parasite examinations in a high-prevalence setting. *J Clin Microbiol* **37:**2408–2411.

Federal Register. 1991. Occupational exposure to bloodborne pathogens. *Fed Regist* 29CFR1910.1030.

Garcia LS. 2016. *Diagnostic Medical Parasitology*, 6th ed. ASM Press, Washington, DC.

Garcia LS, Johnston SP, Linscott AJ, Shimizu RY. 2008. *Cumitech 46, Laboratory procedures for diagnosis of blood-borne parasitic diseases.* Coordinating ed, Garcia LS. ASM Press, Washington, DC.

Garcia LS, Smith JW, Thomas KB, Fritsche R. 2003. *Cumitech 30A, Selection and use of laboratory procedures for diagnosis of parasitic infections of the gastrointestinal tract.* Coordinating ed, Garcia LS. ASM Press, Washington, DC.

Garcia LS, Shimizu RY, Brewer TC, Bruckner DA. 1983. Evaluation of intestinal parasite morphology in polyvinyl alcohol preservative: comparison of copper sulfate and mercuric chloride bases for use in Schaudinn fixative. *J Clin Microbiol* **17:**1092–1095.

Garcia LS, Shimizu RY, Bruckner DA. 1986. Blood parasites: problems in diagnosis using automated differential instrumentation. *Diagn Microbiol Infect Dis* **4:**173–176.

Garcia LS, Shimizu RY, Shum A, Bruckner DA. 1993. Evaluation of intestinal protozoan morphology in polyvinyl alcohol preservative: comparison of zinc sulfate- and mercuric chloride-based compounds for use in Schaudinn's fixative. *J Clin Microbiol* **31:**307–310.

Hiatt RA, Markell EK, Ng E. 1995. How many stool examinations are necessary to detect pathogenic intestinal protozoa? *Am J Trop Med Hyg* **53:**36–39.

Horen WP. 1981. Modification of Schaudinn fixative. *J Clin Microbiol* **13:**204–205.

Kehl KSC. 1996. Screening stools for *Giardia* and *Cryptosporidium:* are antigen tests enough? *Clin Microbiol Newsl* **18:**133–135.

Melvin DM, Brooke MM. 1982. Laboratory procedures for the diagnosis of intestinal parasites, 3rd ed. U.S. Department of Health, Education, and Welfare publication no. (CDC) 82-8282. Government Printing Office, Washington, DC.

NCCLS. 2000. Laboratory diagnosis of blood-borne parasitic diseases. *Approved guideline M15-A*. NCCLS, Wayne, PA.

Scholten TH, Yang J. 1974. Evaluation of unpreserved and preserved stools for the detection and identification of intestinal parasites. *Am J Clin Pathol* **62:**563–567.

Yang J, Scholten T. 1977. A fixative for intestinal parasites permitting the use of concentration and permanent staining procedures. *Am J Clin Pathol* **67:**300–304.

Table 3.1 Fecal specimens for parasites: options for collection and processing[a]

Option	Pros	Cons
Rejection of stools from inpatients who have been in hospital for >3 days	Data suggest that patients who begin to have diarrhea after they have been inpatients for a few days are symptomatic not from parasitic infections but generally from other causes.	There is always the chance that the problem is related to a health care-associated (nosocomial) parasitic infection (rare), but *Cryptosporidium* and microsporidia may be considerations.
Examination of a single stool specimen (O&P examination); data suggest that 40–50% of organisms present will be found with only a single stool exam; two O&P exams (concentration and permanent stained smear) are acceptable but are not always as good as three specimens (this may be a relatively cost-effective approach); any patient remaining symptomatic would require additional testing.	Some think that most intestinal parasitic infections can be diagnosed from examination of a single stool; if the patient becomes asymptomatic after collection of the first stool, subsequent specimens may not be necessary.	Diagnosis from a single stool examination depends on experience of the microscopist, proper collection, and the parasite load in the specimen; in a series of three stool specimens, it is often the case that not all three specimens are positive and/or they may be positive for different organisms.
Examination of a second stool specimen only after the first is negative and the patient is still symptomatic	With additional examinations, the yield of protozoa increases (*Entamoeba histolytica,* 22.7%; *Giardia lamblia,* 11.3%; and *Dientamoeba fragilis,* 31.1%).	Assumes the second (or third) stool specimen is collected within the recommended 10-day time frame for a series of stools (protozoa are shed periodically); may be inconvenient for patient

Table 3.1 *(continued)*

Option	Pros	Cons
Examination of a single stool and an immunoassay (EIA, FA, and lateral or vertical flow cartridge). This approach is a mix. One immunoassay may be acceptable; however, immunoassay testing of two separate specimens may be required to confirm the presence of *Giardia* antigen. One O&P exam is not the best approach (review last option below).	If the examinations are negative and the patient's symptoms subside, probably no further testing is required.	Patients may exhibit symptoms (off and on), so it may be difficult to rule out parasitic infections with only a single stool and one fecal immunoassay. If the patient remains symptomatic, then even if two *Giardia* immunoassays are negative, other protozoa may be missed (the *Entamoeba histolytica/E. dispar* group, *Entamoeba histolytica, Dientamoeba fragilis, Cryptosporidium* spp., microsporidia). Normally, there are specific situations where fecal immunoassays *or* O&P exams should be ordered. It is not recommended to automatically perform both the O&P and the fecal immunoassay as a stool exam for parasites. *Depending on the patient's history and clinical symptoms, either the O&P exam* ***or*** *a fecal immunoassay may be recommended,* ***but generally not both.***
The laboratory pools three specimens for examination, performs *one* concentration, and examines *one* permanent stain.	Three specimens are collected by the patient (three separate collection vials) over 7–10 days; pooling by the laboratory may save time and expense.	Organisms present in low numbers may be missed due to the dilution factor once the specimens are pooled.
The laboratory pools three specimens for examination, performs *one* concentration, and examines *three* permanent stained smears.	Three specimens are collected by the patient (three separate collection vials) over 7–10 days; pooling by the laboratory for the concentration would probably be sufficient for the identification of helminth eggs. Examination of the three separate permanent stained smears (one from each vial) would maximize recovery of intestinal protozoa in areas of the country where these organisms are most common.	Might miss light helminth infection (eggs and larvae) due to the pooling of the three specimens for the concentration; however, with a permanent stain performed on each of the three specimens, this approach would probably be the next best option in lieu of the standard approach (concentration and permanent stained smear performed on every stool). Coding and billing would have to match the work performed; this may present some problems where work performed does not match existing codes.
The patient collects three stools but puts a sample of stool from all three specimens into a single vial (patient is given a single vial only).	Pooling of the specimens requires only a single vial.	This would complicate patient collection and very likely result in poorly preserved specimens, especially regarding the recommended ratio of stool to preservative and the lack of proper mixing of specimen and fixative.

Table continues on next page

Table 3.1 Fecal specimens for parasites: options for collection and processing[a] (*continued*)

Option	Pros	Cons
The lab performs immunoassays on selected patients' samples by FA, EIA, or rapid cartridge methods for *Giardia lamblia* (*G. duodenalis, G. intestinalis*), *Cryptosporidium* spp., and/or *Entamoeba histolytica/E. dispar* or *E. histolytica*).	More cost-effective than performing immuno-assay procedures on all specimens; however, the information needed to group patients is often not received with specimens. *This approach assumes that the physicians have guidance in terms of correct ordering options (see Table 3.2, below).*	Laboratories rarely receive information that would allow them to place a patient in a particular risk group, such as children <5 yr old, children from day care centers (who may or may not be symptomatic), patients with immunodeficiencies, and patients from outbreaks; performance of immunoassay procedures on every stool is not cost-effective, and the positivity rate will be low unless an outbreak situation is involved.
Perform immunoassays and O&P examinations on request: Giardia lamblia (G. duodenalis, G. intestinalis), Cryptosporidium spp., and/or *Entamoeba histolytica/E. dispar* group or *Entamoeba histolytica.* A number of variables will determine the approach to immunoassay testing and the O&P examination (geography, parasites recovered, positivity rate, and physician requests). Immunoassays and/or O&P examinations should be separately ordered, reported, and billed.	This approach will limit the number of stools on which immunoassay procedures for parasites are performed. Immunoassay results do not have to be confirmed by any other testing (such as O&P examinations or modified acid-fast stains). If specific kit performance problems have been identified, individual laboratories may prefer to do additional testing. *However, the fecal immunoassays are more sensitive than the O&P examination and special stains (modified acid-fast stains).* Also, this may be considered duplicate testing and may not be approved for reimbursement unless specifically ordered by the physician.	This approach will require education of the physician clients regarding appropriate times and patients for whom fecal immunoassays should be ordered. Educational initiatives must also include information on the test report indicating the pathogenic parasites that will *not* be detected using these methods. It is critical to make sure clients know that if patients have become asymptomatic, further testing may not be required. However, if the patient remains symptomatic, then further testing (O&P exam) is required. Remember, a single O&P exam may not reveal all organisms present. Present plan to physicians for approval: immunoassays or O&P examinations, procedure discussion, report formats, clinical relevance, and limitations of each approach.
Perform testing using multiplex panel options (parasites available in the panel may vary, depending on the system).	Depending on the parasites included in the panel, testing sensitivity and specificity will be better than results obtained with the examination and/or special stains.	Most of the current multiplex organism panels contain very limited parasite options; thus, many pathogens may not be considered in testing; other methods will need to be used in order to cover other pathogenic parasites; new panels are being developed to include more parasite options.

[a] O&P, ova and parasite; EIA, enzyme immunoassay; FA, fluorescent-antibody immunoassay. See the suggested reading for this section, in particular the articles by Cartwright, Hiatt et al., and Kehl.

Table 3.2 Approaches to stool parasitology: test ordering

Patient and/or situation	Test ordered[a]	Follow-up test ordered
Patient with diarrhea and AIDS or other cause of immune deficiency OR Patient with diarrhea involved in a potential waterborne outbreak (municipal water supply)	*Cryptosporidium* or *Giardia/Cryptosporidium* immunoassay	If immunoassays are negative and symptoms continue, special stains for microsporidia (modified trichrome stain) and other apicomplexans[b] (modified acid-fast stain) and O&P exam should be performed.
Patient with diarrhea (nursery school attendee, day care center attendee, camper, backpacker) OR Patient with diarrhea involved in a potential waterborne outbreak (resort setting) OR Patient from areas within the United States where *Giardia* is the most common parasite found	*Giardia* or *Giardia/Cryptosporidium* immunoassay (*Giardia*: perform testing on two stools before reporting as negative; not necessary for *Cryptosporidium*) **Particularly relevant for areas of the United States where *Giardia* is the most common organism found**	If immunoassays are negative and symptoms continue, special stains for microsporidia and other apicomplexans (see above) and an O&P exam should be performed.
Patient with diarrhea and relevant travel history outside of the United States OR Patient with diarrhea who is a past or present resident of a developing country OR Patient in an area of the United States where parasites other than *Giardia* are found (large metropolitan centers such as New York, Los Angeles, Washington, DC, and Miami)	O&P exam, *Entamoeba histolytica/E. dispar* immunoassay; immunoassay for confirmation of *E. histolytica*; various tests for *Strongyloides* may be relevant (even in the absence of eosinophilia), particularly if there is any history of pneumonia (migrating larvae in lungs) or of sepsis or meningitis (fecal bacteria carried by migrating larvae); agar culture plate is the most sensitive diagnostic approach for *Strongyloides stercoralis*, other than molecular methods.	The O&P exam is designed to detect and identify a broad range of parasites (amebae, flagellates, ciliates, *Cystoisospora belli*, and helminths); if exams are negative and symptoms continue, special tests for apicomplexans (fecal immunoassays, modified acid-fast stains, autofluorescence) and microsporidia (modified trichrome stains and calcofluor white stains) should be performed; fluorescent stains are also options.
Patient with unexplained eosinophilia	Although the O&P exam is recommended, the agar plate culture for *Strongyloides stercoralis* (more sensitive than the O&P exam) or molecular methods are also recommended, particularly if there is any history of pneumonia (migrating larvae in lungs) or of sepsis or meningitis (fecal bacteria carried by migrating larvae).	If tests are negative and symptoms continue, additional O&P exams and special tests for microsporidia (modified trichrome stains, calcofluor white stains, or fluorescent stains) and other apicomplexans (modified acid-fast stains, autofluorescence, or fluorescent stains) should be performed.
Patient with diarrhea (suspected foodborne outbreak)	Test for *Cyclospora cayetanensis* (modified acid-fast stain, autofluorescence, or fluorescent stains).	If tests are negative and symptoms continue, special procedures for microsporidia and other apicomplexans and O&P exam should be performed.

Table continues on next page

Table 3.2 Approaches to stool parasitology: test ordering (*continued*)

Patient and/or situation	Test ordered[a]	Follow-up test ordered
Patient with moderate to severe symptoms with acute diarrhea OR Patient at high risk of spreading disease to others and during known or suspected outbreaks OR Patient with dysentery (blood in the stool) OR Patients with acute diarrhea for >7 days	If available, gastroenterology automated multiplex molecular panel containing bacterial, viral, and limited parasitic organisms (*Cryptosporidium*, *Giardia*, *Entamoeba histolytica*, and *Cyclospora cayetanensis*)	A number of important parasitic pathogens may not be included in the currently available panels (*Blastocystis*, *Dientamoeba*, and microsporidia [*Enterocytozoon bieneusi* and *Encephalitozoon intestinalis*]); panels being developed may also contain *Strongyloides stercoralis*. If the panel is negative for limited parasite testing, additional testing will need to be performed.

[a] Depending on the particular immunoassay kit used, various single or multiple organisms may be included. Selection of a particular kit depends on many variables: clinical relevance, cost, ease of performance, training, personnel availability, number of test orders, training of physician clients, sensitivity, specificity, equipment, time to result, etc. Very few laboratories handle this type of testing in exactly the same way. Many options are clinically relevant and acceptable for good patient care. It is critical that the laboratory report indicate specifically which organisms could be identified by the kit; a negative report should list the organisms relevant to that particular kit. **It is important to remember that sensitivity and specificity data for all of these fecal immunoassay kits (FA, EIA, and cartridge formats) are comparable.** With the continued development and improvement of multiplex organism panels (bacterial, viral, and parasitic), the number of parasites included in the panels may increase. It will be important to verify organisms in the panels from the manufacturer of each system.

[b] The apicomplexans include *Cryptosporidium* spp. and the coccidians (*Cystoisospora belli* and *Cyclospora cayetanensis*).

Table 3.3 Preservatives and procedures commonly used in diagnostic parasitology (stool specimens)[a]

Preservative	Concentration	Permanent stained smear, Wheatley's trichrome, iron hematoxylin, special stains for coccidia and microsporidia	Immunoassays[b] (*Giardia lamblia [G. duodenalis, G. intestinalis], Cryptosporidium* spp.)	Comments
5% or 10% formalin	Yes	No	Yes	EIA, FA, and cartridge
5% or 10% buffered formalin	Yes	No	Yes	EIA, FA, and cartridge
MIF	Yes	Polychrome IV stain	ND	No published data
SAF	Yes	Iron hematoxylin (best) or trichrome (modified for SAF)	Yes	EIA, FA, and cartridge
Schaudinn's fixative[c] (mercury base) with or without PVA[d]	Yes, but rarely used	Trichrome or iron hematoxylin	No	PVA interferes with immunoassays
Modified Schaudinn's fluid[e] (copper base) with PVA	Yes, but rarely used	Trichrome or iron hematoxylin	No	PVA interferes with immunoassays
Modified Schaudinn's fluid[f] (zinc base) with PVA	Yes	Trichrome or iron hematoxylin	No	PVA interferes with immunoassays

Table 3.3 (*continued*)

Preservative	Concentration	Permanent stained smear, Wheatley's trichrome, iron hematoxylin, special stains for coccidia and microsporidia	Immunoassays[b] (*Giardia lamblia [G. duodenalis, G. intestinalis], Cryptosporidium* spp.)	Comments
Single-vial systems[g] (with PVA) (Eco-FIX; Meridian Bioscience)	Rare	Trichrome or iron hematoxylin; works best with EcoStain (Meridian Bioscience)	Some, but not all (No if contains PVA)	Check with the manufacturer
Universal fixative[h] (single-vial system) (Total-Fix; Medical Chemical Corporation)	Yes	Trichrome or iron hematoxylin	Yes	Contains no mercury, no formalin, and no PVA; does not require albumin for adhesion (no PVA and no albumin)

[a] Very detailed information on all fixative options can be found in *Diagnostic Medical Parasitology* (see the suggested reading). EIA, enzyme immunoassay; FA, fluorescent antibody; MIF, merthiolate-iodine-formalin; ND, no data; PVA, polyvinyl alcohol; SAF, sodium acetate-acetic acid-formalin. The apicomplexans include *Cryptosporidium* spp. and the coccidians (*Cystoisospora belli* and *Cyclospora cayetanensis*).

[b] Fecal immunoassays for the *Entamoeba histolytica/E. dispar* complex or group or the true pathogen, *Entamoeba histolytica*, require fresh or frozen specimens; in some cases, Cary-Blair transport medium may be acceptable (check with the manufacturer).

[c] These two fixatives use the mercuric chloride base in the Schaudinn's fluid; this formulation is still considered the gold standard, against which all other fixatives are evaluated (organism morphology after permanent staining). Additional fixatives prepared with non-mercuric chloride-based compounds are used; however, the overall organism morphology is not as good (zinc-based options are much better than copper-based options).

[d] Polyvinyl alcohol (PVA) is a water-soluble synthetic polymer used as a viscosity-increasing agent in pharmaceuticals, as an adhesive in parasitology fecal fixatives, and as a lubricant and protectant in ophthalmic preparations. PVA is also defined as a water-soluble polymer made by hydrolysis of a polyvinyl ester (such as polyvinyl acetate) used in adhesives, as textile and paper sizes, and for emulsifying, suspending, and thickening of solutions. **PVA IS NOT A FIXATIVE BUT AN ADHESIVE TO HELP GLUE THE STOOL MATERIAL ONTO THE SLIDE; this is the only purpose of PVA as an additive to parasitology fecal fixative formulations.** PVA is a plastic resin that is normally incorporated into Schaudinn's fixative; normally, LV-PVA (low-viscosity PVA) is preferred. Although some laboratories may perform a fecal concentration from a preserved specimen containing PVA, some parasites will not concentrate well, nor will some exhibit the typical morphology that would be seen in concentration sediment from a formalin-based fixative. Fixatives containing PVA are highly recommended as a means of preserving cysts and trophozoites for later examination as permanent stained smears. The use of fixatives containing PVA also permits specimens to be shipped (by regular mail service) from any location in the world to a laboratory for subsequent examination. Fixatives containing PVA are particularly useful for liquid specimens and should be used at a ratio of 3 parts fixative to 1 part fecal specimen.

[e] This modification uses a copper sulfate base rather than mercuric chloride; organism morphology is marginal.

[f] This modification uses a zinc base rather than mercuric chloride and works well with both trichrome and iron hematoxylin stains.

[g] These modifications use a combination of ingredients (including zinc) but are prepared from proprietary formulas. The aim is to provide a fixative that can be used for the fecal concentration, permanent stained smear, and available immunoassays for *Giardia lamblia* (*G. duodenalis, G. intestinalis*), *Cryptosporidium* spp., and *Entamoeba histolytica* (or the *Entamoeba histolytica/E. dispar* group).

[h] Universal fixatives contain no mercury, no formalin, and no PVA. The complete O&P (concentration and permanent stained smear), fecal immunoassays for *Giardia lamblia* (*G. duodenalis, G. intestinalis*) and *Cryptosporidium* spp., and special stains for the Apicomplexa, coccidia, and microsporidia can be performed using these options. Fecal immunoassays for the *Entamoeba histolytica/E. dispar* group or the true pathogen, *Entamoeba histolytica*, require fresh or frozen specimens; in some cases, Cary-Blair transport medium may be acceptable (check with the manufacturer). An example is Total-Fix (Medical Chemical Corporation, Torrance, CA). Total-Fix has also been found to be compatible with a number of molecular test options, including PCR.

Table 3.4 Advantages of thin and thick blood films[a]

Advantages of thin blood films	Advantages of thick blood films
RBC morphology (size, shape, and stippling) can be seen after fixation with methanol prior to staining (Giemsa) or as a part of the staining process (Wright's). RBCs are laked in the thick film. Other blood stains, including rapid stains, are acceptable; WBCs serve as the QC organism for the smear being stained. If the WBCs look normal, any parasites will exhibit the same morphologic and staining characteristics. A positive malaria film is not required.	Larger number of parasites are seen per field than in the thin blood film. RBCs are laked (ruptured), so WBCs, platelets, and parasites are visible after staining.
Identification of *Plasmodium* spp. is easier, since the parasite can be seen within the RBC. The size of the parasites within the RBCs can provide information necessary for identification to species level.	Phagocytized malaria pigment can be seen within the WBCs, even with a low parasitemia; however, the pigment may not be recognized or identified as such.
Parasitemia (%) can be calculated from the thin film; determination of parasitemia is mandatory for all *Plasmodium* spp. and is particularly important to monitor therapy for *P. falciparum* and/or *P. knowlesi*. Parasitemia should be reported with every set of blood films positive for *Plasmodium* or *Babesia*.	Stippling may be seen in a well-stained thick film. However, this depends on how long the blood has been in contact with anticoagulant if not collected as a fresh specimen (finger stick).

[a] Although for many years Giemsa stain has been the stain of choice, the parasites can also be seen on blood films stained with Wright's stain, a Wright-Giemsa combination stain, or one of the more rapid stains, such as Diff-Quik (various manufacturers), Wright's Dip Stat stain (Medical Chemical Corp., Torrance, CA), or Field's stain. **IT IS MANDATORY THAT BOTH THIN AND THICK FILMS BE EXAMINED PRIOR TO PROVIDING THE FINAL REPORT (TESTING IS ALWAYS STAT).** RBC, red blood cell.

Table 3.5 Advantages and disadvantages of buffy coat films

Advantages of buffy coat films	Disadvantages of buffy coat films
Volume of blood cells is considerably larger than with both thick and thin films prepared from whole blood.	Practice is needed to remove the correct blood layers from the centrifuged blood to prepare thick and thin films.
More sensitive than thick films for diagnosis of malaria; detection of infected RBCs is more likely.	Same potential problems may occur as with traditional thick or thin blood films.
Malaria pigment, which is phagocytized by the WBCs, may be seen more easily in the concentrated WBCs.	Some organisms and/or stages of parasites might be damaged due to high-speed centrifugation; however, centrifugation at 500 × *g* for 15–20 min should not cause damage.
Detection of parasitemia is easier for larger stages, such as schizonts and gametocytes.	Increased number of platelets may be confusing in terms of parasite differentiation.

Table 3.6 Potential problems of using EDTA anticoagulant for the preparation of thin and thick blood films

Potential problem	Comments
Adhesion to the slide; blood falls off slide during staining.	Incorrect ratio of anticoagulant to blood; fill tube completely with blood (7 ml or pediatric draw tube). Mix well and allow to fix for 30 min prior to blood film preparation.
Distortion of parasites; same type of distortion can also be seen after blood is refrigerated (not recommended).	Prolonged storage of blood in EDTA may lead to distortion (>1 h) and/or loss of parasites (4–6 h). Trophozoites (*P. vivax*) and gametocytes (*P. falciparum*) tend to round up, thus mimicking *P. malariae*.
Change in ring form size	Ring forms of *P. falciparum* continue to grow and enlarge, thus resembling rings of the other species. Typical small rings appear larger than usual.
Use of incorrect techniques. EDTA is the anticoagulant used by the hematology laboratory because the cellular components and morphology of the blood cells are preserved. 1. Blood smears for differentials from acceptable specimens should be prepared within 2 h of collection. 2. Blood counts from acceptable venipuncture specimens should be performed within 6 h of collection. 3. Underfilling the EDTA blood collection tube can lead to erroneously low blood cell counts and hematocrits, morphologic changes to RBCs, and staining alteration. Excess EDTA can shrink RBCs. 4. Conversely, overfilling the blood collection tube does not allow the tube to be properly mixed and may lead to platelet clumping and clotting.	EDTA prevents coagulation of blood by chelating calcium. Calcium is necessary in the coagulation cascade, and its removal inhibits and stops a series of events, both intrinsic and extrinsic, which cause clotting. In some individuals' samples, EDTA may cause inaccurate platelet results. These anomalies, platelet clumping and the formation of platelet satellites, may be the result of changes in the membrane structure occurring when the calcium ion is removed by the chelating agent, allowing the binding of preformed antibodies. Proper mixing of the whole-blood specimen ensures that EDTA is dispersed throughout the sample. Evacuated blood collection tubes with EDTA should be mixed by 8–10 end-over-end inversions immediately following venipuncture collection. Microcollection tubes with EDTA should be mixed by 10 complete end-over-end inversions immediately following collection. They should then be inverted an additional 20 times prior to analysis.
Loss of Schüffner's dots (stippling) in *P. vivax* and *P. ovale*	Schüffner's dots (true stippling) occur in both *P. vivax* and *P. ovale*; in the absence of stippling, identification to the species level may be much more difficult.
Prolonged storage of EDTA-blood (room temperature with stopper removed) Exflagellation of male gametocyte	The pH, CO_2, and temperature changes may reflect conditions within the mosquito. Thus, exflagellation of the male gametocyte may occur while still in the tube of blood prior to thin and thick blood film preparation. Microgametes may be confused with *Borrelia* or may be ignored as debris. If fertilization occurs between the male and female gametocytes, the crescent-shaped ookinete may resemble a *P. falciparum* gametocyte. Exflagellation can occur in any of the five species of human malaria pathogens.

Table continues on next page

Table 3.6 Potential problems of using EDTA anticoagulant for the preparation of thin and thick blood films (*continued*)

Potential problem	Comments
Release of merozoites from the schizonts into the blood. With the exception of heavy parasite loads, merozoites are normally not found outside of the RBCs, in contrast to *Babesia* spp., where rings may be seen outside of the RBCs.	Small rings may be seen outside of the RBCs or appear to be appliqué forms, thus suggesting *P. falciparum*. It is important to differentiate these true rings (both cytoplasmic and nuclear colors) from platelets (uniform color).
Incorrect submission of blood in heparin. Do not reject this specimen, since it is a stat order. Process the heparin tube and request another specimen in EDTA.	EDTA has less impact on parasite morphology than does heparin. Since these requests are stat, process the heparin tube and request a redraw using EDTA.

Table 3.7 Body sites and possible parasites recovered (trophozoites, cysts, oocysts, spores, adults, larvae, eggs, amastigotes, and trypomastigotes)[a]

Site	Parasites
Blood	
Red cells	*Plasmodium* spp.
	Babesia spp.
White cells	*Leishmania* spp.
	Toxoplasma gondii
Buffy coat	All blood parasites
Whole blood or plasma	*Trypanosoma* spp.
	Microfilariae
Bone marrow	*Leishmania* spp.
	Trypanosoma cruzi
	Plasmodium spp.
	Toxoplasma gondii
Central nervous system	*Taenia solium* (cysticerci)
	Echinococcus spp.
	Naegleria fowleri
	Acanthamoeba and *Hartmannella* spp.
	Balamuthia mandrillaris
	Sappinia diploidea
	Toxoplasma gondii
	Microsporidia
	Trypanosoma spp.
	Baylisascaris procyonis
	Angiostrongylus cantonensis
	Toxocara spp. (visceral larva migrans)
Cutaneous ulcers	*Leishmania* spp.
	Acanthamoeba spp.
	Entamoeba histolytica
Eyes	*Acanthamoeba* spp.
	Toxoplasma gondii
	Loa loa
	Microsporidia
	Dirofilaria spp.

Table 3.7 (*continued*)

Site	Parasites
	Thelazia spp.
	Toxocara spp. (ocular larva migrans)
Intestinal tract	*Entamoeba histolytica*
	Entamoeba dispar, E. moshkovskii, E. bangladeshi
	Entamoeba coli
	Entamoeba hartmanni
	Endolimax nana
	Iodamoeba bütschlii
	Blastocystis spp.
	Giardia lamblia
	Chilomastix mesnili
	Dientamoeba fragilis
	Pentatrichomonas hominis
	Balantidium coli
	Cryptosporidium parvum
	Cryptosporidium hominis
	Cryptosporidium spp.
	Cyclospora cayetanensis
	Cystoisospora belli
	Enterocytozoon bieneusi
	Encephalitozoon intestinalis
	Ascaris lumbricoides
	Enterobius vermicularis
	Hookworm
	Strongyloides stercoralis
	Trichuris trichiura
	Hymenolepis nana
	Hymenolepis diminuta
	Taenia saginata
	Taenia solium
	Taenia asiatica
	Diphyllobothrium latum
	Clonorchis (Opisthorchis) sinensis
	Paragonimus spp.
	Schistosoma spp.
	Fasciolopsis buski
	Fasciola hepatica
	Metagonimus yokogawai
	Heterophyes heterophyes
	Angiostrongylus costaricensis
Liver and spleen	*Echinococcus* spp.
	Entamoeba histolytica
	Leishmania spp.
	Microsporidia
	Capillaria hepatica

Table continues on next page

Table 3.7 Body sites and possible parasites recovered (trophozoites, cysts, oocysts, spores, adults, larvae, eggs, amastigotes, and trypomastigotes)[a] (*continued*)

Site	Parasites
	Clonorchis (Opisthorchis) sinensis
Lungs	*Cryptosporidium* spp.[b]
	Echinococcus spp.
	Paragonimus spp.
	Toxoplasma gondii
	Helminth larvae
Muscles	*Taenia solium, T. asiatica* (cysticerci)
	Trichinella spp.
	Onchocerca volvulus (nodules)
	Trypanosoma cruzi
	Microsporidia[c]
Skin	*Leishmania* spp.
	Onchocerca volvulus
	Microfilariae
	Ancylostoma spp. (cutaneous larva migrans)
Urogenital system	*Trichomonas vaginalis*
	Schistosoma spp.
	Microsporidia
	Microfilariae

[a] This table does not include every possible parasite that could be found in a body site. However, the most likely organisms have been listed.
[b] Disseminated in severely immunosuppressed individuals.
[c] The genera *Pleistophora* and *Trachipleistophora* have been documented to occur in muscles.

SECTION **4**

Specimen Test Options: Routine Diagnostic Methods and Body Sites

Practical Guide to Diagnostic Parasitology, Third Edition. Lynne S. Garcia
 DOI: 10.1128/9781683673637.ch04

Diagnostic parasitology includes laboratory procedures that are designed to detect organisms within clinical specimens by using morphologic criteria and visual identification, rather than culture, biochemical tests, and/or physical growth characteristics (see Table 4.1). Many clinical specimens, such as those from the intestinal tract, contain numerous artifacts that complicate the differentiation of parasites from surrounding debris. Final identification is usually based on microscopic examination of stained preparations, often at high magnification, such as with oil immersion (×1,000).

Specimen preparation often requires one of several concentration methods, all of which are designed to increase the chances of finding the organism(s). Microscopic examination requires review of the prepared clinical specimen by using multiple magnifications and different time frames; organism identification also depends on the skill of the microbiologist.

Protozoa are quite small, ranging from 1.5 μm (microsporidia) to ~80 μm (*Balantidium coli* [ciliate]). Some are intracellular and require multiple isolation and staining methods for identification. Helminth infections are usually diagnosed by finding eggs, larvae, and/or adult worms in various clinical specimens, primarily those from the intestinal tract. Identification to the species level may require microscopic examination of the specimen. The recovery and identification of blood parasites can require concentration, culture, and microscopy. Confirmation of suspected parasitic infections depends on the proper collection, processing, and examination of clinical specimens; multiple specimens must often be submitted and examined before the suspected organism(s) is detected and confirmed (see Table 4.1).

Ova and Parasite Examination of Stool Specimens

The most common specimen submitted to the diagnostic laboratory is the stool specimen, and the most commonly performed procedure in parasitology is the ova and parasite examination (O&P exam), which consists of three separate protocols: the direct wet mount (Figure 4.1), the concentration, and the permanent stained smear (Figure 4.2). The direct wet mount requires fresh stool, is designed to allow detection of motile protozoan trophozoites, and is examined microscopically at low and high dry magnifications (×100, entire 22- by 22-mm coverslip; ×400, one-third to one-half of a 22- by 22-mm coverslip) (Figure 4.1). However, due to potential problems with lag time between the time of specimen passage, receipt in the laboratory, and specimen fixation, the direct wet examination has been eliminated from the routine O&P exam in favor of receipt of specimens collected in stool preservatives; **if specimens are received in the laboratory in stool collection preservatives, the direct wet preparation is not performed, since no**

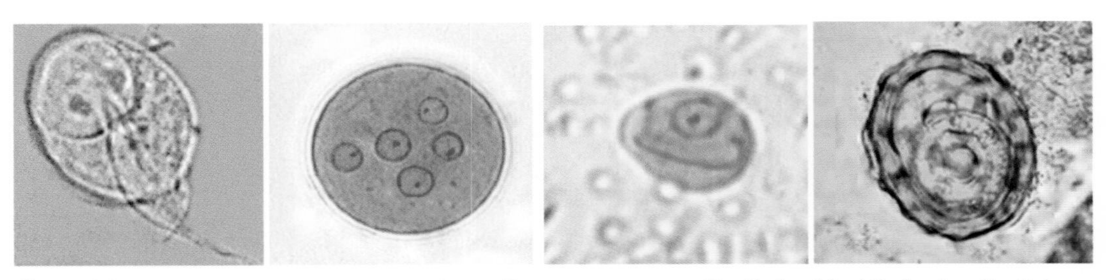

Figure 4.1 Direct wet mount or concentration sediment wet mount. *Giardia lamblia* (*G. duodenalis, G. intestinalis*) trophozoite; *Entamoeba coli* cyst (iodine); *Chilomastix mesnili* cyst (iodine); *Ascaris lumbricoides* fertilized egg.

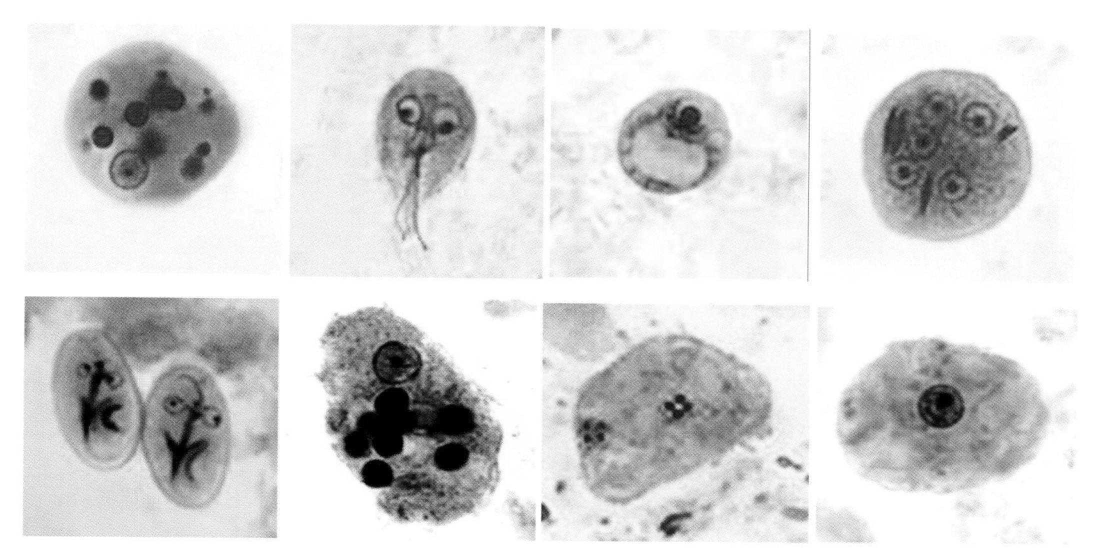

Figure 4.2 Permanent stained smear (trichrome and iron hematoxylin). **(Top row [trichrome])** *Entamoeba histolytica* trophozoite (containing ingested RBCs); *Giardia lamblia* (*G. duodenalis, G. intestinalis*) trophozoite; *Iodamoeba bütschlii* cyst; *Entamoeba coli* cyst (note chromatoidal bars with sharp or spiked ends). **(Bottom row [iron hematoxylin])** *Giardia lamblia* (*G. duodenalis, G. intestinalis*) cysts; *Entamoeba histolytica* trophozoite (containing ingested RBCs); *Dientamoeba fragilis* trophozoite; *Entamoeba coli* trophozoite (note large, eccentric karyosome).

trophozoite motility would be visible. The use of the oil immersion objective on this type of preparation is impractical, especially since morphological detail is more readily seen by oil immersion examination of the permanent stained smear.

The second part of the O&P exam is the concentration, which is designed to facilitate recovery of protozoan cysts, coccidian oocysts, microsporidial spores, and helminth eggs and larvae (Figure 4.1). Both flotation and sedimentation methods are available, the most common procedure being the formalin-ethyl acetate sedimentation method (formerly called the formalin-ether method). The concentrated specimen is examined as a wet preparation, with or without iodine, using low and high dry magnifications (×100 and ×400, respectively) as indicated for the direct wet smear examination.

The third part of the O&P exam is the permanent stained smear, which is designed to facilitate identification of intestinal protozoa. Several staining methods are available, the two most common being the Wheatley modification of the Gomori tissue trichrome and the iron hematoxylin staining methods (Figure 4.2). This part of the O&P exam is critical for the confirmation of suspicious objects seen in the wet examination and for identification of protozoa that might not have been seen in the wet preparation. **The permanent stained smear is the most important procedure performed within the O&P exam for the identification of intestinal protozoan infections (sensitivity and specificity)**; the permanent stained smears are examined using oil immersion objectives (600× for screening and 1,000× for final review of ≥300 oil immersion fields).

Trophozoites (potentially motile forms) of the intestinal protozoa are usually found in liquid specimens; both trophozoites and cysts might be found in soft specimens. The cyst forms are usually found in formed specimens; however, there are always exceptions to these general statements. Apicomplexan and coccidian oocysts and microsporidian spores can be found in any type of fecal specimen; for *Cryptosporidium* spp. (Apicomplexa), the more liquid the stool, the more oocysts are found in the specimen. Helminth eggs may be found in any type of specimen, although the chances of finding eggs in a liquid stool sample are reduced by the dilution factor. Tapeworm proglottids may be found on or beneath the stool on the bottom of the collection container. Adult pinworms and *Ascaris lumbricoides* are occasionally found on the surface or in the stool.

Many laboratories prefer that stool specimens be submitted in some type of preservative. Rapid fixation of the specimen immediately after passage (by the patient) provides an advantage in terms of recovery, morphology preservation, and identification of intestinal protozoa. **This advantage (preservation of organisms before distortion or disintegration) is thought to outweigh the limited motility information that might be gained by examining fresh specimens as direct wet mounts.**

Other Diagnostic Methods for Stool Specimens

Several other diagnostic techniques are available for the recovery and identification of parasitic organisms from the intestinal tract. Most laboratories do not routinely offer all of these techniques, but many of the tests are relatively simple and inexpensive to perform. The clinician should be aware of the possibilities and the clinical relevance of information obtained from using such techniques. Occasionally, it is necessary to examine stool specimens for the presence of scolices and proglottids of cestodes and adult nematodes and trematodes to confirm the diagnosis and/or for identification to the species level. A method for the recovery of these stages is also described in this section.

Culture of Larval-Stage Nematodes

Nematode infections giving rise to larval stages that hatch in soil or in tissues may be diagnosed by using certain fecal culture methods to concentrate the larvae. ***Strongyloides stercoralis*** **larvae are generally the most common larvae found in stool specimens (Figure 4.3).** Depending on the fecal transit time through the intestine and the patient's condition, rhabditiform and, rarely, filariform larvae may be present. Also, if there is a delay in examination of the stool, embryonated eggs as well as larvae of hookworm may be present. Culture of feces for larvae is useful to (i) reveal their presence when they are too scarce to be detected by concentration methods, (ii) distinguish whether the infection is due to *S. stercoralis* or hookworm on the basis of rhabditiform larval morphology

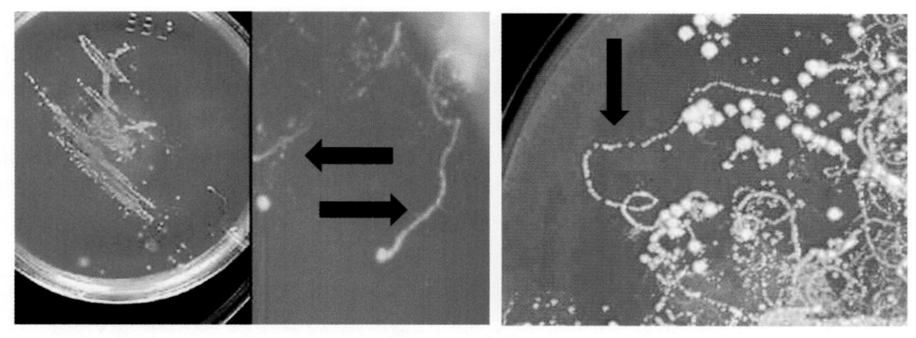

Figure 4.3 *Strongyloides stercoralis* agar plate culture. Agar plate showing tracks on the agar (arrows); specific track on the agar (arrow) (evidence of wandering larvae dragging bacteria along the track).

by allowing hookworm egg hatching to occur, releasing first-stage larvae, and (iii) allow development of larvae into the filariform stage for further differentiation.

The use of certain fecal culture methods (sometimes referred to as coproculture) is especially helpful to detect light infections by hookworm, *S. stercoralis*, and *Trichostrongylus* spp. and for specific identification of parasites. The rearing of infective-stage nematode larvae also helps in the specific diagnosis of hookworm and trichostrongyle infections, because the eggs of many of these species are identical, and specific identifications are based on larval morphology. Additionally, such techniques are useful for obtaining a large number of infective-stage larvae for research purposes. Available diagnostic methods include the Harada-Mori filter paper strip culture, the petri dish filter paper culture, the agar plate method (**the most sensitive method for the recovery of *S. stercoralis***), the charcoal culture, and the Baermann concentration.

Estimation of Worm Burdens through Egg Counts

The only human parasites for which it is reasonably possible to correlate egg production with adult worm burdens are *Ascaris lumbricoides*, *Trichuris trichiura*, and the hookworms (*Necator americanus* and *Ancylostoma duodenale*). The specific instances in which information on approximate worm burdens is useful include determination of the intensity of infection, selection of chemotherapy, and evaluation of the efficacy of the drugs administered. **However, with current therapy, the need for monitoring therapy through egg counts is no longer as relevant, and few laboratories perform this test.** Egg counts are estimates only; count variations occur regardless of how carefully the procedure is followed. If two or more fecal specimens are being compared, it is best to have the same individual perform the technique on both samples and to do multiple counts.

The direct-smear method of Beaver is the easiest to use and is reasonably accurate when performed by an experienced technologist. A direct smear of 2 mg (enough fresh fecal material to form a low cone on the end of a wooden applicator stick) of stool is prepared. Egg counts on the direct smear are reported as eggs per smear, and the appropriate calculations can be made to determine the number of eggs per gram of stool.

The Stoll count is probably the most widely used dilution egg-counting procedure for the purpose of estimating worm burdens. However, because of cost containment and clinical relevance (therapy is often initiated with no egg count data), most laboratories no longer offer this procedure.

Hatching Test for Schistosome Eggs

All fecal and urine specimens used for the hatching test must be collected and processed without using preservatives; any rinse steps must be performed with saline (not water, which may cause premature egg hatching) (Figure 4.4). When schistosome eggs are recovered from either urine or stool, they should be carefully examined to determine viability. The presence of living miracidia within the eggs indicates an active infection that may require therapy. The viability of the miracidia can be determined in two ways: (i) the cilia of the flame cells (primitive excretory cells) may be seen on a wet smear by using high dry power and are usually actively moving, and (ii) the miracidia may be released from the eggs by a hatching procedure. The eggs usually hatch within several hours when placed in 10 volumes of dechlorinated or spring water (hatching may begin soon after contact with the water). The eggs that are recovered in the urine (24-h specimen collected with no preservatives) are easily obtained from the sediment and can be examined under the microscope to determine viability.

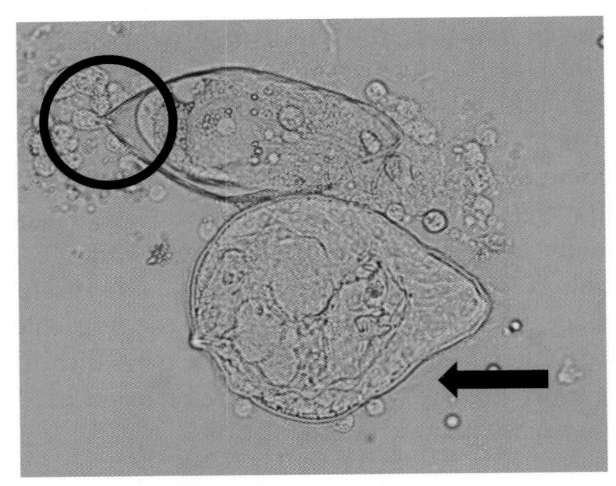

Figure 4.4 Hatched *Schistosoma haematobium* egg found in urine contaminated with water. Note the miracidium larva (arrow) hatched from the eggshell and the terminal spine on the eggshell (circle).

Screening Stool Samples for Recovery of a Tapeworm Scolex

Since the medication used for treatment of tapeworms is usually very effective, **screening for tapeworm scolices is rarely requested and no longer clinically relevant.** However, stool specimens may have to be examined for the presence of scolices and gravid proglottids of cestodes for proper species identification. This procedure requires mixing a small amount of feces with water and straining the mixture through a series of wire screens (graduated from coarse to fine mesh) to look for scolices and proglottids. **Remember to use standard precautions and wear gloves when performing this procedure.** The appearance of scolices after therapy is an indication of successful treatment. If the scolex has not been passed, it may still be attached to the mucosa; the parasite is capable of producing more segments from the neck region of the scolex, and the infection continues. If this occurs, the patient can be re-treated when proglottids begin to reappear in the stool.

After treatment for tapeworm removal, the patient should be instructed to take a saline cathartic and to collect all stool material passed for the next 24 h. The stool material should be immediately placed in 10% formalin, thoroughly broken up, and mixed with the preservative (1-gal [3.8-liter] plastic jars, half full of 10% formalin, are recommended).

Testing of Other Intestinal Tract Specimens

Other specimens from the intestinal tract, such as duodenal aspirates or drainage material, mucus from the Entero-Test Capsule technique, and sigmoidoscopy material, can also be examined as wet preparations and as permanent stained smears after being processed with either trichrome or iron hematoxylin staining. Although not all laboratories perform these procedures, the procedures are included to give some idea of the possibilities for diagnostic testing.

Examination for Pinworm

A roundworm parasite that has worldwide distribution and is commonly found in children is *Enterobius vermicularis*, known as pinworm or seatworm. The adult female worm migrates out of the anus, usually at night, and deposits her eggs on the perianal area. The adult female (8 to 13 mm long) is occasionally found on the surface of a stool specimen

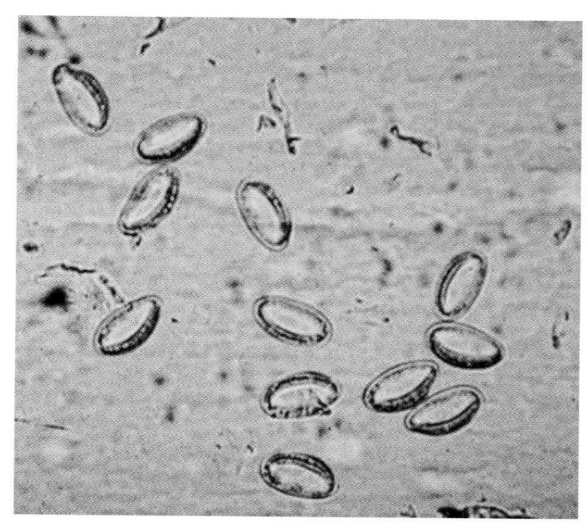

Figure 4.5 *Enterobius vermicularis* (pinworm) eggs collected using cellophane tape.

or on the perianal skin. Since the eggs are usually deposited around the anus, they are not commonly found in feces and must be detected by other diagnostic techniques. Diagnosis of pinworm infection is usually based on the recovery of typical eggs, which are described as thick-shelled, football-shaped eggs with one slightly flattened side (Figure 4.5). Each egg often contains a fully developed embryo and is infective within a few hours after being deposited. Unfortunately, it takes a minimum of four to six consecutive negative tapes or swabs before the infection can be ruled out.

Sigmoidoscopy Material

Material obtained from sigmoidoscopy can be helpful in the diagnosis of amebiasis that has not been detected by routine fecal examinations; however, a series of at least three routine stool examinations for parasites should be performed on each patient before sigmoidoscopy examination is done. Material from the mucosal surface should be aspirated or scraped and should not be obtained with cotton-tipped swabs. At least six representative areas of the mucosa should be sampled and examined (six samples, six slides). The examination of sigmoidoscopy specimens does not take the place of routine O&P exams.

The specimen should be processed immediately. Three methods of examination can be used. All three are recommended; however, depending on the availability of trained personnel, proper fixation fluids, or the amount of specimen obtained, one or two procedures may be used. If the amount of material limits the examination to one procedure, the use of fixative containing polyvinyl alcohol (PVA) is highly recommended. If the material is to be examined using any of the new fluorescent antibody or enzyme immunoassay detection kits (*Cryptosporidium* spp. or *Giardia lamblia*), 5 or 10% formalin or sodium acetate-acetic acid-formalin (SAF) fixative is recommended. Many physicians performing sigmoidoscopy procedures do not realize the importance of selecting the proper fixative for material to be examined for parasites. For this reason, it is recommended that a parasitology specimen tray (containing Schaudinn's fixative, PVA, and 5 or 10% formalin) be provided or a trained technologist be available at the time of sigmoidoscopy to prepare the slides. Even the most thorough examination will be meaningless if the specimen has been improperly prepared.

Duodenal Drainage Material

In infections with *G. lamblia* or *S. stercoralis*, routine stool examinations may not reveal the organisms. Duodenal drainage material can be submitted for examination, a technique that may reveal the parasites. The specimen should be submitted to the laboratory in a tube containing no preservative; the amount may vary from <0.5 ml to several milliliters of fluid. The specimen may be centrifuged (10 min at 500 × *g*) and should be examined immediately as a wet mount for motile organisms (iodine may be added later to facilitate identification of any organisms present). If the specimen cannot be completely examined within 2 h after it is taken, any remaining material should be preserved in 5 to 10% formalin. The "falling-leaf" motility often described for *Giardia* trophozoites is rarely seen in these preparations. The organisms may be caught in mucus strands, and the movement of the flagella on the *Giardia* trophozoites may be the only subtle motility seen for these flagellates. *Strongyloides* larvae are usually very motile. It is important to keep the light intensity low.

The duodenal fluid may contain mucus; this is where the organisms are usually found. Therefore, centrifugation of the specimen is important, and the sedimented mucus should be examined. Fluorescent antibody or enzyme immunoassay detection kits (for *Cryptosporidium* or *Giardia*) can also be used with fresh or formalinized material.

If a presumptive diagnosis of giardiasis is obtained on the basis of the wet preparation examination, the coverslip can be removed and the specimen can be fixed with either Schaudinn's fluid or PVA for subsequent staining with either trichrome or iron hematoxylin. If the amount of duodenal material submitted is very small, one can prepare permanent stains rather than using any of the specimen for a wet smear examination. Some workers think that this approach provides a more permanent record, and the potential problems with unstained organisms, very minimal motility, and a lower-power examination can be avoided by using oil immersion examination of the stained specimen at a magnification of ×1,000.

Duodenal Capsule Technique (Entero-Test)

A simple and convenient method of sampling duodenal contents that eliminates the need for intestinal intubation has been devised. The device consists of a length of nylon yarn coiled inside a gelatin capsule. The yarn protrudes through one end of the capsule; this end of the line is taped to the side of the patient's face. The capsule is then swallowed, the gelatin dissolves in the stomach, and the weighted string is carried by peristalsis into the duodenum (Figure 4.6). The yarn is attached to a weight by a slipping mechanism; the weight is released and passes out in the stool when the line is retrieved after 4 h. Bile-stained mucus clinging to the yarn is then scraped off (mucus can also be removed by pulling the yarn

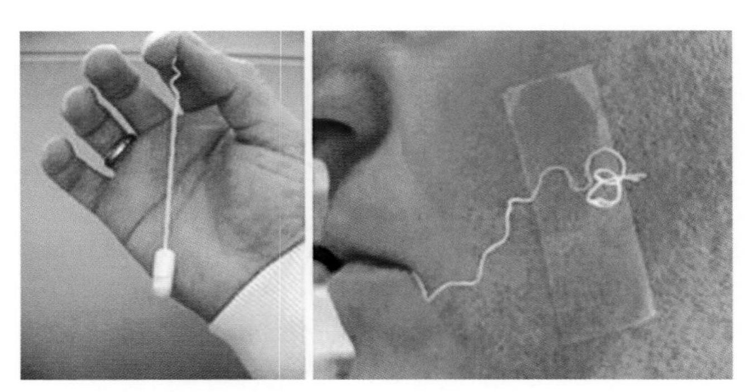

Figure 4.6 Entero-Test capsule for sampling duodenal contents. Left to right: capsule being held by the protruding string; end of the string attached to the cheek.

between thumb and finger) and collected in a small petri dish; disposable gloves are recommended. Usually 4 or 5 drops of material is obtained.

The specimen should be examined immediately as a wet mount for motile organisms (iodine may be added later to facilitate identification of any organisms present). If the specimen cannot be completely examined within an hour after the yarn has been removed, the material should be preserved in 5 to 10% formalin or PVA mucus smears should be prepared. The organism motility is like that described above for duodenal drainage specimens. The pH of the terminal end of the yarn should be checked to ensure adequate passage into the duodenum (a very low pH means that it never left the stomach). The terminal end of the yarn should be yellow-green, indicating that it was in the duodenum (the bile duct drains into the intestine at this point).

Urogenital Tract Specimens

The identification of *Trichomonas vaginalis* is usually based on the examination of wet preparations of vaginal and urethral discharges and prostatic secretions or urine sediment. Multiple specimens may have to be examined to detect the organisms. These specimens are diluted with a drop of saline and examined under low power (×100) and reduced illumination for the presence of actively motile organisms; as the jerky motility begins to diminish, it may be possible to observe the undulating membrane, particularly under high dry power (×400). Stained smears are usually not necessary for the identification of this organism. The large number of false-positive and false-negative results reported on the basis of stained smears strongly suggests the value of confirmation by observation of motile organisms from the direct mount, from appropriate culture media, or from direct detection using more sensitive molecular methods.

Examination of urinary sediment may be indicated for specimens from patients with certain filarial infections. Administration of the drug diethylcarbamazine (Hetrazan) has been reported to enhance the recovery of microfilariae from the urine. The triple-concentration technique is recommended for the recovery of microfilariae. The membrane filtration technique can also be used with urine for the recovery of microfilariae. A membrane filter technique for the recovery of *Schistosoma haematobium* eggs has also been useful.

Sputum

Although not one of the more common specimens, expectorated sputum may be submitted for examination for parasites. Organisms in sputum that may be detected and may cause pneumonia, pneumonitis, or Loeffler's syndrome include the migrating larval stages of *Ascaris lumbricoides*, *Strongyloides stercoralis*, and hookworm; the eggs of *Paragonimus* spp.; *Echinococcus granulosus* hooklets; and the protozoa *Pneumocystis jirovecii* (now classified with the fungi), *Entamoeba histolytica*, *Entamoeba gingivalis*, *Trichomonas tenax*, *Cryptosporidium* spp., and possibly the microsporidia. In a *Paragonimus* sp. infection, the sputum may be viscous and tinged with brownish flecks ("iron filings"), which are clusters of eggs, and may be streaked with blood. Sputum is usually examined as a wet mount (saline or iodine), using low and high dry power (×100 and ×400). The specimen is not concentrated before preparation of the wet mount. If the sputum is thick, an equal amount of 3% sodium hydroxide (NaOH) (or undiluted chlorine bleach) can be added; the specimen is thoroughly mixed and then centrifuged. NaOH should not be used if one is looking for *Entamoeba* spp. or *T. tenax*. After centrifugation, the supernatant fluid is discarded and the sediment can be examined as a wet mount with saline or iodine. If examination has to be delayed for any reason, the sputum should be fixed in 5 or 10% formalin to preserve helminth eggs or larvae or in PVA-containing fixative to be stained later for protozoa.

Concentrated stained preparations of induced sputa are commonly used to detect *P. jirovecii* and differentiate trophozoite and cyst forms of other possible agents of pneumonia, particularly in AIDS patients. Organisms must be differentiated from other fungi, such as *Candida* spp. and *Histoplasma capsulatum*. If the clinical evaluation of a patient suggests *P. jirovecii* pneumonia and the induced sputum specimen is negative, a bronchoalveolar lavage specimen should be evaluated by using appropriate stains.

After patients have used appropriate cleansing procedures to reduce oral contamination, induced sputa are collected by pulmonary or respiration therapy staff. The induction protocol is critical for the success of the procedure, and well-trained individuals are mandatory if organisms are to be recovered.

Aspirates

The examination of aspirated material for the diagnosis of parasitic infections may be extremely valuable, particularly when routine testing methods have failed to demonstrate the organisms. These specimens should be transported to the laboratory immediately after collection. Aspirates include liquid specimens collected from a variety of sites where organisms might be found. Those most commonly processed in the parasitology laboratory include fine-needle aspirates and duodenal aspirates. Fluid specimens collected by bronchoscopy include bronchoalveolar lavage fluid and bronchial washings.

Fine-needle aspirates may be submitted for slide preparation and/or culture. Aspirates of cysts and abscesses for amebae may require concentration by centrifugation, digestion, microscopic examination for motile organisms in direct preparations, and cultures and microscopic evaluation of stained preparations.

Bone marrow aspirates to be tested for *Leishmania* amastigotes, *Trypanosoma cruzi* amastigotes, or *Plasmodium* spp. require staining with any of the blood stains. Examination of these specimens may confirm an infection that has been missed by examination of routine blood films.

Biopsy Specimens

Biopsy specimens are recommended for the diagnosis of tissue parasites. The following procedures may be used for this purpose in addition to standard histologic preparations: impression smears and teased and squash preparations of biopsy tissue from skin, muscle, cornea, intestine, liver, lungs, and brain. Tissue to be examined by permanent sections or electron microscopy should be fixed as specified by the laboratories that will process the tissue. In certain cases, a biopsy may be the only means of confirming a suspected parasitic problem. Specimens to be examined as fresh material rather than as tissue sections should be kept moist in saline and submitted to the laboratory immediately.

Detection of parasites in tissue depends in part on specimen collection and having sufficient material to perform the recommended diagnostic procedures. Biopsy specimens are usually quite small and may not be representative of the diseased tissue. The use of multiple tissue samples often improves diagnostic results. To optimize the yield from any tissue specimen, all areas should be examined and as many procedures as possible should be used. Tissues are obtained by invasive procedures, many of which are very expensive and lengthy; consequently, these specimens deserve the most comprehensive examination possible.

Tissue submitted in a sterile container on a sterile sponge dampened with saline may be used for cultures of protozoa after mounts for direct examination or impression smears for staining have been prepared. If cultures for parasites are to be made, sterile slides should be used for smear and mount preparation.

Blood

Depending on the life cycle, a number of parasites may be recovered in a blood specimen, which can be whole blood, buffy coat preparations, or various types of concentrations. Although some organisms may be motile in fresh whole blood, species are usually identified from the examination of permanent stained blood films, both thick and thin films. Blood films can be prepared from fresh whole blood collected with no anticoagulants, anticoagulated blood, or sediment from the various concentration procedures. The recommended stain of choice is Giemsa stain; however, the parasites can also be seen on blood films stained with Wright's or other blood stains. Delafield's hematoxylin stain is often used to stain the microfilarial sheath; in some cases, Giemsa stain may not provide sufficient stain quality to allow differentiation of the microfilariae (e.g., it may fail to stain the sheath of *Wuchereria bancrofti*).

Thin Blood Films

In any examination of thin blood films for parasitic organisms, the initial screen should be carried out with the low-power objective (10×) of a microscope. Microfilariae may be missed if the entire thin film is not examined. Microfilariae are rarely present in large numbers, and frequently only a few organisms occur in each thin-film preparation. Microfilariae are commonly found at the edges of the thin film or at the feathered end of the film because they are carried to these sites during the process of spreading the blood. The feathered end of the film where the erythrocytes (RBCs) are drawn out into one single, distinctive layer of cells should be examined for the presence of malaria parasites and trypanosomes. In these areas, the morphology and size of the infected RBCs are most clearly seen.

Depending on the training and experience of the microscopist, examination of the thin film usually takes 15 to 20 min (≥300 oil immersion fields) at a magnification of ×1,000. Although some people use a 50× or 60× oil immersion objective to screen stained blood films, there is some concern that small parasites, such as plasmodia, *Babesia* spp., or *Leishmania donovani*, may be missed at this lower total magnification (×500 or ×600) compared with the ×1,000 total magnification obtained when the more traditional 100× oil immersion objective is used. Because people tend to scan blood films at different rates, it is important to examine a minimum number of fields. If something suspicious has been seen in the thick film, considerably more than 300 fields are often examined on the thin film. The request for blood film examination should always be considered a stat procedure, with all reports (negative as well as positive) being reported by telephone to the physician as soon as possible. If the results are positive, appropriate governmental agencies (local, state, and federal) should be notified within a reasonable time frame in accordance with guidelines and laws.

Both malaria and *Babesia* infections have been missed when automated differential instruments were used, and therapy was delayed. Although these instruments are not designed to detect intracellular blood parasites, the inability of the automated systems to discriminate between uninfected RBCs and those infected with parasites may pose serious diagnostic problems.

Thick Blood Films

In the preparation of a thick blood film, the greatest concentration of blood cells is in the center of the film. The examination should be performed at low magnification to detect microfilariae more readily. Examination of a thick film usually requires 5 min (approximately 100 fields at low and high dry magnification). The search for malarial organisms and trypanosomes is best done under oil immersion (total magnification of ×1,000 [approximately 300 oil immersion fields]). Intact RBCs are frequently seen at the very periphery of the thick film; such cells, if infected, may prove useful in malaria diagnosis, since they may demonstrate the characteristic morphology necessary to identify the organisms to the species level.

Blood Staining Methods

For accurate identification of blood parasites, a laboratory should develop proficiency in the use of at least one good staining method. It is better to select one method that provides reproducible results than to use several on a hit-or-miss basis. Blood films should be stained as soon as possible, since prolonged storage may result in stain retention. Failure to stain positive malarial smears within a month may result in failure to demonstrate typical staining characteristics for individual species.

The most common stains are of two types. Wright's stain has the fixative in combination with the staining solution, so that both fixation and staining occur at the same time; therefore, the thick film must be laked before staining. In Giemsa stain, the fixative and stain are separate; thus, the thin film must be fixed with absolute methanol before being stained. Other blood stains can also be used. Although for many years, Giemsa stain has been the stain of choice, the parasites can also be seen on blood films stained with Wright's stain, a Wright-Giemsa combination stain, or one of the more rapid stains, such as Diff-Quik (various manufacturers), Wright's Dip Stat stain (Medical Chemical Corp., Torrance, CA), or Field's stain.

Buffy Coat Films

L. donovani, trypanosomes, and *Histoplasma capsulatum* (a fungus with intracellular elements resembling those of *L. donovani*) are occasionally detected in peripheral blood. The parasite or fungus is found in the large mononuclear cells in the buffy coat (a layer of white cells resulting from centrifugation of whole citrated blood). The nuclear material stains dark red-purple, and the cytoplasm stains light blue (*L. donovani*). *H. capsulatum* appears as a large dot of nuclear material (dark red-purple) surrounded by a clear halo. Trypanosomes in the peripheral blood also concentrate with the buffy coat cells.

QBC Microhematocrit Centrifugation Method

Microhematocrit centrifugation with use of the QBC malaria tube (a glass capillary tube and closely fitting plastic insert [QBC malaria blood tubes; Becton Dickinson, Tropical Disease Diagnostics, Sparks, MD]) can be used for the detection of blood parasites. At the end of centrifugation of 50 to 60 μl of capillary or venous blood (5 min in a QBC centrifuge; 14,387 × *g*), parasites or RBCs containing parasites are concentrated into a small, 1- to 2-mm region near the top of the RBC column and are held close to the wall of the tube by the plastic float, thereby making them readily visible by microscopy. Tubes precoated with acridine orange provide a stain which induces fluorescence in the parasites. This method automatically prepares a concentrated smear which represents the distance between the float and the walls of the tube. Once the tube is placed into the plastic holder (Paraviewer) and immersion oil is applied to the top of the hematocrit tube (no coverslip is necessary), the tube is examined with a 40× to 60× oil immersion objective (the working distance must be 0.3 mm or greater).

Knott Concentration

The Knott concentration procedure is used primarily to detect the presence of microfilariae in the blood, especially when a light infection is suspected. The disadvantage of the procedure is that the microfilariae are killed by the formalin and are therefore not seen as motile organisms.

Membrane Filtration Technique

The membrane filtration technique using Nuclepore filters has proved highly efficient in demonstrating filarial infections when microfilaremias are of low density. It has also been successfully used in field surveys (Figure 4.7).

Culture Methods

Very few clinical laboratories offer specific culture techniques for parasites. The methods for *in vitro* culture are often complex, while quality control is difficult and not really feasible for the routine diagnostic laboratory. In certain institutions, some techniques may be available, particularly where consultative services are provided and for research purposes.

Few parasites can be routinely cultured, and the only procedures in general use are for *Entamoeba histolytica*, *Naegleria fowleri*, *Acanthamoeba* spp., *Trichomonas vaginalis*, *Toxoplasma gondii*, *Trypanosoma cruzi*, and the leishmanias. These procedures are usually available only after consultation with the laboratory and on special request.

Cultures of parasites grown in association with an unknown microbiota are referred to as xenic cultures. A good example of this type of culture is stool specimens cultured for *E. histolytica*. If the parasites are grown with a single known bacterium, the culture is referred to as monoxenic. An example of this type of culture is clinical specimens (corneal biopsy) cultured with *Escherichia coli* as a means of recovering species of *Acanthamoeba* and *Naegleria*. If parasites are grown as pure culture without any bacterial associate, the culture is referred to as axenic. An example of this type of culture is the use of medium for the isolation of *Leishmania* spp. or *T. cruzi*.

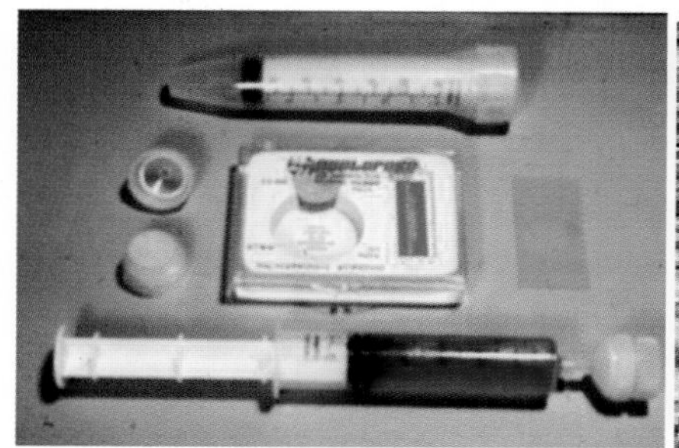
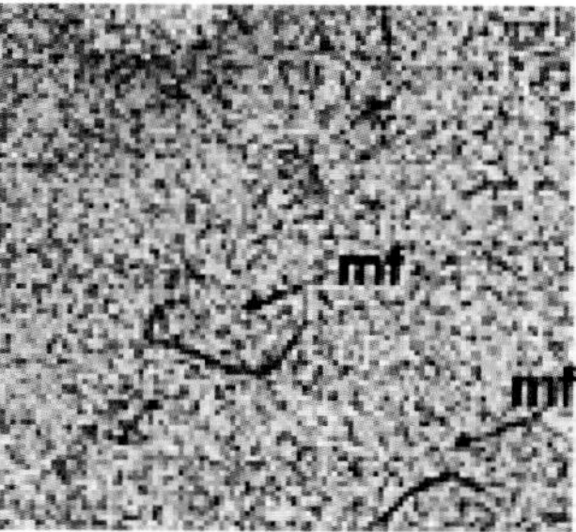

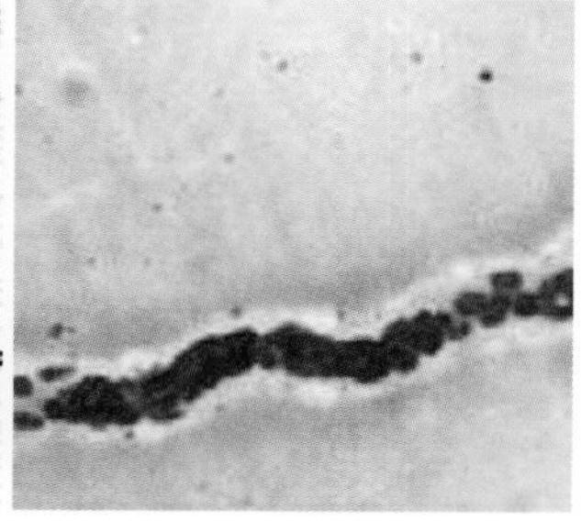

Figure 4.7 Nuclepore membrane filtration for recovery of microfilariae. Filtration apparatus (syringe, filters, and syringe containing diluted blood with filter holder attached); low-power magnification showing microfilariae (mf) on the stained filter pad; high magnification showing a single microfilaria on the stained slide (material taken from the filter pad).

Animal Inoculation and Xenodiagnosis

Most routine clinical laboratories do not have the animal care facilities necessary to provide animal inoculation capabilities for the diagnosis of parasitic infections. Host specificity for many animal parasite species is a well-known fact and limits the types of animals available for these procedures. For certain suspected infections, animal inoculation may be requested and can be very helpful in making the diagnosis, although animal inoculation certainly does not take the place of other, more routine procedures.

Xenodiagnosis uses the arthropod host as an indicator of infection (Figure 4.8). Uninfected reduviid bugs are allowed to feed on the blood of a patient who is suspected of having Chagas' disease (*T. cruzi* infection). After 30 to 60 days, feces from the bugs are examined over a 3-month time frame for the presence of developmental stages of the parasite, which are found in the hindgut of the vector. This type of procedure is used primarily in South America for field work, and the appropriate bugs are raised specifically for this purpose in various laboratories.

Antibody and Antigen Detection

Antibody Detection

In certain parasitic infections, the standard diagnostic laboratory procedures may not be sufficient to confirm infection, or specimen collection may not be practical or cost-effective. In these circumstances, alternative methods may be helpful; these include antibody, antigen, and nucleic acid detection (see Table 4.2). In some cases, serologic methods might be clinically indicated and may be very helpful, particularly if a parasitic infection is suspected and routine results are negative. However, even with the most sophisticated technology, few serologic tests for parasitic infections can be used to confirm an infection or predict the disease outcome.

Although parasites and their by-products are immunogenic for the host, the host immune response is usually not protective. Any immunity that does develop is usually species specific and may even be strain or stage specific. Human parasites are generally divided into two groups: (i) those that multiply within the host (e.g., protozoa) and (ii) those that mature within the host but never multiply, e.g., schistosomes and *Ascaris*.

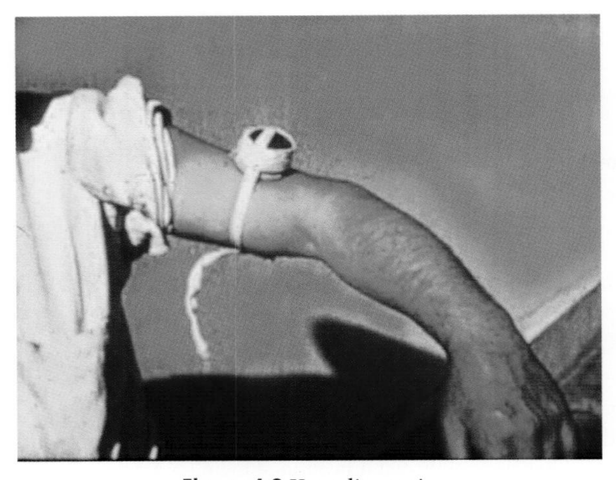

Figure 4.8 Xenodiagnosis.

In infections caused by protozoa that multiply within the host, there is continuous antigenic stimulation of the host's immune system as the infection progresses. In these instances, there is usually a positive correlation between clinical symptoms and serologic test results.

In contrast to the protozoa, helminths often migrate through the body and pass through a number of developmental stages before becoming mature adults. Helminth infections are often difficult to confirm serologically, probably due to a limited antigenic response by the host or failure to use the appropriate antigen in the test system. Most parasitic antigens used in serologic procedures are heterogeneous mixtures that are not well defined. Results of tests performed with such antigens may represent cross-reactions or poor sensitivity.

Interpretation of test results may also present problems, particularly when one is dealing with patients from areas where infection is endemic; such individuals may have higher baseline titers than do patients from other areas, in whom a low titer may actually be significant. Antibody detection generally indicates exposure to the parasite at some time in the past and may not necessarily reflect a current infection. This is particularly true for patients who have lived in an area of endemicity for some time; their current clinical presentation may have no relationship to a positive antibody titer for a particular parasite. Although antibody levels generally decline over a period of months to years, serologic test results neither confirm nor rule out current infection or cure.

However, a positive serologic titer of antibody to a particular parasite in a patient who has had no previous exposure to the organism is clinically relevant and probably indicates recent exposure. The importance of a complete history, including both residence and travel information, is critical for accurate interpretation of serologic results.

Although serologic procedures have been available for many years, they are not routinely offered by most clinical laboratories for a number of reasons (cost, trained personnel, number of test orders, sensitivity, specificity, and interpretation). Standard techniques that are used include complement fixation, indirect hemagglutination, indirect fluorescent-antibody assay, soluble-antigen fluorescent-antibody assay, bentonite flocculation, latex agglutination, double diffusion, counterelectrophoresis, immunoelectrophoresis, radioimmunoassay, and intradermal tests.

The Centers for Disease Control and Prevention (CDC) offers a number of serologic procedures for diagnostic purposes, some of which are not available elsewhere. Because regulations for submission of specimens may vary from state to state, each laboratory should check with its own county or state department of public health for the appropriate instructions. Additional information on procedures, availability of skin test antigens, and interpretation of test results can be obtained directly from CDC. The Infectious Diseases Specimen Submission (IDSS) Help Desk at CDC serves as the central point of contact for submitters of specimens who have questions or issues related to submitting specimens to CDC.

Website: https://www.cdc.gov/laboratory/specimen-submission/help-faqs.html

Email: CDC_ID_lab_info@cdc.gov

Phone (toll free): 1-855-612-7575

The IDSS Help Desk hours of operation are 8:00 a.m. to 5:00 p.m. EST, Monday–Friday; closed on federal holidays.

Antigen Detection, Nucleic Acid-Based Tests, and Molecular Panels

Progress has been made in the development and application of molecular methods for diagnostic purposes, including the use of purified or recombinant antigens and nucleic acid probes. The detection of parasite-specific antigen is more indicative of current disease. Many of the assays were originally developed with polyclonal antibodies which were targeted to unpurified antigens that markedly decreased the sensitivity and specificity of the tests. Fecal immunoassays are generally simple to perform and allow a large number of tests to be performed at one time, thereby reducing overall costs. **A major disadvantage of antigen detection in stool specimens (fecal immunoassays) is that the method can detect only one or two pathogens at a time. A routine O&P exam must be performed to detect other parasitic pathogens.** The current commercially available antigen tests (direct and indirect fluorescent-antibody assays and enzyme immunoassay) are more sensitive and more specific than is routine microscopy. Current testing is available for *E. histolytica*, the *E. histolytica/E. dispar* group, *G. lamblia*, and *Cryptosporidium* spp. Diagnostic reagents are also in development for some of the other intestinal protozoa.

Nucleic acid-based diagnostic tests for parasitology are primarily available only in specialized research or reference centers. PCR and other nucleic acid probe tests have been developed for almost all species of parasites. The only nucleic acid-based probe test commercially available is for the detection of *T. vaginalis*. As the costs of these tests decrease and the various steps necessary to perform the tests become automated, there will be increasing demand for commercially available reagents.

There are also several molecular tests that are in clinical trials for the detection of panels of selected gastrointestinal parasites. These tests are referred to as molecular gastrointestinal panels and target the most commonly occurring parasitic stool pathogens. Although laboratory-developed tests for most parasites have been developed, these are not commercially available or are available only in specialized testing centers. **A parasitic panel currently in development (VERIGENE II GI panel, acquired from Luminex) will test for *Blastocystis* spp., *Cryptosporidium* spp., *Cyclospora cayetanensis*, *Dientamoeba fragilis*, *Entamoeba histolytica*, *Giardia lamblia*, microsporidia (most likely *Encephalitozoon* and *Enterocytozoon*), and *Strongyloides stercoralis*.** FDA clearance is anticipated in 2021.

Intradermal Tests

In the absence of reliable serologic diagnostic tests, skin tests have been used to provide indirect evidence of infection. However, most skin tests have been used primarily for research and epidemiologic purposes. Some of the more commonly used skin tests are the Casoni (hydatid disease) and Montenegro (*L. donovani*) tests. In many cases, the antigens used are difficult to obtain and are not commercially available (Figure 4.9). The antigens are usually crude extracts that have not been standardized and are neither highly sensitive nor specific. They may provoke an immune response that complicates further serologic testing, and there is always the danger of provoking an anaphylactic reaction. In addition, there are ethical questions related to giving patients injections of nonstandardized foreign protein, particularly if the antigens were derived from *in vivo* materials.

UV Autofluorescence

The oocysts of *Cyclospora*, *Cystoisospora*, and *Cryptosporidium* autofluoresce. Strong autofluorescence of *Cyclospora* oocysts is useful for microscopic screening; if autofluorescence is seen, the modified acid-fast stain can be performed for organism confirmation. *Cyclospora* appears blue when exposed to 365 nm UV light and looks green under

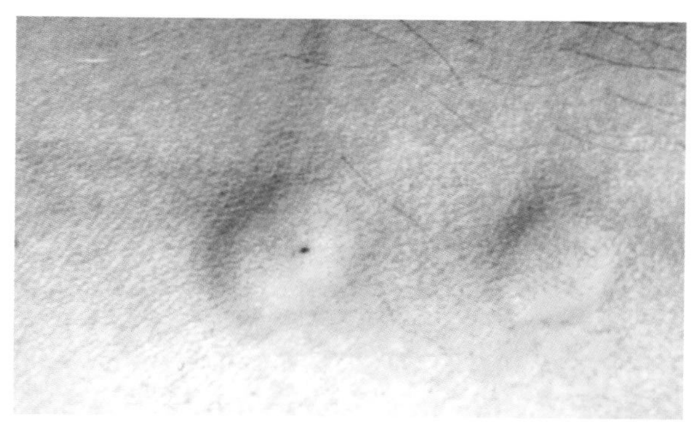

Figure 4.9 Intradermal test showing a positive result.

450–490 nm excitation. *Cryptosporidium* appears violet when exposed to 365 nm UV light and green under 405–436 nm excitation. Fluorescence intensity varies from about 1+ to 2+; it is uncommon to see stronger autofluorescence with these oocysts. Due to potential errors in identification, organisms should be measured to confirm accuracy. Other examples of parasite autofluorescence include *Acanthamoeba* cysts, *Blastocystis* spp. central vacuoles, *Schistosoma mansoni* eggshells, *Ascaris lumbricoides* eggshells, and *Toxocara canis* eggshells.

Suggested Reading

Arakaki T, Iwanaga M, Kinjo F, Saito A, Asato R, Ikeshiro T. 1990. Efficacy of agar-plate culture in detection of *Strongyloides stercoralis* infection. *J Parasitol* **76:**425–428.

Beal C, Goldsmith R, Kotby M, Sherif M, el-Tagi A, Farid A, Zakaria S, Eapen J. 1992. The plastic envelope method, a simplified technique for culture diagnosis of trichomoniasis. *J Clin Microbiol* **30:**2265–2268.

Beal CB, Viens P, Grant RGL, Hughes JM. 1970. A new technique for sampling duodenal contents: demonstration of upper small-bowel pathogens. *Am J Trop Med Hyg* **19:**349–352.

Beaver PC. 1949. A nephelometric method of calibrating the photoelectric meter for making egg-counts by direct fecal smear. *J Parasitol* **35:**13.

Beaver PC. 1950. The standardization of fecal smears for estimating egg production and worm burden. *J Parasitol* **36:**451–456.

Borchardt KA, Smith RF. 1991. An evaluation of an InPouch TV culture method for diagnosing *Trichomonas vaginalis* infection. *Genitourin Med* **67:**149–152.

CLSI. 2005, Procedures for the recovery and identification of parasites from the intestinal tract. *Approved guideline M28-2A*. CLSI, Wayne, PA.

Garcia LS (ed). 2010. *Clinical Microbiology Procedures Handbook*, 3rd ed. ASM Press, Washington, DC.

Garcia LS. 2016. *Diagnostic Medical Parasitology*, 6th ed. ASM Press, Washington, DC.

Garcia LS, Smith JW, Thomas KB, Fritsche R. 2003. *Cumitech 30A, Selection and use of laboratory procedures for diagnosis of parasitic infections of the gastrointestinal tract*. Coordinating ed, Garcia LS. ASM Press, Washington, DC.

Garcia LS, Johnston SP, Linscott AJ, Shimizu RY. 2008. *Cumitech 46, Laboratory procedures for diagnosis of blood-borne parasitic diseases*. Coordinating ed, Garcia LS. ASM Press, Washington, DC.

Harada U, Mori O. 1955. A new method for culturing hookworm. *Yonago Acta Med* **1:**177–179.

Hsieh HC. 1962. A test-tube filter-paper method for the diagnosis of *Ancylostoma duodenale, Necator americanus*, and *Strongyloides stercoralis*. *WHO Tech Rep Ser* **255:**27–30.

Koga K, Kasuya S, Khamboonruang C, Sukhavat K, Ieda M, Takatsuka N, Kita K, Ohtomo H. 1991. A modified agar plate method for detection of *Strongyloides stercoralis*. *Am J Trop Med Hyg* **45:**518–521.

Leber AL (ed). 2016. *Clinical Microbiology Procedures Handbook*, 4th ed. ASM Press, Washington, DC.

Markell EK, Voge M, John DT. 1992. *Medical Parasitology*, 7th ed. The WB Saunders Co, Philadelphia, PA.

Melvin DM, Brooke MM. 1985. *Laboratory Procedures for the Diagnosis of Intestinal Parasites*, p 163–189. US Department of Health, Education, and Welfare publication no. (CDC) 85-8282. US Government Printing Office, Washington, DC.

Murray PR, Baron EJ, Pfaller MA, Tenover FC, Yolken RH (ed). 1995. Section X: Parasitology. *In Manual of Clinical Microbiology*, 6th ed. ASM Press, Washington, DC.

Nagel R, Gray D, Bielefeldt-Ohmann H, Traub RJ. 2015. Features of *Blastocystis* sp. in xenic culture revealed by deconvolutional microscopy. *Parasitol Res.* **114:**3237–3245.

NCCLS. 2000. Laboratory diagnosis of blood-borne parasitic diseases. *Approved guideline M15-A.* NCCLS, Wayne, PA.

Qazi F, Khalid A, Poddar A, Titienne JP, Nadarajah A, Aburto-Medina A, Shahsavari E, Shukla R, Prawer S, Ball AS, Tomljenovic-Hanic S. 2020. Real-time detection and identification of nematode eggs genus and species through optical imaging. *Sci Rep* **10:**7219.

Stoll NR, Hausheer WC. 1926. Concerning two options in dilution egg counting: small drop and displacement. *Am J Hyg* **6**(Suppl):134–145.

Wells KE, Cordingley JS. 1991. *Schistosoma mansoni*: eggshell formation is regulated by pH and calcium. *Exp Parasitol* **73:**295–310.

Wilson M, Schantz P, Pieniazek N. 1995. Diagnosis of parasitic infections: immunologic and molecular methods, p 1159–1170. *In* Murray PR, Baron EJ, Pfaller MA, Tenover FC, Yolken RH (ed), *Manual of Clinical Microbiology*, 6th ed. ASM Press, Washington, DC.

Table 4.1 Body sites, procedures and specimens, recommended methods and relevant parasites, and comments[a]

Body site or sample type	Procedures and specimens	Recommended methods and relevant parasites	Comments
Blood	Microscopy[b]: thin and thick blood films; fresh blood or EDTA-blood (fill EDTA tube completely with blood, then mix well). Since these are STAT procedures, if a heparin tube is received, process immediately and request EDTA tube be drawn ASAP. Morphology not as good with heparin.	Giemsa stain (all blood parasites); hematoxylin-based stain (sheathed microfilariae). For malaria, thick and thin blood films are definitely recommended and should be prepared within 30–60 min of blood collection via venipuncture (other tests may be used as well). Although for many years, Giemsa stain has been the stain of choice, the parasites can also be seen on blood films stained with Wright's stain, a Wright-Giemsa combination stain, or one of the more rapid stains, such as Diff-Quik (various manufacturers), Wright's Dip Stat stain (Medical Chemical Corp., Torrance, CA), or Field's stain.	Most drawings and descriptions of blood parasites are based on Giemsa-stained blood films. Although Wright's stain (or Wright-Giemsa combination stain) works, stippling in samples from malaria patients may not be visible and the organism colors do not match the descriptions. However, with other stains (those listed above, in addition to some of the "rapid" blood stains), the organisms should be detectable on the blood films. The use of blood collected with anticoagulant (rather than fresh) has direct relevance to the morphology of malaria organisms seen in peripheral blood films. If the blood smears are prepared after more than 1 h, stippling may not be visible, even if the correct pH buffers are used. Also, if blood is kept at room temperature (with the stopper removed), the male microgametocyte may exflagellate and fertilize the female macrogametocyte; development continues within the tube of blood (as it would in the mosquito host). The ookinete may actually resemble a *P. falciparum* gametocyte. The microgamete may resemble spirochetes.

Table 4.1 (*continued*)			
Body site or sample type	**Procedures and specimens**	**Recommended methods and relevant parasites**	**Comments**
Blood (*continued*)	Concentration methods: EDTA-blood	Buffy coat, fresh blood films for detection of moving microfilariae or trypanosomes. QBC, a screening method for blood parasites (hematocrit tube contains acridine orange), has been used for malaria parasites, *Babesia*, trypanosomes, and microfilariae. It is usually impossible to identify malaria parasites to the species level; this requires high levels of training.	
	Antigen detection: EDTA-blood for malaria, serum or plasma for circulating antigens (hemolyzed blood can interact in some tests)	Commercial test kits for malaria and some microfilariae	
	Molecular methods (PCR): EDTA-blood, ethanol-fixed or unfixed thin and thick blood films, coagulated blood; possible with hemolyzed or frozen blood samples (other options may be possible; check with individual manufacturers)	Sensitivity is not higher than that of thick films for *Plasmodium* spp., much more sensitive for *Leishmania* (peripheral blood is used from immunodeficient patients only). Sequencing of PCR product is often used for species or genotype identification.	Currently there are two commercial tests available (check for FDA status). Two malaria LAMP kits are now CE-marked and commercially available: the Loopamp Malaria kit (Eiken Chemical Co., Ltd.) and the Illumigene malaria LAMP (Meridian Bioscience). A valid CE mark is a legal requirement for supplying products to Europe[c]. High laboratory standards are needed; check for specimen (may work with finger-prick blood, heparin blood, whole blood or dried blood spots, frozen, coagulated, or hemolyzed blood samples).
	Specific antibody detection: serum or plasma, anticoagulated or coagulated blood (hemolyzed blood can cause problems in some tests)	Most commonly used are EIA (many test kits commercially available), EITB (commercially available for some parasites), and IFA.	Many labs are using in-house tests; only a few fully defined antigens are available; sensitivities and specificities of the tests should be documented by the lab.

Table continues on next page

Table 4.1 Body sites, procedures and specimens, recommended methods and relevant parasites, and comments[a] (*continued*)			
Body site or sample type	**Procedures and specimens**	**Recommended methods and relevant parasites**	**Comments**
Bone marrow	Biopsy specimens or aspirates Microscopy: thin and thick films with aspirate collected in EDTA	Giemsa stain (all blood parasites); other blood stains also acceptable	*Leishmania* amastigotes are recovered in cells of the reticuloendothelial system. If films are not prepared directly after sample collection, infected cells may disintegrate. Sensitivity of microscopy is low; use only in combination with other methods.
	Cultures: sterile material in EDTA or culture medium	Culture for *Leishmania* (or *Trypanosoma cruzi*)	
	PCR: aspirate in EDTA	PCR for blood parasites including *Leishmania*, *Toxoplasma*, and rare other parasites	
Central nervous system	Microscopy: spinal fluid and CSF (wet examination, stained smears), brain biopsy specimen (touch or squash preparations, stained)	Stains: Giemsa or other blood stains (trypanosomes, *Toxoplasma*); blood stains, trichrome, or calcofluor white (amebae [*Naegleria* or *Sappinia* PAM, *Acanthamoeba* or *Balamuthia* GAE]); Giemsa, acid-fast, PAS, modified trichrome, silver methenamine (microsporidia) (tissue Gram stains also recommended for microsporidia in routine histologic preparations); H&E, routine histology (larval cestodes, *Taenia solium* cysticerci, *Echinococcus* spp.)	If CSF is received (with no suspected organism suggested), Giemsa is the best choice; however, modified trichrome or calcofluor is also recommended as a second stain (amebic cysts and microsporidia). If brain biopsy material is received (particularly from an immunocompromised patient), cultivation is recommended for microsporidium isolation and PCR for identification to the species or genotype level. A small amount of the sample should always be stored frozen for PCR analyses in case the results of the other methods are inconclusive.
	Culture: sterile aspirate or biopsy material (in physiologic NaCl)	Free-living amebae (exception: *Balamuthia* does not grow in the routine agar/bacterial overlay method), microsporidia, and *Toxoplasma*	
	PCR: aspirate or biopsy material (fresh, frozen, or fixed in ethanol)	Protozoa and helminths, species and genotype characterization	
Cutaneous ulcers	Microscopy: aspirate, biopsy (smears, touch or squash preparations, histological sections)	Giemsa (*Leishmania*); H&E, routine histology (*Acanthamoeba* spp., *Balamuthia mandrillaris*, *Entamoeba histolytica*). Most likely causative parasites would be *Leishmania* spp., which would stain with Giemsa. PAS could be used to differentiate *Histoplasma capsulatum* from *Leishmania* in tissue.	It is important to remove contaminating organisms in center of lesion (especially important for cultures; see the following section); take clinical specimen from the advancing margins of the ulcer; punch biopsy recommended: one sterile specimen for culture, one for pathology/routine histology, two for microbiology/touch/squash preparations.

Table 4.1 (*continued*)

Body site or sample type	Procedures and specimens	Recommended methods and relevant parasites	Comments
Cutaneous ulcers (*continued*)	Cultures (less commonly used)	*Leishmania*, free-living amebae (often bacterial contaminations)	In immunocompromised patients, skin ulcers have been documented to have amebae as causative agents.
	PCR: aspirate, biopsy material (fresh, frozen, or fixed in ethanol)	*Leishmania* (species identification), free-living amebae	
Eyes	Microscopy: smears; touch or squash preparations; biopsy, scrapings, contact lens, sediment of lens solution	Calcofluor white, cyst only (amebae [*Acanthamoeba*]); Giemsa, trophozoites, cysts (amebae); modified trichrome (preferred) or silver methenamine stain, PAS, acid-fast stains (microsporidial spores); H&E, routine histology (cysticerci, *Loa loa*, *Toxoplasma*)	Some free-living amebae (most commonly *Acanthamoeba*) have been implicated as a cause of keratitis. Although calcofluor white stains the cyst walls, it does not stain the trophozoites. Therefore, for suspected cases of amebic keratitis, both stains (Giemsa, calcofluor white) should be used. H&E (routine histology) can be used to detect and confirm cysticercosis. The adult worm of *Loa loa*, when removed from the eye, can be stained with a hematoxylin-based stain (Delafield's) or can be stained and examined by routine histology.
	Culture: fresh material (see above) in PBS supplemented with antibiotics, if possible, to suppress bacterial growth	Cultures: free-living amebae, *Toxoplasma*, microsporidia	
	PCR: fresh material in physiological NaCl or PBS, or ethanol or frozen	Free-living amebae, *Toxoplasma*, microsporidia species and genotype identification	Microsporidia: confirmation to the species or genotype level may be done by PCR and sequence analyses; however, the spores could be found by routine light microscopy with modified trichrome, calcofluor, and/or tissue Gram stains.
Intestinal tract	Stool and other intestinal material	Concentration methods: ethyl acetate sedimentation of fixed stool samples (no PVA) (most protozoa); flotation or combined sedimentation flotation methods (helminth ova); agar plate culture or Baermann concentration (larvae of *Strongyloides* spp., fresh stool required)	Stool fixation with fecal fixatives preserves parasite morphology, allows prolonged storage (room temperature) and long transportation, and prevents hatching of *Schistosoma* eggs, but it makes *Strongyloides* larval concentration difficult and impedes further PCR analyses; taeniid eggs cannot be identified to the species level.
	Microscopy: stool, sigmoidoscopy material, duodenal contents (all fresh or preserved), direct wet smear, concentration methods	Direct wet smear (direct examination of unpreserved fresh material is also used) (motile protozoan trophozoites; helminth eggs and protozoan cysts may also be detected)	

Table continues on next page

Table 4.1 Body sites, procedures and specimens, recommended methods and relevant parasites, and comments[a] *(continued)*			
Body site or sample type	**Procedures and specimens**	**Recommended methods and relevant parasites**	**Comments**
Intestinal tract (*continued*)		Stains: trichrome or iron hematoxylin (intestinal protozoa); modified trichrome (microsporidia); modified acid-fast stain (*Cryptosporidium*, *Cyclospora*, and *Cystoisospora*)	Microsporidia: confirmation to the species or genotype level requires PCR or electron microscopy; however, modified trichrome and/or calcofluor stains can be used to confirm the presence of spores.
	Anal impression preparation	Adhesive cellulose tape, no stain (*Enterobius vermicularis*)	Four to six consecutive negative tapes are required to rule out infection with pinworm (*E. vermicularis*); now considered impractical for clinical diagnosis; symptomatic patients are often treated without confirmation of diagnosis.
	Adult worms or tapeworm segments (proglottids)	Carmine stains (rarely used for adult worms or cestode segments). Proglottids can usually be identified to the genus level (*Taenia*, *Diphyllobothrium*, or *Hymenolepis*) without using tissue stains.	Worm segments can be stained with special stains. However, after dehydration through alcohols and xylenes (or xylene substitutes) without prior staining, the sexual organs and the branched uterine structure are visible, allowing identification of the proglottid to the species level.
	Antigen detection (fresh or frozen material; suitability of fixation is test dependent)	Commercial immunoassays, e.g., EIA, FA, cartridge formats (*E. histolytica*, the *E. histolytica*/*E. dispar* group, *G. lamblia*, *Cryptosporidium* spp.); in-house tests for *T. solium* and *T. saginata*	Coproantigens can be detected in the prepatent period and independently from egg excretion.
	PCR: fresh, frozen, or ethanol-fixed material	No commercial tests available. Primers for genus or species identification of most helminths and protozoa are published.	Due to potential inhibition after DNA extraction from stool samples, concentration or isolation methods may be required prior to DNA extraction. However, new DNA isolation kits facilitate the isolation of high-quality DNA from stool. Sequence analyses may be required for species or genotype identification.
	Biopsy specimens Microscopy: fixed for histology or touch or squash preparations for staining PCR: see above	H&E, routine histology (*E. histolytica*, *Cryptosporidium*, *Cyclospora*, *Cystoisospora belli*, *Giardia*, microsporidia); less common findings include *Schistosoma* spp., hookworm, and *Trichuris*.	Special stains may be helpful for the identification of microsporidia: tissue Gram stains, silver stains, PAS, and Giemsa or modified acid-fast stains for the coccidia.

Table 4.1 *(continued)*			
Body site or sample type	**Procedures and specimens**	**Recommended methods and relevant parasites**	**Comments**
Liver and spleen	Biopsy specimens or aspirates Microscopy: unfixed material in physiological NaCl; fixed for histology	Examination of wet smears for *E. histolytica* (trophozoites), protoscolices of *Echinococcus* spp., or eggs of *Capillaria hepatica*; Giemsa (*Leishmania*, other protozoa, and microsporidia); H&E (routine histology)	There are definite risks associated with punctures (aspirates and/or biopsy) of spleen or liver lesions (*Echinococcus*). Always keep a small amount of material frozen for PCR.
	Culture: sterile preparation of fresh material	For *Leishmania* (not common)	
	Animal inoculation: sterile preparation of fresh material	Intraperitoneal inoculation of *Echinococcus multilocularis* cyst material for viability test after long-term chemotherapy	
	PCR: fresh, frozen, or ethanol-fixed samples; species or genotype identification (e.g., *Echinococcus* spp.)		
Respiratory tract	Sputum, induced sputum, nasal and sinus discharge, bronchoalveolar lavage fluid, transbronchial aspirate, tracheobronchial aspirate, brush biopsy specimen, open lung biopsy specimen Microscopy: unfixed material, treated for smear preparation PCR: fresh, frozen, or fixed in ethanol	Helminth larvae (*Ascaris*, *Strongyloides*), eggs (*Paragonimus*, *Capillaria*), or hooklets (*Echinococcus*) can be recovered in unstained respiratory specimens. Stains: Giemsa for many protozoa including *Toxoplasma* tachyzoites, modified acid-fast stains (*Cryptosporidium*); modified trichrome (microsporidia) Routine histology (H&E; silver methenamine stain, PAS, acid-fast stains, tissue Gram stains for helminths, protozoa, and microsporidia)	Immunoassay reagents (FA) are available for the diagnosis of pulmonary cryptosporidiosis. Routine histologic procedures allow the identification of any of the helminths or helminth eggs present in the lungs. Disseminated toxoplasmosis or microsporidiosis is well documented, with organisms being found in many different respiratory specimens.

Table continues on next page

Table 4.1 Body sites, procedures and specimens, recommended methods and relevant parasites, and comments[a] *(continued)*			
Body site or sample type	**Procedures and specimens**	**Recommended methods and relevant parasites**	**Comments**
Muscle	Biopsy material Microscopy: touch and squash preparations, unfixed or fixed for histology and EM	Larvae of *Trichinella* spp. can be identified unstained (species identification with a single larva by PCR). H&E, routine histology (*Trichinella* spp., cysticerci); silver methenamine stain, PAS, acid-fast stains, tissue Gram stains, EM (rare microsporidia)	If *Trypanosoma cruzi* is present in the striated muscle, the organisms could be identified by routine histology preparations. Modified trichrome and/or calcofluor stains can be used to confirm the presence of microsporidial spores.
	PCR: fresh, frozen, or ethanol fixed	Microsporidium identification to the species level requires subsequent sequencing.	
Skin	Aspirates, skin snips, scrapings, biopsy specimens	See cutaneous ulcer (above).	Any of the potential parasites present can be identified by routine histology procedures.
	Microscopy: wet examination, stained smear (or fixed for histology or EM)	Wet preparations (microfilariae), Giemsa-stained smears or H&E, routine histology (*Onchocerca volvulus*, *Dipetalonema streptocerca*, *Dirofilaria repens*, other larvae causing cutaneous larva migrans, zoonotic *Strongyloides* spp., hookworms, *Leishmania*, *Acanthamoeba* spp., *Entamoeba histolytica*, microsporidia and arthropods [*Sarcoptes* and other mites])	The larger the amount of material received, the better the chance of finding parasites. Visual identification may require several tissue samples and additional cut sections from multiple tissue blocks in order to find the organisms (parasites tend to be isolated in tissue, not disseminated throughout the specimen like bacteria or fungi).
	PCR: fresh, frozen, or fixed in ethanol	Primers for most parasite species are available.	
Amniotic fluid	PCR (and/or culture): fresh material	PCR based on the detection of highly repetitive gene sequences is the method of choice.	Only applicable to confirm suspected prenatal *Toxoplasma* infections
	Animal inoculation (toxoplasmosis)		

Table 4.1 (*continued*)

Body site or sample type	Procedures and specimens	Recommended methods and relevant parasites	Comments
Urogenital system	Vaginal discharge, saline swab, transport swab (no charcoal), air-dried smear for FA, urethral discharge, prostatic secretions, urine (single unpreserved, 24-h unpreserved, or early-morning specimens) Microscopy: wet smears, smears of urine sediment, stained smears	Giemsa, immunoassay reagents (FA, rapid lateral-flow test) (*Trichomonas vaginalis*); Delafield's hematoxylin (microfilariae); modified trichrome (microsporidia); H&E, routine histology PAS, acid-fast stains, tissue Gram stains (microsporidia); direct examination of urine sediment for *Schistosoma haematobium* eggs or microfilariae	Although *T. vaginalis* is probably the most common parasite identified, there are others to consider, the most recently implicated organisms being in the microsporidian group. Microfilariae could also be recovered and stained. Fixation of urine with formalin prevents hatching of *Schistosoma* eggs.
	Cultivation: vaginal or urethral discharge or swab preparations PCR: fresh, frozen, or fixed in ethanol	Identification and propagation of *T. vaginalis* (commercialized plastic envelope culture systems available); moving trophozoites can be detected using microscopy (or in Giemsa-stained smears).	Material must be put into culture medium immediately after collection; it should not be cooled or frozen.

[a] CSF, cerebrospinal fluid; EIA, enzyme immunoassay; EM, electron microscopy; EITB, enzyme-linked immunoelectrotransfer blot (Western blot); FA, fluorescent antibody; GAE, granulomatous amebic encephalitis; GI, gastrointestinal; H&E, hematoxylin and eosin; PAM, primary amebic encephalitis; PAS, periodic acid-Schiff stain; PBS, phosphate-buffered saline; QBC, quantitative buffy coat; SAF, sodium acetate-acetic acid-formalin. Although Giemsa stain is mentioned, any blood stain could be used in these circumstances.

[b] Many parasites or parasite stages may be detected in standard histologic sections of tissue material. However, species identification is difficult, and additional examinations may be required. Usually, these techniques are not considered first-line methods. Additional methods such as electron microscopy are carried out only by specialized laboratories and are not available for standard diagnostic purposes. Electron microscopy examination for species identification has largely been replaced by PCR assays.

[c] The differences between the two approaches stem from the following: the U.S. FDA approach assesses the device's effectiveness as well as its risk of harm; the CE mark, on the other hand, affirms simply that the product "meets high safety, health and environmental protection requirements" (European Commission 2015). The U.S. FDA approval would ensure not only that the product poses no harm to consumers but also that it does what it claims to do. Critics of the FDA system suggest that this goal adds time and unpredictability to the approval process without in fact establishing the effectiveness of the device. Thus, while the CE mark is less difficult to obtain, it is a less powerful certification. FDA approval means that the device is approved for use in all parts of the world, while the CE mark has restrictions, sometimes even within the European Union.

NOTE: Molecular testing has become more widely used in diagnostic parasitology. Often these procedures provide better sensitivity and specificity than are found in routine non-molecular methods. A number of molecular diagnostic panels are available (primarily for GI parasites), and newer, more comprehensive panels are pending FDA approval (*Entamoeba histolytica*, *Giardia lamblia*, *Dientamoeba fragilis*, *Blastocystis* spp., *Cryptosporidium* spp., *Cyclospora cayetanensis*, microsporidia [at least two species], and *Strongyloides stercoralis*). Other panel options may include *Trichomonas vaginalis*. Although these molecular panel options are becoming more widespread, cost remains a factor for many clinical laboratories.

Table 4.2 Serologic, antigen, and probe tests used in the diagnosis of parasitic infections

Disease or organism	Routine antibody test(s)[a]	Antigen, probe test(s), PCR, NAT, other[b]
Protozoa		
Amebiasis	EIA	EIA, rapid, multiplex PCR
Babesiosis	IFA (CDC)	AST and ALT enzymatic assays; activity is significantly higher in plasma of patients with *B. microti*.
Chagas' disease[c]	EIA, IFA (CDC); automated system/ Elecsys Chagas assay	PCR
Cryptosporidiosis	Currently research use only	DFA, EIA, IFA, rapid, multiplex NAT, multiplex PCR
Cryptosporidium/Giardia		DFA, IFA, EIA, rapid, multiplex PCR
Cryptosporidium, Giardia, Entamoeba histolytica/ E. dispar		Rapid, multiplex PCR
Cyclosporiasis		Multiplex PCR
Giardiasis		DFA, EIA, rapid, multiplex NAT, multiplex PCR
Leishmaniasis	EIA, IFA (CDC), visceral/EIA (PDC)	PCR, rapid, PCR (Walter Reed Army Medical Center)
Malaria	IFA (4 species), IFA (PDC)	Rapid, PCR (CDC), PCR (PDC)
Microsporidiosis		IFA[d]
Toxoplasmosis	EIA IgG, EIA IgM immunocapture, ISAGA IgM, IFA, LA	PCR
Trichomoniasis		DFA, LA, DNA probe, rapid
Trypanosomiasis (African)	CATT, IFA, ELISA, IHA	PCR, PCR (PDC)
Helminths		
Angiostrongyliasis	EIA (PDC)	
Ascariasis	EIA (PDC)	
Baylisascariasis	IB (serum/CSF) (CDC)	
Cysticercosis	EIA, IB (serum/CSF) (CDC), serum/CSF (PDC)	
Echinococcosis	EIA, IB, EIA (PDC)	
Fascioliasis	EIA, IB, EIA (PDC)	
Filariasis	EIA (CDC)	Rapid, EIA
Gnathostomiasis	IB (Mahidol University, Thailand)	
Hookworm	EIA (PDC)	
Paragonimiasis	EIA, IB, WB (CDC), EIA (PDC)	
Schistosomiasis	EIA, IB, FAST-ELISA (CDC), EIA (PDC)	EIA
Strongyloidiasis	EIA (CDC), IgG_4 (PDC)	
Toxocariasis	EIA (CDC), EIA (PDC)	
Trichinellosis	BF, EIA (CDC), EIA (PDC)	

[a] BF, bentonite flocculation; CA, card agglutination; CATT, card agglutination test for trypanosomiasis; CDC, Centers for Disease Control and Prevention; CF, complement fixation; EIA, enzyme immunoassay; IB, immunoblot; IHA, indirect hemagglutination; IFA, indirect fluorescent antibody.

[b] ALT, alanine aminotransferase; AST, aspartate aminotransferase; DFA, direct fluorescent antibody; EIA, enzyme immunoassay; IFA, indirect fluorescent antibody; LA, latex agglutination; multiplex, molecular panels with organism panel targets (bacterial, viral, parasitic); NAT, nucleic acid test; PDC, Parasitic Disease Consultants (Dr. Irving G. Kagan); rapid, lateral-flow cartridge.

[c] Confirmation of the presence of *T. cruzi* antibodies for donor testing commonly requires the use of at least two tests that are based on different methods/antigens.

[d] Reagents are not commercially available (includes all PCR tests and IFA for microsporidia only).

SECTION 5

Specific Test Procedures and Algorithms

Practical Guide to Diagnostic Parasitology, Third Edition. Lynne S. Garcia
 DOI: 10.1128/9781683673637.ch05

Microscopy

CALIBRATION OF THE MICROSCOPE

Description

The identification of protozoa and other parasites depends on several factors, one of which is size. Any laboratory doing diagnostic work in parasitology should have a calibrated microscope available for precise measurements. Measurements are made using a micrometer disk that is placed in the ocular of the microscope; the disk is usually calibrated as a line divided into 50 units. Depending on the objective magnification used, the divisions in the disk represent different measurements. The ocular disk division must be compared with a known calibrated scale, usually a stage micrometer with a scale with 0.1- and 0.01-mm divisions.

Supplies

1. Ocular micrometer disk (line divided into 50 units) (any laboratory supply distributor: Fisher Scientific [Waltham, MA], Baxter Scientific [Leesburg, VA], Scientific Products [Warminster, PA], VWR [Radnor, PA], etc.)
2. Stage micrometer with a scale with 0.1- and 0.01-mm divisions (Fisher, Baxter, Scientific Products, VWR, etc.)
3. Immersion oil
4. Lens paper

Equipment

1. Binocular microscope with 10×, 40×, and 100× objectives. Other objective magnifications may also be used (50× oil or 60× oil immersion lenses).
2. Oculars should be 10×. Some may prefer 5×; however, lower magnification may make final identifications more difficult.
3. A single 10× ocular should be used to calibrate all laboratory microscopes (the same ocular needs to be used when any organism is being measured; do not interchange microscope parts).

Quality Control

1. If the microscope receives heavy use or is moved throughout the laboratory, it should probably be recalibrated once each year. It is up to each laboratory to set the recalibration schedule.
2. Often, the measurement of red blood cells (RBCs) (approximately 7.5 µm) is used to check the calibrations of the three magnifications (×100, ×400, and ×1,000).
3. Latex or polystyrene beads of a standardized diameter can be used to check the calculations and measurements (Sigma [St. Louis, MO] , J. T. Baker [Phillipsburg, NJ], etc.). Beads of 10 and 90 µm are recommended.
4. All measurements should be recorded in quality control records.

Detailed Procedure

1. Unscrew the eye lens of a 10× ocular and place the micrometer disk (engraved side down) within the ocular. Use lens paper (at least double thickness) to handle the disk; keep all surfaces free of dust and lint.

2. Place the calibrated micrometer on the stage and focus on the scale. You should be able to distinguish the difference between the 0.1- and 0.01-mm divisions. Make sure you understand the divisions on the scale before proceeding.
3. Adjust the stage micrometer so that the "0" line on the ocular micrometer is exactly lined up on top of the "0" line on the stage micrometer.
4. After these two "0" lines are lined up, do not move the stage micrometer any farther. Look to the right of the "0" lines for another set of lines that is superimposed. The second set of lines should be as far to the right of the "0" lines as possible; however, the distance varies with the objectives being used (Figure 5.1 The calculation will increase in accuracy the farther to the right the second set of lines is selected.
5. Count the number of ocular divisions between the "0" lines and the point where the second set of lines is superimposed. Then, on the stage micrometer, count the number of 0.1-mm divisions between the "0" lines and the second set of superimposed lines.
6. Calculate the portion of a millimeter that is measured by a single small ocular unit.
7. When the high dry and oil immersion objectives are used, the "0" line of the stage micrometer increases in size, whereas the ocular "0" line remains the same size. The thin ocular "0" line should be lined up in the center or at one edge of the broad stage micrometer "0" line. Thus, when the second set of superimposed lines is found, the thin ocular line should be lined up in the center or at the corresponding edge of the broad stage micrometer line.

EXAMPLE

A. $$\frac{\text{Stage reading (mm)}}{\text{Ocular reading}} \times \frac{1{,}000\ \mu\text{m}}{1\ \text{mm}} = \text{Ocular reading } (\mu\text{m})$$

B. Low power (10×):

$$\frac{0.8\ \text{mm}}{100\ \text{units}} \times \frac{1{,}000\ \mu\text{m}}{1\ \text{mm}} = 8.0\ \mu\text{m (factor)}$$

C. High dry power (40×):

$$\frac{0.1\ \text{mm}}{50\ \text{units}} \times \frac{1{,}000\ \mu\text{m}}{1\ \text{mm}} = 2.0\ \mu\text{m (factor)}$$

D. Oil immersion (100×):

$$\frac{0.05\ \text{mm}}{62\ \text{units}} \times \frac{1{,}000\ \mu\text{m}}{1\ \text{mm}} = 0.8\ \mu\text{m (factor)}$$

Example: If a helminth egg measures 15 ocular units by 7 ocular units (high dry objective), using the factor of 2.0 μm, the egg measures 30 by 14 μm and is probably *Clonorchis sinensis*.

Example: If a protozoan cyst measures 27 ocular units (oil immersion objective), using the factor of 0.8 μm, the cyst measures 21.6 μm.

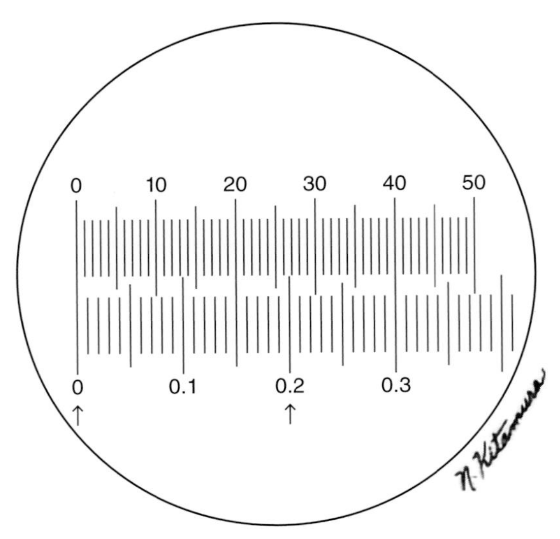

Figure 5.1 Ocular micrometer (top scale) and stage micrometer (bottom scale). (Reprinted from *Diagnostic Medical Parasitology*, 6th ed.)

Reporting

1. For each objective magnification, a factor will be generated (1 ocular unit = certain number of micrometers).
2. If standardized latex or polystyrene beads or an RBC is measured using various objectives, the size for the object measured should be the same (or very close), regardless of the objective magnification.
3. The multiplication factor for each objective should be posted (either on the base of the microscope or on a nearby wall or bulletin board) for easy reference.
4. Once the number of ocular lines per width and length of the organism is measured, then, depending on the objective magnification, the factor (1 ocular unit = certain number of micrometers) can be applied to the number of lines to obtain the width and length of the organism.
5. Comparison of these measurements with reference measurements in various books and manuals should confirm the organism identification.

Procedure Notes/Reminders

1. The final multiplication factors are only as good as your visual comparison of the ocular "0" and stage micrometer "0" lines.
2. As a rule of thumb, the high dry objective (40×) factor should be approximately 2.5 times the factor obtained from the oil immersion objective (100×). The low-power objective (10×) factor should be approximately 10 times that seen using the oil immersion objective (100×).

Procedure Limitations

1. After each objective has been calibrated, the oculars containing the disk and/or these objectives cannot be interchanged with corresponding objectives or oculars on another microscope.

2. Each microscope used to measure organisms must be calibrated as a unit. The original oculars and objectives that were used to calibrate the microscope must also be used when an organism is measured.
3. The objective containing the ocular micrometer can be stored until needed. This single ocular can be inserted when measurements are taken. However, this particular ocular containing the ocular micrometer disk must have also been used as the ocular during microscope calibration.

Ova and Parasite Examination

DIRECT WET FECAL SMEAR

Description

Normal mixing in the intestinal tract usually ensures an even distribution of organisms. However, depending on the level of infection, examination of the fecal material as a direct smear may or may not reveal organisms. The direct wet smear is prepared by mixing a small amount of stool (about 2 mg) with a drop of 0.85% NaCl; this mixture provides a uniform suspension under a 22- by 22-mm coverslip. Some workers prefer a 1.5- by 3-in. (1 in. = 2.54 cm) slide for the wet preparations rather than the standard 1- by 3-in. slide, which is routinely used for the permanent stained smear. A 2-mg sample of stool forms a low cone on the end of a wooden applicator stick. If more material is used for the direct mount, the suspension is usually too thick for an accurate examination; any sample of <2 mg results in the examination of too thin a suspension, thus decreasing the chances of finding organisms. If present, blood or mucus should always be examined as a direct mount. The entire 22- by 22-mm coverslip should be systematically examined with the low-power objective (10×) under low light intensity; any suspicious objects may then be examined with the high dry objective (40×) (Figure 5.2). Use of an oil immersion objective (100×) on mounts of this kind is not routinely recommended unless the coverslip is sealed to the slide (a no. 1 thickness coverslip is recommended for oil immersion). For a temporary seal, use a cotton-tipped applicator stick dipped in equal parts of heated paraffin and petroleum jelly. Nail polish can also be used to seal the coverslip. **Most workers think that the use of the oil immersion objective on this type of preparation is impractical, especially since morphological detail is more readily seen by oil immersion examination of the permanent stained smear. This is particularly true in a busy clinical laboratory situation.**

Figure 5.2 Method of scanning a direct wet film preparation with the 10× objective. (Illustration by Nobuko Kitamura; reprinted from *Diagnostic Medical Parasitology*, 6th ed.)

Reagents

▶ **Saline (0.85% NaCl)**

NaCl . . . . 0.85 g
Distilled water . . . . 100.0 ml

1. Dissolve the sodium chloride in distilled water in a flask or bottle, using a magnetic stirrer.
2. Distribute 10 ml into each of 10 screw-cap tubes.
3. Label as 0.85% NaCl with an expiration date of 1 year.
4. Sterilize by autoclaving at 121°C for 15 min.
5. When cool, store at 4°C.

▶ **D'Antoni's Iodine**

Potassium iodide . . . . 1.0 g
Powdered iodine crystals . . . . 1.5 g
Distilled water . . . . 100.0 ml

1. Dissolve the potassium iodide and iodine crystals in distilled water in a flask or bottle, using a magnetic stirrer.
2. The potassium iodide solution should be saturated with iodine, with some excess crystals left on the bottom of the bottle.
3. Store in a brown, glass-stoppered bottle at room temperature and in the dark.
4. This stock solution is ready for immediate use. Label as D'Antoni's iodine with an expiration date of 1 year (the stock solution remains good as long as an excess of iodine crystals remains on the bottom of the bottle).
5. Aliquot some of the iodine into a brown dropper bottle. The working solution should have a strong-tea color and should be discarded when it lightens (usually within 10 to 14 days). Note: The stock and working solution formulas are identical, but the stock solution is held in the dark and will retain the strong-tea color, while the working solution will fade and have to be periodically replaced.

▶ **Lugol's Iodine**

Potassium iodide . . . . 10.0 g
Powdered iodine crystals . . . . 5 g
Distilled water . . . . 100.0 ml

1. Follow the directions listed above for D'Antoni's iodine, including the expiration date of 1 year.
2. Dilute a portion 1:5 with distilled water for routine use (working solution).
3. Place this working solution into a brown dropper bottle. The working solution should have a **strong-tea color** and should be discarded when it lightens in color (usually within 10 to 14 days).

Quality Control

1. Check the working iodine solution each time it is used or periodically (once a week). The iodine solution should be free of any signs of bacterial or fungal contamination.

2. The color of the iodine should be that of strong tea (discard if it is too light).
3. Protozoan cysts stained with iodine should contain yellow-gold cytoplasm, brown glycogen material, and paler refractile nuclei. The chromatoidal bodies may not be as clearly visible as they were in a saline mount. Human white blood cells (buffy coat cells) mixed with negative stool can be used as a quality control (QC) specimen. The human cells when mixed with negative stool can be used as a QC specimen. The human cells will stain with the same color as that seen in the protozoa.
4. Protozoan trophozoite cytoplasm should stain pale blue and the nuclei should stain a darker blue with the methylene blue stain. Human leukocytes mixed with negative stool should stain the same colors as seen with the protozoa.
5. The microscope should be calibrated, and the original optics used for the calibration should be in place on the microscope when objects are measured. Although some feel that calibration is not required on a yearly basis, if the microscope receives heavy use, is in a position where it can be bumped, or does not receive routine maintenance, yearly calibration is recommended. The calibration factors for all objectives should be posted on the microscope or close by for easy access.
6. All QC results should be appropriately recorded; the laboratory should also have an action plan for results outside the parameters ("out-of-control" results).

Detailed Procedure

1. Place 1 drop of 0.85% NaCl on the left side of the slide and 1 drop of iodine (working solution) on the right side of the slide. If preferred, two slides can be used instead of one. One drop of Nair's methylene blue can also be placed on a separate slide, although this technique is less commonly used.
2. Take a small amount of fecal specimen (the amount picked up on the end of an applicator stick when introduced into the specimen) and thoroughly emulsify the stool in the saline and iodine preparations (use separate sticks for each).
3. Place a 22-mm coverslip (no. 1) on each suspension.
4. Systematically scan both suspensions with the 10× objective. The entire coverslip area should be examined under low power (total magnification, ×100).
5. If something suspicious is seen, the 40× objective can be used for more detailed study. At least one-third of the coverslip should be examined under high dry power (total magnification, ×400) even if nothing suspicious has been seen.
6. Another approach is to prepare and examine the saline mount and then add iodine at the side of the coverslip. The iodine will diffuse into the stool-saline mixture, providing some stain for a second examination. Remember, the iodine kills any organisms present; thus, no motility is seen after the iodine is added to the preparation.

Reporting

Protozoan trophozoites and/or cysts and helminth eggs and larvae may be seen and identified. In a heavy infection with *Cryptosporidium* spp., oocysts may be seen in a direct smear; however, some type of modified acid-fast stain or monoclonal antibody kit is normally used to detect these organisms, particularly when few oocysts are present. Oocysts of *Cystoisospora belli* can also be seen in a direct smear. Spores of the microsporidia are too small, and the shape resembles other debris within the stool; therefore, they are not readily visible in a direct smear.

1. Motile trophozoites and protozoan cysts may or may not be identified to the species level (depending on the clarity of the morphology).

   Examples: *Giardia lamblia* trophozoites
   Entamoeba coli cysts

2. Helminth eggs and/or larvae may be identified.

   Examples: *Ascaris lumbricoides* eggs
   Strongyloides stercoralis larvae

3. Coccidian oocysts may be identified.

   Example: *Cystoisospora belli* oocysts

4. Artifacts and/or other structures may also be seen and reported as follows (note: these crystals and cells are quantitated; however, the quantity is usually assessed when the permanent stained smear is examined under oil immersion):

   Examples: Moderate Charcot-Leyden crystals
   Few RBCs
   Moderate polymorphonuclear leukocytes (PMNs)

Procedure Notes/Reminders

1. In preserved specimens, the formalin replaces the saline and can be used as a direct smear; however, you will not be able to see any organism motility (organisms are killed by 5 or 10% formalin). Consequently, the direct wet smear is usually not performed when the specimen (already preserved) arrives in the laboratory. The technical time is better spent performing the concentration and permanent stained smear.
2. As mentioned above, some workers prefer to make the saline and iodine mounts on separate slides and on 2- by 3-in. slides. Often there is less chance of getting fluids on the microscope stage if separate slides are used (less total fluid on the slide and under the coverslip) or if larger slides are used (Figure 5.3).
3. The microscope light should be reduced for low-power observations, since most organisms are overlooked with bright light. This is particularly true when the preparation is being examined without the use of iodine. Illumination should be regulated so that some of the cellular elements in the feces show refraction. Most protozoan cysts and some coccidian oocysts are refractile under these light conditions.

Procedure Limitations

1. As mentioned above, because motility is lost when specimens are placed in preservatives, many laboratories no longer perform the direct wet smear (whose primary purpose is to see motility) but proceed directly to the concentration and permanent stained smear procedures as a better, more cost-effective use of personnel time.
2. Results obtained from wet smear examinations should be confirmed by permanent stained smears. Some protozoa are very small and difficult to identify to species level with just the direct wet smear technique. Confirmation is particularly important in the case of *Entamoeba histolytica*/*E. dispar* versus *E. coli*. Findings from the direct wet smear examination can be reported as "preliminary," and the final report can be submitted after the concentration and permanent stain procedures are completed.

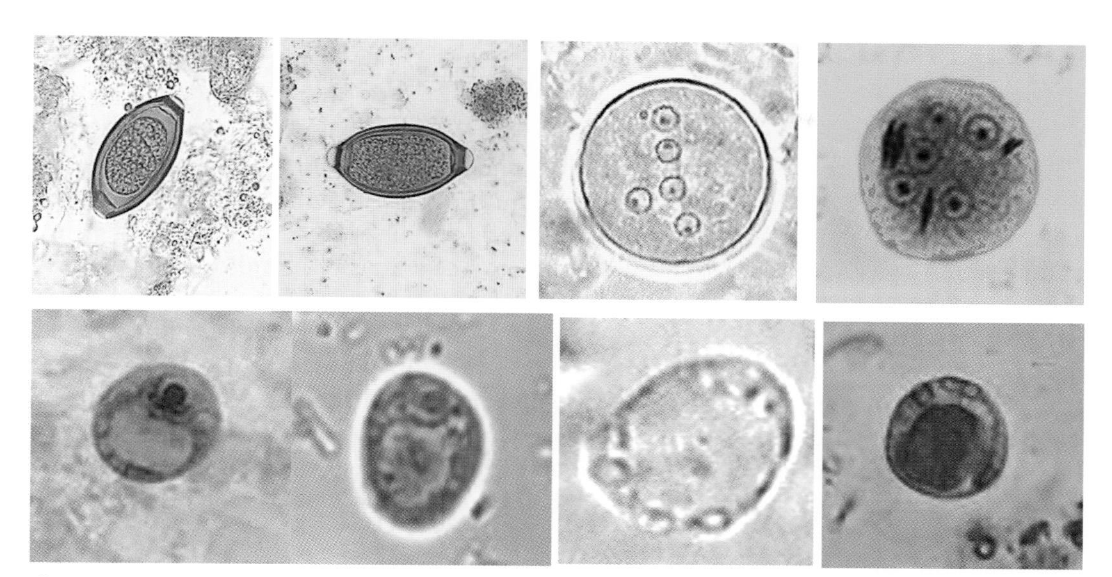

Figure 5.3 Direct wet mount and/or concentration wet mount organism images (with and without iodine). **(Top row)** *Trichuris trichiura* egg in saline; *T. trichiura* egg with iodine; *Entamoeba coli* cyst in saline; *E. coli* cyst with iodine (note the chromatoidal bars with sharp ends). **(Bottom row)** *Iodamoeba bütschlii* cyst in saline; *I. bütschlii* cyst with iodine (note the staining of glycogen vacuole); *Blastocystis* spp. central-body forms in saline; *Blastocystis* spp. with iodine.

REVIEW: DIRECT WET FECAL SMEAR	
Clinical relevance	To assess the parasite burden of the patient (protozoa, helminths), to provide a quick diagnosis of heavily infected specimens, to check organism motility (protozoan trophozoites, helminth larvae), and to diagnose motile organisms that might not be seen by using concentration or permanent stain methods
Specimen	Any fresh stool specimen that has not been refrigerated. Motile organisms are much more likely to be seen in liquid or very soft stools; formed stool often contains only the nonmotile cyst forms. However, if the specimen is not received in the laboratory within an hour, motility may not be visible, even if trophozoites are present.
Reagents	0.85% NaCl; Lugol's or D'Antoni's iodine. **DO NOT USE GRAM'S IODINE.**
Examination requirements	Low-power examination (100×) of entire 22- by 22-mm coverslip preparation (both saline and iodine); high dry power examination (400×) of at least one-third to one-half of the coverslip area (both saline and iodine). **DO NOT USE OIL IMMERSION OBJECTIVES ON THIS TYPE OF PREPARATION.**
Results and laboratory reports	Results from the direct smear examination should often be considered presumptive; however, some organisms can be definitively identified (*Giardia lamblia* cysts and *Entamoeba coli* cysts, helminth eggs and larvae, *Cystoisospora belli* oocysts). The report would be available after the results of both the concentration and permanent stained smear were available.
Procedure notes and limitations	Once iodine is added to the preparation, the organisms are killed and motility is lost. **Specimens that arrive in the laboratory already preserved do not require a direct smear examination; proceed to the concentration and permanent stained smear (consistent with the College of American Pathologists checklist).** Direct smears are normally examined at low (100×) and high dry (400×) power; oil immersion examination (1,000×) is not recommended (organism morphology is not that clear). Do not make the smears too thick.

CONCENTRATION (Sedimentation and Flotation)

Description

Fecal concentration is a routine part of the complete ova and parasite examination (O&P exam) for parasites and allows the detection of small numbers of organisms that may be missed by using only a direct wet smear. There are two types of concentration procedures, sedimentation and flotation, both of which are designed to separate protozoan organisms and helminth eggs and larvae from fecal debris by centrifugation and/or differences in specific gravity (Figure 5.4).

Sedimentation methods (using centrifugation) lead to the recovery of all protozoa, oocysts, microsporidial spores, eggs, and larvae present; however, since the sediment will be examined, the preparation contains more debris. Although some workers recommend using flotation and sedimentation procedures, this approach is impractical for the majority of laboratories. If one technique is to be selected for routine use, the sedimentation procedure is recommended as being easier to perform and less subject to technical error.

A **flotation procedure** permits the separation of protozoan cysts, coccidian oocysts, and certain helminth eggs and larvae through the use of a liquid with a high specific gravity. The parasitic elements are recovered in the surface film, and the debris remains in the bottom of the tube. Since the surface film is examined, this technique yields a cleaner preparation than does the sedimentation procedure; however, some helminth eggs (operculated eggs and/or very dense eggs, such as unfertilized *Ascaris* eggs) do not concentrate well with the flotation method. **Laboratories that use only flotation procedures may fail to recover all of the parasites present; to ensure detection of all organisms in the sample, both the surface film and the sediment should be carefully examined.** Directions for any flotation technique must be followed exactly to produce reliable results.

Commercial Devices

There are a number of commercially available fecal concentration devices which may help a laboratory to standardize the concentration technique. Standardization is particularly important when personnel rotate throughout the laboratory, as they may not be

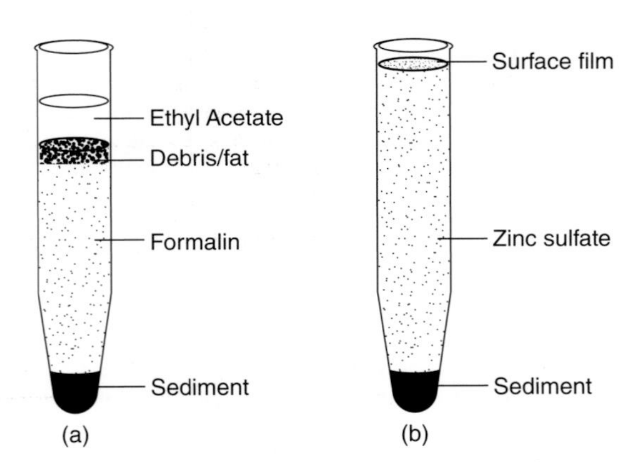

Figure 5.4 Fecal concentration procedures. Various layers are seen in tubes after centrifugation. (A) Formalin-ether (or ethyl acetate). (B) Zinc sulfate (the surface film should be within 2 to 3 mm of the tube rim). (Illustration by Sharon Belkin.)

familiar with parasitology techniques. These devices help ensure consistency, thus leading to improved parasite recovery and subsequent identification. Some of the systems are enclosed and provide a clean, odor-free approach to stool processing, features that may be important to nonmicrobiology personnel processing such specimens. Both 15- and 50-ml systems are available. It is important to remember that you want a maximum of 1.0 ml of sediment in the bottom of the tube. **The amount of sediment in the bottom of the tube is often excessive when 50-ml systems are used.** You can remedy this problem by adding less of the fecal specimen to the concentration system prior to centrifugation. Since the sediment is normally mixed thoroughly and 1 drop is transferred to a coverslip for examination, good mixing may not occur if too much sediment is used. There also appears to be layering in the bottom of the tubes; again, adding less material to the concentrator in the beginning should help eliminate this problem. It is important to review the package insert directions for any of the commercial concentration systems and/or particular fixatives in use.

SEDIMENTATION CONCENTRATION (Formalin-Ethyl Acetate)

By centrifugation, this concentration procedure leads to the recovery of all protozoa, eggs, and larvae present; however, the preparation contains more debris than is found with the flotation procedure. Ethyl acetate is used to extract debris and fat from the feces and leave the parasites at the bottom of the suspension. The formalin-ethyl acetate sedimentation concentration is recommended because it is the easiest to perform, allows recovery of the broadest range of organisms, and is least subject to technical error.

The specimen must be fresh or formalinized stool (5 or 10% buffered or nonbuffered formalin or sodium acetate-acetic acid-formalin [SAF]). Specimens from the single-vial collection systems can also be used (with the universal fixative Total-Fix), as can preservatives containing polyvinyl alcohol (PVA) (the sediment is not as good, and some organisms are difficult to see; **preservatives that do NOT contain PVA are preferred**).

Reagents

▶ **5 or 10% Formalin**

Formaldehyde (USP) . . . . 100 ml (for 10%) or 50 ml (for 5%)

▶ **Saline Solution**

0.85% NaCl . . . . 900 ml (for 10%) or 950 ml (for 5%)

Note: Formaldehyde is normally purchased as a 37 to 40% HCHO solution; however, for dilution it should be considered to be 100%.

Dilute 100 ml of formaldehyde with 900 ml of saline solution. (Distilled water may be used instead of saline solution.)

Quality Control

1. Check the liquid reagents each time they are used; the formalin and saline should appear clear, without any visible contamination.
2. The microscope should be calibrated (within the last 12 months), and the objectives and oculars used for the calibration procedure should be in place on the microscope when objects are measured. The calibration factors for all objectives should be posted on the microscope or close by for easy access. Although some feel that a microscope

does not require calibration every 12 months, if the microscope is moved periodically, can be easily bumped, or does not receive adequate maintenance, it should be rechecked yearly for calibration accuracy.

3. Known positive specimens should be concentrated and organism recovery should be verified at least quarterly, and particularly after the centrifuge has been recalibrated.
4. All QC results should be appropriately recorded; the laboratory should also have an action plan for "out-of-control" results.

Detailed Procedure

1. Transfer 0.5 teaspoon (about 4 g) of fresh stool into 10 ml of 5 or 10% formalin in a shell vial, unwaxed paper cup, or round-bottom tube (the container may be modified to suit individual laboratory preferences). Mix the stool and formalin thoroughly, and let the mixture stand for a minimum of 30 min for fixation. If the specimen is already in 5 or 10% formalin (or SAF or another single-vial fixative), restir the stool-fixative mixture.
2. Depending on the size and viscosity of the specimen, strain a sufficient quantity through wet gauze (no more than two layers of gauze and one layer if the new "pressed" gauze [e.g., nonsterile three-ply gauze; product 7636, Johnson & Johnson, New Brunswick, NJ] is used) into a conical 15-ml centrifuge tube to give the desired amount of sediment (0.5 to 1 ml) for step 3 below. Usually 8 ml of the stool-formalin mixture prepared in step 1 is sufficient. If the specimen is received in vials of preservative (5 or 10% formalin, SAF, or other single-vial systems), unless the specimen has very little stool in the vial, approximately 3 to 4 ml is sufficient. If the specimen contains a lot of mucus, do not strain through gauze but immediately fix in 5 or 10% formalin for 30 min and centrifuge for 10 min at 500 × *g*. Other fixatives are also acceptable (preferably those without PVA). Then proceed directly to step 10.
3. Add 0.85% NaCl or 5 or 10% formalin (some workers prefer to use formalin for all rinses) almost to the top of the tube and centrifuge for 10 min at 500 × *g*. The amount of sediment obtained should be approximately 0.5 to 1 ml.
4. Decant the supernatant fluid, and resuspend the sediment in saline or formalin; add saline or formalin almost to the top of the tube, and centrifuge again for 10 min at 500 × *g*. This second wash may be eliminated if the supernatant fluid after the first wash is light tan or clear. Some prefer to limit the wash to one step (regardless of the clarity or color of the supernatant fluid after centrifugation) to eliminate additional manipulation of the specimen prior to centrifugation **(some organisms may be lost with each centrifugation and wash step).**
5. Decant the supernatant fluid and resuspend the sediment on the bottom of the tube in 5 or 10% formalin. Fill the tube half full only. If the amount of sediment left in the bottom of the tube is very small or the original specimen contained a lot of mucus, do not add ethyl acetate in step 6; merely add the formalin, spin, decant, and examine the remaining sediment.
6. Add 4 to 5 ml of ethyl acetate. Stopper the tube, and shake vigorously for at least 30 s. Hold the tube so that the stopper is directed away from your face.
7. After a 15- to 30-s wait, carefully remove the stopper.
8. Centrifuge for 10 min at 500 × *g*. Four layers should result: a small amount of sediment (containing the parasites) in the bottom of the tube, a layer of formalin, a plug of fecal debris on top of the formalin layer, and a layer of ethyl acetate at the top.

9. Free the plug of debris by ringing the plug with an applicator stick; decant all of the supernatant fluid. After proper decanting, a drop or two of fluid remaining on the side of the tube may run down into the sediment. Mix this fluid with the sediment.
10. If the sediment is still somewhat solid, add 1 or 2 drops of saline or formalin to the sediment, mix, place a small amount of material on a slide, add a coverslip (22 by 22 mm, no. 1), and examine.
11. Systematically scan with the 10× objective. The entire coverslip area should be examined under low power (total magnification, ×100).
12. If something suspicious is seen, the 40× objective can be used for more detailed study. At least one-third of the coverslip should be examined under high dry power (total magnification, ×100) even if nothing suspicious has been seen. As with the direct wet smear, iodine can be added to enhance morphologic detail, and the coverslip can be tapped in an attempt to see objects move and turn over.

Reporting

Protozoan trophozoites and/or cysts and helminth eggs and larvae may be seen and identified. Protozoan trophozoites are less likely to be seen. In a heavy infection with *Cryptosporidium* spp., oocysts may be seen in the concentrate sediment; oocysts of *C. belli* can also be seen. Spores of the microsporidia are too small, and the shape resembles other debris within the stool; therefore, they are not readily visible in the concentration sediment.

1. Protozoan cysts may or may not be identified to the species level (depending on the clarity of the morphology).

   Example: *Entamoeba coli* cysts

2. Helminth eggs and/or larvae may be identified.

   Examples: *Ascaris lumbricoides* eggs
   Strongyloides stercoralis larvae

3. Coccidian oocysts may be identified.

   Example: *Cystoisospora belli* oocysts

4. Artifacts and/or other structures may also be seen and reported as follows (note: these crystals and cells are quantitated; however, the quantity is usually assessed when the permanent stained smear is examined under oil immersion).

   Examples: Moderate Charcot-Leyden crystals
   Few RBCs
   Moderate PMNs

Procedure Notes/Reminders

1. The gauze should never be more than one (pressed gauze) or two (woven gauze) layers thick; more gauze may trap mucus (containing *Cryptosporidium* oocysts and/or microsporidial spores).
2. Tap water may be substituted for 0.85% NaCl throughout this procedure, although the addition of water to fresh stool may cause *Blastocystis* spp. cyst (central-body) forms to rupture and is not recommended. In addition to the original 5 or 10% formalin fixation, some workers prefer to use 5 or 10% formalin for all rinses throughout the procedure.

3. Ethyl acetate is widely recommended as a substitute for ether. It can be used in the same way in the procedure and is much safer. Hemo-De can also be used and is thought to be safer than ethyl acetate.
   a. After the plug of debris is rimmed and excess fluid is decanted, while the tube is still upside down, the sides of the tube can be swabbed with a cotton-tipped applicator stick to remove excess ethyl acetate. This is particularly important if you are working with plastic centrifuge tubes or plastic commercial concentrators. If the sediment is too dry after the tube has been swabbed, add several drops of saline before preparing the wet smear for examination.
   b. If there is excess ethyl acetate in the smear of the sediment prepared for examination, bubbles will be present and will obscure the material that you are trying to see.
4. If specimens are received in SAF, begin the procedure at step 2.
5. If specimens are received in PVA, the first two steps of the procedure should be modified as follows.
   a. Immediately after stirring the stool-PVA mixture with applicator sticks, pour approximately half of the mixture into a tube (container optional) and add 0.85% NaCl (or 5 or 10% formalin) almost to the top of the tube.
   b. Filter the stool-PVA-saline (or formalin) mixture through wet gauze into a 15-ml centrifuge tube. Follow the standard procedure from here to completion, beginning with step 3.
6. Too much or too little sediment results in an ineffective concentration.
7. The centrifuge should reach the recommended speed before the centrifugation time is monitored. However, since most laboratories have their centrifuges on automatic timers, the centrifugation time in this protocol takes into account the fact that some time is spent coming up to speed prior to full-speed centrifugation. If the centrifugation time at the proper speed is reduced, some of the organisms (*Cryptosporidium* oocysts or microsporidial spores) may not be recovered in the sediment.

Procedure Limitations

1. Results obtained with wet smears (direct wet smears or concentrated specimens) should always be confirmed by permanent stained smears. Some protozoa are very small and difficult to identify to the species level with just the direct wet smears. Also, special stains are sometimes necessary for organism identification.
2. Confirmation is particularly important in the case of *E. histolytica*/*E. dispar* versus *E. coli*.
3. Certain organisms (*G. lamblia*, hookworm eggs, and occasionally *Trichuris* eggs) may not concentrate as well from specimens in preservative containing PVA as they do from those preserved in formalin. However, if enough *G. lamblia* organisms are present to concentrate from formalin, the PVA should contain enough for detection on the permanent stained smear. In clinically important infections, the number of helminth eggs present ensures detection regardless of the type of preservative used. Also, the morphology of *S. stercoralis* larvae is not as clear when PVA is used as when specimens are fixed in formalin.
4. For unknown reasons, *Cystoisospora belli* oocysts are routinely missed in the sediment when concentrated from specimens in preservative with PVA. The oocysts would be found if the same specimen were preserved in formalin rather than in preservative containing PVA.
5. In past studies, recommended centrifugation times have not taken into account potential problems with the recovery of *Cryptosporidium* oocysts. There is evidence

(unpublished) to strongly indicate that *Cryptosporidium* oocysts may be missed unless the centrifugation speed is 500 × *g* for a minimum of 10 min.

6. Adequate centrifugation time and speed have become very important for recovery of microsporidial spores. In some earlier studies, use of uncentrifuged material was recommended. However, the UCLA Clinical Microbiology Laboratory has found that centrifugation for 10 min at 500 × *g* definitely increases the number of microsporidial spores available for staining and subsequent examination.

SEDIMENTATION CONCENTRATION USING THE UNIVERSAL FIXATIVE (Total-Fix)

Description

The Total-Fix stool collection kit is a single-vial system that provides a standardized method for untrained personnel to properly collect and preserve stool specimens for the detection of helminth larvae and eggs, protozoan trophozoites and cysts, coccidian oocysts, and microsporidian spores. Concentrations, permanent stains, most fecal immunoassays, special stains, and some molecular testing can be performed from a Total-Fix-preserved specimen. Total-Fix is a mercury-, formalin-, and PVA-free fixative that preserves parasite morphology and helps with disposal and monitoring problems encountered by laboratories. The product contains zinc sulfate, acetic acid, and alcohols in water; however, the formula is proprietary and is patented (see directions below).

Reagents (same as for formalin-ethyl acetate concentration)

Quality Control (same as for formalin-ethyl acetate concentration)

Detailed Procedure (Total-Fix)

1. Add Total-Fix and fecal specimen to a specimen vial (stool-fixative ratio of 1:2); MIX WELL. (Allow to fix 30 min before processing.)
2. Use the fluid at the top of the vial to perform the fecal immunoassays. (Some kit directions recommend shaking the vial. **After shaking, allow particulate matter to settle, and then use the fluid at the top for the immunoassays.**) Use the centrifuged specimen for direct fluorescent-antibody assay (DFA) and special stains (coccidia, microsporidia) (step 7).
3. Tube or concentration device; add approximately 1 to 2 ml of fixative/specimen mix to the tube or device (mix well prior to pouring into the tube or device). (DO NOT ADD ANY RINSE FLUIDS, such as saline, water, or formalin; rinse fluids will prevent routine permanent staining.)
4. Centrifuge for 10 min at 500 × *g*.
5. Pour off most of the excess fixative.
6. Mix sediment with remaining fixative.
7. **Note: STEP 7 IS FOR SMEAR PREPARATION FOR STAINING—THIS STEP COMES FIRST IN THIS PROCEDURE PRIOR TO THE ACTUAL ROUTINE SEDIMENTATION CONCENTRATION. PREPARE SLIDES FOR PERMANENT STAINS and/or DFA; STAIN SLIDES.** Allow slides to dry for ~30 to 45 min at 37°C. Slides can be placed on trays and put in the incubator; room temperature drying requires ~60 min. **ADEQUATE DRYING OF SMEARS IS MANDATORY PRIOR TO STAINING (TRICHROME, IRON HEMATOXYLIN, OR SPECIAL STAINS).**

8. Add rinse fluid to remaining stool sediment (from step 5); mix sediment and rinse fluid well before centrifugation. **PROCEED WITH ROUTINE CONCENTRATION PROCEDURE.** (A single rinse including the ethyl acetate step is recommended, but DO NOT ADD ethyl acetate if the specimen contains a lot of mucus.)
9. Ring debris layer and pour off excess fluid; mix and examine sediment as concentration sediment wet mount (entire 22- by 22-mm coverslip: low power [10× objective]; one-third to one-half of the coverslip: high dry power [40× objective]).

FLOTATION CONCENTRATION (Zinc Sulfate)

Description

The flotation procedure permits the separation of protozoan cysts and eggs of certain helminths from excess debris through the use of a liquid (zinc sulfate) with a high specific gravity. The parasitic elements are recovered in the surface film, and the debris remains in the bottom of the tube. This technique yields a cleaner preparation than does the sedimentation procedure; however, some helminth eggs (operculated eggs and/or very dense eggs, such as unfertilized *Ascaris* eggs) do not concentrate well with the flotation method; a sedimentation technique is recommended to detect these infections. Due to this limitation, most laboratories perform the sedimentation concentration procedure, as it is more practical and yielding more complete information.

When the zinc sulfate solution is prepared, the specific gravity should be 1.18 for fresh stool specimens; it must be checked with a hydrometer. This procedure may be used on formalin-preserved specimens if the specific gravity of the zinc sulfate is increased to 1.20; however, this may cause more distortion in the organisms present. To ensure detection of all possible organisms, both the surface film and the sediment must be examined. For most laboratories, this is not a practical approach.

The specimen must be fresh or formalinized stool (5 or 10% buffered or nonbuffered formalin or SAF). Specimens in PVA-containing fixative can also be used.

Reagents

▶ **Zinc Sulfate (33% aqueous solution)**

Zinc sulfate . . . . 330 g
Distilled water . . . . 670 ml

1. Dissolve the zinc sulfate in distilled water in an appropriate flask or beaker, using a magnetic stirrer.
2. Adjust the specific gravity to 1.20 by the addition of more zinc sulfate or distilled water. Use specific gravity of 1.18 when using fresh stool (nonformalinized).
3. Store in a glass-stoppered bottle with an expiration date of 24 months.

Quality Control

1. Check the reagents each time they are used. The formalin, saline, and zinc sulfate should appear clear, without any visible contamination.
2. The microscope should be calibrated, and the objectives and oculars used for the calibration procedure should be used for all measurements on the microscope. The calibration factors for all objectives should be posted on the microscope or close by for easy access. As mentioned earlier, some feel that recalibration of the microscope is not necessary each year; however, this would depend on the use and maintenance of the equipment.

3. Known positive specimens should be concentrated and organism recovery should be verified at least quarterly and particularly after the centrifuge has been recalibrated.
4. All QC results should be appropriately recorded; the laboratory should also have an action plan for "out-of-control" results.

Detailed Procedure

1. Transfer 0.5 teaspoon (about 4 g) of fresh stool into 10 ml of 5 or 10% formalin in a shell vial, unwaxed paper cup, or round-bottom tube (the container may be modified to suit individual laboratory preferences). Mix the stool and formalin thoroughly, and let the mixture stand for a minimum of 30 min for fixation. If the specimen is already in 5 or 10% formalin (or SAF), restir the stool-formalin mixture.
2. Depending on the size and density of the specimen, strain a sufficient quantity through wet gauze (no more than two layers of gauze or one layer if the new "pressed" gauze [e.g., nonsterile three-ply gauze; product 7636; Johnson & Johnson] is used) into a conical 15-ml centrifuge tube to give the desired amount of sediment (0.5 to 1 ml) for step 3 below. Usually, 8 ml of the stool-formalin mixture prepared in step 1 is sufficient. If the specimen is received in vials of preservative (5 or 10% formalin or SAF), unless the specimen has very little stool in the vial, approximately 3 to 4 ml is sufficient. If the specimen contains a lot of mucus, do not strain through gauze but immediately fix in 5 or 10% formalin for 30 min and centrifuge for 10 min at 500 × *g*. Then proceed directly to step 5.
3. Add 0.85% NaCl almost to the top of the tube, and centrifuge for 10 min at 500 × *g*. The amount of sediment obtained should be approximately 0.5 to 1 ml. Too much or too little sediment will result in an ineffective concentration.
4. Decant the supernatant fluid, resuspend the sediment in 0.85% NaCl almost to the top of the tube, and centrifuge for 10 min at 500 × *g*. This second wash may be eliminated if the supernatant fluid after the first wash is light tan or clear. Some prefer to limit the wash to one step (regardless of the color and clarity of the supernatant fluid) to eliminate additional manipulation of the specimen prior to centrifugation.
5. Decant the supernatant fluid, and resuspend the sediment on the bottom of the tube in 1 to 2 ml of zinc sulfate. Fill the tube within 2 to 3 mm of the rim with additional zinc sulfate.
6. Centrifuge for 2 min at 500 × *g*. Allow the centrifuge to come to a stop without interference or vibration. Two layers should result: a small amount of sediment in the bottom of the tube and a layer of zinc sulfate. The protozoan cysts and some helminth eggs are found in the surface film; some operculated and/or heavy eggs are found in the sediment.
7. Without removing the tube from the centrifuge, remove 1 or 2 drops of the surface film with a Pasteur pipette or a freshly flamed (and allowed to cool) wire loop and place it on a slide. Do not use the loop as a dipper; simply touch the surface (bend the loop portion of the wire 90° so that the loop is parallel with the surface of the fluid) (Figure 5.5). Make sure that the pipette tip or wire loop is not below the surface film.
8. Add a coverslip (22 by 22 mm; no. 1) to the preparation. Iodine may also be added.
9. Systematically scan with the 10× objective. The entire coverslip area should be examined under low power (total magnification, ×100).

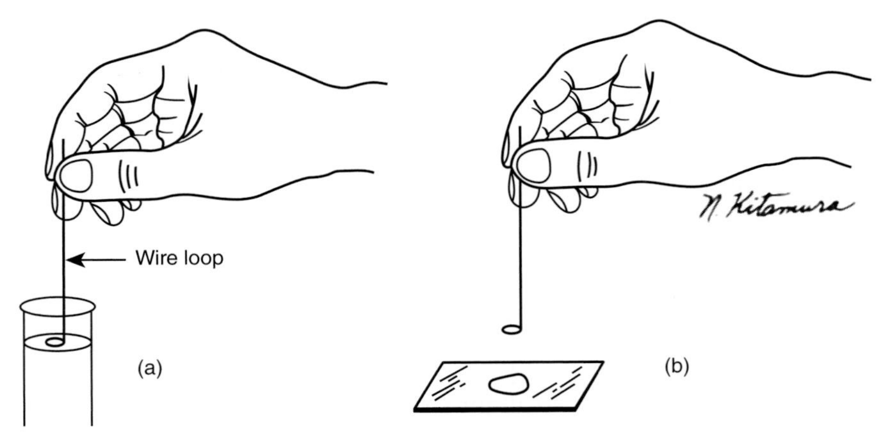

Figure 5.5 Method used to remove surface film in the zinc sulfate flotation concentration procedure. (a) A wire loop is gently placed on (not under) the surface film. (b) The loop is then placed on a glass slide. (Illustration by Nobuko Kitamura; reprinted from *Diagnostic Medical Parasitology*, 6th ed.)

10. If something suspicious is seen, the 40× objective can be used for more detailed study. At least one-third of the coverslip should be examined under high dry power (total magnification, ×400), even if nothing suspicious has been seen. As in the direct wet smear, iodine can be added to enhance morphologic detail, and the coverslip can be tapped gently in an attempt to observe objects moving and turning over.

Reporting

Protozoan trophozoites and/or cysts and some helminth eggs and larvae may be seen and identified. Heavy helminth eggs and operculated eggs do not float in zinc sulfate; they may be seen in the sediment within the tube. The high specific gravity of the zinc sulfate causes the operculum to pop open; the egg fills with fluid and sinks to the bottom. Protozoan trophozoites are less likely to be seen. In a heavy infection with *Cryptosporidium* spp., oocysts may be seen in the concentrate sediment; oocysts of *C. belli* can also be seen. Spores of the microsporidia are too small, and the shape resembles other debris within the stool; therefore, they are not readily visible in the concentration sediment.

1. Protozoan cysts may or may not be identified to the species level (depending on the clarity of the morphology).

   Example: *Giardia lamblia* cysts

2. Helminth eggs and/or larvae may be identified.

   Example: Hookworm eggs

3. Coccidian oocysts may be identified.

   Example: *Cystoisospora belli* oocysts

4. Artifacts and/or other structures may also be seen and reported as follows (note: these cells are quantitated; however, the quantity is usually assessed when the permanent stained smear is examined).

   Examples: Few macrophages
Moderate PMNs

Procedure Notes/Reminders

1. The gauze should never be more than one or two layers thick; more gauze may trap mucus (containing *Cryptosporidium* oocysts and/or microsporidial spores). A round-bottom tube is recommended rather than a centrifuge tube.
2. Tap water may be substituted for 0.85% NaCl throughout this procedure, although the addition of water to fresh stool may cause *Blastocystis* (central-body forms) to rupture and is not recommended. In addition to the original 5 or 10% formalin fixation, some workers prefer to use 5 or 10% formalin for all rinses throughout the procedure.
3. If fresh stool is used (nonformalin preservatives), the zinc sulfate should be prepared with a specific gravity of 1.18. If formalinized specimens are to be concentrated, the zinc sulfate should have a specific gravity of 1.20.
4. If specimens are received in SAF, begin the procedure at step 2.
5. If fresh specimens are received, the standardized procedure requires the stool to be rinsed in distilled water prior to the addition of zinc sulfate in step 4. However, the addition of fresh stool to distilled water may destroy some *Blastocystis* spp. cyst (central-body) forms present and is not a recommended approach.
6. Some workers prefer to remove the tubes from the centrifuge prior to sampling the surface film. This is acceptable; however, there is more chance that the surface film will be disturbed prior to sampling.
7. Some workers prefer to add a small amount of zinc sulfate to the tube so that the fluid forms a slightly convex meniscus. A coverslip is then placed on top of the tube so that the undersurface touches the meniscus. The tube is then left undisturbed for 5 min, after which the coverslip is carefully removed and placed on a slide for examination. This approach tends to be somewhat messy, particularly if too much zinc sulfate has been added.
8. When using the hydrometer (solution at room temperature), mix the solution well. Float the hydrometer in the solution, giving it a slight twist to ensure that it is completely free from the sides of the container. Read the bottom meniscus, and correct the figure for temperature if necessary. Most hydrometers are calibrated at 20°C. A difference of 3°C between the solution temperature (room temperature) and the hydrometer calibration temperature requires a correction of 0.001 to be added if above and subtracted if below 20°C.

Procedure Limitations

1. Results obtained with wet smears (direct wet smears or concentrated specimens) should be confirmed by permanent stained smears. Some protozoa are very small and difficult to identify to the species level with just the direct wet smears. Also, special stains are sometimes necessary for organism identification.
2. Confirmation is particularly important in the case of *E. histolytica*/*E. dispar* versus *E. coli*.
3. Protozoan cysts and thin-shelled helminth eggs are subject to collapse and distortion when left for more than a few minutes in contact with the high-specific-gravity zinc sulfate. **The surface film should be removed for examination within 5 min of the time the centrifuge comes to a stop.** The longer the organisms are in contact with the zinc sulfate, the more distortion will be seen on microscopic examination of the surface film.

4. Since most laboratories have their centrifuges on automatic timers, the centrifugation time in this protocol takes into account the fact that some time will be spent coming up to speed prior to full-speed centrifugation.
5. If zinc sulfate is the only concentration method used, both the surface film and the sediment should be examined to ensure detection of all possible organisms.

REVIEW: CONCENTRATION	
Clinical relevance	To concentrate the parasites present, through either sedimentation or flotation. The concentration is specifically designed to allow recovery of protozoan cysts, apicomplexan oocysts, microsporidian spores, and helminth eggs and larvae.
Specimen	Any stool specimen that is fresh or preserved in formalin, in PVA-containing preservative (mercury or non-mercury based), SAF, merthiolate-iodine-formalin (MIF), or the newer single-vial-system fixatives, such as Total-Fix
Reagents	5 or 10% formalin, ethyl acetate, or zinc sulfate (specific gravity of 1.18 for fresh stool or 1.20 for preserved stool); 0.85% NaCl; Lugol's or D'Antoni's iodine
Examination requirements	Low-power examination (×100) of entire 22- by 22-mm coverslip preparation (iodine is recommended but is optional); high dry power examination (×400) of at least one-third to one-half of the coverslip area (both saline and iodine)
Results and laboratory reports	Results from the concentration examination should often be considered presumptive; however, some organisms can be definitively identified (*Giardia lamblia* cysts and *Entamoeba coli* cysts, helminth eggs and larvae, *Cystoisospora belli* oocysts). The final report is available after the results of the concentration and permanent stained smear are available.
Procedure notes and limitations	Formalin-ethyl acetate sedimentation concentration is the most commonly used. **Zinc sulfate flotation does not detect operculated or heavy eggs; both the surface film and sediment must be examined before a negative result is reported.** Smears prepared from concentrated stool are normally examined at low (×100) and high dry (×400) power; oil immersion examination (×1,000) is not recommended (organism morphology is not that clear). The addition of too much iodine may obscure helminth eggs (which mimic debris).

PERMANENT STAINED SMEAR

Description

Detection and correct identification of many intestinal protozoa frequently depend on examination of the permanent stained smear with the oil immersion lens (100× objective). **Although an experienced microscopist can occasionally identify certain organisms on a wet preparation, most identifications should be considered tentative until confirmed by the permanent stained smear.** The smaller protozoan organisms are frequently seen on the stained smear when they are easily missed with only the direct smear and concentration methods. For these reasons, the permanent stain is recommended for every stool sample submitted for a routine parasite examination.

A number of staining techniques are available; selection of a particular method may depend on the degree of difficulty of the procedure and the amount of time necessary to complete the stain. The older classical method is the long Heidenhain's iron hematoxylin method; however, for routine diagnostic work most laboratories select one of the shorter procedures, such as the trichrome method or one of the modified methods involving iron hematoxylin.

Most problems encountered in the staining of protozoan trophozoites and cysts in fecal smears occur because the specimen is too old, the smears are too dense, the smears are allowed to dry before fixation, or fixation is inadequate. There is variability in fixation in that immature cysts fix more easily than mature cysts, and *E. coli* cysts require a longer fixation time than do those of other species.

Preparation of Material for Staining

Fresh Material

1. When the specimen arrives, prepare two slides with applicator sticks or brushes and immediately (without drying) place them in Schaudinn's fixative. Allow the slides to fix for a minimum of 30 min; overnight fixation is acceptable. The amount of fecal material smeared on the slide should be thin enough that newsprint can be read through the smear. Smears preserved in liquid Schaudinn's fixative should be placed in 70% alcohol to remove the excess fixative before being placed in iodine-alcohol (used for mercury-based fixatives).
2. If the fresh specimen is liquid, place 3 or 4 drops of PVA-containing fixative on the slide, mix several drops of fecal material with the PVA, spread the mixture, and allow it to dry for several hours in a 37°C incubator or overnight at room temperature. This approach works for liquid specimens or those containing a lot of mucus; it is not recommended for more formed specimens.
3. Proceed with the trichrome staining procedure by placing the slides in iodine-alcohol.

Preserved Material Containing PVA

1. Stool specimens that are preserved in fixatives containing PVA should be allowed to fix for at least 30 min. Thoroughly mix the contents of the PVA bottle with two applicator sticks.
2. Pour some of the well-mixed preservative-stool mixture onto a paper towel, and allow it to stand for 3 min to absorb out PVA. **Do not eliminate this step.**
3. With an applicator stick (or brush), apply some of the stool material from the paper towel to two slides and allow them to dry for several hours in a 37°C incubator or overnight at room temperature. Note: The preservative-stool mixture should be spread to the edges of the glass slide; this will cause the film to adhere to the slide during staining. It is also important to thoroughly dry the slides to prevent the material from washing off during staining.
4. Place the dry slides into iodine-alcohol. There is no need to give them a 70% alcohol rinse first, because the smears are already dry (unlike the wet smears coming out of the Schaudinn's fixative).

SAF-Preserved Material

1. Mix the SAF-stool mixture thoroughly, and strain it through gauze into a 15-ml centrifuge tube.
2. After centrifugation (1 min at 500 × *g*), decant the supernatant fluid. Although stains for the coccidia are not recommended with preservatives containing PVA, remember that the centrifuge speed indicated here is probably not sufficient to recover the oocysts. Recommended centrifugation parameters are 10 min at 500 × *g*. The final sediment should be about 0.5 to 1.0 ml. If necessary, adjust by repeating step 1 or by resuspending the sediment in saline (0.85% NaCl) and removing part of the suspension.

3. Prepare a smear from the sediment for later staining by placing 1 drop of Mayer's albumin on the slide, to which is added 1 drop of SAF-preserved fecal sediment. Allow the smear to air dry at room temperature for 30 min prior to staining. The SAF stool smear can also be postfixed in Schaudinn's fixative prior to staining (begin the trichrome stain protocol with the 70% alcohol rinse prior to the iodine-alcohol step).
4. After being dried, the smear can be placed directly in 70% alcohol (step 4) of the staining procedure (the iodine-alcohol step can be eliminated).

Universal Fixative (Total-Fix)-Preserved Material

1. Mix the Total-Fix–stool mixture contained in the collection vial thoroughly; add approximately 1 to 2 ml of fixative-specimen mix to the tube for centrifugation. **DO NOT ADD ANY RINSE FLUIDS.**
2. Centrifuge 10 min at 500 × *g*.
3. Pour off most of the excess fixative.
4. Mix sediment with remaining fixative in the bottom of the tube.
5. Prepare smears for permanent stains and/or DFA.
6. After the smears are thoroughly dry (~30 to 45 min at 37°C or ~60 min at room temperature), they can be placed directly in the 70% alcohol (step 4) of the trichrome staining procedure (the iodine-alcohol step can be eliminated).

Alternate Method for Smear Preparation Directly from Vial

Allow stool to settle to the bottom of the vial. Using a pipette (without creating bubbles), remove fecal material from the bottom of the vial. Smear this fecal material onto the glass slide for subsequent permanent staining (after the slides are allowed to dry for ~30 to 45 min at 37°C or ~60 min at room temperature). Although this is an option, it is recommended that the smear be prepared from the centrifuged material (directions immediately above). See also Algorithm 5.5.

METHOD PROS AND CONS: Smear Preparation Directly from Total-Fix Vial	
PROS	**CONS**
Eliminates one centrifugation step from the algorithm; however, this centrifugation step (step 4) ensures sufficient stool and concentrates the stool prior to smear preparation.	If the specimen is thin or runny or contains very little stool, there is a chance the slide may not contain sufficient stool material for staining and/or organism identification.
	If bubbles are created with the pipette, the fecal material will have to settle out again before the sample is taken.

Note: In order to ensure maximum organism recovery and the best overall results, the procedure should be used as originally written and recommended. However, with practice, sufficient fecal material can be taken from the vial and used to prepare smears for permanent staining.

Stains Used in the Permanent Stained Smear

TRICHROME STAIN (Wheatley's Method)

Description

The trichrome technique of Wheatley for fecal specimens is a modification of Gomori's original staining procedure for tissue. It is a rapid, simple procedure which produces uniformly well-stained smears of the intestinal protozoa, human cells, yeast cells, and artifact material in about 45 min or less.

The specimen usually consists either of fresh stool smeared on a microscope slide which is immediately fixed in Schaudinn's fixative or of stool in fixative containing PVA smeared on a slide and allowed to air dry. Although SAF-preserved specimens can be stained with trichrome, there are other stains which are recommended for better overall results. Trichrome works quite well with specimens preserved in Total-Fix.

Reagents

▶ Trichrome Stain

Chromotrope 2R	0.6 g
Light green SF	0.3 g
Phosphotungstic acid	0.7 g
Acetic acid (glacial)	1.0 ml
Distilled water	100 ml

1. Prepare the stain by adding 1.0 ml of acetic acid to the dry components. Allow the mixture to stand (ripen) for 15 to 30 min at room temperature.
2. Add 100 ml of distilled water. Properly prepared stain is purple.
3. Store in a glass or plastic bottle at room temperature. The shelf life is 24 months.

▶ 70% Ethanol plus Iodine

1. Prepare a stock solution by adding iodine crystals to 70% alcohol until a dark solution is obtained (1 to 2 g/100 ml).
2. To use, dilute the stock solution with 70% alcohol until a dark reddish brown (port wine) or strong-tea color is obtained. As long as the color is acceptable, the new working solution does not have to be replaced. The replacement time depends on the number of smears stained and the size of the container (1 week to several weeks).

▶ 90% Ethanol, Acidified

90% Ethanol	99.5 ml
Acetic acid (glacial)	0.5 ml

Prepare by combining.

▶ 70% Isopropyl or Ethyl Alcohol (100% ethyl alcohol [recommended] or 95% ethyl alcohol [second choice])

▶ Xylene or Xylene Substitute

Quality Control

1. Stool samples used for QC can be fixed stool specimens known to contain protozoa or preserved negative stools containing PVA to which buffy coat cells (PMNs and macrophages) have been added. A QC smear prepared from a positive specimen or a fixative containing buffy coat cells should be used when new stain is prepared or at least once each week. Cultured protozoa can also be used.
2. A QC slide should be included with each new lot number of stain and once a month, particularly if the staining setup is used infrequently.
3. If the xylene becomes cloudy or there is an accumulation of water in the bottom of the staining dish, discard the old reagents, clean the dishes, dry thoroughly, and replace with fresh 100% ethanol and xylene.
4. All staining dishes should be covered to prevent evaporation of reagents (use screw-cap Coplin jars or glass lids).

5. Depending on the volume of slides stained, staining solutions should be changed on an as-needed basis.
6. When the smear is thoroughly fixed and the stain is performed correctly, the cytoplasm of protozoan trophozoites is blue-green, sometimes with a tinge of purple. Cysts tend to be slightly more purple. Nuclei and inclusions (chromatoidal bars, RBCs, bacteria, and Charcot-Leyden crystals) are red, sometimes tinged with purple. The background material usually stains green, providing a nice color contrast with the protozoa. This contrast is more distinct than that obtained with the hematoxylin stain, which tends to stain everything with shades of gray-blue. Color variations among background and organisms are normal (Figure 5.6).
7. The microscope should be calibrated, and the objectives and oculars used for the calibration procedure should be used for all measurements on the microscope. The calibration factors for all objectives should be posted on the microscope for easy access (multiplication factors can be pasted on the body of the microscope).
8. Known positive microscope slides, Kodachrome 2-by-2 projection slides, and photographs (reference books) should be available at the workstation.
9. Record all QC results; the laboratory should also have an action plan for "out-of-control" results.

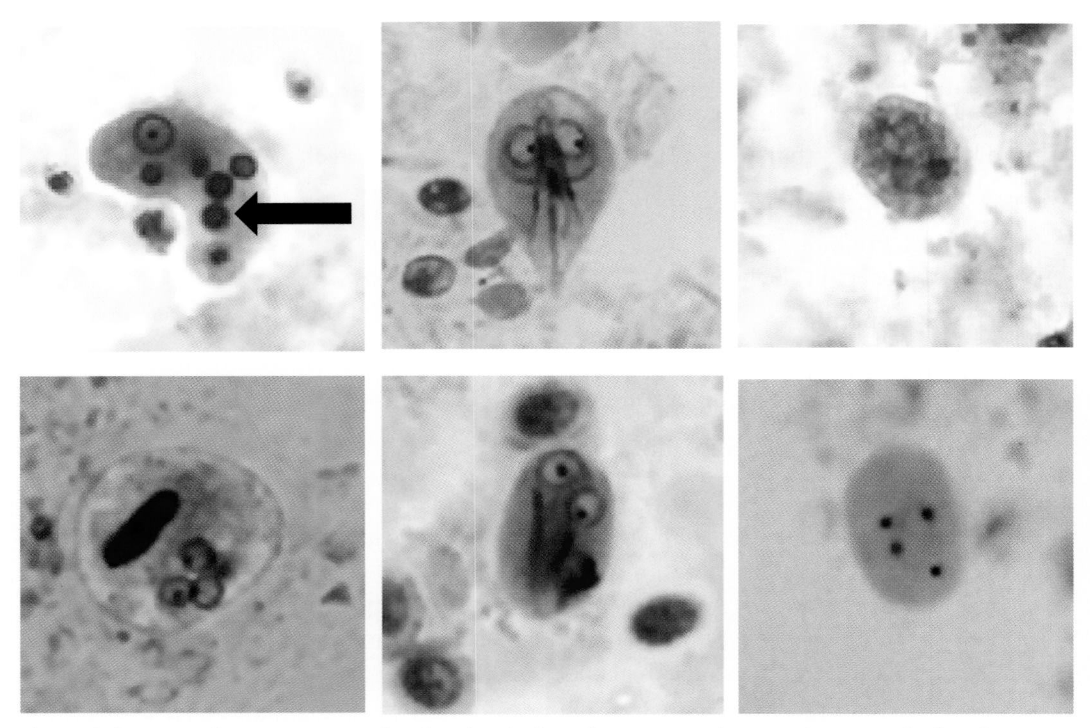

Figure 5.6 Intestinal protozoa stained with Wheatley's trichrome stain. **(Top row)** *Entamoeba histolytica* trophozoite (with ingested RBCs [arrow]); *Giardia lamblia* (*G. duodenalis, G. intestinalis*) trophozoite; *Endolimax nana* trophozoite. **(Bottom row)** *Entamoeba histolytica/E. dispar* complex/group cyst (note the chromatoidal bar with rounded ends); *Giardia lamblia* (*G. duodenalis, G. intestinalis*) cyst; *Endolimax nana* cyst (note the four karyosomes).

Detailed Procedure for Using Mercury-Based Fixatives

Note: In all staining procedures for fecal and gastrointestinal tract specimens, the term "xylene" is used in the generic sense. Xylene substitutes are recommended for the safety of all personnel performing these procedures.

1. Prepare a slide for staining as described above.
2. Remove the slide from Schaudinn's fixative, and place it in 70% ethanol for 5 min.
3. Place the slide in 70% ethanol plus iodine for 1 min for fresh specimens or 5 to 10 min for PVA air-dried smears.
4. Place the slide in 70% ethanol for 5 min.*
5. Place the slide in a second container of 70% ethanol for 3 min.*
6. Place the slide in trichrome stain for 10 min.
7. Place the slide in 90% ethanol plus acetic acid for 1 to 3 s. Immediately drain the rack (see Procedure Notes/Reminders) and proceed to the next step. Do not allow slides to remain in this solution.
8. Dip the slide several times in 100% ethanol. Use this step as a rinse.
9. Place the slide in two changes of 100% ethanol for 3 min each.*
10. Place the slide in xylene for 5 to 10 min.*
11. Place the slide in a second container of xylene for 5 to 10 min.*
12. Mount with a coverslip (no. 1 thickness), using mounting medium (e.g., Permount). (See below this list for an alternative method that does not require the use of mounting medium.)
13. Allow the smear to dry overnight or after 1 h at 37°C.
14. Examine the smear microscopically with the 100× objective. Examine at least 300 oil immersion fields before reporting a negative result.

*Slides may be held for up to 24 h in these solutions without harming the quality of the smear or stainability of organisms.

Alternative Method That Does Not Require Use of Mounting Medium

1. Remove the slide from the last xylene container, place it on a paper towel (flat position), and allow it to air dry. Remember that some of the xylene substitutes may take a bit longer to dry.
2. Approximately 5 to 10 min before you want to examine the slide, place a drop of immersion oil on the dry fecal film. Allow the oil to sink into the film for a minimum of 10 to 15 min. If the smear appears to be very refractile on examination, you have not waited long enough for the oil to sink into the film or you need to add a bit more oil to the film.
3. Once you are ready to examine the slide, place a no. 1 (22- by 22-mm) coverslip onto the oiled smear, add another drop of immersion oil to the top of the coverslip (as you would normally do for any coverslipped slide), and examine with the oil immersion lens (100× objective).
4. Make sure that you do not eliminate adding the coverslip; the dry fecal material on the slide often becomes very brittle after dehydration. Without the addition of the protective coverslip, you might scratch the surface of the oil immersion lens. Coverslips are much cheaper than oil immersion objectives!

Detailed Procedure for Using Non-Mercury-Based Fixatives (Iodine-Alcohol Step and Subsequent Alcohol Rinse Not Required)

1. Prepare the slide for staining as described above.
2. Place the slide in 70% ethanol for 5 min.*
3. Place the slide in trichrome stain for 10 min. Some people prefer to place the dry smear directly in the stain and eliminate step 2.
4. Place the slide in 90% ethanol plus acetic acid for 1 to 3 s. Immediately drain the rack (see Procedure Notes/Reminders) and proceed to the next step. Do not allow slides to remain in this solution.
5. Dip the slide several times in 100% ethanol. Use this step as a rinse.
6. Place the slide in two changes of 100% ethanol for 3 min each.*
7. Place the slide in xylene for 5 to 10 min.*
8. Place the slide in a second container of xylene for 5 to 10 min.*
9. Mount with a coverslip (no. 1 thickness), using mounting medium (e.g., Permount). An alternative method to using mounting medium is given above.
10. Allow the smear to dry overnight or after 1 h at 37°C.
11. Examine the smear microscopically with the 100× objective. Examine at least 300 oil immersion fields before reporting a negative result.

*Slides may be held for up to 24 h in these solutions without harming the quality of the smear or stainability of organisms.

Reporting

Protozoan trophozoites and cysts are readily seen, although helminth eggs and larvae may not be easily identified because of excess stain retention (wet smears from the concentration procedure[s] are recommended for detection of these organisms). Yeasts (single and budding cells and pseudohyphae) and human cells (macrophages, PMNs, and RBCs) can be identified. The following quantitation chart can be used for examination of permanent stained smears with the oil immersion lens (100× objective; total magnification, ×1,000). With very rare exceptions (very heavy infections), coccidia and microsporidia (which normally require specialized stains) are not seen on a routine trichrome-stained fecal smear.

Quantitation of parasites, cells, yeasts, and artifacts	
Quantity	**No./10 oil immersion fields (×1,000)**
Few	2
Moderate	3–9
Many	10

1. Report the organism and stage (do not use abbreviations).

   Examples: *Entamoeba histolytica/E. dispar* group trophozoites
   Giardia lamblia trophozoites

2. Quantitate the number of *Blastocystis* organisms seen (rare, few, moderate, many). Do not quantitate other protozoa.

   Example: Moderate *Blastocystis* spp.

3. Note and quantitate the presence of human cells.

   Example: Moderate WBCs, many RBCs, few macrophages, rare Charcot-Leyden crystals

4. Report and quantitate yeast cells.

   Example: Moderate budding yeast cells and few pseudohyphae

   Reporting yeasts is a bit different. In order to report anything about yeasts, you must know that the stool was fresh or immediately put in preservative. If there are lots of yeast organisms, budding yeasts, and/or pseudohyphae, then this provides some additional information for the physician, often depending on the patient's general condition and whether or not he or she is immunosuppressed. **However, if you do not know whether the collection criteria were met, then reporting anything about yeast is NOT recommended, since this type of report may be misleading.** Certainly, in these circumstances, the physician should be notified directly.

5. Save positive slides for future reference. Label prior to storage (name, patient number, organisms present).

Procedure Notes/Reminders

1. The single most important step in the preparation of a well-stained fecal smear is good fixation (rapid fixation after stool passage, proper ratio between fixative and stool, and adequate mixing). If this has not been done, the protozoa may be distorted or shrunk, may not be stained, or may exhibit an overall pink or red color with poor internal morphology.
2. Slides should always be drained between solutions. Touch the end of the slide to a paper towel for 2 s to remove excess fluid before proceeding to the next step. This will maintain the staining solutions for a longer period.
3. Incomplete removal of mercuric chloride (Schaudinn's fixative and PVA) may cause the smear to contain highly refractive crystals or granules, which may prevent finding or identifying any organisms present. Since the 70% ethanol-iodine solution removes the mercury complex, it should be changed at least weekly to maintain the strong-tea color. It is usually sufficient to keep the slides in the iodine-alcohol for a few minutes; too long a time in this solution may also adversely affect the staining of the organisms.
4. **When non-mercury-based fixatives are used, the iodine-alcohol step (used for the removal of mercury) and the subsequent alcohol rinse can be eliminated from the procedure. The smears for staining can be prerinsed with 70% alcohol and then placed in the trichrome stain, or they can be placed directly into the trichrome stain as the first step in the staining protocol.**
5. Smears that are predominantly green may reflect inadequate removal of iodine by the 70% ethanol (steps 4 and 5). Lengthening the time of these steps or more frequent changing of the 70% ethanol will help. However, remember that there is a large color range seen with this stain; stain color differences are not clinically significant and generally do not impact the identification.
6. To restore weakened trichrome stain, remove the cap and allow the ethanol (carried over on the staining rack from a previous dish) to evaporate. After a few hours, fresh stock stain may be added to restore lost volume. Older, more concentrated stain produces more intense colors and may require slightly longer destaining times (an extra dip). Remember that PVA smears usually require a slightly longer staining time.
7. **Although the trichrome stain is used essentially as a "progressive" stain (that is, no destaining is necessary), the best results are obtained by using the stain "regressively" (destaining the smears briefly in acidified alcohol). Good differentiation is**

obtained by destaining for a very short time (two dips only, approximately 2 to 3 s total); prolonged destaining results in poor differentiation.

8. It is essential to rinse the smears free of acid to prevent continued destaining. Since 90% alcohol continues to leach trichrome stain from the smears, it is recommended that after the acid-alcohol is used, the slides be quickly rinsed in 100% alcohol and then dehydrated through two additional changes of 100% alcohol.
9. In the final stages of dehydration (steps 9 to 11), the 100% ethanol and the xylenes (or xylene substitute) should be kept as free from water as possible. Coplin jars must have tightly fitting caps to prevent both evaporation of reagents and absorption of moisture. If the xylene becomes cloudy after addition of slides from the 100% ethanol, return the slides to fresh 100% ethanol and replace the xylene with fresh stock.
10. If the smears peel or flake off, the specimen might have been inadequately dried on the slide (for specimens in fixative with PVA), the smear may have been too thick, or the slide may have been greasy (fingerprints). However, slides generally do not have to be cleaned with alcohol prior to use.
11. If the stain appears unsatisfactory upon examination and it is not possible to obtain another slide to stain, the slide may be restained. Place the slide in xylene to remove the coverslip, and reverse the dehydration steps, adding 50% ethanol as the last step. Destain the slide in 10% acetic acid for several hours, and wash it thoroughly first in water and then in 50 and 70% ethanol. Place the slide in the trichrome stain for 8 min, and complete the staining procedure.

Procedure Limitations

1. The permanent stained smear is not recommended for staining helminth eggs or larvae; they are often too dark (due to excess stain retention) or distorted. However, they are occasionally recognized and identified. The wet smear preparation from the concentrate is the recommended approach for identification of helminth eggs and larvae.
2. The smear should be examined with the oil immersion lens (100×) for the identification of protozoa, human cells, Charcot-Leyden crystals, yeast cells, and artifact material. These cells and other structures are normally quantitated from the examination of the permanent stained smear, not the wet smear preparations (direct wet smear or concentration wet smear).
3. This high-magnification (oil immersion; total magnification, ×1,000) examination is recommended for protozoa, particularly for confirming species identification.
4. With low magnification (10× objective), one might see eggs or larvae; however, this is not recommended as a routine approach.
5. In addition to helminth eggs and larvae, *C. belli* oocysts are best seen in wet preparations (concentration wet smears prepared from formalin-preserved material, not material in preservative with PVA).
6. *Cryptosporidium* oocysts are generally not recognized on a trichrome-stained smear (acid-fast stains or immunoassay reagent kits are recommended).

IRON HEMATOXYLIN STAIN (Spencer-Monroe Method)

Description

The iron hematoxylin stain can be used to make a permanent stained slide for detecting and quantitating parasitic organisms. Iron hematoxylin was the stain used for most of the original morphologic descriptions of intestinal protozoa found in humans. On oil immersion power (magnification, ×1,000), one can examine the diagnostic features used to identify the protozoan parasite. Although there are many modifications of iron hematoxylin techniques, only two methods are outlined here. Both methods can be used with fresh or preserved specimens, including SAF-preserved specimens and those treated with PVA-containing preservatives.

The specimen usually consists of fresh stool smeared on a microscope slide which is immediately fixed in Schaudinn's fixative, stool in PVA-containing fixative smeared on a slide and allowed to air dry, or SAF-preserved stool smeared on an albumin-coated slide and allowed to air dry.

Reagents

▶ Solution 1

Hematoxylin (crystal or powder) . . . . 10 g
Ethanol (absolute) . . . . 1,000 ml

Place the solution in a stoppered clear flask or bottle and allow it to ripen in a lighted room for at least 1 week at room temperature.

▶ Solution 2

Ferrous ammonium sulfate [$Fe(NH_4)_2(SO_4)_2 \cdot 6H_2O$] . . . . . . 10 g
Ferric ammonium sulfate [$FeNH_4(SO_4)_2 \cdot 12H_2O$] . . . . 10 g
Hydrochloric acid (concentrated) . . . . 10 ml
Distilled water . . . . 1,000 ml

▶ Working Solution

Mix equal volumes of solutions 1 and 2. The working solution should be made fresh every week.

▶ 70% Ethanol plus Iodine

1. Prepare a stock solution by adding iodine crystals to 70% alcohol until a dark solution is obtained (1 to 2 g/100 ml).
2. To use, dilute the stock solution with 70% alcohol until a dark reddish brown or strong-tea color is obtained. As long as the color is acceptable, the new working solution does not have to be replaced. Replacement time depends on the number of smears stained and the size of the container (1 week to several weeks).

▶ 90% Ethanol, Acidified

90% ethanol . . . . 99.5 ml
Acetic acid (glacial) . . . . 0.5 ml

Prepare by combining.

▶ 70% Isopropyl or Ethyl Alcohol (100% ethyl alcohol [recommended] or 95% ethyl alcohol [second choice])

▶ Xylene or Xylene Substitute

Quality Control

1. Stool samples used for quality control can be fixed stool specimens known to contain protozoa or preserved negative stools to which buffy coat cells (PMNs and

macrophages) have been added. A QC smear prepared from a positive specimen or a fixative sample containing buffy coat cells should be used when new stain is prepared or at least once each week. Cultured protozoa can also be used.

2. A QC slide should be included with each run of stained slides, particularly if the staining setup is used infrequently.
3. If the xylene becomes cloudy or there is an accumulation of water in the bottom of the staining dish, discard the old reagents, clean the dishes, dry them thoroughly, and replace with fresh 100% ethanol and xylene.
4. All staining dishes should be covered to prevent evaporation of reagents (use screw-cap Coplin jars or glass lids).
5. Depending on the volume of slides stained, staining solutions should be changed on an as-needed basis.
6. When the smear is thoroughly fixed and the stain is performed correctly, the cytoplasm of protozoan trophozoites is blue-gray, sometimes with a tinge of black. Cysts tend to be slightly darker. Nuclei and inclusions (chromatoidal bars, RBCs, bacteria, and Charcot- Leyden crystals) are dark gray-blue, sometimes almost black. The background material usually stains pale gray or blue, providing some color intensity contrast with the protozoa (Figure 5.7). This contrast is less distinct than that obtained with the trichrome stain, which tends to stain everything with multiple colors (pink, red, purple, green, and blue).
7. The microscope should be calibrated, and the objectives and oculars used for the calibration procedure should be used for all measurements on the microscope. The calibration factors for all objectives should be posted on the microscope for easy access (multiplication factors can be pasted on the body of the microscope).
8. Known positive microscope slides, Kodachrome 2-by-2 projection slides, and photographs (reference books) should be available at the workstation.
9. Record all QC results; the laboratory should also have an action plan for "out-of-control" results.

Detailed Procedure for Using Mercury-Based Fixatives

Note: In all staining procedures for fecal and gastrointestinal tract specimens, the term "xylene" is used in the generic sense. Xylene substitutes are recommended for the safety of all personnel performing these procedures.

1. Prepare the slide for staining as described above (for SAF smears, proceed to step 4).
2. Place the slide in 70% ethanol for 5 min.
3. Place the slide in the iodine–70% ethanol (70% ethanol to which is added enough D'Antoni's iodine to obtain a strong-tea color) solution for 2 to 5 min.
4. Place the slide in 70% ethanol for 5 min. Begin the procedure for SAF-fixed slides at this point.*
5. Wash the slide in running tap water (make sure there is a constant stream of water into the container) for 10 min.
6. Place the slide in iron hematoxylin working solution for 4 to 5 min.
7. Wash the slide in running tap water (make sure there is a constant stream of water into the container) for 10 min.
8. Place the slide in 70% ethanol for 5 min.*
9. Place the slide in 95% ethanol for 5 min.*

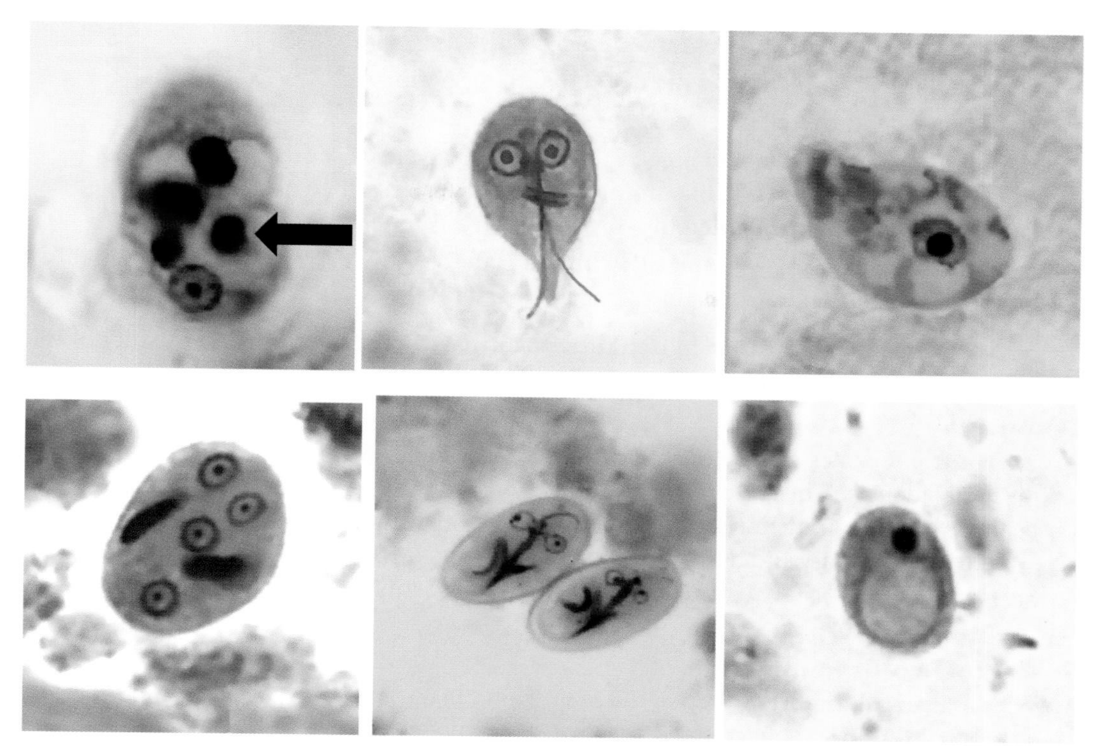

Figure 5.7 Intestinal protozoa stained with iron hematoxylin stain. **Top row**, left to right: *Entamoeba histolytica* trophozoite (with ingested RBCs [arrow]); *Giardia lamblia* (*G. duodenalis*, *G. intestinalis*) trophozoite; *Iodamoeba bütschlii* trophozoite (note the large karyosome). (**Bottom row**) *Entamoeba histolytica*/*E. dispar* complex/group cyst (note the chromatoidal bar with rounded ends); *Giardia lamblia* (*G. duodenalis*, *G. intestinalis*) cysts; *Iodamoeba bütschlii* cyst (note the large glycogen vacuole/clear space).

10. Place the slide in two changes of 100% ethanol for 5 min each.*
11. Place the slide in two changes of xylene for 5 min each.*
12. Add Permount to the stained area of the slide, and cover the slide with a coverslip. An alternative method to using mounting medium is described above (in "Detailed Procedure for Using Mercury-Based Fixatives" under "Trichrome Stain").
13. Examine the smear microscopically with the 100× objective. Examine at least 300 oil immersion fields before reporting a negative result.

*Slides may be held for up to 24 h in these solutions without harming the quality of the smear or stainability of organisms.

Detailed Procedure for Using Non-Mercury-Based Fixatives

1. Prepare the slide for staining as described above.
2. Place the slide in 70% ethanol for 5 min.*
3. Wash the slide in running tap water (make sure there is a constant stream of water into the container) for 10 min.
4. Place the slide in iron hematoxylin working solution for 4 to 5 min.
5. Wash the slide in running tap water (make sure there is a constant stream of water into the container) for 10 min.
6. Place the slide in 70% ethanol for 5 min.*

7. Place the slide in 95% ethanol for 5 min.*
8. Place the slide in two changes of 100% ethanol for 5 min each.*
9. Place the slide in two changes of xylene for 5 min each.*
10. Add Permount to the stained area of the slide and cover the slide with a coverslip. An alternative method to using mounting medium is described above (in "Detailed Procedure for Using Non-Mercury-Based Fixatives" under "Trichrome Stain").
11. Examine the smear microscopically with the 100× objective. Examine at least 300 oil immersion fields before reporting a negative result.

*Slides may be held for up to 24 h in these solutions without harming the quality of the smear or stainability of organisms.

Reporting

Protozoan trophozoites and cysts are readily seen, although helminth eggs and larvae may not be easily identified because of excess stain retention (wet smears from the concentration procedure[s] are recommended for detection of these organisms). Yeasts (single and budding cells and pseudohyphae) and human cells (macrophages, PMNs, and RBCs) can be identified.

For the quantitation of parasites, cells, yeasts, and artifacts, refer to the discussion of trichrome stain (above).

Procedure Notes/Reminders

1. The single most important step in the preparation of a well-stained fecal smear is good fixation. If this has not been done, the protozoa may be distorted or shrunk, may not be stained, or may exhibit an overall gray or blue-gray color with poor internal morphology.
2. Slides should always be drained between solutions. Touch the end of the slide to a paper towel for 2 s to remove excess fluid before proceeding to the next step. This maintains the staining solutions for a longer period.
3. Incomplete removal of mercuric chloride (Schaudinn's fixative and PVA) may cause the smear to contain highly refractive crystals or granules which may prevent detection or identification of any parasites and/or human cells present. Since the 70% ethanol–iodine solution removes the mercury complex, it should be changed at least weekly to maintain the strong-tea color. It is usually sufficient to keep the slides in the iodine-alcohol for a few minutes; too long a time in this solution may also adversely affect the staining of the organisms.
4. **When non-mercury-based fixatives are used, the iodine-alcohol step (used for the removal of mercury) and the subsequent alcohol rinse can be eliminated from the procedure. The smears for staining can be prerinsed with 70% alcohol and then placed in the trichrome stain, or they can be placed directly into the trichrome stain as the first step in the staining protocol.**
5. When large numbers of slides are being stained, the working hematoxylin solution may be diluted; this affects the quality of the stain. If dilution occurs, discard the working solution and prepare a fresh working solution.
6. The shelf life of the stock hematoxylin solutions may be extended by keeping the solutions in the refrigerator at 4°C. Because of crystal formation in the stock solutions, it may be necessary to filter them before preparing a new working solution.
7. In the final stages of dehydration (steps 9 to 11), the 100% ethanol and the xylenes should be kept as free from water as possible. Coplin jars must have tightly fitting caps to prevent both evaporation of reagents and absorption of moisture. If the xylene

becomes cloudy after addition of slides from the 100% ethanol, return the slides to fresh 100% ethanol and replace the xylene with fresh stock.

8. If the smears peel or flake off, the specimen might have been inadequately dried on the slide (in the case of specimens in fixative containing PVA), the smear may have been too thick, or the slide may have been greasy (fingerprints). However, slides generally do not have to be cleaned with alcohol prior to use.
9. If the stain, upon examination, appears unsatisfactory and it is not possible to obtain another slide to stain, the slide may be restained. Place the slide in xylene to remove the coverslip, and reverse the dehydration steps, adding 50% ethanol as the last step. Destain the slide in 10% acetic acid for several hours, and then wash it thoroughly in water and then in 50 and 70% ethanol. Place the slide in the iron hematoxylin stain for 8 min, and complete the staining procedure.

Procedure Limitations

1. The permanent stained smear is not recommended for staining helminth eggs or larvae; they are often too dark (due to excess stain retention) or distorted. However, they are occasionally recognized and identified. The wet smear preparation from the concentrate is the recommended approach for identification of helminth eggs and larvae.
2. The smear should be examined with the oil immersion lens (100× objective) for the identification of protozoa, human cells, Charcot-Leyden crystals, yeast cells, and artifact material. These cells and other structures are normally quantitated from the examination of the permanent stained smear, not the wet smear preparations (direct wet smear, concentration wet smear).
3. This high-magnification (oil immersion; total magnification, ×1,000) examination is recommended for protozoa, particularly for confirming species identification.
4. If the viewer wants to screen the smear with low magnification (10× objective), eggs or larvae might be visible; however, this is not recommended as a routine approach.
5. In addition to helminth eggs and larvae, *C. belli* oocysts are best seen in wet preparations (concentration wet smears prepared from formalin-preserved material, not material in PVA-containing preservative).
6. *Cryptosporidium* oocysts are generally not recognized on an iron hematoxylin-stained smear (acid-fast stains or immunoassay reagent kits are recommended).

IRON HEMATOXYLIN STAIN (Tompkins-Miller Method)

Description

A longer iron hematoxylin method was described by Tompkins and Miller. Since differentiation of overstained slides is critical in most iron hematoxylin staining procedures, Tompkins and Miller have described a method in which phosphotungstic acid is used to destain the protozoa and which gives excellent results, even in unskilled hands.

Detailed Procedure

1. Prepare the slide for staining as described above (for SAF or non-mercury-based smears, proceed to step 4).
2. Place the slide in 70% ethanol for 5 min.
3. Place the slide in the iodine–70% ethanol (70% alcohol to which is added enough D'Antoni's iodine to obtain a strong-tea color) solution for 2 to 5 min.

4. Place the slide in 50% ethanol for 5 min. Begin the procedure for SAF- or non-mercury-fixed slides at this point.*
5. Wash the slide in running tap water (make sure there is a constant stream of water into the container) for 3 min.
6. Place the slide in 4% ferric ammonium sulfate mordant for 5 min.
7. Wash the slide in running tap water (make sure there is a constant stream of water into the container) for 1 min.
8. Place the slide in 0.5% aqueous hematoxylin for 2 min.
9. Wash the slide in tap water (in a container) for 1 min.
10. Place the slide in 2% phosphotungstic acid for 2 to 5 min.
11. Wash the slide in running tap water for 10 min.
12. Place the slide in 70% ethanol (plus a few drops of saturated aqueous lithium carbonate) for 3 min.
13. Place the slide in 95% ethanol for 5 min.*
14. Place the slide in two changes of 100% ethanol for 5 min each.*
15. Place the slide in two changes of xylene for 5 min each.*
16. Add Permount to the stained area of the slide and cover the slide with a coverslip. An alternative method that does not involve mounting medium is as follows.
    a. Remove the slide from the last xylene container, place it on a paper towel (flat position), and allow it to air dry. Remember that some xylene substitutes may take a bit longer to dry.
    b. Approximately 5 to 10 min before you want to examine the slide, place a drop of immersion oil on the dry fecal film. Allow the oil to sink into the film for a minimum of 10 to 15 min. If the smear appears to be very refractile on examination, you have not waited long enough for the oil to sink into the film or you need to add a bit more oil to the film.
    c. Once you are ready to examine the slide, place a no. 1 (22- by 22-mm) coverslip onto the oiled smear, add another drop of immersion oil to the top of the coverslip (as you would normally do for any coverslipped slide), and examine with the oil immersion lens (100× objective).
    d. Make sure that you do not eliminate adding the coverslip; the dry fecal material on the slide often becomes very brittle after dehydration. Without the addition of the protective coverslip, you might scratch the surface of your oil immersion lens. Coverslips are much cheaper than oil immersion objectives!
17. Examine the smear microscopically with the 100× objective. Examine at least 300 oil immersion fields before reporting a negative result.

*Slides may be held up to 24 h in these solutions without harming the quality of the smear or stainability of organisms.

MODIFIED IRON HEMATOXYLIN STAIN (Incorporating the Carbol Fuchsin Step)

Description

The following combination staining method for SAF-preserved fecal specimens was developed to allow the microscopist to screen for acid-fast organisms in addition to other

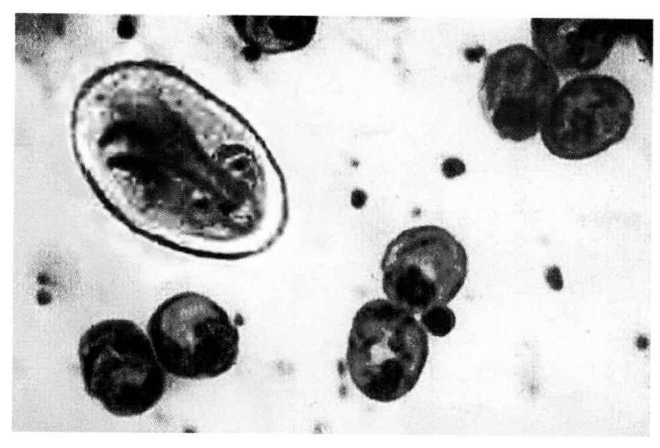

Figure 5.8 Iron hematoxylin stain incorporating the carbol fuchsin step. Note the modified acid-fast-stained *Cryptosporidium* sp. oocysts (4 to 6 μm) and the *Giardia lamblia* cyst.

intestinal parasites. For laboratories using iron hematoxylin stains in combination with SAF-fixed material and modified acid-fast stains for *Cryptosporidium*, *Cyclospora*, and *Cystoisospora*, this modification represents an improved approach to current staining methods. This combination stain provides a saving in both time and personnel use. Some labs prefer to use separate stains (iron hematoxylin and modified acid-fast stains) rather than a combination approach (Figure 5.8).

Any fecal specimen submitted in SAF fixative can be used. Fresh fecal specimens after fixation in SAF for 30 min can also be used. This combination stain approach is not recommended for specimens preserved in Schaudinn's fixative or any fixative containing PVA.

Reagents

▶ Mayer's Albumin

Add an equal quantity of glycerin to a fresh egg white. Mix gently and thoroughly. Store at 4°C, and indicate an expiration date of 3 months. Mayer's albumin from commercial suppliers can normally be stored at 25°C for 1 year (e.g., product Z69; Hardy Diagnostics, 1430 West McCoy Lane, Santa Maria, CA 93455; https://catalog.hardydiagnostics.com/cp_prod/Content/hugo/MayersAlbumin.htm).

▶ Stock Solution of Hematoxylin Stain

Hematoxylin powder . . . . 10 g
Ethanol (95% or 100%) . . . . 1,000 ml

1. Mix well until dissolved.
2. Store in a clear glass bottle in a light area. Allow to ripen for 14 days before use.
3. Store at room temperature with an expiration date of 1 year.

▶ Mordant

Ferrous ammonium sulfate [$Fe(NH_4)_2(SO_4)_2 \cdot 6H_2O$] . . . . . . 10 g
Ferric ammonium sulfate [$FeNH_4(SO_4)_2 \cdot 12H_2O$] . . . . 10 g
Hydrochloric acid (concentrated) . . . . 10 ml
Distilled water to . . . . 1,000 ml

▶ **Working Solution of Hematoxylin Stain**

1. Mix equal quantities of stock solution of stain and mordant.
2. Allow the mixture to cool thoroughly before use (prepare at least 2 h prior to use). The working solution should be made fresh every week.

▶ **Picric Acid**

Mix equal quantities of distilled water and an aqueous saturated solution of picric acid to make a 50% saturated solution.

▶ **Acid-Alcohol Decolorizer**

Hydrochloric acid (concentrated) .... 30 ml
Alcohol to .... 1,000 ml

▶ **70% Alcohol and Ammonia**

70% alcohol .... 50 ml
Ammonia .... 0.5–1.0 ml

Add enough ammonia to bring the pH to approximately 8.0.

▶ **Carbol Fuchsin**

1. To make basic fuchsin (solution A), dissolve 0.3 g of basic fuchsin in 10 ml of 95% ethanol.
2. To make phenol (solution B), dissolve 5 g of phenol crystals in 100 ml of distilled water (gentle heat may be needed).
3. Mix solution A with solution B.
4. Store at room temperature. Solution is stable for 1 year.

Detailed Procedure

1. Prepare the slide.
   a. Place 1 drop of Mayer's albumin on a labeled slide.
   b. Mix the sediment from the SAF concentration well with an applicator stick.
   c. Add approximately 1 drop of the fecal concentrate to the albumin, and spread the mixture over the slide.
2. Allow the slide to air dry at room temperature (the smear appears opaque when dry).
3. Place the slide in 70% alcohol for 5 min.
4. Wash the slide in a container of tap water (not running water) for 2 min.
5. Place the slide in Kinyoun's stain for 5 min.
6. Wash the slide in running tap water (make sure there is a constant stream of water into the container) for 1 min.
7. Place the slide in acid-alcohol decolorizer for 4 min.*
8. Wash the slide in running tap water (constant stream of water into the container) for 1 min.*
9. Place the slide in iron hematoxylin working solution for 8 min.
10. Wash the slide in distilled water (in a container) for 1 min.

11. Place the slide in picric acid solution for 3 to 5 min.
12. Wash the slide in running tap water (make sure there is a constant stream of water into the container) for 10 min.
13. Place the slide in 70% alcohol plus ammonia for 3 min.
14. Place the slide in 95% alcohol for 5 min.
15. Place the slide in 100% alcohol for 5 min.
16. Place the slide in two changes of xylene for 5 min.

*Steps 7 and 8 can also be performed as follows.

a. Place the slide in acid-alcohol decolorizer for 2 min.

b. Wash the slide in running tap water (make sure there is a constant stream of water into the container) for 1 min.

c. Place the slide in acid-alcohol decolorizer for 2 min.

d. Wash the slide in running tap water (make sure there is a constant stream of water into the container) for 1 min.

e. Continue the staining sequence with step 9 (iron hematoxylin working solution).

Procedure Notes/Reminders

1. The first 70% alcohol step acts with the Mayer's albumin to "glue" the specimen to the glass slide. The specimen may wash off if insufficient albumin is used or if the slides are not completely dry prior to staining.
2. The working hematoxylin stain should be checked each day of use by adding a drop of stain to alkaline tap water. If a blue color does not develop, prepare fresh working stain solution.
3. The picric acid differentiates the hematoxylin stain by removing more stain from fecal debris than from the protozoa and removing more stain from the organism cytoplasm than from the nucleus. When properly stained, the background should be various shades of gray-blue and protozoa should be easily seen, with medium blue cytoplasm and dark blue-black nuclei. Color variations are common.

CHLORAZOL BLACK E STAIN

Chlorazol black E staining is a method in which both fixation and staining occur in a single solution. This approach is used for fresh specimens but is not recommended for material in preservative containing PVA. This recommendation is based on the fact that the chlorazol black E staining method does not include an iodine-alcohol step, which is used to remove the mercuric chloride compound found in both Schaudinn's fixative and PVA. The optimal staining time must be determined for each batch of fixative-stain. The length of time for which the fixative-stain can be used depends on the number of slides run through the solution within a 30-day period. If the slides appear red, the solution must be changed. Although this stain is rarely used and does not provide the best staining, it is another option to consider. It is available from Sigma Aldrich (https://www.sigmaaldrich.com/catalog/product/sial/c1144?lang=en®ion=US).

REVIEW: PERMANENT STAINED SMEARS	
Clinical relevance	To provide contrasting colors for both the background debris and parasites present; designed to allow examination and recognition of detailed organism morphology under oil immersion examination (100× objective for a total magnification of ×1,000). Designed primarily to allow recovery and identification of the intestinal protozoa.
Specimen	Any stool specimen that is fresh or preserved in formalin, PVA (mercury based or non-mercury based), SAF, MIF, or the newer single-vial-system fixatives, such as Total-Fix
Reagents	Trichrome, iron hematoxylin, modified iron hematoxylin, polychrome IV, or chlorazol black E stains and their associated solutions; dehydrating solutions (alcohols and xylenes or xylene substitutes); mounting fluid optional
Examination requirements	Oil immersion examination of at least 300 fields; additional fields may be required if suspect organisms have been seen in the wet preparations from the concentrated specimen.
Results and laboratory reports	The majority of the suspected protozoa and/or human cells should be confirmed using the permanent stained smear. These reports should be categorized as "final" and would be signed out as such (along with the results from the concentration examination).
Procedure notes and limitations	The most commonly used stains include trichrome and iron hematoxylin. Unfortunately, helminth eggs and larvae take up too much stain, resemble debris, and usually cannot be identified from the permanent stained smear. Also, coccidian oocysts and microsporidian spores usually require specialized staining methods for identification (modified acid-fast [coccidia] and modified trichrome [microsporidia]). Permanent stained smears are normally examined under oil immersion examination (magnification, ×1,000), and low or high dry power is not recommended. Confirmation of the intestinal protozoa (both trophozoites and cysts) is the primary purpose of this technique.

Specialized Stains for Coccidia and Microsporidia

KINYOUN'S ACID-FAST STAIN (Cold Method)

Description

Cryptosporidium and *Cystoisospora* have been recognized as causes of severe diarrhea in immunocompromised hosts but can also cause diarrhea in immunocompetent hosts. Oocysts in clinical specimens may be difficult to detect without special staining. Modified acid-fast stains are recommended to demonstrate these organisms (Figure 5.9). Unlike the Ziehl-Neelsen modified acid-fast stain, Kinyoun's stain does not require the heating of reagents for staining. With additional reports of diarrheal outbreaks due to *Cyclospora*, it is also important to remember that these organisms are acid fast and can also be identified using this staining approach. Although the microsporidial spores are also acid fast, their size (1 to 2 μm) makes identification very difficult without special stains or the use of molecular biology-based reagents.

It is recommended that concentrated sediment of fresh or formalin-preserved stool be used with this staining method. Other types of clinical specimens, such as duodenal fluid, bile, and pulmonary samples (induced sputum, bronchial wash, or biopsy samples), may also be stained.

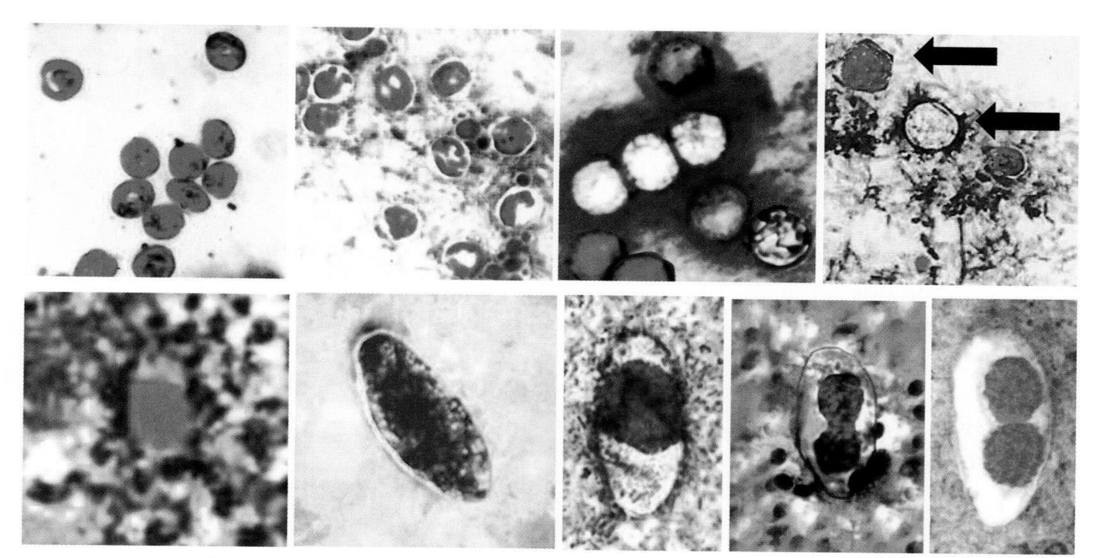

Figure 5.9 Modified acid-fast stains. **(Top row)** *Cryptosporidium* sp. oocysts; *Cryptosporidium* sp. oocysts (note sporozoites in some of the oocysts); *Cyclospora cayetanensis* oocysts with variable staining in modified acid-fast stain; *C. cayetanensis* oocysts (arrows) (note that not all oocysts stain). **(Bottom row)** Immature *Cystoisospora belli* oocyst; very immature *C. belli* oocyst; immature *C. belli* oocyst; more mature *C. belli* oocyst; more mature *C. belli* oocyst (note that internal structures stain).

Reagents

▶ 50% Ethanol

1. Add 50 ml of absolute ethanol to 50 ml of distilled water.
2. Store at room temperature. The solution is stable for 1 year. Note the expiration date on the label.

▶ Kinyoun's Carbol Fuchsin

1. Dissolve 4 g of basic fuchsin in 20 ml of 95% ethanol (solution A).
2. Dissolve 8 g of phenol crystals in 100 ml of distilled water (solution B).
3. Mix solutions A and B.
4. Store at room temperature. The solution is stable for 1 year. Note the expiration date on the label.

▶ 1% Sulfuric Acid

1. Add 1 ml of concentrated sulfuric acid to 99 ml of distilled water.
2. Store at room temperature. The solution is stable for 1 year. Note the expiration date on the label.

▶ Loeffler's Alkaline Methylene Blue

1. Dissolve 0.3 g of methylene blue in 30 ml of 95% ethanol.
2. Add 100 ml of dilute (0.01%) potassium hydroxide.
3. Store at room temperature. The solution is stable for 1 year. Note the expiration date on the label.

Quality Control

1. A control slide of *Cryptosporidium* from a 10% formalin-preserved specimen is included with each staining batch run. If the *Cryptosporidium* slide stains well, any *Cystoisospora* or *Cyclospora* oocysts present will also take up the stain.
2. *Cryptosporidium* stains pink-red. Oocysts measure 4 to 6 μm, and four sporozoites may be present internally. The background should stain uniformly blue.
3. The specimen is also checked for adherence to the slide (macroscopically).
4. The microscope should be calibrated (within the last 12 months), and the objectives and oculars used for the calibration procedure should be used for all measurements on the microscope. The calibration factors for all objectives should be posted on the microscope for easy access (multiplication factors can be pasted on the body of the microscope). If the microscopes receive adequate maintenance and are not moved frequently, yearly recalibration may not be necessary.
5. Known positive microscope slides, Kodachrome 2-by-2 projection slides, and photographs (reference books) should be available at the workstation.
6. Record all QC results; the laboratory should also have an action plan for "out-of-control" results.

Detailed Procedure

1. Smear 1 or 2 drops of specimen on the slide and allow it to air dry. **Do not make the smears too thick (you should be able to see through the wet material before it dries).** Prepare two smears.
2. Fix with absolute methanol for 1 min.
3. Flood the slide with Kinyoun's carbol fuchsin, and stain it for 5 min.
4. Rinse the slide briefly (3 to 5 s) with 50% ethanol.
5. Rinse the slide thoroughly with water.
6. Decolorize by using 1% sulfuric acid for 2 min or until no more color runs from the slide.
7. Rinse the slide with water (it may take less than 2 min; do not destain too much). Drain.
8. Counterstain with methylene blue for 1 min.
9. Rinse the slide with water. Air dry.
10. Examine with the low or high dry objective. To see internal morphology, use the oil immersion objective (100×).

Reporting

Cryptosporidium and *Cystoisospora* oocysts stain pink to red to deep purple. Some of the four sporozoites may be visible in the *Cryptosporidium* oocysts. Some of the immature *Cystoisospora* oocysts (entire oocyst) stain, while mature oocysts usually appear with the two sporocysts within the oocyst wall stained pink to purple and with a clear area between the stained sporocysts and the oocyst wall. The background stains blue but may be pale, depending on the density of the smear. *Cyclospora* oocysts tend to be approximately 10 μm, they resemble *Cryptosporidium* oocysts but are approximately twice as large, and they have no definite internal morphology; the acid-fast staining tends to be more variable than that seen with *Cryptosporidium* or *Cystoisospora* spp. If the patient has a heavy infection with microsporidia (immunocompromised patient), small (1- to 2-μm) spores may be seen but may not be recognized as anything other than bacteria or small yeast cells. There is usually a range of color intensity in the organisms present; not every oocyst appears deep pink to purple. The greatest variation in staining is seen with *Cyclospora*.

1. Report the organism and stage (oocyst). Do not use abbreviations.

   Examples: *Cryptosporidium* sp. oocysts
   Cystoisospora belli oocysts

2. Call the physician when these organisms are identified.
3. Save positive slides for future reference. Label prior to storage (name, patient number, and organisms present).

Procedure Notes/Reminders

1. Routine stool examination stains (trichrome and iron hematoxylin) are not recommended; however, sedimentation concentration (500 × *g* for 10 min) is acceptable for the recovery and identification of *Cryptosporidium* spp., particularly after staining with one of the modified acid-fast stains. The routine concentration (formalin-ethyl acetate) can be used to recover *Cystoisospora* oocysts (wet sediment examination and/or modified acid-fast stains), but routine permanent stains (trichrome and iron hematoxylin) are not reliable for this purpose.
2. Preservatives containing PVA are not acceptable for staining with the modified acid-fast stain. However, specimens preserved in SAF or Total-Fix (two examples) are perfectly acceptable.
3. Avoid the use of wet-gauze filtration (an old, standardized method of filtering stool prior to centrifugation) with too many layers of gauze that may trap organisms and prevent them from flowing into the fluid to be concentrated. It is recommended that no more than two layers of gauze be used; another option is to use the commercially available concentration systems that use no gauze but instead use plastic or metal screens.
4. Other organisms that stain positive include acid-fast bacteria, *Nocardia* spp., and the microsporidia (which are very difficult to find and identify even when they appear to be acid fast).
5. **It is very important that smears not be too thick.** Thicker smears may not adequately destain.
6. **Concentration of the specimen is essential to demonstrate organisms (500 × *g* for 10 min); this approach enhances the sensitivity of the test.** The number of organisms seen in the specimen may vary from many to very few.
7. Some specimens require treatment with 10% KOH because of their mucoid consistency. Add 10 drops of 10% KOH to the sediment, and vortex until it is homogeneous. Rinse with 10% formalin, and centrifuge (500 × *g* for 10 min). Without decanting the supernatant, take 1 drop of the sediment and smear it thinly on a slide.
8. Commercial concentrators and reagents are available.
9. Sulfuric acid at 1.0 to 3.0% is normally used (most laboratories currently use a 1.0% acid rinse). Concentrations higher than 3% remove too much stain, particularly for *Cyclospora*. The use of acid-alcohol (routinely used in the Ziehl-Neelsen acid-fast staining method for mycobacteria) decolorizes all organisms; therefore, one must use the modified decolorizer (1 to 3% H_2SO_4) for good results.
10. There is some debate as to whether organisms lose their ability to take up the acid-fast stain after long-term storage in 10% formalin. Some laboratories have reported this diminished staining.
11. Specimens should be centrifuged in capped tubes, and gloves should be worn during all phases of specimen processing.

Procedure Limitations

1. Light *Cryptosporidium* infections may be missed (small number of oocysts). Immunoassay methods are more sensitive.
2. Multiple specimens must be examined, since the numbers of oocysts present in the stool vary from day to day. A series of three specimens submitted on alternate days is recommended.
3. The identification of both *Cyclospora* and microsporidia is difficult at best. *Cyclospora* may be suspected if the organisms appear to be *Cryptosporidium* but are about twice the size (about 10 μm). The microsporidial spores are extremely small (1 to 2 μm) and will probably not be recognized unless they are very numerous and appear to have a somewhat different morphology than do the bacteria in the preparation.

MODIFIED ZIEHL-NEELSEN ACID-FAST STAIN (Hot Method)

Description

Cryptosporidium and *Cystoisospora* have been recognized as causes of severe diarrhea in immunocompromised hosts but can also cause diarrhea in immunocompetent hosts. Oocysts in clinical specimens may be difficult to detect without special staining. Modified acid-fast stains are recommended to demonstrate these organisms. Application of heat to the carbol fuchsin assists in the staining, and the use of a milder decolorizer allows the organisms to retain more of their pink-red color. With continued reports of diarrheal outbreaks due to *Cyclospora*, it is also important to remember that these organisms are acid fast and can be identified by using this staining approach. Although the microsporidial spores are also acid fast, their size (1 to 2 μm) makes identification very difficult without special stains or the use of molecular assay reagents. Concentrated sediment of fresh or formalin-preserved stool may be used. Other types of clinical specimens, such as duodenal fluid, bile, and pulmonary samples (induced sputum, bronchial washings, and biopsy specimens), may also be stained.

Reagents

▶ Carbol Fuchsin

1. To make basic fuchsin (solution A), dissolve 0.3 g of basic fuchsin in 10 ml of 95% ethanol.
2. To make phenol (solution B), dissolve 5 g of phenol crystals in 100 ml of distilled water (gentle heat may be needed).
3. Mix solution A with solution B.
4. Store at room temperature. The solution is stable for 1 year. Note the expiration date on the label.

▶ 5% Sulfuric Acid

1. Add 5 ml of concentrated sulfuric acid to 95 ml of distilled water.
2. Store at room temperature. The solution is stable for 1 year. Note the expiration date on the label.

▶ Methylene Blue

1. Dissolve 0.3 g of methylene blue chloride in 100 ml of distilled water.
2. Store at room temperature. The solution is stable for 1 year. Note the expiration date on the label.

Quality Control

QC guidelines are the same as those described above for Kinyoun's acid-fast stain.

Detailed Procedure

1. Smear 1 or 2 drops of specimen on the slide, and allow it to air dry. Do not make the smears too thick (you should be able to see through the wet material before it dries). Prepare two smears.
2. Dry on a heating block (70°C) for 5 min.
3. Place the slide on a staining rack, and flood it with carbol fuchsin.
4. With an alcohol lamp or Bunsen burner, gently heat the slide to steaming by passing the flame under the slide. Discontinue heating once the stain begins to steam. Do not boil.
5. Allow the specimen to stain for 5 min. If the slide dries, add more stain without additional heating.
6. Rinse the slide thoroughly with water. Drain.
7. Decolorize with 5% sulfuric acid for 30 s. (Thicker slides may require a longer destaining step; however, do not destain too long.)
8. Rinse the slide with water. Drain.
9. Flood the slide with methylene blue for 1 min.
10. Rinse the slide with water, drain, and air dry.
11. Examine with the low or high dry objective. To see internal morphology, use the oil immersion objective (100×).

Reporting

Cryptosporidium and *Cystoisospora* oocysts stain pink to red to deep purple. Some of the four sporozoites may be visible in *Cryptosporidium* oocysts. Some immature *Cystoisospora* oocysts (entire oocyst) stain, while mature oocysts usually appear with the two sporocysts within the oocyst wall stained pink to purple and with a clear area between the stained sporocysts and the oocyst wall. The background stains blue. If *Cyclospora* oocysts are present (uncommon), they tend to be approximately 10 μm, they resemble *Cryptosporidium* oocysts but are larger, and they have no definite internal morphology; the acid-fast staining tends to be more variable than that seen with *Cryptosporidium* or *Cystoisospora* spp. If the patient has a heavy infection with microsporidia (immunocompromised patient), small (1- to 2-μm) spores may be seen but may not be recognized as anything other than bacteria or small yeast cells (modified trichrome recommended). A range of color intensity is usually seen in the oocysts present; not every oocyst appears deep pink to purple. The greatest variation in staining is seen with *Cyclospora*.

1. Report the organism and stage (oocyst). Do not use abbreviations.

   Examples: *Cryptosporidium* sp. oocysts
   Cyclospora cayetanensis oocysts

2. Call the physician when these organisms are identified.
3. Save positive slides for future reference. Label prior to storage (name, patient number, organisms present).

Procedure Notes/Reminders

1. Routine stool examination stains (trichrome and iron hematoxylin) are not recommended; however, sedimentation concentration (500 × *g* for 10 min) is acceptable for

the recovery and identification of *Cryptosporidium* spp. and *Cyclospora*, particularly after staining with one of the modified acid-fast stains. The routine concentration (formalin-ethyl acetate) can be used to recover *Cystoisospora* oocysts (wet sediment examination and/or modified acid-fast stains), but routine permanent stains (trichrome and iron hematoxylin) are not reliable for this purpose.

2. Specimens in preservative containing PVA are not acceptable for staining with the modified acid-fast stain. However, specimens preserved in SAF are perfectly acceptable.
3. Avoid the use of wet-gauze filtration (an old, standardized method of filtering stool prior to centrifugation) with too many layers of gauze that may trap organisms and prevent them from flowing into the fluid to be concentrated. It is recommended that no more than two layers of gauze be used. Another option is to use the commercially available concentration systems that use no gauze but instead use metal or plastic screens for filtration.
4. Other organisms that stain positive include acid-fast bacteria, *Nocardia* spp., and the microsporidia (which are very difficult to find and identify even when they appear to be acid fast).
5. **It is very important that smears not be too thick.** Thicker smears may not adequately destain.
6. **Concentration of the specimen is essential to demonstrate organisms (500 × *g* for 10 min); this approach enhances the sensitivity of the test.** The number of organisms seen in the specimen may vary from many to very few.
7. Some specimens require treatment with 10% KOH because of their mucoid consistency. Add 10 drops of 10% KOH to the sediment, and vortex until it is homogeneous. Rinse with 10% formalin, and centrifuge (500 × *g* for 10 min). Without decanting the supernatant, take 1 drop of the sediment and smear it thinly on a slide.
8. Commercial concentrators and reagents are available.
9. Do not boil the stain. Gently heat until steam rises from the slide. Do not allow the stain to dry on the slide.
10. Various concentrations of sulfuric acid (0.25 to 10%) may be used; however, the destaining time varies according to the concentration used. Generally, a 1 or 5% solution is used. The use of acid-alcohol (routinely used in the Ziehl-Neelsen acid-fast staining method for the mycobacteria) decolorizes all organisms; therefore, the modified decolorizer (1.0 to 3% H_2SO_4) must be used for good results.
11. There is some debate whether organisms lose their ability to take up the acid-fast stain after long-term storage in 10% formalin. Some laboratories have reported this diminished staining.
12. Specimens should be centrifuged in capped tubes, and gloves should be worn during all phases of specimen processing.

Procedure Limitations

1. Light *Cryptosporidium* or *Cyclospora* infections may be missed (small number of oocysts). The immunoassay methods for *Cryptosporidium* are more sensitive.
2. Multiple specimens must be examined, since the numbers of oocysts present in the stool vary from day to day. A series of three specimens submitted on alternate days is recommended.

3. The identification of both *Cyclospora* and microsporidia may be difficult. *Cyclospora* may be suspected if the organisms appear to be *Cryptosporidium* but are about twice the size (about 10 μm). The microsporidial spores are extremely small (1 to 2 μm) and will probably not be recognized unless they are very numerous and appear to have a somewhat different morphology than do the other bacteria in the preparation (modified trichrome stain is recommended for the microsporidia).

CARBOL FUCHSIN NEGATIVE STAIN FOR *CRYPTOSPORIDIUM* (W. L. Current)

1. Mix thoroughly an equal volume (3 to 10 μl) of fresh or formalin-fixed stool and Kinyoun's carbol fuchsin on a slide.
2. Spread out as a thin film, and allow to air dry at room temperature.
3. Add immersion oil directly to the stained smear, and then cover with a coverslip.
4. Observe with bright-field microscopy (×400). Everything but the oocysts stains darkly. The oocysts are bright and refractile because they contain water whereas everything else is oil soluble.

RAPID SAFRANIN METHOD FOR *CRYPTOSPORIDIUM* (D. Baxby)

1. Smear fresh or formalin-fixed feces on a slide, and allow the film to air dry at room temperature.
2. Fix briefly by one pass through the Bunsen burner flame.
3. Fix for 3 to 5 min with 3% HCl in methanol.
4. Wash the slide with a brief rinse in tap water.
5. Stain with 1% aqueous safranin for 1 min (heat until steam appears) (the authors indicate by personal communication that boiling may be beneficial).
6. Rinse the slide in tap water.
7. Counterstain with 1% methylene blue for 30 s (the authors report that 0.1% aqueous crystal violet was almost as good, but malachite green was unsatisfactory).

RAPID SAFRANIN METHOD FOR *CYCLOSPORA*, USING A MICROWAVE OVEN (Govinda Visvesvara)

Another rapid safranin method uniformly stains *Cyclospora* oocysts a brilliant reddish orange. In this method, the fecal smears must be heated in a microwave oven before being stained. This stain is fast, reliable, and easy to perform (Figure 5.10).

1. Using a 10-μl aliquot of concentrated stool, prepare the smear by spreading the material thinly across the slide.
2. Allow the smear to dry on a 60°C slide warmer.
3. Cool the slide to room temperature before staining.
4. Place the slide in a Coplin jar containing acidic alcohol (3% [vol/vol] HCl in methanol), and let it stand for 5 min.
5. Wash off the excess acidic alcohol with cold tap water.
6. Place the slide in the Coplin jar containing 1% safranin solution in acidified water (pH 6.5), and microwave on full power (650 W) for 1 min. (Place the staining jar in another container to catch any overflow of stain because of boiling.)
7. Wash off excess stain with cold tap water.
8. Place the slide in a Coplin jar containing 1% methylene blue for 1 min.

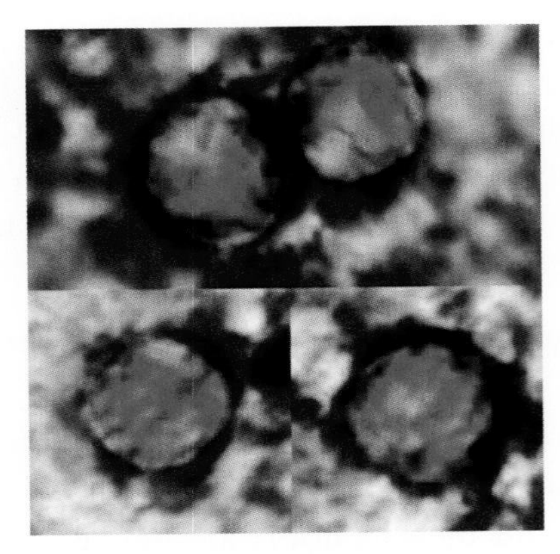

Figure 5.10 Rapid hot safranin stain of *Cyclospora cayetanensis* oocysts.

9. Rinse gently with cold tap water.
10. Air dry.
11. Coverslip the slide using Cytoseal 60 or other mounting medium; the immersion oil mounting method can also be used.
12. Examine the smear under low power or high dry power objectives. To see additional morphology, use the oil immersion objective (100×).

AURAMINE O STAIN FOR APICOMPLEXA (INCLUDING COCCIDIA) (Thomas Hänscheid)

Coccidia are acid-fast organisms and also stain well with auramine O (phenolized auramine O). The size and typical appearance of *Cryptosporidium* (Apicomplexa, but no longer grouped with the coccidia), *Cyclospora*, and *Cystoisospora* oocysts enable auramine O-stained slides to be examined at low power under the 10× objective. The entire sample area can usually be examined in less than 30 s. The low cost of the reagents, the simple staining protocol, and the rapid microscopic examination also make this staining method suitable for screening unconcentrated fecal specimens.

Concentrated sediment from fresh stool or stool preserved with non-PVA-containing fixative may be used. Other stool samples may also be used, such as unconcentrated stool submitted for culture in a bacteriology transport medium. However, to increase the sensitivity of the test, small numbers of oocysts are more easily detected in concentrated stools.

▶ Auramine O Stain

1. **Auramine O**

   Dissolve 0.1 g of auramine O in 10 ml of 95% ethanol.
2. **Phenol**

   Dissolve 3.0 g of phenol crystals in 87 ml of distilled water.

   Combine solutions 1 and 2. Store in a dark bottle at room temperature for up to 3 months.
3. **Destaining agent:** 0.5% acid alcohol

   Add 0.5 ml of concentrated HCl to 100 ml of 70% ethanol. Store at room temperature for 3 months.

4. **Counterstain:** 0.5% potassium permanganate
   Dissolve 0.5 g of potassium permanganate in 100 ml of distilled water.

Quality Control

QC guidelines are the same as those for Kinyoun's acid-fast stain and are given on p. 134.

Detailed Procedure

1. Using a 10- to 20-μl aliquot of concentrated stool, prepare the smear by spreading the material across the slide.
2. Heat fix the slides either on a 65 to 75°C heat block for at least 2 h or using the flame of a Bunsen burner. Do not overheat. Another fixation option would be to fix the slide in absolute methanol for 1 min, air dry, and then proceed with staining.
3. Cool the slide to room temperature before staining.
4. Flood the slide with the phenolized auramine O solution.
5. Allow the smear to stain for ca. 15 min. Do not heat.
6. Rinse the slide in water. Drain excess water from the slide.
7. Flood the slide with the destaining solution (0.5% acid-alcohol).
8. Allow the specimen to decolorize for 2 min.
9. Flood the slide with counterstain (potassium permanganate) solution.
10. Stain for 2 min. The timing of this step is critical.
11. Rinse the slide in water. Drain excess water from the slide.
12. Allow the smear to air dry. Do not blot.
13. Examine the smear under a fluorescence microscope with a 10× objective and fluorescein isothiocyanate (FITC) optical filters (auramine O: excitation maximum, ~435 nm in water; emission maximum, ~510 nm). Screen the whole sample area for the presence of fluorescent oocysts. Suspicious objects can be reexamined with a 20× or 100× objective.
14. Smears can be restained by any of the carbol fuchsin (modified acid-fast) staining procedures to allow examination by light microscopy.

Reporting

Cryptosporidium and *Cyclospora* oocysts fluoresce brightly and have a regular round appearance ("starry-sky" appearance with the 20× objective [*Cryptosporidium*] and 10× objective [*Cystoisospora*]) (Figure 5.11, bottom row). In contrast to the large majority of fluorescent artifacts, the oocysts do not stain homogeneously. Thus, the fluorescence is heterogeneously distributed in the interior of the oocyst; no staining of the cyst walls is observed (Figure 5.11). *Cyclospora* oocysts also stain well; however, no sporozoites are seen. *Cystoisospora* oocysts fluoresce brightly with three patterns: (i) a fairly brightly but heterogeneously stained interior of the whole oocyst, (ii) one brightly staining sporocyst, or (iii) two brightly staining sporocysts within the oocyst wall.

1. Report the organism and stage (oocyst). Do not use abbreviations.

   Examples: *Cryptosporidium* sp. oocysts
   Cystoisospora belli oocysts
   Cyclospora cayetanensis oocysts

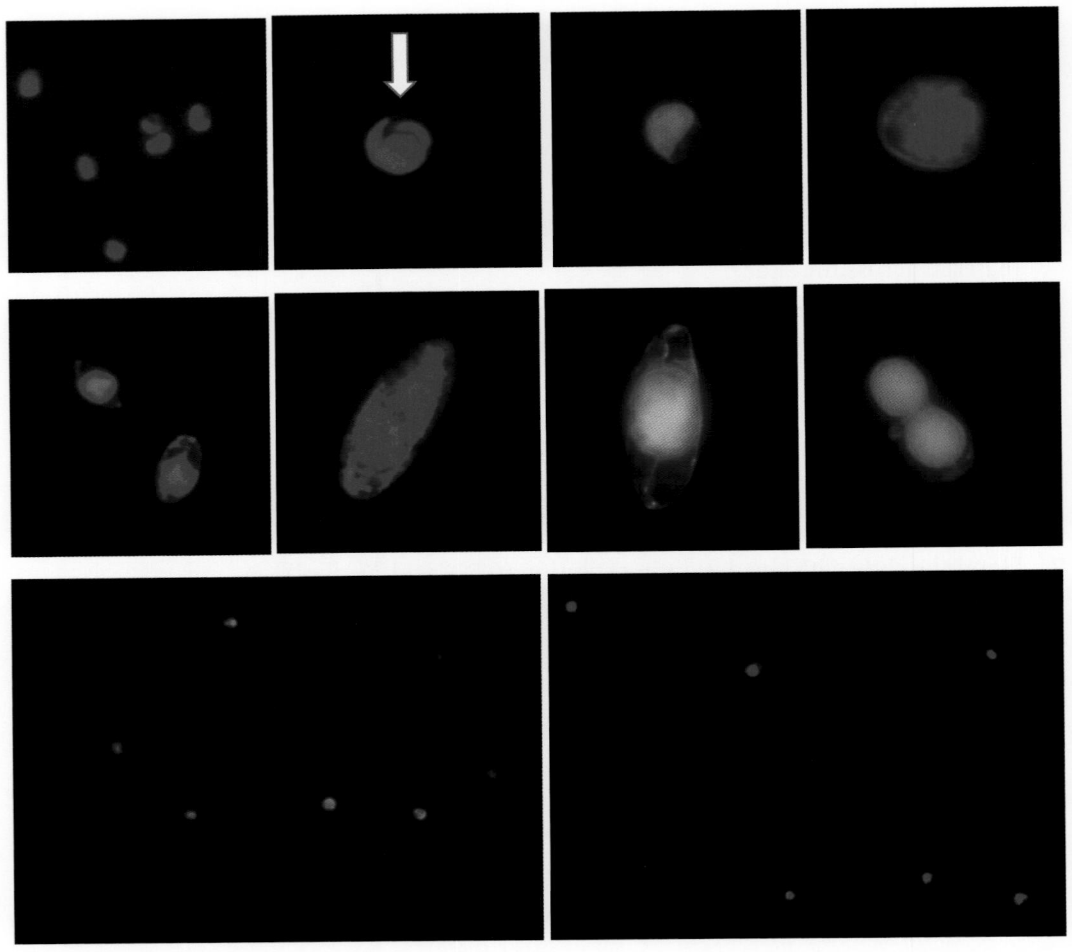

Figure 5.11 Auramine O staining for Apicomplexa. **(Top row)** *Cryptosporidium* sp. oocysts, ×400; *Cryptosporidium* oocyst showing sporozoites (arrow), ×1,000 (note that the *Cryptosporidium* sporozoites stain, but the cyst wall does not stain); *Cryptosporidium* oocyst, ×1,000; *Cyclospora cayetanensis* oocyst, ×1,000. **(Middle row)** Immature *Cystoisospora belli* oocysts, ×200; immature *C. belli* oocyst, ×1,000; immature *C. belli* oocyst, ×1,000; mature *C. belli* oocyst, ×1,000. **(Bottom row)** *Cryptosporidium* sp. oocysts, ×200 (both images). (Images courtesy of Soraia Vieira, Pedro Marinho, and Thomas Hänscheid, Faculdade de Medicina, Universidade de Lisboa, Lisbon, Portugal.)

2. Call the physician when these organisms are identified.
3. Save positive slides for future reference. Label prior to storage (name, patient number, and organisms present). These slides can be kept at room temperature in the dark, and the fluorescence remains stable for up to 4 weeks.

Procedure Notes

1. It is mandatory that positive-control smears be stained and examined each time patient specimens are stained and examined.
2. For best results, examine the auramine O solution for deposits and remove them by filtration or centrifugation. This problem can also be avoided by preparing smaller volumes more frequently.
3. Slides should be observed as soon as possible after staining. However, they can be kept at room temperature in the dark, and fluorescence remains stable for up to 4 weeks.

Procedure Limitations

1. Light infections might be missed, particularly if unconcentrated stool is used; it is always recommended that concentrated stool sediment be used for staining (500 × *g* for 10 min).
2. Using the 40× high dry objective often causes a blurred image (a fluorescent "halo" around the image and hazy contours), which appears to be the effect of interfering fluorescence from the auramine O stain located outside the plane of focus. Using the 100× oil immersion objective gives higher-quality images. Immersion oils used for light microscopy may be autofluorescent; special low-fluorescence immersion oil should be used.
3. If the fluorescence is not clear or definitive, a suspicious slide can be restained with a modified acid-fast stain and reexamined by light microscopy with the 100× oil immersion objective.
4. If protected from sunlight, auramine O slides can be kept on the bench at room temperature for up to 3 weeks, with only minor loss of fluorescence (photobleaching).

REVIEW: MODIFIED ACID-FAST SMEARS AND OTHER STAINS FOR APICOMPLEXA[a]	
Clinical relevance	To provide contrasting colors for both the background debris and parasites present; designed to allow examination and recognition of the acid-fast characteristic of the organisms under high dry examination (40× objective for a total magnification of ×400). Designed primarily to allow recovery and identification of intestinal Apicomplexan oocysts (*Cryptosporidium*, *Cyclospora*, and *Cystoisospora*). Internal morphology (sporozoites) is seen in some *Cryptosporidium* oocysts under oil immersion (magnification, ×1,000).
Specimen	Any stool specimen that is fresh or preserved in formalin, SAF, or the newer single-vial-system fixatives
Reagents	Kinyoun's acid-fast stain, modified Ziehl-Neelsen stain, and their associated solutions; dehydrating solutions (alcohols and xylenes); mounting fluid optional; remember that the decolorizing agents are less intense than the routine acid-alcohol used in routine acid-fast staining (this is what makes these procedures "modified" acid-fast procedures). Safranin and auramine O stains and associated solutions.
Examination requirements	High dry examination of at least 300 fields; additional fields may be required if suspect organisms have been seen but are not clearly acid fast.
Results and laboratory reports	The identification of *Cryptosporidium* and *Cystoisospora* oocysts should be possible; *Cyclospora* oocysts, which are twice the size of *Cryptosporidium* oocysts, should be visible but tend to be more acid-fast variable. Although microsporidia are acid fast, their small size makes recognition very difficult. Final laboratory results would depend heavily on the appearance of the QC slides and comparison with patient specimens.
Procedure notes and limitations	Both the cold and hot modified acid-fast methods are excellent for the staining of Apicomplexan oocysts. There is some feeling that the hot method may result in better stain penetration, but the differences are probably minimal. Procedure limitations are related to specimen handling (proper centrifugation speeds and times; use of no more than two layers of wet gauze for filtration) and a complete understanding of the difficulties in recognizing microsporidial spores. There is also some controversy concerning whether the organisms lose the ability to take up acid-fast stains after long-term storage in 10% formalin. The organisms are more difficult to find in specimens from patients who do not have the typical, watery diarrhea (more formed stool = more artifact material). Both the safranin and auramine O stains are also good options for staining oocysts.

[a] *Cryptosporidium* spp. are no longer considered coccidia; however, they remain in the Apicomplexa along with the other coccidia (*Cyclospora cayetanensis, Cystoisospora belli*).

MODIFIED TRICHROME STAIN FOR MICROSPORIDIA (Weber, Green Counterstain)

Description

A few years ago, the diagnosis of intestinal microsporidiosis (*Enterocytozoon bieneusi* and *Encephalitozoon intestinalis*) depended on the use of invasive procedures and subsequent examination of biopsy specimens, often by electron microscopy methods. Slides prepared from fresh or formalin-fixed stool specimens can be stained using chromotrope-based techniques and can be examined by light microscopy (Figure 5.12, left). Other single-vial fixative options such as Total-Fix can also be used. This staining method is based on the fact that stain penetration of the microsporidial spore is very difficult; thus, more dye is used in the chromotrope 2R than that routinely used to prepare Wheatley's modification of Gomori's trichrome method, and the staining time is much longer (90 min). At least several of these stains are available commercially from a number of suppliers. The specimen can be fresh stool or stool that has been preserved in 5 or 10% formalin, SAF, or one of the newer single-vial-system fixatives. Actually, any specimen other than tissue thought to contain microsporidia could be stained by this method.

Reagents

▶ **Stain**

Chromotrope 2R	6.0 g*
Fast green	0.15 g
Phosphotungstic acid	0.7 g
Acetic acid (glacial)	3.0 ml
Distilled water	100.0 ml

*This is 10 times the amount used in the normal trichrome stain formula.

1. Prepare the stain by adding 3.0 ml of acetic acid to the dry ingredients. Allow the mixture to stand (ripen) for 30 min at room temperature.
2. Add 100 ml of distilled water. Properly prepared stain is dark purple.
3. Store in a glass or plastic bottle at room temperature. The shelf life is at least 24 months.

▶ **Acid-Alcohol**

90% ethyl alcohol	995.5 ml
Acetic acid (glacial)	4.5 ml

Prepare by combining the two solutions.

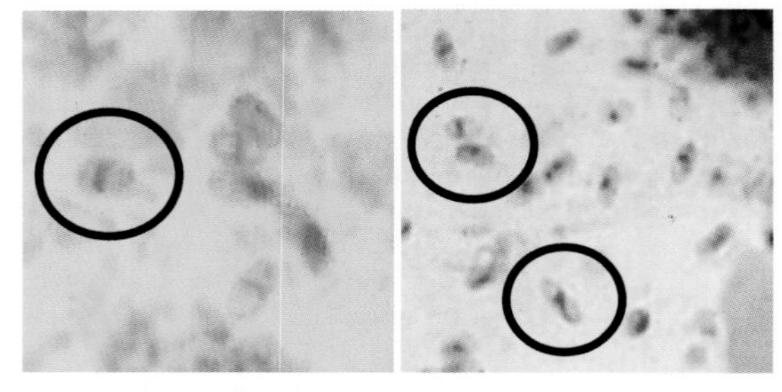

Figure 5.12 Modified trichrome stain for microsporidian spores. Weber green counterstain; Ryan blue counterstain. Note the spores with internal evidence of polar tubules, which appear as horizontal or diagonal lines (circles).

Quality Control

1. Unfortunately, the only way to perform acceptable QC procedures for this method is to use actual microsporidial spores as the control organisms. Obtaining these positive controls may be somewhat difficult. It is particularly important to use the actual organisms because the spores are very small (1 to 1.5 µm) and difficult to stain.
2. **A QC slide should be included with each run of stained slides, particularly if the staining setup is used infrequently.**
3. All staining dishes should be covered to prevent evaporation of reagents (screw-cap Coplin jars or glass lids should be used).
4. Depending on the volume of slides stained, staining solutions should be changed on an as-needed basis.
5. When the smear is thoroughly fixed and the stain is performed correctly, the spores are seen to be ovoid and refractile, with the spore wall being bright pinkish red. Occasionally, the polar tube can be seen either as a stripe or as a diagonal line across the spore. The majority of the bacteria and other debris tend to stain green. However, some bacteria and debris stain red.
6. The specimen should also be checked (macroscopically) for adherence to the slide.
7. The microscope should be calibrated (within the last 12 months), and the objectives and oculars used for the calibration procedure should be used for all measurements on the microscope. The calibration factors for all objectives should be posted on the microscope for easy access (multiplication factors can be pasted on the body of the microscope). Although recalibration every 12 months may not be necessary, this varies from laboratory to laboratory, depending on equipment care and use.
8. Known positive microscope slides, Kodachrome 2-by-2 projection slides, and photographs (reference books) should be available at the workstation.
9. Record all QC results; the laboratory should also have an action plan for "out-of-control" results.

Detailed Procedure

1. Using a 10-µl aliquot of unconcentrated, preserved liquid stool (5 or 10% formalin, SAF, or other single-vial option), prepare the smear by spreading the material over an area 45 by 25 mm. **Although this original procedure specifies unconcentrated specimen, organism recovery can be dramatically enhanced by centrifuging the specimen for 10 min at 500 × *g* prior to smear preparation.**
2. Allow the smear to air dry.
3. Place the smear in absolute methanol for 5 min.
4. Allow the smear to air dry.
5. Place in trichrome stain for 90 min.
6. Rinse in acid-alcohol for no more than 10 s.
7. Dip slides several times in 95% alcohol. Use this step as a rinse.
8. Place in 95% alcohol for 5 min.
9. Place in 100% alcohol for 10 min.
10. Place in xylene substitute for 10 min.
11. Mount with a coverslip (no. 1 thickness), using mounting medium.
12. Examine smears under oil immersion (1,000×) and read at least 300 fields; the examination time will probably be at least 10 min per slide.

Reporting

Results are reported as for the Ryan stain, described below (see the following protocol).

Procedure Notes/Reminders

The procedure is the same as for the Ryan stain (see below).

Procedure Limitations

The procedure limitations are the same as for the Ryan stain (see below).

MODIFIED TRICHROME STAIN FOR MICROSPORIDIA (Ryan, Blue Counterstain)

Description

A number of variations to the modified trichrome stain (Ryan blue) were tried in an attempt to improve the contrast between the color of the spores and the background staining (Figure 5.12, right). Optimal staining was achieved by modifying the composition of the trichrome solution. This stain is also available commercially from a number of suppliers. The specimen can be fresh stool or stool that has been preserved in 5 or 10% formalin, SAF, or one of the newer single-vial-system fixatives. Actually, any specimen other than tissue thought to contain microsporidia could be stained using this method.

Reagents

▶ **Stain**

Chromotrope 2R	6.0 g*
Aniline blue	0.5 g
Phosphotungstic acid	0.25 g
Acetic acid (glacial)	3.0 ml
Distilled water	100.0 ml

*This is 10 times the amount used in the normal trichrome stain formula.

1. Prepare the stain by adding 3.0 ml of acetic acid to the dry ingredients. Allow the mixture to stand (ripen) for 30 min at room temperature.
2. Add 100 ml of distilled water and adjust the pH to 2.5 with 1.0 M HCl. Correctly prepared stain is dark purple. The staining solution should be protected from light.
3. Store in a glass or plastic bottle at room temperature. The shelf life is at least 24 months.

▶ **Acid-Alcohol**

90% ethyl alcohol	995.5 ml
Acetic acid (glacial)	4.5 ml

Prepare by combining the two solutions.

Quality Control

1. Unfortunately, the only way to perform acceptable QC procedures for this method is to use actual microsporidial spores as the control organisms. Obtaining these positive controls may be somewhat difficult. It is particularly important to use the actual organisms because the spores are very small (1 to 1.5 μm) and difficult to stain.
2. **A QC slide should be included with each run of stained slides, particularly if the staining setup is used infrequently.**
3. All staining dishes should be covered to prevent evaporation of reagents (screw-cap Coplin jars or glass lids should be used).

4. Depending on the volume of slides stained, staining solutions should be changed on an as-needed basis.
5. When the smear is thoroughly fixed and the stain is performed correctly, the spores are seen to be ovoid and refractile, with the spore wall being bright pinkish red. Occasionally, the polar tube can be seen either as a stripe or as a diagonal line across the spore. The majority of the bacteria and other debris tend to stain blue. However, some bacteria and debris stain red.
6. The specimen should also be checked (macroscopically) for adherence to the slide.
7. The microscope should be calibrated (within the last 12 months), and the objectives and oculars used for the calibration procedure should be used for all measurements on the microscope. The calibration factors for all objectives should be posted on the microscope for easy access (multiplication factors can be pasted on the body of the microscope). Although recalibration every 12 months may not be necessary, this varies from laboratory to laboratory, depending on equipment care and use.
8. Known positive microscope slides, Kodachrome 2-by-2 projection slides, and photographs (reference books) should be available at the workstation.
9. Record all QC results; the laboratory should also have an action plan for "out-of-control" results.

Detailed Procedure

1. Using a 10-µl aliquot of concentrated (10 min at 500 × *g*), preserved stool (5 or 10% formalin, SAF, or one of the zinc-based single-vial preservatives, including Total-Fix), prepare the smear by spreading the material over an area 45 by 25 mm.
2. Allow the smear to air dry.
3. Place the slide in absolute methanol for 5 or 10 min.
4. Allow the smear to air dry.
5. Place the slide in trichrome stain for 90 min.
6. Rinse in acid-alcohol for no more than 10 s.
7. Dip the slide several times in 95% alcohol. Use this step as a rinse (no more than 10 s).
8. Place the slide in 95% alcohol for 5 min.
9. Place the slide in 95% alcohol for 5 min.
10. Place the slide in 100% alcohol for 10 min.
11. Place the slide in xylene substitute for 10 min.
12. Mount with a coverslip (no. 1 thickness), using mounting medium.
13. Examine the smear under oil immersion (magnification, ×1,000), and read at least 300 fields; the examination time will probably be at least 10 min per slide.

Reporting

The microsporidial spore wall should stain pinkish to red, with the interior of the spore being clear or perhaps showing a horizontal or diagonal stripe, which represents the polar tube. The background should appear green or blue, depending on the method. Some bacteria, some yeast cells, and some debris stain pink to red; the shapes and sizes of the various components may be helpful in differentiating the spores from other structures. The results of this staining procedure should be reported only if the positive-control smears are acceptable. The use of immunoassay reagents should provide a more specific and sensitive approach to the identification of the microsporidia in fecal specimens.

1. Report the organism. Do not use abbreviations.

Examples: Microsporidia present

Most likely *Enterocytozoon bieneusi* or *Encephalitozoon intestinalis* present (if fecal specimens are used)

Encephalitozoon intestinalis present (identification to species level is highly likely; this is generally the organism involved in disseminated cases from the gastrointestinal tract to the kidneys, with the organisms being recovered in urine)

Procedure Notes/Reminders

1. It is mandatory that positive-control smears be stained and examined each time patient specimens are stained and examined.
2. Because of the difficulty in getting stain to penetrate the spore wall, prepare thin smears and do not reduce the staining time in trichrome. Also, make sure that the slides are not left too long in the decolorizing agent (acid-alcohol). If the control organisms are too light, leave them in the trichrome longer and shorten the time to two dips in the acid-alcohol solution. Also, remember that the 95% alcohol rinse after the acid-alcohol step should be performed quickly to prevent additional destaining from the acid-alcohol reagent.
3. When you purchase the chromotrope 2R, obtain the highest dye content available. Two sources are Harleco (Gibbstown, NJ) and Sigma Chemical Co. (St. Louis, MO) (the dye content is among the highest [85%]). Fast green and aniline blue can be obtained from Allied Chemical and Dye (New York, NY).
4. In the final stages of dehydration, the 100% ethanol and the xylenes (or xylene substitutes) should be kept as free from water as possible. Coplin jars must have tightly fitting caps to prevent both evaporation of reagents and absorption of moisture. If the xylene becomes cloudy after addition of slides from 100% alcohol, return the slides to 100% alcohol and replace the xylene with fresh stock.

Procedure Limitations

1. **Although this staining method stains the microsporidia, the range of stain intensity and the small size of the spores tend to cause some difficulty in identifying these organisms. Since this procedure results in many other organisms or objects staining in stool specimens, differentiation of the microsporidia from surrounding material is still very difficult. There also tends to be some slight size variation among the spores. Identification to the genus and species levels cannot be made based on spore morphology in stained smears; molecular testing and/or electron microscopy is required to determine the exact identification.**
2. If the patient has severe watery diarrhea, there will be less artifact material in the stool to confuse with the microsporidial spores; however, if the stool is semiformed or formed, much more artifact material will be present and the spores will be much harder to detect and identify. Also, remember that the number of spores varies according to the stool consistency (the more diarrhetic the stool, the more spores will be present).
3. The investigators who developed some of these procedures feel that concentration procedures result in an actual loss of microsporidial spores; thus, there is a strong recommendation to use unconcentrated, formalinized stool. However, there are no data indicating what centrifugation speeds, etc., were used in the study.

4. **In the UCLA Clinical Microbiology Laboratory, we have generated data (unpublished) to indicate that centrifugation at 500 × *g* for 10 min dramatically increases the number of microsporidial spores available for staining (from the concentrate sediment).** This is the method we use for centrifugation of all stool specimens, regardless of the suspected organism.
5. Avoid the use of wet gauze filtration (an old, standardized method of filtering stool prior to centrifugation) with too many layers of gauze that may trap organisms and allow them to flow into the fluid to be concentrated. It is recommended that no more than two layers of gauze be used. Another option is to use the commercially available concentration systems that use metal or plastic screens for filtration.

MODIFIED TRICHROME STAIN FOR MICROSPORIDIA (Evelyn Kokoskin, Hot Method)

Description

Changes in temperature from room temperature to 50°C and in the staining time from 90 to 10 min have been recommended as improvements for the modified trichrome staining methods. The procedure is as follows.

1. Using a 10-µl aliquot of unconcentrated, preserved liquid stool (5 or 10% formalin or SAF), prepare the smear by spreading the material over an area 45 by 25 mm.
2. Allow the smear to air dry.
3. Place the slide in absolute methanol for 5 min.
4. Allow the smear to air dry.
5. Place the slide in trichrome stain for 10 min at 50°C.
6. Rinse in acid-alcohol for no more than 10 s.
7. Dip the slide several times in 95% alcohol. Use this step as a rinse (no more than 10 s).
8. Place the slide in 95% alcohol for 5 min.
9. Place the slide in 100% alcohol for 10 min.
10. Place the slide in xylene substitute for 10 min.
11. Mount with a coverslip (no. 1 thickness), using mounting medium.
12. Examine smears under oil immersion (magnification, ×1,000), and read at least 100 fields; the examination time will probably be at least 10 min per slide.

REVIEW: MODIFIED TRICHROME-STAINED SMEARS	
Clinical relevance	To provide contrasting colors for both the background debris and parasites present; designed to allow examination and recognition of organism morphology under oil immersion (100× objective for a total magnification of ×1,000). Designed primarily to allow recovery and identification of microsporidial spores. Internal morphology (**horizontal or diagonal stripes**) may be seen in some spores under oil immersion (magnification, ×1,000). Exact identification to genus/species requires molecular testing and/or electron microscopy.
Specimen	Any stool specimen that is fresh or preserved in formalin, SAF, or one of the single-vial-system fixatives
Reagents	Modified trichrome stain (using the high-dye-content chromotrope 2R) and associated solutions; dehydrating solutions (alcohols and xylenes); mounting fluid optional

REVIEW: MODIFIED TRICHROME-STAINED SMEARS *(continued)*	
Examination requirements	Oil immersion examination of at least 300 fields; additional fields may be required if suspect organisms have been seen but are not clearly identified. Identification of microsporidial spores may be possible; however, their small size makes recognition difficult. Final laboratory results depend heavily on the appearance of the QC slides and comparison with patient specimens. It is mandatory that some spores which contain either horizontal or diagonal stripes (polar tubule) be seen.
Procedure notes and limitations	Because of the difficulty in getting dye to penetrate the spore wall, this staining approach can be very helpful. Procedure limitations are related to specimen handling (proper centrifugation speeds and times, use of no more than two layers of wet gauze for filtration, and a complete understanding of the difficulties in recognizing microsporidial spores due to their small size [1 to 2.5 μm]).
Important questions for commercial suppliers	Make sure to ask about specific fixatives and whether the fecal material can be stained with the modified trichrome stains or modified acid-fast stains. Also, ask if the fixatives prevent the use of any of the immunoassay methods now available for several of the intestinal amebae, flagellates, coccidia, and microsporidia.

Fecal Immunoassays for Intestinal Protozoa

Fecal immunoassays are generally simple to perform and allow a large number of tests to be performed at one time, thereby reducing overall costs. A major disadvantage of antigen detection in stool specimens is that the method can detect only one or two pathogens at a time. One still must perform a routine O&P examination to detect other parasitic pathogens. The current commercially available antigen tests have superior sensitivity and specificity compared with routine microscopy. Current formats include the enzyme-linked immunosorbent assay, the fluorescent-antibody (FA) test, and rapid membrane flow cartridges (Figure 5.13). Sensitivity and specificity are comparable among the various formats and kits currently available. Selection of any particular format immunoassay often depends on the workflow options within the laboratory, based primarily on test menu orders received from the physician. The methods are different, but the results are comparable.

ENTAMOEBA HISTOLYTICA

Antigen-based fecal immunoassays have several significant advantages over other methods currently used for diagnosis of amebiasis: (i) some of the assays differentiate the true pathogen *E. histolytica* from nonpathogenic *E. dispar*, (ii) they have excellent sensitivity and

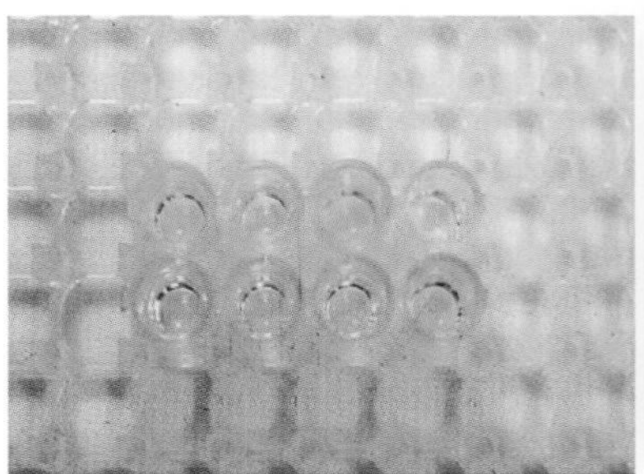

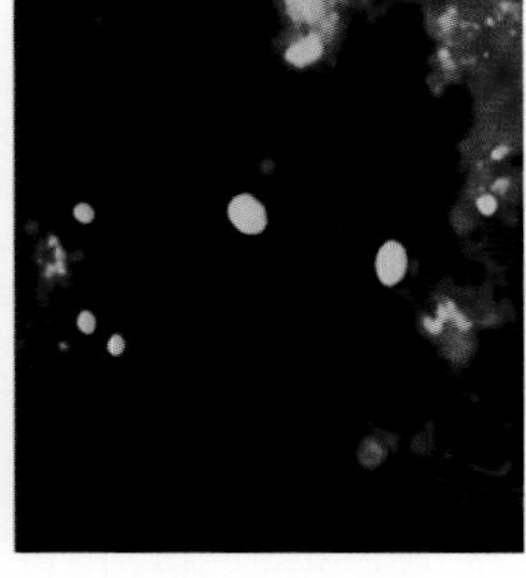

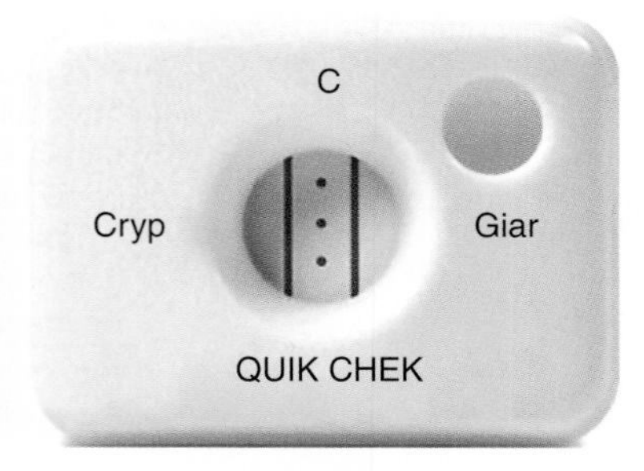

Figure 5.13 Fecal immunoassay formats. EIA plate for *Giardia* with positive yellow wells; *Cryptosporidium*/*Giardia* combination FA procedure with counterstain, positive test for both; rapid membrane flow cartridge for *Cryptosporidium*/*Giardia* combination test (demonstrating positive results for both organisms).

specificity, (iii) they are readily usable by most laboratory personnel, and (iv) they have potential use in situations such as waterborne outbreaks. Because there are distinct genetic differences between *E. dispar* (a nonpathogen) and *E. histolytica* (a true pathogen), commercial kits have been developed to detect their presence and differentiate them in clinical samples. However, current antigen detection tests require the examination of fresh or frozen (not preserved) stool specimens, while many laboratories have switched to stool collection methods using various preservatives. A microplate ELISA and a rapid enzyme immunoassay (EIA) cartridge are available formats for testing for the true pathogen, *E. histolytica* (available through TECHLAB/Alere, Orlando, FL).

CRYPTOSPORIDIUM SPP.

A number of commercially available immunoassay kits are available for detection of *Cryptosporidium* spp. and are more sensitive and specific than routine microscopic examination of modified acid-fast stained smears. Stool specimens may be fresh, frozen, or fixed; however, preserved specimens containing PVA are currently unacceptable for use in the fecal immunoassays.

GIARDIA LAMBLIA

Detection of *Giardia* in stool specimens by various immunoassay methods has been reported. These tests are reliable and more sensitive and specific than routine O&P exams. Commercial immunoassay kits are readily available. Users will have to evaluate which kit format will be most useful for their own laboratories. Some of the methods may require fresh specimens, and stools fixed in preservatives may not be suitable. Also, the kits are designed to detect cysts of *G. lamblia;* however, since some antigenic sites are shared between cysts and trophozoites, pale trophozoites may occasionally be seen using the FA test format. Data indicate that when fecal immunoassays are used, two negative specimens must be found in order to report a negative result (this approach is limited to testing for *Giardia* and does not apply to *Cryptosporidium*).

KITS UNDER DEVELOPMENT

Although not currently available commercially, several immunoassays are in various developmental phases. These include antigen detection kits for *Dientamoeba fragilis*, *Blastocystis* spp., *Cyclospora cayetanensis*, and various species of the microsporidia.

COMMENTS ON THE PERFORMANCE OF FECAL IMMUNOASSAYS

Some comments about various immunoassay formats are provided to assist you in evaluating test performance and/or result interpretation. **It is very important to read the kit information sheet before use.** Currently, fecal immunoassays are available for *G. lamblia*, the *E. histolytica*/*E. dispar* group, *E. histolytica*, and *Cryptosporidium* spp. Based on the published literature, fecal immunoassays are more sensitive and specific than the routine O&P exam; this is particularly true for *G. lamblia*. However, unlike the O&P exam, which facilitates the recovery of many different parasites, the fecal immunoassays are limited to one or two organisms only. The fecal immunoassays are also more sensitive than the special stains (modified acid-fast stains) for *Cryptosporidium* spp. Fresh specimens can be stored at 2 to 8°C and should be tested within 48 h, or they should be frozen at −20 to −70°C (freezing is not acceptable for the FA method; the freeze-thaw cycle damages the organisms). Stool specimens preserved in 10% formalin, MF, SAF, or some of the other single-vial fixatives may be refrigerated at 2 to 8°C or stored at room temperature (20 to 25°C) and should be tested within 2 months. Stool specimens submitted in Cary-Blair transport medium (or equivalent) should be refrigerated or frozen and tested within 1 week after collection. Fecal specimens that have been preserved in fixatives containing PVA are not acceptable for testing. With the FA procedure, the actual organisms (*G. lamblia* and/or *Cryptosporidium* spp.), not antigens, are seen via a color change. To enhance the sensitivity of the FA procedure, it is recommended that testing be performed on centrifuged (500 × *g* for 10 min) stool specimen.

ENZYME IMMUNOASSAYS (Antigen Detection, No Centrifugation Recommended)

In enzyme immunoassays, the antigen is found in the top fluid layer of the stool collection vial.

1. Remember to thoroughly rinse the wells according to the instructions; do not eliminate any of the rinse steps. Make sure that each well receives the total number of rinses required.
2. Make sure the stream of buffer goes directly into the wells. Use a wash bottle with a small opening, so you have to squeeze the bottle to get the fluid to squirt directly into the wells.
3. When the directions tell you to "slap" the tray down onto some paper towels to remove the last of the rinse fluid, make sure that you slap it several times. Don't be too gentle; the cups will not fall out of the tray. If they do come loose, just push them back into the tray.
4. Before the last reagents are added, the wells should be empty of rinse buffer (not dry, but empty of excess fluid).
5. **Note: If you shake the specimen vial prior to testing, allow the vial contents to settle out for several minutes. Addition of too much particulate stool to the wells interferes with testing.**

FLUORESCENCE (Visual Identification of the Organisms, Centrifugation Recommended)

1. Since you will be looking for the actual organisms (cysts of *Giardia* and/or oocysts of *Cryptosporidium*), this test should be performed on centrifuged (500 × *g* for 10 min) stool to increase the sensitivity of the test.
2. Remember to thin out the smear; it is important to make sure the slides are thoroughly dry before adding reagents. The slides can be placed in a 35°C incubator for about 30 min to 1 h to make sure they are dry before being processed. If the material on the wells is too thick, it may not dry thoroughly and may fall off of the glass. It is better to let the slides dry longer rather than for too short a time. A heat block is **not** recommended for this purpose.
3. **Very gently** rinse the reagents from the wells; do not squirt directly into the wells, but allow the rinse fluid to flow over the wells.
4. Remember that not all clinical specimens will provide the 3+ to 4+ fluorescence that is often seen in the positive control. Also, from time to time, you may see fluorescing bacteria and/or some yeasts in certain patient specimens. This is not common, but the shapes can be distinguished from *Giardia* cysts and *Cryptosporidium* oocysts.

5. The intensity of the fluorescence may vary, depending on the filters. If the fluorescence microscope dual-filter system is used, it demonstrates both the yellow-green fluorescence and the red-orange counterstain, and neither *Giardia* nor *Cryptosporidium* may appear quite as bright as when the yellow-green filter only is used. Both approaches are acceptable and may reflect laboratory preferences. However, remember that when the single FITC (yellow-green) filter is used alone, some artifact materials may also appear to fluoresce more brightly, while the artifact material might not be seen when both filters (FITC and counterstain) are used. Artifact material may fluoresce a dull color without the bright outlines seen around the *Giardia* cysts and *Cryptosporidium* oocysts that can be seen when both filter systems are used.

   Both filters = lower fluorescence intensity, less visible artifacts
   Single FITC filter = brighter fluorescence, more visible artifacts
6. Make sure to examine the edges of the wells. Sometimes in a light infection, the edges may contain organisms while in the middle of the well (thick area), the organisms may be a bit more difficult to detect.
7. *Giardia* trophozoites may also fluoresce with a very pale fluorescent outline (teardrop shape); **this can be important if the patient has diarrhea and no cysts are present in the specimen.** The fluorescence level will usually be no stronger than a 1+.

LATERAL-FLOW CARTRIDGES (Antigen Detection, No Centrifugation Recommended)

In the lateral-flow cartridge system, the antigen is found in the top fluid layer of the stool collection vial.

1. If the stool is too thick, the addition of reagents will not thin it out enough. If the specimen poured into the well remains too thick, the fluid does not flow up the membrane. If your specimens arrive in fixative and there is no fluid at the top of the vial overlying the stool, this means that the vial may have been overfilled with stool. These specimens will have to be diluted with the appropriate diluent before being tested.
2. It is always important to see the control line indicated as positive all the way across the membrane, not just at the edges.
3. **A positive test result may be much lighter than the control line; this is normal.**
4. At the cutoff time to read the result, the presence of any acceptable color intensity visible in the test area should be interpreted as a positive result (often the color will be pale). Note: some kits indicate that certain color lines are not acceptable. Check the package insert.
5. **Do not read/interpret the results after the time indicated in the directions; you may get a false-positive result.**
6. If you shake the specimen vial prior to testing, allow the vial contents to settle out for several minutes. Adding too much particulate stool to the wells interferes with testing.

Larval Nematode Culture

HARADA-MORI FILTER PAPER STRIP CULTURE

Description

To detect light infections with hookworm, *S. stercoralis*, and *Trichostrongylus* spp., as well as to facilitate specific identification, the Harada-Mori filter paper strip culture technique is recommended (Figure 5.14). The technique requires filter paper to which

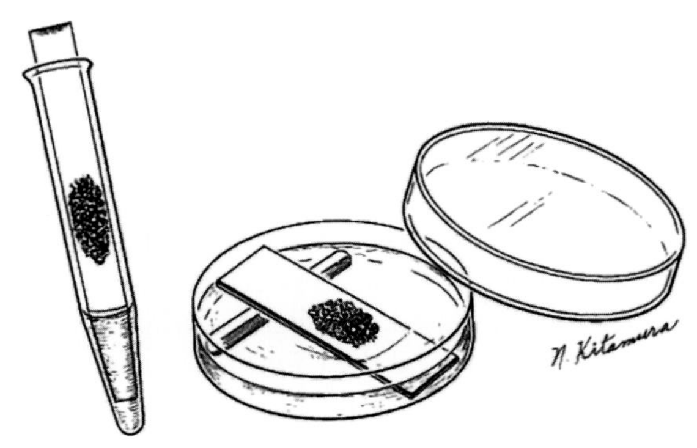

Figure 5.14 Culture methods for the recovery of larval-stage nematodes, including the Harada-Mori tube method and the petri dish culture method. (Illustration by Nobuko Kitamura; reprinted from *Diagnostic Medical Parasitology*, 6th ed.)

fresh fecal material is added and a test tube into which the filter paper is inserted. Moisture is provided by adding water to the tube, which continuously soaks the filter paper by capillary action. Incubation under suitable conditions favors hatching of ova and/or development of larvae. Fecal specimens to be cultured should not be refrigerated, since some parasites (especially *Necator americanus*) are susceptible to cold and may fail to develop after refrigeration. Also, caution must be exercised in handling the filter paper strip itself, since infective *Strongyloides* larvae may migrate upward as well as downward on the paper strip. Always observe standard precautions and wear gloves when performing these procedures.

Quality Control

1. Follow routine procedures for optimal collection and handling of fresh fecal specimens for parasitologic examination.
2. Examine known positive and negative samples of stools (from laboratory animals), if available, to make sure that the procedure works.
3. Review larval diagrams and descriptions for confirmation of larval identification.
4. The microscope should be calibrated, and the objectives and oculars used for the calibration procedure should be used for all measurements on the microscope. The calibration factors for all objectives should be posted on the microscope for easy access (multiplication factors can be pasted on the body of the microscope).
5. Record all QC results.

Detailed Procedure

1. Smear 0.5 to 1 g of feces in the center of a narrow strip of filter paper (3/8 by 5 in. [1 in. = 2.54 cm], slightly tapered at one end).
2. Add 3 to 4 ml of distilled water to a 15-ml conical centrifuge tube; identify the specimen on the tube.
3. Insert the filter paper strip into the tube so that the tapered end is near the bottom of the tube. The water level should be approximately 0.5 in. below the fecal spot. It is not necessary to cap the tube. However, a cork stopper or a cotton plug may be used.
4. Maintain the tube upright in a rack at 25 to 28°C. Add distilled water to maintain the original level (usually evaporation takes place over the first 2 days, and then the culture becomes stabilized).

5. Keep the tube for 10 days, and check it daily by withdrawing a small amount of fluid from the bottom of the tube. Prepare a smear on a glass slide, cover the slide with a coverslip, and examine the smear with the 10× objective.
6. Examine the larvae for motility and typical morphologic features to reveal whether hookworm, *Strongyloides*, or *Trichostrongylus* larvae are present.

Reporting

Larval nematodes of hookworm, *S. stercoralis*, or *Trichostrongylus* spp. may be recovered. If *Strongyloides* organisms are present, free-living stages and larvae may be found after several days in culture.

1. Report "No larvae detected" if no larvae could be detected at the end of the incubation.
2. Report larvae detected by fecal culture.

   Example: *Strongyloides stercoralis* larvae detected by fecal culture

Procedure Notes/Reminders

1. If the larvae are too active to observe under the microscope and morphologic details are difficult to see, the larvae can be heat killed within the tube or after removal to the slide; iodine can also be used to kill larvae.
2. Infective larvae may be found any time after the fourth day or even on the first day in a heavy infection. Since infective larvae may migrate upward as well as downward on the filter paper strip, caution must be exercised in handling the fluid and the paper strip itself to prevent infection. Handle the filter paper with forceps, and wear gloves when handling the cultures.
3. It is important to maintain the original water level to maintain optimum humidity.
4. Fresh stool is required for this procedure; preserved fecal specimens or specimens obtained after a barium meal are not suitable for processing by this method.

Procedure Limitations

1. This technique allows both parasitic and free-living forms of nematodes to develop. If specimens have been contaminated with soil or water containing these forms, it may be necessary to distinguish parasitic from free-living forms. This distinction is possible, since parasitic forms are more resistant to slight acidity than are free-living forms. Proceed as follows. Add 0.3 ml of concentrated hydrochloric acid per 10 ml of water containing the larvae (adjust the volume accordingly to achieve a 1:30 dilution of acid). Free-living nematodes are killed, while parasitic species live for about 24 h.
2. Specimens that have been refrigerated or preserved are not suitable for culture. Larvae of certain species are susceptible to cold environments.

BAERMANN CONCENTRATION

Description

Another method of examining a stool specimen suspected of having small numbers of *Strongyloides* larvae is the use of a modified Baermann apparatus (Figure 5.15). The Baermann technique, which involves using a funnel apparatus, relies on the principle that active larvae migrate from a fresh fecal specimen that has been placed on a wire mesh with several layers of gauze which are in contact with tap water. Larvae migrate through the gauze into the water and settle to the bottom of the funnel, where they can be collected and examined. The main difference between this method and the Harada-Mori and petri dish methods is the greater amount of fresh stool used, possibly providing a better chance of larval recovery

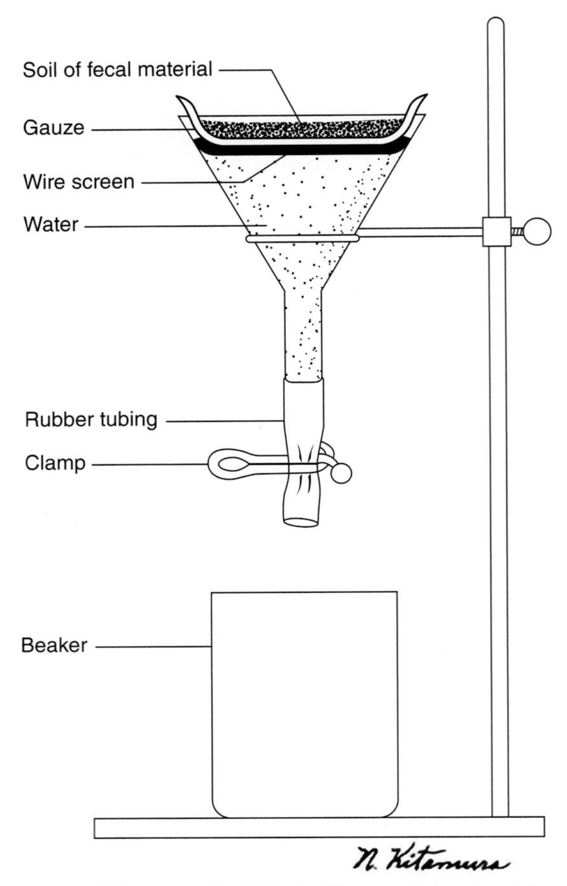

Figure 5.15 Baermann apparatus. (Illustration by Nobuko Kitamura; reprinted from *Diagnostic Medical Parasitology*, 6th ed.)

in a light infection. Besides being used for patient fecal specimens, this technique can be used to examine soil specimens for the presence of larvae.

Quality Control

1. Follow routine procedures for optimal collection and handling of fresh specimens for parasitologic examination.
2. Examine known positive and negative samples of stools (from laboratory animals), if available, to make sure that the procedure is precise.
3. Review larval diagrams for confirmation of larval identification.
4. The microscope should be calibrated, and the objectives and oculars used for the calibration procedure should be used for all measurements on the microscope. The calibration factors for all objectives should be posted on the microscope for easy access (multiplication factors can be pasted on the body of the microscope).
5. Record all QC results.

Detailed Procedure

1. If possible, use a fresh fecal specimen obtained after administration of a mild saline cathartic, not a stool softener. Soft stool is recommended; however, any fresh fecal specimen is acceptable.

2. Set up a clamp supporting a 6-in. glass funnel. Attach rubber tubing and a pinch clamp to the bottom of the funnel. Place a collection beaker underneath.
3. Place a wire gauze or nylon filter over the top of the funnel, followed by a pad consisting of two layers of gauze.
4. Close the pinch clamp at the bottom of the tubing, and fill the funnel with tap water until it just soaks the gauze padding.
5. Spread a large amount of fecal material on the gauze padding so that it is covered with water. If the fecal material is very firm, first emulsify it in water.
6. Allow the apparatus to stand for 2 h or longer; then draw off 10 ml of fluid into the beaker by releasing the pinch clamp, centrifuge for 2 min at 500 × *g*, and examine the sediment under the microscope (magnifications, ×100 and ×400) for the presence of motile larvae. Make sure that the end of the tubing is well inside the beaker before slowly releasing the pinch clamp. Infective larvae may be present; wear gloves when performing this procedure.

Reporting

Larval nematodes (hookworm, *S. stercoralis*, or *Trichostrongylus* spp.) may be recovered. Both infective and noninfective *Strongyloides* larvae may be recovered, particularly in a heavy infection.

1. Report "No larvae detected" if no larvae could be detected at the end of incubation.
2. Report larvae detected by fecal culture.

   Example: *Strongyloides stercoralis* larvae detected by fecal culture

Procedure Notes/Reminders

1. It may be difficult to observe morphological details in rapidly moving larvae; a drop of iodine or formalin or slight heating can be used to kill the larvae.
2. Infective larvae may be found any time after the fourth day and occasionally after the first day in heavy infections. Caution must be exercised in handling the fluid, gauze pad, and beaker to prevent infection. Wear gloves when using this technique.
3. Remember to make sure that the pinch clamp is tight until you want to release some of the water.
4. Preserved fecal specimens or specimens obtained after a barium meal are not suitable for processing by this method; fresh stool specimens must be obtained.

Procedure Limitations

1. This technique allows both parasitic and free-living forms of nematodes to develop. If specimens have been contaminated with soil or water containing these forms, it may be necessary to distinguish parasitic from free-living forms. This distinction is possible, since parasitic forms are more resistant to slight acidity than are free-living forms. Proceed as follows. Add 0.3 ml of concentrated hydrochloric acid per 10 ml of water containing the larvae (adjust the volume accordingly to achieve a 1:30 dilution of acid). Free-living nematodes are killed, while parasitic species live for about 24 h.
2. Specimens that have been refrigerated or preserved are not suitable for culture. Larvae of certain species are susceptible to cold environments.
3. Gloves should be worn when this procedure is performed.
4. Release the pinch clamp slowly to prevent splashing; have the end of the tubing close to the bottom of the beaker for the same reason.

AGAR PLATE CULTURE FOR *STRONGYLOIDES STERCORALIS*

Description

Agar plate cultures are also recommended for the recovery of *S. stercoralis* larvae and **tend to be more sensitive than some of the other diagnostic methods.** Stool is placed on agar plates, and the plates are sealed with cellulose tape to prevent accidental infections and held for 2 days at room temperature. As the larvae crawl over the agar, they carry bacteria with them, thus creating visible tracks over the agar. The plates are examined under the microscope for confirmation of larvae, the surface of the agar is then washed with 10% formalin, and final confirmation of larval identification is made via wet examination of the sediment from the formalin washings (Figure 5.16).

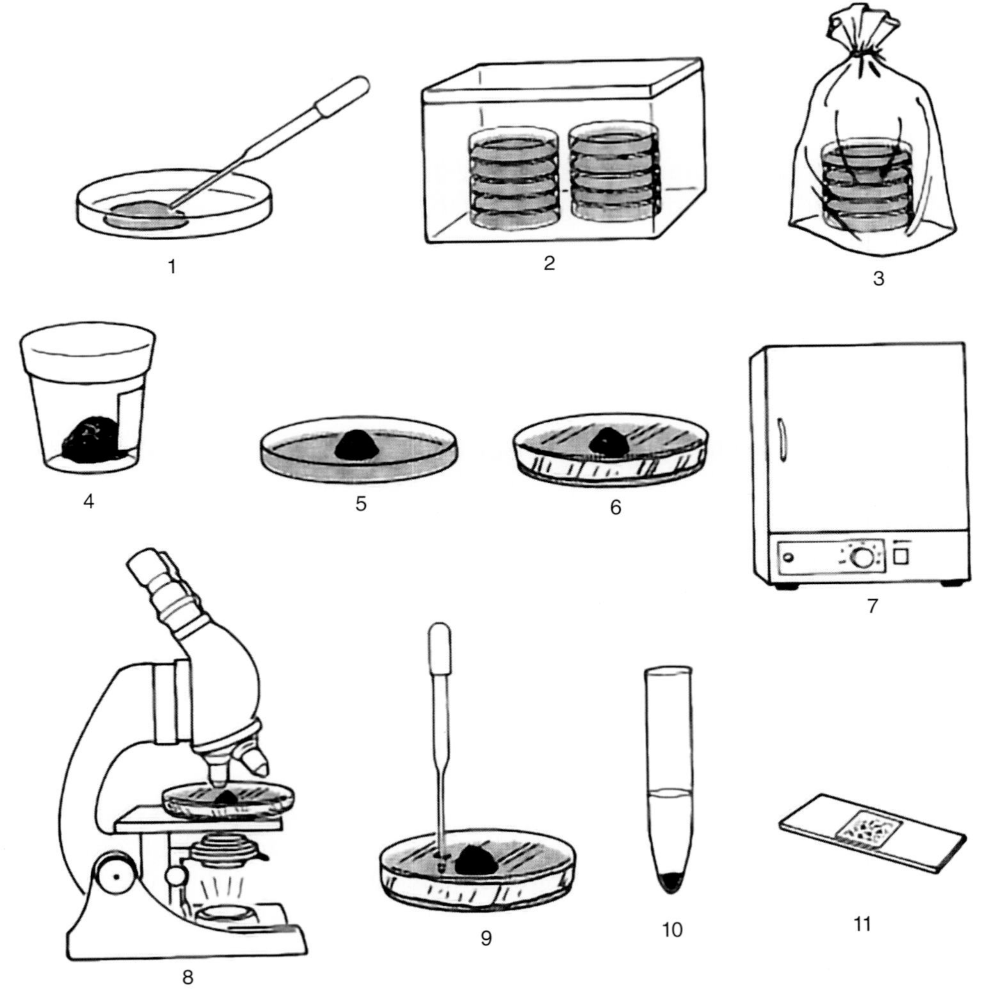

Figure 5.16 Agar culture method for *Strongyloides stercoralis*. (1) Agar plates are prepared; (2) agar is dried for 4 to 5 days on the bench top; (3) plates are stored in plastic bags; (4) fresh stool is submitted to the laboratory; (5) approximately 2 g of stool is placed on an agar plate; (6) the plate is sealed with tape; (7) the culture plate is incubated at 26 to 33°C for 2 days; (8) the plate is examined microscopically for the presence of tracks (indicating bacteria carried over the agar by migrating larvae); (9) 10% formalin is placed onto the agar through a hole made in the plastic by hot forceps; (10) material from the agar plate is centrifuged; (11) the material is examined as a wet preparation for rhabditiform or filariform larvae (high dry power; magnification, ×400). (Illustration by Sharon Belkin; reprinted from *Diagnostic Medical Parasitology*, 6th ed.)

Reagents

▶ **Agar**
1.5% Agar
0.5% Meat extract
1.0% Peptone
0.5% NaCl

Note: Positive tracking on agar plates has been seen on a number of different types of agar. However, the most appropriate agar formula is that seen above.

Quality Control

1. Follow routine procedures for optimal collection and handling of fresh fecal specimens for parasitologic examination.
2. Examine agar plates to ensure that there is no cracking and that the agar pour is sufficient to prevent drying. Also, make sure that there is no excess water on the surface of the plates.
3. Review larval diagrams and descriptions for confirmation of larval identification.
4. The microscope should be calibrated, and the objectives and oculars used for the calibration procedure should be used for all measurements on the microscope. The calibration factors for all objectives should be posted on the microscope for easy access (multiplication factors can be pasted on the body of the microscope).
5. Record all QC results (condition of agar plates).

Detailed Procedure

1. Place approximately 2 g of fresh stool (approximately 1 in. in diameter) in the center of the agar plate.
2. Replace the lid, and seal the plate with cellulose tape.
3. Maintain the plate (right side up) at room temperature for 2 days.
4. After 2 days, examine the sealed plate through the plastic lid under the microscope for microscopic colonies that develop as random tracks on the agar and evidence of larvae at the ends of the tracks away from the stool.
5. With the end of hot forceps, make a hole in the top of the plastic petri dish.
6. Gently add 10 ml of 10% formalin through the hole onto the agar surface, swirl to cover the surface, and rinse the agar plate. Allow the plate to stand for 30 min.
7. Remove the tape and lid of the agar plate. Pour the 10% formalin through a funnel into a centrifuge tube. Do not pour the formalin off directly into the centrifuge tube, because the tube opening is too small and formalin will be spilled onto the counter.
8. Centrifuge the formalin rinse fluid for 5 min at 500 × *g*.
9. Prepare a wet smear preparation from sediment, and examine it with the 10× objective (low power) for presence of larvae. If larvae are found, confirm the identification with the 40× objective (high dry power).

Reporting

Larval nematodes of hookworm, *S. stercoralis*, or *Trichostrongylus* spp. may be recovered. If *Strongyloides* organisms are present, free-living stages and larvae may be found after several days on the agar plates.

1. Report "No larvae detected" if no larvae could be detected at the end of the incubation and rinse procedure.

2. Report larvae detected by agar plate culture.

   Example: *Strongyloides stercoralis* larvae detected by agar plate culture

Procedure Notes/Reminders

1. If the larvae are too difficult to observe under the microscope and morphologic details are difficult to see, the larvae can be formalin killed within the plate and examined in the formalin-concentrated sediment.
2. Infective larvae may be found any time after the first or second day or even on the first day in a heavy infection. Since infective larvae may be present on the agar, caution must be exercised in handling the plates once the cellulose tape is removed. Wear gloves when handling the cultures.
3. It is important to maintain the plates upright at room temperature. Do not incubate or refrigerate them at any time; this also applies to the fresh stool specimen.
4. Fresh stool is required for this procedure; preserved fecal specimens or specimens obtained after a barium meal are not suitable for processing by this method.

Procedure Limitations

1. This technique is successful if any larvae present are viable. If the fresh stool specimen is too old, larvae may not survive and a negative result will be reported.
2. Specimens that have been refrigerated or preserved are not suitable for culture. Larvae of certain species are susceptible to cold environments.
3. The overall sensitivity of this method does not equal that seen with molecular methods. Multiplex organism panels containing several parasites including *Strongyloides* are currently under development within the United States. A number of molecular tests are also available elsewhere in the world; however, they are not FDA cleared for use within the United States.

REVIEW: LARVAL NEMATODE CULTURE	
Clinical relevance	To (i) reveal their presence when they are too scarce to be detected by concentration methods; (ii) distinguish whether the infection is due to *S. stercoralis* or hookworm on the basis of rhabditiform larval morphology by allowing hookworm egg hatching to occur, thus releasing first-stage larvae; and (iii) allow the development of larvae into the filariform stage for further differentiation.
Specimen	Any stool specimen that is fresh and has not been refrigerated
Supplies	Appropriate tubes, plates, funnels, gauze, and agar formula
Examination requirements	Daily checking of the fluid for the presence of larvae; **hold the cultures for 10 days prior to making a final report.**
Results and laboratory reports	The failure to recover larvae does not completely rule out the possibility of infection; however, the probability of infection is lower when results are negative.
Procedure notes and limitations	There is always the prospect of recovering infective larvae; gloves must be worn at all times when performing these procedures and examining fluid. Make sure that the culture systems are kept hydrated; a certain amount of water will evaporate and be lost as a result of culture equilibration, particularly during the first couple of days. The agar plate culture for *S. stercoralis* is considered to be more sensitive than most other nonmolecular diagnostic methods for this particular parasite.

Other Methods for Gastrointestinal Tract Specimens

EXAMINATION FOR PINWORM (Cellulose Tape Preparations)

Description

Enterobius vermicularis, known as pinworm, is a roundworm parasite that has worldwide distribution and is commonly found in children. The adult female worm migrates out of the anus, usually at night, and deposits her eggs on the perianal area. The adult female (8 to 13 mm long) may be found on the surface of a stool specimen or on the perianal skin. Since the eggs are usually deposited around the anus, they are not commonly found in feces and must be detected by other diagnostic techniques. Diagnosis of pinworm infection is based on the recovery of typical eggs, which are described as thick-shelled, football-shaped eggs with one slightly flattened side. Each egg often contains a fully developed embryo and is infective within a few hours after being deposited (Figure 5.17).

The clear-cellulose-tape preparation is the most widely used procedure for the detection of human pinworm infections (Figure 5.18). Tapes for egg collection are also available commercially. The eggs, and occasionally the adult female worms, stick to the sticky surface of the cellulose tape. These cellulose tape preparations are submitted to the laboratory, where they are examined microscopically. Several commercial collection procedures are also available. Specimens should be obtained in the morning before the patient bathes or goes to the bathroom. At least four to six consecutive negative slides should be observed before the patient is considered free of infection.

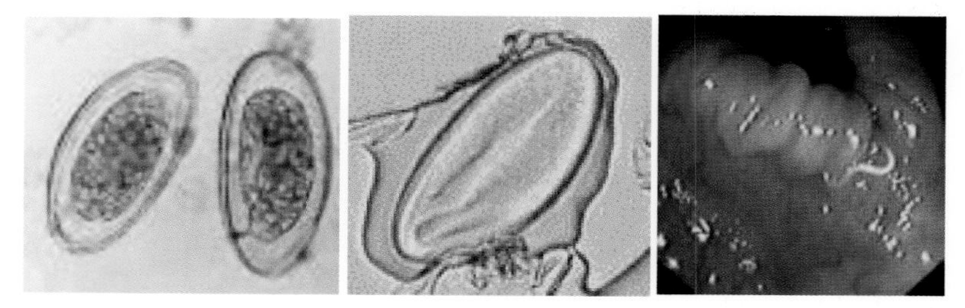

Figure 5.17 *Enterobius vermicularis* (pinworm). *E. vermicularis* eggs; *E. vermicularis* egg showing internal developing larva; adult *E. vermicularis* female on perianal skin.

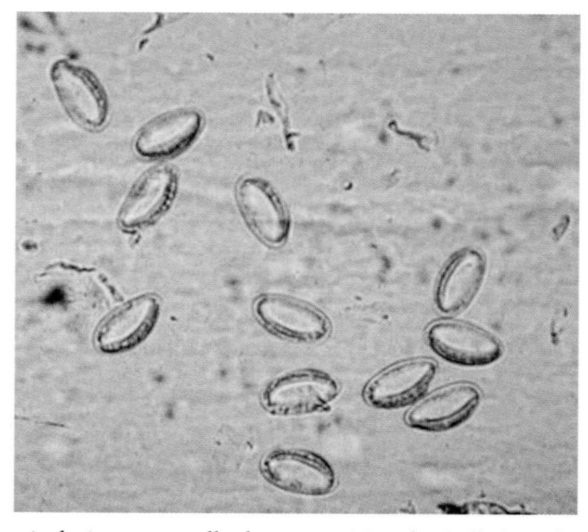

Figure 5.18 *Enterobius vermicularis* eggs on cellophane tape. Note football-shaped eggs with one flattened side.

Quality Control

The microscope should be calibrated, and the objectives and oculars used for the calibration procedure should be used for all measurements on the microscope. The calibration factors for all objectives should be posted on the microscope for easy access (multiplication factors can be pasted right on the body of the microscope). Pictures of *Enterobius* eggs (with measurements) should be available for comparison with the clinical specimen.

Detailed Procedure

1. Place a strip of cellulose tape on a microscope slide, starting 0.5 in. (1 in. = 2.54 cm) from one end and running toward the same end, continuing around this end lengthwise; tear off the strip evenly with the other end. Place a strip of paper, 0.5 by 1 in., between the slide and the tape at the end where the tape is torn flush.
2. To obtain the sample from the perianal area, peel back the tape by gripping the label, and with the tape looped (adhesive side outward) over a wooden tongue depressor held against the slide and extended about 1 in. beyond it, press the tape firmly against the right and left perianal folds.
3. Spread the tape back on the slide, adhesive side down.
4. Write the name and date on the label.

   Note: Do not use Magic transparent tape; use regular clear cellulose tape. If Magic tape is submitted, a drop of immersion oil can be placed on top of the tape to facilitate clearing.
5. Lift one side of the tape, apply 1 small drop of toluene or xylene, and press the tape down on the glass slide. The preparation will then be cleared, and the eggs will be visible.
6. Examine the slide at low power and under low illumination.

Reporting

Typical pinworm eggs are thick-shelled and football shaped with one flattened side and may contain a partially or fully developed larva. Occasionally, adult worms are seen on the cellulose tape preparation.

1. Report the organism and stage (do not use abbreviations).

   Example: *Enterobius vermicularis* eggs present
2. Report adult worms.

   Example: *Enterobius vermicularis* adult worm present
3. Report "No *Enterobius vermicularis* eggs or adults seen" for a negative result.

Procedure Notes/Reminders

1. Pinworm eggs are usually infectious. The use of glass slides and tapes may expose laboratory personnel to these eggs.
2. Some investigators recommend the use of the Swube (a paddle with a sticky adhesive coat [Becton Dickinson]) as a safer alternative (Figure 5.19).
3. If opaque tape is submitted by mistake, a drop of immersion oil on top of the tape will clear it enough to proceed with the microscopic examination.

Procedure Limitations

1. The female pinworm deposits eggs on the perianal skin only sporadically. Therefore, without multiple tapes (taken consecutively, one each morning), it is not possible to determine if the patient is positive or negative for the infection.

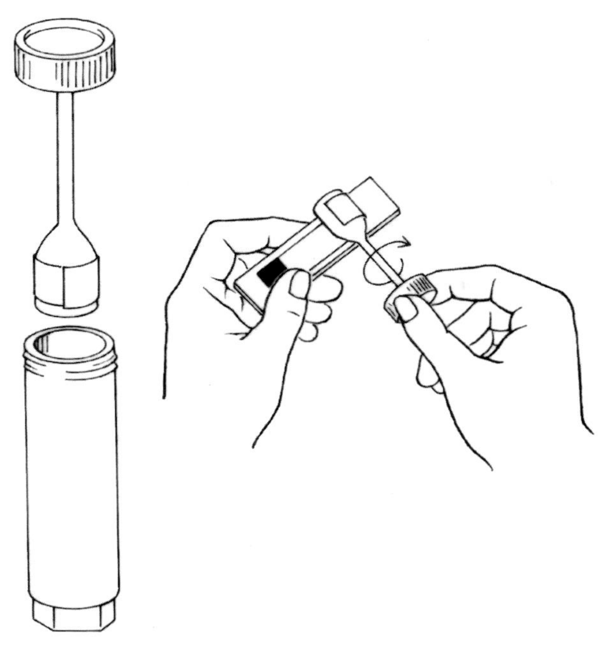

Figure 5.19 Diagram of a commercial kit (Evergreen Scientific) for use in sampling the perianal area for the presence of pinworm (*E. vermicularis*) eggs. On the left is the vial containing the sampler, which has sticky tape around the end. Once this is applied to the perianal area and eggs are picked up on the tape, the label area is placed at one end of the slide. The sticky tape is rolled down the slide and attaches to the glass. This device is easy to use and provides an area sufficient for adequate sampling. A minimum of four to six consecutive negative tapes are required to rule out a pinworm infection; most laboratories are accepting four rather than requesting the full six. (Illustration by Sharon Belkin.)

2. Occasionally, a parent will bring in an adult worm collected from the perianal skin or from the surface of the stool. The identification of the adult worm (almost always the female) confirms the infection (Figure 5.17, right).

SIGMOIDOSCOPY SPECIMENS (Direct Wet Smear)

Description

The direct smear is used primarily to detect motile parasites that are found in the colon (the organism in question is usually *Entamoeba histolytica*). Specific ulcerated areas should always be sampled; in the absence of specific lesions, the mucosa is randomly sampled. On low-power (magnification, ×100) examination of the smear, motility of trophozoites and/or human cells might be detected. At high dry power (magnification, ×430), organisms might be tentatively identified based on size, nucleus/cytoplasm ratio, appearance of the cytoplasm, and motility (saline only). The direct smear can be prepared with either 0.85% NaCl or the addition of iodine (Lugol's or D'Antoni's). Presumptive findings from this procedure must be confirmed by using some type of permanent stained smear. The specimen may consist of mucosal lining, mucus, stool, or a combination of the three. The specimen is taken by the physician and either prepared at the bedside for immediate review or submitted to the laboratory for subsequent examination. Prepare direct wet mounts on clean, new 1- by 3-in. glass slides. Depending on the specimen type, the following amounts should be used.

1. For mounts of mucus or similar material, place approximately 1 to 2 drops onto the slide.
2. For mounts of stool, place approximately 1 to 2 drops onto the slide.
3. If the material is very wet (watery), add 1 to 2 drops onto the slide.

If the specimen must be transported to the laboratory, the material can be placed in a small amount of 0.85% NaCl (0.5 to 1.0 ml) to keep the specimen from drying out. These specimens should be transported to the laboratory within no more than 30 min from the collection time and should be examined immediately.

Reagents

▶ 0.85% NaCl

Sodium chloride (NaCl) . . . . 850 mg
Distilled water . . . . 100 ml

1. Dissolve the sodium chloride in distilled water in an appropriate glass flask, using a magnetic stirrer.
2. Store in a glass bottle.
3. Label as 0.85% saline with the preparation date and an expiration date of 6 months. Store at room temperature.

▶ Modified D'Antoni's Stock Iodine

Potassium iodide (KI) . . . . 1.0 g
Powdered iodine crystals . . . . 1.5 g
Distilled water . . . . 100 ml

1. Dissolve the ingredients in distilled water in an appropriate glass flask, using a magnetic stirrer.
2. The D'Antoni's solution should be saturated with iodine, with some excess crystals left in the bottle. Store the solution in a brown bottle at room temperature. The stock solution remains good as long as an excess of iodine crystals remains on the bottom of the bottle.
3. Label as D'Antoni's stock iodine with the preparation date and an expiration date of 1 year.
4. Small amounts of stock iodine solution can be aliquoted into brown dropper bottles and used for routine daily use. The expiration date will be 30 to 60 days, depending on the amount of fading of the solution from the normal strong-tea color. The use of a small and/or clear-glass dropper bottle will result in a shorter time to expiration, whereas the use of a brown bottle will lengthen the expiration time.

▶ Lugol's Iodine Solution

Potassium iodide (KI) . . . . 10.0 g
Iodine crystals . . . . 5.0 g
Distilled water . . . . 100 ml

1. Dissolve the ingredients in distilled water in an appropriate glass flask, using a magnetic stirrer.
2. The Lugol's iodine solution should be saturated with iodine, with some excess crystals left in the bottle. Store the solution in a brown bottle at room temperature. The stock solution remains good as long as an excess of iodine crystals remains on the bottom of the bottle.
3. Label as Lugol's stock iodine with the preparation date and an expiration date of 1 year.
4. Small amounts of stock iodine solution can be aliquoted into brown dropper bottles and used for routine daily use. The expiration date will be 30 to 60 days, depending on the amount of fading of the solution from the normal strong-tea color. The use of a small and/or clear-glass dropper bottle will result in a shorter time to expiration, whereas the use of a brown bottle will lengthen the expiration time.

Quality Control

1. Check the direct-mount reagents each time they are used.
   a. The saline should appear clear, without any visible contamination.
   b. The iodine should be a strong-tea color, and there should be crystals in the bottom of the bottle. Small aliquots of the stock solution should always be a strong-tea color. If not, discard them and aliquot some stock solution into your dropper bottle.
2. The microscope should be calibrated, and the objectives and oculars used for the calibration procedure should be used for all measurements on the microscope. The calibration factors for all objectives should be posted on the microscope for easy access (multiplication factors can be pasted right on the body of the microscope).
3. Record all QC results.

Detailed Procedure

1. To 1 or 2 drops of patient material on the slide, add 1 or 2 drops of 0.85% NaCl, mix with the corner of the coverslip or an applicator stick, and mount with a no. 1 coverslip (22 by 22 mm). The amount of saline is determined by the specimen (less saline is needed if the material is very liquid).
2. Examine the smear with the low-power objective (10×) under low light. View each field for a few seconds, looking for any organism motility. Any suspicious objects can be examined using the high dry objective (40×) under low light.
3. Prepare a wet mount using Lugol's or D'Antoni's iodine (working solution) rather than saline. Another option would be to add a small drop of iodine at the side of the coverslip on the saline wet preparation. The iodine will diffuse into the saline suspension under the coverslip. However, if the specimen is thick or contains mucus, capillary action pulling the iodine under the coverslip and into the saline may not occur and a separate iodine mount may be required. Addition of the iodine gives the material some color (organisms may be easier to see); however, motility is lost.

Reporting

1. The organism and stage (trophozoite, cyst, oocyst, etc.) should be reported (do not use abbreviations); however, confirmation of species may require some type of permanent stained smear.

   Example: *Entamoeba histolytica* trophozoites (RBCs seen within the organism cytoplasm—confirmatory for the true pathogen, *E. histolytica*)

2. The presence of human cells should be noted and quantitated.

   Example: Moderate WBCs, many RBCs, few macrophages

3. The physician should be called if pathogenic organisms are identified.
4. If the results are negative, this should be reported as a presumptive report (based on wet examination only) prior to the examination of the permanent stained smear.

Quantitation of parasites, cells, yeasts, and artifacts

Quantity	Protozoa, cells, yeasts, and artifacts		Helminths
	PVA smears (no. per 10 oil immersion fields) (×1,000)	Wet preparations (no. per 10 40× fields) (×400)	Wet preparations (no. per 22-mm coverslip)
Few	≤2	≤2	≤2
Moderate	3–9	3–9	3–9
Many	≥10	≥10	≥10

This is a general chart for the quantitation of parasites, cells, yeast, and artifacts found in specimens from the intestinal tract. In general, protozoa are not quantitated on the laboratory slip (*Blastocystis* spp. are an exception); however, human cells, yeasts, and artifacts like Charcot-Leyden crystals are normally reported and quantitated.

Procedure Notes/Reminders

1. Remember that the iodine working solution should be a strong-tea color; if it is not, discard it and prepare a new working solution.
2. Final identification of some of the intestinal protozoa may be difficult (because of small size or confusion between organisms and human cells); a permanent stained smear must be used as a confirmatory method and examined at ×1,000 to see morphologic details.
3. In saline, human cells and/or protozoan trophozoites may exhibit some motility.
4. In iodine, human cells and/or protozoan trophozoites may be seen (but exhibit no motility).
5. Presumptive findings (either positive or negative) must be confirmed using a permanent stained smear.
6. Protozoan trophozoites may be confused with human cells (macrophages), so any identification should be reported as "presumptive" until the permanent stained smears have been examined.
7. The presumptive identification and quantitation of the human cells (macrophages, PMNs, eosinophils, and RBCs) could be obtained from the wet preparations. However, this information should also be considered presumptive until the permanent stained smears have been examined.

Procedure Limitations

1. Multiple areas of the mucosa should be examined (six smears are often recommended); this technique should not take the place of the routine O&P exam.
2. Wet preparations are normally not examined by using oil immersion power (magnification, ×1,000). Consequently, permanent stained smears should be used to confirm morphology and organism (or human cell) identification.
3. If the specimen amount is limited, then the wet preparation should not be performed and the specimen that is available should be processed by the permanent stained smear method to maximize the amount and clinical relevance of the information obtained.
4. Microsporidial spores are too small to be seen on a wet preparation; they will mimic bacteria or very small yeast. Permanent stains are mandatory for the identification of these organisms.

SIGMOIDOSCOPY SPECIMENS (Permanent Stained Smear)

Description

Most of the material obtained at sigmoidoscopy can be smeared (gently) onto a slide and immediately immersed in Schaudinn's or other fixatives. These slides can then be stained with trichrome stain and examined for specific cell morphology, either protozoan or otherwise. The procedure and staining times are identical to those for routine fecal smears. Specific ulcerated areas are always sampled; in the absence of specific lesions, the mucosa is randomly sampled. On oil immersion power (magnification, ×1,000) examination of the smear, protozoan trophozoites and/or cysts might be detected. Coccidian oocysts, helminth eggs or larvae, and/or human cells are also detected by this procedure. The permanent smear can be stained with trichrome or iron hematoxylin stains. These permanent stained smears usually confirm the identity of structures that might have been seen on the wet specimen examinations. The specimen may consist of mucosal lining, mucus, stool, and/or a combination of the three.

1. Prepare smears on clean, new 1- by 3-in. glass slides.
2. For mounts of mucus or similar material, place approximately 1 to 2 drops on the slide.
3. For mounts of stool, place approximately 1 drop on the slide.
4. If the material is very wet (watery), you can add 1 to 2 drops to the slide.
5. If Total-Fix fixative is being used, make sure the smear is thoroughly dry (1 h at room temperature) prior to staining.

Reagents

1. Schaudinn's fixative (see Section 3)
2. Fixative containing PVA (see Section 3)
3. Single-vial fixatives (see Section 3)

Quality Control

1. Check the fixatives weekly or when a new lot is used. Fresh stool containing protozoa or negative stool seeded with human buffy coat cells can be used to evaluate the efficacy of the fixatives. Cultured protozoa can also be used.
2. The fixative should appear clear, without floating debris or crystals. It is acceptable to have some crystal sediment on the bottom of the Coplin jar or dish.
3. The fixative containing PVA should be clear (a slight milky or smoky color is acceptable). There may be a slight precipitate on the bottom of the container (acceptable). The fluid should easily move in the bottle when it is inverted, and the viscosity of many of the available formulations actually approaches that of water.

Detailed Procedure

1. Gently smear 1 or 2 drops of patient material onto the slide and immediately immerse the slide in Schaudinn's fixative. The fixation and staining times are identical to those for routine fecal smears. Refer to the trichrome stain procedure above.
2. If the material is bloody, contains a lot of mucus, or is a "wet" specimen, gently mix 1 or 2 drops of patient material with 3 to 4 drops of PVA fixative directly on the slide. The smear should be allowed to air dry for at least 2 h before being stained. The fixation and staining times are identical to those for routine fecal smears.
3. Examine the stained smear using the oil immersion lens (97 to 100× objective) under maximum light. At least 300 oil immersion fields of the smear should be examined.
4. If the sigmoidoscopy material is collected in Total-Fix, centrifuge the specimen (10 min at 500 × *g*) and place 1 or 2 drops of sediment onto the slide, spread into a smear, and allow to thoroughly dry (1 h at room temperature) prior to staining.

Reporting

1. With either the trichrome or iron hematoxylin stains, the protozoan trophozoites and cysts can easily be seen. Oocysts are not clearly delineated; if suspect organisms are seen, additional procedures should be used for confirmation (see the procedure for modified acid-fast stains for Apicomplexans including the coccidia, above).
2. Helminth eggs or larvae may not be easily identified on the permanent stained smear, and wet mount examinations may have to be performed.

3. Human cells are readily identified (macrophages, PMNs, RBCs, etc.). Yeast cells (single cells, budding, and presence of pseudohyphae) are also readily identified.
4. The organism and stage should be reported (do not use abbreviations).

   Example: *Entamoeba histolytica* trophozoites (visible ingested RBCs—confirmation of the true pathogen, *E. histolytica*)

5. The presence of human cells should be noted and quantitated.

   Example: Moderate WBCs, many RBCs, few macrophages, etc.

6. Yeast cells should also be reported and quantitated. Refer to the quantitation table (in the section on sigmoidoscopy specimens [direct wet preparation], above).

   Example: Moderate budding yeast cells and few pseudohyphae

Procedure Notes/Reminders

1. Sigmoidoscopy specimens are submitted to help differentiate between inflammatory bowel disease and amebiasis. It is critical that the specimens be preserved immediately after being taken. Any delay could result in the disintegration of amebic trophozoites or distortion of human cells.
2. It is critical that permanent stained smears of this material be carefully examined using the oil immersion lens (magnification, ×1,000).

Procedure Limitations

1. The more areas of the mucosa are sampled, the more likely it is that the organisms might be found (*Entamoeba histolytica*). If only one or two smears are submitted for examination, the physician must be informed (the recommendation is to submit six smears from representative areas of the mucosa).
2. The examination of smears prepared at sigmoidoscopy does not take the place of routine O&P exams but serves as a supplemental procedure. Stool samples for routine examinations should also be submitted (a minimum of three specimens collected every other day or within no more than 10 days).
3. Specimens submitted for the identification of microsporidial spores must be stained using a modified trichrome stain; routine trichrome (Wheatley's trichrome) will not reveal the spores.

DUODENAL ASPIRATES

Description

Some organisms may be more difficult to recover in the stool, particularly those normally found in the duodenum. An alternative approach to routine stool examinations would be to sample the duodenal contents. Samples are obtained through the use of nasogastric intubation or the Entero-Test capsule (string test). Fluid from the duodenum is examined for the presence of *Strongyloides stercoralis*, *Giardia lamblia* trophozoites, *Cryptosporidium* spp., and *Cystoisospora belli* oocysts. The specimen can be examined as a wet preparation or as a permanent stained smear. In rare instances, *Clonorchis sinensis* eggs may be recovered. Duodenal fluid must be transported stat in a securely covered container placed in a plastic bag. A screw-cap urine container or plastic centrifuge tube with no preservatives is practical for this purpose. It is important to remember that specimens submitted for the identification of microsporidial spores must be stained using modified trichrome stain (not Wheatley's trichrome).

Reagents

1. Schaudinn's fixative (see Section 3)
2. PVA fixative (see Section 3)
3. 10% Formalin (see Section 3)
4. Other fixatives (see Section 3)

Quality Control

1. Check the fixatives weekly or when a new lot is used. Fresh stool containing protozoa or negative stool seeded with human buffy coat cells can be used to evaluate the efficacy of the fixatives. Cultured protozoa can also be used.
2. The Schaudinn's fixative should appear clear, without floating debris or crystals. It is acceptable to have some crystal sediment on the bottom of the Coplin jar or dish. This fixative (which contains mercuric chloride) is no longer routinely used. Other fecal fixatives are acceptable, including the universal fixative Total-Fix.
3. If the fixative being used contains PVA, the fluid should be clear (a slight milky or smoky color is acceptable). There may be a slight precipitate on the bottom of the container (this is acceptable). The fluid should easily move in the bottle when it is inverted, and the viscosity of many of the available formulations actually approaches that of water.
4. The 10% formalin and/or other single-vial fixatives should appear clear, without any visible contamination.
5. Record all QC results.

Detailed Procedure

1. Gloves must be worn when handling this specimen; infectious *Strongyloides* larvae can penetrate intact skin.
2. Examine the specimen within 1 h after it is taken. Note the amount of yellow color present, which indicates bile staining and confirms that it is actually from the duodenum.
3. Centrifugation (500 × *g* for 10 min) may be necessary to concentrate the mucus and any organisms present. Centrifugation should be routinely performed if the volume of fluid is ≥2 ml.
4. Place 1 drop of fluid on a clean slide, and cover with a 22- by 22-mm coverslip. If the specimen is very viscous, a drop of saline may be added.
5. Examine the entire coverslip under low power (magnification, ×100) for larvae or motile trophozoites, looking especially carefully around the mucus, where *Giardia* may be entangled.
6. Examine the mucus present under high dry power (magnification, ×400), since *Giardia* may be detectable only by the flutter of the flagella rather than by the motility of the organism.
7. Immediately after reading, place the slide in a Coplin jar containing Schaudinn's solution or other fixative so that permanent stained smears can be prepared. **Do not dry the slide,** or the coverslip will float off and sink to the bottom. If you have enough material, gently smear 1 or 2 drops of specimen on the slide and immediately immerse the slide in Schaudinn's fixative. The fixation and staining times are identical to those for routine fecal smears.
8. If the material contains a lot of mucus or is a "wet" specimen, gently mix 1 or 2 drops of the specimen with 3 or 4 drops of PVA fixative directly on the slide. The smear

should be allowed to air dry for at least 2 h before being stained. The fixation and staining times are identical to those for routine fecal smears.

9. Place 1 drop of the duodenal fluid on one or more slides to be stained for *Cryptosporidium* and *Cystoisospora*, and then repeat the wet mount procedure (steps 4 to 7) until all the remaining mucus (after centrifugation) or sediment is gone.
10. Stain the *Cryptosporidium*/*Cystoisospora* slide(s) with modified acid-fast stain, and examine as usual.
11. Examine the permanent stained smear, using the oil immersion lens (97 to 100× objective) under maximum light. At least 300 oil immersion fields on each smear should be examined.
12. If *Strongyloides* larvae are found, the rest of the specimen can be preserved in 10% formalin for teaching purposes.

Reporting

1. With either the trichrome or iron hematoxylin stain, the protozoan trophozoites and cysts can easily be seen. Oocysts are not clearly delineated; if suspect organisms are seen, additional procedures should be used for confirmation.
2. Oocysts of *Cryptosporidium* and *Cystoisospora* are visible on permanent stained smears (modified acid-fast procedures).
3. Helminth eggs or larvae may not be easily identified on the permanent stained smear but are visible in the wet preparations.
4. Report the organism and stage (trophozoite, cyst, oocyst, etc.); do not use abbreviations. However, confirmation of species may require some type of permanent stained smear.

   Examples: *Giardia lamblia* trophozoites
   Strongyloides stercoralis larvae

5. Call the physician if pathogenic organisms are identified.
6. Quantitate *Clonorchis sinensis* eggs if they are recovered (this is optional).
7. If the results are negative on the wet smear examination, report them as a presumptive (based on the wet examination only) prior to examination of the permanent stained smear.

Procedure Notes/Reminders

1. If you receive more than 2 ml of specimen, you will have to centrifuge the specimen (500 × *g* for 10 min) and examine the mucus or material in the bottom of the tube.
2. Modified acid-fast methods (or monoclonal antibody direct-detection methods) should be used for the identification of *Cryptosporidium* spp. *Cystoisospora belli* can be identified on the basis of the wet examination or from smears stained by the modified acid-fast methods.
3. If the specimen is submitted in Total-Fix preservative, centrifuge the specimen for 10 min at 500 × *g*. Place one or two drops of the sediment on the slide, prepare as a smear, and thoroughly dry the smear (1 h at room temperature) prior to staining. If microsporidial spores are suspected, the smear must be stained using modified trichrome stain (not Wheatley's trichrome stain).

Procedure Limitations

1. Many of the parasites will be caught up in the mucus; therefore, it is very important to centrifuge the specimen, concentrating this material for examination. Centrifugation is mandatory if there is more than 2 ml of specimen.

2. Although duodenal aspirate specimens are normally examined as wet preparations, it is important to remember that some of the organisms may be missed without the use of additional permanent stains (*Cryptosporidium*, microsporidia).

Methods for Urogenital Tract Specimens

RECEIPT OF DRY SMEARS

If dry smears prepared from urogenital specimens are received in the laboratory, several options can be used to salvage the smears, as follows.

1. If you have a fluorescence microscope (blue-light excitation as for fluorescent antibody work and/or auramine acid-fast staining) and can obtain the low-pH acridine orange stain, you can perform a very sensitive acridine orange stain (twice the sensitivity of the wet preparation) for *Trichomonas* on air-dried smears made when and where the specimen is collected.
2. You can treat the dry smear like a thin blood film. Fix the smear with absolute methanol, allow the slide to dry, and then stain it with a blood stain.
3. In addition to salvaging the smear, these approaches are more sensitive than a simple direct saline mount (see below). These approaches also avoid the problem of delays in transport to the laboratory.

DIRECT SALINE MOUNT

Description

Trichomonas vaginalis infections are diagnosed primarily by detecting live motile flagellates from direct saline (wet) mounts. Microscope slides containing patient specimens can be examined under low and high power for the presence of actively moving organisms. The specimens may include vaginal discharge, urethral discharge, penile discharge, urethral mucosa scrapings, and first-voided urine with or without prostatic massage. The specimens may be collected with a platinum loop, cotton or Dacron swab, or speculum. They can be placed in a small amount (<1.0 ml) of 0.85% NaCl in a test tube or placed on a microscopic slide and diluted with a drop of 0.85% NaCl. If the specimen cannot be examined immediately, the swab may be placed in Amies transport medium, which keeps the organisms viable for approximately 24 h. Urine specimens should be collected in a clean-catch urine collection container. The urine should be centrifuged at 500 × *g*, and the sediment should be examined for *Trichomonas*. All specimens should be held at room temperature, because refrigeration temperatures inhibit motility and have a deleterious effect on the organisms. Returning the specimen to room temperature does not reverse these deleterious morphologic changes. Specimens more than 24 h old should be rejected.

If microsporidia are suspected, one or two drops of the centrifuged sediment can be placed on a slide, prepared as a smear, and allowed to thoroughly dry (1 h at room temperature) prior to staining with modified trichrome stain (not Wheatley's trichrome stain). The stained smear must be examined using the ×100 oil immersion objective.

Reagents

▶ **0.85% NaCl**

Sodium chloride (NaCl) . . . . 850 mg
Distilled water . . . . 100 ml

1. Dissolve the NaCl in distilled water in an appropriate glass flask, using a magnetic stirrer.
2. Store in a glass bottle.
3. Label as 0.85% NaCl with the preparation date and an expiration date of 6 months. Store at room temperature.

▶ **Modified D'Antoni's Stock Iodine**

Potassium iodide (KI)	1.0 g
Powdered iodine crystals	1.5 g
Distilled water	100 ml

1. Dissolve in distilled water in an appropriate glass flask, using a magnetic stirrer.
2. The D'Antoni's iodine solution should be saturated with iodine, with some excess crystals left in the bottle. Store in a brown bottle at room temperature. The stock solution remains good as long as an excess of iodine crystals remains on the bottom of the bottle.
3. Label as D'Antoni's stock iodine with the preparation date and an expiration date of 1 year.
4. Small amounts of stock iodine solution can be aliquoted into brown dropper bottles and used for routine daily use. The expiration date is 30 to 60 days, depending on the amount of fading of the solution from the normal strong-tea color. The use of a small and/or clear-glass dropper bottle results in a shorter time to expiration. The use of a brown bottle lengthens the expiration time.

Quality Control

1. Check the direct-mount reagents each time they are used.
   a. The saline should appear clear, without any visible contamination.
   b. The iodine should be a strong-tea color, and there should be crystals in the bottom of the bottle. Small aliquots of the stock solution should always be a strong-tea color. If not, discard them and aliquot some stock solution into the dropper bottle.
2. The microscope should be calibrated, and the objectives and oculars used for the calibration procedure should be used for all measurements on the microscope. The calibration factors for all objectives should be posted on the microscope for easy access (multiplication factors can be pasted right on the body of the microscope).
3. Record all QC results.

Detailed Procedure

1. Apply the specimen to a small area of a clean microscope slide.
2. Immediately before the specimen dries, add 1 or 2 drops of saline with a pipette. If urine sediment is used, the addition of saline may not be necessary.
3. Mix the saline and specimen with the pipette tip or the corner of the coverslip.
4. Cover the specimen with a no. 1 coverslip.
5. Examine the wet mount with the low-power objective (10×) under low light.
6. Examine the entire coverslip for motile flagellates. Suspicious objects can be examined with the high-power objective (40×).
7. The organism is usually slightly larger than a polymorphonuclear leukocyte, and flagellar movement should be detected.

Reporting

1. If motile flagellates are seen (axostyle and undulating membrane), trophozoites of *T. vaginalis* are present.
2. If the nonmotile organisms are visible after staining with D'Antoni's iodine (axostyle), trophozoites of *T. vaginalis* are present.
3. Report the presence of the organism. The organism stage is not necessary since there is no known cyst stage for the trichomonads. The organisms do not have to be quantitated.

   Example: *Trichomonas vaginalis* present

4. If no flagellated organisms are seen, report the specimen as negative for *T. vaginalis*.

   Example: No *Trichomonas vaginalis* seen

Procedure Notes/Reminders

1. It is very important that specimens to be examined for *T. vaginalis* be delivered to the laboratory within 1 h after collection.
2. The organisms will lose their motility, particularly when they begin to dry out.
3. If a dry smear is delivered to the laboratory, it can be salvaged by being fixed as for a thin blood film (absolute methanol) and stained with Giemsa for at least 20 min at a 1:20 dilution (see the Giemsa staining procedure in the section on blood film preparation). The stained organisms may be difficult to see, but the specimen may provide some clinically relevant information if you can actually see and identify the organisms as *T. vaginalis*.
4. Calgiswabs are not recommended (because of tight adherence of the specimen to the swab), and the specimen should be rejected if submitted on this type of swab.
5. When the specimen is examined microscopically, always confirm that no fecal contamination is present (artifacts, vegetable debris, etc.). This type of contamination is rare and would probably be limited to a urine specimen.

Procedure Limitations

1. If the specimen is left at room temperature or held at refrigeration temperature for a prolonged period (usually >1 h), the organisms round up, lose their motility, and eventually die. Motility is occasionally enhanced by warming the specimen to 37°C, but this does not revive dying organisms.
2. Wet mounts have been reported to detect *T. vaginalis* in 75 to 85% of infected patients. Alternative diagnostic methods include culture, monoclonal antigen detection kits, permanent stained slides, and collection of a second sample for examination.
3. If the patient has a *Pentatrichomonas hominis* intestinal infection and the urogenital specimen becomes contaminated with fecal material, a false-positive *T. vaginalis* result may be reported because *P. hominis* and *T. vaginalis* are similar in shape. **This false-positive report can result in risk management issues (including an erroneous interpretation as indicating a possible sexual abuse case and, in instances where children are being tested, serious legal charges and incorrect consequences).**
4. If microsporidial spores are present, they will not be detected on a wet mount; they are too small and will resemble bacteria and/or small yeast cells.

PERMANENT STAINED SMEAR

Description

Trichomonas vaginalis infections are diagnosed primarily from direct saline (wet) mounts by detecting live motile flagellates. Permanent stained smears can be made from patient specimens for specific identification of the organism. Although a number of stains can be used, Giemsa and Papanicolaou stains are the ones most frequently used to diagnose *T. vaginalis* infections. If microsporidial spores are suspected, modified trichrome stain should be used; routine Wheatley's trichrome stain is not acceptable. Acceptable specimens include vaginal discharge, urethral discharge, penile discharge, urethral mucosa scrapings, and first-voided urine with or without prostatic massage. The specimens may be collected with a platinum loop, cotton or Dacron swab, or speculum. They can be placed in a small amount (<1.0 ml) of 0.85% NaCl in a test tube or smeared directly on a microscope slide. A drop of 0.85% NaCl may be used to dilute the direct smear when it is placed on the slide. Slides prepared in this manner should be air dried before they are transported to the laboratory. Specimens collected with a cotton or Dacron swab may also be placed in Amies transport medium if the specimen cannot be immediately processed. Organisms remain viable for approximately 24 h in Amies transport medium. Urine specimens should be collected in a clean-catch urine collection container. The urine should be centrifuged at 500 × *g* for 5 min, and the sediment should be examined for *Trichomonas*. All specimens should be held at room temperature because refrigeration temperatures have a deleterious effect on the organisms. Returning the specimen to room temperature does not reverse these deleterious morphologic changes. Specimens more than 24 h old should be rejected.

Reagents

▶ **0.85% NaCl**

Sodium chloride (NaCl) . . . . 850 mg
Distilled water . . . . 100 ml

1. Dissolve the NaCl in distilled water in an appropriate glass flask, using a magnetic stirrer.
2. Store in a glass bottle.
3. Label as 0.85% NaCl with a preparation date and an expiration date of 6 months. Store at room temperature.

▶ **Giemsa Stain**

See elsewhere in this section for preparation of Giemsa stain (p. 190), modified trichrome stain (p. 144), and phosphate buffer solutions (p. 191).

▶ **Absolute Methanol**

Quality Control

1. Check the direct mount reagents each time they are used.
   a. The saline should appear clear, without any visible contamination.
   b. A peripheral blood film may be used as quality control for the Giemsa stain. For staining characteristics, see Section 3.
   c. Check the phosphate buffer each time it is used. The buffer should appear clear, with no signs of visible contamination or precipitates. The pH should be between 6.8 and 7.2.

2. The Giemsa-stained control slide should be reviewed before the patient's specimen is searched for the organism. If there was potential fecal contamination of the specimen, one may have to differentiate *T. vaginalis* from *P. hominis*.
3. The microscope should be calibrated, and the objectives and oculars used for the calibration procedure should be used for all measurements on the microscope. The calibration factors for all objectives should be posted on the microscope for easy access (multiplication factors can be pasted right on the body of the microscope).
4. Record all QC results.

Detailed Procedure

1. Apply the specimen to a small area of a clean microscope slide.
2. Immediately before the specimen dries, add 1 or 2 drops of saline with a pipette. If urine sediment is used, the addition of saline may not be necessary.
3. Mix the saline and specimen with the pipette tip.
4. Air dry the slide.
5. Fix in absolute methanol for 1 min (use a Coplin jar or slide staining rack).
6. Place the slide in the Giemsa solution and stain for the desired time depending on the stain dilution used (20 min at 1:20 dilution).
7. Rinse the slide with tap water to remove excess stain solution (use gently running water or a Coplin jar).
8. Air dry the slide; do not apply a coverslip.
9. Examine the slide with the oil immersion objective (100×).
10. Examine the entire smear for flagellates.
11. The organism is usually slightly larger than a PMN: 7 to 23 μm long and 5 to 15 μm wide. Differential characteristics to be observed include anterior flagella, an undulating membrane, an axostyle, and nucleus.

Reporting

1. If motile flagellates are seen (axostyle and undulating membrane), trophozoites of *T. vaginalis* are present.
2. If nonmotile organisms are visible after staining with D'Antoni's iodine (axostyle), trophozoites of *T. vaginalis* are present.
3. Report the organism. The organism stage is not necessary since there is no known cyst stage for the trichomonads. The organisms do not have to be quantitated.

   Example: *Trichomonas vaginalis* present

4. If no flagellated organisms are seen, report the specimen as negative for *T. vaginalis*.

   Example: No *Trichomonas vaginalis* seen

Procedure Notes/Reminders

1. It is very important that specimens to be examined for *T. vaginalis* be delivered to the laboratory within 1 h after collection.
2. The organisms will lose their motility, particularly when they begin to dry out.
3. If a dry smear is delivered to the laboratory, you can salvage it by fixing it as you would a thin blood film (absolute methanol) and staining it with Giemsa for at least 20 min at a 1:20 dilution. The stained organisms may be difficult to see but may

provide some clinically relevant information if you can actually see and identify them as *T. vaginalis*.

4. Calgiswabs are not recommended (because of tight adherence of the specimen to the swab), but a specimen should not be rejected if submitted on this type of swab.
5. When the specimen is examined microscopically, always confirm that no fecal contamination is present (artifacts, vegetable debris, etc.). This type of contamination is rare and would probably be limited to a urine specimen.
6. Specimens for microsporidial spores can be submitted in various fixatives. Centrifuged sediment should be used for smear preparation prior to staining with modified trichrome stain.

Procedure Limitations

1. If the specimen is left at room temperature or held at refrigeration temperature for a prolonged period (usually >1 h), the trichomonads will round up and eventually die.
2. If the patient has a *Pentatrichomonas hominis* intestinal infection and the urogenital specimen becomes contaminated with fecal material, a false-positive *T. vaginalis* result may be reported, because *P. hominis* and *T. vaginalis* are similar in shape. The position of the undulating membrane will allow differentiation between *T. vaginalis* and *P. hominis*.

Diagnostic differential characteristics

Characteristic	*T. vaginalis*	*P. hominis*
Size and shape	7–23 µm long	5–15 µm long
	5–15 µm wide	7–10 µm wide
	Pear shaped	Pear shaped
Flagella	Four anterior	Three to five anterior
Undulating membrane	**Extends half the length of the organism**	**Extends the entire length of the organism, with a free trailing flagellum**
Axostyle	Present	Present
Nucleus	Anterior end, oval	Anterior end, oval

URINE CONCENTRATION (Centrifugation)

Description

Helminthic larval stages and eggs and some protozoa infecting humans may be found in the urine whether or not they cause pathologic sequelae in the urinary tract. Infections diagnosed or organisms which can be detected in the urine include *Trichomonas vaginalis*, *Schistosoma haematobium*, and filariasis. Microfilariae may be detected in the urine in heavily infected patients or in patients recently treated with diethylcarbamazine. The urine specimen should be collected as follows.

1. For *T. vaginalis*, collection of first-voided urine, particularly after prostatic massage in male patients, is useful for the diagnosis.
2. For *S. haematobium*, collection of a midday urine specimen or a 24-h collection in a container without preservatives is recommended. Peak egg excretion occurs between noon and 3 p.m. In patients with hematuria, eggs may be found trapped in the blood and mucus in the terminal portion (last portion voided) of the urine specimen.

3. For filariasis, microfilariae may be detected in urine of patients with chyluria, those who have very heavy filarial infections, and those treated with diethylcarbamazine.
4. For microsporidial infections, the spores can be detected in urine using fluorescent stains (calcofluor white or auramine) or modified trichrome stain (examined using the 100× oil immersion objective). Although these stains are nonspecific, spores can be seen inside and outside of the cells. Horizontal and/or diagonal lines within some of the spores are an indication of the polar tubule and provide a general diagnosis of microsporidial spores present (seen using modified trichrome stain). The spores cannot be identified to genus or species level based on the stained morphology. Electron microscopy or molecular means will be required for generic identification.

Reagents

▶ **0.85% NaCl**

Sodium chloride (NaCl) .... 850 mg
Distilled water .... 100 ml

1. Dissolve the NaCl in distilled water in an appropriate glass flask, using a magnetic stirrer.
2. Store in a glass bottle.
3. Label as 0.85% NaCl with the preparation date and an expiration date of 6 months. Store at room temperature.

Quality Control

1. Check the direct-mount reagents each time they are used. The saline should appear clear, without any visible contamination.
2. The microscope should be calibrated, and the objectives and oculars used for the calibration procedure should be used for all measurements on the microscope. The calibration factors for all objectives should be posted on the microscope for easy access (multiplication factors can be pasted right on the body of the microscope).
3. Record all QC results.

Detailed Procedure

1. If a 24-h urine specimen was collected, allow it to sediment for 2 h and decant the major portion of the supernatant. There may be 100 to 200 ml of sediment left. If a first-voided urine specimen is received, use the entire specimen.
2. Place the remaining urine specimen (sediment) in centrifuge tubes.
3. Centrifuge the specimen at 500 × *g* for 5 min.
4. Decant the supernatant fluid.
5. With a pipette, mix and aspirate the sediment.
6. Place 1 drop of the sediment on a microscope slide.
7. Place a coverslip on top of the sediment.
8. Observe the specimen under the coverslip at magnifications of ×100 and ×400. Examine the entire coverslip at ×100 and at least half the coverslip at ×400.

Reporting

1. If motile flagellates are seen (axostyle and undulating membrane), trophozoites of *T. vaginalis* are present.
2. If live microfilariae are seen, species identification can be confirmed or accomplished with permanent stains.

3. If eggs of *S. haematobium* are seen, observe the eggs for live miracidia. If flame cell activity (motile cilia) is detected inside the miracidium larva, the miracidium is viable. A hatching test may also be used to determine if the eggs are viable.
4. For *T. vaginalis*, report the organism. The stage is not necessary since there is no known cyst stage for the trichomonads. The organisms do not have to be quantitated.

   Example: *Trichomonas vaginalis* present

5. For filariae, report the presence of microfilariae. Genus and species should be reported if possible. The organisms do not have to be quantitated.

   Example: *Wuchereria bancrofti* microfilariae present

6. For *S. haematobium*, if eggs are present, report the genus and species and whether the eggs are viable or nonviable.

   Examples: *Schistosoma haematobium* eggs present (some viable, some nonviable)
   Schistosoma haematobium eggs present (nonviable, eggshells only)

7. For microsporidial spores, report as follows: Microsporidial spores seen (unable to identify to genus or species level).

Procedure Notes/Reminders

1. Specimens to be examined for *T. vaginalis* should be delivered to the laboratory as soon as possible after collection. Hold all specimens at room temperature, because refrigeration temperatures have a deleterious impact on *T. vaginalis.*
2. Species identification of the microfilariae may not be possible if unstained preparations (urine sediment) are used; permanent stains may be necessary for further identification.
3. It is very important that all urine specimens (24-h and single-void specimens) be collected with no preservatives. It is clinically important to determine whether the eggs are viable. This can be accomplished by examining eggs in the wet preparations at a magnification of ×400.
4. Reject specimens more than 24 h old for *T. vaginalis* and filariae, and reject midday urine specimens for schistosomes.
5. Reject all 24-h urine specimens more than 48 h old.
6. When the specimen is examined microscopically, always confirm that no fecal contamination is present (artifacts, vegetable debris, etc.). This type of contamination is rare and would probably be limited to a urine specimen.

Procedure Limitations

1. If the urine specimen is left at room temperature or held at a low temperature for a prolonged period, *T. vaginalis* may round up, become nonmotile, and eventually die.
2. **If the patient has a *P. hominis* intestinal infection and the urogenital specimen becomes contaminated with fecal material, a false-positive *T. vaginalis* result may be reported because *P. hominis* and *T. vaginalis* are similar in shape. This incorrect identification as *T. vaginalis* can be interpreted as indicating a possible sexual abuse case and, in instances where children are being tested, can lead to serious legal charges and incorrect consequences.**
3. Microfilariae can be identified to the species level only by making a permanent stained slide from the specimen.

URINE CONCENTRATION (Nuclepore Membrane Filter)

Description

Microfilariae may be detected in the urine of heavily infected patients or in patients recently treated with diethylcarbamazine. Eggs of *Schistosoma haematobium* can also be recovered in urine specimens. Microfilariae and *S. haematobium* eggs can be easily concentrated by passing the specimen through a membrane filter and then observing the filter through a microscope (Figure 5.20). Specimens should be collected as indicated below.

1. For filariasis, microfilariae may be detected in urine of patients with chyluria, those who have very heavy filarial infections, and those treated with diethylcarbamazine. Specimens should be collected as first-voided specimens or a 24-h collection in a container without preservatives.
2. For *S. haematobium*, collection of a midday urine specimen or a 24-h collection in a container without preservatives is recommended. Peak egg excretion occurs between noon and 3 p.m. In patients with hematuria, eggs may be found trapped in the blood and mucus in the terminal portion (last-voided portion) of the urine specimen.

Reagents

▶ **0.85% NaCl**

Sodium chloride (NaCl) . . . . 850 mg
Distilled water . . . . 100 ml

1. Dissolve the NaCl in distilled water in an appropriate glass flask, using a magnetic stirrer.
2. Store in a glass bottle.
3. Label as 0.85% NaCl with the preparation date and an expiration date of 6 months. Store at room temperature.

Quality Control

1. Check the direct-mount reagents each time they are used. The saline should appear clear, without any visible contamination.

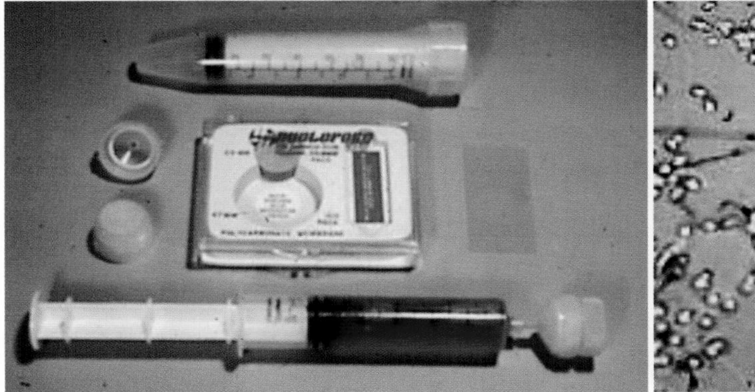

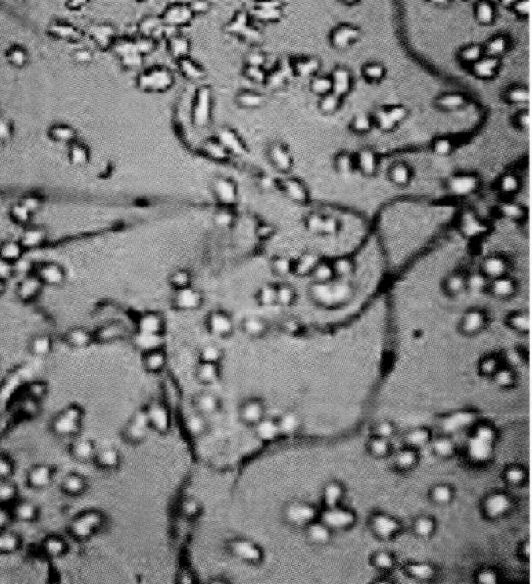

Figure 5.20 Nuclepore filtration system for various human parasites. **(Left)** Nuclepore filtration system showing syringe, filtration pads, and filter pad holder attached to syringe holding diluted blood; **(right)** microfilariae on wet filtration pad (note that identification to genus or species level is not possible without permanent staining of the pad and/or blood films).

2. The microscope should be calibrated, and the objectives and oculars used for the calibration procedure should be used for all measurements on the microscope. The calibration factors for all objectives should be posted on the microscope for easy access (multiplication factors can be pasted right on the body of the microscope).
3. Record all QC results.

Detailed Procedure

1. If a 24-h urine sample was collected for *S. haematobium* diagnosis, allow the specimen to sediment for 2 h and decant the major portion of the supernatant. There may be 100 to 200 ml of sediment left.
2. Thoroughly mix the urine specimen.
3. Draw up 10 ml of urine into the syringe.
4. Attach the filter holder containing the filter to the syringe. For *S. haematobium*, use the 8-μm filters; for *Wuchereria bancrofti*, *Brugia malayi*, and *Loa loa*, use the 5-μm filters; and for *Mansonella* species, use the 3-μm filters.
5. Express the urine through the filter.
6. The membrane may be washed with physiological saline by removing the filter holder, drawing up 10 ml of saline into the syringe, reattaching the filter holder, and expressing the saline through the filter.
7. Repeat step 6, but fill the syringe with air rather than saline and express the air through the filter.
8. Remove the filter holder from the syringe.
9. Disassemble the filter holder to expose the filter.
10. Remove the filter from the holder with forceps.
11. Place the filter upside down on a microscope slide.
12. With a Pasteur pipette, add 1 drop of saline to moisten the filter.
13. Examine the filter for microfilariae and eggs under ×100 magnification.

Reporting

1. If live microfilariae are seen, species identification could be confirmed or accomplished with permanent stains.
2. If eggs of *S. haematobium* are seen, observe them for live miracidia. If flame cell activity (motile cilia) is detected inside the miracidium larva, the miracidium is viable. A hatching test may also be used to determine if the eggs are viable.
3. For filariae, report the presence of microfilariae. Genus and species should be reported if possible. The organisms do not have to be quantitated.

   Example: *Wuchereria bancrofti* microfilariae present

4. For *S. haematobium*, if eggs are present, report the genus and species and whether the eggs are viable or nonviable.

   Examples: *Schistosoma haematobium* eggs present (some viable, some nonviable)
   Schistosoma haematobium eggs present (nonviable, eggshells only)

Procedure Notes/Reminders

1. Species identification of the microfilariae may not be possible from unstained preparations (urine sediment), and permanent stains may be necessary for further

identification. They measure 3 to 10 µm wide by 160 to 330 µm long. Depending on the species and stain used, a sheath may or may not be present.

2. It is very important that all urine specimens (24-h and single-void specimen) be collected with no preservatives. It is clinically important to determine whether the eggs are viable. This can be accomplished by examining eggs in the wet preparations at a magnification of ×400.
3. Reject specimens more than 24 h old for filariae and midday urine for schistosomes.
4. Reject all 24-h urine specimens more than 48 h old.
5. If you accidentally put the filter right side up, don't add more than 1 drop of saline (the organisms may accidentally float off the filter and onto the glass slide).

Procedure Limitations

1. Microfilariae can be identified to the species level only by making a permanent stained slide from the specimen. For a method using the membrane filter, refer to *Diagnostic Medical Parasitology*, 6th ed.
2. A hatching test may also be used to determine if the eggs are viable. For the hatching method, refer to *Diagnostic Medical Parasitology*, 6th ed.
3. Infrequently, eggs of other *Schistosoma* species may be recovered in the urine.

Species	Egg	
	Size	Shape
S. haematobium	112–170 by 40–70 µm	Elongate, terminal spine
S. japonicum	55–85 by 40–60 µm	Oval, minute lateral spine
S. mansoni	114–180 by 45–73 µm	Elongate, prominent lateral spine

4. Because microsporidial spores are so small (1 to 2.5 µm), the membrane filtration system is not used for recovery of these organisms. The membrane pore size would have to be extremely small; debris plugging of the membrane pad would be very likely.

Preparation of Blood Films

Although some parasites (microfilariae and trypanosomes) can be detected in fresh blood on the basis of their characteristic shape and motility, specific identification of the organisms is best done with a permanent stain. Two types of blood films are recommended for detection of all blood-borne parasites. Thick films allow a larger amount of blood to be examined and provide an increased possibility of detecting light infections. However, species identification from the thick film, particularly for malaria parasites, can be difficult and can usually be made only by experienced workers. The morphologic characteristics of blood parasites are best seen in thin films, in which the RBC morphology is preserved and the size relationship between infected and uninfected RBCs can be determined after staining. Examination of the thin film is often valuable in determining the *Plasmodium* species present.

Accurate examination of thick and thin blood films and identification of parasites depend on the use of clean, grease-free slides. Old (unscratched) slides should be cleaned first with detergent and then with 70% ethyl alcohol. New, unused slides are coated with a greasy substance that allows them to be pulled apart; therefore, these slides must be cleaned with alcohol prior to use for preparation of blood films.

Blood films should be ordered and prepared immediately as stat procedures (ordered, collected, processed, examined, and reported within 1 h if possible). Blood collection should not be delayed for any evidence of periodic fevers, since periodicity is often not yet evident in travelers who present to the clinic or emergency room. When malaria is a possible diagnosis, after the first set of negative smears, samples should be taken at intervals of 6 to 8 h for at least 36 h. If the finger-stick method is used, the blood should flow freely; blood that has to be "milked" from the finger is diluted with tissue fluids, decreasing the number of parasites per field. If the blood specimen from a venipuncture is sent directly to the laboratory, the following approach can be used. Request a tube of fresh blood (EDTA anticoagulant is preferred over heparin), and prepare the smears.

THIN BLOOD FILMS

Description

The thin film is prepared like that for a differential blood cell count and provides an area for examination where the RBCs are neither overlapping nor distorted (Figure 5.21). Here, the morphologies of parasites and infected RBCs are most typical. The specimen usually consists of fresh whole blood collected by finger puncture or whole blood containing EDTA (0.020 g per 10 ml of blood) collected by venipuncture that is less than 1 h postcollection. Occasionally a buffy coat (for leishmaniasis) or the sediment from a special concentration procedure (triple centrifugation for trypanosomes) is spread into a thin film.

Reagents

▶ **Absolute Methanol (for smear fixation/Giemsa stain)**

Quality Control

1. Visually, the thin film should be rounded, feathered, and progressively thinner toward the middle of the slide.
2. There should not be any clear areas or smudges in the film itself (indicating that grease or fingerprints were on the glass).
3. The WBCs on the blood film being stained serve as the quality control organisms. If the WBCs look typical (color, morphology), any parasites present will look acceptable. It is not necessary to use an actual positive malaria smear for quality control.

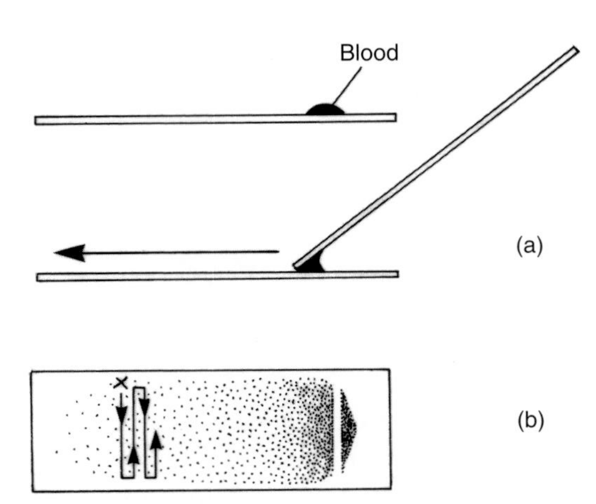

Figure 5.21 Method of thin blood film preparation. (a) Position of spreader slide; (b) well-prepared thin film. Arrows indicate the area of the slide used to observe accurate cell morphology. (Illustration by Nobuko Kitamura; reprinted from *Diagnostic Medical Parasitology*, 6th ed.)

Detailed Procedure

1. Wear gloves when performing this procedure.
2. The procedure depends on the source of the specimen.
   a. For blood from finger puncture, after wiping off the first drop of blood, touch a clean 1- by 3-in. glass microscope slide, about 0.5 in. from the end, to a small drop of blood (10 to 15 μl) standing on the finger, remove the slide from the finger, turn it blood side up, and place it on a horizontal surface.
   b. For blood from venipuncture, place a clean 1- by 3-in. glass microscope slide on a horizontal surface. Place a small drop (10 to 15 μl) of specimen on the center of the slide about 0.5 in. from the end.
3. Holding a second clean glass slide at a 40° angle, touch the angled end to the mid-length area of the specimen slide.
4. Pull the angled slide back into the blood, and allow the blood to almost fill the end area of the angled slide.
5. Continuing contact with the blood under the lower edge, quickly and steadily move the angled slide toward the opposite end of the specimen slide until the blood is used up.
6. The result will be a thin film that is rounded, feathered, and progressively thinner toward the center of the slide.
7. Label the slide appropriately, and allow it to air dry for at least 10 min while protected from dust.
8. The most common stains are of two types. Wright's stain has the fixative in combination with the staining solution, so that both fixation and staining occur at the same time; therefore, the thick film must be laked before staining. Giemsa stain has the fixative and stain separate; therefore, the thin film must be fixed with absolute methanol before staining. If you are using one of the rapid blood stains, read the package insert to confirm laking and/or fixation requirements for the thin and thick blood films. If the film will be stained with Giemsa stain, after the film is completely dry, fix it by dipping the slide into absolute methanol and allow the film to air dry in a vertical position. If the film will be stained with Wright's stain, it does not have to be fixed, because Wright's stain contains the fixative and stain in one solution. There are a number of stains that can be used. Although for many years Giemsa stain has been the stain of choice, the parasites can also be seen on blood films stained with Wright's stain, a Wright-Giemsa combination stain, or one of the more rapid stains, such as Diff-Quik (various manufacturers), Wright's Dip Stat stain (Medical Chemical Corp., Torrance, CA), or Field's stain. Delafield's hematoxylin stain is often used to stain the microfilarial sheath; in some cases, Giemsa stain does not provide sufficient stain quality to allow differentiation of the microfilariae.

Procedure Notes/Reminders

1. A diamond marking pen is recommended.
2. An indelible ink pen can be used.
3. Pencil can be used if the information is actually written in the thick part of the smear (where the original drop of blood was placed).
4. Do not use wax pencils; the material may fall off during the staining procedure.
5. Make sure the films are protected from dust (while drying).

6. The last few drops of blood remaining in the needle after a venipuncture can also be used to prepare thin blood films. However, if you are preparing thick films, remember that the blood has not been in contact with the anticoagulant, and so you will have to follow directions given in the protocol for thick blood films (from finger puncture). This approach carries the risk of a needlestick injury.

Procedure Limitations

1. A light infection may be missed in a thin film, whereas the increased volume of blood present on a thick film may allow it to be detected, even with a low parasitemia.
2. If the smears are prepared from anticoagulated blood which is more than 1 h old, the morphology of both parasites and infected RBCs may not be typical. Approximately 6 h after collection, the parasites begin to disappear.
3. *Plasmodium vivax* and *P. ovale* should be able to be identified, even in the absence of Schüffner's dots (stippling).

THICK BLOOD FILMS

Description

The thick film contains more blood than the thin film, and therefore, this method is more likely to detect a low-level parasitemia. The RBCs are lysed during staining, making the preparation more or less transparent and leaving only parasites, platelets, and WBCs for examination. The specimen usually consists of fresh whole blood collected by finger puncture or whole blood containing EDTA (0.02 g per 10 ml of blood) collected by venipuncture that is less than 1 h postcollection. Heparin (2 mg per 10 ml of blood) or sodium citrate (0.05 g per 10 ml of blood) may be used as an anticoagulant if trypanosomes or microfilariae are suspected. The sediment from a concentration procedure for trypanosomes or microfilariae is frequently spread into a thick film that is stained, examined, and kept as a permanent record.

Quality Control

1. Visually, the thick smear should be round to oval, approximately 2.0 cm across.
2. One should be able to barely read newsprint through the wet or dry film.
3. The film itself should not have any clear areas or smudges (indicating that grease or fingerprints were on the glass).

Detailed Procedure

1. Wear gloves when performing this procedure.
2. The procedure depends on the source of the specimen.
   a. For blood from finger puncture, after wiping off the first drop of blood, touch a clean 1- by 3-in. glass microscope slide to a large drop of blood standing on the finger, and rotate the slide on the finger until the circle of blood is nearly the size of a dime or nickel (1.8 to 2.0 cm). Without breaking contact with the blood, rotate the slide back to the center of the circle. Remove the slide from the finger, quickly turn it blood side up, and place it on a horizontal surface. The blood should spread out evenly over the surface of the circle and be sufficiently thin that fine print (newsprint size) can be just barely read through it. If not, take the corner of a second clean slide or an applicator stick and expand the circle until the print is just readable.
   The final thickness of the film is important. If it is too thick, it might flake off while

drying or wash off while staining. If it is too thin, the amount of blood available for examination is insufficient to detect a low-level parasitemia. Continue stirring for 30 s to prevent the formation of fibrin strands.

b. For blood from venipuncture, place a clean 1- by 3-in. glass microscope slide on a horizontal surface. Place a drop (30 to 40 µl) of blood onto the center of the slide about 0.5 in. from the end. Using either the corner of another clean glass slide or an applicator stick, spread the blood into a circle about the size of a dime or nickel (1.8 to 2.0 cm). Immediately place the thick film over some small print and be sure that the print can just barely be read through it. If not, expand the film until the print can be read. Three or four small drops of blood may be used in place of the larger drop, and the small ones can be pooled into a thick film by using the corner of a clean slide or an applicator stick. Be sure that small print can be read through it.

3. Allow the film to air dry in a horizontal position and protected from dust for several hours (6 to 8 h) or overnight. Do not attempt to speed the drying process by applying any type of heat, because heat fixes the RBCs and they subsequently do not lyse in the staining process.
4. **Do not fix the thick film.** If thin and thick films are made on the same slide, do not allow the methanol or its vapors to contact the thick film during fixing of the thin film.
5. Label the slide appropriately.
6. If staining with Giemsa stain is delayed for more than 3 days or the film is to be stained with Wright's stain, lyse the RBCs in the thick film by placing it in buffered water (pH 7.0 to 7.2) for 10 min, removing it from the water, and placing it in a vertical position (thick film down) to air dry.

Procedure Notes/Reminders

1. A diamond marking pen is recommended.
2. An indelible ink pen can be used.
3. Do not use wax pencils; the material may fall off during the staining procedure.
4. Make sure the films are protected from dust (while drying).
5. The last few drops of blood remaining in the needle after a venipuncture can also be used to prepare thick blood films. However, when you are preparing thick films, remember that the blood has not been in contact with the anticoagulant, and so you will have to follow directions given in the protocol for thick blood films (from finger puncture). Remember the possibility of a needlestick injury.

Procedure Limitations

1. If the smears are prepared from anticoagulated blood that is more than 1 h old, the morphology of the parasites may not be typical and the film may wash off the slide during the staining procedure. After 6 h postcollection, the organisms will begin to disappear.
2. Identification to the species level, particularly between *P. ovale* and *P. vivax* and between the ring forms of *P. falciparum* and *Babesia* spp., may be impossible without examining the stained thin blood film. Also, *Trypanosoma cruzi* trypomastigotes are frequently distorted in thick films.
3. Excess stain deposition on the film may be confusing and make the detection of organisms difficult.

COMBINATION THICK-THIN BLOOD FILMS

Description

To prepare a slide containing blood that can be stained for either a thick or thin blood film, the method described here was developed (Figure 5.22). The specimen usually consists of fresh whole blood collected by finger puncture or whole blood containing EDTA (0.02 g per 10 ml of blood) collected by venipuncture that is less than 1 h postcollection.

Quality Control

1. Visually, the smear should alternate, with thick and thin portions throughout the length of the glass slide.
2. One should barely be able to read newsprint through the wet or dry film.
3. The film itself should not have any clear areas or smudges (indicating that grease or fingerprints were on the glass).

Detailed Procedure

1. Wear gloves when performing this procedure.
2. The procedure depends on the source of the specimen.
   a. Finger puncture blood is not recommended; the procedure is best performed with blood that contains EDTA anticoagulant. The procedure does not lend itself to "stirring" to prevent fibrin strands.
   b. For blood from venipuncture, place a clean 1- by 3-in. glass microscope slide on a horizontal surface. Place 1 drop (30 to 40 μl) of blood at one end of the slide about 0.5 in. from the end. Using an applicator stick lying across the glass slide and keeping the applicator stick in contact with the blood and glass, rotate the stick in a circular motion while moving it down the glass slide to the opposite end. The appearance of the smear should be alternating thick and thin areas of blood that

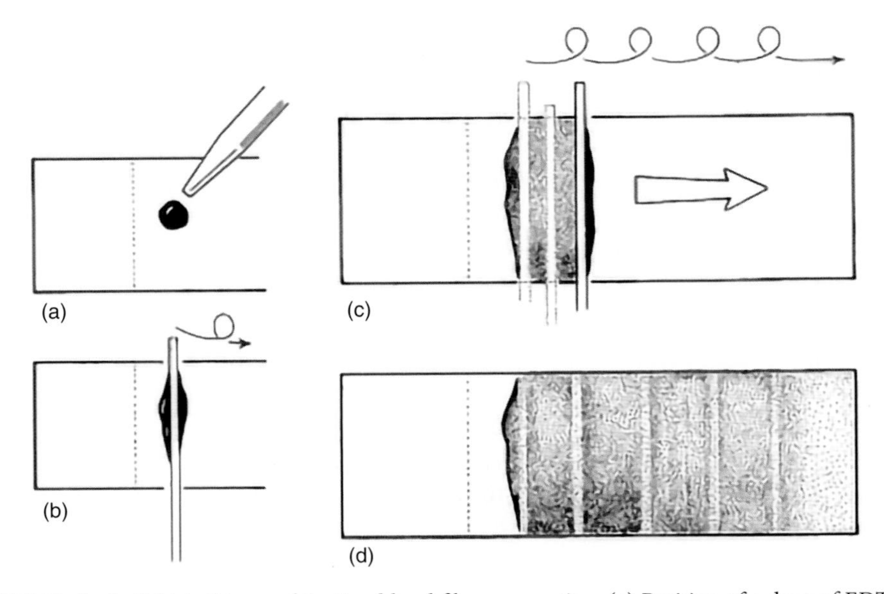

Figure 5.22 Method of thick-thin combination blood film preparation. (a) Position of a drop of EDTA-blood; (b) position of the applicator stick in contact with blood and glass slide; (c) rotation of the applicator stick; and (d) completed thick-thin combination blood film prior to staining. (Illustration by Sharon Belkin; reprinted from *Diagnostic Medical Parasitology*, 6th ed.)

cover the entire slide. Immediately place the film over some small print and be sure that the print can just barely be read through it.

3. Allow the film to air dry in a horizontal position and protected from dust for at least 30 min to 1 h. Do not attempt to speed the drying process by applying any type of heat, because heat fixes the RBCs, and they subsequently will not lyse in the staining process.
4. This slide can be stained as **either** a thick or thin blood film.
5. Label the slide appropriately.
6. If staining with Giemsa is delayed for more than 3 days or the film is to be stained with Wright's stain, lyse the RBCs in the thick film by placing it in buffered water (pH 7.0 to 7.2) for 10 min, and then remove it from the water and place it in a vertical position (thick film down) to air dry. Other blood stains are also possible.
7. Although thick blood films are not fixed with absolute methanol, after the thick films are thoroughly dry, they can be dipped twice in acetone and allowed to dry before being stained. This extra step does not interfere with RBC lysis that occurs either prior to or during staining. The acetone "quick dip" makes the thick film less likely to fall off during staining and provides a cleaner background for microscopic examination.

Procedure Notes/Reminders

1. A diamond marking pen is recommended.
2. An indelible ink pen can be used.
3. Do not use wax pencils; the material may fall off during the staining procedure.
4. Make sure that the films are protected from dust while drying.
5. Anticoagulated blood is recommended for this procedure (EDTA).

Procedure Limitations

1. If the smears are prepared from anticoagulated blood that is over 1 h old, the morphology of the parasites may not be typical and the film may wash off the slide during the staining procedure.
2. Identification to the species level, particularly distinguishing between *P. ovale* and *P. vivax* and between the ring forms of *P. falciparum* and *Babesia* spp., may be impossible without examining the stained thin blood film. Also, *Trypanosoma cruzi* trypomastigotes are frequently distorted in thick films.
3. Excess stain deposition on the film may be confusing and make the detection of organisms difficult.

RISK MANAGEMENT ISSUES ASSOCIATED WITH BLOOD FILMS

It is important for both physicians and laboratorians in areas where malaria is not endemic to be aware of the difficulties related to malaria diagnosis and the fact that patient symptoms are often nonspecific and may mimic other problems. It is important for physicians to recognize that travelers are susceptible to malarial infection when they return from a country where malaria is endemic, and they should receive prophylactic medication. It is also important to remember that 80% of the *P. falciparum* cases acquired by American civilians are contracted in sub-Saharan Africa; currently there are 40 cities in Africa with over 1 million inhabitants, and by 2025 over 800 million people will live in urban areas. Malaria has always been a rural disease in Africa; however, it appears that the urban poor are at a much higher risk of contracting malaria than was previously recognized. These changes also have the potential to affect travelers to Africa.

Patient presentations may be very different between those living in an area of endemicity with possible multiple exposures and travelers who are immunologically naive with no prior exposure to the infection. Patients from areas of endemicity often present with a higher parasitemia and typical symptoms: periodic fevers and a history suggesting probable malaria. Travelers may present with symptoms that do not necessarily suggest malaria: continuous low fever, no periodicity, malaise, and possible diarrhea (which can be misdiagnosed as gastrointestinal infections). It is very important to recognize the differences between these groups of patients. In immunologically naive patients, parasitemia can be extremely low, even in patients who may be quite ill; these parasitemia levels are often missed when hematology automation is used.

Potential factors contributing to malaria-related fatalities in U.S. travelers include (i) failure to seek pretravel advice, (ii) failure to prescribe correct chemoprophylaxis, (iii) failure to obtain prescribed chemoprophylaxis, (iv) failure to adhere to the chemoprophylaxis regimen, (v) failure to seek medical care promptly for their illness, (vi) failure to obtain an adequate patient history, (vii) delay in diagnosis of malaria, and (viii) delay in initiating treatment of malaria. One or more of these potential problems can lead to the death of the patient.

For the above reasons, any order requesting blood films for parasites must be considered a stat request. Any delay could result in serious sequelae for the patient.

USE OF A REFERENCE LABORATORY FOR PARASITE BLOOD DIAGNOSTIC TESTING

There are several important considerations when a reference laboratory is used for blood parasite diagnostic testing. Different laboratory situations are presented below, with recommendations for reference laboratory use.

1. If your test timing does not consider stat testing time limits for blood parasite diagnostic work, this test option should not be listed as an available test on your test menu.
2. If you make the test available to your physicians, the following guidelines should be considered.
   - **a.** Blood film examination for parasites should be ordered on a stat basis (ordering, collection, processing, examination, and reporting).
   - **b.** Even if the blood films will be sent to a reference laboratory, the laboratory handling blood collection should do the following:
     - **i.** Examine the blood films on site with a report sent to the physician (stat basis). At the same time, send the original lavender (EDTA) blood tube with thick and thin blood films prepared at the time the blood was received in the collecting laboratory (stained or unstained). These thick and thin blood films eliminate the potential problems with lag time between when the blood was collected and the blood films were made. If no blood films are prepared prior to shipment to the reference laboratory, organism distortion (after 1 h) and disintegration (after 6 h) tend to occur. If the blood films are unstained, the thin films should be fixed (absolute methanol) prior to shipment.
     - **ii.** If the collecting laboratory performs a stat dipstick malarial rapid test (BinaxNOW malaria test kit; Abbott), the following steps are relevant. The rapid test should be performed immediately on receipt of the blood (stat). If positive, the result and interpretation should be sent to the physician. If the BinaxNOW test is negative, the result should be reported to the physician. HOWEVER, at the same time, the laboratory is responsible for examining and reporting the

results of thick and thin blood films (stat basis), even if the blood is being sent to a reference laboratory.

iii. It is critical to understand the BinaxNOW limitations in patients with low-level parasitemia; these patients' samples may yield a false-negative rapid test. These patients are often travelers who are symptomatic at a very low-level parasitemia. THIS NEGATIVE BinaxNOW RESULT DOES NOT RULE OUT A BLOOD PARASITE (*Plasmodium/Babesia*). THESE LOW-LEVEL-PARASITEMIA CASES WILL ALSO NOT BE DETECTED USING AUTOMATED HEMATOLOGY SYSTEMS.

BLOOD FILM REPORTING WITH ADDITIONAL REPORT COMMENTS

Additional report comments can be extremely valuable for the physician. Various examples (underlined), reports, and report comments are presented below.

1. **Negative thick and thin blood films:** No blood parasites seen. One set of negative blood films will not rule out *Plasmodium* or *Babesia*.
2. ***Plasmodium* seen, not identified to species level:** *Plasmodium* spp. seen. Unable to rule out *Plasmodium falciparum* or *Plasmodium knowlesi*. Note: In this circumstance the presence or absence of gametocytes could not be confirmed (this is another comment that could be added); the answer (yes/no) will influence therapy.
3. ***Plasmodium* seen, good morphology, unable to identify to species level:** *Plasmodium* spp. seen. Unable to rule out *Plasmodium falciparum* or *Plasmodium knowlesi*. No gametocytes seen. Note: If morphology allows identification to *Plasmodium* genus, provide parasitemia for the physician.
4. ***Plasmodium vivax* seen (including gametocytes); possible other unidentified species present:** *Plasmodium vivax* (including gametocytes) seen; possible mixed infection. Unable to rule out *Plasmodium falciparum* or *Plasmodium knowlesi*.
5. ***Plasmodium falciparum* seen, rings only:** *Plasmodium falciparum* seen (no gametocytes seen); make sure you report parasitemia.

BUFFY COAT BLOOD FILMS

In suspected infections with malaria (negative thick and thin blood films), trypanosomiasis, filariasis, and leishmaniasis, concentration procedures are designed to increase the number of organisms recovered from blood specimens. The buffy coat (a layer of WBCs resulting from centrifugation of whole anticoagulated blood) containing WBCs and platelets, and the layer of RBCs just below the buffy coat layer, can be used to prepare thick and thin blood films (Figure 5.23). The sensitivity of this approach is much higher than that of the routine thick film (see "Blood Concentration" below for the complete protocol).

Leishmania donovani, trypanosomes, and *Histoplasma capsulatum* (a fungus with intracellular elements resembling those of *L. donovani*) are occasionally detected in peripheral blood. The parasites or fungi are found in the large mononuclear cells in the buffy coat. Depending on the stain used, the nuclear material and cytoplasm stain with colors similar to that of the WBCs. *H. capsulatum* appears as a large dot of nuclear material (dark stain) surrounded by a clear halo area. Trypanosomes in the peripheral blood also concentrate with the buffy coat cells.

After centrifugation and removal of the relevant layers, some of the material can be examined as a wet mount; trypomastigotes and microfilariae can be seen as motile objects in the wet mount. After staining, *L. donovani* amastigotes can be found in the monocytes and *Plasmodium* parasites can be seen in the thick and thin films.

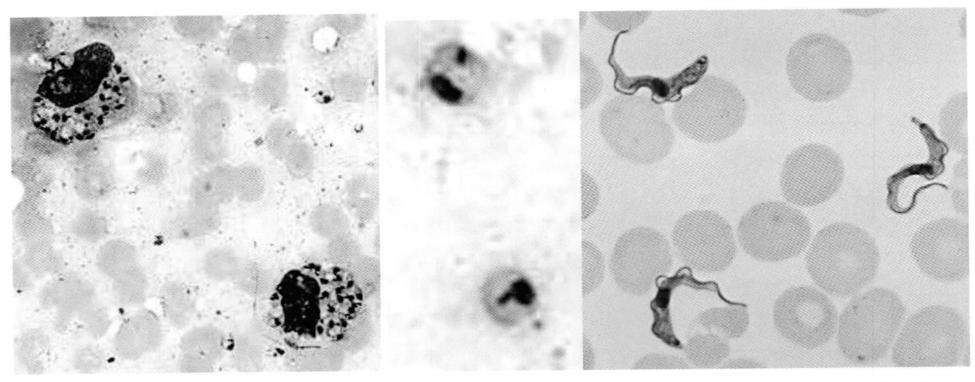

Figure 5.23 Buffy coat stained blood films. *Leishmania donovani* (note the small amastigotes within the monocytes); enlarged amastigotes showing the dot nucleus and kinetoplast bar (primitive flagellum); African trypomastigotes (*Trypanosoma brucei gambiense* or *T. b. rhodesiense*).

Blood Stains

STAIN OPTIONS

For accurate identification of blood parasites, a laboratory should develop proficiency in the use of at least one good staining method. It is better to select one method that provides reproducible results than to use several on a hit-or-miss basis. Blood films should be stained as soon as possible, since prolonged storage may result in stain retention. Failure to stain positive malarial smears within a month may result in failure to demonstrate typical staining characteristics for individual species.

Although in the past the recommended stain of choice has been Giemsa stain, the parasites can also be seen on blood films stained with Wright's stain, a Wright-Giemsa combination stain, or one of the more rapid stains. Delafield's hematoxylin stain is often used to stain the microfilarial sheath; in some cases, Giemsa stain does not provide sufficient stain quality to allow differentiation of the microfilariae.

The most common stains are of two types. Wright's stain has the fixative in combination with the staining solution, so that both fixation and staining occur at the same time; therefore, the thick film must be laked (i.e., the RBCs must be ruptured) before staining. Giemsa stain has the fixative and stain separate; therefore, the thin film must be fixed with absolute methanol before staining. Rapid blood stains are also good options. **The WBCs on the blood film will serve as the QC "organisms"—if they look good, any parasites will also exhibit good morphology and will stain the same colors as the WBCs.**

GIEMSA STAIN

Description

Giemsa stain is used to differentiate nuclear and/or cytoplasmic morphology of platelets, RBCs, WBCs, and parasites. The most dependable staining for blood parasites, particularly in thick films, is obtained with Giemsa stain containing azure B. Liquid stock is available commercially or can be made from dry stain powder. Either must be diluted for use with water buffered to pH 6.8 (more red color) or 7.0 to 7.2 (more blue tones), depending on the specific technique used. Either should be tested for proper staining reaction before use. The stock is stable for years but must be protected from moisture, because the staining reaction is oxidative; therefore, the oxygen in water initiates the reaction and ruins the stock stain. The aqueous working dilution of stain is good only for 1 day.

Although not essential, the addition of Triton X-100, a nonionic surface-active agent, to the buffered water used to dilute the stain enhances the staining properties of Giemsa stain and helps to eliminate possible transfer of parasites from one slide to another. For routine staining of thin films and combination thin-thick films, a 0.01% (vol/vol) final concentration of Triton X-100 is best. For staining thick films for microfilariae, use a 0.1% (vol/vol) concentration. The specimen may consist of a thin blood film that has been fixed in absolute methanol and allowed to dry, a thick blood film that has been allowed to dry thoroughly and is not fixed, or a combination of a fixed thin film and an adequately dried thick film (not fixed) on the same slide.

Reagents

▶ Stock Giemsa Stain

▶ Stock Solution of Triton X-100 (10% aqueous solution)

▶ Buffered Water, pH 7.0 to 7.2 (for diluting stain and washing films); some prefer a pH of 6.8

▶ Buffered Water, pH 6.8 (called for by some commercial stains for diluting stain and washing films)

▶ Triton-Buffered Water Solutions (optional)

1. For combination thin and thick blood films, after determining the pH of the buffered water, add 1 ml of the stock 10% aqueous dilution of Triton X-100 to 1 liter of buffered water (pH 7.0 to 7.2; 0.01% final concentration).
2. Label appropriately, and store in a tightly stoppered bottle. The solution can be used as long as the pH is within limits listed for the procedure.

▶ Methyl Alcohol, Absolute

Quality Control

1. The stock buffer solutions and buffered water should appear clear, without any visible contamination.
2. Check the Giemsa stain reagents, including the pH of the buffered water, before each use. If Triton X-100 has been added to the buffered water, do not use a colorimetric method to determine the pH, because Triton X-100 interferes with the color indicators. Use a pH meter to test buffered water that contains Triton X-100. The buffered water is usable as long as the pH is within the limits listed for the procedure.
3. Prepare and stain films from "normal" blood, and microscopically evaluate the staining reactions of the RBCs, platelets, and WBCs.
   a. Macroscopically, blood films appear purplish. If they are blue, the buffered water was too alkaline; if they are pink to red, the buffered water was too acidic. HOWEVER, color variations rarely prevent identifications; if the WBCs look good, any parasites present will also look good (Figure 5.24).

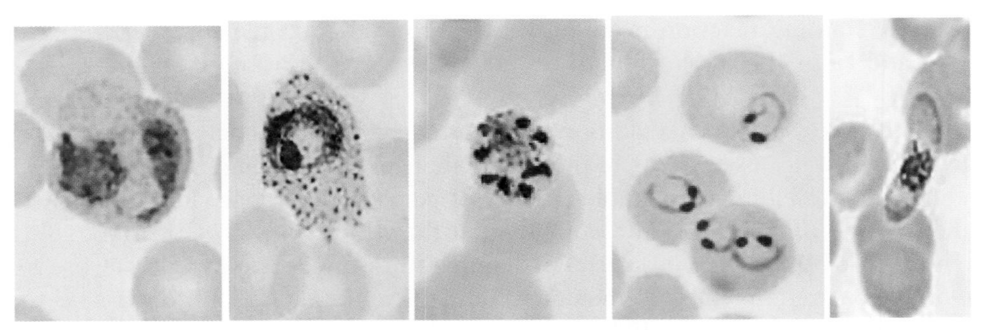

Figure 5.24 Stained examples of *Plasmodium* spp. parasites. *Plasmodium vivax*, developing trophozoite; *Plasmodium ovale*, developing trophozoite (note fimbriated RBC edges); *Plasmodium malariae* mature schizont; *Plasmodium falciparum* ring forms; *P. falciparum* gametocyte. Note the normal color variations; these may depend on the buffer pH used in the staining process. Variation may also occur due to blood stain being used.

b. Microscopically, the RBCs appear pinkish gray, the platelets appear deep pink, and the WBCs have purple-blue nuclei and lighter cytoplasm. Eosinophilic granules are bright purple-red, and neutrophilic granules are purple. Basophilic stippling within uninfected RBCs is blue.

c. Variations may appear in the colors described above, depending on the batch of stain used, the buffer pH, and the character of the blood itself, but if the various morphological structures are distinct, the stain is satisfactory.

4. The microscope should be calibrated (within the last 12 months), and the objectives and oculars used for the calibration procedure should be used for all measurements on the microscope. The calibration factors for all objectives should be posted on the microscope for easy access (multiplication factors can be pasted right on the body of the microscope).
5. Record the QC results on appropriate sheets.

Detailed Procedure

Thin Blood Films (Only)

1. Fix the air-dried film in absolute methyl alcohol by dipping the film briefly (two dips) in a Coplin jar containing methyl alcohol.
2. Remove the film and let it air dry.
3. Stain the film with diluted Giemsa stain (1:20 [vol/vol]) for 20 min. (Add 2 ml of stock Giemsa stain to 40 ml of buffered water containing 0.01% Triton X-100 in a Coplin jar.)
4. Wash by briefly dipping the slide in and out of a Coplin jar of buffered water (one or two dips). Note: Excessive washing will decolorize the film.
5. Let the film air dry in a vertical position.

Thick Blood Films (Only)

1. Allow the film to air dry thoroughly for several hours or overnight. Do not dry films in an incubator (>30 min) or by heat, because this will fix the blood and interfere with the lysing of the RBCs. Note: If a rapid diagnosis of malaria is needed, thick films can be made slightly thinner than usual, allowed to dry for 1 h, and then stained.
2. **Do not fix.**
3. Stain with diluted Giemsa stain (1:50 [vol/vol]) for 50 min. Add 1 ml of stock Giemsa stain to 50 ml of buffered water containing 0.01% Triton X-100 (if staining microfilariae, use 0.1% Triton X-100) in a Coplin jar.

4. Wash by placing the film in buffered water for 3 to 5 min.
5. Let the film air dry in a vertical position.

Combination Thin and Thick Blood Films

1. Stain some of these slides as thin blood films (see above).
2. Stain some of these slides as thick blood films (see above).

Reporting

1. If *Plasmodium* spp. are present, the cytoplasm of the organisms stains blue and the nuclear material stains red to purple-red.
2. Schüffner's stippling and other inclusions in the RBCs infected by *Plasmodium* spp. stain red.
3. Nuclear and cytoplasmic colors that are seen in the malarial parasites also apply to the trypanosomes and any intracellular leishmaniae that might be present.
4. The sheath of microfilariae may or may not stain with Giemsa stain, while the body usually appears blue to purple.
5. Any parasite, including the stage(s) seen, should be reported (do not use abbreviations).

   Examples: *Plasmodium falciparum* rings and gametocytes, or rings only
   Plasmodium vivax rings, trophozoites, schizonts, and gametocytes
   Wuchereria bancrofti microfilariae
   Trypanosoma brucei gambiense/rhodesiense trypomastigotes
   Trypanosoma cruzi trypomastigotes
   Leishmania donovani amastigotes

6. Any laboratory providing malaria diagnoses should be able to identify *P. vivax* and *P. ovale*, even in the absence of Schüffner's stippling.

Procedure Notes/Reminders

1. If blood films are to be prepared from venipuncture blood (use of anticoagulant), they must be prepared within 1 h of collection. Otherwise, certain morphologic characteristics of both parasites and infected RBCs may be atypical. Also, thick blood films may wash off the slide during the staining procedure.
2. It is important to use the correct pH for all buffered water and staining solutions. Solutions with an incorrect pH may prevent certain morphologic characteristics from being visible (e.g., stippling) and do not give typical nuclear and cytoplasmic colors on the stained film.
3. A QC slide can be stained each time patient blood films are stained. If several patient specimens are stained on the same day (using the same reagents), only one control slide needs to be stained and examined. Also, the patient slide serves as its own control (if the WBCs look good, then any parasites will look good).
4. The slide you are staining can also serve as the QC slide. If the WBCs look good, then any parasites will also be acceptable.

Procedure Limitations

1. **Finding no parasites in one set of blood films does not rule out a parasitic infection.**
2. A minimum of 300 oil immersion (magnification, ×1,000) fields should be examined before reporting "no parasites found."
3. The entire smear should be examined under low power (×100) for the presence of microfilariae. Remember that the sheath may not be visible (*W. bancrofti*).

Blood Concentration

Blood concentration procedures increase the number of organisms recovered from blood specimens submitted for diagnosis of trypanosomiasis, filariasis, and leishmaniasis. A concentration procedure should be performed routinely on all blood specimens submitted for examination for trypanosomes or microfilariae when the suspected organisms are not found in thick blood films or so few organisms are present that more are needed to make a positive identification of the species.

BUFFY COAT CONCENTRATION

Description

Leishmania donovani amastigotes are difficult to detect in blood specimens but are occasionally found within monocytes by fractional centrifugation of comparatively large amounts of blood. The procedure may also be used to recover trypanosomes and microfilariae, both of which are found in the plasma. The specimen of choice is whole blood, collected using EDTA, heparin, or sodium citrate anticoagulant.

Reagents

- ▶ **Methyl Alcohol, Absolute**
- ▶ **Giemsa Stain (see above)**

Quality Control

1. Check the calibration of the centrifuge.
2. Perform the procedure on "normal" blood. The film should be composed almost exclusively of WBCs, which stain characteristically with Giemsa or other blood stains. If parasites are present, they should also stain like the WBCs (**morphology seen, stain intensity consistent**). More RBCs may be present if part of that layer has also been sampled.
3. The microscope should be calibrated, and the objectives and oculars used for the calibration procedure should be used for all measurements on the microscope. The calibration factors for all objectives should be posted on the microscope for easy access (multiplication factors can be pasted right on the body of the microscope).
4. Record QC results on appropriate sheets.

Detailed Procedure

1. Wear gloves when performing this procedure.
2. Centrifuge the anticoagulated blood specimen in a sealed cup at 100 × *g* for 15 min.
3. Remove the thin creamy layer (buffy coat) between the RBCs and plasma with a capillary pipette, or transfer the buffy coat and plasma to another tube and centrifuge in a sealed cup at 300 × *g* for 15 min.
4. Examine the buffy coat directly for motile trypomastigotes and microfilariae.
   - **a.** Place 0.5 drop of saline on a clean microscope slide.
   - **b.** Remove 1 drop of sediment and mix it in the saline.
   - **c.** Add a coverslip and examine for organism motility with the low-power (10×) and high dry (40×) objectives.
5. Prepare thin films, dry, fix, and stain with a blood stain.

Reporting

1. If present, *Leishmania donovani* amastigotes are found within the monocytes on a Giemsa-stained film. Nuclear material stains dark purple-red, the cytoplasm is light blue, and the kinetoplast may or may not be visible as a dark bluish-purple structure (Figure 5.23).
2. Trypomastigotes are found extracellularly (they are motile in the wet smear). Morphologic detail is seen in the Giemsa-stained film. The stain reaction is like that of *Leishmania*; the kinetoplast is visible (Figure 5.23).
3. Microfilariae may be found in the wet smear. Morphologic detail is seen in a Giemsa- or hematoxylin-stained film. The stain reaction is typical for each stain.
4. Report the presence of organisms from the wet smear.

   Examples: Trypomastigotes present
   Microfilariae present

5. Report the genus and species of organisms from the Giemsa-stained film.

   Examples: *Trypanosoma cruzi* trypomastigotes present
   Leishmania donovani amastigotes present

Procedure Notes/Reminders

1. If you need to add anticoagulant to blood, mix 9 ml of blood and 1 ml of 5% sodium citrate in a glass centrifuge tube. Then proceed with the centrifugation.
2. This procedure can be performed in a microhematocrit tube if the tube is carefully scored and broken at the buffy coat interface and the WBCs are prepared and stained as a thin blood film.
3. Also, the tube can be examined microscopically (high dry magnification) at the buffy coat layer for motile trypomastigotes and microfilariae, before the tube is scored and broken.

Procedure Limitations

1. When examined as a wet smear, the intracellular leishmaniae are very difficult to see.
2. Although trypomastigote and microfilarial motility may be visible on the wet smear, specific identification may be difficult without a permanent stain.

KNOTT CONCENTRATION

Description

This technique is used to recover low numbers of microfilariae from blood. A solution is used to lyse the red blood cells in a large blood sample, and the organisms are concentrated from the supernatant fluid by centrifugation. The disadvantage of this technique is that the microfilariae are killed and immobilized and are therefore not readily revealed by their motility. The specimen of choice is whole blood, collected using EDTA, heparin, or sodium citrate anticoagulant.

Reagents

▶ **2% Aqueous Formalin**
Formaldehyde, liquid .... 2 ml
Distilled water .... 98 ml

Mix thoroughly. Store in a stoppered bottle. Label appropriately. The shelf life is 24 months.

Quality Control

1. Check the calibration of the centrifuge.
2. If possible, check the procedure with human or canine blood containing microfilariae, with or without a sheath (often not practical).
3. If positive blood is not available, follow the procedure carefully in testing the specimen submitted for diagnosis. Examine the sediment thoroughly using low- and high-power magnification.
4. The microscope should be calibrated, and the objectives and oculars used for the calibration procedure should be used for all measurements on the microscope. The calibration factors for all objectives should be posted on the microscope for easy access (multiplication factors can be pasted right on the body of the microscope).
5. Record QC results on appropriate sheets.

Detailed Procedure

1. Wear gloves when performing this procedure.
2. Place 1 ml of fresh whole blood or anticoagulated blood in a centrifuge tube containing 10 ml of 2% formalin. Mix thoroughly.
3. Centrifuge for 5 min at 300 × *g*.
4. Pour off the supernatant fluid without disturbing the sediment.
5. Using a capillary pipette, transfer a portion of the sediment to a slide.
6. Apply a coverslip, and examine microscopically under low-power (×100) and high-power (×400) magnification.
7. If microfilariae are present, prepare a thick film from the remainder of the sediment, air dry, fix in absolute methanol for 5 min, air dry again, and stain with Giemsa or Delafield's hematoxylin.

Reporting

1. If present in the sample, microfilariae are concentrated and appear nonmotile in the wet smear.
2. After staining with Giemsa or Delafield's hematoxylin, the microfilariae exhibit diagnostic morphology and typical staining characteristics.
3. Report the presence of organisms from the wet smear.

   Example: Microfilariae present

4. Report the genus and species of organisms from the Giemsa- or hematoxylin-stained film.

   Example: *Wuchereria bancrofti* microfilariae present

Procedure Notes/Reminders

1. If you need to add anticoagulant to blood, mix 9 ml of blood and 1 ml of 5% sodium citrate in a glass centrifuge tube. Then proceed with the centrifugation.
2. Morphologic details may not be visible prior to Giemsa or hematoxylin staining.

Procedure Limitations

1. Motility is not visible after formalin fixation.
2. Species determination may be difficult for most laboratorians without additional staining.
3. The blood-formalin mixture can be sent to a reference laboratory for staining and identification of microfilariae.

MEMBRANE FILTRATION CONCENTRATION

Description

Membrane filtration methods have been developed for recovering microfilariae in light infections (Figure 5.20). They have an advantage over simple centrifugation methods in that large samples of blood (20 ml or more) can be used if necessary. The technique described here is one of the most efficient for the clinical laboratory when other procedures used to recover microfilariae are unsatisfactory. Membrane filtration recovers most species of microfilariae; however, because of their small size, *Mansonella perstans* and *Mansonella ozzardi* may not be recovered. Membranes with smaller pores (3 µm) have been suggested to recover these two species. The specimen of choice is whole blood, collected using EDTA, heparin, or sodium citrate anticoagulant.

Reagents

- **Distilled Water**
- **Methyl Alcohol, Absolute**
- **Giemsa Stain (see above)**
- **Toluene**

Quality Control

1. If possible, check the procedure using human or canine blood containing microfilariae (often not practical).
2. If positive blood is not available, follow the procedure carefully in testing the specimen submitted for diagnosis. Examine sediment thoroughly under low- and high-power magnification.
3. The microscope should be calibrated, and the objectives and oculars used for the calibration procedure should be used for all measurements on the microscope. The calibration factors for all objectives should be posted on the microscope for easy access (multiplication factors can be pasted right on the body of the microscope).
4. Record QC results on appropriate sheets.

Detailed Procedure

1. Wear gloves when performing this procedure.
2. Draw 1 ml of fresh whole blood or anticoagulated blood into a 15-ml syringe containing 10 ml of distilled water.
3. **Gently** shake the mixture for 2 to 3 min to ensure that all blood cells are lysed.
4. Place a 25-mm Nuclepore filter, 5-µm porosity, over a moist 25-mm filter paper pad and place in a Swinney filter adapter.
5. Attach the Swinney filter adapter to the syringe containing the lysed blood.
6. With gentle but steady pressure on the piston, push the lysed blood through the filter.
7. Without disturbing the filter, remove the Swinney adapter from the syringe and draw approximately 10 ml of distilled water into the syringe. Replace the adapter, and gently push the water through the filter to wash the debris from the filter.

8. Remove the adapter again, draw the piston of the syringe to about half the length of the barrel, replace the adapter, and push the air in the barrel through the filter to expel excess water.
9. To prepare the filter for staining, remove the adapter, draw the piston about half the length of the barrel, and then draw 3 ml of absolute methanol into the syringe. Holding the syringe vertically, replace the adapter and push the methanol followed by the air through the filter to fix the microfilariae and expel the excess methanol, respectively.
10. To stain, remove the filter from the adapter, place it on a slide, and allow it to air dry thoroughly. Stain with Giemsa or other blood stain as for a thick film.
11. To cover the stained filter, dip the slide in toluene **before** mounting with neutral mounting medium and a coverslip. This will lessen the formation of bubbles in or under the filter.

Reporting

1. If present in the sample, microfilariae are concentrated and appear on the wet membrane.
2. The microfilariae stain characteristically with Giemsa or Delafield's hematoxylin. The sheath, if present, may or may not stain with Giemsa (*Wuchereria bancrofti* sheath will not stain with Giemsa).
3. Report the presence of organisms from the wet Nuclepore membrane.

   Example: Microfilariae present

4. Report the genus and species of organisms from the Giemsa- or hematoxylin-stained membrane.

   Example: *Wuchereria bancrofti* microfilariae present

Note: stained with hematoxylin-based stain. The sheath of *W. bancrofti* may not be visible if stained with Giemsa.

Procedure Notes/Reminders

1. Gently shake the water-blood mixture to ensure total lysis of blood cells. Some parasitologists prefer to use an aqueous solution of 10% Teepol (Shell Oil Co., Houston, TX) to lyse the blood cells.
2. Motile microfilariae may be seen on the membrane filter; however, low light intensity is needed.
3. The membrane filter must be supported by the moistened filter pad to prevent rupture when the water is expelled through the membrane.
4. If you need to add anticoagulant to the blood, mix 9 ml of blood and 1 ml of 5% sodium citrate in a glass centrifuge tube. Then proceed with centrifugation.

Procedure Limitations

1. Giemsa or hematoxylin staining may be necessary to identify the organisms to the species level. The sheath of *W. bancrofti* may not be visible if stained with Giemsa.
2. Species identification of microfilariae on filters may be difficult.

Suggested Reading

Arakaki T, Iwanaga M, Kinjo F, Saito A, Asato R, Ikeshiro T. 1990. Efficacy of agar-plate culture in detection of *Strongyloides stercoralis* infection. *J Parasitol* **76:**425–428.

Baermann G. 1917. Eine einfache Methode zur Auffindung vor Ankylostomum (Nematoden) Larven in Erdproben. *Geneeskd Tijdschr Ned Indie* **57:**131–137.

Baxby D, Blundell N, Hart CA. 1984. The development and performance of a simple, sensitive method for the detection of *Cryptosporidium* oocysts in faeces. *J Hyg (Lond)* **93:**317–323.

Didier ES, Orenstein JM, Aldras A, Bertucci D, Rogers LB, Janney FA. 1995. Comparison of three staining methods for detecting microsporidia in fluids. *J Clin Microbiol* **33:**3138–3145.

Garcia LS. 2016. *Diagnostic Medical Parasitology*, 6th ed. ASM Press, Washington, DC.

Garcia LS, Arrowood M, Kokoskin E, Paltridge GP, Pillai DR, Procop GW, Ryan N, Shimizu RY, Visvesvara G. 2018. Practical guidance for clinical microbiology laboratories: laboratory diagnosis of parasites from the gastrointestinal tract. *Clin Microbiol Rev* **31:**e00025–17. doi:10.1128/CMR.00025-17

Garcia LS, Brewer TC, Bruckner DA. 1979. A comparison of the formalin-ether concentration and trichrome-stained smear methods for the recovery and identification of intestinal protozoa. *Am J Med Technol* **45:**932–935.

Garcia LS, Bruckner DA, Brewer TC, Shimizu RY. 1983. Techniques for the recovery and identification of *Cryptosporidium* oocysts from stool specimens. *J Clin Microbiol* **18:**185–190.

Garcia LS, Johnston SP, Linscott AJ, Shimizu RY. 2008. *Cumitech 46, Laboratory procedures for diagnosis of blood-borne parasitic diseases.* Coordinating ed, Garcia LS. ASM Press, Washington, DC.

Garcia LS, Shimizu RY. 1997. Evaluation of nine immunoassay kits (enzyme immunoassay and direct fluorescence) for detection of *Giardia lamblia* and *Cryptosporidium parvum* in human fecal specimens. *J Clin Microbiol* **35:**1526–1529.

Garcia LS, Shimizu RY. 2000. Detection of *Giardia lamblia* and *Cryptosporidium parvum* antigens in human fecal specimens using the ColorPAC combination rapid solid-phase qualitative immunochromatographic assay. *J Clin Microbiol* **38:**1267–1268.

Hanson KL, Cartwright CP. 2001. Use of an enzyme immunoassay does not eliminate the need to analyze multiple stool specimens for sensitive detection of *Giardia lamblia. J Clin Microbiol* **39:**474–477.

Harada U, Mori O. 1955. A new method for culturing hookworm. *Yonago Acta Med* **1:**177–179.

Hsieh HC. 1962. A test-tube filter-paper method for the diagnosis of *Ancylostoma duodenale, Necator americanus,* and *Strongyloides stercoralis. WHO Tech Rep Ser* **255:**27–30.

Koga K, Kasuya S, Khamboonruang C, Sukhavat K, Ieda M, Takatsuka N, Kita K, Ohtomo H. 1991. A modified agar plate method for detection of *Strongyloides stercoralis. Am J Trop Med Hyg* **45:**518–521.

Kokoskin E, Gyorkos TW, Camus A, Cedilotte L, Purtill T, Ward B. 1994. Modified technique for efficient detection of microsporidia. *J Clin Microbiol* **32:**1074–1075.

Leber AL (ed). 2016. Section 9: Parasitology. *In Clinical Microbiology Procedures Handbook*, 4th ed. ASM Press, Washington, DC.

Markell EK, Voge M. 1981. *Medical Parasitology*, 5th ed. The WB Saunders Co, Philadelphia, PA.

Melvin DM, Brooke MM. 1982. *Laboratory Procedures for the Diagnosis of Intestinal Parasites*, 3rd ed. U.S. Department of Health, Education, and Welfare publication no. (CDC) 82-8282. Government Printing Office, Washington, DC.

National Committee for Clinical Laboratory Standards. 2000. Use of blood film examination for parasites. *Approved guideline M15-A.* National Committee for Clinical Laboratory Standards, Wayne, PA.

National Committee for Clinical Laboratory Standards. 1997. Procedures for the recovery and identification of parasites from the intestinal tract, 2nd ed. *Approved guideline M28-A.* National Committee for Clinical Laboratory Standards, Wayne, PA.

Neimeister R, Logan AL, Gerber B, Egleton JH, Kleger B. 1987. Hemo-De as substitute for ethyl acetate in formalin-ethyl acetate concentration technique. *J Clin Microbiol* **25:**425–426.

Palmer J. 1991. Modified iron hematoxylin/Kinyoun stain. *Clin Microbiol Newsl* **13:**39–40. (Letter.)

Ritchie LS. 1948. An ether sedimentation technique for routine stool examinations. *Bull US Army Med Dep* **8:**326.

Ryan NJ, Sutherland G, Coughlan K, Globan M, Doultree J, Marshall J, Baird RW, Pedersen J, Dwyer B. 1993. A new trichrome-blue stain for detection of microsporidial species in urine, stool, and nasopharyngeal specimens. *J Clin Microbiol* **31:**3264–3269.

Spencer FM, Monroe LS. 1976. *The Color Atlas of Intestinal Parasites*, 2nd ed. Charles C Thomas, Springfield, IL.

Tompkins VN, Miller JK. 1947. Staining intestinal protozoa with iron-hematoxylin-phosphotungstic acid. *Am J Clin Pathol* **17:**755–758.

Visvesvara GS, Moura H, Kovacs-Nace E, Wallace S, Eberhard ML. 1997. Uniform staining of *Cyclospora* oocysts in fecal smears by a modified safranin technique with microwave heating. *J Clin Microbiol* **35:**730–733.

Weber R, Bryan RT, Owen RL, Wilcox CM, Gorelkin L, Visvesvara GS, The Enteric Opportunistic Infections Working Group. 1992. Improved light-microscopical detection of microsporidia spores in stool and duodenal aspirates. *N Engl J Med* **326:**161–166.

Wheatley WB. 1951. A rapid staining procedure for intestinal amoebae and flagellates. *Am J Clin Pathol* **21:**990–991.

Wilson M, Schantz PM. 2000. Parasitic immunodiagnosis, p. 1117–1122. *In* Strickland GT (ed), *Hunter's Tropical Medicine and Emerging Infectious Diseases*, 8th ed. The WB Saunders Co, Philadelphia, PA.

Yang J, Scholten T. 1977. A fixative for intestinal parasites permitting the use of concentration and permanent staining procedures. *Am J Clin Pathol* **67:**300–304.

Algorithm 5.1 Procedure for processing fresh stool for the O&P examination

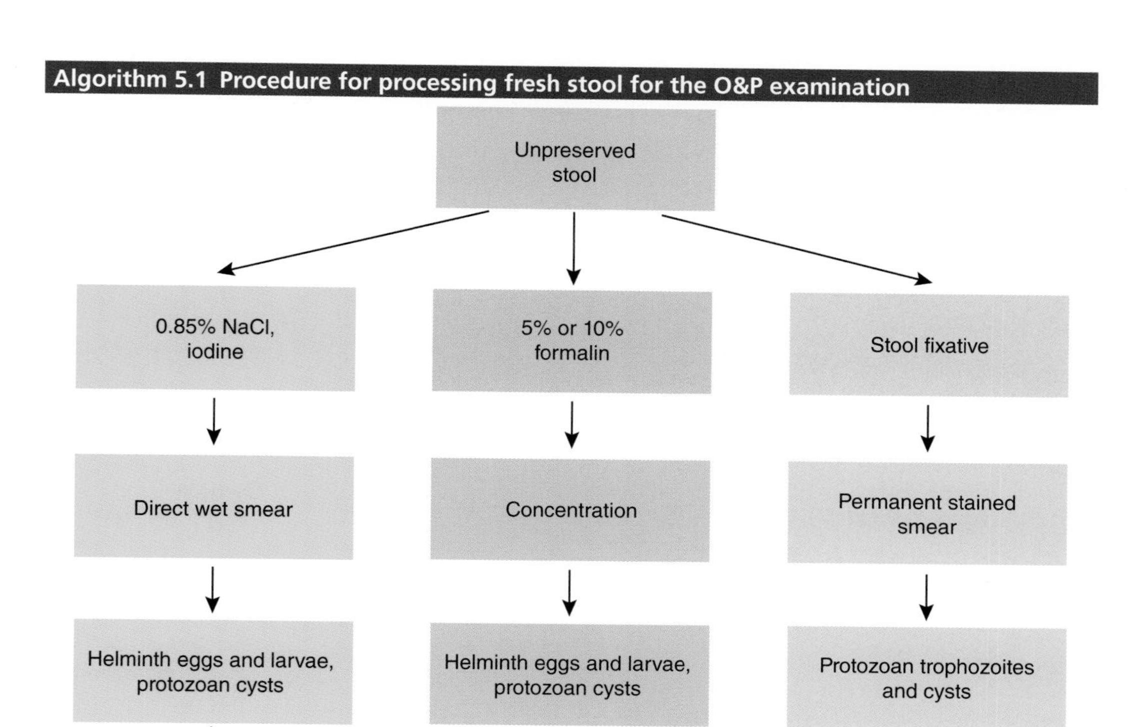

Notes: Special stains are necessary for *Cryptosporidium* and *Cyclospora* (modified acid-fast stains) and the microsporidia (modified trichrome stain or the nonspecific calcofluor white). If the iron hematoxylin permanent staining method (containing the carbol fuchsin step) is used for the O&P examination, members of the Apicomplexa (including the coccidia) will stain pink. Also, fecal immunoassay kits are now available for *Giardia lamblia*, *Cryptosporidium* spp., *Entamoeba histolytica*, and the *Entamoeba histolytica/E. dispar* group.

Algorithm 5.2 Procedure for processing liquid specimens for the O&P examination

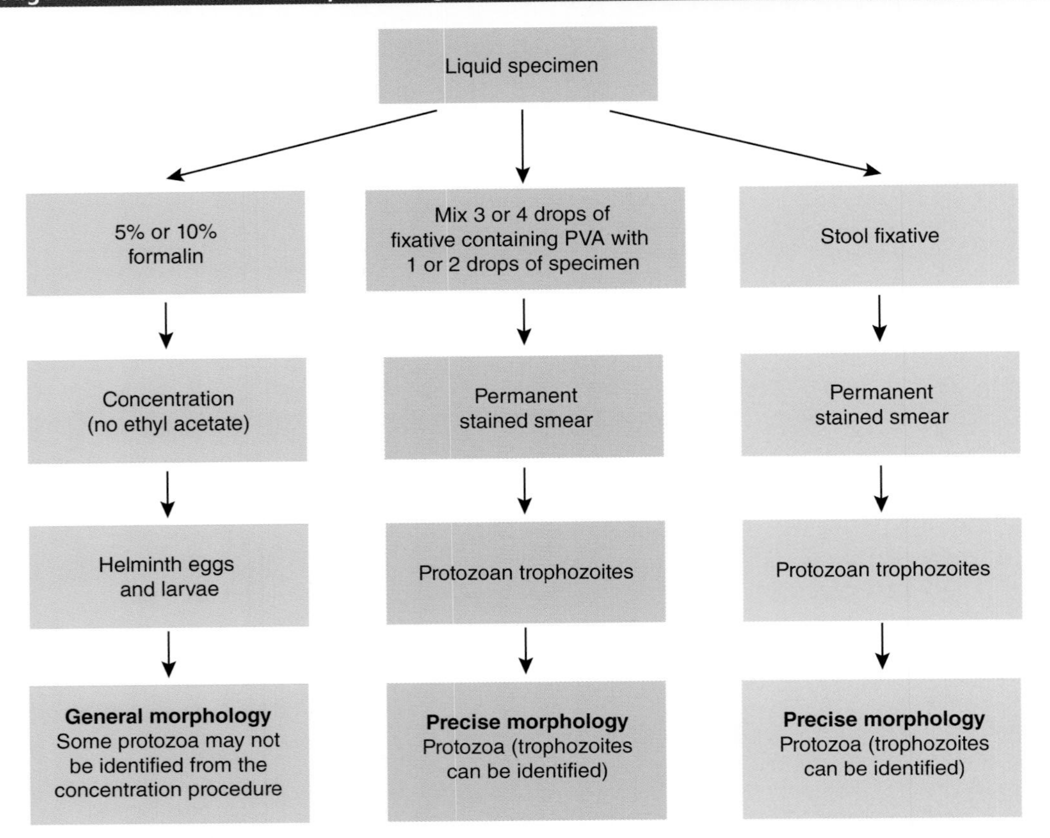

Notes: PVA (polyvinyl alcohol)/fixative and specimen are mixed on the slide, allowed to air dry, and then stained. This approach is appropriate only for liquid specimens and those containing a lot of mucus; semiformed or formed specimens will not be adequately fixed using this method. Not all fixatives will require PVA or albumin as adhesives.

Algorithm 5.3 Procedure for processing preserved stool for the O&P examination—two-vial collection kit

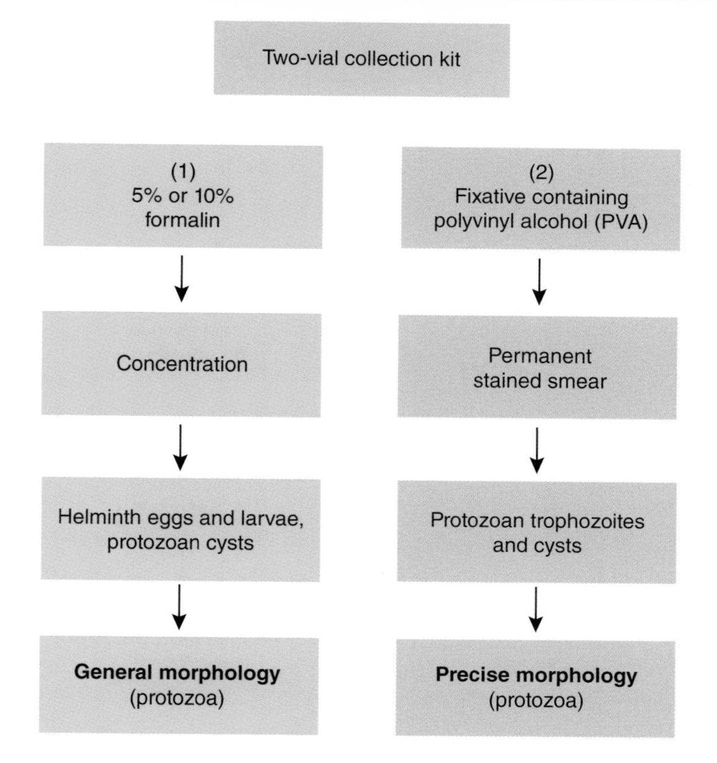

Notes: The following fixatives are listed and compared with those containing the mercuric chloride base and plastic powder (used to glue the stool onto the slide) polyvinyl alcohol (PVA).

1. Mercuric chloride base (Schaudinn's) containing PVA (gold standard) (trichrome, iron hematoxylin)
2. Zinc sulfate base containing PVA (good substitute) (trichrome, iron hematoxylin)
3. Copper sulfate base containing PVA (poor substitute) (trichrome, iron hematoxylin)
4. SAF (no PVA) (good substitute) (single-vial system) (iron hematoxylin best, trichrome OK)
5. Unifix (no PVA) (good substitute) (single-vial system) (trichrome probably best)
6. Ecofix (no PVA) (acceptable substitute) (good when used with Ecostain; trichrome OK)
7. Total-Fix, new universal fixative (see Algorithm 5.5).

Algorithm 5.4 Procedure for processing SAF-preserved stool for the O&P examination

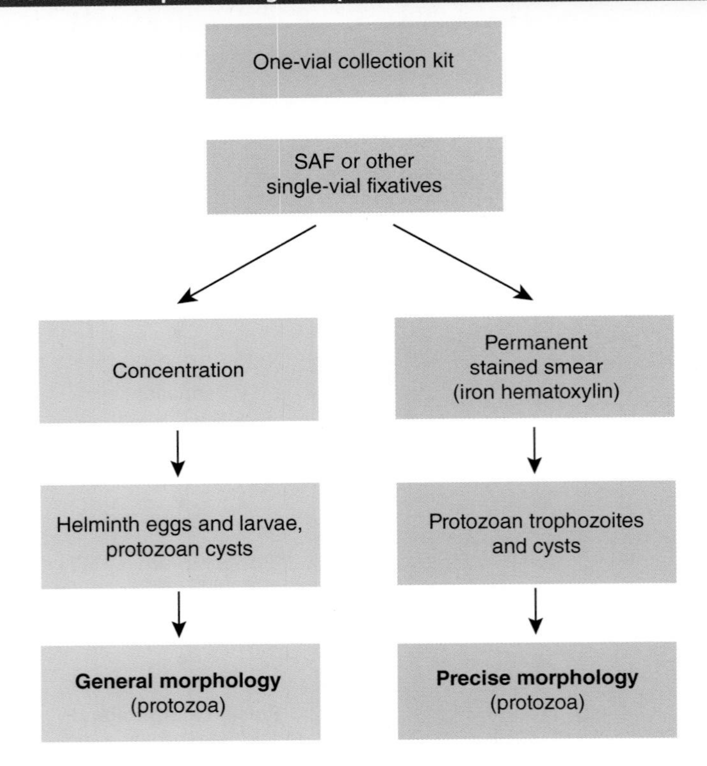

Notes: SAF and some of the other single-vial systems can also be used with EIA, FA, and cartridge rapid fecal immunoassays for the detection of *Giardia* and *Cryptosporidium*. Centrifuged material can also be used for the modified acid-fast and/or modified trichrome stains.

Algorithm 5.5 Procedure for the use of Total-Fix (universal fixative, single-vial system) (this fixative contains no mercury, no PVA, and no formalin)*

Step 1 **Specimen vial containing Total-Fix and fecal specimen**	Ratio of 1/3 stool, 2/3 fixative – **MIX WELL**. Allow to fix 30 min before processing. *See end of algorithm for optional Smear Preparation Directly from Vial
Step 2 **Use the fluid at the top of the vial to run the immunoassays**	Some kit directions recommend shaking the vial. After shaking, allow particulate matter to settle, and then use the fluid at the top for the immunoassays. Use centrifuged specimen for DFA and special stains (Apicomplexa, microsporidia) (Step 7).
Step 3 **Tube or concentration device**	Add approximately 1 to 2 ml of fixative/specimen mix to the tube or device. Mix well prior to pouring into the tube or device. **DO NOT ADD ANY RINSE FLUIDS:** Saline, water, formalin. Rinse fluids will prevent routine permanent staining.
Step 4 **Centrifuge**	10 min at 500 × *g*
Step 5 **Pour off most of excess fixative**	
Step 6 **Mix sediment with remaining fixative**	
Step 7 **Prepare slides for permanent stains and/or DFA; stain slides**	Allow slides to dry for ~30 to 45 min at 37°C. Slides can be placed on trays and put in the incubator. Room temperature drying requires ~60 min.
Step 8 **Add rinse fluid to remaining stool sediment (from Step 5). Proceed with routine concentration procedure.**	Mix sediment and rinse fluid well before centrifugation. Single rinse including ethyl acetate step is recommended. **DO NOT ADD** ethyl acetate if the specimen contains a lot of mucus.
Step 9 **Ring debris layer, pour off excess fluid. Mix and examine sediment as concentration sediment slide.**	Entire 22- by 22-mm coverslip: low power (10× objective) 1/3 to 1/2 coverslip: high dry power (40× objective)

(continued on next page)

Algorithm 5.5 Procedure for the use of Total-Fix (universal fixative, single-vial system) (this fixative contains no mercury, no PVA, and no formalin)* (*continued*)

*ALTERNATE METHOD FOR SMEAR PREPARATION DIRECTLY FROM VIAL

1. Allow stool to settle to the bottom of the vial.
2. Using a pipette (without creating bubbles), remove fecal material from the bottom of the vial.
3. Smear this fecal material onto the glass slide for subsequent permanent staining (after the slides are allowed to dry for ~30 to 45 min at 37°C or ~60 min at room temperature).

METHOD PROS AND CONS (Smear Preparation Directly from Vial)	
PROS	**CONS**
Eliminates one centrifugation step from the algorithm; however, this centrifugation step (step 4) ensures sufficient stool and concentrates the stool prior to smear preparation.	If the specimen is thin or runny or contains very little stool, there is a chance the slide may not contain sufficient stool material for staining and/or organism identification.
	If bubbles are created with the pipette, the fecal material will have to settle out again before the sample is taken.

Note: In order to ensure maximum organism recovery and the best overall results, the algorithm should be used as originally written and recommended. However, with practice, sufficient fecal material can be taken from the vial and used to prepare smears for permanent staining.

Algorithm 5.6 Use of various fixatives and their recommended stains

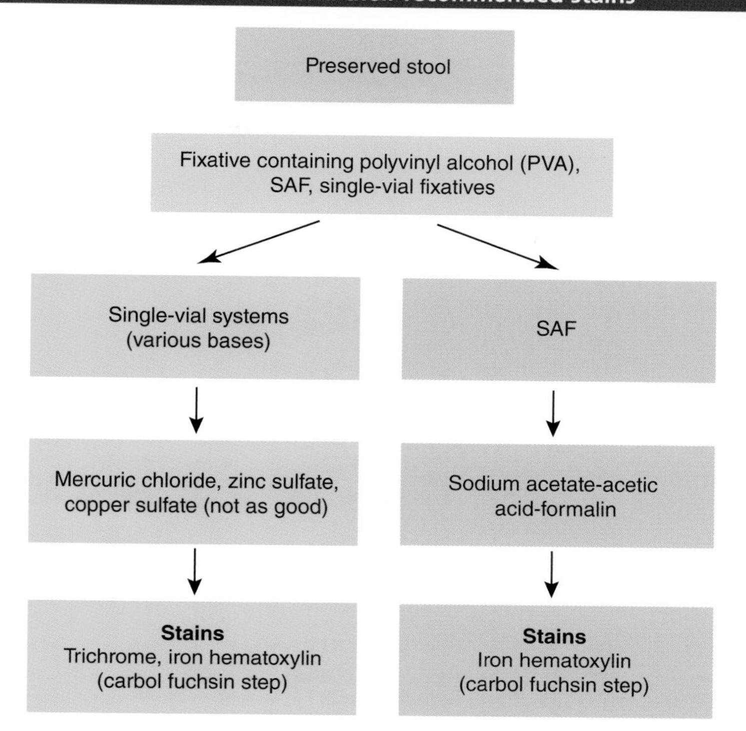

Notes: 5 or 10% formalin, SAF, Unifix, Ecofix, and some of the other single-vial systems can also be used with EIA, FA, and rapid cartridge fecal immunoassay kits, as well as modified acid-fast (coccidia) and modified trichrome (microsporidia) stains. Check with the manufacturer and/or published results. Although both the concentration and permanent stained smear can be performed using these fixatives, the concentration performed from PVA-containing fixatives is an uncommon approach.

Algorithm 5.7 Ordering algorithm for laboratory examination for intestinal parasites

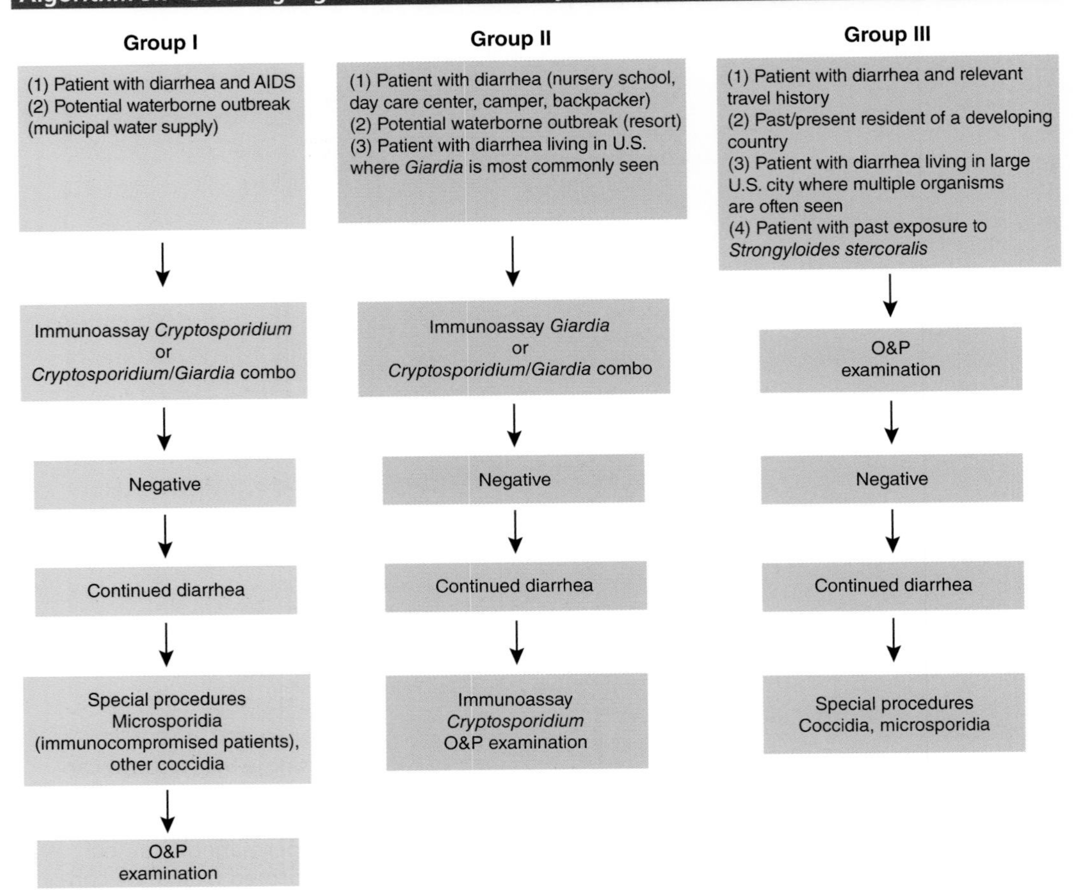

Notes: The first group of patients, for which the *Cryptosporidium* fecal immunoassay is recommended as the first test order, includes (1) patients with diarrhea and AIDS (or other compromised states) and (2) patients with diarrhea from a potential municipal waterborne outbreak. Group 2, for which the first test order would be the *Giardia* fecal immunoassay,* includes (1) patients with diarrhea from a nursery school or day care center and campers or backpackers, (2) patients with diarrhea from a potential waterborne outbreak at a resort, and (3) patients with diarrhea who live in areas within the United States where *Giardia* is the most common organism seen. Group 3, for which the first test order would be the O&P examination, includes (1) patients with diarrhea and relevant travel history, (2) patients with diarrhea who are past or present residents of a developing country, (3) patients with diarrhea who live in large metropolitan areas within the United States where multiple parasites are often seen, and (4) patients who may have a history of exposure to *Strongyloides stercoralis*. For group 3, it is also very important to remember that there may be symptomatic patients who are immunocompromised (either from underlying disease or from therapy, etc.) who may have a history of possible exposure to *S. stercoralis*. Remember, this past exposure could have occurred as long as 40+ years ago, after the patient has left the area of endemicity. With the internal autoinfection cycle, the patient can carry this infection without symptoms until becoming debilitated or immunocompromised. In these cases, it is very important to perform the O&P examination, as well as the agar plate culture for *Strongyloides* infection. In the event of disseminated disease with strongyloidiasis, this type of hyperinfection can be fatal if not recognized and treated.

*It is important to remember that *Blastocystis* spp. is the most common organism seen in stool worldwide; *Dientamoeba fragilis* is as common as or more common than *Giardia*. As multiplex organism panels that include parasites continue to be developed, *Blastocystis*, *Dientamoeba*, and *Strongyloides* may be included in the parasite panel.

Algorithm 5.8 Procedure for processing blood specimens for examination

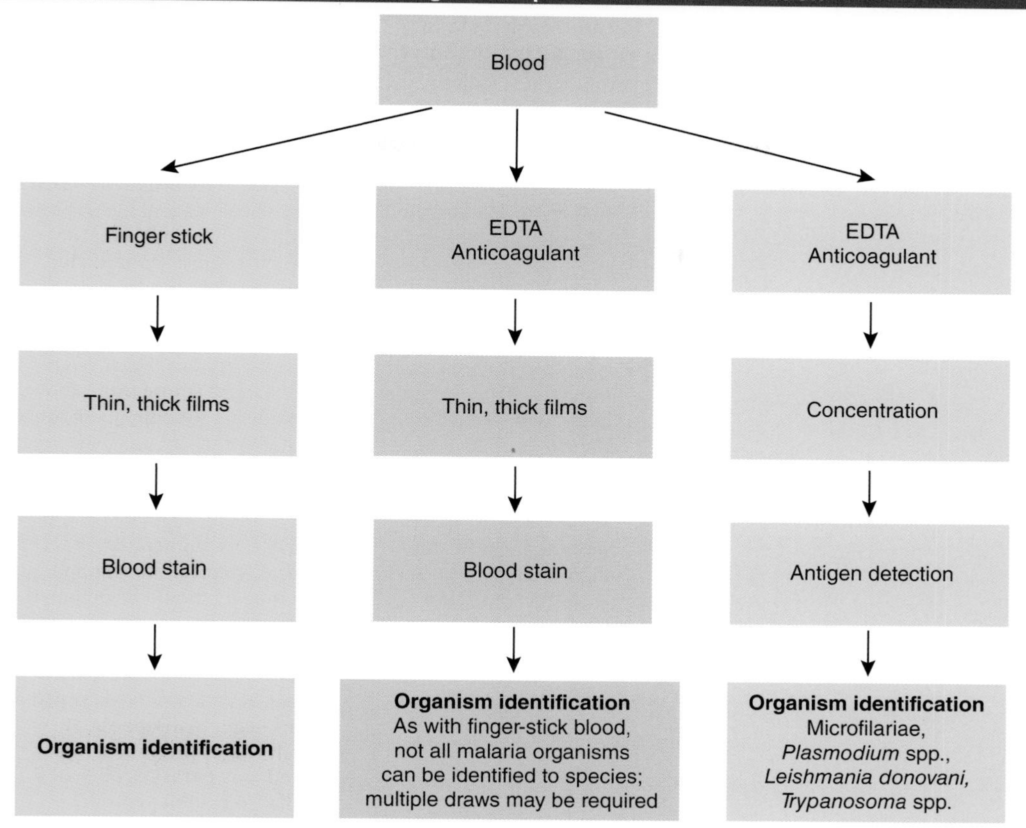

Notes: Special stains may be necessary to identify some of the microfilariae to the species level (Delafield's or other hematoxylin stains). Many laboratories use automated instruments for hematology. Parasite detection with these instruments has not proven to be acceptable; manual examination of stained blood films is recommended for the detection of all blood parasites. One negative set of blood films does not rule out a possible infection. Although for many years, Giemsa stain has been the stain of choice, the parasites can also be seen on blood films stained with Wright's stain, a Wright-Giemsa combination stain, or one of the more rapid stains, such as Diff-Quik (American Scientific Products, McGaw Park, IL), Wright's Dip Stat stain (Medical Chemical Corp., Torrance, CA), or Field's stain.

Table 5.1 Body sites, specimens, and recommended stains

Body site or sample type	Specimen	Recommended stain(s) for suspected organism	Comments
Blood	Whole or anticoagulated blood	Blood stains (Giemsa, Wright's, Wright-Giemsa combination, Field's, rapid blood stains) All blood parasites Hematoxylin-based stain Microfilariae (sheathed) Note: The QBC tube is also available as a screening tool for blood parasites (hematocrit tube contains acridine orange) and has been used for malaria, *Babesia*, trypanosomes, leishmania, and microfilariae; **however, the sensitivity and specificity of the QBC tube do not equal those seen with thick and thin blood films.**	**Stat request (collection, processing, staining, examination, and reporting).** Most drawings and organism descriptions of blood parasites were originally based on Giemsa-stained blood films. Although Wright's stain (or Wright-Giemsa combination stain) works, stippling in malaria parasites may not be visible and the organism colors do not match the descriptions. However, if other stains (those listed above, in addition to some of the "quick" blood stains) are used, the organisms should be visible on the blood film. The use of anticoagulated blood (EDTA recommended) is acceptable. For detection of stippling, the smears should be prepared within 1 h after the specimen is drawn. After that time, stippling may not be visible on stained films, and after 4 to 6 h, organisms begin to disappear. **Overall QC: If the WBCs look good, any parasites present will also exhibit good morphology (both thick and thin blood films).**
Bone marrow	Aspirate	Blood stains (Giemsa, Wright's, Wright-Giemsa combination, Field's, rapid blood stains) All blood parasites	See comments above for blood.
Central nervous system	Spinal fluid, brain biopsy specimen	Blood stains Trypanosomes, *Toxoplasma gondii* Blood stains, trichrome, or calcofluor white Amebae (*Naegleria*, *Acanthamoeba*, *Balamuthia*, *Sappinia*) Blood stains, acid-fast, periodic acid-Schiff (PAS), modified trichrome, silver methenamine Microsporidia Hematoxylin and eosin (routine histology) Larval cestodes	**Stat request (collection, processing, staining, examination, and reporting).** If cerebrospinal fluid is received (with no suspected organism suggested), any of the blood stains would be acceptable; however, calcofluor white is also recommended as a second stain. If brain biopsy material is received (particularly from an immunocompromised patient), electron microscopy studies may be required in order to identify microsporidia to the genus or species level. Immunofluorescent methods (along with routine histologic morphology) may be required to identify free-living amebae to the correct genus.

Table 5.1 (continued)

Body site or sample type	Specimen	Recommended stain(s) for suspected organism	Comments
Cutaneous ulcers	Aspirate, biopsy specimen	Blood stains Leishmaniae Hematoxylin and eosin (routine histology) *Acanthamoeba* spp. *Balamuthia mandrillaris* *Entamoeba histolytica*	Most likely causative agent would be leishmaniae, all of which would stain with any of the blood stains. Hematoxylin and eosin (routine histology) could also be used to identify these organisms.
Eye	Biopsy specimen, scrapings, contact lens, lens solution if opened; unopened lens care solutions subject to FDA regulations are not tested by routine laboratories	Calcofluor white (cysts only) Amebae (*Acanthamoeba*) Blood stains (trophozoites, cysts) Amebae Hematoxylin and eosin (routine histology) Cysticerci *Loa loa* *Toxoplasma gondii* Silver methenamine stain, PAS, acid-fast, electron microscopy Microsporidia	Some free-living amebae (most commonly *Acanthamoeba*) have been implicated as a cause of keratitis. Although calcofluor white stains the cyst walls, it does not stain the trophozoites. Therefore, in suspected cases of amebic keratitis, both stains should be used. Hematoxylin and eosin (routine histology) can be used to detect and confirm cysticercosis. The adult worm of *Loa loa*, when removed from the eye, can be stained with a hematoxylin-based stain (Delafield's) or can be stained and examined via routine histology. *Toxoplasma* infection could be diagnosed using routine histology and/or serology results. Microsporidium confirmation to the genus or species level may require electron microscopy or molecular studies.
Intestinal tract	Stool, sigmoidoscopy material, duodenal contents	Trichrome or iron hematoxylin Intestinal protozoa Modified trichrome Microsporidia Modified acid fast *Cryptosporidium* spp. *Cyclospora cayetanensis* *Cystoisospora belli* Fecal immunoassays (enzyme immunoassay, FA, rapid cartridges) *E. histolytica*, *G. lamblia* *Cryptosporidium* spp. Microsporidia (experimental)	Although trichrome or iron hematoxylin stains can be used on almost all specimens from the intestinal tract, actual worm segments (tapeworm proglottids) can be stained with special stains. However, after routine dehydration through alcohols and xylenes (or xylene substitutes), the branched uterine structure is visible, thus allowing identification of the proglottid to the species level. Fecal immunoassay detection kits are also available for the identification of *Giardia lamblia*, *Entamoeba histolytica*, *Entamoeba dispar*, and *Cryptosporidium* spp. Microsporidium confirmation to the genus or species level may require electron microscopy studies.
	Anal impression smear	No stain, cellulose tape	Four to six consecutive negative tapes are required to rule out infection.

Table continues on next page

Table 5.1 Body sites, specimens, and recommended stains *(continued)*			
Body site or sample type	**Specimen**	**Recommended stain(s) for suspected organism**	**Comments**
Intestinal tract *(continued)*	Adult worm or worm segments	Carmine stains (rarely used)	Proglottids can usually be identified to the species level without using tissue stains.
	Biopsy specimen	Hematoxylin and eosin (routine histology) *Entamoeba histolytica* *Cryptosporidium* spp. *Cyclospora cayetanensis* *Cystoisospora belli* *Giardia lamblia* Microsporidia	Special stains may be helpful in the identification of microsporidia: tissue Gram stains, silver stains, PAS, and Giemsa.
Liver, spleen	Aspirates, biopsy specimen	Blood stains Leishmaniae Hematoxylin and eosin (routine histology) Leishmaniae, microsporidia, trematodes (adults, eggs)	Aspirates and/or touch preparations from biopsy material can be routinely stained with blood stains. This allows identification of the leishmania. There are definite risks associated with spleen aspirates and/or biopsy specimens. Other parasites, such as larval cestodes, trematodes, amebae, or microsporidia, can be seen and identified from routine histologic staining.
Lungs	Sputum, induced sputum, bronchoalveolar lavage fluid, transbronchial aspirate, tracheobronchial aspirate, brush biopsy specimen, open-lung biopsy specimen	Silver methenamine stain, calcofluor white (cysts only) *Pneumocystis jirovecii* Giemsa (trophozoites only) *Pneumocystis jirovecii* Modified acid-fast stains *Cryptosporidium* spp. Hematoxylin and eosin (routine histology) *Strongyloides stercoralis* *Paragonimus* spp. Amebae Silver methenamine stain PAS, acid-fast stains, tissue Gram stains, modified trichrome, electron microscopy Microsporidia	*Pneumocystis jirovecii* is the most common parasite (now classified with the fungi) recovered and identified from the lungs by using silver or Giemsa stains or monoclonal reagents (FA). Monoclonal reagents (FA) are also available for the diagnosis of pulmonary cryptosporidiosis. Routine histology procedures would allow the identification of any helminths or helminth eggs present in the lungs.
Muscle	Biopsy specimen	Hematoxylin and eosin (routine histology) *Trichinella* spp. Cysticerci *Sarcocystis* spp. Silver methenamine stain PAS, acid-fast stains, tissue Gram stains, electron microscopy Microsporidia	If *Trypanosoma cruzi* is present in the striated muscle, it can be identified from routine histology preparations. Microsporidium confirmation to the genus or species level may require electron microscopy studies.

Table 5.1 *(continued)*			
Body site or sample type	**Specimen**	**Recommended stain(s) for suspected organism**	**Comments**
Nasopharynx, sinus cavities	Nasal swabs, cytospin deposits from nasopharyngeal aspirates	Modified trichrome, acid-fast stain, Giemsa, optical brightening agent (calcofluor white), methenamine silver, electron microscopy Microsporidia Giemsa, trichrome *Acanthamoeba* sp. *Naegleria* sp.	Microsporidium identification to the genus or species level may require electron microscopy studies. Some free-living amebae (most commonly *Acanthamoeba* or *Naegleria*) have been found to colonize the nasopharynx and/or sinus cavities. Although calcofluor white stains the cyst walls, it does not stain the trophozoites. Therefore, in suspected cases, both calcofluor white and one of the routine stains should be used.
Skin	Aspirates, skin snip, scrapings, biopsy specimens	See "Cutaneous Ulcer" above Hematoxylin and eosin (routine histology) *Onchocerca volvulus* *Dipetalonema streptocerca* *Acanthamoeba* spp. *Entamoeba histolytica*	Any of the potential parasites present could be identified using routine histology procedures and stains.
Urogenital system	Vaginal discharge, urethral discharge, prostatic secretions, urine	Giemsa stain, immunoassay reagents (FA) *Trichomonas vaginalis* Delafield's hematoxylin Microfilariae Modified trichrome Microsporidia	Although *T. vaginalis* is probably the most common parasite identified, there are others to consider, the most recently implicated organisms being in the microsporidian group. Microfilariae could also be recovered and stained.
	Biopsy specimen	Hematoxylin and eosin (routine histology) *Schistosoma haematobium* Microfilariae PAS, acid-fast stain, tissue Gram stains, electron microscopy Microsporidia	

Table 5.2 Approaches to stool parasitology: test ordering		
Patient and/or situation	**Test ordered[a]**	**Follow-up test ordered**
Patient with diarrhea and AIDS or other cause of immune deficiency OR Patient with diarrhea involved in a potential waterborne outbreak (municipal water supply)	*Cryptosporidium* or *Giardia/Cryptosporidium* immunoassay	If immunoassays are negative and symptoms continue, special stains for microsporidia (modified trichrome stain) and the coccidia (modified acid-fast stain) and O&P exam should be performed.
Patient with diarrhea (nursery school, day care center, camper, backpacker) OR Patient with diarrhea involved in a potential waterborne outbreak (resort setting) OR Patient from areas within the United States where *Giardia* is the most common parasite found	*Giardia* or *Giardia/Cryptosporidium* immunoassay	If immunoassays are negative and symptoms continue, special stains for microsporidia and the coccidia (see above) and O&P exam should be performed.
Patient with diarrhea and relevant travel history outside the United States OR Patient with diarrhea who is a past or present resident of a developing country OR Patient in an area of the United States where parasites other than *Giardia* are found (large metropolitan centers such as New York, Los Angeles, Washington, DC, Miami)	O&P exam, *Entamoeba histolytica/E. dispar* immunoassay; immunoassay for confirmation of *E. histolytica*; various tests for *Strongyloides* may be relevant (even in the absence of eosinophilia), particularly if there is any history of pneumonia (migrating larvae in lungs), sepsis, or meningitis (fecal bacteria carried by migrating larvae); the agar culture plate is the most sensitive diagnostic approach for *Strongyloides*.	The O&P exam is designed to detect and identify a broad range of parasites (amebae, flagellates, ciliates, *Cystoisospora belli*, and helminths). If exams are negative and symptoms continue, special tests for coccidia (fecal immunoassays, modified acid-fast stains, autofluorescence) and microsporidia (modified trichrome stains, calcofluor white stains) should be performed; fluorescent stains are also options.
Patient with unexplained eosinophilia	Although the O&P exam is recommended, the agar plate culture for *Strongyloides stercoralis* (it is more sensitive than the O&P exam) is also recommended, particularly if there is any history of pneumonia (migrating larvae in lungs), sepsis, or meningitis (fecal bacteria carried by migrating larvae).	If tests are negative and symptoms continue, additional O&P exams and special tests for microsporidia (modified trichrome stains, calcofluor white stains, fluorescent stains) and other coccidia (modified acid-fast stains, autofluorescence, fluorescent stains) should be performed.
Patient with diarrhea (suspected foodborne outbreak)	Test for *Cyclospora ayetanensis* (modified acid-fast stain, autofluorescence, fluorescent stains).	If tests are negative and symptoms continue, special procedures for microsporidia and other coccidia and O&P exam should be performed.

[a] Depending on the particular immunoassay kit used, various single or multiple organisms may be included. Selection of a particular kit depends on many variables: clinical relevance, cost, ease of performance, training, personnel availability, number of test orders, training of physician clients, sensitivity, specificity, equipment, time to result, etc. Very few laboratories handle this type of testing exactly the same way. Many options are clinically relevant and acceptable for good patient care. It is critical that the laboratory report indicate specifically which organisms could be identified by the kit; a negative report should list the organisms relevant to that particular kit. **It is important to remember that sensitivity and specificity data for all of these fecal immunoassay kits (FA, enzyme immunoassay, and cartridge formats) are comparable.**

Table 5.3 Laboratory test reports: notes and optional comments[a,b]

***Entamoeba histolytica/E. dispar* group or complex**

Unless you see trophozoites containing ingested RBCs (these are from the true pathogen, *E. histolytica*), you cannot tell from the organism morphology whether you have actual pathogenic *E. histolytica* organisms or nonpathogenic *E. dispar* present.

Report as *Entamoeba histolytica/E. dispar* group or complex.

Additional computer comments:

A. Unable to determine pathogenicity from organism morphology
B. Depending on patient's clinical condition, treatment may be appropriate.

If you have the kit reagents to differentiate the two organisms, comments could also be added:

A. If you wish to determine which of the two organisms is present, please submit a fresh stool specimen.
B. To determine the presence or absence of pathogenic *E. histolytica*, please submit a fresh stool specimen.

Identification of nonpathogens

Comments that can be used for reporting nonpathogens include the following. However, these statements assume that a complete stool exam was performed on multiple stool samples; you may detect nonpathogens in the first examination or an incomplete examination but miss a pathogen (example: *Dientamoeba fragilis* requires the permanent stained smear for identification).

Additional computer comments:

A. Considered nonpathogenic; treatment not recommended
B. Nonpathogen; indication that patient has ingested something contaminated with fecal material (same method for acquiring pathogens).

Reporting *Blastocystis* spp.

Several comments are optional for reporting *Blastocystis* spp.

A. Clinical significance is confirmed; it appears that approximately half of the human subtypes are pathogenic.
B. Status as a pathogen is confirmed; it appears that approximately half of the human subtypes are pathogenic.

You may want to add a second comment:

Other organisms capable of causing diarrhea should also be ruled out.

Negative stool examination (O&P exam)[c]

An additional comment may be helpful:

A. Certain antibiotics such as metronidazole or tetracycline may interfere with the recovery of intestinal parasites, particularly the protozoa.
B. Antibiotics such as metronidazole or tetracycline may prevent recovery and identification of the intestinal protozoa.

Positive *Plasmodium* spp. report (ALWAYS a stat request and report)[d]

If *Plasmodium* spp. are detected but you are unable to identify to the species level, report as follows.

A. *Plasmodium* spp. seen; unable to rule out *Plasmodium falciparum* or *Plasmodium knowlesi*
B. (for a mixed infection) *Plasmodium* mixed infection possible; unable to rule out *Plasmodium falciparum* or *Plasmodium knowlesi*

Quantitate all positive malarial smears (percentage of infected RBCs within 100 total RBCs); quantitate from thin blood film. **Always add quantitation to the report (this applies to *Babesia* spp. as well).**

Always report presence or absence of gametocytes; this will influence therapy.

Send additional blood samples approximately every 6 h for 3 days (unless malaria is no longer a consideration).

[a] It is important to remember that educational information for your clients is critical to the success of your test reporting formats. The information in the table should be shared with your clients prior to changing your actual report formats. Your physician group may have a preference regarding additional comments. Information updates or newsletters are appropriate for this purpose.

[b] All of these comments are optional, and wording can be changed to fit your circumstances. However, it is recommended that you select specific comments and try not to use "free text" so that everyone reports test results the same way each time.

[c] This information could also be conveyed to your clients via the newsletter or update format.

[d] It is very important to add the comments about *Plasmodium falciparum* and *Plasmodium knowlesi*; this is critical information for patient care.

Table 5.4 Parasitemia determined from conventional light microscopy: clinical correlation

Parasitemia (%)	No. of parasites/µl	Clinical correlation
0.0001–0.0004	5–20	Number of organisms required for positive thick film (sensitivity) Examination of 100 thick-blood-film fields (0.25 µl) may miss up to 20% of infections (sensitivity). At least 300 fields should be examined before a negative result is reported (80–90%). Examination of 100 thin-blood-film fields (0.005 µl); at least 300 fields should be examined before a negative result is reported; **both** thick and thin blood films should be examined for every specimen submitted for a suspected malaria case (report final results with the 100× oil immersion objective). **One set (thick plus thin blood films) of negative blood films does not rule out a malaria infection.**
0.002	100	**Patients may be symptomatic below this level, particularly if they are immunologically naive** (with no prior exposure to malaria).
0.02	1,000	Level seen in travelers (immunologically naive); results may also be lower than this.
0.2	10,000	Level above which immune patients exhibit symptoms; **0.1% (5,000) is the BinaxNOW level of sensitivity for rapid malaria test (dipstick)[a].**
2	100,000	Maximum parasitemia of *P. vivax* and *P. ovale* (which infect young RBCs only)
2–5	100,000–250,000	Hyperparasitemia, severe malaria[b]; increased mortality
10	500,000	Exchange transfusion may be considered; high mortality

[a] It is not uncommon for immunologically naive travelers to have a false-negative BinaxNOW result; they will have an extremely low parasitemia. If the result is negative, stat thick and thin blood films are required. Often the thin blood film may appear to be negative; however, the thick film must also be examined and may reveal the presence of organisms.

[b] World Health Organization criteria for severe malaria are parasitemia of >10,000/µl and severe anemia (hemoglobin, <5 g/liter). The prognosis is poor if >20% of parasites are pigment-containing trophozoites and schizonts and/or if >5% of neutrophils contain visible pigment.

SECTION 6

Commonly Asked Questions about Diagnostic Parasitology

Practical Guide to Diagnostic Parasitology, Third Edition. Lynne S. Garcia
 DOI: 10.1128/9781683673637.ch06

As laboratory personnel become more widely cross-trained, the availability of people who have expertise in diagnostic parasitology will become more limited. Over the years, many questions have been asked about various aspects of diagnostic medical parasitology. Answers to these questions may be helpful for those working in this area of microbiology and may provide some "tips of the trade" that are learned through many years of bench experience.

Stool Parasitology

Specimen Collection

Intestinal Tract

1. **What is the gold standard for stool collection systems?**
The standard two-vial collection set has been used for many years and consists of one vial of 10% formalin from which the concentration is performed and one vial of fixative containing polyvinyl alcohol (PVA) (base fixative can be mercury or nonmercury to which has been added the plastic resin powder PVA) from which the permanent stained smear is prepared.

2. **What is the basis for the recommendation that three stools should be collected on alternate days, rather than three days in a row or three in one day?**
Intestinal protozoa are shed in the stool on a cyclic basis. It is generally accepted that this time frame is approximately 10 days, so if specimens are collected too close together, sampling may occur at a low point in the cycle and the organisms may be missed. By collecting specimens over a wider time frame, days with heavier shedding may be sampled as well.

3. **Is it acceptable to accept three stools collected on three consecutive days?**
Yes, but sampling during the total time frame for shedding (10 days) might be more likely to include days of heavy shedding.

4. **Is it ever appropriate to accept three stools obtained in one day from the same patient?**
With rare exceptions, this is unacceptable. If a patient has severe diarrhea or dysentery, there is a large dilution factor that may make finding any organisms more difficult. Therefore, after consultation with the physician, more than one stool sample could be examined within any one day.

5. **Is it good laboratory practice to accept two (rather than three) stools for the O&P examination?**
Although the examination of three stool specimens provides a more statistically accurate result, many laboratories feel that the percentage of organism recovery with the examination of two stool samples is acceptable. The recovery rate varies from ~80% to over 95% with the submission and examination of two stool specimens from a single patient. Therefore, two specimens are acceptable, but three are recommended. The examination of a single stool specimen is only about 50% accurate in the detection of parasites, especially intestinal protozoa. In one study (Hiatt et al.; see Suggested Reading), the authors found the following increases when examining the third stool:

- Yield increased 22.7%: *Entamoeba histolytica*
- Yield increased 11.3%: *Giardia lamblia*
- Yield increased 31.1%: *Dientamoeba fragilis*

6. **Do you have to collect the stool specimen any differently if you are looking for *Cryptosporidium* spp., *Cyclospora cayetanensis*, or *Cystoisospora belli*?**
   Fresh stool samples or specimens preserved in routine stool fixatives (formalin based or non-formalin based) can be used for diagnostic procedures for the identification of oocysts (special modified acid-fast stains for all three apicomplexan organisms [*Cryptosporidium*, *Cyclospora*, and *Cystoisospora*] or fecal immunoassays for *Cryptosporidium* spp.). Some of the single-vial collection options (universal fixatives; e.g., Total-Fix) are also recommended (no mercury, no formalin, no PVA).

7. **Do you have to collect the stool specimen any differently if you are looking for microsporidia?**
   Fresh stool or specimens preserved in routine stool fixatives (formalin based or non-formalin based) can be used for diagnostic procedures (modified trichrome stains) for the identification of the microsporidial spores. Some of the single-vial collection options (Total-Fix) are also recommended.

Fixatives

1. **What are the pros and cons of using stool fixatives?**
   Although it would be helpful to see motile organisms in the direct saline wet preparation, the trophozoite forms have often already begun to disintegrate by the time the stool specimen reaches the laboratory. To reduce the effect of the lag time between passage of the specimen and arrival in the laboratory, stool fixatives are recommended. **The benefits of fixation and preservation of organism morphology far outweigh the benefits of receiving a fresh, unpreserved stool sample (looking for motile organisms).**

2. **When a single-vial stool fixative is being selected, what questions should be asked?**
   The following are two such questions:
   - Can one perform a concentration, permanent stained smear, special stains for the apicomplexans and microsporidia, fecal immunoassay procedures, and/or molecular testing from the specimen received in that vial (universal fixative)?
   - What stains work best with that particular fixative?

   Another issue for consideration involves the types and numbers of parasites that might be missed using this approach (specific tests that require fresh or frozen specimens).

3. **What are the advantages and disadvantages of using 5% versus 10% formalin?**
   Although 5% formalin is thought to provide a more "gentle" fixation of protozoa, this concentration does not always kill all helminth eggs. The 10% formalin will kill most helminth eggs (exceptions: *Ascaris lumbricoides*, *Toxocara*, and *Baylisascaris*), but it may be more damaging to protozoa. In reality, most people probably cannot tell the difference by examining clinical specimens preserved in either 5 or 10% formalin, with the exception of helminth egg viability (containing motile larvae: *Ascaris* and *Toxocara*).

4. **What is the difference between buffered and nonbuffered formalin?**
Buffered formalin tends to produce fewer osmotic changes in the organisms during fixation; however, on a day-to-day basis, it may be difficult to detect morphologic differences arising from buffered versus unbuffered formalin when you are viewing direct wet mounts or concentrate sediment wet mounts microscopically.

5. **Should the laboratory be concerned about the amount of formalin used within the parasitology lab? *(See a complete discussion in "Concentrations" under "Diagnostic Methods," below.)***
The amount of formalin used in diagnostic parasitology testing is minimal. The regulations governing formalin use were developed for industry, where large amounts of formalin are sometimes used. However, the regulations do indicate that any place using formalin must be monitored. Once you have been monitored and your results have been found not to exceed the stated limits and your records are on file, there is no need to be rechecked (unless something dramatically changes in terms of your formalin volume use). **We have never heard of any microbiology laboratory coming even close to the limits, so formalin use within diagnostic parasitology laboratories is perfectly acceptable. However, many laboratories have elected to eliminate formalin use in parasitology testing (this may be due to formalin interference seen in some molecular testing and/or safety considerations).**

6. **What is PVA?**
PVA stands for polyvinyl alcohol, a plastic powder/resin that is incorporated into the fixative (Schaudinn's or other fixatives) and serves as an adhesive to "glue" the stool material to the slide. **PVA itself has no preservation capability and is inert in terms of fixation. The term "PVA fixative" is somewhat misleading; PVA is merely added to the fixative as an adhesive only.**

7. **What is albumin (stool adhesive)?**
Egg albumin has been used as a stool adhesive by coating the glass slide prior to smear preparation, particularly when sodium acetate-acetic acid-formalin (SAF), which does not contain PVA plastic resin (also an adhesive), is being used. However, with the development of universal fixatives such as Total-Fix, smears can be prepared without using PVA or albumin.

8. **What are some good stool fixative-stain combinations?**
The gold standard has been to use one vial of 10% formalin, from which the concentration is performed, and a second vial containing mercuric chloride-based PVA, from which the permanent stained smear is prepared and then stained (trichrome or iron hematoxylin). Other options in terms of overall quality are as follows:
    - Total-Fix and trichrome stain
    - SAF and iron hematoxylin stain
    - Unifix or Z-PVA (zinc-based fixative containing PVA) and trichrome stain
    - EcoFix and EcoStain
    - SAF and trichrome stain (not as good as SAF and iron hematoxylin)
    - Other options may be available; check the published literature.

9. **What are universal fixatives?**
Examples of universal fixatives are Total-Fix (no mercury, no PVA, no formalin); SAF (no mercury, no PVA; contains formalin); and EcoFix (no mercury, no formalin;

contains PVA). **Currently, Total-Fix is the only fixative that contains NO formalin, NO PVA, and NO mercury. Total-Fix can be used without the addition of the PVA to the fixative; drying smears for an adequate time prior to staining is the most important step (a minimum of 30 min in 37°C incubator is needed, and more time is required for thicker fecal smears). No "glue" (PVA or albumin) is required with Total-Fix.**

Specimen Processing

O&P Exam

1. **What procedures constitute the ova and parasite examination (O&P exam)?**
The direct wet smear, concentration, and permanent stained smear constitute the routine O&P exam on fresh stool specimens. **If the specimens are submitted to the laboratory in stool preservatives, the concentration and permanent stained smear should be performed.** If a laboratory indicates that they provide an O&P exam, the College of American Pathologists (CAP) checklist indicates that the O&P exam must include the concentration and permanent stained smear. Also, the CAP checklist requires that the direct wet smear be performed on **fresh** liquid or soft stool only (looking for motile trophozoites). There is no need to perform a direct wet mount on fresh formed stool; the likelihood of seeing motile trophozoites is low, since formed stools tend to contain only the cyst forms.

2. **Why is it necessary to perform a permanent stained smear examination on every stool specimen submitted for an O&P exam?**
Intestinal protozoa are not always seen and identified from the concentration examination. **Since the permanent stained smear is designed to facilitate identification and/or confirmation of the intestinal protozoa, it is important that this procedure be performed on all stools for which the O&P exam has been ordered.** Also, since trophozoite stages are not visible on the concentration examination (with rare exceptions; e.g., trophozoites can occasionally be seen in concentration sediments prepared from preserved specimens), it is especially important to examine the permanent stained smear. Even if organisms (trophozoites and/or cysts) are seen in the concentration wet mount, they might not be identified accurately and will require confirmation from examination of the permanent stained smear. Patients who are symptomatic with diarrhea are more likely to have protozoan trophozoites in the stool, not the more resistant cyst form. This approach is consistent with the O&P exam (as described in the CAP checklist for laboratory inspections).

3. **Why do you need to pour out some PVA onto paper towels prior to preparation of slides for permanent staining?**
The reason this step has been included in processing directions is that some people were taking the PVA/stool right out of the vial and onto the slide, and too much PVA was being carried over onto the slide. It takes quite a bit of time for PVA to dry; thus, the material often falls off during staining because the thick PVA is not yet dry. **Also, the amount of PVA (plastic powder) it takes to glue the stool onto the slide is extremely small.** So, if you eliminate the excess PVA prior to making the smears, your slides will require less drying time, the stool will adhere to the glass, and you will get a better stained smear. However, if the material (PVA-stool mixture) is taken right out

of the vial onto the glass slide and this approach is working (no excess PVA), there is no need to change protocols. Note: If Total-Fix is being used, no PVA is used; however, drying prior to staining is still very important.

4. **Can you use concentrated sediment to prepare slides for permanent staining?**
   The main thing to remember is that **a routine permanent stain (e.g., a trichrome stain) cannot be performed from the concentration sediment if the specimen was originally preserved in formalin or has been rinsed using formalin, saline, or water.** One can perform a trichrome stain on SAF-preserved material, but not if the specimen has been rinsed with formalin, saline, or water. If you centrifuge the feces-fixative material (without adding any rinse reagents), then provided that your fixative is compatible with the permanent stain you are using, you can use some of the sediment for smear preparation prior to staining. However, once you continue with the rinse steps, the final concentration sediment will often (again depending on the rinse fluids) not be compatible with trichrome staining. **If you want to use a single-vial fixative system (Total-Fix), then you can spin down the stool-fixative mixture (first spin using no additional rinse reagents), prepare your slide for permanent staining from the sediment, and then proceed with the regular rinses called for in the concentration procedure.**

Diagnostic Methods

Direct Wet Examinations

1. **What is the purpose of the direct wet examination?**
   This procedure is designed to allow the viewer to detect motile trophozoites (using saline); this procedure should not be performed on preserved specimens and should be reserved for fresh stool specimens that are very soft or liquid. Often, the organism identification will be presumptive; permanent stained smears will also need to be examined. If iodine is added to the direct wet mount, organism morphology may be more easily seen, but the organisms will be killed, and thus, no motility will be visible.

2. **How should the direct wet preparation be examined?**
   The entire coverslip preparation (22 by 22 mm) should be examined under low magnification (×100); approximately one-third to one-half of the coverslip preparation should be examined under high dry magnification (×400). **It is not practical to examine this preparation using oil immersion magnification (×1,000).** Saline and/or iodine mounts can be examined; however, iodine kills any organisms present, so trophozoite motility will no longer be visible.

3. **What do you expect to see during a wet preparation examination?**
   Helminth eggs and/or larvae can be seen, as well as some protozoan cysts, white blood cells (WBCs), some yeasts, and fecal debris. Many of the intestinal protozoa will need to be confirmed using the oil immersion magnification (×1,000) for the permanent stained smear.

Concentrations

1. **What are the recommended time and speed for centrifugation for the concentration method? Why is this important?**
   The current recommendation is for every centrifugation step in the concentration method (sedimentation) to be performed for **10 min at 500 × g.** If this recommendation

is not followed, small coccidian oocysts and microsporidial spores may not be recovered in the sediment. The number of organisms obtained after centrifugation is much greater than that obtained by taking the sample from the uncentrifuged specimen.

2. **What is the purpose of the concentration procedure?**
   The purpose of the concentration is to concentrate the parasites present, through either sedimentation or flotation. The concentration is specifically designed to allow recovery of protozoan cysts, apicomplexan/coccidian oocysts, microsporidian spores, and helminth eggs and larvae.

3. **Why is the flotation concentration used less frequently than the sedimentation concentration?**
   There are several reasons. First, **not all parasites float; therefore, you need to examine both the surface film and the sediment before indicating that the concentration examination is negative.** Second, the organisms must not be left in contact with the high-specific-gravity zinc sulfate for too long or the protozoa will tend to become distorted, so the timing of the examination is more critical. Also, the specific gravity of the fluid must be checked periodically.

4. **What specific gravity zinc sulfate should be used for the flotation concentration procedure?**
   If the concentration is being performed on fresh stool, the specific gravity of the solution should be 1.18. However, if the concentration is being performed on stool preserved in a formalin-based fixative, the specific gravity of the zinc sulfate should be 1.20.

5. **How should the concentration wet preparation be examined?**
   Formalin-ethyl acetate (substitute for ether) sedimentation concentration is the most commonly used concentration procedure. Zinc sulfate flotation will not detect operculated or heavy eggs (e.g., unfertilized *Ascaris* eggs and operculated cestode and/or trematode eggs); both the surface film and sediment will need to be examined before a negative result is reported if the flotation method is used. Smears prepared from concentrated stool are normally examined as for the direct wet mount using the low-power objective (10×) and the high dry power objective (40×); use of the oil immersion objective (100×) is not recommended (organism morphology is not sufficiently clear). The addition of too much iodine may obscure helminth eggs (i.e., they will mimic debris).

6. **What semiautomated methods are available to read the concentration sediments?**
   The traditional sampling approach using pipettes and the preparation of wet smears using glass slides and coverslips can be replaced with a semiautomated sampling and viewing system like the ParaSys parasitology instrument (Apacor Limited, Wokingham, United Kingdom). The specimen is drawn through tubing from the mixed concentrated stool sediment into the ParaSlide chamber. Organism morphology can be seen within the viewing chambers at a magnification of ×10. Selection of such a system often depends on the laboratory workload.

7. **Are there any tips for specimen processing for detection of the microsporidia?**
   The earlier studies of microsporidia were performed mainly at the Centers for Disease Control and Prevention (CDC). When early comparisons of methods were performed, the authors compared slow centrifugation with using uncentrifuged material. Of these methods, they felt that using uncentrifuged material was better for recovery

of microsporidial spores. **However, at the University of California Los Angeles, when we looked carefully at this approach compared with centrifugation at 500 × *g* for 10 min (the standard centrifugation speed and time), we found considerably more spores in the sediment after centrifugation.** If the stool contains a lot of mucus or is runny, formalin should be added before centrifugation. It is not necessary to use ethyl acetate, since it may pull much of the material you want to examine up into the mucus layer. If the stool is not particularly watery or does not contain a lot of mucus, it can be treated just like a regular concentration (but every centrifugation should be at 500 × *g* for 10 min). Also, the more a stool specimen is manipulated, the more likely it is that some organisms will be lost (this applies to all parasites in stool). Therefore, you may want to eliminate the wash steps and work with the sediment you obtain from the first centrifugation step. The smears should be fairly thin to help visualize the spores, but also remember not to decolorize too much. Filtration is acceptable using any of the commercial concentration systems. If you are using gauze, make sure you use woven gauze and use only two layers (do not use pressed gauze, which is too thick).

8. How safe is the use of formalin within the microbiology laboratory?

The formalin regulations were originally developed for industry (plywood, fabrication, etc.), where large amounts of formalin are used in the manufacturing process. The amount of formalin we are exposed to in the laboratory is minimal; I have never heard of any microbiology laboratory (including a full-service parasitology service) even coming close to the limits.

The Occupational Safety and Health Administration (OSHA) amended the original regulations for occupational exposure to formaldehyde in May 1992 (*Fed Regist* 57:22290–22328, 1992; see Suggested Reading). The final amendments lower the permissible exposure level for formaldehyde from an 8-hour time-weighted average (TWA) of 1 ppm (part per million) to a TWA of 0.75 ppm. The amendments also add medical removal protection provisions to supplement the existing medical surveillance requirements for employees suffering significant eye, nose, or throat irritation and for those suffering from dermal irritation or sensitization from occupational exposure to formaldehyde. In addition, changes have been made to the standard's hazard communication and employee training requirements. These amendments establish specific hazard labeling for all forms of formaldehyde, including mixtures and solutions composed of 0.1% or more formaldehyde in excess of 0.1 ppm. Additional hazard labeling, including a warning that formaldehyde presents a potential cancer hazard, is required where formaldehyde levels, under reasonably foreseeable conditions of use, may potentially exceed 0.5 ppm. The final amendments also provide for annual training of all employees exposed to formaldehyde at levels of 0.1 ppm or higher.

Laboratories that have been monitored have not come close to either measurement. Once a laboratory has been measured and the results (below thresholds for regulatory requirements) are on file, this information does not have to be generated again. No badges are required. Even without a fume hood (many labs do not use a fume hood), performing the routine formalin-ethyl acetate concentration does not seem to be a problem. A number of people who have indicated that they want to remove formalin from the laboratory probably do not really understand the history of the regulation or the actual issues. The only possible problem seen in the clinical and pathology laboratory setting might be in a routine anatomical pathology laboratory where very large amounts of formalin are used, and where use is sloppy. However, within the microbiology laboratory (even large laboratories), it does not seem to be a problem. Appendix B to OSHA standard 1910.1048 (see United States Department of Labor, Suggested Reading) is titled "Sampling Strategy and Analytical Methods for Formaldehyde." **Ultimately, the removal of formalin from the laboratory is an individual laboratory decision** (keeping in mind the exposure limits indicated below).

> TWA: The employer shall assure that no employee is exposed to an airborne concentration of formaldehyde which exceeds 0.75 parts formaldehyde per million parts of air (0.75 ppm) as an 8-hour TWA (time weighted average).
>
> STEL: The employer shall assure that no employee is exposed to an airborne concentration of formaldehyde which exceeds two parts formaldehyde per million parts of air (2 ppm) as a 15-minute STEL (short-term exposure limit) (https://www.osha.gov/pls/oshaweb/owadisp.show_document?p_id=10075&p_table=STANDARDS#:~:text=Short%20Term%20Exposure%20Limit%20(STEL,as%20a%2015%2Dminute%20STEL).

Formalin often interferes with molecular testing. Thus, as multiplexed organism panels continue to become available, laboratories are eliminating formalin for testing compatibility with the newer methods.

Permanent Stains

1. What is the purpose of the permanent stained smear?

The purpose of the permanent stained smear is to provide contrasting colors for the background debris and parasites present; it is designed to allow examination and recognition of detailed organism morphology during oil immersion examination (100× objective, for a total magnification of ×1,000). This examination is designed primarily to allow recovery and identification of the intestinal protozoa.

2. How long should the permanent stained smear be examined?

Rather than a specific number of minutes, the recommendation is to examine at least 300 oil immersion (×1,000 total magnification) fields; additional fields may be required if suspect organisms have been seen in the wet preparations from the concentrated specimen.

3. What recent changes have influenced the overall quality of the permanent stained smear?

The use of mercury substitutes in PVA generally leads to diminished quality of the overall morphology of the intestinal protozoa. However, some of the mercury substitutes provide a morphologic quality close to that of mercury, allowing identification of most of the intestinal protozoa. Differences in detection and identification are usually minor unless very few organisms are present or the organisms are quite small. In these circumstances, some organisms may be missed using mercury substitutes. **The recently developed Total-Fix (no mercury, no formalin, and no PVA) provides a fixative that results in excellent overall protozoan morphology.**

4. What is the purpose of the iodine dish in the Wheatley's trichrome stain protocol?

Mercury is removed from the smear when it is placed in the iodine dish; there is a chemical substitution of iodine for mercury. The iodine is removed during the next two alcohol rinses. Therefore, at the point at which the slide is placed in trichrome stain, neither mercury nor iodine is left on the smear.

5. **Why don't you need to use the iodine dish when staining fecal smears prepared from specimens preserved in the newer single-vial non-mercury-based systems (zinc sulfate-based PVA)?**
The zinc sulfate-based PVA is water soluble, and so the dry smears can be placed directly into the trichrome dish without having to go through the iodine and subsequent rinse steps. The zinc sulfate is removed by the water in the trichrome stain.

6. **Why might you have to use the iodine dish and subsequent rinses in your staining setup when staining slides from the proficiency testing agencies (American Association of Bioanalysts, various states, etc.)?**
Some smears used for proficiency testing are prepared from fecal specimens that have been preserved in mercury-based fixatives, so the iodine dish and subsequent rinse steps are required to remove mercury and iodine prior to staining with either trichrome or iron hematoxylin stains. For several years, CAP proficiency testing specimens have been preserved in nonmercury fixatives; therefore, the iodine dish is not required. If you're not sure if the smears came from mercury-based or non-mercury-based fixatives, the iodine dish and subsequent rinse steps can be retained in the staining procedure.

7. **What role does the acetic acid play in the trichrome stain?**
Both the trichrome and iron hematoxylin stains are considered regressive stains; the fecal smears are overstained and then destained. The acetic acid in the 90% alcohol rinse step of the trichrome staining procedure removes some of the stain and provides better contrast. However, in some cases, differences in the quality of staining between stained protozoa that have been subjected to the 90% alcohol rinse with and without the acetic acid may be difficult to detect. It is important not to destain the smears too much; overall organism morphology will be diminished.

8. **What causes the xylene (or xylene substitute) dehydration solutions to turn cloudy when a slide from the previous alcohol dish is moved forward into the xylene dish?**
If there is too much water carryover from the last alcohol dish, the xylene solution may turn cloudy. When this occurs, replace the 100% alcohol dishes, back up the slide into 70% alcohol (you can also use a series of steps, 95% and then 70%), allow it to stand for 15 min, and then move the slide forward through the replacement 100% alcohol and xylene dishes.

9. **Why is absolute ethanol (100%) recommended as the best approach?**
Although many laboratories use the commercially available 95/5% denatured alcohol mix as their "absolute alcohol," the dehydration of stained fecal smears is not as good as that obtained with 100% ethanol. You may want to add an additional dish of the denatured alcohol (absolute alcohol) to your staining setup to obtain better dehydration.

10. **What is the difference between xylene and xylene substitutes?**
There are several differences. Most laboratories have made the decision to eliminate xylene from their laboratories as a safety measure. However, xylene substitutes generally do not dehydrate as well. Also, after the slides are removed from the last dish of xylene substitute, they take longer to dry. You may want to add an additional dish of the xylene substitute to your staining setup to obtain better overall dehydration.

11. **How do stained fecal smears differ when the fixative DOES and DOES NOT contain PVA?**
Although PVA is inert, the material does stain; there will be some background "haze" associated with PVA on the permanent stain. When specimens in Total-Fix (no PVA

present) are stained, the background will be quite clean, and the organisms tend to stain a bit paler. It is important to remember not to destain too long (two quick dips in the 90% alcohol containing 1% acetic acid are sufficient) and to immediately move into the first dish of absolute alcohol. Use this dish as a wash and begin dehydration in the second absolute alcohol dish; then move smears into the xylene substitute dishes.

Stool Immunoassay Options

1. **What immunoassay options are available for stool protozoa?**
Currently, immunoassays are available for *Giardia lamblia*, *Cryptosporidium* spp., the *Entamoeba histolytica/E. dispar* group, and the true pathogen *E. histolytica*. Reagents for the detection of *Dientamoeba fragilis* and the microsporidia are under development. **All of the fecal immunoassay test formats are comparable in terms of sensitivity and specificity. Selection of a particular method depends on work flow, the number of test requests, and the preference of the individual laboratory.** Outside the United States (non-FDA cleared), there are also reagents available for some of the microsporidia, as well as other organisms.

2. **What methods are available commercially?**
Direct fluorescence assay (DFA), enzyme immunoassay (EIA), and cartridge formats (membrane flow or a solid-phase qualitative immunochromatographic procedure) are currently available.

3. **Why might someone want to use a fecal immunoassay option?**
If the most common organisms found in the area are *Giardia*, *Cryptosporidium*, and/or the *E. histolytica/E. dispar* group, then fecal immunoassays are certainly options. Specific patient histories and symptoms may suggest the use of fecal immunoassays. **Both the O&P exams and fecal immunoassays are recommended for a laboratory test menu; both would be orderable, billable procedures.** Refer to Table 5.2, which contains order recommendations based on the patient's history; sharing this table with your physician clients is highly recommended.

4. **How would stool immunoassay requests fit into a laboratory that also performs O&P exams?**
Any diagnostic laboratory performing routine parasitology testing should offer both routine O&P exams and fecal immunoassays. **Both tests are recommended for a laboratory test menu; both would be orderable, billable procedures** (see Table 5.2).

5. **What do you mean by the "routine O&P exam"?**
The routine O&P exam includes (for liquid or semiliquid fresh stools) a direct wet smear, concentration, and the permanent stained smear; preserved specimens would require a concentration and permanent stained smear (no direct wet mount is required).

6. **How would you fit the fecal immunoassays into your laboratory?**
You could offer the routine O&P exam, and you could also offer on request the stool immunoassay as a separate option. Both options should be in the laboratory ordering test menu.

7. **What type of educational initiatives would have to be undertaken prior to offering these options?**
Physicians would need to know the pros and cons of ordering either the immunoassay or the routine O&P examination (see Table 5.2).

8. What are some of the pros and cons of the fecal immunoassays?

Refer to Section 5, "Comments on the Performance of Fecal Immunoassays" under "Fecal Immunoassays for Intestinal Pathogens."

Some pros are as follows.

- Depending on the format selected, the immunoassays are fast and relatively simple to perform.
- The result can rule in or out some very specific organisms.
- If the patient becomes asymptomatic at the time the immunoassay is negative, additional testing may not be necessary.
- They may help reduce personnel costs (time to perform procedures).
- The fecal immunoassays are more sensitive than the routine O&P exams and/or the special stains for the coccidia or microsporidia.

Some cons are as follows.

- The fecal immunoassay kits test only for selected organisms.
- Depending on the format, they might be somewhat complex to perform.
- Test requests may not justify certain formats (cost, equipment, or training).
- It is critical that the physician realize that a negative immunoassay does not rule out all possible parasitic etiologic agents causing diarrhea.
- **In tests for *Giardia lamblia*, fecal immunoassays on two different stool specimens may be required to get a positive result. Consequently, it will require two negative stools to rule out giardiasis. This is not the case for other protozoa.**

9. Can fecal immunoassays be used for duodenal fluid (giardiasis)?

Fecal immunoassays have not been cleared or validated for this type of specimen. Also, the duodenal fluid or aspirate would contain primarily the trophozoite form, not the cyst stage for which the reagents have been designed. As an example, if one is using the fluorescent assay (FA) test (rather than EIA or cartridge), the trophozoites may appear to be fluorescent at a very pale 1+ while the cysts are a strong 3+ to 4+. So, while a few antigenic sites may be shared by the cysts and trophozoites, the commercial tests for *Giardia* detect the cyst antigens. You can try using the reagents on these specimens, but if the result is negative, the result in no way has ruled giardiasis out. I recommend testing stool only; if the result is positive, this approach might avoid the need for duodenal aspirate testing altogether.

10. In the *Giardia* fecal immunoassay, how many specimens should be tested before it can be assumed that the patient is not infected? *(Also see question 8 above on the pros and cons of fecal immunoassays).*

Since the evidence indicates that *Giardia* is shed sporadically and that more than one immunoassay might be required to diagnose the infection, the recommendation is similar to that seen for stool collection for the routine O&P examination: **for *Giardia*, perform the immunoassay on two different stools (assuming the first specimen is negative)**, collected within no more than a 10-day period; a good collection schedule would be day 1, then day 3 or 4. That way, one would assume that one of those collections would yield a positive result if the organisms (in sufficient numbers) are present. However, you also have to remember that **if the patient is a carrier with a low organism load, even the second immunoassay might be negative. Note: One stool is sufficient for immunoassay testing for *Cryptosporidium* spp.**

11. Although an FA test for *Cryptosporidium/Giardia* was negative, the O&P concentrate showed *Giardia* trophozoites on the wet mount. Might we miss a *Giardia*-positive sample if only the FA test is ordered? Also, do any of the antigen EIAs detect both trophozoites and cysts, and should we switch to one of those?

Although the antibody in the immunoassay kits is to the cyst antigen (primarily, but it is a polyclonal reagent), in some of the kits the trophozoites do fluoresce, but much less intensely (around a 1+ or even a 2+, thus indicating some shared epitopes). In almost all patients, there is a combination of cysts and trophozoites, unless the patient has active diarrhea and is passing only trophozoites (there is no time for cysts to form with rapid passage through the gastrointestinal [GI] tract). However, most patients harbor cysts as well, and the results (if they are above the test limits for sensitivity) will be positive. The situation described in the question can happen, but it is probably not that common. However, with only trophozoites present, you may get a negative result. Some of the kits tend to provide a bit higher fluorescence with the trophozoite, but it varies.

12. After the patient has been treated, how long does the *Giardia* antigen test remain positive?

It has been recommended that patients be tested about 7 days after therapy, in hopes to avoid picking up residual antigen. However, if you wait too long (several weeks), you always run the risk of picking up antigen from a possible reinfection. Some also feel that low antigen levels can be found for up to about 2 weeks. A good time frame for retesting would probably be about 7 to 10 days after therapy. If the first specimen at 7 days is still positive, then retesting at 10 to 12 days would be appropriate. Also, we know that the immunoassays (for diagnosis) may not pick up low antigen loads (organism shedding issues); therefore, the recommendation for diagnostic immunoassay testing for giardiasis has been changed and now involves performing immunoassay testing on one additional stool specimen (if the first one is negative). The testing on two different stool specimens should be performed within about 3 to 5 days.

13. Why do fecal immunoassay kits that test for either *Entamoeba histolytica* or the *E. histolytica/E. dispar* complex/group require fresh or frozen stools?

Unfortunately, at present these reagents do not function properly on preserved fecal specimens. Although the manufacturers are trying to develop such kits, they are not yet available.

14. How long does antigen survive in fresh stools?

It is recommended that fresh stools be tested within 24 h of collection; they can be stored overnight in the refrigerator. They can also be frozen or preserved in 10% formalin before being tested (both freezing and formalin preservation methods preserve antigen for long periods, even years).

MOLECULAR TEST PANELS (FDA CLEARED)

The reagents used in molecular test panels contain ingredients that could be irritating or caustic if they come in contact with skin, eyes, or mucous membranes. Wear gloves, safety glasses, and laboratory coat, and use standard laboratory precautions when handling reagents. If a reagent is swallowed, call a physician. In case of skin or eye contact, flush with copious amounts of water. Pathogenic microorganisms, including hepatitis viruses and human immunodeficiency virus, may be present in clinical specimens. Proper handling and disposal methods should be established. Wipe up spillage of patient specimens immediately and disinfect with an appropriate disinfectant. Treat the cleaning materials as biohazardous waste.

The test options listed below are examples of FDA-cleared molecular diagnostic tests for parasites. In the future, we expect to see more test options, since many test parasite panels are currently under development.

A. APTIMA *Trichomonas vaginalis* Assay

Although there are many molecular methods published in the literature for parasites, most of these tests were developed in-house and have not been cleared by the FDA. However, there are several FDA-cleared molecular diagnostic tests for parasites, including the APTIMA *Trichomonas vaginalis* assay (GenProbe, San Diego, CA). The APTIMA *Trichomonas vaginalis* assay is a nucleic acid test that utilizes target capture, transcription-mediated amplification (TMA), and hybridization protection assay (HPA) technologies that have previously been used for detection of *Neisseria gonorrhoeae* and *Chlamydia trachomatis*. It also accepts the same specimen types (i.e., clinician-collected vaginal and endocervical swabs, female urine, and PreservCyt solution). However, urine, vaginal swab, and PreservCyt solution liquid Pap specimen sampling is not designed to replace cervical exams and endocervical specimens for diagnosis of female urogenital infections. Patients may have cervicitis, urethritis, urinary tract infections, or vaginal infections due to other causes or concurrent infections with other agents. Reliable results depend on adequate specimen collection. Because the transport system used for this assay does not permit microscopic assessment of specimen adequacy, training of clinicians in proper specimen collection techniques is necessary.

The APTIMA provides high sensitivity and specificity and does not require organism viability or motility. There are approximately 7 million new cases of *T. vaginalis* infection in the United States per year. This figure probably represents an underestimate of the disease burden, considering that there are no recommended screening guidelines and the most commonly used testing methods, such as wet mounts and culture, are insensitive. Clinical infection is associated with cervical neoplasia, cervical carcinoma, tubal infertility, posthysterectomy infection, pelvic inflammatory disease, preterm rupture of membranes, preterm birth, low-birth-weight infants, neonatal infection, and increased transmission of HIV. It has been estimated that 10 to 50% of *T. vaginalis* infections in women are asymptomatic, and in men, the proportion may be even higher. It is important to remember that therapeutic failure or success cannot be determined with the APTIMA *Trichomonas vaginalis* assay, since nucleic acid may persist following appropriate antimicrobial therapy.

KEY POINTS

1. A negative result does not exclude a possible infection, since results depend on adequate specimen collection. Test results may be affected by improper specimen collection, technical error, specimen mix-up, or target levels below the assay limit of detection.

2. A negative result does not exclude a possible infection, because the presence of *Trichomonas tenax* or *Pentatrichomonas hominis* in a specimen may affect the ability to detect *T. vaginalis* rRNA. However, these organisms are relatively rare and may not present a problem. Also, performance of the APTIMA *T. vaginalis* assay has not been evaluated in the presence of *Dientamoeba fragilis*.

3. The APTIMA *T. vaginalis* assay provides qualitative results. Therefore, a correlation cannot be drawn between the magnitude of a positive assay signal and the number of organisms in a specimen.

4. If a specimen contains few *T. vaginalis* organisms, uneven distribution of the trichomonads may occur, which may affect the ability to detect *T. vaginalis* rRNA in the collected material. If negative results from the specimen do not fit with the clinical impression, a new specimen may be necessary.

5. This test has not been FDA cleared for male clinical specimens.

B. Affirm VPIII Microbial Identification Test

The Affirm VPIII DNA probe technology (Becton, Dickinson and Co., Franklin Lakes, NJ) provides a dependable, rapid means for the differential detection and identification of three organisms causing vaginitis: *Candida* species, *Gardnerella vaginalis*, and *Trichomonas vaginalis*. The test features an easy-to-read color reaction that is more accurate than current microscopic methods such as wet mounts and culture for the detection of the causative agents of vaginitis. Also, the test performance is not affected by "difficult" specimens (collected during menstruation or after intercourse), self-medication, or the presence of mixed infections.

The test uses two distinct single-stranded nucleic acid probes for each organism, a capture probe and a color development probe, that are complementary to unique genetic sequences of the target organisms. The capture probes are immobilized on a bead embedded in a probe analysis card (PAC), which contains a separate bead for each target organism. The color development probes are contained in a multiwell reagent cassette (RC). During the procedure, the assay aligns complementary nucleic acid strands to form specific, double-stranded complexes called hybrids. For each of the three organisms, this hybrid is composed of capture and color development single-stranded nucleic acid probes, complementary to the released target nucleic acid analyte. Enzyme conjugate then binds to the captured analyte, thus providing the test results. A positive result is a visible blue color on the organism-specific bead embedded in the PAC. Each PAC contains three test beads, one for each organism, plus positive- and negative-control beads to ensure test quality. The total time to results is under 45 minutes, with less than 2 minutes of hands-on time needed. The simple, automated procedure can be performed with minimal training.

KEY POINTS

1. With the Affirm VPIII ambient-temperature transport system, the total time between sample collection and proceeding with sample preparation should be no longer than 72 h when the specimen is stored under ambient conditions (15 to 30°C). The system has also been qualified for transport use at refrigerated conditions (2 to 8°C).

2. The Affirm VPIII microbial identification test includes two internal controls on each PAC: a positive-control bead and a negative-control bead. These control beads are tested simultaneously with each patient specimen, ensuring the proper performance of PAC, RC, and processor. The positive control also ensures the absence of specimen interference. The negative control also ensures the absence of nonspecific binding from the specimen.

3. The assay is intended to be used with the Affirm VPIII ambient temperature transport system, the Affirm VPIII sample collection set, or the swabs provided in the Affirm VPIII microbial identification kit. Other methods of collection have not been evaluated. Optimal test results require appropriate specimen collection. Test results may be affected by improper specimen collection, handling, and/or storage conditions. A negative test result does not exclude the possibility of vaginitis or vaginosis.

4. Vaginitis/vaginosis is most frequently caused by *G. vaginalis*, *Candida* species, and *T. vaginalis*. Vaginitis symptoms may also be seen in toxic shock syndrome (caused by *Staphylococcus aureus*) or may be caused by nonspecific factors or by specific organisms. Mixed infections may occur. Therefore, a test indicating the presence of *Candida* species, *G. vaginalis*, and/or *T. vaginalis* does not rule out the presence of other organisms. Bacterial vaginosis is not caused by a single organism but is thought to reflect a shift in the vaginal flora.

5. The performance of this test on patient specimens collected during or immediately after antimicrobial therapy is unknown. The presence or absence of *Candida* species, *G. vaginalis*, or *T. vaginalis* cannot be used as a test for therapeutic success or failure.

C. Cepheid Xpert TV Assay for *Trichomonas vaginalis* from Men and Women

The Cepheid Xpert TV assay is a nucleic acid-based test using real-time PCR (Cepheid, Sunnyvale, CA). *T. vaginalis* is detected through the use of real-time PCR to amplify and detect *T. vaginalis* DNA. Genomic *T. vaginalis* DNA is amplified and detected using a sequence-specific probe that is cleaved during PCR amplification, resulting in a signal that occurs when the fluorescent reporter dye is released from the quencher.

Nucleic acid amplification tests (NAAT) have been FDA cleared in the United States for detection of *T. vaginalis* in specimens from both women and men. In specimens from women, the performance of the Xpert TV assay was compared to the patient infection status (PIS) derived from the results of InPouch TV broth culture and Aptima NAAT for *T. vaginalis*. The diagnostic sensitivities and specificities of the Xpert TV assay for combined female specimens (urine samples, self-collected vaginal swabs, and endocervical swabs) range from 99.5 to 100% and 99.4 to 99.9%, respectively. For male urine samples, the diagnostic sensitivity and specificity were 97.2% and 99.9%, respectively, compared to PIS results derived from the results of broth culture for *T. vaginalis* and bidirectional gene sequencing of amplicons. The CDC recommends NAAT as the preferred diagnostic modality due to their superior sensitivity over those of direct microscopy or culture-based methods for detecting *T. vaginalis*. Diagnostic testing is recommended for symptomatic women, and screening should be considered for those with multiple sex partners, persons who exchange sex for payment, use drugs, and/or have a history of sexually transmitted diseases (STDs), and women in high-prevalence settings, such as STD clinics and correctional facilities.

The Xpert TV Assay is performed on the Cepheid GeneXpert instrument systems (GeneXpert Dx, GeneXpert Infinity-48, GeneXpert Infinity-48s, and GeneXpert Infinity-80 systems). The GeneXpert instrument system consists of an instrument, personal computer, and preloaded software for running tests and viewing the results. The GeneXpert instrument system requires single-use, disposable cartridges (the Xpert TV cartridges) that hold the PCR reagents and host the PCR process. The cartridges are self-contained, so specimens never come into contact with the working parts of the instrument.

A sample processing control (SPC), sample adequacy control (SAC), and probe check control (PCC) are controls utilized by the GeneXpert instrument system. The SPC is present to control for adequate processing of the target trichomonads and to monitor the presence of inhibitors in the real-time PCR to reduce the possibility of false-negative results. The

SAC reagents detect the presence of a single-copy human gene and monitor whether the specimen contains human cells. The PCC verifies reagent rehydration, real-time PCR tube filling in the cartridge, probe integrity, and dye stability.

KEY POINTS

1. This is the first NAAT for male urine specimens.

2. Xpert TV is FDA cleared for symptomatic and asymptomatic patients.

3. Results are available in about 40 min.

4. This assay delivers high-sensitivity results for samples from both men and women.

D. BD MAX Enteric Parasite Panel

The BD MAX enteric parasite panel (Becton, Dickinson and Co., Sparks, MD; see Suggested Reading) is a multiplex qualitative *in vitro* diagnostic test that simultaneously and differentially detects DNA from ***Giardia lamblia*, *Cryptosporidium* (*C. hominis* and *C. parvum*), and *Entamoeba histolytica*** in unpreserved and preserved (10% formalin) stool specimens. These pathogens represent commonly isolated and highly pathogenic organisms. Results from this assay are available in approximately 4 hours, require minimal technologist time, and may replace traditional methods such as microscopy and immunoassay.

Positive results do not rule out coinfection with other organisms that are not detected by this test, and organisms detected by this test may not be the sole cause of patient illness. Negative results in the setting of clinical illness compatible with gastroenteritis and/or colitis may indicate that the symptoms are due to infection by pathogens that are not detected by this test or noninfectious causes, such as ulcerative colitis, irritable bowel syndrome, or Crohn's disease.

A soft diarrheal stool is collected, homogenized, and looped into a sample buffer tube (SBT). The SBT is closed with a septum cap and heated on the BD prewarming heater to lyse the parasite organisms. The SBT is then vortexed and transferred to the BD MAX system. The cells are lysed, nucleic acids are extracted on magnetic beads and concentrated, and then an aliquot of the eluted nucleic acids is added to PCR reagents which contain the target-specific primers used to amplify the genetic targets. The assay also includes an SPC, which is present in the extraction tube and undergoes the extraction, concentration, and amplification steps to monitor for inhibitory substances or instrument or reagent failure. The BD MAX system automates sample lysis, nucleic acid extraction and concentration, reagent rehydration, nucleic acid amplification, and detection of the target nucleic acid sequence using real-time PCR. Amplified targets are detected with hydrolysis probes labeled with quenched fluorophores. The amplification, detection, and interpretation of the signals are done automatically by the BD MAX system.

KEY POINTS

1. To avoid amplicon contamination, do not break apart the BD MAX PCR cartridges after use. The seals of the BD MAX PCR cartridges are designed to prevent contamination.

2. In cases where open-tube PCR tests are conducted in the laboratory, care must be taken to ensure that the BD MAX enteric parasite panel, any additional reagents required for testing, and the BD MAX system are not contaminated. Gloves must be changed before manipulation of reagents and cartridges.

3. BD MAX enteric parasite panel components are stable at 2 to 25°C through the stated expiration date. Do not use expired components.

4. A positive result with the BD MAX enteric parasite panel does not necessarily indicate the presence of viable organisms. It indicates the presence of DNA from *E. histolytica*, *G. lamblia*, *Cryptosporidium parvum*, or *Cryptosporidium hominis*, allowing identification of the BD MAX enteric parasite panel organisms.

5. As with all PCR-based *in vitro* diagnostic tests, extremely low levels of target below the analytical sensitivity of the assay may be detected, but results may not be reproducible. False negatives may also occur if the level of organism nucleic acid is below the threshold of detection.

6. False-negative results may occur due to loss of nucleic acid from inadequate collection, transport, or storage or due to inadequate organism lysis. The SPC has been added to the test to aid in the identification of specimens that contain inhibitors of PCR amplification. The SPC does not indicate if nucleic acid has been lost due to inadequate collection, transport, or storage or whether cells have been inadequately lysed.

7. BD MAX enteric parasite panel results may or may not be affected by concurrent antimicrobial therapy, which may reduce the amount of target present.

8. The performance of this test has not been established for monitoring treatment of *Giardia lamblia*, *Cryptosporidium hominis*, *C. parvum*, or *Entamoeba histolytica*. This test is a qualitative test and does not provide quantitative values or indicate the quantity of organisms present.

9. The performance of this test has not been evaluated for immunocompromised individuals or for patients without symptoms of gastrointestinal infection.

10. The effect of interfering substances has been evaluated only for those listed in the product information sheet (https://www.bd.com/resource.aspx?IDX=31265). Potential interference has not been evaluated for substances other than those described in the "Interference" section of the product information sheet.

11. Cross-reactivity with organisms other than those listed in the "Analytical Specificity" section of the package insert has not been evaluated.

E. BioFire FilmArray Gastrointestinal Panel

The FilmArray gastrointestinal panel (bioMérieux, Marcy l'Etoile, France; see Suggested Reading) is designed for the simultaneous detection of 22 different enteric pathogens directly from stool specimens: *Campylobacter* spp. (*C. jejuni*, *C. coli*, and *C. upsaliensis*), *Clostridioides* (formerly *Clostridium*) *difficile* (toxin A/B), *Plesiomonas shigelloides*, *Salmonella*, *Vibrio* spp. (*V. parahaemolyticus*, *V. vulnificus*, and *V. cholerae*), *Yersinia enterocolitica*, enteroaggregative *Escherichia coli*, enteropathogenic *E. coli*, enterotoxigenic *E. coli*, Shiga-like toxin-producing *E. coli* (stx_1 and stx_2), STEC O157, *Shigella*/enteroinvasive *E. coli*, ***Cryptosporidium* spp., *Cyclospora cayetanensis*, *Entamoeba histolytica* (may cross-react with nonpathogenic *Entamoeba dispar* when present at higher levels), *Giardia lamblia* (also called *G. intestinalis* or *G. duodenalis*)**, adenovirus group F serotypes 40 and 41, astrovirus, norovirus GI and GII, rotavirus A, and sapovirus (genogroups I, II, IV, and V). This test is a qualitative test and does not provide a quantitative value for the organism(s) in the sample.

The FilmArray GI pouch is a closed disposable system that houses all the chemistry required to isolate, amplify, and detect nucleic acid from multiple gastrointestinal pathogens in a single stool specimen. The rigid plastic component (fitment) of the FilmArray GI pouch contains reagents in freeze-dried form. The flexible plastic portion of the pouch is divided into discrete segments (blisters) where the required chemical processes are carried out. The user of the FilmArray GI panel loads the sample into the FilmArray GI pouch, places the pouch into the FilmArray instrument, and starts the run. All other operations are automated.

The sample volume requirement is 200 μl. The FilmArray GI panel test consists of automated nucleic acid extraction, reverse transcription, amplification, and analysis, with results available in 1 h per run (i.e., per specimen). Each FilmArray pouch contains an internal nucleic acid extraction control and a PCR control. FilmArray GI panel runs are considered valid if the run completes normally and internal controls pass. The FilmArray software performs automated result analysis with each target in a valid run reported as "detected" or "not detected." If either internal control fails, the software automatically provides a result of "invalid" for all panel analytes.

KEY POINTS

1. Stool samples may contain a high concentration of organisms. To avoid possible contamination, samples should be processed in a biosafety cabinet. If a biosafety cabinet is not used, a dead-air box, a splash shield, or a face shield should be used when samples are being prepared. A biosafety cabinet or work station that is used for performing stool pathogen testing (e.g., culture or EIA) should not be used for sample preparation or pouch loading.

2. Prior to processing a sample, thoroughly clean both the work area and the FilmArray pouch loading station using a suitable cleaner such as freshly prepared 10% bleach or a similar disinfectant. To avoid residue build-up and potential PCR inhibition, wipe disinfected surfaces with water. Samples and pouches should be handled one at a time. Change gloves and clean the work area between samples.

3. Stool specimens should be collected in Cary-Blair transport medium according to the manufacturer's instructions. Two hundred microliters of sample is required for testing. Specimens should be processed and tested as soon as possible, though they may be stored at room temperature or under refrigeration for up to 4 days.

4. The RNA process control assay targets an RNA transcript from the yeast *Schizosaccharomyces pombe*. The freeze-dried yeast is present in the pouch and becomes rehydrated when the sample is loaded. The control material is carried through lysis, nucleic acid purification, reverse transcription, first-stage PCR, dilution, second-stage PCR, and DNA melting. A positive control result indicates that all steps carried out in the FilmArray GI pouch were successful.

5. The second-stage PCR (PCR2) control assay detects a DNA target that is dried in the wells of the array along with the corresponding primers. A positive result indicates that second-stage PCR was successful.

6. Good laboratory practice recommends running external positive and negative controls regularly. Enteric transport medium can be used as an external negative control. Previously characterized positive stool samples or negative samples spiked with well-characterized organisms can be used as external positive controls. External controls should be used in accordance with requirements of the appropriate accrediting organization, as applicable.

7. Once melting curves have been identified, the software evaluates the three replicates for each assay to determine the assay result. For an assay to be called positive, at least two of the three associated melting curves must be called positive, and the melting temperatures for at least two of the three positive melting curves must be similar (within 1°C). Assays that do not meet these criteria are called negative.

8. If four or more distinct organisms are detected in a specimen, retesting is recommended to confirm the polymicrobial result.

9. The FilmArray GI panel contains two assays (Crypt 1 and Crypt 2) for detection of *Cryptosporidium* spp. Approximately 23 different *Cryptosporidium* species are detected, including the most common species of human clinical relevance (i.e., *C. hominis* and *C. parvum*), as well as several less common species (e.g., *C. meleagridis*, *C. felis*, *C. canis*, *C. cuniculus*, *C. muris*, and *C. suis*). The assays do not differentiate between these *Cryptosporidium* spp. and the very rare species *C. bovis*, *C. ryanae*, and *C. xiaoi*, which may not be detected. A positive result for either or both assays will give a "*Cryptosporidium* detected" test result.

10. The FilmArray GI panel contains a single assay (Ehist) for the detection of *E. histolytica*, the only *Entamoeba* species implicated in gastroenteritis. This assay may cross-react with the closely related *E. dispar* when it is present at higher levels (approximately 10^5 oocysts/ml or more).

11. The Result Summary section of the test report lists the result for each target tested by the panel. Possible results for each organism are "detected," "not detected," "not applicable," and "invalid." See the Results Summary section in the instruction booklet for detailed information about interpretation of test results and appropriate follow-up for invalid results.

12. The performance of this test has been validated only with human stool collected in Cary-Blair transport medium, according to the medium manufacturer's instructions. It has not been validated for use with other stool transport media, raw stool, rectal swabs, endoscopy stool aspirates, or vomitus. Test performance has not been established for patients without signs and symptoms of gastrointestinal illness.

13. Virus, bacteria, and parasite nucleic acid may persist *in vivo* independently of organism viability. Additionally, some organisms may be carried asymptomatically. Detection of organism targets does not imply that the corresponding organisms are infectious or are the causative agents of clinical symptoms. The performance of this test has not been established for monitoring treatment of infection with any of the panel organisms.

F. Luminex (Verigene II GI Flex Assay; Includes Parasites)

In January 2013, the FDA cleared the first test that can simultaneously detect 11 common viral, bacterial, and parasitic (*Cryptosporidium, Giardia*) causes of infectious gastroenteritis from a single patient fecal specimen (xTAG gastrointestinal pathogen panel [GPP]; Luminex, Inc., Austin, TX). The xTAG GPP is a multiplexed nucleic acid test for the simultaneous qualitative detection and identification of multiple viral, parasitic, and bacterial nucleic acids in human stool specimens from individuals with signs and symptoms of infectious colitis or gastroenteritis.

The target is amplified using a PCR or reverse transcription-PCR and then analyzed with Luminex xTAG technology to detect the presence or absence of each pathogen in the panel. The test detects 11 gastrointestinal pathogen targets: *Campylobacter* (*C. jejuni*, *C. coli*, and *C. lari* only), *Clostridioides difficile* toxins A and B, *Cryptosporidium* (*C. parvum* and *C. hominis* only), *Escherichia coli* O157, enterotoxigenic *Escherichia coli* (ETEC) producing heat-labile and heat-stable enterotoxins, *Giardia* (*G. lamblia* only, also known as *G. intestinalis* and *G. duodenalis*), norovirus GI and GII, rotavirus A, *Salmonella*, Shiga toxin-producing *Escherichia coli* (STEC) stx_1 and stx_2, and *Shigella* (*S. boydii*, *S. sonnei*, *S. flexneri*, and *S. dysenteriae*). The xTAG GPP is capable of delivering multiple results within 5 hours. The assay is cleared for use on the widely available Luminex 100/200 system. Additionally, simultaneous molecular testing on a single sample within a single shift provides significant benefit to laboratories in terms of workflow and resource utilization.

Human stool samples are pretreated and then subjected to nucleic acid extraction. For each sample, 10 µl of extracted nucleic acid is amplified in a single multiplex RT-PCR/PCR. Each target or internal control in the sample results in PCR amplimers ranging from 58 to 202 bp (not including the 24-mer tag). A 5-µl aliquot of the RT-PCR product is then added to a hybridization/detection reaction containing bead populations coupled to sequences from the universal array ("antitags"), streptavidin, and R-phycoerythrin conjugate. Each Luminex bead population detects a specific microbial target or control through a specific tag-antitag hybridization reaction. Following the incubation of the RT-PCR products with the xTAG GPP bead mix and xTAG reporter buffer, the Luminex instrument sorts and reads the hybridization/detection reactions. A signal or median fluorescence intensity is generated for each bead population. These fluorescence values are analyzed to establish the presence or absence of bacterial, viral, or parasitic targets and/or controls in each sample. A single multiplex reaction identifies all targets.

KEY POINTS

1. Due to the open system design of the platform, there is a potential for contamination, the "intended use" portion of the package insert for this device states that all positive results obtained with the xTAG GPP assay are presumptive and need to be confirmed by another FDA-cleared assay or acceptable reference method. All results should be used and interpreted in the context of a full clinical evaluation as an aid in the diagnosis of gastrointestinal infection. Confirmed positive results do not rule out coinfection with other organisms that are not detected by this test, and the organism identified may not be the sole or definitive cause of patient illness. Negative xTAG GPP results in the setting of clinical illness compatible with gastroenteritis may reflect infection by pathogens that are not detected by this test or noninfectious causes, such as ulcerative colitis, irritable bowel syndrome, or Crohn's disease.

2. Analyte targets (virus, bacterial, or parasite nucleic acid sequences) may persist *in vivo*, independent of virus, bacterial, or parasite viability. Detection of an analyte target(s) does not guarantee that the corresponding live organism(s) is present, or that the corresponding organism(s) is the causative agent of clinical symptoms.

3. Regarding *Cryptosporidium*, the xTAG GPP assay detects *C. parvum* and *C. hominis* only.

4. Regarding *Giardia*, the xTAG GPP assay detects *G. lamblia* (*G. intestinalis*, *G. duodenalis*) only.

5. There is a risk of false-negative results due to strain and species sequence variability in the targets of the assay, procedural errors, amplification inhibitors in specimens, or inadequate numbers of organisms for amplification.

6. The performance of this test has not been established for monitoring treatment of infection with any of the panel organisms.

7. Positive and negative predictive values are highly dependent on prevalence. False-negative test results are more likely when prevalence is high. False-positive test results are more likely during periods when prevalence is low.

8. This test is a qualitative test and does not provide the quantitative value of detected organism present.

9. Known strains or positive clinical samples for the targeted viruses, bacteria, or parasites should be included in routine quality control procedures (external controls) as positive controls for the assay. One or more of these external controls are analyte-positive controls and should be included with each batch of patient specimens. Controls positive for different targets should be rotated from batch to batch. External controls should be prepared, extracted, and tested in the same manner as patient samples (http://www.accessdata.fda.gov/cdrh_docs/reviews/K121454.pdf; accessed 24 August 2019).

10. Failure to correctly interpret test results in the context of other clinical and laboratory findings may lead to inappropriate or delayed treatment. For example, a microorganism present as a colonizer may be correctly detected but not be the true cause of illness. Although the same risk would be present in the use of any microbiological assay in this setting, simultaneous testing of multiple analytes in a multiplex assay may be more likely to detect an unanticipated colonizer that might not be tested for individually.

11. It should be recognized that the device is intended for use as an aid in the diagnosis of gastroenteritis in conjunction with clinical presentation and the results of other laboratory tests. Both clinical presentation and other results will likely substantially mitigate concerns with both false-positive and false-negative test results.

12. A parasitic diagnostic panel currently in development (Verigene II GI panel) will test for *Blastocystis* spp., *Cryptosporidium* group, *Cyclospora cayetanensis*, *Dientamoeba fragilis*, *Entamoeba histolytica*, *Giardia lamblia*, microsporidia, and *Strongyloides stercoralis*. FDA clearance is anticipated in 2021.

G. Other Pending Molecular Tests

There are also several molecular tests that are in clinical trials for the detection of selected gastrointestinal parasites. These tests are molecular gastrointestinal panels and target the most commonly occurring parasitic stool pathogens. Although there are laboratory-developed tests for most parasites, these are not commercially available or are available only in specialized testing centers. When such tests are used, attention should be given to the use of internal amplification controls to detect inhibition, since common specimens, such as blood and stool, contain PCR inhibitors. Thorough validation is required before these are implemented for clinical testing. Specific organisms under consideration include *Entamoeba histolytica*, *Blastocystis* spp. (*Blastocystis hominis*), and *Dientamoeba fragilis*. Blood parasites include *Plasmodium* spp. (including *P. knowlesi*), *Trypanosoma* spp., and *Leishmania* spp.

Organism Identification

Protozoa

1. **What is the most effective technique for the identification of the intestinal protozoa?**

   Although some protozoan cysts can be seen and identified on the wet preparation smear (direct mount and concentration sediment wet mount), the permanent stained smear is recommended as the most relevant and accurate procedure for identification of the intestinal protozoa. These preparations are examined using the oil immersion objective (100×) for a total magnification of ×1,000. At least 300 oil immersion fields should be examined prior to reporting the result of the permanent stained smear.

2. **Are trophozoites ever seen in the wet mounts of stool?**
Usually the trophozoites are not seen in the concentration sediment wet mount preparations unless they are prepared from stool preserved in a fixative containing no formalin. Motile trophozoites can occasionally be seen in the direct wet smear prepared from fresh soft or liquid stool that is more likely to contain trophozoites, but this is rare.

3. **What are some tips to consider when reporting *Entamoeba hartmanni*?**
The measurements of 10 µm or less for the *Entamoeba hartmanni* cyst refer to wet preparation measurements, so the measurement should be decreased ~1 µm on the permanent stained smear. When you see a cyst on the permanent stained smear, there is often a halo representing shrinkage. **The cyst measurement must include that halo.** On the bench, the measurements for this cyst generally run from around 9.5 down to about 8 µm, and the cysts are morphologically definitely *E. hartmanni*. The *E. hartmanni* cyst generally contains more chromatoidal bars than are seen in *E. histolytica*/*E. dispar*. Also, the *E. histolytica*/*E. dispar* cysts tend to measure routinely on the bench (trichrome slides) from about 10.5 up to about 13 µm. Also, on the bench, the *E. hartmanni* nucleus, particularly in the trophozoite, tends to look like a bulls'-eye, consisting of a very sharp ring of nuclear chromatin with the karyosome right in the middle; often the cysts contain two nuclei only.

4. **Do nonpathogenic protozoa ever cause symptoms?**
Endolimax nana, *Iodamoeba bütschlii*, *Chilomastix mesnili*, and *Pentatrichomonas hominis* (as examples) have been categorized as nonpathogens. Patients have been documented to be symptomatic when infected with a nonpathogen, although this is rare. However, it is sometimes difficult to determine from the case history the extent of the workup, including Apicomplexa/coccidia and the microsporidia. Before ascribing symptoms to nonpathogenic protozoa, a comprehensive search for other proven pathogens should be performed.

5. **What are some considerations when reporting *Entamoeba histolytica* versus the *Entamoeba histolytica*/*E. dispar* complex/group?**
Morphologically, unless the trophozoite measures >12 µm and contains ingested red blood cells (RBCs), the identification must be *Entamoeba histolytica*/*E. dispar* complex/group, indicating that the true pathogen (*E. histolytica*) cannot be differentiated from the nonpathogen (*E. dispar*). One can also report the true pathogen if the fecal immunoassay specific for *E. histolytica* is positive. Since *Entamoeba moshkovskii* and *Entamoeba bangladeshi* also look very much like others in the genus *Entamoeba*, some have suggested that such organisms should be reported as the "*Entamoeba histolytica* complex" or "*Entamoeba histolytica*/*E. dispar* complex or group."

6. **What color is the autofluorescence seen with *Cyclospora cayetanensis*?**
The color depends on the particular FA filters used. If the filters are used for calcofluor white, the oocysts appear as pale blue rings; if the filters are used for fluorescein isothiocyanate, the oocysts appear to be more yellow green. Fluorescence intensity varies from about 1+ to 2+; it is rare to see stronger autofluorescence with these oocysts.

7. **Why are *Blastocystis* spp. (*Blastocystis hominis*) so controversial regarding pathogenicity?**
What we currently call *Blastocystis hominis* **(morphologically) appears to be approximately 10 different strains, subtypes, or species, some of which are pathogenic and some of which are nonpathogenic.** Currently, the term most commonly used to refer to the various types is "subtype." As molecular studies continue to be published, changes in the classification of this group are probable. As all subtypes currently make up what is called "*Blastocystis hominis*," this could explain the controversy regarding pathogenicity and the fact that some patients are symptomatic, while some are asymptomatic. Unfortunately, all of the subtypes are morphologically identical; therefore, **the correct identification report continues to be *Blastocystis* spp.** (*Blastocystis hominis*), and the number should be quantitated from the permanent stained smear (rare, few, moderate, many, or packed). More laboratories have begun reporting *Blastocystis* spp., with a report comment explaining the controversy regarding pathogenicity. This is the recommended approach. It is important to remember that this organism is the most common parasite found in stool worldwide. In some populations, the percentage can reach >80%. Laboratories that do not report this organism should reevaluate their diagnostic reporting practices. Another potential problem would be the failure to recognize, identify, and report this organism.

Helminths

1. **Why can't all helminth eggs be recovered using the flotation concentration rather than the sedimentation concentration?**
Some helminth eggs are quite heavy and will not float, even when zinc sulfate with a specific gravity of 1.20 is used. Other helminth eggs are operculated; when the egg is placed in a high-specific-gravity solution, the operculum "pops" open and the egg fills with fluid and sinks to the bottom of the tube. Thus, both the surface film and the sediment should be examined before the specimen is reported as negative.

2. **Why must helminth larvae be identified to the species level? Shouldn't all larvae recovered in stool be those of *Strongyloides stercoralis*?**
Although helminth larvae in stool are normally *Strongyloides stercoralis*, there is always the possibility that hookworm eggs have continued to mature in fresh stool and may hatch before the stool is processed and/or placed in fixatives. It is important to make sure that the larvae seen are, in fact, the rhabditiform (noninfectious) larvae of *S. stercoralis* rather than larvae of hookworm. The agar plate culture is the most sensitive method for the recovery of *S. stercoralis* larvae. Also, migrating larvae could also be recovered from respiratory specimens (sputum, bronchoalveolar lavage fluid, etc.). Note: *S. stercoralis* larval morphology can be quite poor if the larvae have been preserved in a fixative containing PVA. Concentrations are recommended for fixatives that do not contain PVA.

3. **What is the most sensitive test for the diagnosis of strongyloidiasis?**
Agar plate cultures are recommended for the recovery of *S. stercoralis* larvae and tend to be more sensitive than some of the other diagnostic methods, such as the O&P exam. Stool is placed on agar plates, and the plates are sealed to prevent accidental infections and held for 2 days at room temperature. As the larvae crawl over the agar, they carry bacteria with them, thus creating visible tracks over the agar. The plates are examined under the microscope for confirmation of larvae, the surface of the agar is

then washed with 10% formalin, and final confirmation of larval identification is made via wet examination of the sediment from the formalin washings (see Section 5).

In a study of the prevalence of *S. stercoralis* in three areas of Brazil (see Kobayashi et al., Suggested Reading), the diagnostic efficacy of the agar plate culture method was as high as 93.9%, compared to only 28.5% and 26.5% by the Harada-Mori filter paper culture and fecal concentration methods, respectively, when fecal specimens were processed using all three methods. Among the 49 positive samples, about 60% were confirmed as positive by using only the agar plate method. These results indicate that the agar plate approach is probably a much more sensitive diagnostic method and is recommended for the diagnosis of strongyloidiasis.

It is important to remember that more than half of *S. stercoralis*-infected individuals tend to have low-level infections. The agar plate method continues to be documented as a more sensitive method than the usual direct smear and fecal concentration methods. A daily search for furrows on agar plates for up to six consecutive days results in increased sensitivity for diagnosis of both *S. stercoralis* and hookworm infections. Also, a careful search for *S. stercoralis* should be made for all patients with comparable clinical findings before a diagnosis of idiopathic eosinophilic colitis is made, because consequent steroid treatment may have a fatal outcome by inducing widespread dissemination of the parasite.

4. Why might pulmonary specimens be important in the diagnosis of nematode larvae?

In some nematode life cycles, larvae migrate through various parts of the body, including the lungs. Therefore, occasionally migrating larvae (e.g., *Ascaris* and *Strongyloides*) may be found. The examination of pulmonary specimens is not necessarily recommended; often the larvae are accidentally found when pulmonary specimens are submitted for routine cultures and/or stains.

5. Are there any specific recommendations for the detection of schistosome eggs?

When trying to diagnose schistosomiasis, regardless of the species suspected, you should examine both stool (several different stool specimens) and urine (spot urine samples plus a 24-h specimen), collected with no preservatives. Occasionally adult worms get into blood vessels, where they are not normally found (for example, *S. mansoni* eggs might be found in urine only rather than stool, the more common specimen). When you perform a sedimentation concentration, you should use saline to prevent premature egg hatching. Once you are ready to try a hatching procedure, you can put the sediment into spring water (dechlorinated water) to stimulate hatching. When examining wet mounts under the microscope, you need to look for the movement of cilia on the larvae within the eggshell (hence the need to collect the specimens with no preservatives). You should be able to tell the physician whether the eggs, if present, are viable or whether you see only dead eggshells. If you suspect schistosomiasis, it is recommended that you examine a number of wet mounts, particularly if you are not going to perform a hatching test.

6. Where can one get serologies for a *Baylisascaris procyonis* infection?

CDC

(404) 718-4745 (M–F, 7:30 a.m. to 4 p.m. EST); https://www.cdc.gov/parasites/contact.html; https://www.cdc.gov/dpdx/diagnosticprocedures/serum/tests.html

After-hours emergencies: (770) 488-7100

Reporting

Organism Identification

1. **Should common or scientific names be used when reporting the presence of parasites?**
   The scientific (genus and species) names should be used on the final report that goes to the physician and on the patient's chart. It is also recommended that the stage of the organism be included (trophozoite, cyst, oocyst, spore, egg, larvae, and adult worm), as this information may impact therapy. **Various stages of the malaria parasites also determine recommended therapy (presence or absence of gametocytes). This information should be included on the test result report.**

2. **What happens if several different names are used in the literature (e.g., *Giardia lamblia, G. intestinalis,* and *G. duodenalis*)?**
   It is appropriate to use the most commonly accepted name (*Giardia lamblia*); you can also let the proficiency testing organism lists be your guide. When a replacement name begins to appear on the proficiency test list, then it may be appropriate to consider the name change. Remember, it is very important to notify all clients prior to making any name changes. The general recommendation for nomenclature changes is to follow the *International Code for Zoological Nomenclature* published by International Trust of Zoological Nomenclature (ITZN; https://www.iczn.org/the-code/the-international-code-of-zoological-nomenclature/the-code-online/) (approved in 1961). Binomial nomenclature is the formal naming system for living things that all scientists use. The names lose much of their usefulness if they are changed frequently and arbitrarily. Therefore, it is generally recommended that name changes be thoroughly reviewed and changed only according to accepted guidelines. According to the Integrated Taxonomic Information System, the current valid taxonomy for *Giardia lamblia* is found under taxonomic serial number 553109 (*Giardia lamblia* Kofoid and Christiansen, 1915) (https://www.itis.gov/servlet/SingleRpt/SingleRpt?search_topic=TSN&search_value=553109#null; accessed 25 August 2019). **Any nomenclatural acts (e.g., publications that create or change a taxonomic name) need to be registered with ZooBank to be "officially" recognized by the ITZN.**

3. **How has the reporting of *Cryptosporidium parvum* changed and why?**
   It is now known that more than one species of *Cryptosporidium* can cause disease in humans (*C. parvum* infects humans and other mammals, while *C. hominis* infects only humans). However, the different species cannot be differentiated on the basis of morphology. It is now recommended that the more correct reporting format would be *Cryptosporidium* spp. (rather than *Cryptosporidium parvum*).

4. **Should WBCs and/or other cells or yeasts be reported, and why?**
   The reporting and quantitation (rare, few, moderate, or many) of WBCs (polymorphonuclear leukocytes [PMNs], macrophages, and eosinophils) provide some additional information for the physician. If the patient continues to have diarrhea, this may give the physician something more to consider, particularly if a stool culture has not been ordered. Conditions related to noninfectious diarrheas may also result in WBCs and macrophages in the stool. We always include this type of information/explanation in our educational process regarding reporting formats. Physicians can then decide on the relevance of the information. When reporting WBCs, we can also identify eosinophils from the permanent stained smear (this may or may not be related to parasitic

infection and may or may not correlate with peripheral eosinophilia); this information may also be helpful.

Reporting yeast infections is a bit different. To report anything about yeasts, it is important to know that the stool was fresh or immediately put in preservative. If there are large numbers of yeasts, budding yeasts, and/or pseudohyphae, this provides some additional information for the physician, often depending on the patient's general condition and whether the patient is immunosuppressed. However, if you do not know whether the collection criteria were met, reporting anything about yeasts is NOT recommended, since this type of report may be misleading. The physician can be contacted for additional information regarding the patient's condition; at that time, any possible relevant yeast information can be conveyed to the physician. One can also report yeast and add a clinical comment:

Moderate to many budding yeast cells seen.

Reports of yeasts (budding and/or pseudohyphae) might be misleading due to a lag time between stool passage and specimen fixation, during which the yeasts continue to grow and multiply. Therefore, reporting the presence of yeasts may or may not be clinically relevant.

5. How should *Blastocystis* spp. (*Blastocystis hominis*) be reported?

In the past, there was some agreement that the number of organisms present (moderate, many, or packed) was more likely to be associated with symptoms. However, in the past few years, there have been anecdotal reports of patients being symptomatic after infection with rare or few organisms. The current recommendation is to report the organism and quantitate as rare, few, moderate, many, or packed (using your proficiency testing quantitation scheme). It is important to confirm that the physicians know what the report means and understand the controversial issues surrounding this organism; they can then correlate the numbers with symptoms. For cases where the organisms are rare or few, there are no solid data on the relevance of numbers other than the anecdotal case reports (many of which are word of mouth). Many physicians treat patients who are symptomatic and in whom no other organisms are found (including coccidia and/or microsporidia).

Blastocystis hominis comprises a mix of several subtypes, some of which are pathogenic and some of which are nonpathogenic. Based on continuing molecular biology studies, the classification of this group could be updated. If pathogenic and nonpathogenic subtypes currently make up what is called "*Blastocystis* spp.," this could explain the controversy regarding pathogenicity and the fact that some patients are symptomatic while some are asymptomatic. It is also recommended that the organism be reported as *Blastocystis* spp. and a report comment be included regarding the presence of pathogenic and/or nonpathogenic subtypes, none of which can be differentiated on the basis of microscopic morphology.

6. How should intestinal protozoa be reported?

Reporting trophozoites and cysts has been the accepted way of reporting intestinal protozoa for a couple of reasons. (i) Different drugs are used to treat *E. histolytica* cysts and trophozoites. (ii) Since the cyst form is the infective stage for the protozoa (except for the trichomonads), the report does convey some epidemiological information. For these reasons, the current recommendation is to continue to report all protozoa (genus, species, and stage).

7. What comments should be added to the reporting of the O&P examination and the fecal immunoassays?

It is important to add the following comment to the O&P examination: "The O&P examination is not designed to detect *Cryptosporidium* spp., *Cyclospora cayetanensis*, or the microsporidia. *Cystoisospora belli* oocysts (coccidia) can be detected from the concentration sediment examination."

The results of the fecal immunoassays should be reported based on the specific organism(s) relevant to the kit:

"No *Giardia lamblia* antigen detected." OR "*Giardia lamblia* antigen present."

"No *Giardia lamblia* or *Cryptosporidium* spp. antigen detected."

OR

"Negative for *Giardia lamblia*."

"Negative for *Giardia lamblia* and *Cryptosporidium* spp."

The report of *Blastocystis* spp. should be accompanied by the following report comment: "*Blastocystis* spp. (morphologically) are **composed of a number of different subtypes, some of which are pathogenic and some of which are nonpathogenic (they can't be differentiated on the basis of morphology).** This explains why some patients are symptomatic and some are asymptomatic."

Quantitation

1. What organisms should be quantitated in the final report to the physician and patient's chart?

Organisms and nonorganism cells and other structures (stool blood cells, Charcot-Leyden crystals) that are recommended for quantitation include the following:

- **a.** Intestinal protozoa: *Blastocystis* spp.
- **b.** Helminths: *Trichuris trichiura* eggs, *Clonorchis sinensis* eggs, schistosome eggs (also report the viability of *Schistosoma* eggs). Unless the eggs mentioned above are described as "moderate" to "many," it is probably neither practical nor clinically relevant to quantitate the eggs on the report.
- **c.** Blood parasites: all malaria organisms and *Babesia* spp.
- **d.** Blood cells (PMNs, macrophages, RBCs)
- **e.** Charcot-Leyden crystals

2. Why are all proficiency testing specimen answers reported and quantitated, while clinical specimen reports are rarely quantitated in terms of organism numbers?

There are two issues to consider.

- When reporting proficiency testing specimen results, you are asked to quantitate organisms in both the formalin wet mounts and permanent stained smears. This information serves as a **quality control check for the proficiency testing agency to ensure that organism numbers are consistent throughout the challenge vials/slides.**
- Most laboratories (with few organism exceptions) do not quantitate organisms on the concentration or permanent stained smear reports. One exception is *Trichuris trichiura* eggs in a concentration wet mount (light infections might not be treated). Since many organisms are shed on a random basis, quantitation may change dramatically from day to day and generally has little clinical relevance.

Proficiency Testing

Wet Preparations

1. **How should you examine a wet preparation for proficiency testing?**
   Most directions recommend that you shake the vial and take a sample from the mixed vial contents. However, if only a few eggs are present, you may be better off taking a very small drop from the settled material (allow the vial to stand on the counter for 5 min prior to sampling sediment). If you take too large a drop, it will be too thick to examine properly. Make sure you do not add too much iodine; very darkly stained helminth eggs can resemble debris. Also, if you accidentally create a bubble while pipetting fluid and the vial contents are remixed, wait another 5 minutes prior to sampling the settled vial contents.

2. **Are there any tips regarding the microscope setup for the examination of an unstained wet preparation?**
   The microscope should be aligned properly using Köhler illumination. Make sure the light is not too bright; otherwise, you may shine right through the organisms and miss the parasites. Also, you may want to close the diaphragm a bit to provide a bit more contrast. Extra contrast is particularly important when reading with saline only (no iodine added).

Permanent Stained Smears

1. **How should you examine a permanent stained fecal smear for proficiency testing?**
   Make sure the light is very bright (with the condenser all the way up) and the diaphragm is open. Review both thin and thick areas of the smear; examine at least 300 oil immersion fields (using the 100× oil immersion objective) before you report the specimen as a negative. If you use the 50× or 60× scanning oil immersion lens, you still need to review 300 oil fields as indicated above (using the 100× oil immersion objective).

2. **How should you examine a permanent stained blood film for proficiency testing?**
   Make sure the light is very bright (with the condenser all the way up) and the diaphragm is open. Review both thin and thick areas of the smear; examine at least 300 oil immersion fields (using the 100× oil immersion objective) before you report the specimen as negative. If you use the 50× or 60× scanning oil immersion lens, you still need to review 300 oil fields as indicated above (using the 100× oil immersion objective). Because microfilaria may be present, remember to examine the entire stained blood film (thick and/or thin) using the 10× low-power objective. Be sure to cover all areas of the film, including the drop from which the smear was pulled as well as thicker sections and the feather end of the thin blood film. There may be only 3 or 4 larvae per stained smear, so it is critical that the entire blood film be screened prior to using oil immersion objectives.

Tissues or Fluids

1. **How should tissues be submitted to the laboratory?**
   Tissue specimens should be immediately sent to the laboratory and kept moist during transit. If the tissue is to be processed for culture (*Acanthamoeba*, *Naegleria*, *Leishmania* spp., *Trypanosoma* spp., or *Toxoplasma gondii*), the specimen should be kept

sterile and submitted in a sterile container. If the tissue is to be processed for wet examinations and permanent stained smears, any type of container is acceptable. Remember that the specimen must be kept moist; if it dries out in transit, neither culture nor stained smears will be acceptable.

2. **How should eye specimens be submitted for *Acanthamoeba* culture?**
If you are sending the specimen in a screw cap tube or vial (or other capped container), make sure that the container is filled to the top with transport fluid (saline is acceptable). This will prevent the small tissue specimen from drying out if it is shaken onto the side of the container during transit. If the container is not full of fluid, the tissue may dry out and will be unacceptable for culture. If trophozoites are present, they will be able to encyst within the fluid, so disintegration of the trophozoite forms is not critical. This situation is unlike that seen with fecal protozoan trophozoites, which begin to disintegrate within hours if they are not placed in an appropriate fixative (trophozoites will not encyst once the stool is passed).

3. **How should duodenal drainage specimens be handled?**
Duodenal drainage specimens must be submitted to the laboratory as quickly as possible for processing. Delays may prevent organism recovery and identification. These specimens may be very liquid and may contain a lot of mucus. It is best to centrifuge the tube, discard any supernatant fluid left, and examine the mucus only as wet preparations (low light) and/or permanent stained smears (trichrome, iron hematoxylin). It is recommended that a single centrifugation be used; do not rinse the material prior to examination. Do not use ethyl acetate in the concentration procedure; a single spin is recommended (do not rinse the specimen).

Blood

Many questions have been asked regarding the examination of blood specimens in diagnostic medical parasitology. The information below may be helpful for those working in this area of microbiology and/or hematology and provide some professional hints learned through many years of bench experience.

Specimen Collection

1. **If *Plasmodium falciparum* parasites are sequestered in the capillaries, why not do a finger-stick rather than a venipuncture?**
The capillaries are generally in the deep tissues (spleen, liver, and bone marrow), so a finger-stick blood sample is no more likely to be positive than a venipuncture blood sample. There may be some very minor differences with *Plasmodium falciparum*, but these are not sufficient to warrant eliminating venipunctures. The anticoagulant tube also provides additional blood for multiple thick and thin blood films and/or buffy coat films. Many laboratorians are unfamiliar with the preparation of finger-stick blood films; thus, this practice within the United States is much less common than in the past.

2. **What is the best anticoagulant to use for blood specimens?**
Although heparin (green-top tube) or EDTA (lavender-top tube) can be used, EDTA is recommended as providing better organism morphology, particularly for *Plasmodium* spp. Blood collected using EDTA anticoagulant is acceptable; however, if the blood remains in the tube for >1 to 2 h, true stippling may not be visible within the infected

RBCs (for *Plasmodium vivax* and *P. ovale*). Also, when anticoagulants are used, the proper ratio between blood and anticoagulant is necessary for good organism morphology. The lavender-top tube should be filled to the top with blood to provide the proper blood/anticoagulant ratio. Although heparin can be used, EDTA is preferred. Finger-stick blood is recommended when the volume of blood required is minimal (i.e., when no other hematologic procedures have been ordered). The blood should be free-flowing when taken for smear preparation and should not be contaminated with alcohol used to clean the finger prior to the stick. However, finger-stick blood is no longer commonly used in many parts of the world. When blood is collected in EDTA, specimens should be processed immediately after blood collection. **Parasite numbers may decrease if processing is delayed, even 4 to 6 h.** Adhesion of the blood to the slide can be a problem if the ratio of anticoagulant to blood is high, the patient is anemic, or the blood was held in EDTA too long. If the laboratory receives a heparin-blood specimen, do not reject it; process it and request that a lavender-top EDTA sample be submitted on a stat basis.

Table 6.1

Organism	Morphology (normal)	Morphology (contact with EDTA for >1–2 h)
Plasmodium falciparum		
Rings	Typical, small to medium rings; double rings/cell	Numbers tend to remain constant; all parasites continue to grow in EDTA.
Gametocytes	Crescent-shaped	Gametocytes may round up and be confused with other species.
Plasmodium knowlesi	Rings resemble *P. falciparum*; rest of stages resemble *P. malariae*	As with all species, parasites may actually disappear after 4–6 h in EDTA. This infection is often reported as a *P. falciparum*/*P. malariae* mix.
Plasmodium vivax		
Trophozoites	Ameboid trophozoites, for both early and late rings	May round up and lose their characteristic shape
Schüffner's dots	Typical dots appear in late rings; present throughout the rest of the life cycle stages	Stippling (Schüffner's dots) may not be visible at all, in any of the life cycle stages (regardless of buffer pH used). Note: If slides are prepared soon after collection in EDTA, the dots will be seen after staining; however, if slides are prepared after the blood has been in contact with EDTA for several hours, the dots may not be visible after staining.
***Plasmodium* spp.**		
Male gametocytes	Exflagellation	As blood cools and becomes oxygenated (cap removed from the EDTA tube; tube is allowed to stand at room temperature), the parasites "think" they are now in the mosquito (this cycle may continue). The male gametocyte may exflagellate; microgametes may resemble *Borrelia* (this is also related to pH and pCO_2, e.g., when the lid is left off the tube).

3. Why do new slides have to be cleaned with alcohol prior to use?

Even new slides are coated with a very fine layer of oil (to allow the slides to be pulled apart from each other). If the coating is removed, the blood will flow more smoothly over the glass during blood film preparation. "Holes" in the blood film are evidence of oil or grease on the slide. **Use standard precautions;** remember that both slides and spreaders must be held on the edges and not on any part of the slide that will come in contact with blood.

4. When should blood specimens be drawn for a suspected malaria diagnosis?

The majority of patients with malaria in the United States have never been exposed to the organism before; therefore, they have no antibody and when they present, they do NOT have a synchronized fever cycle. These immunologically naive patients may present with nonspecific symptoms that can mimic many other diseases. The rule of thumb is to **draw immediately**; do not wait for some "magic" periodic cycle that may never appear. Patients with a very low parasitemia with *P. falciparum* or *P. knowlesi* can become quite ill before they have any type of fever cycle or gametocytes. For patients who are suspected of having malaria or who have a fever of unknown origin, blood should be drawn and both thick and thin blood films should be prepared and examined immediately. **This request is always considered a stat request.** In areas of the world where malaria is endemic and people have been exposed to the parasite before (and therefore have some antibody), they may not become symptomatic until they actually have some sort of periodic fever cycle and a higher parasitemia. However, you should always use the general stat guideline to draw immediately. Also, remember that one set of negative blood films does not rule out malaria. If the first set (both thick and thin films) is negative, you can recommend that additional blood films be drawn in about 4 to 6 h. Also, any decision to delay treatment should be left to the physician, not the laboratory.

Specimen Processing

1. Why is it important that the EDTA-blood be processed as quickly as possible?

If a tube of blood containing EDTA cools to room temperature and the cap has been removed, several parasite changes can occur. The parasites within the RBCs will respond as if they were now in the mosquito after being taken in with a blood meal. The morphology of these changes in the life cycle and within the RBCs can cause confusion when blood films prepared from this blood are examined.

- Stippling (Schüffner's dots) may not be visible.
- The male gametocyte (if present) may exflagellate.
- The ookinetes of *Plasmodium* spp. other than *P. falciparum* may develop as if they were in the mosquito and may mimic the crescent-shaped gametocytes of *P. falciparum*.
- Smears left longer than 24 h can autofix; if this occurs, lysis of the RBCs will be difficult, if not impossible, to achieve. This autofixation tends to occur more quickly in warm, humid climates.
- Thick smears can be dried in a 37°C incubator for 10 to 15 min without fixation of the RBCs; **do not go beyond 15 min when using this method. Do not use a heat block to dry thick films.**

2. **Why is it important to keep thick films from getting hot (heat fixation)?**
Heat fixes the RBCs, and they subsequently do not lyse in the staining process (see the last item in the question above).

3. **What should you do if you have blood films that can't be stained for several days?**
If staining with Giemsa is delayed for more than 3 days or if the film is to be stained with Wright's stain, lyse the RBCs in the thick film by placing the slide in buffered water (pH 7.0 to 7.2) for 10 min, remove it from the water, and place it in a vertical position to air dry. The laked thick films and thin films should be dipped in absolute methanol and placed in a vertical position to air dry. This is particularly important if the blood films are to be stored for days or weeks before being stained.

Diagnostic Methods

1. **Why is it important to always examine both thick and thin blood films prior to reporting the specimen as negative for blood parasites?**
Thick blood films allow a larger amount of blood to be examined, which increases the possibility of detecting light infections. However, only experienced workers can usually make a species identification from a thick film, particularly for malaria parasites. The morphologic characteristics of blood parasites are best seen in thin films. However, it is mandatory that both types of films be examined prior to reporting negative findings.

2. **What are the advantages of the thick blood film?**
The purpose of using a thick film is to have a drop of blood with 20 or 30 layers of RBCs on the slide and then to lyse the RBCs, wash off the hemoglobin, and stain the parasites, which remain intact in the process. As a consequence, RBCs are normally not visible, but WBCs and parasites may be seen. It is essential to become familiar with the characteristics of malaria parasites and know what to look for on such a preparation. An experienced microscopist should be able to detect 20 parasites per μl of blood (i.e., a parasitemia of 0.0001%) after examining 100 fields in 10 min. A greater volume of blood can be examined in the same time as that taken for examination of the thin blood film. The presence of phagocytized malaria pigment within WBCs (particularly in cases with low levels of parasitemia) can be very helpful. Occasionally, one can see Schüffner's dots in a thick film. However, recognition of the malarial pigment within WBCs and Schüffner's dots in a thick blood film may be difficult.

3. **What are the disadvantages of the thick blood film?**
Due to lysis of the RBCs during processing of thick blood films, the thick film cannot be used to compare the sizes of the parasite within the RBCs or to compare the sizes of the infected RBCs to those of the uninfected RBCs. Recognition of organism distortion and identification to the species level are generally more difficult from the thick film.

4. **What are some of the problems associated with thick blood films?**
The thick film flakes off during the staining process.
    - The film was not dry.
    - The film did not dry evenly.
    - The film was too thick (refer to the combination thick-thin blood film protocol).
    - The blood was too diluted with anticoagulant or the patient was anemic (in this case, centrifuge blood at 500 × *g* for 1 min and repeat the thick film).

- The slides were greasy or dirty.
- The blood was in EDTA too long (prepare thick films thinner than normal; dry for a longer period).
- The thick film does not stain adequately.
- The film was too thick.
- The staining solution was too dilute or the staining time was too short.

5. What are some tips for improving thick blood films?

After the thick films are dry, fix the smear with acetone for a few seconds **(dip only twice)** and then dry the film. This approach improves the durability of the thick film and does **not** interfere with the lysis of the RBCs. These films tend to have a clean background, making the parasites easier to see. Overall adherence to the slide is enhanced by using this approach.

6. How do you prepare thin blood films?

Keep the spreader at about a 40 to 45° angle to allow blood to flow to almost the edge of the slide. Lower the angle to about 30 to 35° to pull the film in one smooth motion (Figure 6.1). The finished blood film should appear as a feathered edge in the center of the slide, **with a free margin on either side.** Before the thin blood films are fixed in absolute methanol, the film must be completely dry. If slides must be stored unfixed, they should be frozen. If fixation lasts too long (>30 to 45 s), stippling may be reduced. If fixation is too short, RBCs may be distorted (crescent shaped) or partially lysed. During fixation, it is better to use a dispensing bottle for methanol rather than a Coplin jar, in which the methanol will pick up water from the air. Methanol used on one day (Coplin jar) should not be reused the next day; it is important to begin with fresh stock.

7. What are the advantages of the thin blood film?

The RBC morphology can be seen, as well as the size of the infected RBC in which the parasite resides compared to the uninfected RBC size. It is much easier to identify malaria organisms to the species level by using the thin blood film. The parasitemia can be calculated much more easily from the thin blood film than the thick film (as percent infected RBCs within 100 RBCs). In a low-parasitemia case, many RBCs will need to be counted to obtain an accurate percentage.

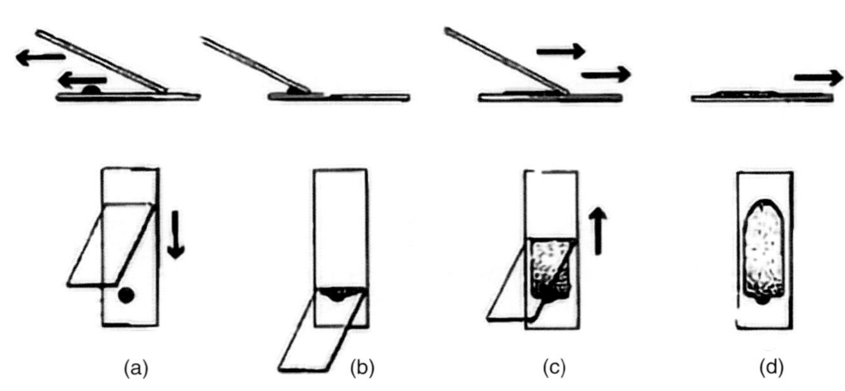

Figure 6.1 Traditional method for preparing a thin blood film. The blood can be either "pushed" or "pulled" by the spreader slide. Image (d) illustrates a thin film with a good feather edge. (Illustration by Sharon Belkin.)

8. **What are the disadvantages of the thin blood film?**
The thin blood film has a much lower sensitivity than does the thick blood film; thus, infections with a low parasitemia may be missed. The thin blood film is used to identify the organism(s) to the species level, and the thick blood film is used to determine whether organisms are present and the patient is infected.

9. **What are some of the problems associated with thin blood films?**
Films are too thick.
   - Too large a drop of blood was used.
   - The RBCs may be crowded together with no thin feathered edge.
   - The film was prepared too quickly or with too large an angle, >30 to 35°.
   - Films are too thin.
   - Blood from anemic patients does not spread well; it should be spread more quickly. Allow the blood to settle or centrifuge, remove some plasma, and repeat the thin-film preparation procedure.
   - The angle between the spreader and slide was too small (<30°).
   - Smears have ragged edges.
   - The drop of blood may have been in front of the spreader.
   - The slides were greasy or dirty.
   - The spreader slide was chipped.
   - The spreader slide was reused (a new slide should be used each time).
   - If the blood contained no anticoagulant, a delay in spreading may have led to fibrin formation, causing streaking.
   - The problem may be due to a high plasma fibrinogen level and not to the technique.

Poorly prepared thick and thin blood films can be seen in Figure 6.2.

10. **How does one prepare a combination thick-thin blood film?**
The combination thick-thin blood film provides both options on one glass slide, and the slide can be stained as either a thick or a thin blood film (Figure 6.3). If the film is fixed prior to being stained, the smear will be read as a thin blood film; if RBCs are lysed during staining, the preparation will be read as a thick blood film (parasites, platelets, and WBCs). This combination blood film dries more rapidly than the

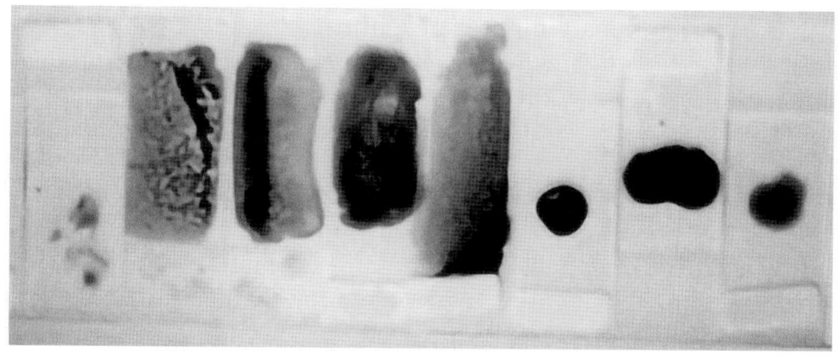

Figure 6.2 Poorly prepared thin and thick blood films (dirty slides, oil on slides, too-thick preparations, poor spreading of the blood); organism morphology will be very poor on the stained films.

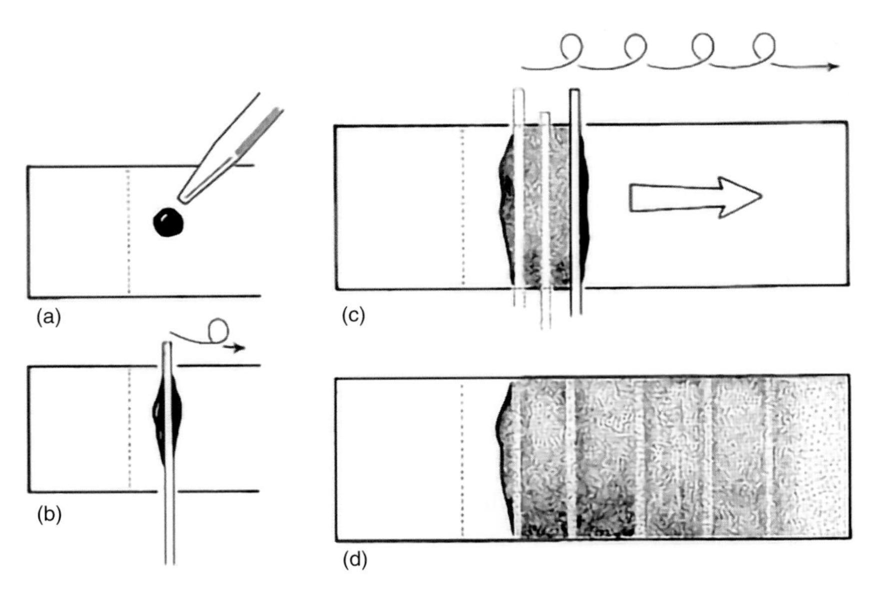

Figure 6.3 Method of thick-thin combination blood film preparation. (a) Position of drop of EDTA-containing blood. (b) Position of the applicator stick in contact with blood and glass slide. (c) Rotation of the applicator stick. (d) Completed thick-thin combination blood film prior to staining. (Illustration by Sharon Belkin.)

traditional thick blood film, thus allowing staining and examination to proceed with very little waiting time for the slide(s) to dry.

- Place a drop (30 to 40 μl) of blood onto one end of a clean slide about 0.5 in. from the end.
- Using an applicator stick lying across the glass slide and keeping the applicator in contact with the blood and glass, rotate (do not "roll") the stick in a circular motion while moving the stick down the glass slide to the opposite end.
- The appearance of the blood smear should be alternate thick and thin areas of blood.
- Immediately place the film over some small print and be sure that the print is just barely readable.
- Allow the film to air dry horizontally and protected from dust for at least 30 min to 1 h. Do not attempt to speed the drying process by applying any type of heat, because the heat will fix the RBCs and they subsequently will not lyse in the staining process.
- This slide can be stained as either a thick or thin blood film.

11. How should malaria smears be stained with Giemsa stain?

Giemsa stain is a mixture of eosin and methylene blue. Stock solutions of Giemsa stain can be purchased commercially. Some brands are better than others. The stock solution of Giemsa stain is easily prepared from commercially available Giemsa powder.

Stock solutions of Giemsa stain must always be diluted by mixing an appropriate amount of stain with distilled neutral or slightly alkaline water; buffered saline is preferred because it provides a cleaner background and better preservation of parasite morphology. Although most people do not filter the working stain prior to use (if using a Coplin jar), results are better overall if the working stain is filtered through Whatman no.1 filter paper immediately before use. The working stain is stable for 1 day. Make sure to use absolute methanol (acetone free) for fixing thin blood films. Stock buffered water is stable for 1 year at room temperature. Working stock buffered water is stable for 1 month at room temperature. Stock Giemsa stain is stable at room temperature indefinitely; stock stain appears to improve with age (similar to iron hematoxylin stains). A 45- to 60-min staining time appears to work better than 15 min; staining times depend on stain dilution. Some workers feel that the use of a 10% Triton solution is helpful, while some feel that it is detrimental to the overall quality.

12. How should you handle a delay between thick-film preparation and staining?

Thick films can be preserved, particularly if there will be a delay prior to staining. Dip the slides in a buffered 0.65% methylene blue solution for 1 to 2 s and allow them to dry.

Methylene blue	0.65 g
Disodium hydrogen phosphate (Na_2HPO_4)	2.0 g
Potassium dihydrogen phosphate (KH_2PO_4)	0.65 g
Distilled water	1.0 liter

Mix and store in a stoppered brown bottle.
The benefits of such an approach are as follows.

- The RBCs are hemolyzed.
- Color is introduced into the cytoplasm of the parasites.
- Prestaining with methylene blue helps preserve the organisms on the smear.

13. Can blood stains other than Giemsa stain be used to stain the blood films?

Detection of parasites in the blood has been made possible by the use of Romanovsky-type differential stains which selectively color the nuclear material red and the cytoplasm blue. This reaction takes place under optimal pH conditions: rather than using a neutral pH, optically perfect results are obtained using the slightly basic pH of 7.2 (due to the optical dominance of red pigmentation). However, some like the color modifications seen using water at pH 6.8. Although Giemsa stain has been the stain of choice for many years, the parasites can also be seen on blood films stained with Wright's stain, a Wright-Giemsa combination stain, or one of the more rapid stains, such as Diff-Quik (various manufacturers), Wright's Dip Stat stain (Medical Chemical Corp., Torrance, CA), or Field's stain. It is more appropriate to use a stain with which you are familiar than to use Giemsa, which is somewhat more complicated to use, if you are not familiar with it. WBCs serve as the quality control organism for any of the blood stains. Any parasites present stain like the WBCs, regardless of the stain used. Also, the CAP checklist does not mandate the use of Giemsa stain.

14. How should malaria blood films (both thick and thin films) be examined?

A minimum of 300 oil immersion fields should be examined using the 100× objective. The blood film can be scanned using a 50× or 60× oil immersion lens, but final reporting of the results should be based on the use of the 100× oil immersion lens for a total magnification of ×1,000. In order to cover the possibility of microfilariae, the blood films should be thoroughly screened using the 10× low-power objective prior to using oil immersion objectives.

15. What type of quality control (QC) slides should be used for blood parasite work?

Regardless of the stain you are using (Giemsa, Wright's, Wright-Giemsa, or rapid stains), **your QC slide is the actual slide you are staining.** This approach to QC is acceptable to CAP as well. Any parasites present stain like WBCs, so your QC is built into the system. Another good source of teaching slides is Meridian Bioscience, Inc. (Cincinnati, OH). This company currently provides proficiency testing specimens and has extra slides and specimens available for sale. Since these have been reviewed a number of times before being accepted for proficiency testing, the quality is good. Contact them for a brochure: (800) 543-1980 ext. 335 (Proficiency and Controls Sales Coordinator) or jross@meridianbioscience.com.

16. What problems can occur when automated hematology instruments are used for the examination of blood films for parasites?

Although the use of automated blood cell analyzers is not yet clinically relevant for the diagnosis of blood parasites, improvements in the systems may offer future information that will supplement the routine microscopy procedures currently in use. Unfortunately, when the parasitemia is light, automation tends to have some of the problems seen with other alternative procedures. In many cases, the changes seen using analyzers are not specific to malaria infection and could occur in many other diseases (increases in the number of large, unstained cells and thrombocytopenia). The accuracy for malaria diagnosis appears to vary according to the *Plasmodium* species, parasitemia, immunity, and clinical context. **In patients with a very light parasitemia (*Plasmodium* spp. and/or *Babesia* spp.), organisms may be missed, resulting in a false-negative report. Even if a positive result is obtained using automation, microscopy is still required for species determination and parasite quantitation.**

Organism Identification

1. Why is it so important to rule out infections with *P. falciparum* or *P. knowlesi*?

P. falciparum and *P. knowlesi* cause more serious disease than the other species (*P. vivax*, *P. ovale*, and *P. malariae*). Both species tend to invade all ages of RBCs, and the proportion of infected cells may exceed 50%. Schizogony occurs in the internal organs (spleen, liver, bone marrow, etc.) rather than in the circulating blood. Particularly in *P. falciparum* infections, ischemia caused by the plugging of vessels within these organs by masses of parasitized RBCs causes various symptoms, depending on the organ involved.

The onset of a *P. falciparum* malaria attack occurs 8 to 12 days after infection and is preceded by 3 to 4 days of vague symptoms, such as aches, pains, headache, fatigue, anorexia, or nausea. The onset is characterized by fever, a more severe headache, and nausea and vomiting, with occasional severe epigastric pain. There may be only a feeling of chilliness at the onset of fever. Periodicity of the cycle is not established during the early stages, and the presumptive diagnosis may be totally unrelated to a possible malaria infection. If the fever does develop a synchronous cycle, it is usually a cycle of somewhat less than 48 h. In the case of *P. knowlesi*, the cycle is approximately 24 h, the most rapid of all five species infecting humans. An untreated primary attack of *P. falciparum* malaria usually ends within 2 to 3 weeks. True relapses from the liver do not occur, and after a year, recrudescences are rare. Severe or fatal complications of *P. falciparum* and *P. knowlesi* malaria can occur at any time during the infection and are related to the plugging of vessels in the internal organs; the symptoms depend on the organ(s) involved. **Note: The primary objective when performing a blood film examination for parasites is to rule out *P. falciparum* or *P. knowlesi*. It is also becoming more widely recognized that *P. vivax* can cause severe sequelae; this infection is not nearly as benign as previously thought.**

2. What are some of the problems associated with the differentiation between *Plasmodium* spp. and *Babesia* spp.? Can *Babesia* spp. cause serious infections?

The ring forms of all five species of *Plasmodium* can mimic the ring forms of *Babesia* spp. Multiple rings per cell are more typical of *P. falciparum* than of the other species causing human malaria. Babesia rings are often numerous and smaller and tend to be very pleomorphic, while those of *P. falciparum* tend to be fewer and more consistent in size and shape. Remember that *P. falciparum* rings may appear somewhat larger if the blood has been drawn in EDTA and there is any lag time prior to thick- and thin-film preparation. Differentiation between *Plasmodium* spp. and *Babesia* spp. may be impossible without examining a thin blood film (rather than a thick blood film). Often, the parasitemia in a *Babesia* infection is heavier than that in a *P. falciparum* infection, particularly when a patient presents early in the infection. It is also important to remember that some *Babesia* spp. can cause severe illness (*Babesia divergens* from Europe, etc. [42% mortality], and *Babesia* from California, Oregon, Washington, and Missouri), while some infections, such as those due to *Babesia microti*, cause less serious or subclinical infections. **However, any of the *Babesia* spp. can cause severe disease in immunocompromised patients, particularly patients who have undergone splenectomy. Rings seen outside the RBCs generally suggest *Babesia* (Figure 6.4, square); however, in heavy malaria infections, rings can occasionally be seen outside the RBCs.**

It is also important to remember that not all *Babesia* spp. exhibit the ring configuration called the Maltese cross (Figure 6.4, circles).

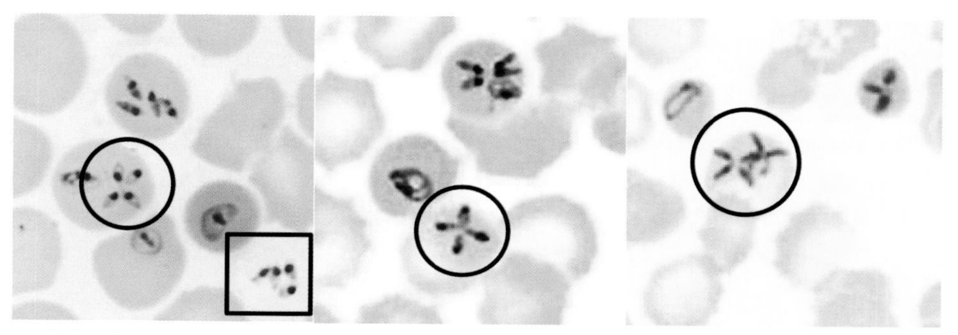

Figure 6.4 *Babesia* spp. The ring configuration called the Maltese cross (circles) may not be seen in all *Babesia* spp. Rings seen outside the RBCs (square) generally suggest *Babesia*; however, in heavy malaria infections, rings can occasionally be seen outside the RBCs.

3. **What is the significance of finding only ring forms on two sets of blood films drawn 6 hours apart?**

   Remember that all of the life cycle stages (rings, developing trophozoites, early schizonts, late schizonts, mature schizonts, and gametocytes) can be seen on the blood films in infections with *P. vivax*, *P. ovale*, *P. malariae*, and *P. knowlesi*. Due to the unique characteristics of the life cycle, only rings and gametocytes (and occasional mature schizonts) are seen in the peripheral blood of a patient with a *P. falciparum* infection. **Therefore, if you see two sets of blood films (collected 6 hours apart) that contain ring forms only, there is an excellent chance that the patient is infected with *P. falciparum*, which causes the most serious infections of the five human-pathogenic *Plasmodium* spp.**

4. **Why aren't gametocytes of *P. falciparum* seen in many patients presenting to the emergency room?**

   Many patients present to the emergency room early in the infection, prior to the formation of the gametocytes; thus, the diagnosis must be made on the basis of seeing the ring forms only. As you can imagine, identification of *Plasmodium* organisms to the species level can be very difficult when only rings are present. These patients tend to be travelers who have had no prior exposure to *P. falciparum* (immunologically naive persons) and who become symptomatic very early after being infected. It normally takes approximately 10 days for the crescent-shaped *P. falciparum* gametocytes to form. **The presence or absence of gametocytes must be reported (this will determine therapy).**

5. **What should be considered if the patient has been diagnosed with a *P. falciparum/P. malariae* mixed infection, particularly with a history of travel to Malaysia?**

   P. knowlesi mimics *P. falciparum* early in the infection, while the morphology mimics that of *P. malariae* later in the cycle. Although relatively uncommon, infections with the fifth agent of human malaria, *P. knowlesi*, are being seen in other parts of the world, primarily in travelers who have been to Malaysia.

Reporting

1. **Why is it important to identify malaria organisms to the species level?**

   Since *P. falciparum* and *P. knowlesi* can cause severe disease and death, it is very important for the physician to know whether these infections can be ruled out. It is also important to know if any of the other three species are present, particularly *P. vivax* or *P. ovale*, which would require therapy for both the liver and RBC stages due to potential relapse from the liver stages. It is also important because of potential drug resistance (chloroquine resistance of *P. falciparum* and *P. vivax*; primaquine tolerance or resistance of *P. vivax* [rare but documented]).

2. **How should a positive malaria blood film be reported?**

   - Using the thin-blood-film method, report the percentage of parasite-infected RBCs per 100 RBCs counted. Always report the presence or absence of gametocytes.

     Example: *Plasmodium falciparum*, parasitemia = 0.01%, no gametocytes seen

   - Using the thick/thin-blood-film method, report the number of parasites per microliter of blood.

     Example: *Plasmodium falciparum*, parasitemia = 10,000 per µl of blood, no gametocytes seen

3. **How should results be reported if *Plasmodium* sp. parasites are seen but *P. falciparum* or *P. knowlesi* infection cannot be ruled out?**
It is important to convey to the physician that *P. falciparum* or *P. knowlesi* cannot be ruled out; therapy may be initiated on the assumption that this species might be present. The report should read, "***Plasmodium* spp. present; unable to rule out *Plasmodium falciparum* or *Plasmodium knowlesi*, no gametocytes seen.**"

4. **How often do mixed infections occur, and how should they be reported?**
It is important to remember that mixed infections are much more common than suspected and/or reported. When rings are present, along with other developing stages (*P. vivax*, *P. ovale*, *P. malariae*, or *P. knowlesi*), always look for the presence of two populations of ring forms, one of which might be *P. falciparum*. The report should read, "***Plasmodium* spp. present, possible mixed infection; unable to rule out *P. falciparum* or *P. knowlesi*. No gametocytes seen.**" Another report example is, "***Plasmodium vivax* rings, developing schizonts, and gametocytes; possible mixed infection: unable to rule out *P. falciparum* or *P. knowlesi*.**"

5. **Why is it important to report the *Plasmodium* stages seen in the blood films?**
Patients who have been diagnosed with *Plasmodium* infection and who are not suspected of having drug-resistant malaria (*P. falciparum* or *P. vivax*) are often treated with chloroquine. Chloroquine does not eliminate any gametocytes present, and there are mosquitoes in the United States that can transmit malaria if they take a blood meal from an individual with gametocytes in the blood. Thus, it is important for the physician to know which stages are present in the blood (rings, developing trophozoites, schizonts, and/or gametocytes).

6. **How do parasitemia and malaria severity correlate?**
The following percentages are helpful in interpretation of malaria severity:

    Parasitemia of >10,000/μl, heavy infection

    0.002%, 100/μl, immunologically naive, symptomatic below this level

    0.2%, 10,000/μl, immune patients symptomatic

    0.1%, 5,000/μl, minimum sensitivity of rapid malaria test (BinaxNOW)

    2%, 100,000/μl, maximum parasitemia for *P. vivax* and *P. ovale* (rarely go above 2%)

    2 to 5%, up to 250,000/μl, severe malaria, mortality

    10%, 500,000/μl, exchange transfusion

    Note: Patients with no prior exposure (immunologically naive) to malaria can present with very low parasitemias and nonspecific symptoms. Those living in areas of endemicity with past (often several) exposures to malaria tend to have antibody and will present with a higher parasitemia and more typical symptoms (including periodic fevers).

Proficiency Testing

1. **How should blood films be examined for proficiency testing?**
Since you do not know what organisms might be present, always review the blood films with the 10× objective (review the **entire** slide). This examination is likely to reveal any microfilariae that are present; however, small parasites like *Plasmodium* and

Babesia may be missed. Before reporting the smear as negative, examine at least 300 oil immersion fields with the 100× oil immersion lens.

2. **Do proficiency testing blood films match those seen from actual patients?**
Yes and no! Blood films for proficiency testing (PT) are actual patient specimens; however, they may have a higher parasitemia than is seen in many patients reporting to the emergency room, clinic, etc. Often, smears contain a higher parasitemia than is commonly seen. Therefore, the PT smears represent a mix, some of which are fairly typical and some of which contain a large number of organisms. **As in the case of specimens from travelers, the thin blood film may appear to be negative, while the thick film is positive. In a clinical setting, these blood film results are not unusual, particularly in an immunologically naive patient (e.g., a traveler with no prior exposure to *Plasmodium* spp.).**

3. **What blood parasites might be seen in proficiency testing specimens?**
The following parasites may be seen in proficiency testing specimens: *P. falciparum*, *P. vivax*, *P. ovale*, *P. malariae*, mixed malaria infections, *Babesia* spp., *Trypanosoma cruzi*, *Trypanosoma brucei rhodesiense*/*Trypanosoma brucei gambiense*, and microfilariae.

General Questions

1. **What stains are recommended for staining microfilariae?**
Although Giemsa stain is generally recommended, it does not stain the sheath of *Wuchereria bancrofti*; hematoxylin-based stains (Delafield's stain) are recommended. Also, the sheath of *Brugia malayi* stains pink with Giemsa stain.

2. **What other methods can be used for the identification of blood parasites?**
Concentrations (trypanosomes, microfilariae, leishmaniae), buffy coat thick and thin blood films (all blood parasites, including *Plasmodium* spp.), skin biopsies and bone marrow aspirates (Giemsa or other blood stains) (leishmaniae), and rapid methods/dipstick formats (malaria and microfilariae) can all be used (Figure 6.5). **Note: The Binax (Abbott) rapid test for *Plasmodium falciparum* and *Plasmodium* spp.**

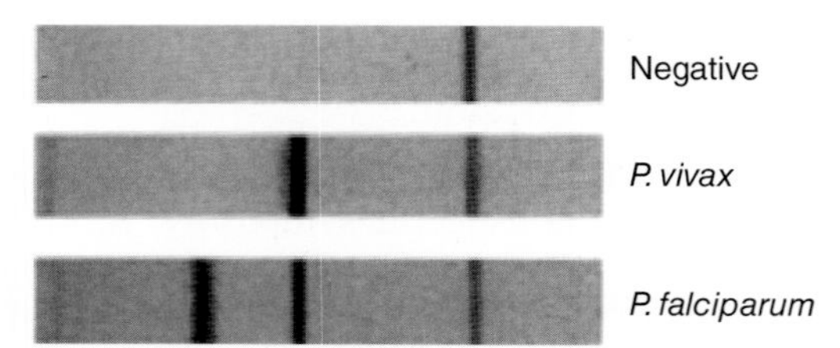

Figure 6.5 General diagram of a rapid malaria test. (Top) The negative test shows the control line only; (middle) a control line plus the *Plasmodium vivax* line indicates the presence of a panspecific antigen (common to all *Plasmodium* spp. but most sensitive for *P. vivax* rather than *P. ovale* and *P. malariae*); (bottom) a control line, a panspecific antigen line, and a line specific for *Plasmodium falciparum* antigen presence are visible. Note: most cartridge rapid malaria tests have not been developed to detect *Plasmodium knowlesi* (they are still in development).

has been cleared by the FDA (June 2007); a positive control is also available (https://www.fishersci.com/shop/products/alere-binaxnow-malaria-tests-positive-control-kit/bnx665010). The Binax rapid test is as sensitive as a good microscopist; however, often these skilled individuals may not be available during off-hour shifts. It is important to recognize the pros and cons of the rapid test compared to microscopic examination of blood films, particularly those from travelers in whom the parasitemia may be considerably less than 0.1%.

3. How good are serologic tests for malaria, and when should they be performed?

Malaria serologies are available through CDC. They always test with antigens of all five species, but the reactions are not species specific—there is a lot of cross-reactivity. *Plasmodium* spp. can be identified in about 80% of NONIMMUNE patients in their PRIMARY infection (as was found among those serving in the U.S. Army during the Vietnam War), but species cannot be determined in patients with long-term exposure and multiple infections (e.g., Africans). CDC does not recommend serology for identification to the species level; PCR is the best tool for this purpose if pretreatment EDTA-blood is available.

4. How do you categorize malaria resistance (seen in both *P. falciparum* and *P. vivax*)?

Table 6.2

Definition	Comments
Sensitive	From initiation of therapy, asexual parasites are cleared by day 6; no evidence of recrudescence up to day 28. • Peripheral blood films appear to go from positive to negative very quickly (can be a change from one draw to the second draw 6 h later).
Resistance type I	From initiation of therapy, asexual parasites have cleared for at least two consecutive days (the latest day being day 6); recrudescence follows. • Parasite count initially drops and blood films appear to be negative; patient should be monitored for a period of days, particularly if drug-resistant *P. falciparum* is suspected.
Resistance type II	Within 48 h of initiation of therapy, marked reduction of asexual parasitemia to <25% of pretreatment count; however, no subsequent disappearance of parasitemia (smear positive on day 6) • Patient appears to be improving; parasite count drops, but blood films always appear positive.
Resistance type III	Modest reduction in parasitemia may be seen; no change or increase in parasitemia is seen during first 48 h after treatment; no clearing of asexual parasites. • In some cases, the parasite count continues to increase with no visible decrease at any time; blood films show overall parasite increase.

5. How soon after initiation of therapy for *P. vivax* malaria would one expect to see clearance in blood smears? How soon should follow-up blood smears be submitted for evaluation after therapy?

Although we certainly see fewer cases in the United States than in areas of endemicity, we begin to see a decrease in parasite numbers very quickly (within a few hours) after the initiation of therapy. If an individual presents in the emergency room (ER), the parasitemia is often below 1% (0.1 to 0.01%); the patient is given therapy and is not admitted to the hospital (in the case of *P. vivax*, *P. malariae*, or *P. ovale*). If the organism is identified as *P. falciparum*, *P. knowlesi*, or a mix (the most common are *P. falciparum-P. vivax* and *P. falciparum-P. malariae*), the patient is admitted. In some cases, we do not receive blood for follow-up examination (from ER patients for whom *P. falciparum* and *P. knowlesi* have been ruled out). When a patient is being treated for *P. falciparum*, we receive blood samples approximately every 4 to 6 hours for routine checks (patients are admitted to the hospital for treatment). Although the number of resistant *P. vivax* cases is small, such resistance has been confirmed, so this is always a consideration, as are mixed infections with *P. falciparum* and *P. vivax*.

6. **Where can one get PCR performed for the malaria diagnosis and identification to the species level?**
CDC performs PCR for malaria, which identifies the organisms to the species level. See also https://www.cdc.gov/malaria/diagnosis_treatment/index.html. CDC requires about 0.5 ml of pretreatment EDTA-blood. Also, if possible, they like to review both thick and thin smears if extra blood films are available. Specimens can be sent overnight to:

   Division of Parasitic Diseases

   Centers for Disease Control and Prevention

   Building 109, Room 1302

   4770 Buford Highway NE

   Atlanta, GA 30341

7. **What is the current status of rapid testing for malaria?**
This issue has been a difficult one to resolve in the United States. The Binax rapid test was approved in June 2007 (Figure 6.5) (see Table 6.3). The company now has a positive control available for customers. Certainly, the use of these rapid tests would be a tremendous advantage for personnel on the night shift. However, a negative test would not rule out a *Plasmodium* infection. If the test was positive, follow-up testing would need to be performed on slides to detect mixed infections. All kits are helpful (according to published data), but they are no better and somewhat less accurate than a good microscopist. However, as we all know, these individuals are often not available during off-hour shifts. Many places have solved the problem by just bringing someone in to prepare and read slides during off hours. Often, the afternoon shifts prepare and stain smears (Wright-Giemsa or one of the rapid stains) and another individual examines the thick and thin stained blood films. We exposed our residents, postdocs, etc., to these kits and found them to be helpful. However, when dealing with immunologically naive patients with very low parasitemias, it would be easy to miss an infection unless careful examination of the thick films was performed. Once the external QC specimen is commercially available, it will be very helpful to have this rapid test available. **However, the rapid test will not take the place of careful examination of both thick and thin blood films. In other parts of the world, there are a number of rapid malaria tests available; however, they are not FDA cleared for sale within the United States.**

Table 6.3

Parasitemia	Comments
>10,000/μl	Heavy infection
0.0001–0.0004%, 5–20/μl	Required for a positive thick blood film
0.002%, 100/μl	Immunologically naive patients may be symptomatic below this level
0.02%, 1,000/μl	Can be seen in travelers in ER (these patients are symptomatic very early with low parasitemia, meaning that they are immunologically naive)
0.1%, 5,000/μl	**BinaxNOW rapid lateral flow method:** level of sensitivity below this percentage drops significantly—problem with travelers (0.01% or less)
0.2%, 10,000/μl	Immune patients are symptomatic
2%, 100,000/μl	Maximum parasitemia for *P. vivax* and *P. ovale*; rarely goes above 2%
2–5%, up to 250,000/μl	Severe malaria, high mortality
10%, 500,000/μl	Exchange transfusion usually required

8. Can mosquitoes be infected with more than one species of *Plasmodium*?
There appears to be no barrier to infection of *Anopheles* with mixed *Plasmodium* species. Mosquitoes doubly infected with *P. falciparum* and *P. vivax* were able to transmit both species to humans following deliberate feeding on volunteers. In natural situations, however, suppressive effects in the human host may lead to overlapping waves of gametocytes of different species, so that there is a tendency for mosquitoes to be infected with only one species. In some cases, there may be specific suppression of sporogony of one species compared to the other. For example, in areas where *P. falciparum* and *P. malariae* are sympatric, the cold temperatures associated with altitude may be disadvantageous to *P. falciparum*, which requires a higher temperature for sporogony than *P. malariae*. This may lead to seasonal changes in the prevalence of the two species.

9. What is the general status of PCR for leishmaniasis?
PCR for leishmaniasis has been through some very rigorous validation at the Leishmaniasis Laboratory at Walter Reed Army Institute of Research and is awaiting FDA clearance. Contact information is available from https://www.wrair.army.mil/collaborate/leishmania-diagnostics-laboratory.

10. Are serologic tests available for the African trypanosomes?
CDC does not perform PCR for African trypanosomiasis. It is unnecessary for diagnosing *T. b. rhodesiense* (these parasites are easily detectable by microscopy). Serologic testing for *T. b. gambiense* is used for screening purposes only and is not available within the United States. The Parasitic Diseases Hotline for Healthcare Providers (non-malaria parasitic diseases) is as follows:
Telephone: (404) 718-4745; after hours: (770) 488-7100
Email: parasites@cdc.gov
Hours: 8 a.m.–4 p.m. EST, Monday–Friday

Suggested Reading

Anderson NW, Buchan BW, Ledeboer NA. 2014. Comparison of the BD MAX enteric bacterial panel to routine culture methods for detection of *Campylobacter*, enterohemorrhagic *Escherichia coli* (O157), *Salmonella*, and *Shigella* isolates in preserved stool specimens. *J Clin Microbiol* **52:**1222–1224.

Becton, Dickinson and Co. 2011. Affirm VPIII microbial identification test package insert. Becton, Dickinson and Co., Franklin Lakes, NJ.

Becton, Dickinson and Co. 2015. BD MAX enteric parasite panel package insert. Becton, Dickinson and Co., Franklin Lakes, NJ. https://moleculardiagnostics.bd.com/wp-content/uploads/2017/08/Enteric-Parasite-Panel-Info-Sheet.pdf

BioFire Diagnostics, LLC. 2017. FilmArray gastrointestinal (GI) panel CE IVD instruction booklet. BioFire Diagnostics, LLC, Salt Lake City, UT. Available from https://www.online-ifu.com/ITI0030/14480/EN.

Biswas JS, Al-Ali A, Rajput P, Smith D, Goldenberg SD. 2014. A parallel diagnostic accuracy study of three molecular panels for the detection of bacterial gastroenteritis. *Eur J Clin Microbiol Infect Dis* **33:**2075–2081.

Brown HL, Fuller DD, Jasper LT, Davis TE, Wright JD. 2004. Clinical evaluation of Affirm VPIII in the detection and identification of *Trichomonas vaginalis*, *Gardnerella vaginalis*, and *Candida* species in vaginitis/vaginosis. *Infect Dis Obstet Gynecol* **12:**17–21.

Buss SN, Leber A, Chapin K, Fey PD, Bankowski MJ, Jones MK, Rogatcheva M, Kanack KJ, Bourzac KM. 2015. Multicenter evaluation of the BioFire FilmArray gastrointestinal panel for etiologic diagnosis of infectious gastroenteritis. *J Clin Microbiol* **53:**915–925.

Code of Federal Regulations. 2015. Title 29, CFR 1910.1048—formaldehyde. US Government Printing Office, Washington, DC.

Federal Register. 1992. Occupational exposure to formaldehyde—OSHA. Response to court remand; final rule. *Fed Regist* **57:**22290–22328.

Garcia LS. 2016. *Diagnostic Medical Parasitology*, 6th ed. ASM Press, Washington, DC.

Garcia LS, Johnston SP, Linscott AJ, Shimizu RY. 2008. *Cumitech 46, Laboratory procedures for diagnosis of blood-borne parasitic diseases*. Coordinating ed, Garcia LS. ASM Press, Washington, DC.

Gen-Probe, Inc. August 2012. *APTIMA Trichomonas vaginalis assay package insert*. Gen-Probe, Inc., San Diego, CA.

Goldenberg SD, Bacelar M, Brazier P, Bisnauthsing K, Edgeworth JD. 2015. A cost benefit analysis of the Luminex xTAG gastrointestinal pathogen panel for detection of infectious gastroenteritis in hospitalised patients. *J Infect* **70:**504–511.

Hanson KL, Cartwright CP. 2001. Use of an enzyme immunoassay does not eliminate the need to analyze multiple stool specimens for sensitive detection of *Giardia lamblia*. *J Clin Microbiol* **39:**474–477.

Hiatt RA, Markell EK, Ng E. 1995. How many stool examinations are necessary to detect pathogenic intestinal protozoa? *Am J Trop Med Hyg* **53:**36–39.

Imwong M, Nakeesathit S, Day NP, White NJ. 2011. A review of mixed malaria species infections in anopheline mosquitoes. *Malar J* **10:**253.

Khare R, Espy MJ, Cebelinski E, Boxrud D, Sloan LM, Cunningham SA, Pritt BS, Patel R, Binnicker MJ. 2014. Comparative evaluation of two commercial multiplex panels for detection of gastrointestinal pathogens by use of clinical stool specimens. *J Clin Microbiol* **52:**3667–3673.

Kobayashi J, Hasegawa H, Soares EC, Toma H, Dacal AR, Brito MC, Yamanaka A, Foli AA, Sato S. 1996. Studies on prevalence of *Strongyloides* infection in Holambra and Maceió, Brazil, by the agar plate faecal culture method. *Rev Inst Med Trop Sao Paulo* **38:**279–284.

Leber AL (ed). 2016. Section 9: Parasitology. *In Clinical Microbiology Procedures Handbook*, 4th **ed.** ASM Press, Washington, DC.

Luminex. 2012. Luminex xTAG gastrointestinal pathogen panel package insert. Luminex Molecular Diagnostics, Toronto, Canada.

Navidad JF, Griswold DJ, Gradus MS, Bhattacharyya S. 2013. Evaluation of Luminex xTAG gastrointestinal pathogen analyte-specific reagents for high-throughput, simultaneous detection of bacteria, viruses, and parasites of clinical and public health importance. *J Clin Microbiol* **51:**3018–3024.

NCCLS. 2000. Use of blood film examination for parasites. *Approved guideline M15-A*. NCCLS, Wayne, PA.

NCCLS. 1997. Procedures for the recovery and identification of parasites from the intestinal tract, 2nd ed. *Approved guideline M28-A*. NCCLS, Wayne, PA.

Nye MB, Schwebke JR, Body BA. 2009. Comparison of APTIMA *Trichomonas vaginalis* transcription-mediated amplification to wet mount microscopy, culture, and polymerase chain reaction for diagnosis of trichomoniasis in men and women. *Am J Obstet Gynecol* **200:**188.e1–188.e7.

Perry MD, Corden SA, Howe RA. 2014. Evaluation of the Luminex xTAG gastrointestinal pathogen panel and the Savyon Diagnostics gastrointestinal infection panel for the detection of enteric pathogens in clinical samples. *J Med Microbiol* **63:**1419–1426.

Radonjic IV, Dzamic AM, Mitrovic SM, Arsic Arsenijevic VS, Popadic DM, Kranjcic Zec IF. 2006. Diagnosis of *Trichomonas vaginalis* infection: the sensitivities and specificities of microscopy, culture and PCR assay. *Eur J Obstet Gynecol Reprod Biol* **126:**116–120.

Schwebke JR, Gaydos CA, Davis T, Marrazzo J, Furgerson D, Taylor SN, Smith B, Bachmann LH, Ackerman R, Spurrell T, Ferris D, Burnham CAD, Reno H, Lebed J, Eisenberg D, Kerndt P, Philip S, Jordan J. 2018. Clinical evaluation of the Cepheid Xpert TV assay for detection of *Trichomonas vaginalis* with prospectively collected specimens from men and women. *J Clin Microbiol* **56:**e01091-17.

United States Department of Labor, Occupational Safety and Health Administration. 2001. Occupational Safety and Health Standards, Toxic and Hazardous Substances: Sampling strategy and analytical methods for formaldehyde. Standard number 1910.1048 App B. https://www.osha.gov/laws-regs/regulations/standardnumber/1910/1910.1048AppB

Wessels E, Rusman LG, van Bussel MJ, Claas EC. 2014. Added value of multiplex Luminex gastrointestinal pathogen panel (xTAG® GPP) testing in the diagnosis of infectious gastroenteritis. *Clin Microbiol Infect* **20:**O182–O187.

World Health Organization. 2018. Malaria rapid diagnostic test performance. Results of WHO product testing of malaria RDTs: Round 8 (2016–2018), https://www.who.int/malaria/publications/atoz/9789241514965/en/.

SECTION 7

Parasite Identification

Practical Guide to Diagnostic Parasitology, Third Edition. Lynne S. Garcia
 DOI: 10.1128/9781683673637.ch07

Entamoeba histolytica

Pathogenic	Yes
Disease	Amebiasis
Acquired	Fecal-oral transmission; contaminated food and water
Body site	Intestine, liver, skin, miscellaneous body sites
Symptoms	*Intestinal*: Diarrhea, dysentery
	Extraintestinal (liver): Right upper quadrant pain, fever
Clinical specimen	*Intestinal*: Stool, sigmoidoscopy specimens
	Extraintestinal: Liver aspirate, biopsy, serum for antibody
Epidemiology	Worldwide, primarily human-to-human transmission
Control	Improved hygiene, adequate disposal of fecal waste, adequate washing of contaminated fruits and vegetables

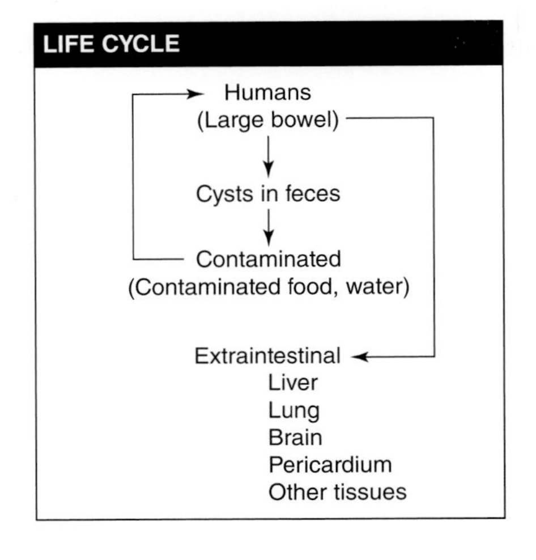

Diagnosis

The standard ova and parasite examination (O&P exam) is recommended for recovery and identification of *Entamoeba histolytica* in stool specimens. Microscopic examination of a direct saline wet mount may reveal motile trophozoites, which may contain red blood cells (RBCs). However, in many patients who do not present with acute dysentery, trophozoites may not contain RBCs. An asymptomatic individual may have few trophozoites or possibly only cysts in the stool. Although the concentration technique is helpful in demonstrating cysts, *the most important technique for recovery and identification of protozoan organisms is the permanent stained smear* (normally stained with trichrome or iron hematoxylin). At least three specimens collected over no more than 10 days is often recommended. Unless one is working in an area of high prevalence, the majority of organisms seen will probably not contain RBCs; thus, the diagnosis would be *Entamoeba histolytica/E. dispar* complex (or group) (pathogenicity cannot be determined from morphology). In many parts of the world, you are more likely to see some of the other pathogens such as *Blastocystis* spp., *Giardia lamblia*, or *Dientamoeba fragilis*. Nonpathogens might include *Endolimax nana* or *Iodamoeba bütschlii*. Confirmation of the true pathogen *Entamoeba histolytica* could also be accomplished by using an organism-specific fecal immunoassay (e.g., TechLab *E. histolytica* Quik Chek) (identifies the true pathogen, not the *E. histolytica/E. dispar* group).

Diagnostic Tips

If the permanent stained smear is quite dark, a delicate *E. histolytica* cell may appear more like *Entamoeba coli*; on a very thin, pale smear, *E. coli* can appear more like *E. histolytica*. Also, when organisms are dying or have been poorly fixed, they appear more highly vacuolated, looking like *E. coli* trophozoites rather than *E. histolytica*. Diagnosis of liver abscess can be made by identification of organisms from liver aspirate material, a procedure which is rarely performed. Serologic tests are recommended (extraintestinal cases, not gastrointestinal infections limited to the gut).

Additional tips include the following:

1. The morphology of the trophozoite karyosome (compact, NOT blot-like) is more important than the position (central or eccentric).
2. As the trophozoite begins to deteriorate, the cytoplasm becomes more highly vacuolated.

3. The overall staining color can vary tremendously (trichrome: blue, green, red tones: iron hematoxylin: blue, gray, dark gray tones).
4. Trophozoites may mimic macrophages; cysts may mimic polymorphonuclear leukocytes (PMNs) (see below).
5. Ingested RBCs appear round as they are digested (yeasts appear oval, both in the background and within the trophozoite cytoplasm); RBCs in the background stain appear to be very pleomorphic. (See Hiatt et al. and Stanley, Suggested Reading.)

Artifact	Resemblance	Differential characteristics of artifacts (permanent stain)		
		Saline mount	Cytoplasm	Nucleus
PMNs (seen in dysentery and other inflammatory bowel diseases)	*E. histolytica*/ *E. dispar* cysts	Usually not a problem if cells are from fresh blood. Granules in cytoplasm. Cell border irregular	Less dense, often frothy. Border less clearly demarcated than that of amebae. May look very similar to that in protozoa	Coarser. Larger, relative to size of organism. Irregular shape and size. Chromatin unevenly distributed. Chromatin strands may link nuclei or may appear to be 4 separate nuclei as the cell ages; note the fragmented nucleus (mimics multiple nuclei)
Macrophages (seen in dysentery and other inflammatory bowel diseases; may be present in purged specimens)	Amebic trophozoites, especially *E. histolytica*/ *E. dispar*	Nuclei larger and irregular in shape, with irregular chromatin distribution. Cytoplasm granular; may contain ingested debris. Cell border irregular and indistinct. Movement irregular; pseudopodia indistinct but may mimic protozoa	Coarse. May contain inclusions. May include RBCs; if RBCs present, will mimic *E. histolytica*	Large and often irregular in shape. Chromatin irregularly distributed. May appear to have karyosome. Will have more nuclear material per cell than protozoan trophozoite (nucleus may be absent)

General Comments

Although many people worldwide are infected, only a few develop clinical symptoms. Morbidity and mortality due to *E. histolytica* vary depending on geographic area, organism strain, and the patient's immune status. For many years, the issue of pathogenicity has been very controversial: some felt that what was called *E. histolytica* was really two separate species, one being pathogenic and causing invasive disease and the other being nonpathogenic and causing mild or asymptomatic infections, and others felt that all organisms designated *E. histolytica* were potentially pathogenic, with symptoms depending on the result of host or environmental factors, including intestinal flora.

Based on current knowledge, pathogenic *E. histolytica* is considered the etiologic agent of amebic colitis and extraintestinal abscesses, while nonpathogenic *E. dispar* produces no intestinal symptoms and is not invasive in humans. The division of *Entamoeba* into invasive *E. histolytica* and noninvasive *E. dispar* by isoenzyme analysis is supported by genetic differences; pathogenesis differences also help explain the epidemiology, clinical syndromes, and pathology of amebiasis. The total body of evidence supports the differentiation of pathogenic *E. histolytica* from nonpathogenic *E. dispar*. Amebiasis has been diagnosed in approximately 12% of travelers returning from the developing world with acute diarrhea.

Although *Entamoeba moshkovskii* and *Entamoeba bangladeshi* are morphologically very similar to *E. histolytica* and *E. dispar*, these two species are not normally reported as a part of the *E. histolytica/E. dispar* group. The addition of these names to the report is not clinically relevant and may be more confusing than helpful.

Fecal immunoassays can identify the *E. histolytica/E. dispar* group in stool, or they can confirm the presence of the true pathogen *E. histolytica*; fresh or frozen stool is required. The two formats available for confirmation of the true pathogen, *E. histolytica*, are the enzyme immunoassay (EIA) tray and the rapid cartridge (TechLab/Abbott).

Description (Trophozoite)

Living trophozoites vary in diameter from about 10 to 60 μm. Organisms recovered from diarrheic or dysenteric stools are generally larger than those in a formed stool from an asymptomatic individual. Motility is rapid and unidirectional, with pseudopods forming quickly in response to the conditions around the organism. The motility may appear sporadic. Although this characteristic motility is often described, it is rare to diagnose amebiasis on the basis of motility seen in a direct mount. The cytoplasm is differentiated into a clear outer ectoplasm and a more granular inner endoplasm.

When the organism is examined on a permanent stained smear (trichrome or iron hematoxylin), the morphologic characteristics are readily seen. The nucleus is characterized by having evenly arranged chromatin on the nuclear membrane and a small, compact, centrally located karyosome. Note: The karyosome appearance (neat, compact) is much more

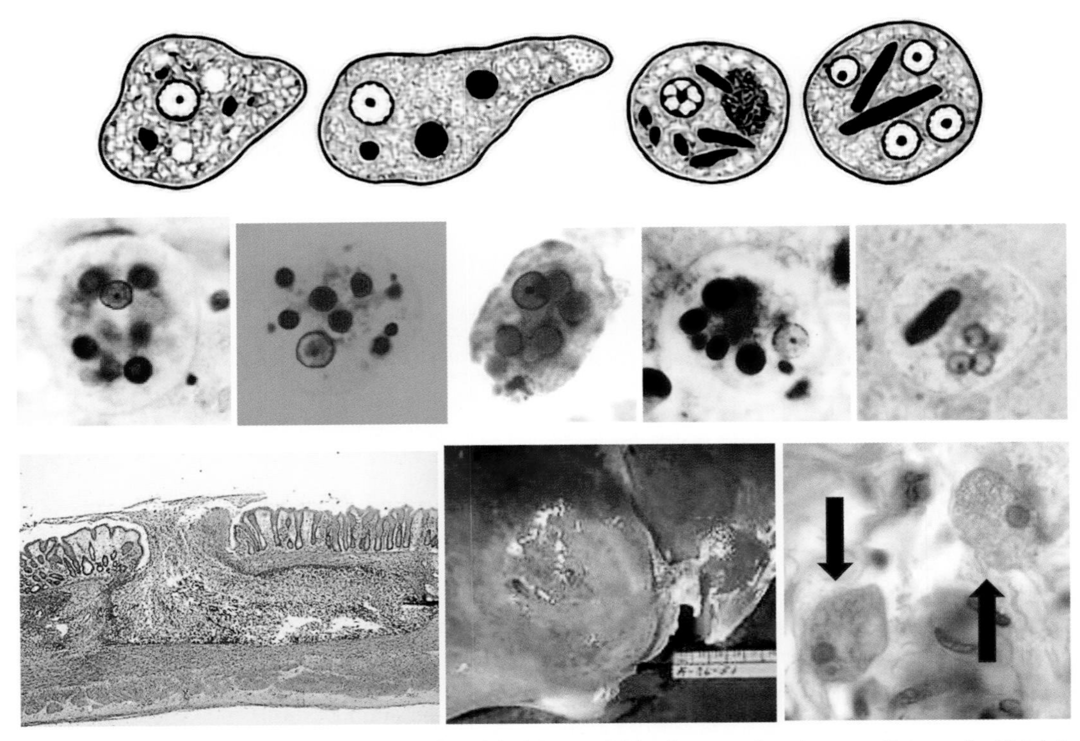

Figure 7.1 ***Entamoeba histolytica*** **(true pathogen). (Top row)** The first two drawings are *Entamoeba histolytica* trophozoites (note ingested RBCs); the third drawing is a precyst with developing chromatoidal bars; the last illustration is a mature cyst containing four nuclei. The precyst and cyst must be identified as *Entamoeba histolytica/E. dispar* complex or group; the presence of the true pathogen, *E. histolytica*, cannot be determined by cyst morphology. **(Middle row)** The first four images are *E. histolytica* trophozoites (note the ingested RBCs). The fifth image is a cyst with a single chromatoidal bar. This organism would need to be identified as *E. histolytica/E. dispar* complex or group (may or may not be the true pathogen, *E. histolytica*). **(Bottom row)** Flask-shaped ulcer in the gastrointestinal tract; abscess in liver; trophozoites (arrows) in histology section.

important than the location (central or eccentric); tremendous variation in location will be seen. The cytoplasm is usually described as finely granular with few ingested bacteria or debris in vacuoles. In a patient with dysentery, RBCs may be visible in the cytoplasm, and this feature is considered microscopically diagnostic for the true pathogen, *E. histolytica*. Most often, infection with *E. histolytica*/*E. dispar* is diagnosed on the basis of trophozoite morphology without the presence of RBCs or cyst morphology (pathogenic *E. histolytica* and nonpathogenic *E. dispar* look identical).

Based on microscopic morphology, the majority of organisms seen cannot be identified as the true pathogen, *E. histolytica*. Therefore, the results would be reported as "*Entamoeba histolytica*/*E. dispar* complex/group trophozoites and/or cysts seen."

Description (Cyst)

For unknown reasons, the trophozoites may condense into a round mass (precyst), and a thin wall is secreted around the immature cyst. There may be two types of inclusions within this immature cyst, a glycogen mass and highly refractile chromatoidal bars with smooth, rounded edges. As the cysts mature (metacysts), there is nuclear division with the production of four nuclei; occasionally, eight nuclei are produced, and the cysts range in size from 10 to 20 μm. Often, as the cyst matures, the glycogen completely disappears; the chromatoidal bars may also be absent in the mature cyst. Cyst formation occurs only within the intestinal tract, not when the stool has left the body. The uni-, bi-, and tetranucleate cysts are infective and represent the mode of transmission from one host to another.

After cyst ingestion, no changes occur in an acid environment; however, once the pH becomes neutral or slightly alkaline, the encysted organism becomes active, with the outcome being four separate trophozoites (small, metacystic trophozoites). These develop into the normal trophozoites when they become established in the large intestine.

E. histolytica trophozoites and cysts can be confused with human macrophages and PMNs, respectively. Also, because there are no ingested RBCs within the cyst forms, the correct diagnosis would be "*Entamoeba histolytica*/*E. dispar* complex or group."

A brief history of *Entamoeba histolytica* is presented below. These studies eventually led to the acceptance of *E. histolytica* as the true pathogen and the designation of *E. dispar* as nonpathogenic (from Pinilla AE, López MC, Viasus DF, *RevMéd Chile* **136:**118–124, 2008, https://scielo.conicyt.cl/scielo.php?script=sci_arttext&pid=S0034-98872008000100015&lng=en&nrm=iso&tlng=en).

1. Feder Losch (1875), in Saint Petersburg, found amebae in fecal samples but only regarded them as responsible for maintaining the inflammatory process, not as a cause of dysentery.
2. Fritz Schaudinn (1903) established the differentiation between *Entamoeba histolytica* and *Endamoeba coli*. Schaudinn decided to call it *E. histolytica* because of its ability to cause tissue lysis.
3. Emile Brumpt (1925), based on experimental studies, pointed out the existence of *E. histolytica* as a species complex comprising two morphologically indistinguishable species: *E. dysenteriae*, the cause of symptomatic infection, and *Entamoeba dispar*, found only in asymptomatic carriers.
4. Louis Diamond et al. (1961) during the 1960s developed an axenic culture medium for *E. histolytica* which allowed *in vivo* and *in vitro* studies.
5. Sargeaunt and Williams (1978) for the first time distinguished *E. histolytica* strains by isoenzyme electrophoresis, thus confirming that *E. histolytica* was a species complex comprising both pathogenic and nonpathogenic species.
6. William Petri et al. (1987) demonstrated that the 170-kDa protein with greater antigenicity was the Gal/GalNac-specific lectin.
7. Diamond and Clark (1993) described again Brumpt's original 1925 hypothesis, concluding definite support for the existence of two morphologically indistinguishable species, pathogenic *E. histolytica* and nonpathogenic *E. dispar*.
8. The World Health Organization accepted this hypothesis in 1997.

Entamoeba histolytica/Entamoeba dispar

Pathogenic	Perhaps
Disease	Same as for *E. histolytica* if the true pathogen is present
Acquired	Fecal-oral transmission; contaminated food and water
Body site	Intestine
Symptoms	Same as for *E. histolytica* if the true pathogen is present
Clinical specimen	Stool
Epidemiology	Worldwide, primarily human-to-human transmission
Control	Improved hygiene, adequate disposal of fecal waste, adequate washing of contaminated fruits and vegetables

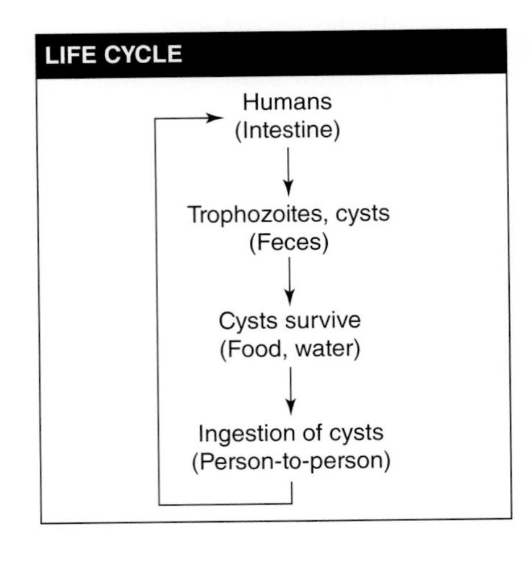

Diagnosis

The separation of pathogenic (*Entamoeba histolytica*) and nonpathogenic (*Entamoeba dispar*) species is not possible based on microscopic morphology, assuming that no ingested RBCs are seen. The standard O&P exam is recommended for recovery and identification of *E. dispar* in stool specimens. Microscopic examination of a direct saline wet mount may reveal motile trophozoites, which do not contain RBCs. Any trophozoite that is the appropriate size but contains ingested RBCs in the cytoplasm should be identified as the true pathogen, *E. histolytica*. Note: Because the true pathogen *E. histolytica* and nonpathogen *E. dispar* look the same if no RBCs are present, the report must be written as follows: "*Entamoeba histolytica/E. dispar* complex or group, trophozoites and/or cysts seen." IT IS INCORRECT TO REPORT *Entamoeba dispar* alone. Except for the mature cyst, the morphologies of *E. histolytica*, *E. dispar*, *E. moshkovskii*, and *E. bangladeshi* (*E. histolytica/E. dispar* group) and *E. coli* are similar. See Additional Information below for comments on the *E. histolytica/dispar/moshkovskii* complex (generally, *E. moshkovskii* is a part of the complex but rarely listed).

Although the concentration technique is helpful in demonstrating cysts, **the most important technique for the recovery and identification of protozoan organisms is the permanent stained smear** (normally stained with trichrome or iron hematoxylin). A minimum of three specimens collected over not more than 10 days is often recommended.

Diagnostic Tips

If the permanent stained smear is quite dark, a delicate *E. histolytica/E. dispar* cell may appear more like *Entamoeba coli*; on a very thin, pale smear, *E. coli* can appear more like *E. histolytica/E. dispar*. Also, when the organisms are dying or have been poorly fixed, they appear more highly vacuolated, looking like *E. coli* trophozoites rather than *E. histolytica/E. dispar*.

Fecal immunoassays can identify the *E. histolytica/E. dispar* group in the stool, or they can confirm the presence of the true pathogen *E. histolytica*; fresh or frozen stool is required. The two formats that are available include the EIA tray and the rapid cartridge (TechLab/Abbott).

General Comments

Although many people worldwide are infected, only a small percentage develop clinical symptoms and actually have *E. histolytica* rather than the nonpathogen, *E. dispar*. For many years, the issue of pathogenicity has been very controversial. Some felt that what was called *E. histolytica* was really two separate species of *Entamoeba*, one being pathogenic and causing invasive disease and the other being nonpathogenic and causing mild or asymptomatic infections.

Based on current knowledge, pathogenic *E. histolytica* is considered the etiologic agent of amebic colitis and extraintestinal abscesses, while nonpathogenic *E. dispar* produces no intestinal symptoms and is not invasive in humans. The division of *Entamoeba* into invasive *E. histolytica* and noninvasive *E. dispar* by isoenzyme analysis is supported by genetic differences; pathogenesis differences also help explain the epidemiology, clinical syndromes, and pathology of amebiasis. **The total body of evidence supports the differentiation of the pathogen *E. histolytica* and the nonpathogen *E. dispar* as two distinct species. However, they cannot be differentiated on the basis of microscopic morphology. Differentiation between the two species is complicated, because not every *E. histolytica* trophozoite contains ingested RBCs.**

Description (Trophozoite)

Living trophozoites vary in diameter from about 10 to 60 μm. Organisms recovered from diarrheic or dysenteric stools are generally larger than those in formed stool from asymptomatic individuals. Motility is rapid and unidirectional, with pseudopods forming quickly in response to the conditions around the organism; it may appear sporadic. Although this characteristic motility is often described, it is rare to diagnose amebiasis on the basis of motility seen in a direct mount. The cytoplasm is differentiated into a clear outer ectoplasm and a more granular inner endoplasm.

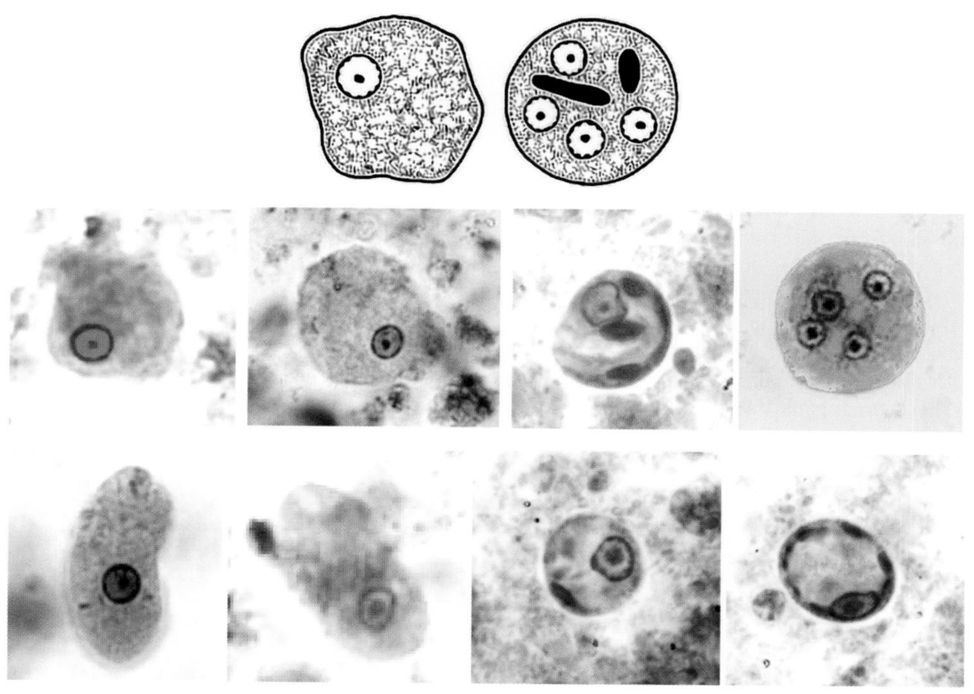

Figure 7.2 ***Entamoeba histolytica/E. dispar.*** **(Top row)** *Entamoeba histolytica/E. dispar* trophozoite (no ingested RBCs); cyst. **(Middle row)** The first two images are *E. histolytica/E. dispar* trophozoites (no ingested RBCs). The third image is a precyst with an enlarged nucleus and developing chromatoidal bars. The last organism is a cyst with four nuclei and no chromatoidal bars. **(Bottom row)** The first two organisms are *E. histolytica/E. dispar* complex/group trophozoites (no ingested RBCs). The third image is a precyst with an enlarged nucleus. The last image is another precyst with an enlarged nucleus and some developing chromatoidal bars (around the inside edges of the cyst). Note: The precysts of the true pathogen, *E. histolytica*, and the *E. histolytica/E. dispar* complex/group tend to have a single enlarged nucleus. The precyst of *Entamoeba coli* tends to have two enlarged nuclei, one at each side of the cyst.

When the organism is examined on a permanent stained smear (trichrome or iron hematoxylin), the morphologic characteristics are readily seen. The nucleus is characterized by having evenly arranged chromatin on the nuclear membrane and the presence of a small, compact, centrally located karyosome. The cytoplasm is usually described as finely granular with few ingested bacteria or debris in vacuoles. In a patient with dysentery, RBCs may be visible in the cytoplasm, and this feature is considered diagnostic for *E. histolytica*. Infection with *E. histolytica*/*E. dispar* group is diagnosed on the basis of organism morphology without the presence of RBCs.

Description (Cyst)

For unknown reasons, the trophozoites may condense into a round mass (precyst), and a thin wall is secreted around the immature cyst. There may be two types of inclusions within this immature cyst, a glycogen mass and highly refractile chromatoidal bars with smooth, rounded edges. As the cysts mature (metacyst), there is nuclear division with the production of four nuclei; occasionally, eight nuclei are produced, and the cysts range in size from 10 to 20 μm. Often, as the cyst matures, the glycogen completely disappears; the chromatoidal bars may also be absent in the mature cyst. Cyst formation occurs only within the intestinal tract, not when the stool has left the body. The uni-, bi-, and tetranucleate cysts are infective and represent the mode of transmission from one host to another.

After cyst ingestion, no changes occur in an acid environment; however, once the pH becomes neutral or slightly alkaline, the encysted organism becomes active, with the outcome being four separate trophozoites (small, metacystic trophozoites). These organisms develop into the normal trophozoites when they become established in the large intestine.

E. histolytica/*E. dispar* trophozoites and cysts can be confused with human macrophages and PMNs, respectively. See the section on *Entamoeba histolytica*, above.

Additional Information

Most of the epidemiological surveys on *E. histolytica* infection were carried out before the characterization of the *E. histolytica/dispar/moshkovskii* complex. Therefore, additional surveys have been conducted to differentiate these species and to determine more correct distribution patterns of *E. histolytica* infection worldwide. In one report, the authors (Soares NM, Azevedo AC, Pacheco FTF, et al, *Biomed Res Int* 2019, https://www.ncbi.nlm.nih.gov/pmc/articles/PMC6681611/) used immunological and molecular tools to determine the prevalence of *E. histolytica*, *E. dispar*, and *E. moshkovskii* infections among 55,218 patients seen at the Clinical Laboratory of Pharmacy College in the city of Salvador (Bahia/Brazil). The prevalence rate of *E. histolytica/dispar/moshkovskii* complex based on fecal examination by optical microscopy was around 0.49% (wet mount identification of four-nucleated cyst); this prevalence rate was lower than those previously recorded (3.2% and 5.0%). In the samples tested using coproantigen and PCR, it was not possible to prove the presence of *E. histolytica* and *E. moshkovskii*. Only *E. dispar* was diagnosed by PCR, although the presence of circulating IgG anti-*E. histolytica* was detected. Apparently, the prevalence of *E. dispar* is 10 times higher than that of *E. histolytica* worldwide; therefore, physicians must decide if treatment is necessary based more on clinical evidence. With the ongoing development of molecular panels, the confirmation or "rule out" of the true pathogen *E. histolytica* may become routine.

Comments on *Entamoeba moshkovskii* and *Entamoeba bangladeshi*

Entamoeba moshkovskii is found worldwide and is generally considered a free-living ameba, although more comprehensive studies on pathogenicity are under way. Based on microscopic morphology, this organism is indistinguishable from *Entamoeba histolytica* and *Entamoeba dispar,* except in cases of invasive disease when *E. histolytica* contains ingested RBCs. Apparently, there are some differences that separate this organism from *E. histolytica* and *E. dispar*. However, these differences pertain to physiology rather than morphology.

Life Cycle and Morphology

The life cycle is essentially identical to that of *E. dispar,* and morphological differences are minimal or nonexistent. In wet preparations, trophozoites usually range in size from 15 to 20 μm and cysts normally range in size from 12 to 15 μm.

Morphology of Trophozoites and Cysts

Trophozoites do not ingest RBCs, and the motility is similar to that of both *E. histolytica* and *E. dispar*. Nuclear and cytoplasmic characteristics are very similar to those seen in *E. histolytica*; however, trophozoites of *E. moshkovskii* do not contain ingested RBCs. Nuclear characteristics and chromatoidal bars are similar to those in *E. histolytica* and *E. dispar*.

Diagnosis

Because *E. moshkovskii* can easily be confused with other amebae, particularly in a wet preparation, it is almost mandatory that the final identification be obtained from the permanent stained smear. However, without the use of molecular tools to specifically identify this organism, the results are usually reported as "*Entamoeba histolytica*/*E. dispar* group" with no mention of *E. moshkovskii.* **The true prevalence of this organism may be much higher than suspected; this may also explain some of the microscopy-positive, antigen-negative results obtained with the *Entamoeba* test kit. In the majority of cases, the actual identification of *E. moshkovskii* is probably not possible on the basis of morphology.**

Epidemiology and Prevention

The mode of transmission is similar to that for other protozoa and is related to ingestion of cysts in contaminated food or water.

Entamoeba bangladeshi

Microscopy and culture for the genus *Entamoeba* have limited specificity, because several species, which vary in their pathogenic potential, have morphologically similar cysts and trophozoites. During analysis of feces positive for *Entamoeba* organisms by microscopy or culture but negative for *E. histolytica*, *E. dispar*, and *E. moshkovskii* by PCR, a new species was identified, which was named *Entamoeba bangladeshi* sp. nov. in recognition of the support of the Bangladesh community for this research. These isolates represent a new species of *Entamoeba* (GenBank accession no. JQ412861 and JQ412862), here referred to as *E. bangladeshi*.

Life Cycle and Morphology

The life cycle is essentially identical to that of *E. dispar and E. moshkovskii*, and morphological differences are minimal or nonexistent. In wet preparations, trophozoites usually range in size from 15 to 20 μm and cysts normally range in size from 12 to 15 μm. It is important to remember that on the permanent stained smear there is a certain amount of artificial shrinkage due to dehydration; therefore, all of the organisms, including pathogenic *E. histolytica,* may be somewhat smaller (from 1 to 1.5 μm) than the sizes quoted for the wet-preparation measurements. By light microscopy, no apparent differences between *E. bangladeshi* and *E. histolytica* were detected, with the exception of ingested RBCs in *E. histolytica*. The physical resemblance between *E. histolytica* and *E. bangladeshi* is a diagnostic problem, because direct microscopic examination of fecal samples is still used as a diagnostic tool to detect *E. histolytica* in areas in which these species are endemic.

Morphology of Trophozoites and Cysts

Trophozoites do not ingest RBCs, and the motility is similar to that of *E. histolytica*, *E. dispar*, and *E. moshkovskii*. Nuclear and cytoplasmic characteristics are very similar to those seen in *E. histolytica*; however, trophozoites of *E. bangladeshi* do not contain ingested RBCs.

In the cyst forms, nuclear characteristics and chromatoidal bars are similar to those in *E. histolytica*, *E. dispar*, and *E. moshkovskii*.

Diagnosis

E. bangladeshi is microscopically indistinguishable from *E. histolytica* in cyst and trophozoite stages. Currently, *E. bangladeshi* is identifiable only by its small-subunit-rRNA gene sequence.

Epidemiology and Prevention

The mode of transmission is similar to that for other protozoa and is related to ingestion of cysts in contaminated food or water. *E. bangladeshi* is grouped with the clade of *Entamoeba* species infecting humans, including *E. histolytica*. *E. bangladeshi*, however, appears to be more distantly related than the noninvasive *E. dispar*, but more closely than *E. moshkovskii*, to *E. histolytica*. Future epidemiologic studies of *E. histolytica* infection should include the capability of diagnosing all four species individually (*E. histolytica*, *E. dispar*, *E. moshkovskii*, and *E. bangladeshi*).

Entamoeba hartmanni

Pathogenic	No
Disease	None
Acquired	Fecal-oral transmission; contaminated food and water
Body site	Intestine
Symptoms	None
Clinical specimen	Stool
Epidemiology	Worldwide, primarily human-to-human transmission
Control	Improved hygiene, adequate disposal of fecal waste, adequate washing of contaminated fruits and vegetables

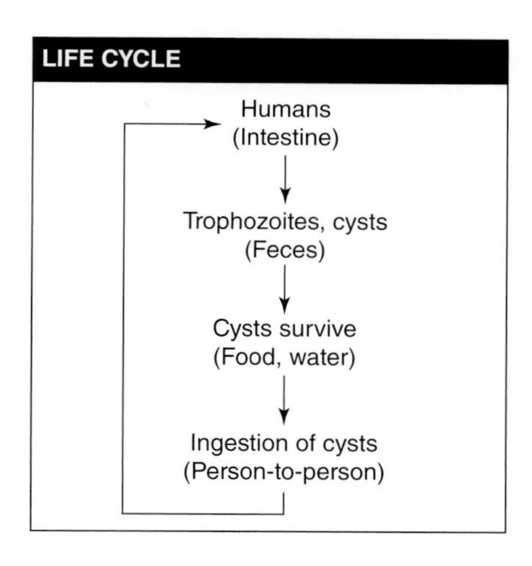

Diagnosis

The standard O&P exam is recommended for recovery and identification of *Entamoeba hartmanni* in stool specimens. Microscopic examination of a direct saline wet mount may reveal small motile trophozoites. An asymptomatic individual may have few trophozoites and possibly only cysts in the stool. Although the concentration technique is helpful in demonstrating cysts, **the most important technique for the recovery and identification of protozoan organisms is the permanent stained smear** (normally stained with trichrome or iron hematoxylin). A minimum of three specimens collected over not more than 10 days is often recommended.

Because *E. hartmanni* can be easily confused with other small amebae, particularly in a wet preparation, it is almost mandatory that the final identification be obtained from the permanent stained smear. Accurate measurement of organisms also confirms the tentative visual diagnosis.

The mode of transmission is similar to that of other protozoa and is related to ingestion of cysts in contaminated food or water. In areas where accurate identifications have been made, the prevalence of *E. hartmanni* is similar to that of *Entamoeba histolytica*.

Note: Depending on the intensity of the stain, the trophozoite stage can mimic *Endolimax nana* or *Dientamoeba fragilis* trophozoites.

Diagnostic Tips

If the permanent stained smear is quite dark, a delicate *E. hartmanni* trophozoite or cyst may appear more like small *Entamoeba coli* organisms; on a very thin, pale smear, *E. coli* can appear more like *E. hartmanni*. Also, when the organisms are dying or have been poorly fixed, they appear more highly vacuolated, looking like *E. coli* trophozoites rather than *E. hartmanni*. Remember, in the case of *E. hartmanni*, the differentiation among the *Entamoeba* species can be determined on the basis of size.

When examining permanent stained smears, it is important to look for populations of organisms. Occasionally, one will see organisms that tend to fall outside the size limits; these are probably outliers unless they look quite different from the original population of organisms. This is an important distinction when reviewing proficiency testing smears. The organism in question tends to be in quantities of "moderate" or "many"; the presence of a rare organism (different morphology and/or size) is almost always an outlier and not a separate species. However, even with multiple reviews prior to selection for proficiency testing, occasionally a challenge smear will contain a rare organism of another genus and/or species.

Fecal immunoassays can identify the *E. histolytica*/*E. dispar* group in the stool or can confirm the presence of the true pathogen

E. histolytica; fresh or frozen stool is required. This approach may be appropriate if a number of organisms overlap between the sizes of *E. hartmanni* and the *E. histolytica*/*E. dispar* group.

General Comments

Although many people worldwide may be infected, a large proportion of them will remain asymptomatic. The main problem is the correct identification and differentiation between this nonpathogen and the *E. histolytica*/*E. dispar* complex/group.

E. histolytica is considered the etiologic agent of amebic colitis and extraintestinal abscesses, while nonpathogenic *E. dispar* produces no intestinal symptoms and is not invasive in humans. Organisms in the *E. histolytica*/*E. dispar* group can be confused morphologically with the nonpathogenic *E. hartmanni*, although the latter is smaller. *E. hartmanni* can also be confused with human cells, including PMNs and macrophages.

Generally, *E. hartmanni* is not as common as *E. coli* or the *E. histolytica*/*E. dispar* group. The morphology is usually very clear. The only problem with identification might occur if one was examining the permanent stained smear with a 50× or 60× oil immersion objective rather than a 100× oil immersion objective.

Although early studies suggested that *E. hartmanni* could transform from small "nonpathogenic" strains to large "pathogenic" ones, none of this information has been confirmed by subsequent studies. Because *E. hartmanni* is a separate and nonpathogenic species, treatment is not recommended. However, remember that this organism is acquired in the same way as pathogenic species.

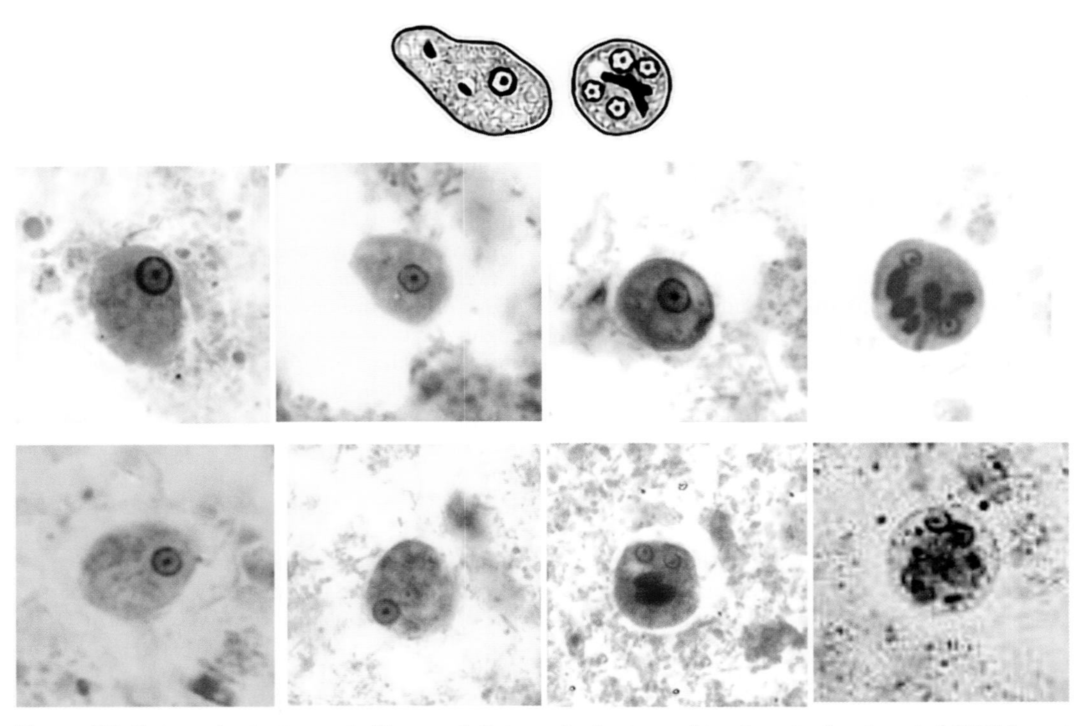

Figure 7.3 ***Entamoeba hartmanni.*** **(Top row)** *Entamoeba hartmanni* trophozoite (no ingested RBCs); cyst (although four nuclei are shown, often the cysts seen contain only two nuclei). **(Middle row)** The first two images are *E. hartmanni* trophozoites (no ingested RBCs); the third image is a precyst with an enlarged nucleus; the last organism is a cyst with two nuclei and many small chromatoidal bars. **(Bottom row)** The first two organisms are *E. hartmanni* trophozoites (no ingested RBCs); the third image is a cyst with an enlarged nucleus and several chromatoidal bars; the last image is another cyst with two nuclei and some chromatoidal bar debris. Although the mature cyst of *E. hartmanni* contains four nuclei, most often the cysts contain only two nuclei and many chromatoidal bars.

Description (Trophozoite)

The life cycle is essentially identical to that of *E. histolytica* and *E. dispar*, and morphologic differences involve size, although there is an overlap in size between the two species. In wet preparations, trophozoites range in size from 4 to 12 μm and cysts range in size from 5 to 10 μm. However, on the permanent stained smear there is a certain amount of artificial shrinkage due to dehydration; thus, all of the organisms, including the *E. histolytica*/*E. dispar* group, may be somewhat smaller on a permanent stained smear (from 1 to 1.5 μm smaller) than the sizes quoted for the wet-preparation measurements.

Trophozoites do not ingest RBCs, and the motility is usually less rapid. Otherwise, nuclear and cytoplasmic characteristics are very similar to those seen in the *E. histolytica*/*E. dispar* group.

In some stained smears, there may be confusion among *E. hartmanni*, *Endolimax nana*, and *Dientamoeba fragilis*, particularly in the trophozoite stage. However, the nucleus in the trophozoite of *E. hartmanni* tends to be very precise and resembles a bull's-eye with very even chromatin and a central karyosome.

Description (Cyst)

Frequently cysts contain only one or two nuclei, even though the mature cyst contains four nuclei. The tetranucleate state is seen in the drawing below. Mature *E. hartmanni* cysts also tend to retain their chromatoidal bars, a characteristic often not seen in the *E. histolytica*/*E. dispar* group.

Chromatoidal bars are similar to those in the *E. histolytica*/*E. dispar* group but are smaller and more numerous, with the same smooth, rounded ends. Because differentiation between *E. hartmanni* and the *E. histolytica*/*E. dispar group* at the species level depends primarily on size, it is mandatory that laboratories use calibrated microscopes which are periodically rechecked. Although the College of American Pathologists (CAP) checklist does not require yearly microscope calibration, it is recommended if the microscope receives heavy use and/or is moved frequently during the year. *E. hartmanni* trophozoites and cysts can be confused with PMNs and macrophages, particularly since these human cells appear somewhat shrunken on the permanent stained smear; they always appear smaller in a permanent stained smear of fecal material than on a stained thin blood film.

Entamoeba coli

Pathogenic	No
Disease	None
Acquired	Fecal-oral transmission; contaminated food and water
Body site	Intestine
Symptoms	None
Clinical specimen	Stool
Epidemiology	Worldwide, primarily human-to-human transmission
Control	Improved hygiene, adequate disposal of fecal waste, adequate washing of contaminated fruits and vegetables

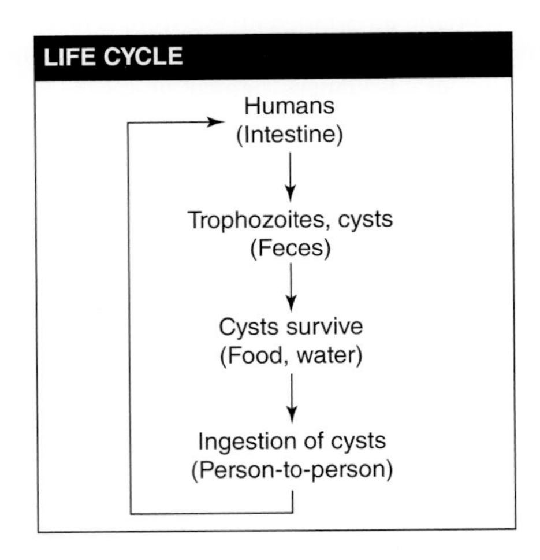

Diagnosis

The standard O&P exam is recommended for recovery and identification of *Entamoeba coli* in stool specimens. Microscopic examination of a direct saline wet mount may reveal motile trophozoites. An asymptomatic individual may have few trophozoites and possibly only cysts in the stool. Although the concentration technique is most helpful in demonstrating *E. coli* cysts, they can also be seen on the permanent stained smear (normally stained with trichrome or iron hematoxylin). A minimum of three specimens collected over not more than 10 days is often recommended.

Diagnostic Tips

If the permanent stained smear is quite light, an *E. coli* cell may appear more like the *Entamoeba histolytica*/*E. dispar* group. Also, when the organisms are dying or have been poorly fixed, they appear more highly vacuolated; the nucleus can be lost within the cytoplasmic debris.

Except for the mature cyst, the morphologies of *E. histolytica*, *E. dispar*, *E. moshkovskii*, and *E. bangladeshi* (*E. histolytica*/*E. dispar* group) are similar to that of *E. coli*. It is very important to examine permanent stained smears, even if a tentative identification has been made from a wet-preparation examination. Correct differentiation is critical to good patient care. Also, multiple species can be found in the same patient. If few *E. histolytica*/*E. dispar* organisms are present among many *E. coli* organisms, additional searching and/or species-specific immunoassay testing may be necessary to correctly identify the different species.

For some reason, as the cyst of *E. coli* matures, it becomes more resistant to fixation with various preservatives. Therefore, the cyst may be seen on the wet preparation but not on the permanent stained smear. Occasionally, on trichrome smears, the cysts appear distorted, shrunken, and somewhat pink. This is not an indication of poor reagents or techniques but rather an indication that the cyst walls do not fix well with the routinely used fecal preservatives. Better fixation and more detailed morphology can be obtained by heating some of the fixatives prior to specimen preservation. Although for most laboratories, this is not a practical approach, it can be helpful in preparing permanent stains for teaching slides.

General Comments

Although a large number of people throughout the world are infected with this organism, many of them will remain asymptomatic. The main problem is the correct identification and differentiation between the *E. coli* nonpathogen and the *E. histolytica*/*E. dispar* group.

E. histolytica is considered the etiologic agent of amebic colitis and extraintestinal abscesses, while nonpathogenic *E. dispar* produces no intestinal symptoms and is not invasive in humans. The *E. histolytica/E. dispar* group can be confused morphologically with nonpathogenic *E. coli*. Specific treatment is not recommended for this nonpathogen.

E. coli can also be confused with human cells, including PMNs and macrophages. *E. coli* is generally described as a "dirty-looking" organism with a large blot-like karyosome. It is more commonly found in human stool specimens than the *E. histolytica/E. dispar* group or the true pathogen, *E. histolytica*.

The mode of transmission is ingestion of cysts from contaminated food or water. Apparently, the infection is readily acquired, and in some warmer

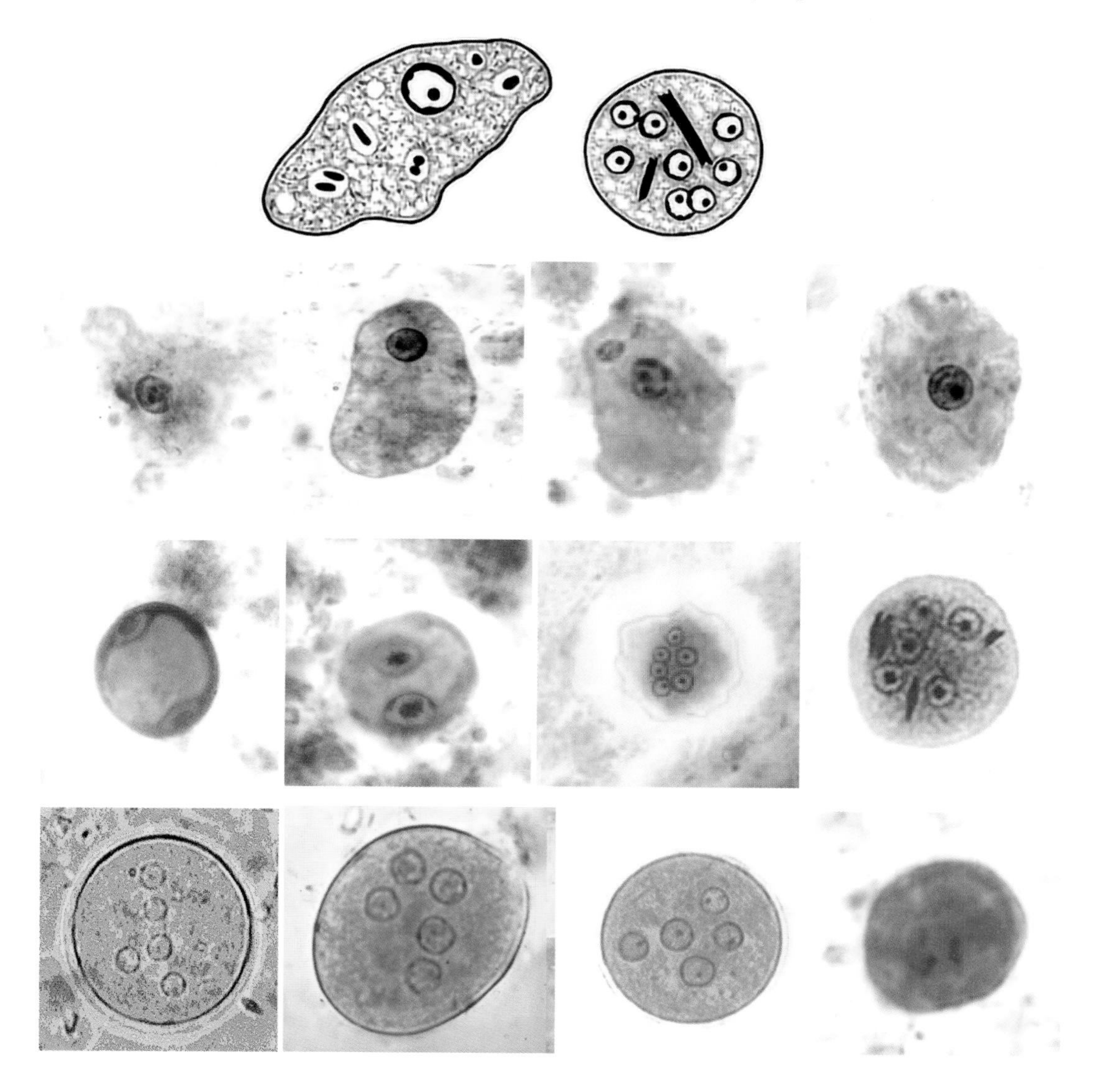

Figure 7.4 ***Entamoeba coli.*** **(Row 1)** *Entamoeba coli* trophozoite (no ingested RBCs); cyst showing eight nuclei and chromatoidal bars with sharp, pointed ends. **(Row 2)** All four images are *E. coli* trophozoites (no ingested RBCs). Note the eccentric nuclear chromatin and the large, blot-like karyosomes. **(Row 3)** The first two organisms are *E. coli* precysts with two nuclei (often one at each side of the precyst); the third image is a shrunken, distorted cyst that indicates poor fixation; the last image is a typical cyst with five nuclei and typical chromatoidal bars with sharp, pointed ends. **(Row 4)** The first two organisms are *E. coli* cysts (left, saline; right, iodine); the third cyst is also stained more lightly with iodine; the last *E. coli* cyst is very shrunken and distorted. This image is typical of cysts that do not fix well using routine fecal fixatives. This pink/purple cyst can often be seen in the trichrome-stained permanent smear.

climates or areas with primitive hygienic conditions, the infection rate can be quite high.

Description (Trophozoite)

Living trophozoites tend to be somewhat larger than those of *E. histolytica* and range from 15 to 50 μm. Motility has been described as sluggish with broad, short pseudopods. In wet preparations, it may be extremely difficult to differentiate nonpathogenic *E. coli* from the potentially pathogenic *E. histolytica*/*E. dispar* group. On the permanent stained smear, the cytoplasm appears granular, with few to numerous vacuoles containing bacteria, yeasts, and other food materials. The nucleus has a moderately large blot-like karyosome that is frequently eccentric. The chromatin on the nuclear membrane is usually clumped and irregular in placement. If there are RBCs in the intestinal tract, *E. coli* may ingest them rather than bacteria; however, this rarely occurs. Occasionally the cytoplasm also contains ingested *Sphaerita* spores and possibly a *Giardia lamblia* cyst.

E. coli trophozoites (which resemble macrophages) and immature cysts with four nuclei (which resemble PMNs) can be confused with human cells.

Description (Cyst)

Trophozoites discharge their undigested food and begin to round up prior to precyst and cyst formation. Early cysts usually contain a dense glycogen mass and may also contain chromatoidal bars, which tend to be splinter shaped and irregular. Eventually the nuclei divide until the mature cyst, containing 8 (occasionally 16) nuclei, is formed. The cysts measure 10 to 35 μm and almost always lose their chromatoidal bars as they mature. It is important to remember that as the cyst of *E. coli* matures, it becomes more difficult to fix using the routine fecal preservatives. Consequently, it may be seen on the wet-smear preparation and not on the permanent stained smear. Occasionally on the permanent stained smear, the cysts appear distorted and somewhat dark, with little or no definition of the internal structures, including the nuclei. This does not mean that there is a problem with poor reagents or techniques; it means that the cyst wall of *E. coli* is harder to fix than that of other protozoan cysts. Better fixation and more detailed morphology can be obtained by heating some of the fixatives prior to specimen preservation.

After cyst ingestion, the metacyst undergoes division of the cytoplasm, thus becoming metacystic trophozoites that will grow and divide within the lumen of the intestine. Usually, fewer than eight trophozoites are formed from the mature cyst.

Entamoeba gingivalis

Pathogenic	No; however, studies on periodontitis are ongoing
Disease	None
Acquired	Droplet spray from the mouth; close contact, contaminated cups
Body site	Mouth, vagina, cervix
Symptoms	None
Clinical specimen	Material from gums or teeth
Epidemiology	Worldwide, primarily human-to-human transmission
Control	Proper care of teeth and gums; removal of intrauterine devices (IUDs)

Comments

Entamoeba gingivalis was the first parasitic ameba of humans to be described. It was recovered from the soft tartar between the teeth. It has also been recovered from the tonsillar crypts and can multiply in bronchial mucus, thus appearing in sputum. Since morphologically it is very similar to *Entamoeba histolytica*, it is important to make the correct identification from a sputum specimen (nonpathogenic *E. gingivalis* rather than *E. histolytica* from a possible pulmonary abscess).

Organisms identified as *E. gingivalis* have been recovered in vaginal and cervical smears from women using intrauterine devices (IUDs); the organisms spontaneously disappeared after removal of the devices. In an unusual case, *E. gingivalis* was identified in a left upper neck nodule by fine-needle aspiration. Apparently, the patient had an increased number of amebae within the oral cavity secondary to radiation therapy, which may have contributed to a fistula tract between the oral cavity and the surgical incision site, resulting in the formation of a small inflammatory nodule in the upper neck. Generally, no treatment is indicated, regardless of the body site from which the parasites are recovered. The infection suggests a need for better oral hygiene and can be prevented by proper care of teeth and gums.

Description

The *E. gingivalis* trophozoite measures approximately 5 to 15 μm, and the **cytoplasm most often contains ingested leukocytes.** On the permanent stained smear, nuclear fragments of the white blood cells (WBCs) can be seen within food vacuoles, which are usually larger than the vacuoles seen in *E. histolytica*. This helps differentiate the two, since *E. gingivalis* is the only species that ingests WBCs. No cysts are formed by this species. Although *E. gingivalis* is most often recovered from patients with gingivitis or periodontitis, it is still considered nonpathogenic. More recent findings suggest that *E. gingivalis* may be asymptomatically present in some sulci and may be associated with the disease after environmental changes, reminiscent of the intestinal pathogen *E. histolytica*. As the only human cells observed in these samples were polymorphonuclear cells, and the literature mentions that neutrophils are predominant in periodontal pockets, the target cells of amoebic phagocytosis may be the latter. The processes leading to the modifications in nuclear and cytoplasmic morphology in these cells remain to be defined and could be linked. Further studies will have to take into consideration the physicochemical and biological characteristics of the periodontal pockets to allow relevant studies of the biology of the parasites, either *in vitro* or in animal models. Further research should determine if *E. gingivalis* merely takes part in the pathophysiology of periodontitis or is an actual marker of the disease. Also, as the clinical picture is becoming more complex and the genetic susceptibility of patients shapes different microbiota, periodontitis might be redefined as a group of related diseases with comparable outcomes. Also, a new *Entamoeba* variant (kamaktli variant), which is closely related to *E. gingivalis*, has been identified from the oral cavity. Work to expand our information on the organism morphology and pathogenesis of the kamaktli variant is in progress. Delineation of subtypes versus species is ongoing.

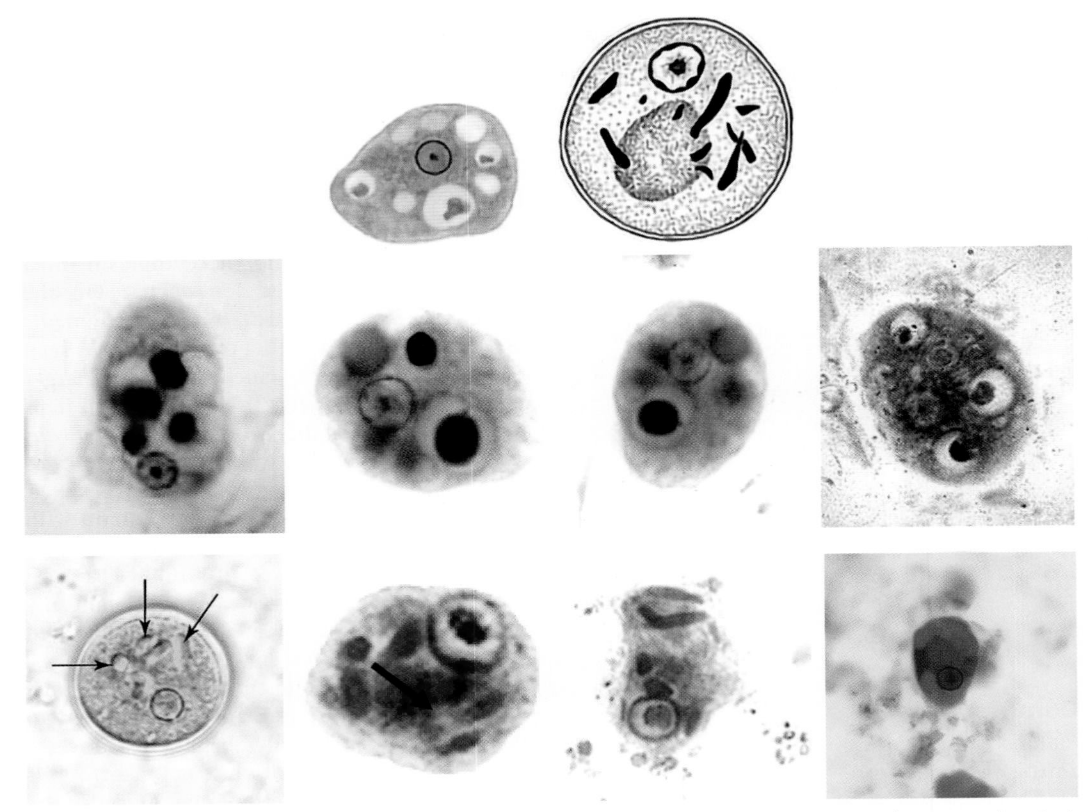

Figure 7.5 ***Entamoeba gingivalis*** **(trophozoites only) and** ***Entamoeba polecki*****.** **(Top row)** *Entamoeba gingivalis* trophozoite containing ingested WBCs within the vacuoles; *Entamoeba polecki* cyst with large inclusion and many multiple-shaped chromatoidal bars. **(Second row)** All four organisms are *E. gingivalis* trophozoites (no cysts are found in the life cycle); note the typical *Entamoeba* nuclei and ingested WBCs within the vacuoles. **(Bottom row)** Cyst of *E. polecki* (wet mount stained with iodine; note the numerous chromatoidal bars [arrows]); *E. polecki* cyst (note the large dark blue inclusion, not seen in other *Entamoeba* species [arrow], and large nucleus/karyosome); *E. polecki* cyst (note the large nucleus/karyosome and numerous purple chromatoidal bars, also seen in the second image); *E. polecki* trophozoite (note the large pale blue inclusion). (Organisms in bottom row courtesy of the CDC Public Health Image Library and CDC DpDx.)

Entamoeba polecki

Pathogenic	No
Disease	None
Acquired	Fecal-oral transmission; contaminated food and water
Body site	Intestine
Symptoms	None
Clinical specimen	Stool
Epidemiology	Worldwide; usually found in pigs and monkeys, less often in humans
Control	Improved hygiene, adequate disposal of fecal waste, adequate washing of contaminated fruits and vegetables

Comments

Entamoeba polecki was originally found in the intestines of pigs and monkeys and has also been found as a human parasite. Where humans and pigs live in close association and where sanitation is poor, pig-to-human transmission is considered the most likely source of human infection (via ingestion of infective cysts). However, human-to-human transmission is very likely if the prevalence rate and intensity of infection are high.

This ameba has been reported infrequently in the Western world; most reported cases have been from the New Guinea region. In most described patients, no definite gastrointestinal symptoms could be directly attributed to *E. polecki* infection.

Due to increased travel and numbers of immigrants to the United States, *E. polecki* may be identified here more frequently than in the past. Physicians and laboratory personnel should be familiar with this organism, because it may be confused with *E. histolytica*, a true pathogen.

In the earlier literature, one patient was reported to experience intermittent episodes of abdominal cramps, diarrhea, nausea, and malaise associated with large numbers of *E. polecki* cysts in the stool. However, this is probably a rare occurrence of symptoms.

Description

In certain areas of the world, such as Papua New Guinea, *E. polecki* is the most common intestinal ameba of humans. Few cases are reported, possibly because it resembles *E. histolytica*/*E. dispar*, *E. moshkovskii*, and *E. coli*. The trophozoites resemble *E. coli* in that the cytoplasm is granular, containing ingested bacteria, and the motility tends to be sluggish. **The nuclear morphology is almost a composite of those of *E. histolytica*/*E. dispar*, *E. moshkovskii*, and *E. coli*. Without some of the cyst stages for comparison, it would be very difficult to identify this organism to the species level on the basis of the trophozoite alone.** The cyst normally has only a single nucleus, chromatoidal material like that in *E. histolytica*, and often an inclusion body. This mass tends to be round or oval and is not sharply defined on the edges. The material, which is not glycogen, remains on the permanent stained smear and stains less intensely than nuclear material or chromatoidal bars. This organism is rarely differentiated from *E. histolytica* and *E. coli* by a wet-preparation examination. The size on the permanent stained smear ranges from 10 to 12 µm for the trophozoite and 5 to 11 µm for the cyst.

Organisms identified as *E. gingivalis* have been recovered in vaginal and cervical smears from women using IUDs; one of the cases also had *E. polecki* reported from the same genital tract specimen.

Endolimax nana

Pathogenic	No
Disease	None
Acquired	Fecal-oral transmission; contaminated food and water
Body site	Intestine
Symptoms	None
Clinical specimen	Stool
Epidemiology	Worldwide, primarily human-to-human transmission
Control	Improved hygiene, adequate disposal of fecal waste, adequate washing of contaminated fruits and vegetables

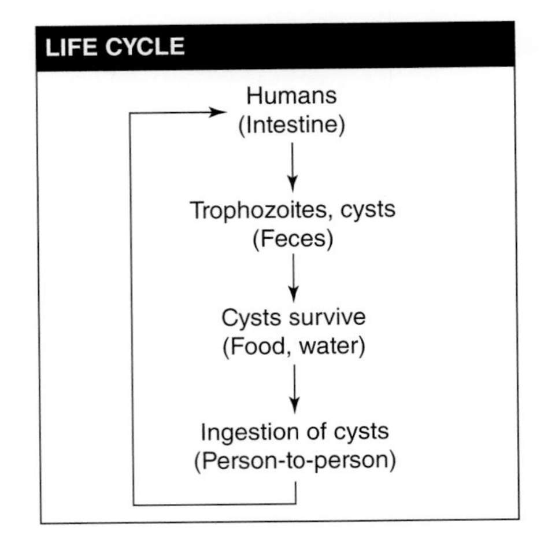

Diagnosis

The standard O&P exam is recommended for recovery and identification of *Endolimax nana* in stool specimens. Microscopic examination of a direct saline wet mount may reveal small, motile trophozoites. An asymptomatic individual may have few trophozoites and possibly only cysts in the stool. Although the concentration technique is helpful in demonstrating cysts, **the most important technique for the recovery and identification of protozoan organisms is the permanent stained smear** (normally stained with trichrome or iron hematoxylin). A minimum of three specimens collected over not more than 10 days is often recommended.

Although *E. nana* is a nonpathogen and no therapy is recommended, it is still important to differentiate it from other amebae, some of which are pathogenic. As indicated above, because these organisms are small, definitive diagnosis of *E. nana* is often based on the permanent stained smear. The four nuclear karyosomes appear very refractile in the wet preparation.

E. nana (particularly the cyst stage) does not fix as well with now commonly used mercury substitute fixatives and may be very difficult to find and identify on the permanent stained smear.

Diagnostic Tips

The trophozoite can often be confused with other protozoa, including *Entamoeba hartmanni* and *Dientamoeba fragilis*. Because the nuclear karyosome is so pleomorphic, the configuration mimics many other organisms. Provided that the organism numbers are sufficient, it is important to look for populations of organisms, rather than basing the identification on the observation of a few single trophozoites. The trophozoites of *E. nana* and *Iodamoeba bütschlii* are almost identical; it is not clinically relevant if they are confused. Finding either indicates that the patient has ingested food and/or water contaminated with fecal material containing infective cysts. Proficiency testing challenges do not require differentiation between these two trophozoite stages; however, the cysts are quite different morphologically and are used for proficiency testing.

Potential problems associated with *E. nana* cysts include failure to stain consistently (due to fixation, not a problem with laboratory procedures) and their small size. Often the cysts are round rather than oval, and the four karyosomes may not be clearly visible.

General Comments

E. nana is one of the smaller nonpathogenic amebae and was distinguished as a separate ameba around 1908. Its distribution is worldwide; it is seen in most populations at least as frequently as *Blastocystis* spp. and *Entamoeba coli*. There might be a need to revise the taxonomy of *Endolimax*. Researchers have recently observed extensive genetic variation in the small-subunit (SSU) rRNA gene among human-derived *Endolimax* (unpublished data). The only complete SSU rRNA sequence of *E. nana* available in GenBank is from a monkey (*Cercocebus albigena*) isolate, which clusters with only some of the unpublished human-derived sequences, indicating that there may be different ribosomal lineages infecting humans and nonhuman primates, as was already observed in the closely related genus *Iodamoeba*. The genetic diversity might also explain the differences in host specificity observed between research groups.

Based on available data, the global prevalence of *E. nana* in healthy individuals is estimated to be 13.9% on average, which, however, is probably an overestimate; still, hundreds of millions are most likely infected. Very little research has been performed on *Endolimax* since the 1920s, '30s, and '40s. With the availability of DNA-based detection methods, resolving major issues, such as host specificity, diversity, and which *Endolimax* species can infect humans, should be straightforward. In addition, the development of diagnostic primers will allow *Endolimax* to be detected with high sensitivity using fecal DNAs and distinguished easily from other amebae.

Although many people worldwide are infected with this organism, a large proportion of them remain asymptomatic. The clinical picture may be subtle, however, and it has been suggested that symptoms may develop if a heavy infection is present or that the pathogenicity might be limited to particularly virulent strains and/or subtypes. The main problem is the correct identification and differentiation between this nonpathogen and the pathogenic *Dientamoeba fragilis* or the nonpathogenic *Entamoeba hartmanni*.

This organism is one of the smaller amebae, and its trophozoite stage and that of *Iodamoeba bütschlii* look very similar. However, misidentification of these two organisms has little clinical significance, since both are considered nonpathogens. In general, *E. nana* is much more common than *I. bütschlii*, and the two organisms can also be found in the same specimen.

Of all the nonpathogenic amebae, *E. nana* is the most common and often seen in mixed protozoan infections.

Although *E. nana* is a nonpathogen, it is acquired via the fecal-oral route just like the pathogenic protozoa. Its presence in a stool specimen should always be reported to the physician.

E. nana cysts tend to be less resistant to desiccation than those of *E. coli*. This organism is also found in the same areas of the world as are the other amebae, that is, in warm, moist climates and in other areas where there is a low standard of personal hygiene and poor sanitary conditions.

Description (Trophozoite)

Motility has been described as sluggish and nonprogressive with blunt, hyaline pseudopods. In the permanent stained smear, the nucleus is more easily seen. There is normally no peripheral chromatin on the nuclear membrane, and the karyosome tends to be large and has a central or eccentric location within the nucleus. The trophozoites measure 6 to 12 μm, with a usual range of 8 to 10 μm.

There is tremendous nuclear variation; occasionally, the overall morphology mimics that of *D. fragilis* and *E. hartmanni*. The more organisms there are on the smear, the more likely it is that some of them will mimic other species of the amebae. The cytoplasm may have small vacuoles containing ingested debris or bacteria.

Nuclear morphology can be described as a Y shape, a band of chromatin across the nucleus, peripheral chromatin, or numerous other variations.

Description (Cyst)

Trophozoites discharge their undigested food and begin to round up prior to precyst and cyst formation. Early cysts may contain very thin, curved chromatoidal bars (often difficult to see). Eventually, the nuclei divide until the mature cyst, containing four nuclei, is formed.

Cysts usually measure 5 to 10 μm, with a normal range of 6 to 8 μm. In some instances, cysts as large as 14 μm have been seen. They are usually oval to round. Occasionally very small, slightly curved chromatoidal bars are present. It is unusual to see the binucleate stage. Clinical specimens often contain both trophozoites and cysts.

Since there tends to be no peripheral nuclear chromatin, the cyst nuclei may be difficult to see, particularly if the fecal specimen was originally fixed in non-mercury-based fixatives.

Cysts can be difficult to fix and may be very difficult to identify, particularly when the non-mercury fixatives are used. If a binucleate cyst is seen, it may mimic the cyst of *Enteromonas hominis*—a small oval cyst with two small nuclei. However, the appearance of binucleate cysts of *E. nana* is quite unusual; almost all cysts contain four nuclei.

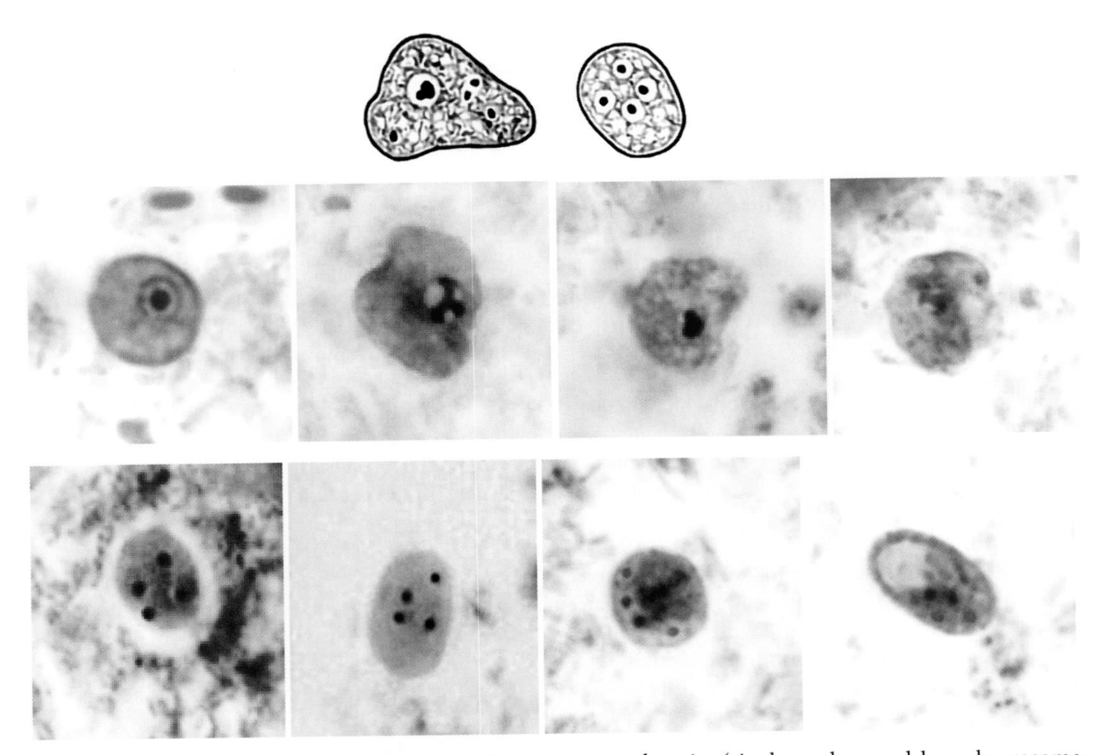

Figure 7.6 ***Endolimax nana.*** **(Top row)** *Endolimax nana* trophozoite (single nucleus and large karyosome; many nuclear variations can be seen in one fecal smear); *E. nana* cyst (note the four nuclei with compact karyosomes). **(Middle row)** All four organisms are trophozoites of *E. nana* (note the tremendous nuclear variation). **(Bottom row)** All four organisms are cysts of *E. nana* (note the four dot-like karyosomes). Note that in the last cyst on the right, there is a vacuole present (this could be confused with *Iodamoeba bütschlii*; however, note that there are four nuclear karyosomes, not just one).

Iodamoeba bütschlii

Pathogenic	No
Disease	None
Acquired	Fecal-oral transmission; contaminated food and water
Body site	Intestine
Symptoms	None
Clinical specimen	Stool
Epidemiology	Worldwide, primarily human-to-human transmission
Control	Improved hygiene, adequate disposal of fecal waste, adequate washing of contaminated fruits and vegetables

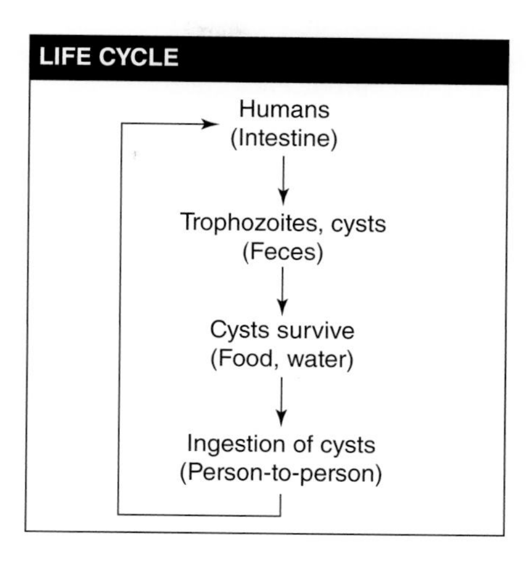

Diagnosis

The standard O&P exam is recommended for recovery and identification of *Iodamoeba bütschlii* in stool specimens. Microscopic examination of a direct saline wet mount may reveal small, motile trophozoites. An asymptomatic individual may have few trophozoites and possibly only cysts in the stool. Although the concentration technique is helpful in demonstrating cysts, **the most important technique for the recovery and identification of protozoan organisms is the permanent stained smear** (normally stained with trichrome or iron hematoxylin). A minimum of three specimens collected over not more than 10 days is often recommended.

This organism can be easily confused with other small amebae, particularly in the trophozoite stage. The cyst form is visible when the organisms are stained with iodine in the wet preparation.

The mode of transmission is ingestion of infective cysts in contaminated food or water. Prevention depends on improved personal hygiene and sanitary conditions.

Diagnostic Tips

In clinical specimens, *Endolimax nana* is much more common than *I. bütschlii*; the latter organism is often seen in mixed infections with other intestinal protozoa (both pathogenic and nonpathogenic). As mentioned in the section above on *E. nana*, the trophozoites of *E. nana* and *I. bütschlii* look almost identical (size, cytoplasmic debris, nuclear karyosome, and single nucleus); proficiency testing challenges do not require the participants to differentiate the two trophozoites. However, since the cyst forms of the two are quite different, they are routinely used for proficiency test challenges.

Occasionally, the cyst of *E. nana* contains a large vacuole that can be confused with the large glycogen vacuole seen in *Iodamoeba* cysts; remember that four nuclear dot-like karyosomes are found in *E. nana*, whereas a single nucleus is seen in the *I. bütschlii* cyst.

General Comments

Although *I. bütschlii* is found in areas where the other amebae have been recovered, its incidence is not as high as that of *Blastocystis* spp. (Stramenopiles), *Entamoeba coli*, or *E. nana*. One of the most striking morphologic features of this organism is the large glycogen vacuole which appears in the cyst and readily stains with iodine on a wet-preparation smear.

Many people are infected with this organism worldwide, but a large proportion of them remain asymptomatic. The main problem is correct identification and differentiation between this nonpathogen and the pathogenic *Dientamoeba fragilis* or the nonpathogens *Entamoeba hartmanni* and *Endolimax nana*.

This organism is one of the smaller amebae; its trophozoite stage looks very similar to that of *E. nana*. However, misidentification of these two organisms has little clinical significance, since both are considered nonpathogens. In general, *E. nana* is much more common than *I. bütschlii*, and they can also be found in the same specimen. In proficiency testing specimens, identification of *E. nana* and *I. bütschlii* cysts is relevant; however, participants will not be asked to differentiate the two trophozoites (the morphology is too similar).

When you identify both *E. nana* and *I. bütschlii* trophozoites in a fecal specimen, look for two different organism populations which differ mainly in size. However, differentiation of the two is most commonly based on cyst morphology. There is no solid evidence explaining why *E. nana* tends to be more common than *I. bütschlii* in human fecal specimens.

Description (Trophozoite)

Motility is active in the wet preparation. In the permanent stained smear, the nucleus has a large karyosome which can be either central or eccentric and may appear to have a halo. Chromatin granules fan out around the karyosome. If the granules are on one side, the nucleus may appear to have the "basket nucleus" arrangement of chromatin, which is often seen in the cyst stage (but can be seen in the trophozoite as well). The cytoplasm is rather granular, containing numerous vacuoles with ingested debris and bacteria. In general, the vacuolated cytoplasm is more obvious than that in *E. nana* trophozoites. The trophozoites measure 8 to 20 μm.

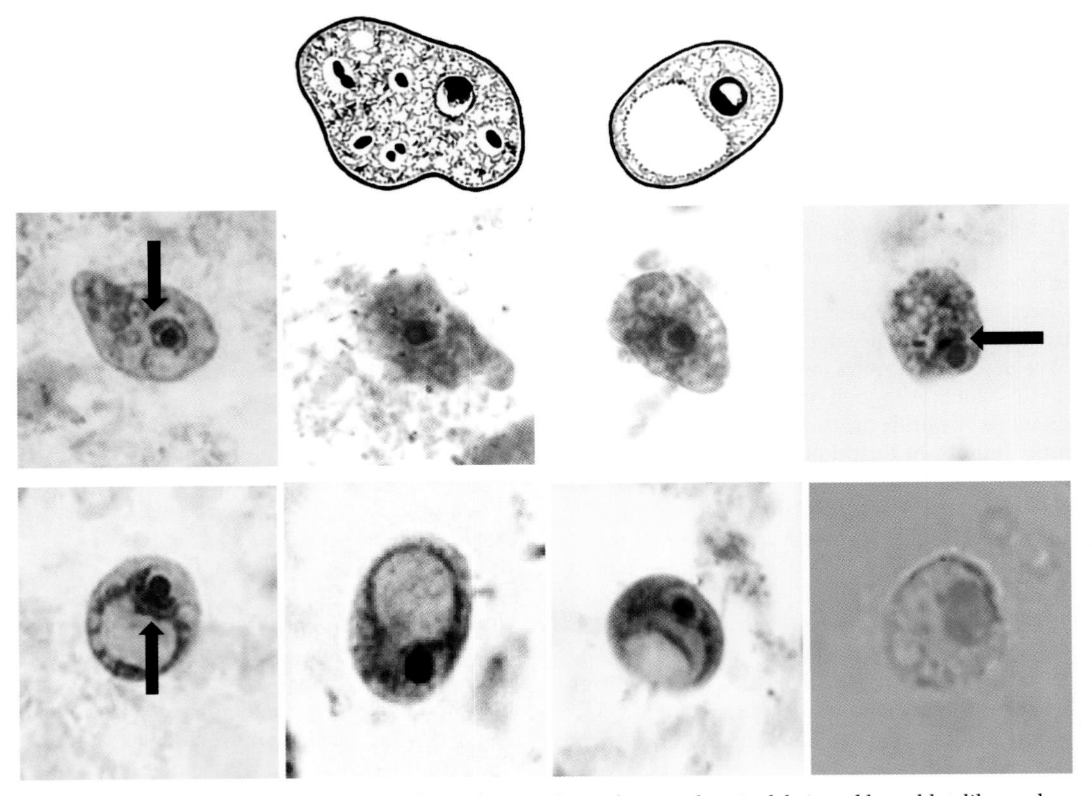

Figure 7.7 ***Iodamoeba bütschlii.*** **(Top row)** Trophozoite (note the cytoplasmic debris and large blot-like nuclear karyosome); cyst (note the large glycogen vacuole and single nucleus). **(Middle row)** All four images are *Iodamoeba* trophozoites. Note the large karyosome, ingested debris, and vacuolated cytoplasm. The arrows point to the extra curved nuclear chromatin above the karyosome (a configuration known as the "basket nucleus"). **(Bottom row)** The first three images are typical cysts with the large glycogen vacuole and large karyosome (the arrow points to excess, curved nuclear chromatin above the karyosome, i.e., the basket nucleus). The last image shows the glycogen vacuole in an iodine wet mount; the iodine stains the glycogen.

The trophozoites of *E. nana* and *I. bütschlii* may be very similar and difficult to differentiate, even on the permanent stained smear. Both are considered nonpathogens, and *E. nana* is recovered more frequently than *I. bütschlii*.

Description (Cyst)

Trophozoites discharge their undigested food and begin to round up prior to precyst and cyst formation. The cyst of *I. bütschlii* contains a single nucleus, like that in the trophozoite.

Cysts usually measure 5 to 20 μm and are rarely confused with other protozoa. Occasionally, cysts may appear collapsed and the typical, large glycogen vacuole may not be clearly visible. The cyst is usually oval to round.

The cyst form is often seen in the concentration wet-sediment preparation, particularly when iodine is added to the smear; the iodine stains the glycogen in the vacuole, making it easier to see the large vacuole within the cyst. However, the typical morphology is usually clear on the permanent stained smear (the glycogen vacuole looks like a clear area within the cytoplasm).

In a specimen containing both *E. nana* and *I. bütschlii* trophozoites, there are usually two populations of organisms, with the main difference usually being size. Although their sizes overlap, *I. bütschlii* is almost always the larger of the two. Also, *I. bütschlii* tends to have more debris and a more vacuolated cytoplasm than *E. nana*.

Blastocystis spp. (formerly *Blastocystis hominis*)

Pathogenic	Yes, depending on the subtype(s) present
Disease	Yes, depending on the subtype(s) present
Acquired	Fecal-oral transmission; contaminated food and water
Body site	Intestine; rare reports of dissemination to lymph nodes in debilitated patients
Symptoms	Yes, depending on the subtype(s) present
Clinical specimen	Stool
Epidemiology	Worldwide; primarily human-to-human transmission; most common intestinal protozoan found in stool specimens worldwide
Control	Improved hygiene, adequate disposal of fecal waste, adequate washing of contaminated fruits and vegetables

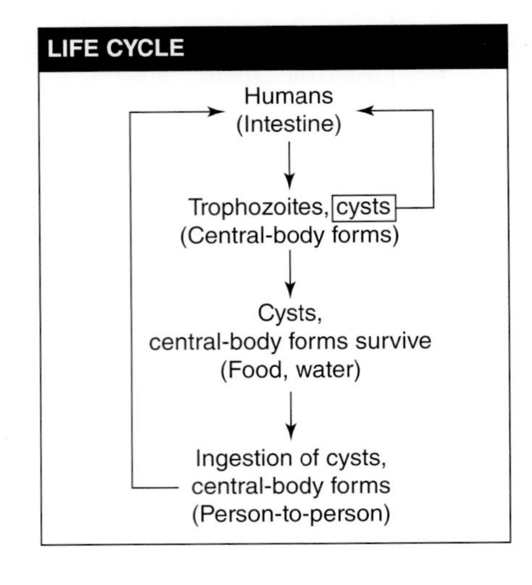

Diagnosis

The standard O&P exam is recommended for recovery and identification of *Blastocystis* spp. (formerly *Blastocystis hominis*) in stool specimens. Microscopic examination of a direct saline wet mount may reveal small to large "central-body" forms. An asymptomatic individual may have few organisms present, all of which are the same central-body forms. Although other forms are present in the gut (primarily in patients with diarrhea), the morphology is difficult to identify; identification relies almost exclusively on the presence of the central-body form. While the concentration technique is helpful in demonstrating these organisms, **the most important technique for the recovery and identification of protozoan organisms is the permanent stained smear** (normally stained with trichrome or iron hematoxylin). A minimum of three specimens collected over not more than 10 days is often recommended. (See Markell and Udow and Cakir et al., Suggested Reading.)

Diagnostic Tips

This organism can be easily confused with other small protozoa and/or yeast cells. There is a tremendous size range in *Blastocystis* organisms, particularly the central-body form. Typical morphology of the peripheral nuclei and the central body area may be difficult to see, especially in the smaller organisms. Also, there tends to be great variation in staining intensity and/or differentiation between the nuclei, as well as limited cytoplasm and central body contents.

Some directions for the stool concentration recommend rinsing the stool specimen with water prior to fixation with formalin. *This is not recommended* for any of the protozoa, but especially not for *Blastocystis*, because the organisms may be ruptured by contact with tap or distilled water. Fecal specimens should be placed in a fecal fixative for 30 min before the concentration procedure and permanent staining are performed. Identification and quantitation (rare, few, moderate, many, or packed) are normally based on the permanent stained smear.

The classic form usually seen in human stool specimens varies tremendously in size, from 6 to 40 µm, and is characterized by a large central body, which may be involved with carbohydrate and lipid storage (visually like a large vacuole). The more amebic form is occasionally seen in diarrheal fluid but may be extremely difficult to recognize. Generally, *Blastocystis* spp. are identified on the basis of the more typical round form with the central body; these forms are consistently present, whether or not some of the other life cycle stages are seen.

When this organism is identified within a stool specimen, it should be quantitated on the report (rare, few, moderate, many, or packed). There may be some relationship between pathogenesis and organism numbers; this association is still being clarified.

General Comments

Blastocystis is a stramenopile (strict anaerobe protist) with worldwide significance due to its capacity to colonize several hosts. It is estimated that more than a billion individuals are infected with *Blastocystis*. Based on its high level of genetic diversity, *Blastocystis* is classified into global ribosomal subtypes (STs). Currently, 17 STs are known, of which ST1 to ST9 and ST12 have been identified in humans. In humans in Europe, ST1, -2, -3, and -4 reportedly occur most commonly, whereas ST1, -2, and -3 commonly occur in South America. More than one ST can reportedly colonize humans, and infections with mixed STs have been reported. *Blastocystis* may cause clinical manifestations such as diarrhea, abdominal pain, irritable bowel syndrome, constipation, and flatulence, along with extraintestinal manifestations such as chronic urticaria. However, these symptoms are nonspecific. Different *Blastocystis* STs could exhibit different growth rates, drug susceptibilities, host ranges, and other biological characteristics. These differences could therefore affect the protist's influence on the gut microbiota. It is possible that microbiota composition in relation to *Blastocystis* may be dependent on the organism's subtype identity. Various studies have confirmed that the subtype differences play an important role in the pathogenesis of *Blastocystis* spp. Approximately half of the human subtypes may cause symptoms; relevant factors include trophozoite adhesion to epithelium and release of cysteine proteases leading to mucosal sloughing, increased intestinal permeability, and proinflammatory cytokine responses. There also tend to be distinct links between parasite colonization and the gut microbial flora profiles.

Although many people worldwide are infected, most probably remain asymptomatic. The main problem is correct identification and differentiation between this organism and pathogenic intestinal protozoa.

The organisms should be quantitated when reported (rare, few, moderate, many, or packed); both large and small numbers of organisms may cause symptoms. However, clinical specimens should be examined thoroughly for the presence of other potential pathogens before pathogenic status is assigned. Therapy is also effective in eradicating other intestinal protozoa (*Giardia*, etc.). However, recent studies indicate that persistence of symptoms in some patients could be due to resistance to metronidazole demonstrated in ST3. Unfortunately, the various subtypes cannot be differentiated on the basis of microscopic morphology.

Description

Blastocystis is capable of pseudopod extension and retraction; it reproduces by binary fission or "sporulation" and has a membrane-bound body that takes up about 90% of the organism. This form is called the central-body form; it is usually seen in the stool specimen and measures from 6 to 40 μm. It is characterized by having a large central body that can be seen in the wet preparation from the concentration and in the permanent stained smear.

The more amebic form can be very difficult to identify and is seen in patients with more severe diarrhea. However, patients with *Blastocystis* infections tend to have the central-body forms in the stool, regardless of whether the amebic forms are seen.

The presence of thin- and thick-walled cysts has been suggested. The thin-walled cysts may be autoinfectious, leading to multiplication of the organism in the intestinal tract. The thick-walled cysts are responsible for external transmission via the fecal-oral route. This life cycle might explain the high percentage of positive *Blastocystis* carriers relative to those infected with other protozoa in many studies.

EIAs and fluorescent-antibody (FA) tests have been used to detect serum antibody to *Blastocystis* spp. A strong antibody response is consistent with the ability to cause symptoms. Also, serum antibody production during and after *Blastocystis* symptomatic disease is immunologic evidence for the pathogenic role of this protozoan.

Additional Information

Terminology for *Blastocystis* may change, depending on the outcome of molecular studies attempting to classify this organism more accurately. Some use the term "cyst" to denote the central-body form, while others continue to use "central-body form." Either term is acceptable at present.

The organisms may resemble yeast cells, debris, and/or amebic cysts. In some preparations, the

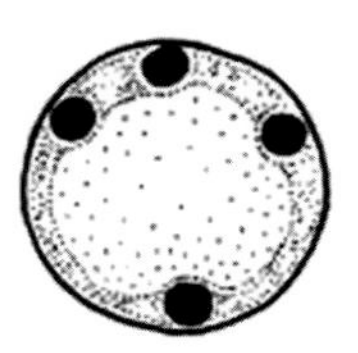

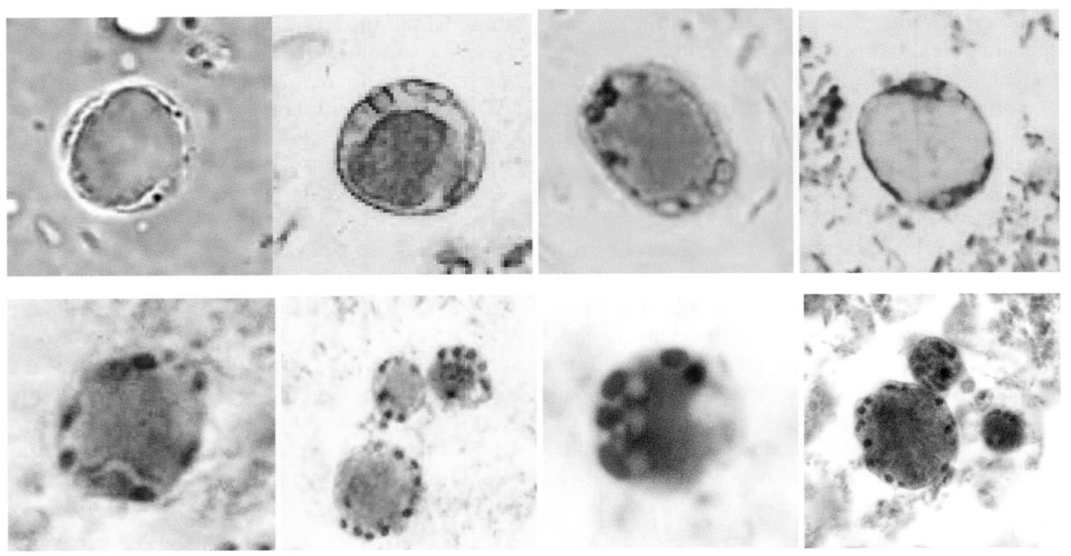

Figure 7.8 *Blastocystis* **spp.** (**Top row**) Drawing of the *Blastocystis* central-body form; note the nuclei around the edges of the central body area. (**Middle row**) Central-body form in saline wet mount (note nuclei around the edges of the central body area); central-body form with D'Antoni's iodine stain; dividing organism with D'Antoni's iodine stain; central-body form with Lugol's iodine stain. (**Bottom row**) All four images show central-body forms stained with Wheatley's trichrome stain (note the peripheral nuclei surrounding the central body area; also note the tremendous size variation even on a single smear).

organisms appear to be dividing, a finding not consistent with a cyst form of the intestinal protozoa.

In the absence of other organisms, *Blastocystis* may cause diarrhea, cramps, nausea, fever, vomiting, and abdominal pain and may require therapy. It has also been linked to infective arthritis. In patients with other underlying conditions, the symptoms may be more pronounced. The incidence of this organism in stools submitted for parasite examination appears to be higher than suspected. In symptomatic patients in whom no other etiologic agent has been identified, *Blastocystis* spp. should be considered the possible pathogen. When a symptomatic *Blastocystis* infection responds to therapy, the improvement may represent elimination of some other undetected pathogenic organism (*E. histolytica*, *G. lamblia*, or *D. fragilis*). However, it is also clear that some organisms within the *Blastocystis* group of subtypes are almost certainly pathogenic. In symptomatic patients in whom no other etiologic agent has been identified, *Blastocystis* spp. should be considered the possible pathogen.

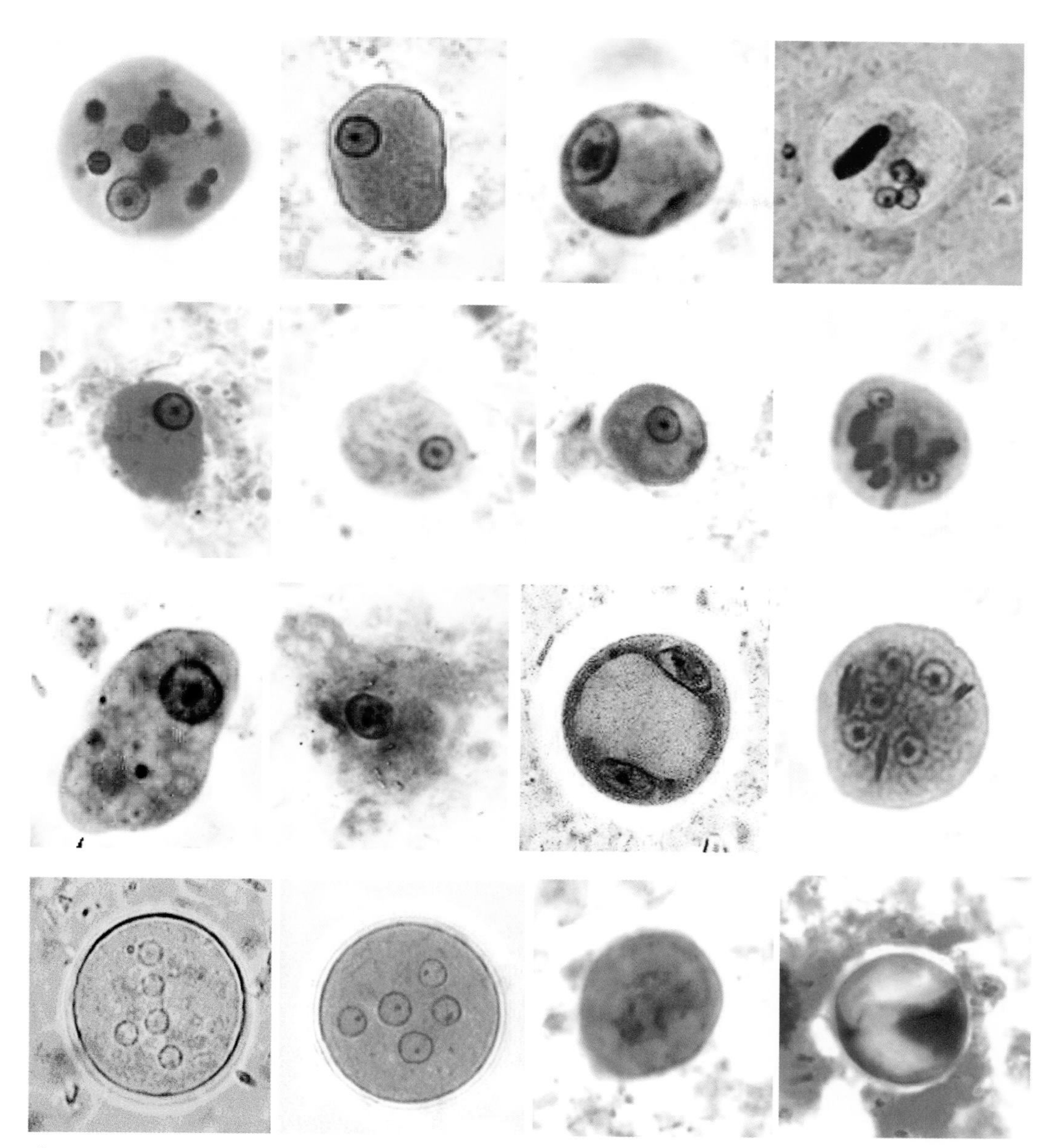

Plate 7.A **(Row 1)** *Entamoeba histolytica* trophozoite; *Entamoeba histolytica/E. dispar* group trophozoite; *E. histolytica/E. dispar* group precyst (note single enlarged nucleus compared with precyst of *Entamoeba coli* with two enlarged nuclei); *E. histolytica/E. dispar* group cyst (note chromatoidal bar with smooth, rounded ends). **(Row 2)** *Entamoeba hartmanni* trophozoites; *E. hartmanni* trophozoites (note "bull's-eye" nuclear configuration); *E. hartmanni* precyst; *E. hartmanni* cyst (note multiple chromatoidal bars and only two nuclei). **(Row 3)** *Entamoeba coli* trophozoite; *E. coli* trophozoite (note large blot-like karyosomes); *E. coli* precyst (note two nuclei, one at each side of the precyst wall); *E. coli* cyst (note five nuclei and chromatoidal bars with sharp, pointed ends). **(Row 4)** *E. coli* cyst (saline wet preparation); *E. coli* cyst (iodine wet mount); *E. coli* cysts; *E. coli* cysts (note that both images are shrunken, some red/some blue; because *E. coli* cyst walls do not fix well using routine fixatives, the cysts often appear shrunken and distorted on the permanent stained smear).

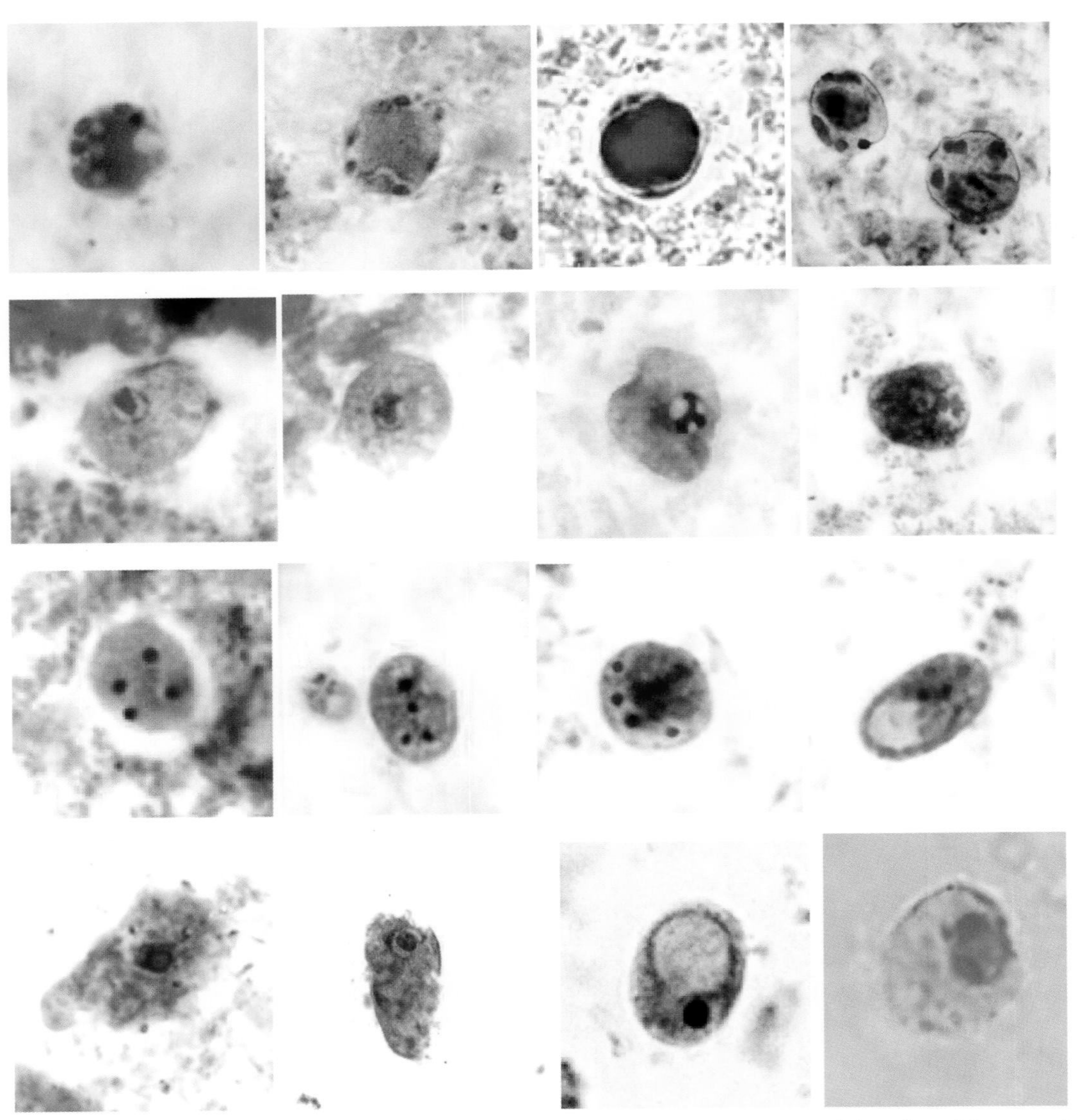

Plate 7.B (**Row 1**) *Blastocystis* central-body forms (note the peripheral nuclei around the edge of the central body area). (**Row 2**) *Endolimax nana* trophozoites (note the tremendous nuclear variation in morphology, which is more common with this organism than the other protozoa; the second organism mimics *Dientamoeba fragilis* with what appears to be a fragmenting karyosome). (**Row 3**) *E. nana* cysts and (on the right) an *E. nana* cyst that contains a vacuole but has four karyosomes (this could be misidentified as *Iodamoeba bütschlii* due to the vacuole). (**Row 4**) *Iodamoeba bütschlii* trophozoite; *I. bütschlii* trophozoite; *I. bütschlii* cyst; *I. bütschlii* cyst (note that the cyst on the right is from an iodine-stained wet mount; the glycogen vacuole stains golden brown).

Giardia lamblia (*G. duodenalis, G. intestinalis*)

Pathogenic	Yes
Disease	Giardiasis
Acquired	Fecal-oral transmission; contaminated food and water
Body site	Intestine, occasionally gallbladder, and rarely bronchoalveolar lavage fluid
Symptoms	Diarrhea, epigastric pain, flatulence, increased fat and mucus in stool; gallbladder colic and jaundice
Clinical specimen	*Intestinal*: Stool
	Extraintestinal: Fluids
Epidemiology	Worldwide, primarily human-to-human transmission
Control	Improved hygiene, adequate disposal of fecal waste, adequate washing of contaminated fruits and vegetables

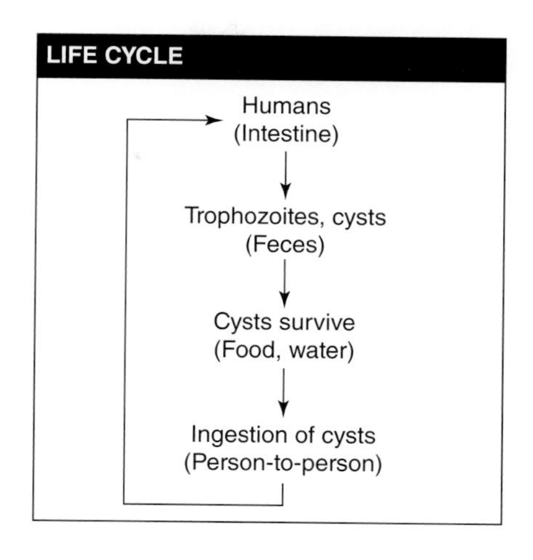

Diagnosis

The standard O&P exam is recommended for recovery and identification of *Giardia lamblia* in stool specimens; however, even after a series of four to six stool examinations, the organisms may not be seen. Duodenal aspirate fluid and/or the Entero-Test capsule may be used. Microscopic examination of a direct saline wet mount or mucus from the Entero-Test capsule may reveal motile trophozoites and/or nonmotile cysts. Although the concentration technique is helpful in demonstrating these organisms, **the most important technique for the recovery and identification of protozoan organisms is the permanent stained smear** (normally stained with trichrome or iron hematoxylin). A minimum of three specimens collected over not more than 10 days is often recommended.

It is important to remember that this organism can easily be missed, even with multiple stool examinations. There is tremendous variation in organism shedding; thus, collection of three stools over a period of 10 days is highly recommended.

Antigenic variation occurs with surface antigen changes during *G. lamblia* infections; although the biological importance of this is not clear, it suggests that this variation may allow the organism to escape the host immune response.

Cysteine proteases (CPs) are major virulence factors in protozoan parasites, influencing disease pathogenesis and parasitic life cycles. *Giardia* CPs are directly involved in the disruption of the intestinal epithelial junctional complex, intestinal epithelial cell apoptosis, and degradation of host immune factors, including chemokines and immunoglobulins. *Giardia* CPs have also been implicated in mucus depletion and microbiota dysbiosis induced by the parasite. Although patients with symptomatic giardiasis usually have no underlying abnormality of serum immunoglobulins, giardiasis is common in patients with immunodeficiency syndromes, particularly in those with common variable hypogammaglobulinemia.

Diagnostic Tips

When performing the concentration method, if the stool contains a lot of mucus, do not use ethyl acetate in the fecal concentration; the organisms can be within the debris layer that is discarded.

A single centrifugation is recommended; excess rinsing will also remove some of the organisms.

Giardia trophozoites tend to stain quite pale; the internal structures (median bodies, axonemes, and nuclei) can be difficult to see, especially if the trophozoites are seen in the thinner part of the permanent stained smear. You may need to adjust the light and condenser for the appropriate light intensity. The trophozoites can occasionally be seen lined up in mucus within the stained smear. When seen head on, the trophozoites have a teardrop shape; when seen from the side, they look like the curved portion of a spoon (see Figure 7.9).

On the permanent stain, the cyst morphology varies from stages that cannot really be identified to ones that appear to be perfect like a drawing. Often the cysts vary from round to oval. They are very three-dimensional and require continuous focusing of the microscope. Occasionally, shrunken cysts mimic large *Cystoisospora belli* oocysts; however, if the organisms are measured, they can be very easily differentiated, since *Cystoisospora* are about twice as large (20 to 33 μm by 10 to 19 μm).

Fecal immunoassays, such as EIA, FA assays, and the rapid cartridges (membrane flow), are more sensitive than the routine O&P exam. Test menus should contain both the O&P exam and one or more of the fecal immunoassays. Because organism shedding is so variable, if the first fecal immunoassay is negative, a second stool should be tested; if both are negative, then the report can indicate that *Giardia* is probably not present. However, a report of negative on the basis of a single negative immunoassay is not acceptable. Appropriate ordering by the physician depends on patient history (travel history,

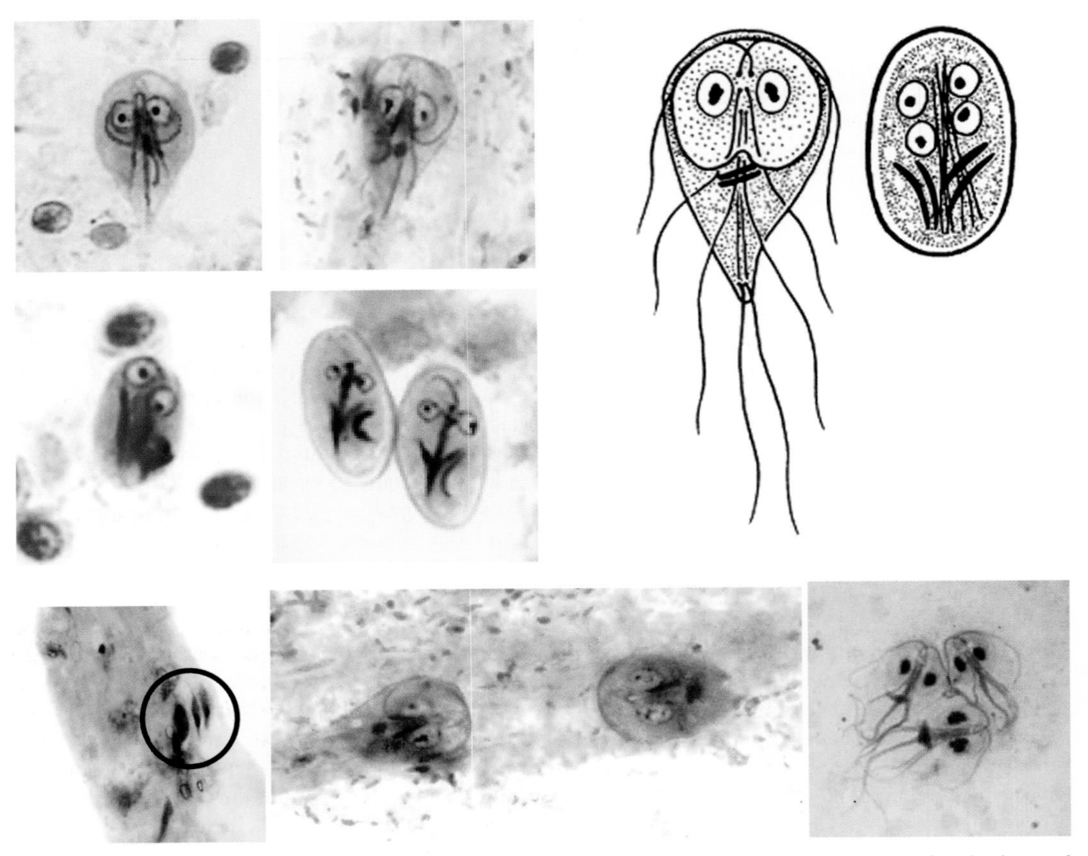

Figure 7.9 *Giardia lamblia*. **(Top row)** Two *Giardia* trophozoites (note the two nuclei, curved median bodies and linear axonemes); drawing of trophozoite and cyst. **(Middle row)** Two *Giardia lamblia* cysts. These cysts are oval; however, they often appear round. Cysts are very three-dimensional; microscopic examination requires constant focusing. **(Bottom row)** *Giardia lamblia* trophozoites in mucus (note that they look like curved spoons when seen from the side [circle]); two trophozoites shown from the front (teardrop shape); three trophozoites from a mucosal imprint/Giemsa stain (courtesy of CDC DpDx, https://www.cdc.gov/dpdx/index.html).

possible outbreaks, etc.). Unfortunately, serodiagnostic procedures for giardiasis do not yet fulfill the criteria necessary for wide clinical use, particularly since they may indicate either past or present infection. (See Hanson and Cartwright, Suggested Reading.)

General Comments

The incidence of giardiasis worldwide may be as high as a billion cases. However, remember that *Blastocystis* spp. are the most common intestinal parasites, and *Dientamoeba fragilis* is seen as often as *G. lamblia*. Although many people worldwide are infected, not all patients present with continuous symptoms. Another challenge in making the correct identification is differentiation between this organism and other intestinal protozoa that may be nonpathogenic.

This organism is currently classified with the protozoa (flagellates), and studies at the molecular level have led to differentiation among *G. lamblia*, *G. duodenalis*, and *G. intestinalis*. Consistent with common terminology used in the United States, the term *G. lamblia* is used throughout this text. The actual term used for identification is not as important as consistent use by both laboratory and clinician. The flagellate trophozoites are often seen in diarrheal specimens but may be very difficult to recognize, even on the permanent stained smear; they tend to be quite pale, and morphology may be difficult to differentiate. *Giardia* trophozoites seen in mucus can be helpful (see Figure 7.9).

Treatment failures have been reported with all the common antigiardial agents, and resistance to these drugs has been demonstrated in the laboratory. Clinical resistance has been reported, including resistance to both metronidazole and albendazole in the same individual.

Potential wild-animal reservoirs have complicated control measures, particularly with waterborne outbreaks. The single most effective practice that prevents the spread of infection in the child care setting is thorough handwashing by the children, staff, and visitors.

Description (Trophozoite)

The standard O&P exam is recommended for recovery and identification of *G. lamblia* in stool specimens. The trophozoite is usually described as being teardrop shaped from the front, with the posterior end being pointed. From the side, it resembles the curved portion of a spoon. The concave portion is the area of the sucking disk, used for attachment to the mucosal lining. There are four pairs of flagella, two nuclei, two linear axonemes, and two curved bodies called the median bodies. Trophozoites usually measure 10 to 20 μm long and 5 to 15 μm wide.

Although the concentration technique is helpful in demonstrating these organisms, **the most important technique for the recovery and identification of protozoan organisms is the permanent stained smear** (normally stained with trichrome or iron hematoxylin). As with most intestinal protozoa, the trophozoite is usually seen in diarrheic or soft stools, while the cyst is seen in the more normal stool or when the patient may be asymptomatic. Trophozoites may remain attached or detach from the mucosal surface. Since the epithelial surface sloughs off the tip of the villus every 72 h, apparently the trophozoites detach at that time.

If mucus from the Entero-Test or fluid from a duodenal aspirate is submitted, the organisms may be trapped in the mucus. Keep the microscope light low, particularly when examining wet preparations. Make sure your eye becomes adjusted to each field before moving the stage.

Description (Cyst)

Cyst formation takes place as the organisms move through the jejunum after exposure to biliary secretions. Trophozoites discharge their undigested food and begin to round up prior to precyst and cyst formation. They retract the flagella into the axonemes, the cytoplasm becomes condensed, and the cyst wall is secreted. As the cyst matures, the internal structures are doubled, so that when excystation occurs, the cytoplasm divides, producing two trophozoites. Excystation normally occurs in the duodenum or appropriate culture medium.

G. lamblia cysts may be round or oval; they contain four nuclei, axonemes, and median bodies. Often some cysts appear to be distorted or shrunken; there may be two halos, one around the cyst wall and one inside the cyst wall around the shrunken organism. The halo around the outside of the cyst wall is often easily seen in the permanent stained fecal smear. Cysts normally measure 11 to 14 μm long and 7 to 10 μm wide.

Dientamoeba fragilis

Pathogenic	Yes
Disease	Dientamoebiasis
Acquired	Fecal-oral transmission; contaminated food and water; may be transmitted within helminth eggs (*Ascaris*, *Enterobius*); cyst form has been confirmed, also impacts transmission
Body site	Intestine
Symptoms	Intermittent diarrhea, abdominal pain, nausea, anorexia, malaise, fatigue, poor weight gain, unexplained eosinophilia
Clinical specimen	Stool
Epidemiology	Worldwide, primarily human-to-human transmission; often as common as *Giardia*
Control	Improved hygiene, adequate disposal of fecal waste, adequate washing of contaminated fruits and vegetables

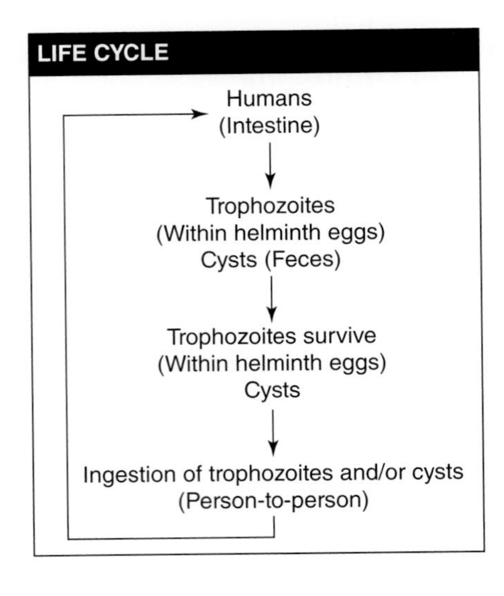

Diagnosis

Dientamoeba fragilis is an important and underestimated cause of gastrointestinal disease in both the autochthonous and immigrant or traveler populations worldwide. The standard O&P exam is recommended for recovery and identification of *D. fragilis* in stool specimens; however, even after a series of several stool examinations, the organisms may not be seen, particularly if wet preparations are the only ones examined. Although the concentration technique is helpful in demonstrating some intestinal protozoa, **the most important technique for the recovery and identification of *D. fragilis* is the permanent stained smear** (normally stained with trichrome or iron hematoxylin). A minimum of three specimens collected over not more than 10 days is often recommended.

D. fragilis has been recovered in formed stool; therefore, a permanent stained smear must be prepared for every stool sample submitted. Organisms seen in direct wet mounts may appear as refractile, round forms; the nuclear structure cannot be seen without examination of the permanent stained smear.

This organism can easily be missed, even with multiple stool examinations. Although a cyst form has been confirmed, very few cysts are found in a stool specimen, and the trophozoites can easily be missed, even with the permanent stained smear.

Fecal immunoassays, such as EIA, FA assay, and the rapid cartridges (membrane flow), would probably be more sensitive than the routine O&P exam. However, these methods are still under development and are not yet commercially available. Some multiplex organism panels are being developed for relevant parasites, including *D. fragilis*.

Diagnostic Tips

Although the survival time for this parasite has been reported as 24 to 48 h, the survival time in terms of morphology is limited, and stool specimens must be examined immediately or preserved in a suitable fixative soon after defecation.

The trophozoite forms have been recovered from formed stool; hence the need to perform the permanent stained smear on specimens other than liquid or soft stools.

Trophozoites seen in direct wet mounts may appear as refractile, round forms; the nuclear structure cannot be seen without examination of the permanent stained smear. With the recent confirmation of the cyst stage, one needs to take into account the more shrunken appearance of this form compared with the trophozoite (see Figure 7.10). However, the number of cysts seen on a fecal smear is very limited; identification is based almost entirely on the trophozoite morphology. The cyst form will appear shrunken, with two cyst walls; often there will be a large clear area surrounding the cyst (permanent stain).

Organisms with a single nucleus can easily be confused with *Endolimax nana* or *Entamoeba hartmanni*, both of which are considered nonpathogens. However, reporting the presence of nonpathogens is important, since it conveys information to the physician that the patient has ingested something contaminated with fecal material. Where nonpathogens are present, there may also be pathogens present; thus the need for multiple stools.

General Comments

Although a large number of people throughout the world are infected with *D. fragilis*, not all patients have continuous symptoms. In some areas of the world, this organism is recovered as often as *Giardia*. It has a cosmopolitan distribution, and surveys show incidence rates of 1.4 to 19%. There is speculation that *D. fragilis* may be infrequently recovered and identified in some areas; low incidence or absence from survey studies may be due to limited laboratory techniques and a general lack of knowledge about the organism.

To date, two *D. fragilis* genotypes (1 and 2) are recognized, with a strong predominance of genotype 1 in both humans and few animal hosts. Recent studies indicate that a very low level of genetic variability characterizes *D. fragilis* isolates collected in various geographic areas and from both symptomatic and asymptomatic cases; thus, it has been suggested that *D. fragilis* may be a clonal organism. The recent availability of transcriptome data should be helpful in the development of markers used to understand genetic diversity of *D. fragilis* at the population level.

On the basis of electron microscopy studies, *D. fragilis* has been reclassified as an ameboflagellate rather than an ameba and is closely related to *Histomonas* and *Trichomonas* spp. *D. fragilis* is currently classified with the protozoa (flagellates); studies at the molecular level may lead to reclassification and differentiation. The flagellate trophozoites are often seen in diarrheal specimens but may be very difficult to recognize, even on the permanent stained smear. The flagella are internal, and the organism resembles a small ameba rather than a flagellate.

Incidence is hard to assess; laboratories that do not routinely examine a permanent stained smear will tend to miss this organism entirely. However, many reports indicate that this infection may be as common as that seen with *Giardia*. These infections have been seen in nursery schools, along with giardiasis and cryptosporidiosis.

The significance of two genetically distinct forms of *D. fragilis* may ultimately clarify the issues of virulence and clinical perceptions regarding pathogenicity. Case reports of children and adults infected with *D. fragilis* reveal a number of symptoms, including intermittent diarrhea, abdominal pain, nausea, anorexia, malaise, fatigue, poor weight gain, and unexplained eosinophilia. The most common symptoms in patients infected with this parasite appear to be intermittent diarrhea and fatigue. In some patients, both the organism and the symptoms persist or reappear until appropriate treatment is initiated.

Description (Trophozoite)

The standard O&P exam is recommended for recovery and identification of *D. fragilis* in stool specimens. Motility is usually nonprogressive. The trophozoite is characterized by having one (20 to 40%) or two (60 to 80%) nuclei. The nuclear chromatin tends to be fragmented into three to five granules, and there is normally no peripheral chromatin on the nuclear membrane. In some organisms, the nuclear chromatin mimics that of *E. nana*, *E. hartmanni*, or even *Chilomastix mesnili*. The cytoplasm is usually vacuolated and may contain ingested debris or bacteria. There may also be some large granules. The cytoplasm can also appear quite clean with few inclusions. The trophozoites usually measure 5 to 15 μm, with the usual range being 9 to 12 μm.

Again, **the most important technique for the recovery and identification of protozoan organisms is the permanent stained smear** (normally stained with trichrome or iron hematoxylin).

Description (Cyst)

The discovery of a protozoan cyst in *D. fragilis*-infected mice was made initially by light microscopy and was confirmed by transmission electron microscopy. Subsequently, a total of five true cysts were detected from four patient samples, giving a prevalence of 0.01% cyst per patient sample. The cysts were detected independently in two different laboratories in different locations (Australia and the United States).

The cysts are composed of a distinct cyst wall (~5 μm in diameter) with a zone of clearance around the cyst. A peritrophic space is present between the cyst wall and the amoebic parasite enclosed within. The nuclear structure is characteristically identical to what is found in *D. fragilis* trophozoites. All cysts seen were binucleate, with each nucleus containing a large central karyosome with a delicate nuclear membrane. No chromatin is visible on the nuclear membrane, and the nucleus is often fragmented into distinct granules of chromatin, often referred to as chromatin packets. According to these findings, these "true" cysts are rarely encountered in clinical samples, which probably accounts for the limited number of reports describing these structures (see Figure 7.10). In contrast, the precystic forms of *D. fragilis* were more frequently encountered, with a prevalence of up to 5% in clinical samples. This precystic stage is characterized by a compact spherical shape with a reduction in size of up to 50% relative to "normal" trophozoites. These forms range in size from 4 to 5 μm. The cytoplasm is darkly staining, indicating a denser structure than what is found in normal trophozoites. The cytoplasm is homogeneous and rarely contains any inclusions.

Additional Information

Although its pathogenic status is still not well defined, *D. fragilis* has been associated with a wide range of symptoms. In one study, 11 pediatric patients, 7 of whom had peripheral eosinophilia, a history of recent travel, and symptoms of anorexia, intermittent vomiting, abdominal pain, and diarrhea, were diagnosed with *D. fragilis* (see Stark et al., Suggested Reading). Based on these and other findings, including bovine protein allergy and eosinophilic colitis, the authors recommend that *D. fragilis* be included in the differential diagnosis of chronic diarrhea and eosinophilic colitis.

Clinical improvement has been observed in adults receiving tetracycline; symptomatic relief was reported in children receiving diiodohydroxyquin, metronidazole, or tetracycline. Current recommendations include iodoquinol, paromomycin, and tetracycline. Since symptomatic relief has been observed to follow appropriate therapy, *D. fragilis* is probably pathogenic in infected individuals who are symptomatic.

With clarification of the cyst form, fecal-oral transmission is probably possible. Preventive measures would coincide with those used for other protozoan infections. Since transmission also occurs from the ingestion of certain helminth eggs (*Enterobius vermicularis* and *Ascaris lumbricoides*), hygienic and sanitary measures to prevent contamination with fecal material would be appropriate.

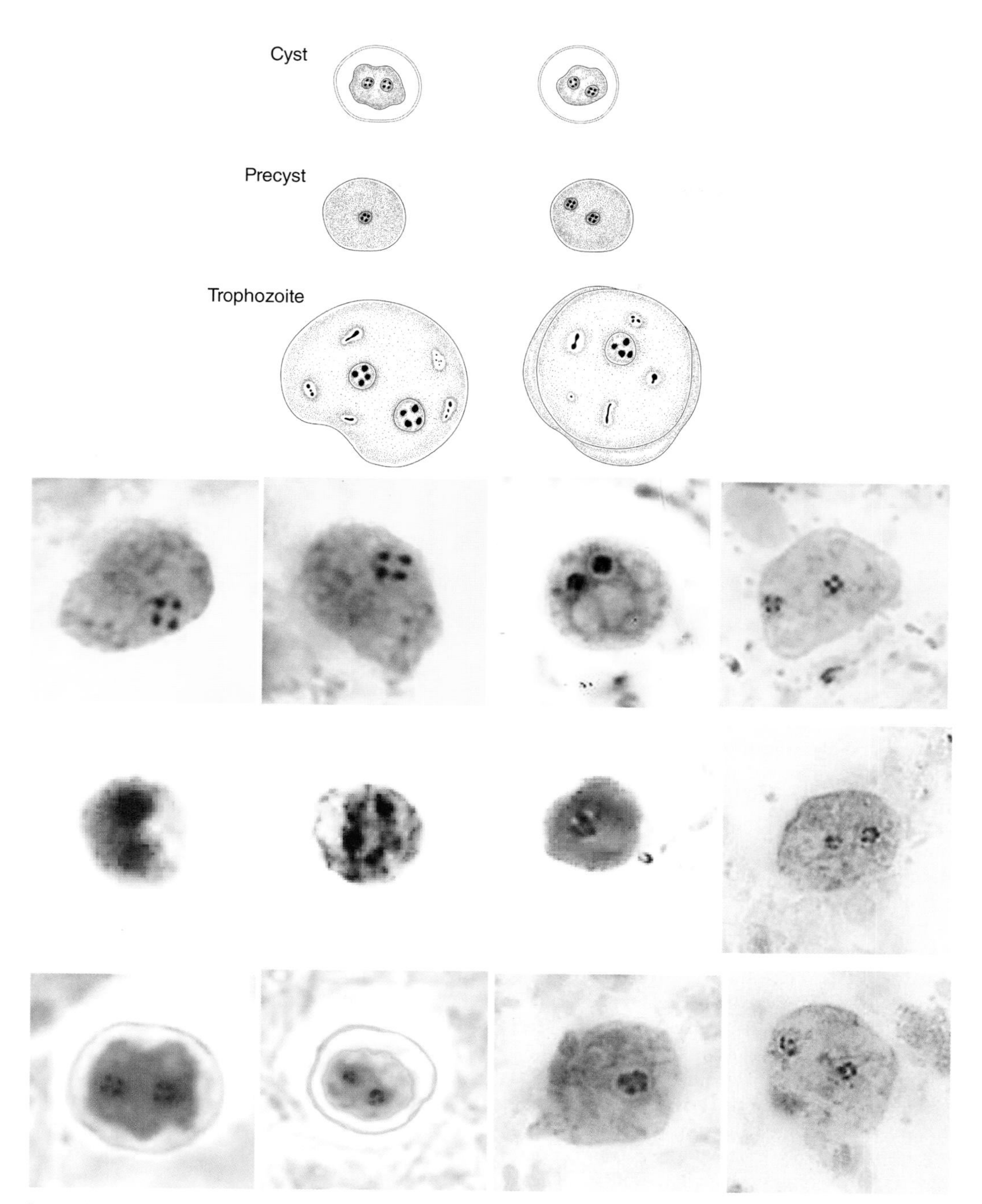

Figure 7.10 ***Dientamoeba fragilis.*** Drawings: Forms of *Dientamoeba fragilis*: cysts, precysts, and trophozoites. **(Row 1)** Photos of *D. fragilis* trophozoites (the first two have a single fragmented karyosome; the last two have two nuclei with fragmented karyosomes). **(Row 2)** The first three images are *D. fragilis* precysts; the fourth image is a trophozoite with two fragmented karyosomes (courtesy of CDC Public Health Image Library). **(Row 3)** The first two images are *D. fragilis* cysts (note the double wall and space between inner and outer wall); the last two images are *D. fragilis* trophozoites, one with a single fragmented karyosome (left) and one with two fragmented karyosomes (right) (last two courtesy of the CDC Public Health Image Library). (Drawings and remainder of images courtesy of Stark D, Garcia LS, Barratt JLN, et al, *J Clin Microbiol* **52**:2680–2863, 2014, doi:10.1128/JCM.00813-14.)

Chilomastix mesnili

Pathogenic	No
Disease	None
Acquired	Fecal-oral transmission; contaminated food and water
Body site	Intestine
Symptoms	None
Clinical specimen	Stool
Epidemiology	Worldwide, primarily human-to-human transmission
Control	Improved hygiene, adequate disposal of fecal waste, adequate washing of contaminated fruits and vegetables

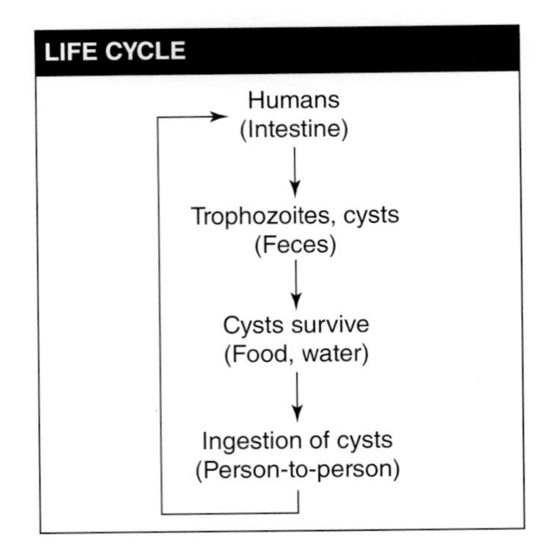

Diagnosis

The standard O&P exam is recommended for recovery and identification of *Chilomastix mesnili* in stool specimens. Microscopic examination of a direct saline wet mount may reveal motile trophozoites and/or nonmotile cysts. Although the concentration technique is helpful in demonstrating these organisms, **the most important technique for the recovery and identification of protozoan organisms is the permanent stained smear** (normally stained with trichrome or iron hematoxylin). A minimum of three specimens collected over not more than 10 days is often recommended. This organism can easily be missed, even with multiple stool examinations.

These organisms normally live in the cecal region of the large intestine, where they feed on bacteria and debris. They are considered nonpathogens, and no treatment is recommended. Since transmission is through ingestion of infective cysts, prevention depends on improved personal hygiene and upgraded sanitary conditions.

Diagnostic Tips

Chilomastix trophozoites resemble many other small protozoa; while the trophozoites are not acceptable for a proficiency testing challenge, the cysts are more appropriate. In drawings of the trophozoite, the shape is elongated; however, in actual clinical specimens, this is not the case. The organism is more round. Also, the flagella are very rarely seen, and the feeding/oral groove, or cytostome, is not clearly visible in most trophozoites. There is tremendous variation among the shapes of the trophozoites, while that of the cyst is much more typical. The cyst is typically pear or lemon shaped with a single nucleus and a curved fibril that delineates the cytostome.

General Comments

C. mesnili is currently classified with the protozoa (flagellates). *C. mesnili* tends to have a cosmopolitan distribution, although it is found more frequently in warm climates. It has both trophozoite and cyst stages and is somewhat more easily identified than are some of the smaller flagellates, such as *Enteromonas hominis* and *Retortamonas intestinalis*. Although many people worldwide are probably infected, the organisms may not always be seen, particularly if a permanent stained smear is not examined. Another challenge to correct identification is differentiation between this organism and other intestinal protozoa that may be pathogenic.

The presence of this organism should always be reported, since nonpathogenic organisms are transmitted in the same way as pathogens. Confirmation of the presence of this organism indicates that the patient has ingested something (food, water, etc.) contaminated with fecal material. If nonpathogenic protozoa are found, pathogens may also be found on additional examination of the specimen.

Description (Trophozoite)

The trophozoite is somewhat elongated, measuring 6 to 24 μm long and 4 to 8 μm wide. There is a single nucleus and a distinct oral groove, or cytostome, close to the nucleus. Flagella are difficult to see unless there is obvious motility in a wet preparation. Morphology can be confirmed by a permanent stained smear, particularly when the cytostome is visible.

Although the concentration technique is helpful in demonstrating these organisms, **the most important technique for the recovery and identification of protozoan organisms is the permanent stained smear** (normally stained with trichrome or iron hematoxylin). As with most intestinal protozoa, the trophozoite form is usually seen in diarrheic or soft stools while the cyst stage is seen in the more normal stool or when the patient may be asymptomatic.

If this trophozoite appears somewhat round to oval and the cytostome is not visible, *C. mesnili* can easily be confused with other small amebae and/or flagellates.

Description (Cyst)

Trophozoites discharge their undigested food and begin to round up prior to precyst and cyst formation. The cysts are pear or lemon shaped and range from 6 to 10 μm long and from 4 to 6 μm wide. There is also a single nucleus in the cyst and the typical curved cytostomal fibril, which is called the shepherd's crook.

These cysts can be confused with other small flagellate cysts, especially *Retortamonas intestinalis* cysts. The cytostomal fibril (shepherd's crook) may not always be visible, and if the cyst is somewhat shrunken, confusion with other small organisms is possible. Often one can see shrinkage (halo) surrounding the cyst on the permanent stained smear; thus, the organism may measure at the small end of the size range due to shrinkage.

Due to the small size of the cyst, identification is not as easy as identification of *Giardia lamblia*. Also, when the cyst is preserved with mercuric-chloride substitutes, the overall morphology, even on the permanent stained smear, is not as clear.

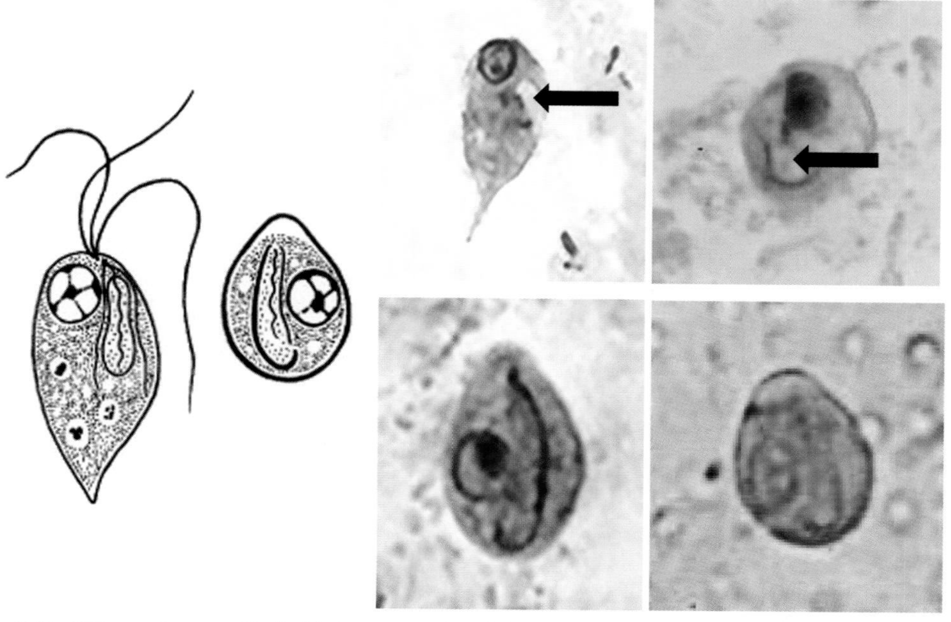

Figure 7.11 *Chilomastix mesnili.* **(Left)** Drawings of a *Chilomastix mesnili* trophozoite and cyst. **(Top photos)** Two trophozoites are visible; note the single nucleus, the more typical shape of the one on the left, and the clear cytostome (arrows). **(Bottom photos)** Two cysts (note the curved fibril, or shepherd's crook); the image on the right is an iodine-stained wet mount.

Pentatrichomonas hominis

Pathogenic	No
Disease	None
Acquired	Fecal-oral transmission; contaminated food and water
Body site	Intestine
Symptoms	None
Clinical specimen	Stool
Epidemiology	Worldwide, primarily human-to-human transmission
Control	Improved hygiene, adequate disposal of fecal waste, adequate washing of contaminated fruits and vegetables

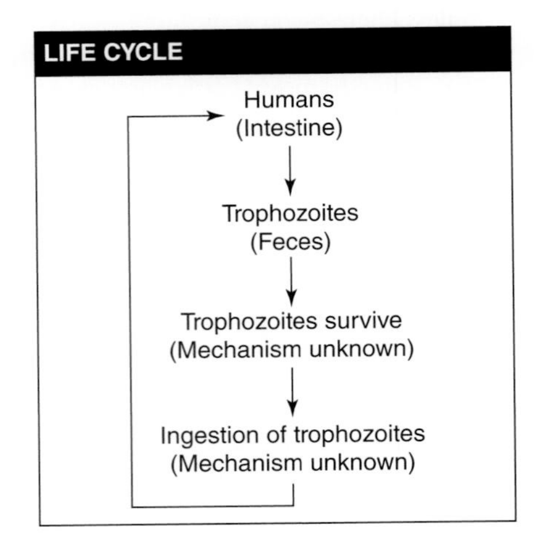

Diagnosis

The standard O&P exam is recommended for recovery and identification of *Pentatrichomonas hominis* in stool specimens. Microscopic examination of a direct saline wet mount may reveal motile trophozoites; there are no known cyst forms. The concentration technique is not very useful; **the most important technique for the recovery and identification of protozoan organisms is the permanent stained smear** (normally stained with trichrome or iron hematoxylin). A minimum of three specimens collected over not more than 10 days is often recommended.

This organism can easily be missed, even with multiple stool examinations and the permanent stained smear.

This organism is found in the intestinal tract only; *Pentatrichomonas* and *Trichomonas* spp. tend to be site specific, and only *Trichomonas vaginalis* is pathogenic (it is found in the genitourinary tract).

Diagnostic Tips

Although the concentration technique is helpful in demonstrating these organisms, the most important technique for the recovery and identification of protozoan organisms is the permanent stained smear (normally stained with trichrome or iron hematoxylin). As with most intestinal protozoa, the trophozoite form is usually seen in diarrheic or soft stools.

Pentatrichomonas trophozoites tend to stain very pale and can easily be missed, even on the permanent stained smear. Often one can see the axostyle protruding from the bottom of the organism. On the permanent stained smear, look for the small granules that may be clustered along the axostyle. These small granules are often typically found in trichomonads.

These trophozoites can easily be confused with those of other small flagellates, such as *Enteromonas hominis* and *Retortamonas intestinalis*.

There are no known cyst forms.

General Comments

The genus name has been changed from *Trichomonas* to *Pentatrichomonas* based on the five anterior flagella and a granular parabasal body. This organism is currently classified with the protozoa (flagellates). *P. hominis* infection is common in dogs, monkeys, and humans, especially in children and young dogs. Given the infection prevalence, *P. hominis* may pose a risk of zoonotic and anthroponotic transmission.

Many people worldwide may be infected; however, the organisms are not always seen, particularly if a permanent stained smear is not examined. Correct identification and differentiation between this organism and other intestinal protozoa that may be pathogenic (*Giardia lamblia*) are difficult. A recent study is the first report presenting the high association between *P. hominis* and gastrointestinal cancers. Nevertheless, whether there is any possible pathological role of *P. hominis* infection in cancer patients needs to be further clarified. Organisms in the study population were identified using nested PCR amplifying the internal transcribed spacer region and partial 18S rRNA gene.

The presence of this nonpathogen should always be reported; confirmation of its presence indicates that the patient has ingested something (food, water, etc.) contaminated with fecal material, which also contains pathogens.

P. hominis is probably the most commonly identified flagellate other than *G. lamblia* and *Dientamoeba fragilis*. It has been recovered worldwide, in both warm and temperate climates.

Description (Trophozoite)

The trophozoite measures 5 to 15 μm long and 7 to 10 μm wide. It is pyriform and has both an axostyle and undulating membrane, which help in identification. The undulating membrane extends the entire length of the body, in contrast to that of *T. vaginalis*, which goes only halfway down the length of the body. Flagella are difficult to see unless there is obvious motility in a wet preparation.

Additional Information

The trophozoites live in the cecal area of the large intestine and feed on bacteria. The organism is not considered invasive.

When urine is examined for *T. vaginalis* (pathogen), it is important not to accidentally misidentify *P. hominis* as *T. vaginalis* if the urine is contaminated with fecal material. *T. vaginalis* is a sexually transmitted disease, and incorrect organism identification could have multiple ramifications, including risk management issues, particularly if the urine specimen is from a child (i.e., it could potentially lead to suspicion of child abuse).

Since there is no known cyst stage, transmission probably occurs in the trophic form. If ingested in a protecting substance, such as milk, these organisms can survive passage through the stomach and small intestine in patients with achlorhydria. *P. hominis* cannot be transplanted into the vagina, the natural habitat of *T. vaginalis*. The incidence of this organism is relatively low, but it tends to be recovered more often than *E. hominis* or *R. intestinalis*. The infection is diagnosed more often in warm climates and in children. Because of the fecal-oral transmission route, preventive measures should emphasize improved hygienic and sanitary conditions.

Trichomonas tenax

Trichomonas tenax was first recovered from tartar from the teeth. Although the pathological role of some specific bacterial strains associated with periodontal diseases is well documented, the impact of parasites in periodontitis pathology is still being clarified. In a very recent study, a link was established between the carriage in patients of *T. tenax* and the severity of the periodontitis. Genotyping demonstrated the presence of strain diversity with three major different clusters and a relation between disease strains and the periodontitis severity. More frequently detected in periodontal cases than from other sites, *T. tenax* is likely to be related to the onset and/or evolution of periodontal diseases. Additional studies are ongoing.

The distribution of this organism is worldwide. There is no known cyst stage. Transmission is assumed to be by direct contact or use of contaminated dishes and glasses. Prevalence rates vary from 0 to 25%. Although *T. tenax* has been considered a harmless commensal in the mouth, there are also reports of respiratory infections and thoracic abscesses, particularly in patients with underlying cancers or other lung diseases.

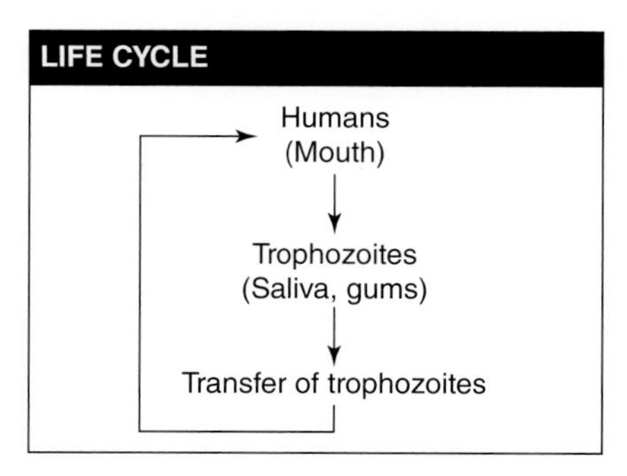

Life cycle for *Trichomonas tenax*

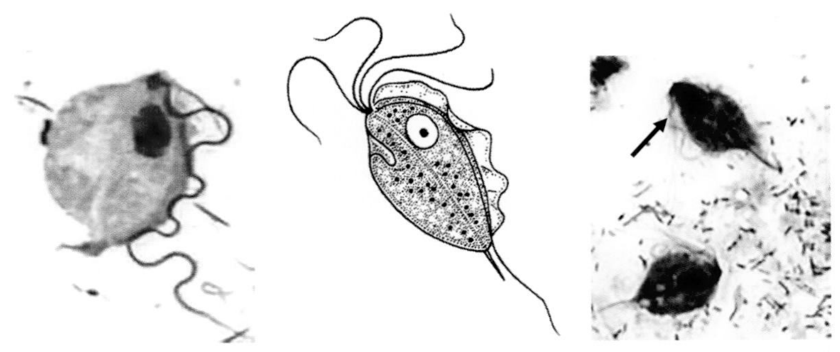

Figure 7.12 ***Pentatrichomonas hominis, Trichomonas tenax.*** Photograph of *P. hominis* trophozoite (note that the undulating membrane extends to the bottom of the organism and the axostyle/supporting rod extends through the bottom); drawing of *P. hominis*. Photograph of *T. tenax* trophozoite (note that the undulating membrane does NOT extend to the bottom of the organism [arrow]; the axostyle is clearly visible protruding from the bottom of the organism) (last image courtesy of Alameda County Medical Center, Oakland, CA).

Enteromonas hominis
Retortamonas intestinalis

Pathogenic	No
Disease	None
Acquired	Fecal-oral transmission; contaminated food and water
Body site	Intestine
Symptoms	None
Clinical specimen	Stool
Epidemiology	Worldwide, primarily human-to-human transmission
Control	Improved hygiene, adequate disposal of fecal waste, adequate washing of contaminated fruits and vegetables

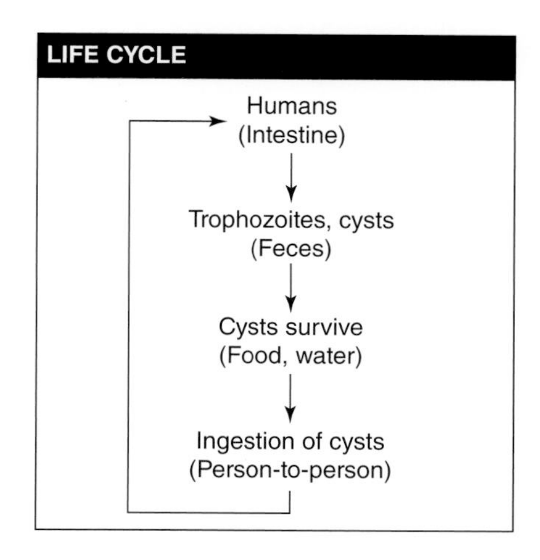

Diagnosis

The standard O&P exam is recommended for recovery and identification of *Enteromonas hominis* and *Retortamonas intestinalis* in stool specimens. Microscopic examination of a direct saline wet mount may reveal motile trophozoites or non-motile cyst forms. The concentration technique is helpful in demonstrating these organisms, but **the most important technique for the recovery and identification of protozoan organisms is the permanent stained smear** (normally stained with trichrome or iron hematoxylin). A minimum of three specimens collected over not more than 10 days is often recommended. These organisms can easily be missed, even with multiple stool examinations and the permanent stained smear. Both flagellates can be easily confused with other small flagellates.

These organisms are rarely found in clinical specimens, probably because of the difficulties in accurate identification of small flagellates, even when the permanent stained smear is used. Infection is through ingestion of the cysts, and improved hygienic conditions would certainly prevent spread of the infection. *E. hominis* and *R. intestinalis* are considered nonpathogens; thus, no therapy is indicated.

General Comments

Although many people worldwide may be infected with *E. hominis* or *R. intestinalis*, the organisms are not always seen, particularly if a permanent stained smear is not examined. *R. intestinalis* has been recovered from both warm and temperate areas of the world. It has also been found in certain groups such as psychiatric hospital patients. Another challenge to correct identification is differentiation between these organisms and other intestinal protozoa that may be pathogenic (*Giardia lamblia*). These organisms are currently classified with the protozoa (flagellates).

The presence of these organisms should always be reported, since nonpathogenic organisms are transmitted the same way as pathogens. The presence of these protozoa indicates that the patient has ingested something (food, water, etc.) contaminated with fecal material. If nonpathogenic protozoa are found, pathogens may also be found on additional examination of the specimen.

Phylogenetic and network analyses of *Retortamonas* spp. have revealed three statistically supported clusters among the vertebrate-isolated haplotypes, while insect-isolated haplotypes independently clustered with *Chilomastix*. In the clade of vertebrate isolates, assemblage A (amphibian genotype), which included the amphibian references, was addressed as an out-group of the other clusters. Assemblage B (mammalian and chicken genotype) included most haplotypes from various mammals, including humans, with the haplotypes isolated from a chicken. Human isolates were all classified into this assemblage; thus, assemblage B might correspond to *R. intestinalis*. Assemblage C (bovine genotype), which included specific haplotypes from water buffalos and cattle, was considered a sister lineage of assemblage B. Among the diversified haplotypes of assemblage B, a specific haplotype, which was identified from multiple host mammals (humans, dogs, pigs, cattle, water buffalos, elks, goats, and rats), indicates the potential zoonotic transmission of the *Retortamonas* among them.

Description (Trophozoite)

The standard O&P exam is recommended for recovery and identification of *E. hominis* and *R. intestinalis* in stool specimens. The

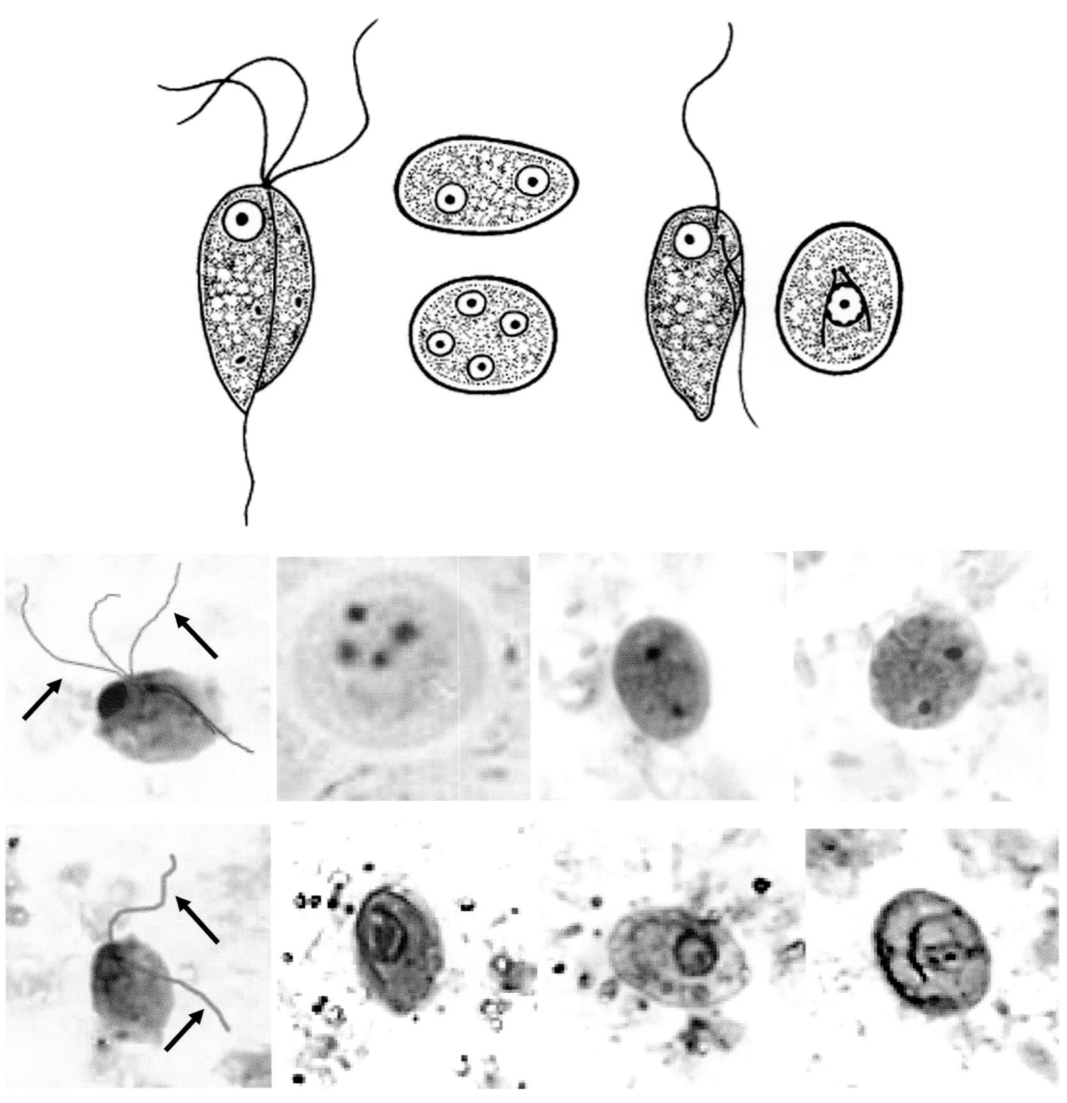

Figure 7.13 ***Enteromonas hominis, Retortamonas intestinalis.*** **(Top row)** Drawings of *E. hominis* trophozoite and cysts; *R. intestinalis* trophozoite and cyst. **(Middle row)** *E. hominis* trophozoite (note the pale flagella [arrows]) and three examples of cysts (numbers of nuclei range from 2 to 4 in these images). **(Bottom row)** *R. intestinalis* trophozoite (note flagella [arrows]) and three examples of cysts (some fibrils are seen). The fibrils in the first cyst image look most like the "bird's beak" configuration. (Photos courtesy of the CDC Public Health Image Library.)

trophozoites of *E. hominis* are somewhat pear shaped, measuring approximately 4 to 10 µm by 3 to 6 µm. There is no cytostome, and the flagella are rarely visible unless motile organisms are seen.

The trophozoites of *R. intestinalis* are elongate pyriform or ovoidal. They measure 4 to 9 µm long by 3 to 4 µm wide. This flagellate has a cytostome, which may be difficult to see, even in a permanent stained smear. Flagella are difficult to see unless there is obvious motility in a wet preparation. Morphology can be confirmed by using a permanent stained smear, particularly when the cytostome is visible.

As with most intestinal protozoa, the trophozoite form is usually seen in diarrheic or soft stools, while the cyst stage is seen in the more normal stool or when the patient is asymptomatic.

These trophozoites can be easily confused with those of *Chilomastix mesnili* and *Pentatrichomonas hominis*.

Description (Cyst)

Trophozoites discharge their undigested food and begin to round up prior to precyst and cyst formation. The cyst of *E. hominis* measures approximately 6 to 10 µm by 4 to 6 µm and tends to be oval. There are two nuclei, and the cyst may mimic a binucleate *Endolimax nana* cyst. However, *E. nana* cysts containing two nuclei are quite rare in most clinical specimens. Also, *E. hominis* is not often reported, perhaps in part because of its small size and the difficulties in identification. Even on a permanent stained smear, this organism is often difficult to identify accurately.

The cysts of *R. intestinalis* are somewhat pear shaped, measuring 4 to 9 µm long and 4 to 6 µm wide. Both trophozoite and cyst have a single nucleus. These cysts can be confused with other small flagellate cysts, especially *E. hominis* cysts. The typical "bird's beak" structure of the fibrils within the cyst may not always be easy to see; these cysts occasionally mimic those of *C. mesnili*.

Balantidium coli

Pathogenic	Yes
Disease	Balantidiasis
Acquired	Fecal-oral transmission; contaminated food and water
Body site	Intestine
Symptoms	Intermittent diarrhea, abdominal pain, nausea, anorexia, malaise, fatigue, poor weight gain
Clinical specimen	Stool
Epidemiology	Worldwide, primarily human-to-human transmission, also pig-to-human transmission
Control	Improved hygiene, adequate disposal of fecal waste, adequate washing of contaminated fruits and vegetables; prevention of human-pig contact, hygienic rearing of pigs

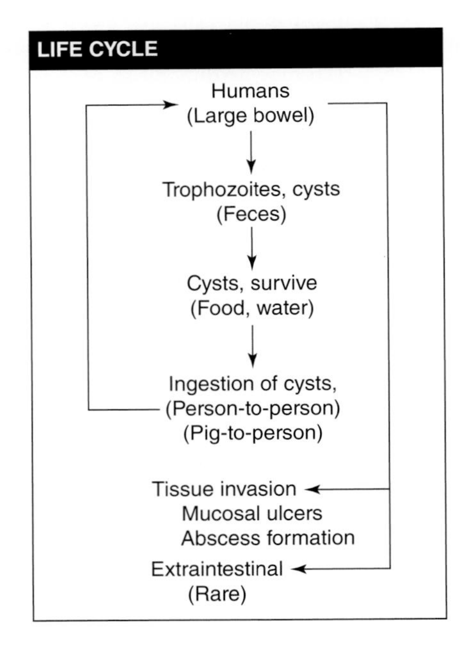

Diagnosis

Balantidium coli is a ciliated protozoan that lives in the large intestine (cecum and colon) of pigs (the natural reservoir host), humans, nonhuman primates, and rodents. *Balantidium coli* is often a neglected pathogen and is now considered an emerging and reemerging protozoan.

The standard O&P exam is recommended for recovery and identification of *B. coli* in stool specimens. Microscopic examination of a direct saline wet mount may reveal motile trophozoites or cyst form. The concentration technique is very useful in demonstrating these organisms, but **the most important technique for the recovery and identification of protozoan organisms is generally the permanent stained smear** (normally stained with trichrome or iron hematoxylin). A minimum of three specimens collected over not more than 10 days is often recommended.

Balantidiasis is considered a neglected zoonotic disease. In areas where pigs are the main domestic animal, the incidence of human infection can be quite high. Particularly susceptible are persons working as pig farmers or in slaughterhouses (28% infection rate in New Guinea). Human infection is fairly rare in temperate areas, although once the infection is established, it can develop into an epidemic, particularly where poor environmental sanitation and personal hygiene are found.

Although several publications mention potential changes in taxonomy, none of these are currently approved. Thus, we retain the name *Balantidium coli* here.

Diagnostic Tips

These organisms are quite large and are generally identified from a concentration wet mount. In a permanent stained smear, these organisms take up so much stain that they can be confused with artifact material and/or helminth eggs. For these reasons, proficiency testing challenges rely on the concentration sediment wet-mount examination identification. Usually *B. coli* can be seen using the low-power objective (10×). Since the permanent stained smear is normally examined using the oil immersion objective (100×), the organisms will appear to be dark-staining debris.

When examining wet mounts with saline alone, remember to set the light and condenser correctly to gain as much contrast as possible. In wet-mount preparations, the trophozoite exhibits a rotary or boring motility. The large macronucleus will be visible, and hopefully, you will be able to see the cilia around the edges. Normally, the micronucleus is not visible. In some of the trophozoites, the cytostome (feeding groove) may be visible (see Figure 7.14).

Because these ciliates are so large, constant focusing will be required to see the internal structures. Excellent morphology can usually be seen using the 10× and 40× objectives.

General Comments

B. coli is widely distributed in hogs, particularly in warm and temperate climates, and in monkeys in the tropics. Human infection is found in warmer climates, sporadically in cooler areas, and in institutionalized groups with low levels of personal hygiene. *B. coli* is rarely recovered in clinical specimens within the United States. This organism is currently classified with the protozoa (ciliates).

With a global prevalence of 0.02 to 1%, infection with *B. coli* is a significant problem in some developed countries, with a high prevalence rate in Latin America, the Philippines, Papua New Guinea, western New Guinea, and the Middle East. The parasite is endemic in Yunnan province in China, with infection rates of up to 4.24% in some villages. Apparently, the highest rate (up to 20% prevalence) has been seen in areas of Indonesia; there appears to be a very close association between pigs and humans in this area. Certainly, human-to-human contact has been widely documented, particularly in psychiatric units within the United States and other countries.

B. coli infection is a common protozoan infection in wild boars in Iran, with prevalence rates of 25 to 70%. Swine balantidiasis is often asymptomatic, yet these asymptomatic pigs serve as the major reservoir hosts of the parasite. Wild boars in rural western Iran are considered a reservoir for human balantidiasis.

Some individuals with *B. coli* infections are asymptomatic, whereas others have symptoms of severe dysentery similar to those seen with amebiasis. Symptoms usually include diarrhea or dysentery with multiple liquid stools per day, tenesmus, nausea, vomiting, anorexia, and headache. Insomnia, muscular weakness, and weight loss have also been reported. The diarrhea may persist for weeks to months prior to the development of dysentery. There may be tremendous fluid loss with a type of diarrhea similar to that seen in cholera or in some coccidial infections. The disease responds to treatment with tetracycline or metronidazole.

B. coli can invade tissue. It may penetrate the mucosa on contact, with cellular infiltration in the area of the developing ulcer. Some of the abscess formations may extend to the muscular layer. The ulcers may vary in shape, and the ulcer bed may be full of pus and necrotic debris. Although the number of cases is small, extraintestinal disease has been reported.

On very rare occasions, the organisms may invade extraintestinal organs, and extraintestinal spread to the peritoneal cavity, appendix, genitourinary tract, and lung has rarely been reported.

Warming of the earth's surface may provide a more favorable environment, even in the now-temperate areas of the world, for survival of trophic and cystic stages of *Balantidium*, and its prevalence may increase.

Description (Trophozoite)

The trophozoite is quite large, oval, and covered with short cilia; it measures ca. 50 to 100 μm long and 40 to 70 μm wide. It can easily be seen in a wet preparation on lower power. The anterior end is somewhat pointed and has a cytostome; in contrast, the posterior end is broadly rounded. The cytoplasm contains many vacuoles with ingested bacteria and debris. There are two nuclei, a very large bean-shaped macronucleus and a smaller round micronucleus. The organisms normally live in the large intestine.

Trophozoites are occasionally found in tissue; however, dissemination from the intestine is rare. Although the number of cases is small, extraintestinal disease has been reported (peritonitis, urinary tract infection, and inflammatory vaginitis).

Description (Cyst)

The cyst is formed as the trophozoite moves down the intestine. Nuclear division does not occur in the cyst; therefore, only two nuclei are present, the macronucleus and the micronucleus. The cysts measure 50 to 70 μm.

Cysts are never found in tissues, and the trophozoite form invades the intestinal tract on occasion. However, in a more formed stool, both forms can be found.

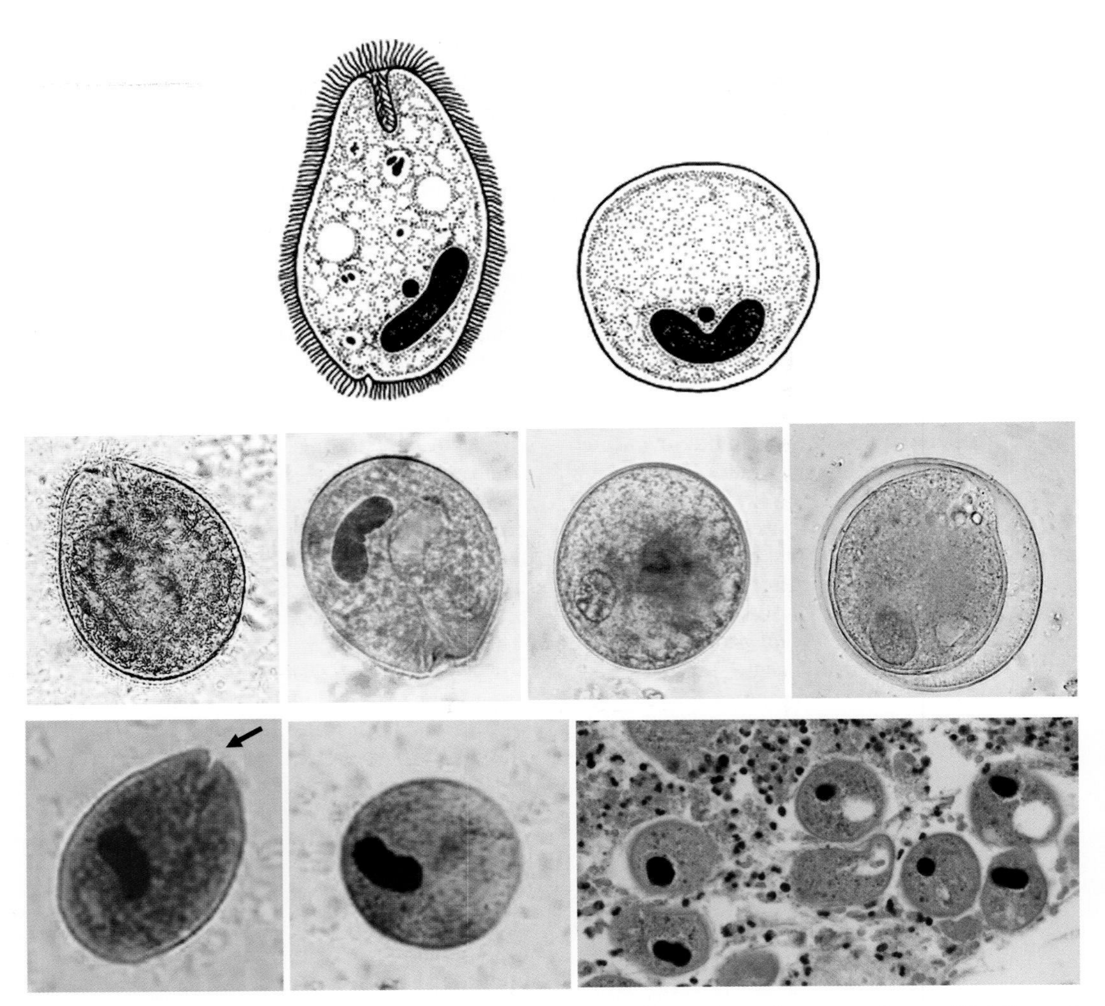

Figure 7.14 ***Balantidium coli.*** **(Top row)** Drawings of *Balantidium coli* trophozoite and cyst. **(Middle row)** Iodine wet mounts of *B. coli* trophozoites (first two images; note the cilia around the edges of the trophozoite on the left and the large bean-shaped nucleus in the organism on the right) and cysts (last two images; note that the internal structures are not as clearly delineated). **(Bottom row)** *B. coli* trophozoite (note the cytostome [arrow]); *B. coli* cyst; trophozoites in intestinal tissue (note the dark macronucleus). It is unusual to see permanent stains this good. If *B. coli* is suspected, the fecal smears can be prepared somewhat thin; this approach should produce better staining results, as seen here.

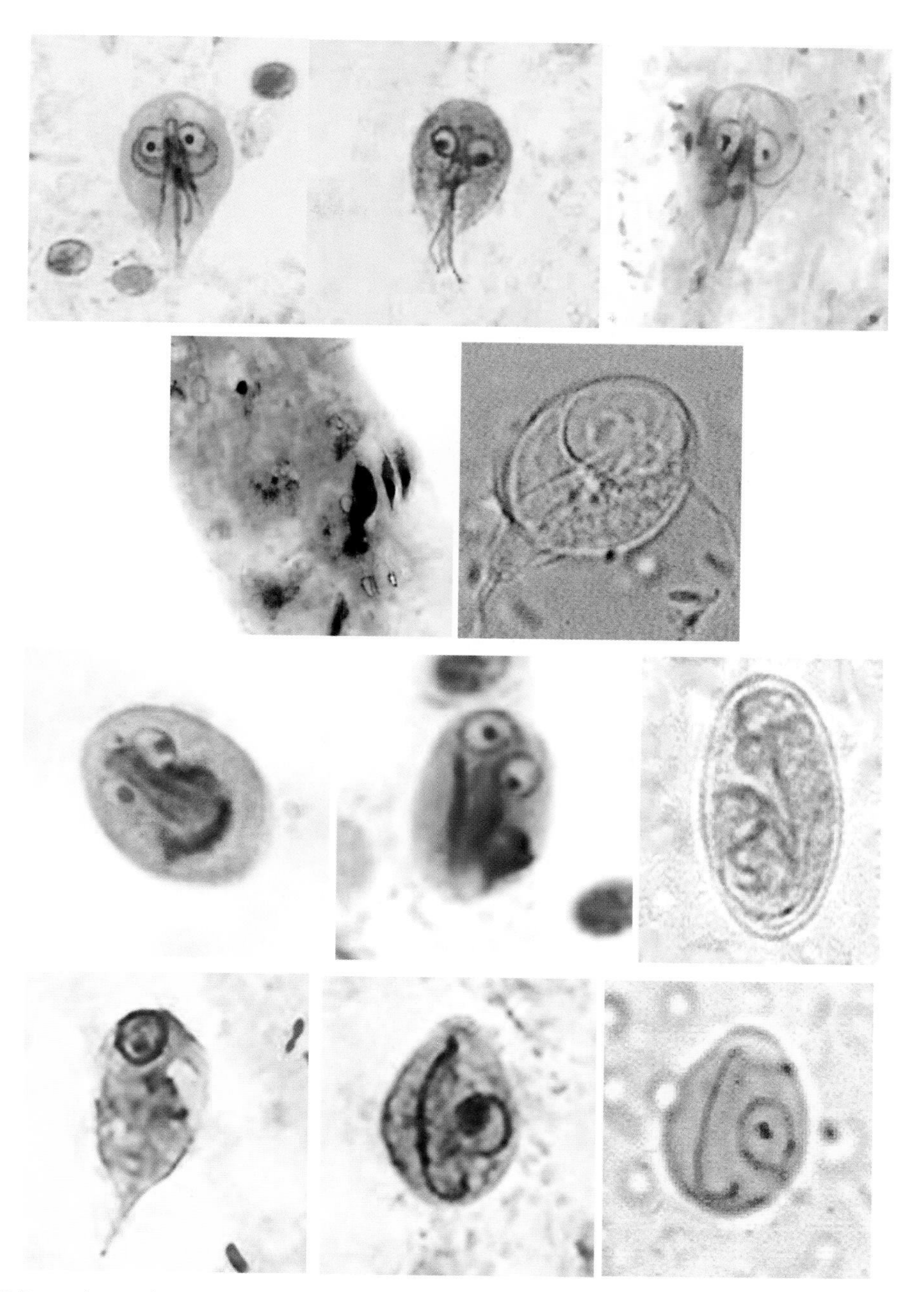

Plate 7.C (**Row 1**) *Giardia lamblia* trophozoites. (**Row 2**) *G. lamblia* trophozoites in mucus; *G. lamblia* trophozoite (wet mount). (**Row 3**) *G. lamblia* cysts; *G. lamblia* cyst (wet mount). (**Row 4**) *Chilomastix mesnili* trophozoite; *C. mesnili* cyst; *C. mesnili* cyst (wet mount).

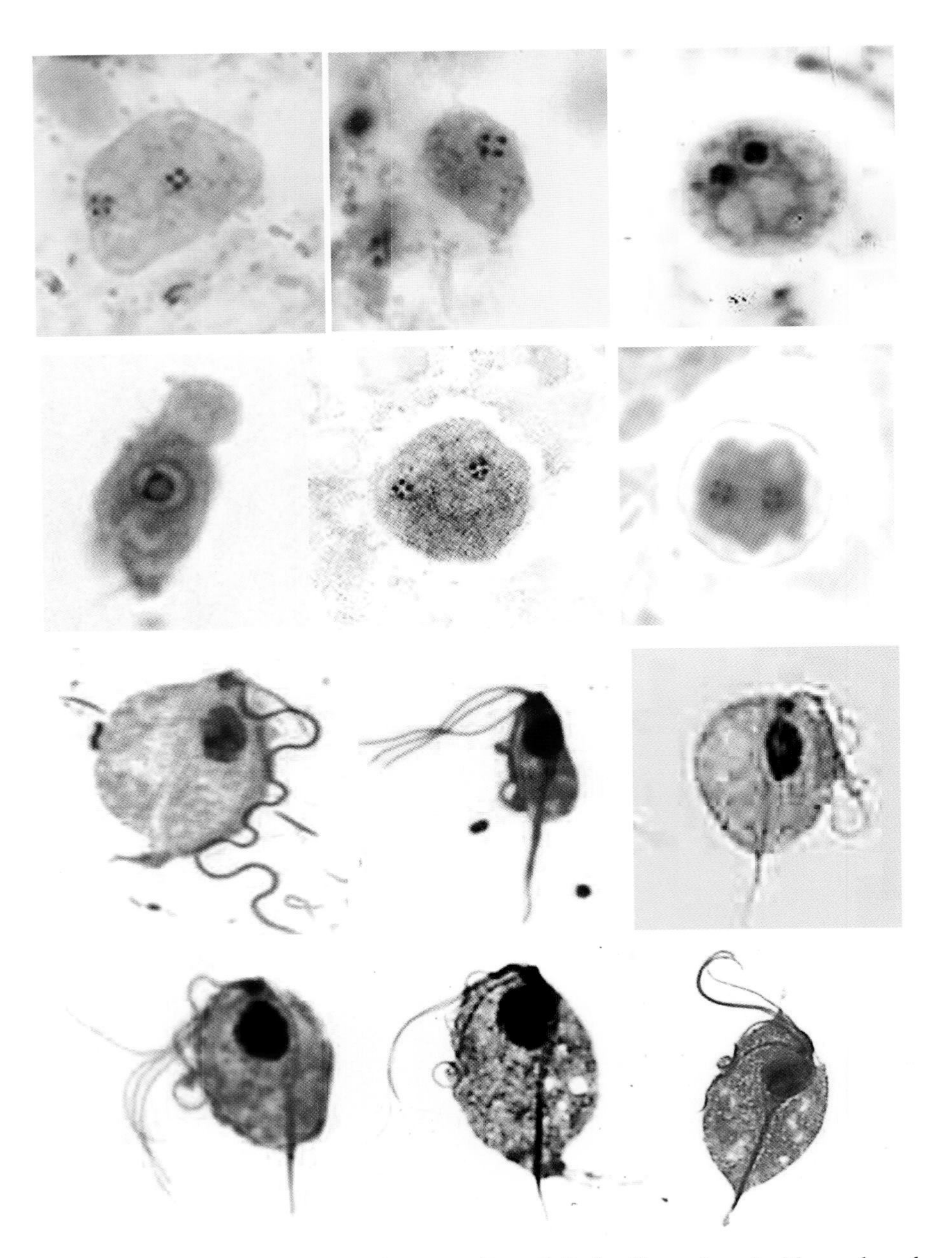

Plate 7.D **(Row 1)** *Dientamoeba fragilis* trophozoites. **(Row 2)** *D. fragilis* trophozoite (the nuclear chromatin has not yet completely fragmented; it can therefore mimic *Endolimax nana*); *D. fragilis* trophozoite; *D. fragilis* cyst (note the double wall). **(Row 3)** *Pentatrichomonas hominis* trophozoite; *Trichomonas tenax* trophozoite; *Trichomonas vaginalis* trophozoite. **(Row 4)** *T. vaginalis* trophozoites.

Cryptosporidium spp.

Pathogenic	Yes
Disease	Cryptosporidiosis (in the Apicomplexa, but no longer coccidian)
Acquired	Fecal-oral transmission; contaminated food and water
Body site	Intestine, disseminated infection in severely compromised patients
Symptoms	Nausea, low-grade fever, abdominal cramps, anorexia, and 5 to 10 watery, frothy bowel movements per day
Clinical specimen	Stool
Epidemiology	Worldwide, primarily human-to-human transmission, also animal-to-human transmission
Control	Improved hygiene, adequate disposal of fecal waste, water testing, adequate washing of contaminated fruits and vegetables

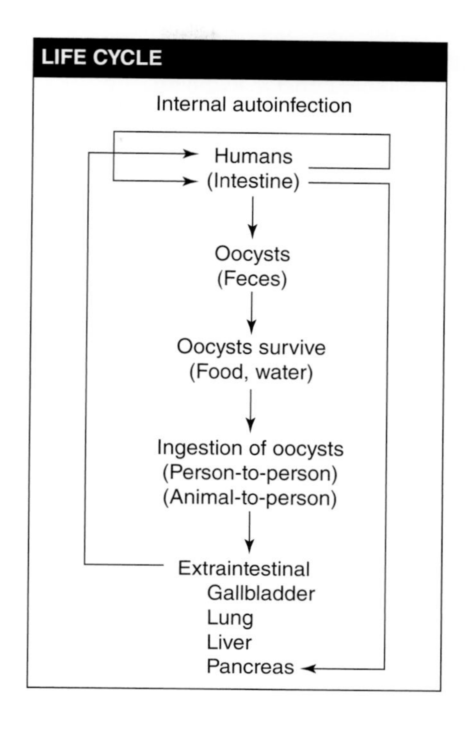

Diagnosis

Cryptosporidiosis is well recognized as a disease in humans, particularly in those who are in some way immunosuppressed or immunodeficient. *Cryptosporidium* has been implicated as one of the more important opportunistic agents seen in patients with AIDS. Also, reports of respiratory tract and biliary tree infections confirm that the developmental stages of this organism are not always confined to the gastrointestinal tract.

Oocysts recovered in clinical specimens are difficult to see without special staining techniques, such as the modified acid-fast, Kinyoun's, and Giemsa methods, or the newer direct FA assay or EIA methods. The four sporozoites may be seen within the oocyst wall in some of the organisms, although in freshly passed specimens they are not always visible. The standard O&P concentration is recommended (500 × *g* for 10 min), with the concentrate sediment being used for all smears and for the FA procedure. Shorter centrifugation at lower speeds (often mentioned in some references) may not guarantee recovery of the oocysts. Multiple stool specimens may have to be examined to diagnose the infection, particularly when formed stool specimens are being tested.

Uncentrifuged fresh, frozen, or fixed fecal material can be used for the antigen detection immunoassays (EIA, FA assay, rapid cartridges); except for FA, these procedures do not rely on visual identification of the oocysts but on antigen detection. The selection of fresh or frozen stool rather than fixed specimens will depend on testing parameters; currently, if the test reagents or kit format includes *Entamoeba histolytica*/*E. dispar* or *E. histolytica* alone, the test format requires fresh or frozen stool (or possibly stool in Cary-Blair medium). If the test format includes *Cryptosporidium* and/or *Giardia*, fresh, frozen, or preserved stools can be used. The fecal immunoassays have excellent specificity and sensitivity and result in a significantly increased detection rate over conventional staining methods (modified acid-fast stains). Some of these reagents, particularly the combination direct fluorescent-antibody assay (DFA) product used to identify both *Giardia lamblia* cysts and *Cryptosporidium* spp. oocysts, are being widely used in water testing and outbreak situations.

Diagnostic Tips

Frozen specimens should not be used for FA; the freeze-thaw cycle destroys the oocysts. When preparing smears for FA, remember to make the smears thinner rather than too thick. When examining the smear, read edges as well as the center area of the smear. Occasionally, small fluorescent objects may be seen; however, these are artifacts and do not really look like oocysts. When the FA method is used, the counterstain can be used; use of the counterstain is not mandatory and depends on user preference.

When removing fluid from a fecal preservative/stool mix vial, use only the relatively clear liquid portion for antigen detection. Too much particulate matter will interfere with the test procedure. This applies to the EIA as well as the membrane flow cartridge format.

When staining is done using the modified acid-fast method, the decolorizer can be 1% H_2SO_4 only; it will provide good results for *Cyclospora*, as well as *Cryptosporidium*. Although sporozoites can be seen in some of the oocysts, all oocysts when passed (regardless of the stool consistency) are infectious.

For *Cryptosporidium* in tissues, hematoxylin-eosin (H&E)-stained developing forms are seen as basophilic spherical bodies, from 2.0 to 7.5 µm, located on the surfaces of epithelial cells in histological sections. Other techniques that may be used are periodic acid-Schiff and silver-based stains. Transmission electron microscopy allows the parasites' ultrastructural morphology to be seen.

General Comments

Cryptosporidium parvum is found in animals and humans; *Cryptosporidium hominis* is found in humans. Differentiation is not possible from routine microscopy. Approximately 17 species have been implicated in human infections.

Cryptosporidium developmental stages occur in an intracellular, extracytoplasmic location. Each stage is within a parasitophorous vacuole of host cell origin; however, the vacuole containing the organism is located at the microvillous surface of the host cell. The thin-walled autoinfective oocyst may explain why a small inoculum can lead to overwhelming infection in a susceptible host and why immunosuppressed patients may have persistent, life-threatening infections in the absence of documentation of repeated exposure to oocysts. The stages found on the microvillous surface are 1 µm, and the oocysts recovered in stool specimens are 4 to 6 µm.

Oocysts undergo sporogony while they are in the host cells and are already infective when passed in the stool. Approximately 20% of *Cryptosporidium* oocysts do not form the thick two-layered, environmentally resistant oocyst wall. The four sporozoites within this autoinfective stage are surrounded by a single-unit membrane. After release from a host cell, this membrane ruptures, and the invasive sporozoites penetrate the microvillous region of other cells within the intestine and reinitiate the life cycle.

The prepatent period from the time of ingestion of infective oocysts to completion of the life cycle with excretion of newly developed oocysts in the human is ca. 4 to 22 days. Characteristic antibody responses develop following infection, and persons with preexisting antibodies may be less likely to develop illness, particularly when infected with low oocyst doses.

Description

In histology preparations, *Cryptosporidium* developmental stages are found at all levels of the intestinal tract, with the jejunum being the most heavily infected site. Routine H&E staining is sufficient to demonstrate these parasites. With regular light microscopy, the organisms are visible as small (1- to 3-µm) round structures aligned along the brush border. They are intracellular but extracytoplasmic and are found in parasitophorous vacuoles. Developmental stages are more difficult to identify without using transmission electron microscopy.

Another very relevant finding is the verification that *Cryptosporidium* spp. can develop their life cycle outside a host cell. The possibility of *Cryptosporidium* spp. multiplying in biofilms and in water and sewage treatment plants reveals a greater risk of outbreaks from waterborne transmission. In addition, microbial biofilms can form in the interior of the human intestine, increasing the host's susceptibility, as well as during the course of the infection.

In severely compromised patients, *Cryptosporidium* occurs in other body sites, primarily the lungs, as disseminated infections. Within tissue, confusion with *Cyclospora cayetanensis* is unlikely, since *Cyclospora* oocysts are approximately 8 to 10 µm and the developmental stages occur within a vacuole at the luminal end of the enterocyte, rather than at the brush border. Developmental stages of *Cystoisospora belli* also occur within the enterocyte, so they should not be confused with *Cryptosporidium*.

Microsporidia are present in ca. 30% of severely immunocompromised patients (<100 $CD4^+$ cells) with cryptosporidiosis. The diagnostic procedures for identification of *Cryptosporidium* spp. are not appropriate for identification of microsporidial spores. Modified trichrome stains and optical brightening agents (calcofluor white) can be used for that purpose.

Additional Information

In immunocompetent individuals, clinical symptoms include nausea, low-grade fever, abdominal cramps, anorexia, and 5 to 10 watery, frothy bowel movements per day, which may be followed by constipation. Some patients present with diarrhea, while others have relatively few symptoms, particularly later in the infection. In patients with the typical watery diarrhea, the stool specimen contains very little fecal material, mainly water and mucus flecks. Often the organisms are trapped in the mucus. A patient with a normal immune system will have a self-limiting infection.

In immunocompromised individuals, the duration and severity of diarrhea depend on the immune status of the patient. Most severely immunocompromised patients cannot self-cure, the illness becomes progressively worse, and the sequelae may be a major factor leading to death. In these patients, *Cryptosporidium* infections are not always confined to the gastrointestinal tract;

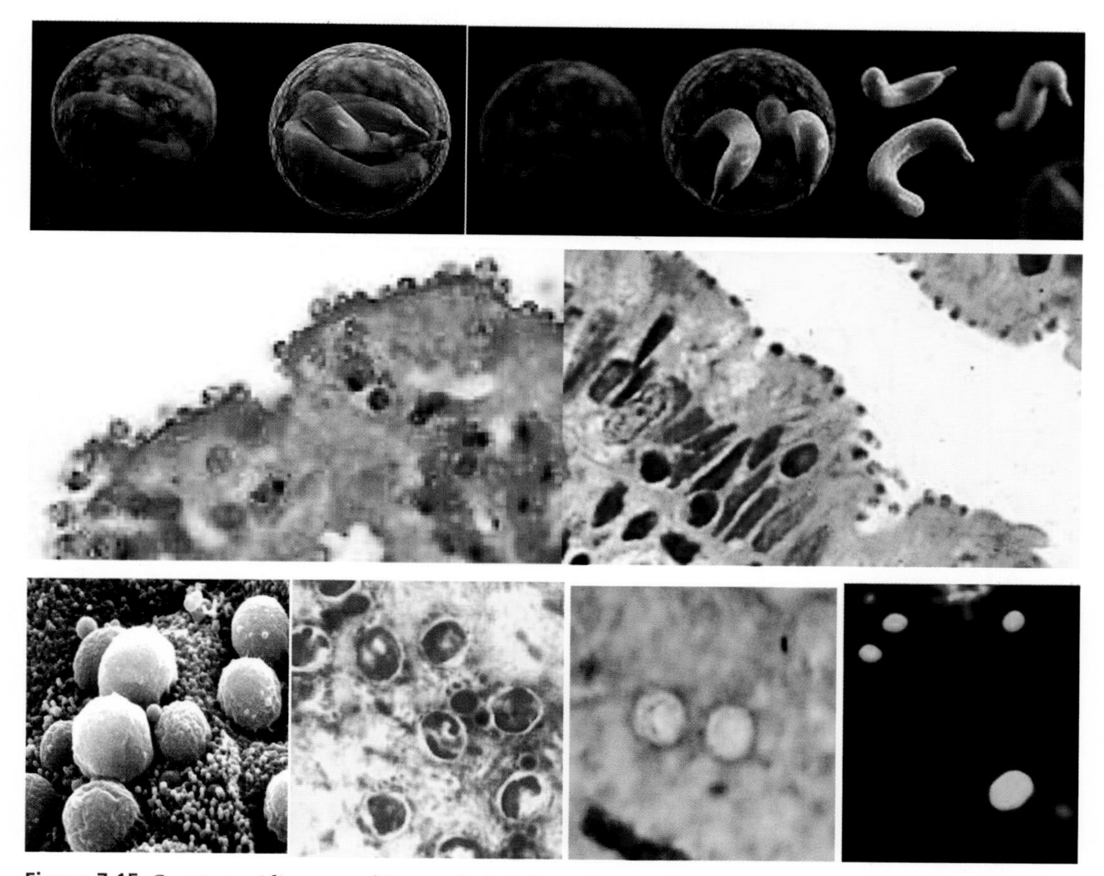

Figure 7.15 ***Cryptosporidium*** **spp. (Top row)** Artist's rendering of *Cryptosporidium* oocysts containing sporozoites; sporozoites being released from the oocyst (courtesy of CDC PHIL [https://www.cdc.gov/parasites/crypto/] and CDC Newsroom [https://www.cdc.gov/media/releases/2017/p0518-cryptosporidium-outbreaks.html]). **(Middle row)** *Cryptosporidium* sp. oocysts along the microvillous surface; developing *Cryptosporidium* at the microvillous surface (courtesy of Armed Forces Institute of Pathology). **(Bottom row)** Scanning electron micrograph of *Cryptosporidium* oocysts (4 to 6 μm); oocysts stained with modified acid-fast stain (these oocysts stain more consistently than those of *Cyclospora cayetanensis*; in some oocysts, the sporozoites can be seen; oocysts are infectious even if the sporozoites are not visible); two "ghost" cells (routine Wheatley's trichrome stain; oocysts do not stain well); results of an FA assay—the large object is a *Giardia lamblia* cyst, and the smaller objects are *Cryptosporidium* sp. oocysts.

additional symptoms (respiratory problems, cholecystitis, hepatitis, and pancreatitis) have been associated with extraintestinal infections.

Cryptosporidiosis tends to be self-limiting in patients who have an intact immune system. In patients who are receiving immunosuppressive agents, one therapeutic option would be to discontinue the regimen. Other approaches with specific therapeutic drugs have been tried, but with inconclusive results. Highly active antiretroviral therapy (HAART) has had a dramatic impact on cryptosporidiosis in AIDS patients, leading to an increased $CD4^+$ count. Resolution of the cryptosporidiosis diarrhea is related to the enhanced $CD4^+$ count rather than any change in the viral load or any therapeutic impact of the drugs.

Cryptosporidiosis is a leading cause of diarrheal disease. There remains a critical need for new approaches to therapy for this infection. There are no vaccines, and nitazoxanide (NTZ) is the only drug approved for the treatment of cryptosporidiosis. Basic research to enhance new molecule development has progressed slowly, mainly due to the lack of genetic tools and appropriate animal models. However, progress is expected to increase in the next decade, especially considering the completion of several *Cryptosporidium* genome sequences and the capacity to genetically engineer *Cryptosporidium*. Since there is no continuous culture system for *Cryptosporidium*, studies have relied on short-term maintenance in cultured cell monolayers to assess the efficacy of potential drugs.

The routes of drug uptake are still poorly understood, and it is unclear whether, for intestinal cryptosporidiosis, the optimal anti-cryptosporidial agent should be absorbed systemically and/or retained in the gastrointestinal tract. Ideally, the treatment of extra-intestinal or severe cryptosporidiosis requires an injectable compound. Several options have been identified. Developing therapeutics to reduce morbidity and mortality due to *Cryptosporidium* spp. among children in low-resource settings is a challenge. Some studies have centered on various plant extracts as anti-diarrheal options, some of which look promising and a few of which appear to be more effective than paromomycin.

Cyclospora cayetanensis

Pathogenic	Yes
Disease	Cyclosporiasis
Acquired	Fecal-oral transmission; contaminated food and water
Body site	Intestine
Symptoms	Nausea, low-grade fever, fatigue, anorexia, and up to seven bowel movements per day; relapses can occur
Clinical specimen	Stool
Epidemiology	Worldwide, primarily human-to-human transmission, although direct person-to-person transmission is unlikely; animal-to-human transmission is not yet documented; ingestion of contaminated food (lettuce, basil, berries, other fresh produce)
Control	Improved hygiene, adequate disposal of fecal waste, adequate washing of contaminated fruits and vegetables

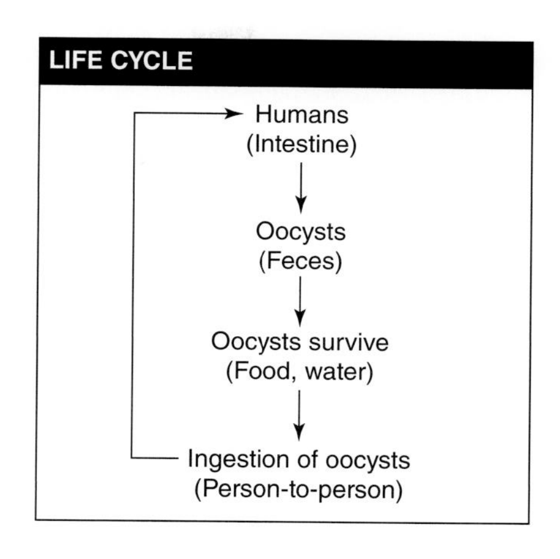

Diagnosis

In recent years, human *Cyclospora cayetanensis* infection has emerged as an agent of important illness, with a number of outbreaks being reported in the United States and Canada. Prior to 1995, only one outbreak of cyclosporiasis had been reported in the United States; however, from May through August 1996, more than 1,400 cases were reported from 20 states, Washington, D.C., and two Canadian provinces. Understanding of the biology and epidemiology of *C. cayetanensis* is limited and has been complicated by a lack of information about the parasite's origins, possible animal reservoir hosts, and relationship to other coccidia.

The organisms stain orange with safranin and are acid-fast variable, with some organisms staining deep red with a mottled appearance but no internal organization visible. Clean wet mounts show nonrefractile spheres that are acid-fast variable with the modified acid-fast stain; those that are unstained appear as glassy, wrinkled spheres. Modified acid-fast stains stain the oocysts light pink to deep red; some contain granules or have a bubbly appearance (like wrinkled cellophane). When the modified acid-fast stain is used for *Cryptosporidium* and detects other similar but larger structures (approximately twice the size of *Cryptosporidium* oocysts), they may be *Cyclospora.* All acid-fast oocysts should be measured, particularly if they appear to be somewhat larger than those of *Cryptosporidium.* Organisms can be seen by using a fluorescent microscope (380- to 420-nm excitation filter), which demonstrates bright green to intense blue autofluorescent oocysts. During concentration (formalin-ethyl acetate) of stool specimens, centrifugation should be for 10 min at 500 × *g* to guarantee recovery of oocysts. The standard O&P concentration is recommended, with the concentrate being used for all smears and for the FA procedure.

Apparently, *Cyclospora* oocysts have a much thicker oocyst wall than *Cryptosporidium* oocysts. Also, the oocyst contents of *Cyclospora* tend to be more granular. (See Almeria et al., Suggested Reading.)

Diagnostic Tips

On modified acid-stained smears, the oocysts stain light pink to deep purple, and some contain granules or have a bubbly appearance. Those that do not stain may have a wrinkled appearance that has been described as looking like wrinkled cellophane. **A strong decolorizer should not be used; 1% sulfuric acid is recommended and also works well for modified acid-fast stains for *Cryptosporidium* spp. and/or *Cryptoisospora belli* acid-fast stains.** The 3 to 5% concentration of sulfuric acid that was originally recommended is usually too strong for *Cyclospora* and removes too much color. Even with the 1% acid decolorizer, some oocysts may appear clear or very pale.

In the modified safranin technique, the oocysts uniformly stain a brilliant reddish orange if fecal smears are heated in a microwave during staining. The stained slide can also be examined by epifluorescence microscopy first, and suspected oocysts can be confirmed by bright-field microscopy.

The oocysts of *Cyclospora*, *Cystoisospora*, and *Cryptosporidium* autofluoresce. Strong autofluorescence of *Cyclospora* oocysts is useful for microscopic identification. *Cyclospora* appears blue when exposed to 365-nm UV light and looks green under 450- to 490-nm excitation. *Cryptosporidium* appears violet when exposed to 365-nm UV light and green under 405- to 436-nm excitation. Due to potential errors in identification, organisms should be measured to confirm accuracy.

The detection of *C. cayetanensis* in produce is difficult. Usually, low numbers of oocysts are detected in contaminated produce, and the methods used for clinical human specimens are not always appropriate for detection in produce. One very important step is efficient recovery of oocysts from fresh produce after careful washing. Of a number of rinse solutions tried for removal of oocysts from produce, a commercial laboratory detergent (Alconox) demonstrated improved recovery compared to other solutions. With the development of advanced molecular methods, as few as five oocysts in raspberries, cilantro, parsley, basil, and carrots were detected by qPCR using an FDA-validated technique. However, overall, this is not a routine procedure.

General Comments

The oocysts that had previously been recovered from human stool were immature, so the structure of the mature oocyst had not been seen. Unsporulated oocysts are passed in the stool; oocyst maturation takes approximately 5 to 13 days, so the mature stage may not have been seen in human specimens. Because oocysts are excreted unsporulated and need to sporulate in the environment, direct person-to-person transmission is unlikely. The oocyst contains two sporocysts, each containing two sporozoites, a pattern which places these organisms in the coccidian genus *Cyclospora*. The species name comes from the university where it was initially studied (Universidad Peruana Cayetano Heredia).

In patients with *Cyclospora* in their stool specimens, parasites with coccidian characteristics have been found within the jejunal enterocytes. The entire life cycle can be completed within a single host. Information on potential reservoir hosts has yet to be obtained.

Developmental stages of *C. cayetanensis* usually occur within epithelial cells of the jejunum and lower portion of the duodenum. *Cyclospora* infection reveals characteristics of a small-bowel pathogen, including upper gastrointestinal symptoms, malabsorption of D-xylose, weight loss, and moderate to marked erythema of the distal duodenum. Histopathology in small-bowel biopsy specimens reveals acute and chronic inflammation, partial villous atrophy, and crypt hyperplasia.

Description

Two types of meronts and sexual stages have been seen in jejunal enterocytes in biopsy specimens from infected patients. These findings confirm that the entire life cycle can be completed within a single host. Unsporulated oocysts are passed in the stool, and sporulation occurs within approximately 7 to 13 days. Complete sporulation produces two sporocysts that rupture to reveal two crescent-shaped sporozoites measuring 1.2 by 9.0 μm.

Additional Information

Individuals of all ages, including both immunocompetent and immunosuppressed persons, can become infected. In Peru, infections have shown some seasonal variation, with peaks during April to June. This pattern is similar to that seen with *Cryptosporidium* infections in Peru.

Transmission of *Cyclospora* is thought to be fecal-oral, although direct person-to-person transmission has not been well documented and may not be a factor, since sporulation takes days. Outbreaks linked to contaminated water and various types of fresh produce (raspberries, basil, and baby lettuce leaves) have been reported. The presence

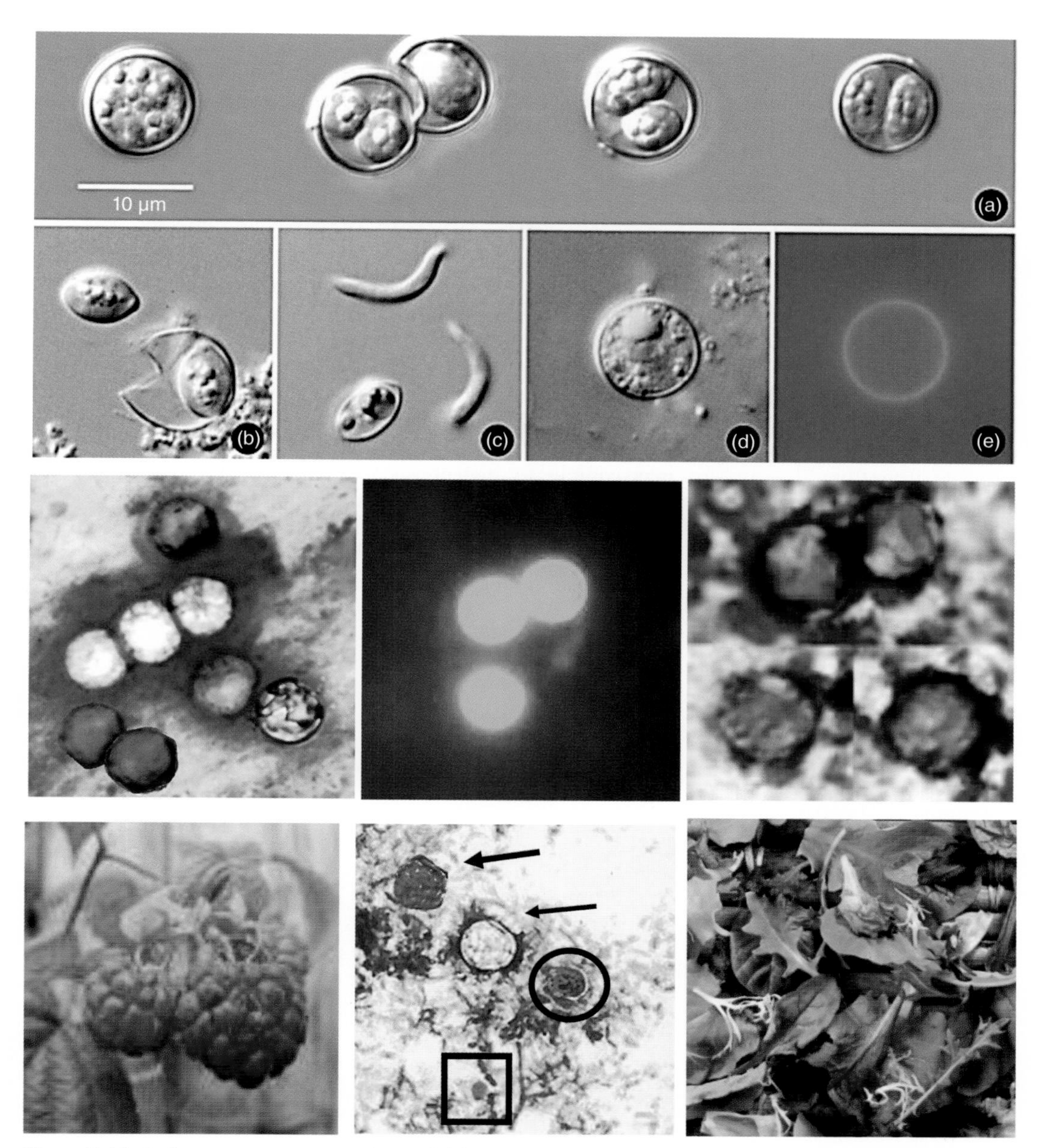

Figure 7.16 ***Cyclospora cayetanensis.*** **(Row 1)** (a) Unsporulated oocyst with undifferentiated cytoplasm, sporulating oocyst that contains two immature sporocysts. **(Row 2)** (b) An oocyst that was mechanically ruptured has released one of its two sporocysts; (c) one free sporocyst is shown as well as two free sporozoites, the infective stage of the parasite; (d) oocyst; (e) the oocyst is autofluorescent when viewed under UV microscopy (courtesy of CDC DpDx, https://www.cdc.gov/dpdx/cyclosporiasis/index.html). **(Row 3)** The first image shows *C. cayetanensis* oocysts (8 to 10 μm) stained with modified acid-fast stain. There is a range of clear to deeply stained oocysts; there is a lot of variation with modified acid-fast staining (modified acid-fast variable). The second image shows autofluorescent oocysts on filters commonly used for calcofluor white staining. The third image is a hot safranin stain (courtesy of CDC/Dr. Govinda Visvesvara). **(Row 4)** Raspberries; a modified acid-fast stain (stained and unstained *Cyclospora* oocysts [arrows], *Cryptosporidium* oocyst [circle], and artifact [box]); mixture of baby lettuce leaves (mesclun).

of oocysts in animals may simply reflect passage through the gastrointestinal tract, since to date there is no evidence of tissue infection in animals.

There is generally 1 day of malaise and low-grade fever, with rapid onset of diarrhea of up to seven stools per day. There may also be fatigue, anorexia, vomiting, myalgia, and weight loss with remission of self-limiting diarrhea in 3 to 4 days, followed by relapses lasting from 4 to 7 weeks.

In patients with AIDS, symptoms may persist for 12 weeks; biliary disease has also been reported, as has diarrhea alternating with constipation (this is not uncommon in a number of protozoal gastrointestinal infections). Clinical clues include unexplained prolonged diarrheal illness during the summer in any patient and in persons returning from tropical areas. Most infected individuals had intermittent diarrhea for 2 to 3 weeks, and many complained of intense fatigue, as well as anorexia and myalgia, during the illness. The clinical presentation is similar to that of patients infected with *Cryptosporidium*.

The disease appears to be self-limiting within a few weeks. Trimethoprim-sulfamethoxazole (TMP-SMX) is the drug of choice; relief of symptoms has been seen 1 to 3 days posttreatment. AIDS patients may need higher doses and long-term maintenance treatment. However, recurrence of symptoms occurs within 1 to 3 months posttreatment in over 40% of patients.

A high prevalence is generally observed in patients immunosuppressed with pathologies such as Hodgkin's lymphoma and acute lymphoblastic leukemia. In a study in cancer patients receiving chemotherapy in Saudi Arabia, a very high prevalence was observed (52% of 54 patients) and a particularly high *C. cayetanensis* prevalence was observed in patients with lymphomas.

Cystoisospora (formerly *Isospora*) *belli*

Pathogenic	Yes
Disease	Cystoisosporiasis
Acquired	Fecal-oral transmission; contaminated food and water
Body site	Intestine
Symptoms	Diarrhea, which may last for months to years, weight loss, abdominal colic, and fever
Clinical specimen	Stool
Epidemiology	Worldwide, primarily human-to-human transmission
Control	Improved hygiene, adequate disposal of fecal waste, adequate washing of contaminated fruits and vegetables

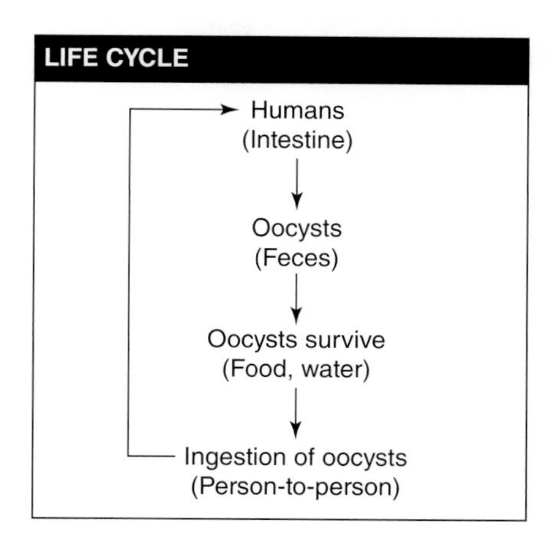

Diagnosis

Although isosporiasis has been found in various parts of the world, certain tropical areas in the Western Hemisphere appear to contain some well-defined locations of endemic infections. These organisms can infect both adults and children, and intestinal involvement and symptoms are generally transient unless the patient is immunocompromised.

Examination of a fecal specimen for the oocysts is recommended. However, wet-preparation examination of fresh material, either as the direct smear or as concentrated material, is recommended rather than the permanent stained smear. These organisms are positive with modified acid-fast staining and can also be demonstrated by using auramine-rhodamine stains. Organisms tentatively identified with auramine-rhodamine stains should be confirmed by wet-smear examination or acid-fast stains, particularly if the stool contains other cells or excess artifact material (more normal stool consistency).

The patent period is not known but may be only 15 days. Chronic infections develop in some patients, and oocysts can be shed for several months to years. In one particular case, an immunocompetent individual had symptoms for 26 years and *Cystoisospora belli* was recovered in stool a number of times over a 10-year period.

Diagnostic Tips

When a wet mount is examined microscopically, the light level should be reduced, and additional contrast should be obtained by dropping the condenser for optimal examination conditions.

The oocysts are very pale and transparent and can easily be overlooked. Oocysts can also be very difficult to see if the concentration sediment is performed from stool specimens in preservative containing polyvinyl alcohol (PVA) (fixatives that do not contain PVA are recommended).

The routine Wheatley's trichrome stain is not recommended; often the oocysts stain very darkly and can be misidentified as distorted helminth eggs. Modified acid-fast stains are recommended. Do not use more than a 1% acid decolorizer; if the fecal smear is thin, a higher-percentage acid may remove most or all of the color.

It is possible to have a positive biopsy specimen and not recover the oocysts in the stool because of the small number of organisms present. Stool microscopy cannot be used to rule out cystoisosporiasis, and even intestinal biopsies can provide false-negative results.

General Comments

Justification for the creation of the genus *Cystoisospora* in 1977 and placement of mammalian *Isospora* species in it were based on morphological, biological, and genetic differences between *Isospora* species from nonmammalian hosts and mammalian hosts.

The oocysts are very resistant to environmental conditions and may remain viable for months if kept cool and moist. Oocysts usually mature within 48 h following stool evacuation and are then infectious.

Unsporulated or partially sporulated oocysts of *C. belli* are excreted in feces. When sporulated oocysts in contaminated water or food are ingested, asexual and sexual stages of *C. belli* are confined to the epithelium of intestines, bile ducts, and gallbladder. Monozoic tissue cysts occur in extraintestinal organs (lamina propria of the small and large intestines, lymph nodes, spleen, and liver) of immunosuppressed humans. However, a paratenic host (not needed for the development of the parasite, but serving to maintain the parasite's life cycle) has not been demonstrated. *Cystoisospora* infections can last for months, and relapses are common. However, the mechanism of relapse has not been defined. Eventually oocysts are passed in the stool; they are long and oval (20 to 33 μm by 10 to 19 μm). Usually the oocyst contains only one immature sporont, but two may be present. Continued development occurs outside the body to form two mature sporocysts, each containing four sporozoites, which can be recovered from the fecal specimen. The sporulated oocyst is the infective stage that will excyst in the small intestine, releasing the sporozoites that penetrate the mucosal cells and initiate the life cycle. The life cycle stages (schizonts, merozoites, gametocytes, gametes, and oocysts) are structurally similar to those seen in the other coccidia.

Additional Information

C. belli is thought to be the only *Cystoisospora* species that infects humans, and no other reservoir hosts are recognized for this infection. Transmission is through ingestion of water or food contaminated with mature, sporulated oocysts. Sexual transmission by direct oral contact with the anus or perineum has been postulated but is probably much less common. The oocysts are very resistant to environmental conditions and can survive for many months. Diagnostic methods for laboratory examinations may tend to miss the organisms. Since transmission is via the infective oocysts, prevention centers on improved personal hygiene measures and sanitary conditions to eliminate possible fecal-oral transmission from contaminated food, water, and possibly environmental surfaces.

Clinical symptoms include diarrhea, which may last for months or years, weight loss, abdominal colic, and fever; diarrhea is the main symptom. Bowel movements (usually 6 to 10 per day) are watery to soft, foamy, and offensive smelling, suggesting a malabsorption process. Eosinophilia is found in many patients, recurrences are quite common, and the disease is more severe in infants and young children.

Patients who are immunosuppressed, particularly those with AIDS, often present with profuse diarrhea associated with weakness, anorexia, and weight loss. In one patient with a well-documented long-term infection, a series of biopsies showed a markedly abnormal mucosa with short villi, hypertrophied crypts, and infiltration of the lamina propria with eosinophils, neutrophils, and round cells. Charcot-Leyden crystals derived from eosinophils have also been found in the stools of infected patients. The diarrhea and other symptoms may continue in compromised patients, even those on immunosuppressive therapy when the regimen of therapy is discontinued. This infection has been found in homosexual men, all of whom were immunosuppressed and had had diarrhea for several months.

Extraintestinal infections have occurred in AIDS patients. Microscopic findings associated with *C. belli* infection have been seen in the walls of both the small and large intestines, mesenteric and mediastinal lymph nodes, liver, and spleen. Finding the merozoites within the lymphatic channels documents a means of dissemination to lymph nodes and other tissues.

Effective eradication has been achieved by using cotrimoxazole, TMP-SMX, pyrimethamine-sulfadiazine, primaquine phosphate-nitrofurantoin, and primaquine phosphate-chloroquine phosphate. Ineffective drugs include dithiazanine, tetracycline, metronidazole, phanquone, and quinacrine hydrochloride. The drug of choice is TMP-SMX, which is classified as investigational for treatment of this infection.

It has been recommended that physicians consider *C. belli* in AIDS patients with diarrhea who have emigrated from or traveled to Latin America, are Hispanics born in the United States, are young adults, or have not received prophylaxis with TMP-SMX for *Pneumocystis jirovecii*. It has also been recommended that AIDS patients who travel to Latin America and other developing countries be advised of the waterborne and foodborne transmission of *C. belli*; they should consider chemoprophylaxis.

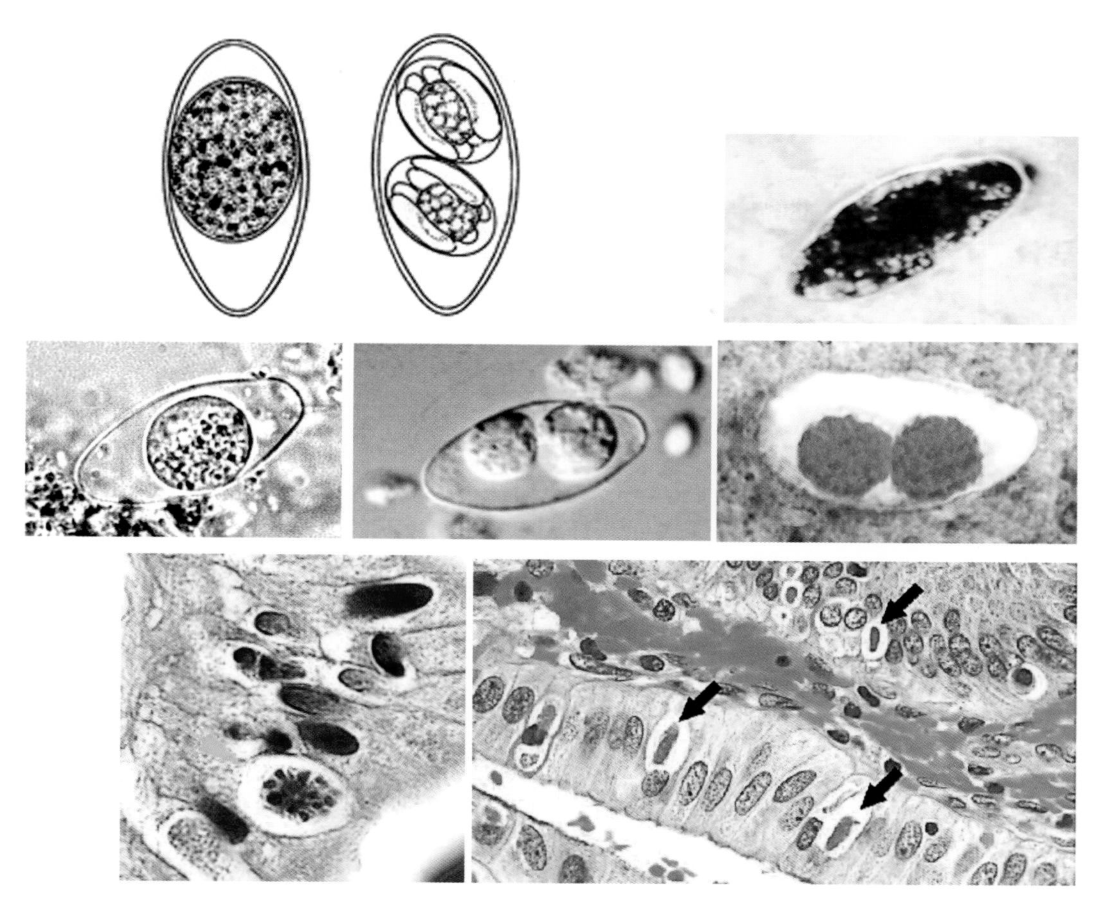

Figure 7.17 *Cystoisospora belli.* **(Top row)** Drawing of an immature oocyst; drawing of a mature oocyst containing two sporocysts; image of an immature oocyst stained with modified acid-fast stain (note that the entire oocyst stains). **(Middle row)** Immature oocyst; mature oocyst (wet mounts); mature oocyst stained with modified acid-fast stain. **(Bottom row)** Oocyst of *C. belli* in the epithelial cells of a mammalian host (yellow arrow) (courtesy of CDC DpDx, https://www.cdc.gov/dpdx/cystoisosporiasis/index.html); donor gallbladder (H&E; magnification, ×1,000) showing characteristic morphology of *C. belli*, including its banana shape and perinuclear parasitophorous vacuoles (arrows), within the gallbladder epithelium (reprinted from: Akateh C, Arnold CA, et al, Case Rep Infect Dis 2018, Article ID 3170238, https://doi.org/10.1155/2018/3170238, © 2018 Clifford Akateh et al, CC-BY 4.0).

Enterocytozoon bieneusi

Pathogenic	Yes
Disease	Microsporidiosis (now classified with the fungi)
Acquired	Fecal-oral transmission; contaminated food and water
Body site	Intestine; dissemination can be seen in compromised patients
Symptoms	Intractable diarrhea, fever, malaise, and weight loss
Clinical specimen	Stool
Epidemiology	Worldwide, primarily human-to-human transmission
Control	Improved hygiene, adequate disposal of fecal waste, adequate washing of contaminated fruits and vegetables

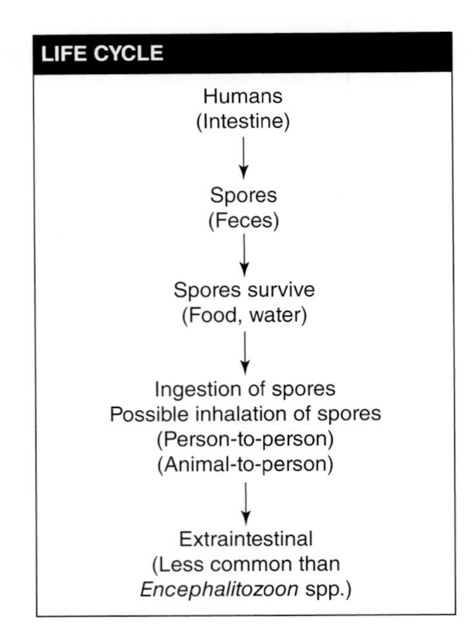

Diagnosis

Microsporidia are in a group of obligate intracellular protist-like fungi with hosts ranging from protists to mammals. Microsporidian species are generally grouped into generalists and specialists according to the range of phenotypic variations. Generalists have broad host ranges and wide cell and tissue specificities, and they often cause opportunistic infections in higher vertebrates, among which *Enterocytozoon bieneusi* and three *Encephalitozoon* species (*Encephalitozoon cuniculi*, *E. hellem*, and *E. intestinalis*) are the best-known representatives with simple developmental cycles. They affect a wide variety of mammal hosts and together represent the most frequently reported causative agents of zoonotic microsporidian infections. In contrast, specialists are closely adapted for infections and developments in a single host species or a very narrow range of closely related host species, most of which would have no or minimal effects on public health.

The standard O&P concentration is recommended (500 × *g* for 10 min), with the concentrate being used for all smears and for FA assays. Recommendations include using modified trichrome stains in which the concentration of the chromotrope 2R component added to the stain is 10 times the concentration normally used in the routine trichrome stain for stool. Stool preparations must be very thin, the staining time is 90 min, and the slide must be examined at ×1,000 (or higher) magnification. Unfortunately, there are many objects within stool material that are oval, stain pinkish with trichrome, and measure ca. 1.5 to 3 μm. If this stain is used to identify microsporidia in stool, positive-control material should be available for comparison. Additional modifications include the use of heat and a shorter staining time. Pretreatment of fecal specimens (1:1) with 10% KOH may provide a better-quality smear with the modified trichrome stains.

Another approach involves chemofluorescent agents (optical brightening agents) such as calcofluor white, Fungi-Fluor, or Uvitex 2B. These reagents are sensitive but nonspecific; objects other than microsporidial spores also fluoresce. This is a particular problem when examining stool specimens; both false-positive and false-negative results have been seen.

Although immunofluorescent antibody tests are available in other countries, none are currently FDA cleared for use in the United States. Relatively high percentages of *E. intestinalis* and *E. bieneusi* have been reported in cancer patients undergoing therapy.

A multiplex panel is being developed that will detect *Enterocytozoon bieneusi* and

Encephalitozoon intestinalis. FDA clearance should be obtained in 2021.

Diagnostic Tips

The use of concave well slides is recommended for preparing smears for staining for microsporidian spores. When fecal smears are prepared for staining, it is important to make the smears thinner rather than thick; spread the material all the way to the outside edges of the well. Microscopic examination using the 100× oil immersion lens is critical; make sure the smear is examined at the edges, as well as the thicker area in the center. When the modified trichrome method is used, the room temperature stain time is 90 min; it can be lengthened considerably without damaging or overstaining the smear.

It is very important to confirm the presence of the polar tubule; horizontal and/or diagonal lines across the spores provide confirmatory evidence. However, not all spores will display evidence of the polar tubule. If several positive spores are seen to contain the polar tubule, this is sufficient to indicate a positive finding (microsporidian spores present).

When using chemofluorescent agents, use a thinner smear rather than one that is too thick. Examination should include all areas of the smear (thinner edges and thicker center). Because this approach provides nonspecific results, if presumptive spores are seen, this should be confirmed using the modified trichrome stain. Note that in a cleaner specimen, such as a urine sediment, spores are more likely to be real (both intra- and extracellular); however, in stool, the large number of artifacts (both color and shape) will make the presumptive positive result less likely. Again, confirmatory staining with the modified trichrome stain is recommended.

If the first set of stained smears is negative and the infection is still suspected, it is recommended that both stool and urine be submitted within the following 2 weeks for additional stained smear examinations.

Microsporidial infections can be misdiagnosed in tissues and can be confused with *Cryptococcus neoformans* infections. Mucus granules in goblet cells can take up stain and be very confusing. Good preservation and thin tissue sections (1 µm) that have been resin embedded enhance the resolution of cellular detail. Demonstration of the coiled polar tube within spores is diagnostic for microsporidial infection.

Routine microscopy is insufficient for the identification of the various genera infecting humans; transmission electron microscopy and/or molecular testing can confirm the various genera and species.

General Comments

Infection occurs with the introduction of infective sporoplasm through the polar tubule into the host cell. The microsporidia multiply extensively within the host cell cytoplasm; the life cycle includes repeated divisions by binary fission (merogony) or multiple fission (schizogony) and spore production (sporogony). Merogony and sporogony can occur in the same cell at the same time. During sporogony, a thick spore wall is formed, providing environmental protection for this infectious stage of the parasite. *E. bieneusi* spores are released into the intestinal lumen and are passed in the stool. They are environmentally resistant and can then be ingested by other hosts. There is also evidence for inhalation of spores and evidence in animals suggesting that human microsporidiosis may also be transmitted via the rectal route.

Eight genera have been isolated from humans: *Anncaliia* (former names, *Nosema*, *Brachiola*), *Encephalitozoon*, *Vittaforma*, *Pleistophora*, *Trachipleistophora*, *Enterocytozoon*, *Microsporidium*, and *Tubulinosema*. Classification criteria include spore size, configuration of nuclei within the spores and developing forms, number of polar tubule coils within the spore, and relationship between the organism and host cell. These generic designations are frequently changed as more information becomes available.

E. bieneusi is the most common microsporidian found in humans, particularly in immunocompromised patients. However, there are very few data regarding infections in immunocompetent hosts.

Additional Information

Prior to the HIV/AIDS pandemic in the mid-1980s, microsporidiosis was rarely reported in human patients. The pandemic brought to light the opportunistic capability of microsporidia to infect humans and produce disease in virtually all organs. Before the common use of antiretroviral therapies, microsporidiosis was reported in at least 15% (and up to 85%) of HIV/AIDS patients. However, although prevalence declined with improved therapy, an increase in newly diagnosed cases of HIV in people over 50 years of age, coupled with an aging population of patients living with HIV, is leading to HIV-associated non-AIDS (HANA) conditions that accelerate the onset of diseases normally seen in the elderly. Reactivation of latent microsporidian infections with age, or with subsequent use of chemotherapy or immunosuppressive treatments, has also been reported. Although at least 10 microsporidian genera have been associated with human patients, the most accepted eight are listed in "General Comments" above; the most frequently detected species is the gut-infecting *Enterocytozoon bieneusi* in patients with HIV/AIDS, in whom it produces chronic diarrhea.

Up to 30% of patients infected with *Cryptosporidium* spp. may have concurrent infections with microsporidia. This emphasizes the importance of considering both organisms in compromised patients, particularly HIV patients, with diarrhea. The use of antiretroviral combination therapy has led to decreases in both cryptosporidiosis and microsporidiosis in this patient group. Microsporidia also cause disease in organ transplant recipients, children, travelers, contact lens wearers, and the elderly. Conjunctival and corneal epithelium infections occur in HIV patients, while corneal stroma and ulceration occur in immunocompetent individuals.

Symptoms in AIDS patients include chronic intractable diarrhea, fever, malaise, and weight loss, similar to those of cryptosporidiosis or cystoisosporiasis. These patients tend to have four to eight watery, nonbloody stools daily which can be accompanied by nausea and anorexia. There may be dehydration with mild hypokalemia and hypomagnesemia, as well as D-xylose and fat malabsorption. The patients tend to be severely immunodeficient, with a $CD4^+$ count always below 200 and often below 100. Dual infections with *E. bieneusi* and *E. intestinalis* have also been reported. Unfortunately, these infections do not respond to therapy with albendazole, unlike infections with *Encephalitozoon* spp.

The microsporidia are obligate intracellular parasites that have been recognized in a variety of animals, particularly invertebrates. Typical sizes range from 1.5 to 5 μm wide and 2 to 7 μm long; organisms found in humans tend to be quite small (1.5 to 2 μm). Until the recent increased understanding of AIDS within the immunosuppressed population, awareness and understanding of human microsporidial infections were marginal. Limited availability of electron microscopy has also affected the ability to recognize and diagnose these infections. However, the introduction of newer diagnostic methods has improved our diagnostic ability. The organisms are characterized by having spores containing a polar tubule, which is an extrusion mechanism for injecting the infective spore contents into host cells.

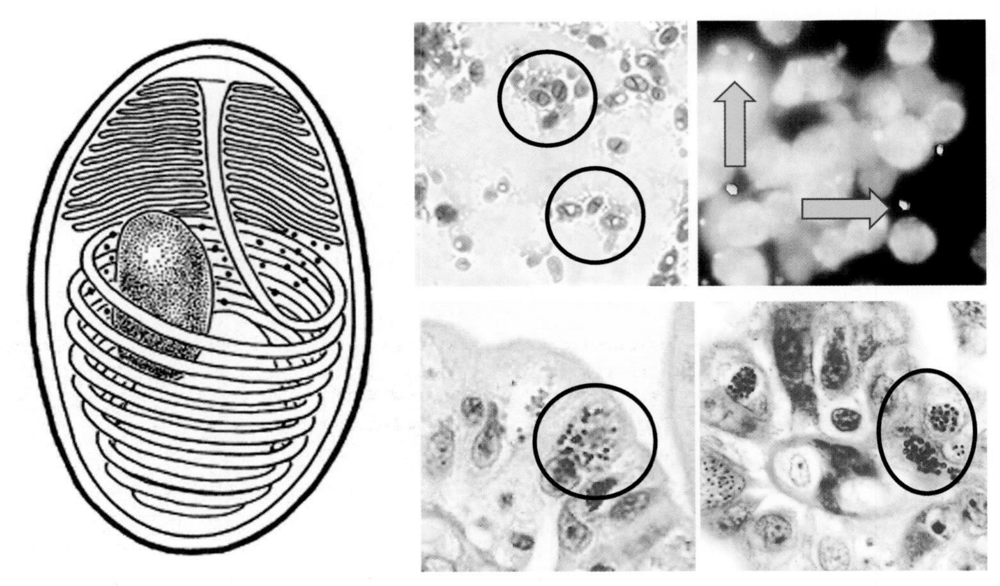

Figure 7.18 ***Enterocytozoon bieneusi*****.** **(Left)** Drawing of an infective spore (1.5 to 2.5 μm) containing the coiled polar tubule. **(Top row)** Spores in stool, stained with Ryan modified trichrome blue (some of the spores show the horizontal "stripe" that indicates the presence of the polar tubule [circles]); calcofluor white staining of spores in a urine sediment (intra- and extracellular) (arrows). **(Bottom row)** Spore development in the human intestine. Note the small size of the spores in tissue (circle and oval).

Encephalitozoon intestinalis *Encephalitozoon* spp.

Pathogenic	Yes
Disease	Microsporidiosis (now classified with the fungi)
Acquired	Fecal-oral transmission, inhalation, fomites and direct inoculation; contaminated food and water
Body site	Intestine; dissemination to kidneys, lower airways, and biliary tract appears to occur via infected macrophages; eye
Symptoms	Intractable diarrhea, fever, malaise, and weight loss
Clinical specimen	Stool
Epidemiology	Worldwide, primarily human-to-human transmission
Control	Improved hygiene, adequate disposal of fecal waste, adequate washing of contaminated fruits and vegetables

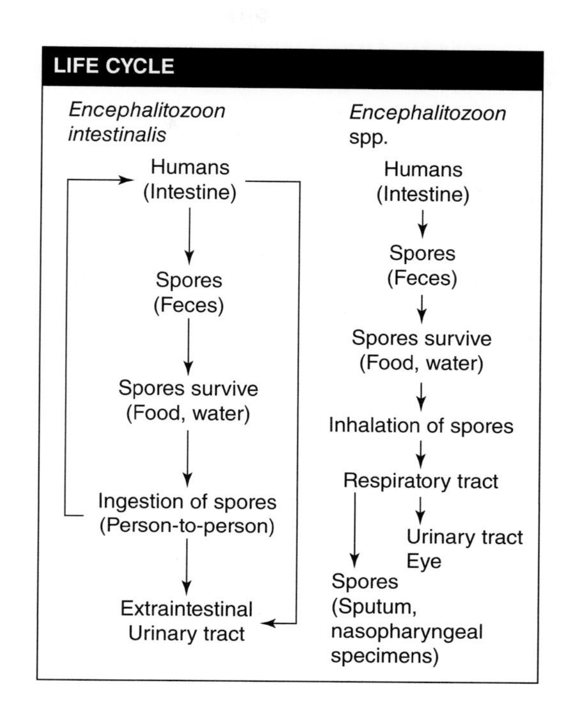

Diagnosis

Microsporidia are in a group of obligate intracellular protist-like fungi with hosts ranging from protists to mammals. Microsporidian species are generally grouped into generalists and specialists according to the range of phenotypic variations. Generalists have broad host ranges and wide cell and tissue specificities, and they often cause opportunistic infections in higher vertebrates, among which *Enterocytozoon bieneusi* and three *Encephalitozoon* species (*Encephalitozoon cuniculi*, *E. hellem*, and *E. intestinalis*) are the best-known representatives with simple developmental cycles. They affect a wide variety of mammal hosts and together represent the most frequently reported causative agents of zoonotic microsporidian infections. In contrast, specialists are closely adapted for infections and developments in a single host species or a very narrow range of closely related host species, most of which would have no or minimal effects on public health.

The standard O&P concentration is recommended (500 × *g* for 10 min), with the concentrate being used for all smears and for FA assays. Recommendations include using modified trichrome stains in which the concentration of the chromotrope 2R component added to the stain is 10 times that normally used in the routine trichrome stain for stool. Stool preparations must be very thin, the staining time is 90 min, and the slide must be examined at ×1,000 (or higher) magnification. Unfortunately, there are many objects in stool material that are oval, stain pinkish with trichrome, and measure ca. 1.5 to 3 µm. If this stain is used to identify microsporidia in stool, positive control material should be available for comparison. Additional modifications include the use of heat and a shorter staining time. Pretreatment of fecal specimens (1:1) with 10% KOH may provide a better-quality smear with the modified trichrome stains.

Another approach involves the use of chemofluorescent agents (optical brightening agents) such as calcofluor white, Fungi-Fluor, or Uvitex 2B. These reagents are sensitive but nonspecific; objects other than microsporidial spores also fluoresce. This is a particular problem when stool specimens are examined; both false-positive and false-negative results have been seen. Although immunofluorescent-antibody tests are available in other countries, none are currently FDA cleared for use in the United States. Relatively high percentages of

E. intestinalis and *E. bieneusi* have been reported in cancer patients undergoing therapy.

Currently a multiplex panel is being developed that will detect *Enterocytozoon bieneusi* and *Encephalitozoon intestinalis*. Information suggests FDA clearance should be obtained in 2021.

Diagnostic Tips

Although most diagnostic testing is on stool and urine, microsporidiosis should definitely be considered in the differential diagnosis of pulmonary infections in immunosuppressed patients; therefore, pulmonary as well as other specimens have to be added to the possible specimens tested.

The use of concave well slides is recommended when smears are prepared for staining for microsporidian spores. When preparing fecal smears for staining, it is important to make the smears thinner rather than thick; spread the material all the way to the outside edges of the well. Microscopic examination using the 100× oil immersion lens is critical; make sure the smear is examined at the edges, as well as the thicker area in the center. When the modified trichrome method is used, the room temperature stain time is 90 min; it can be lengthened considerably without damaging or overstaining the smear.

It is very important to confirm the presence of the polar tubule; horizontal and/or diagonal lines across the spores provide confirmatory evidence. However, not all spores display evidence of the polar tubule. If several positive spores are seen to contain the polar tubule, this is sufficient to indicate a positive finding (microsporidian spores present).

When using chemofluorescent agents, use a thinner smear rather than one that is too thick. Examination should include all areas of the smear (thinner edges and thicker center). Because this approach provides nonspecific results, if presumptive spores are seen, this should be confirmed using the modified trichrome stain. Note that in a cleaner specimen such as a urine sediment, spores are more likely to be real (both intra- and extracellular); however, in stool, the large number of artifacts (both color and shape) will make the presumptive positive result less likely. Again, confirmatory staining with the modified trichrome stain is recommended.

If the first set of stained smears is negative and the infection is still suspected, it is recommended that both stool and urine be submitted within the following 2 weeks for additional stained smear examinations.

Microsporidial infections can be misdiagnosed in tissues and can be confused with *Cryptococcus neoformans* infections. Mucus granules in goblet cells can take up stain and be very confusing. Good preservation and thin tissue sections (1 μm) that have been resin embedded enhance the resolution of cellular detail. Demonstration of the coiled polar tube within spores is diagnostic for microsporidial infection.

Routine microscopy is insufficient for the identification of the various genera infecting humans; transmission electron microscopy and/or molecular testing can confirm the various genera and species.

General Comments

Infection occurs with the introduction of infective sporoplasm through the polar tubule into the host cell. The microsporidia multiply in the host cell cytoplasm; the life cycle includes repeated divisions by binary fission (merogony) or multiple fission (schizogony) and spore production (sporogony). Merogony and sporogony can occur in the same cell at the same time. During sporogony, a thick spore wall is formed, providing environmental protection for this infectious stage of the parasite.

Eight genera have been isolated from humans: *Anncaliia* (formerly *Nosema, Brachiola*), *Encephalitozoon, Vittaforma, Pleistophora, Trachipleistophora, Enterocytozoon, Microsporidium*, and *Tubulinosema*. Classification criteria include spore size, configuration of nuclei within the spores and developing forms, number of polar tubule coils within the spore, and relationship between the organism and host cell.

E. intestinalis infects primarily small intestinal enterocytes, but infection does not remain confined to epithelial cells. It is also found in lamina propria macrophages, fibroblasts, and endothelial cells. Dissemination to the kidneys, lower airways, and biliary tract occurs via infected macrophages. Infections respond to therapy with albendazole, unlike infections with *E. bieneusi*.

Both *E. cuniculi* and *E. hellem* have been isolated from human infections, the first species from the central nervous system and the second from the eye. Conjunctival and corneal epithelium infections occur in HIV patients, while corneal stroma and ulceration occur in immunocompetent individuals.

Additional Information

E. intestinalis is not confined to epithelial cells but is seen in macrophages in the lamina propria. Although the primary site appears to be the small bowel, organisms can disseminate to other sites, including duodenum, jejunum, ileum, colon, kidneys, liver, and gallbladder. They have also been identified in the lower airways. Concurrent infections with *E. bieneusi* have also been seen.

Currently, there are at least three *E. cuniculi* strains which may become more important in the epidemiology of human infections. Several *E. hellem* eye infections have been found in AIDS patients, as well as infections in the sinuses, conjunctivae, and nasal epithelium. In one case of disseminated *E. hellem* infection in an AIDS patient, autopsy revealed organisms in the eyes, urinary tract, and respiratory tract. The presence of numerous organisms within the lining epithelium of almost the entire length of the tracheobronchial tree suggests respiratory acquisition.

The sources and routes of ocular infection are less clear than those for digestive diseases. Although cases of zoonotic, aerial, alimentary, and interhuman infection have been described, water sources are probably the most important route of infection. Most species that are potentially pathogenic for humans have been identified in water and are resistant to chlorine. As reported previously in immunocompetent cases, clinical symptoms include unilateral pain, conjunctival redness, photophobia, the sensation of a foreign body in the eye, and blurred vision. Ocular microsporidiosis affects the cornea, with deep or superficial lesions, and/or the conjunctiva. The most frequently reported forms include punctate stromal keratitis and superficial keratoconjunctivitis. Since early 2000, an increase in stromal keratitis and keratoconjunctivitis cases has been observed in patients with no underlying problems. Although rare, microsporidia are an emerging cause of ocular infection in immunocompetent patients. This etiology should be considered in a patient with keratitis of undetermined origin originating or returning from Asia, particularly during the rainy season.

The microsporidia are obligate intracellular parasites that have been recognized in a variety of animals, particularly invertebrates. Typical sizes range from 1.5 to 5 μm wide and 2 to 7 μm long; organisms found in humans tend to be quite small (1.5 to 2 μm). Until recent increased understanding of AIDS within the immunosuppressed population, awareness and understanding of human infections were marginal. Limited availability of EM has also affected the ability to recognize and diagnose these infections. However, the introduction of newer diagnostic methods has improved our ability to identify these parasites. The organisms are characterized by having spores containing a polar tubule, which is an extrusion mechanism for injecting the infective spore contents into host cells.

Serologic studies show that immunocompetent humans probably have persistent or chronic infections with microsporidia. This is also supported by the fact that microsporidiosis patients who are HIV seronegative show clinical resolution of their infections after a few weeks.

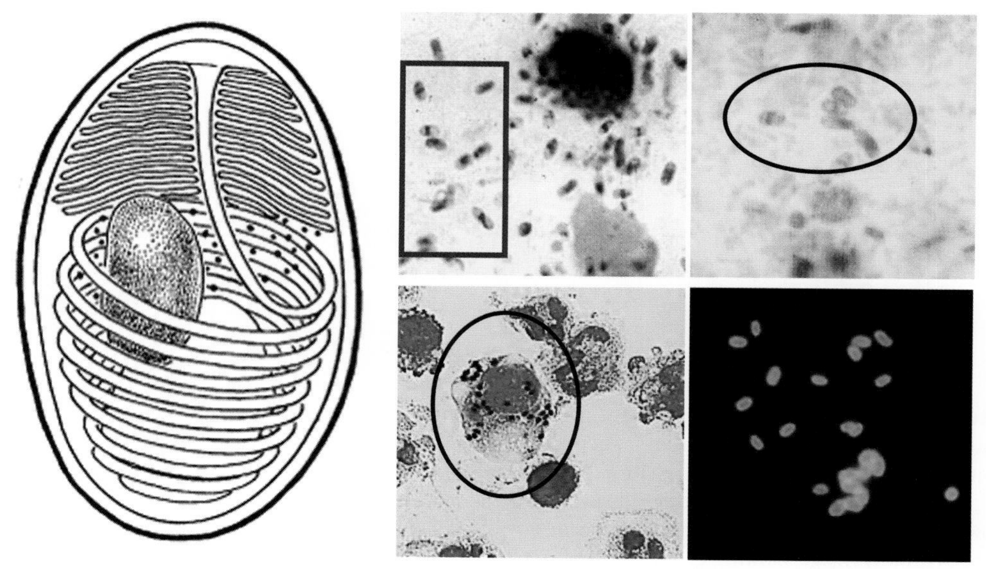

Figure 7.19 ***Encephalitozoon*** **spp. (Left)** Drawing of an infective spore (1.5 to 2.5 μm), containing the coiled polar tubule. **(Top row)** Nasopharyngeal aspirate with spores (Ryan blue-modified trichrome stain; note the spores with horizontal and/or diagonal lines, evidence of the polar tubule [rectangle]); microsporidial spores in fecal specimen (Weber green modified trichrome stain; note the spores with horizontal and/or diagonal lines, evidence of the polar tubules [oval]). **(Bottom row)** Gram stain showing spores within a white blood cell (oval) (these Gram stains can be easily misinterpreted as bacteria, with subsequent inappropriate therapy); urine sediment with direct-FA reagent (to the genus *Encephalitozoon*).

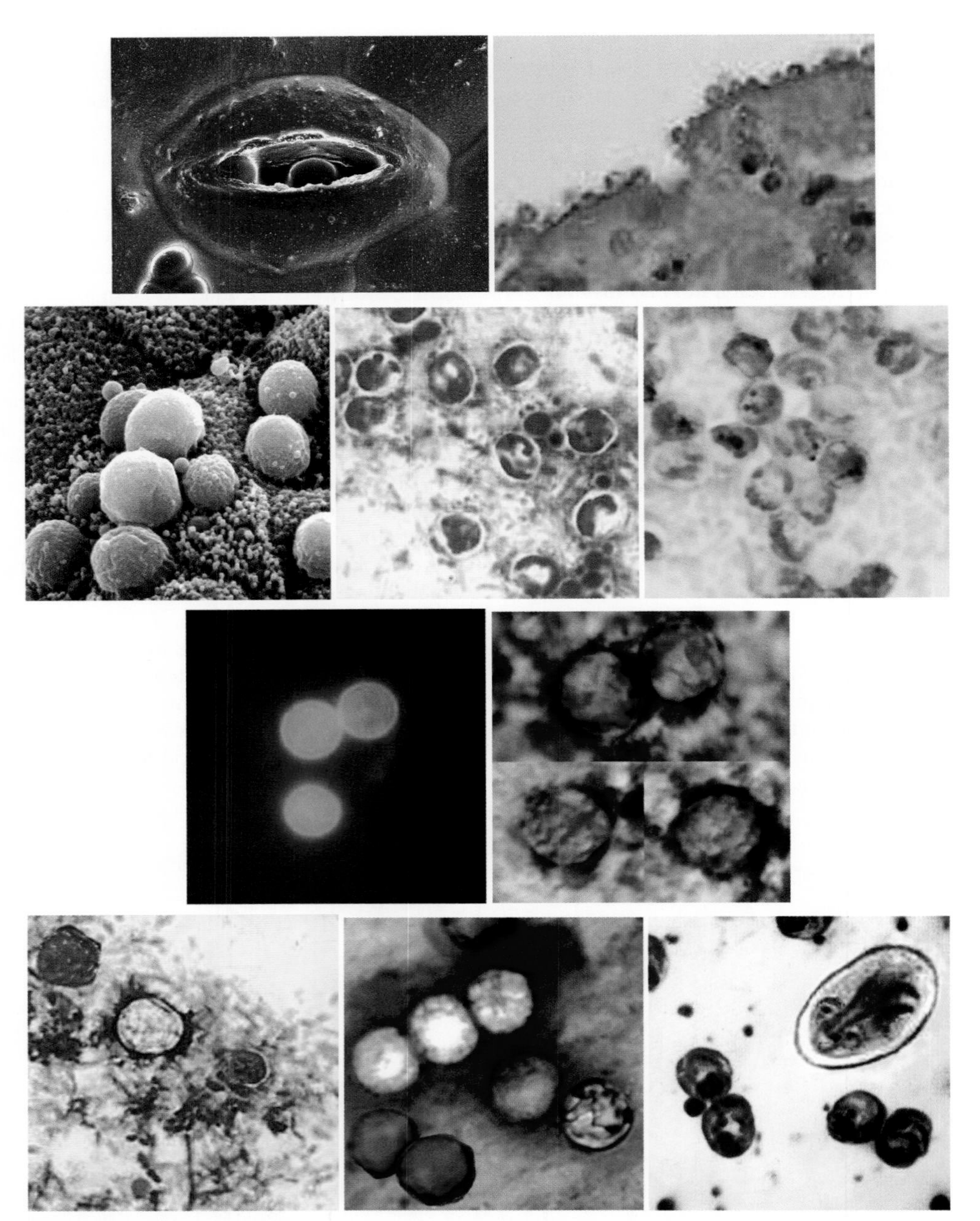

Plate 7.E (**Row 1**) *Cryptosporidium* oocysts in a spinach leaf stoma (SEM, courtesy of US Department of Agriculture, https://www.ars.usda.gov/oc/images/photos/aug13/d2901-1/); *Cryptosporidium* spp. on the surface of the intestinal tract epithelium (right). (**Row 2**) *Cryptosporidium* organisms on the surface of the intestinal tract (electron micrograph) (left); *Cryptosporidium* oocysts stained with modified acid-fast stain (middle and right). (**Row 3**) *Cyclospora cayetanensis* autofluorescent oocysts (left); *C. cayetanensis* oocysts stained with safranin (right). (**Row 4**) Image showing *C. cayetanensis* oocysts (the two large objects, one stained, one clear), a *Cryptosporidium* oocyst (the medium-size object), and an artifact (the small dark-staining object at the bottom of the image) (left); *C. cayetanensis* oocysts (modified acid-fast variable staining) (center); *Cryptosporidium* sp. oocysts and a *Giardia lamblia* cyst (combination modified acid-fast–iron hematoxylin stain) (right).

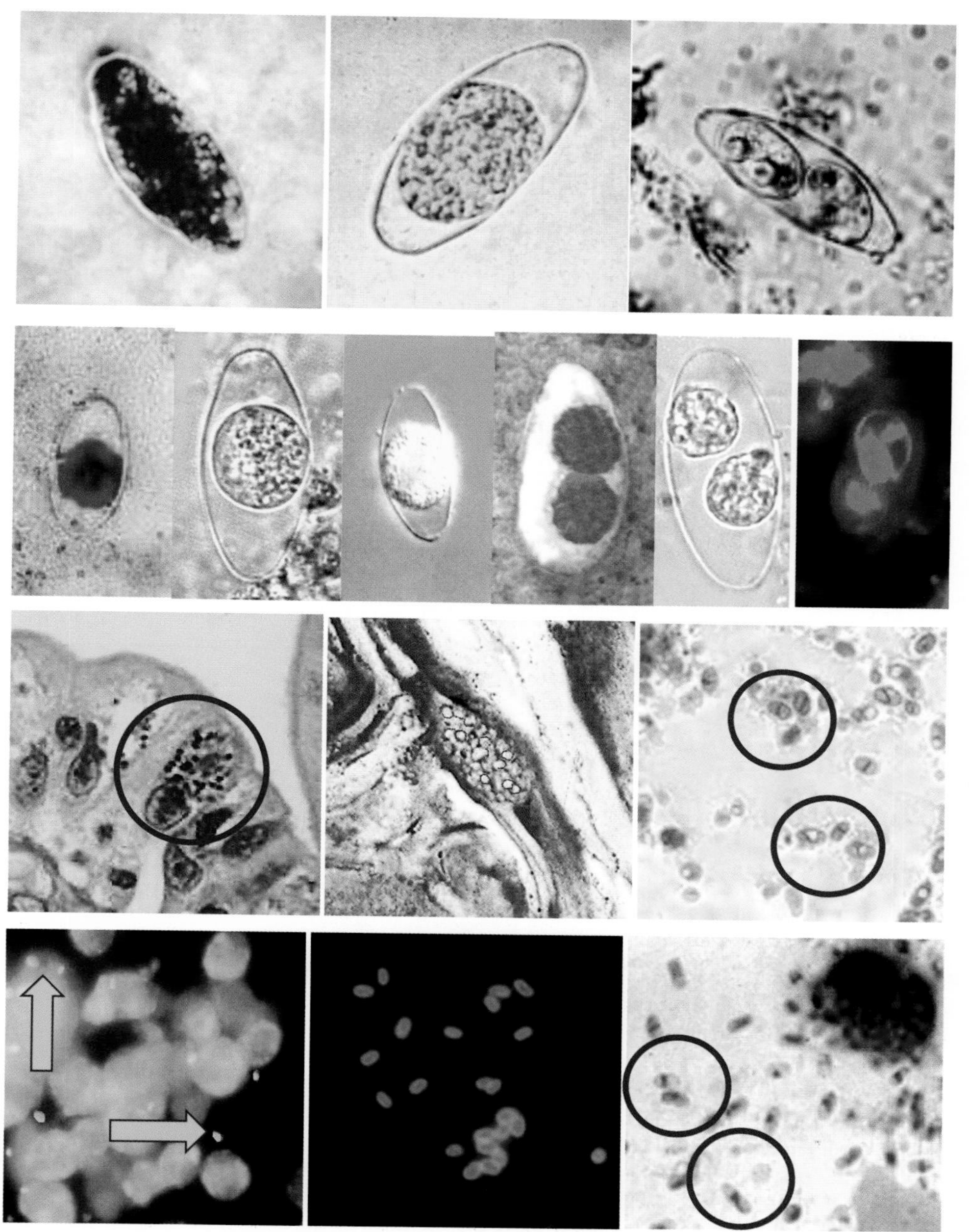

Plate 7.F **(Row 1)** *Cystoisospora belli* oocysts (from left to right: modified acid-fast stain, immature oocysts, and mature oocysts [wet mounts]). **(Row 2)** Mature *C. belli* oocysts (from left to right: modified acid-fast stain, wet mount, wet mount, modified acid-fast stain, wet mount, and calcofluor white) (last image courtesy of CDC Public Health Image Library). **(Row 3)** Microsporidial spores in intestinal tissue (circle) (routine H&E stain); spores from an eye specimen (silver stain); spores (modified trichrome stain; note the horizontal lines representing the polar tubules) (circles). **(Row 4)** Spores in urine sediment, intra- and extracellular (calcofluor white) (arrows); spores visualized using an experimental DFA reagent; spores (circles) (Ryan modified trichrome blue stain; note the horizontal and/or diagonal lines representing the polar tubules.

Plasmodium vivax

Pathogenic	Yes, 2.5 billion at risk with WHO estimates of 13.8 million *P. vivax* infections in 2015
Disease	Tertian malaria, periodicity of approximately 48 h
Acquired	Bite of female anopheline mosquito, blood, shared needles, congenital infections
Body site	Liver, RBCs
Symptoms	Few; anemia, splenomegaly, paroxysm (cold stage, fever, and sweats); may cause severe symptoms (not as benign as previously thought)
Clinical specimen	Blood, multiple draws (EDTA)
Epidemiology	May account for 80% of malaria in tropics, subtropics, and temperate zones; mosquito-to-human and human-to-human transmission
Control	Vector control, avoiding shared needles, checking blood supply; various vaccines are undergoing trials

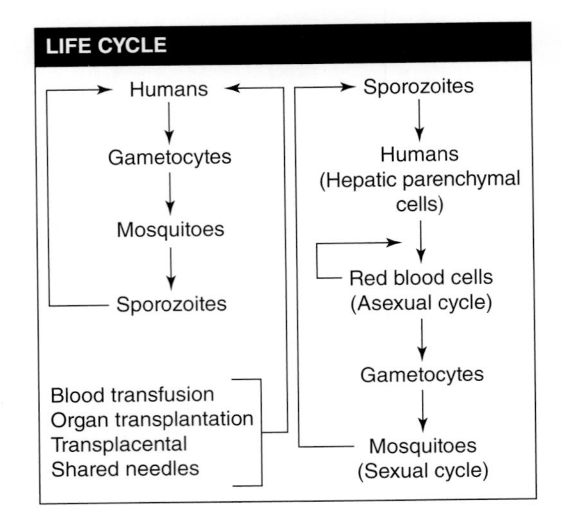

Diagnosis

Although malaria is no longer endemic in the United States, it is life-threatening, and laboratory requests for blood smear examination and organism identification should be treated as stat requests. Frequently, for a number of reasons, organism recovery and identification may be more difficult than the textbooks imply. It is very important that this fact be recognized, particularly when a possibly fatal infection with *Plasmodium falciparum* or *Plasmodium knowlesi* could be involved.

As drug-resistant *Plasmodium vivax* strains emerge and spread and fatality rates increase, the need to implement better control and elimination strategies is becoming urgent. Many of the interventions used for controlling *P. falciparum* malaria are not as effective against *P. vivax*. Consequently, *P. vivax* has become the dominant malaria parasite in several countries where *P. falciparum* transmission has been successfully reduced. As a result of this development, the importance of this species has definitely increased. Both thick and thin blood films should be prepared *on admission of the patient* (to the clinic or emergency room or in hospital), and at least 300 oil immersion fields should be examined on the thick and thin films before a negative report is issued. Any slide that is protected by a cover glass and is going to be examined with an oil immersion lens must be covered by a no. 1 cover glass. If a no. 2 cover glass is used, the extra thickness may prevent the oil immersion lens from focusing properly.

Since one set of negative films does not rule out malaria, additional blood specimens should be examined over 36 h. Although Giemsa stain is often used for parasitic blood work, the organisms can also be seen with other blood stains, such as Wright's stain, a Wright-Giemsa stain, or any of the rapid blood stains. Blood collected with EDTA anticoagulant is acceptable; however, if the blood remains in the tube for any length of time, true stippling may not be visible within the infected

RBCs (e.g., for *P. vivax* and *Plasmodium ovale*); organism morphology changes, and some of the parasites actually disintegrate (after about 4 to 6 h). Also, the proper ratio between blood and anticoagulant is necessary for good organism morphology (fill the EDTA tube completely).

Diagnostic Tips

Any request for blood film identification requires STAT handling (orders, collection, processing of both thick and thin films, examination, and reporting (see Figure 7.20).

The accurate examination of thick and thin blood films and identification of parasites depend on the use of absolutely clean, grease-free slides for preparation of all blood films. Old (unscratched) slides should be cleaned first with detergent and then 70% ethyl alcohol; new slides should also be cleaned with alcohol before use. New slides are coated with a substance that allows them to be pulled apart; these slides should be cleaned before use for preparation of blood films. Do not use cotton; gauze is recommended with 70% alcohol.

Blood should be collected immediately on admission or when the patient is first seen in the emergency room and/or clinic; if the initial blood films are negative, collect daily specimens for two or three additional days (ideally between paroxysms if present; however, there is often no periodicity seen).

Unless you are positive that you will receive well-prepared slides, request a tube of fresh blood (EDTA anticoagulant [lavender top] is preferred) and prepare the smears. In general, the use of finger-stick blood has declined, particularly in areas of the world where automated hematology instruments have become much more widely used. For detection of stippling, the smears should be prepared within 1 h after the specimen is drawn. After that time, stippling may not be visible on stained films; however, the overall organism morphology will still be acceptable.

The time when the specimen was drawn should be clearly indicated on the tube of blood and also on the result report. The physician will then be able to correlate the results with any fever pattern or other symptoms that the patient may have. However, with travelers who are immunologically naive (i.e., they have never come in contact with malaria before), there may not be any fever periodicity at all, and the symptoms may be very general and not specific for a malaria infection. There should also be some indication on the report that is sent back to the physician that one negative specimen does not rule out the possibility of a parasitic infection.

Thick Blood Films—Fresh Blood. To prepare the thick film, place 2 or 3 small drops of fresh blood (no anticoagulant) on an alcohol-cleaned slide. With the corner of another slide and using a circular motion, mix the drops and spread them over an area ~2 cm in diameter. Continue stirring for 30 s to prevent the formation of fibrin strands, which might obscure the parasites after staining.

Thick Blood Films—Anticoagulant. If blood containing an anticoagulant is used, 2 or 3 drops may be spread over an area about 2 cm in diameter; it is not necessary to continue stirring for 30 s, since fibrin strands do not form. If the blood is too thick or any grease remains on the slide, the blood may flake off during staining. It is far better to make the preparation too thin, rather than too thick.

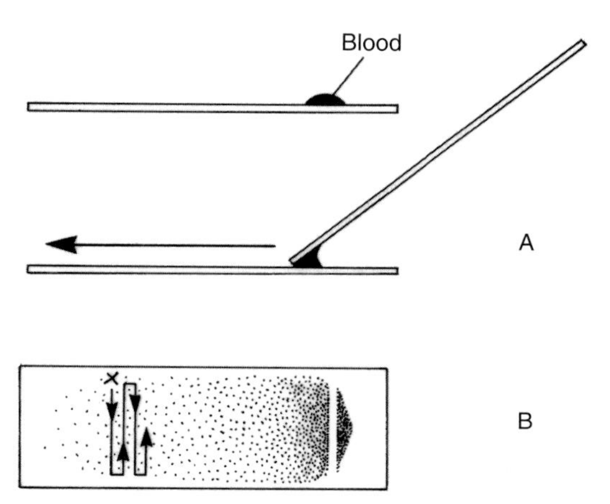

Figure 7.20 Method for preparation of thin blood film. (A) Position of spreader slide; (B) well-prepared thin film. Arrows indicate the area of the slide (feather edge) used to observe accurate cell morphology. (Illustration by Sharon Belkin.)

Allow the thick film to air dry (room temperature) in a dust-free area. Never apply heat to a thick film, since heat will fix the blood, causing the RBCs to remain intact during staining; the result is stain retention and inability to identify the parasites. After the thick films are thoroughly dry, they can be laked to remove the hemoglobin. Rupture of the RBCs during laking removes the RBCs from the final stained blood film; the only structures remaining on the thick film are the WBCs, the platelets, and any parasites present. To lake the films, place them in buffer solution before staining or directly into Giemsa stain, which is an aqueous stain. If thick films are to be stained later, they should be laked before storage.

Thin Blood Films—Fresh Blood or Anticoagulant. The thin blood film is routinely used for specific parasite identification, although the number of organisms per field is much reduced compared with that in the thick film. The thin film is prepared exactly as for a differential count. A well-prepared film is thick at one end and thin at the other (one layer of evenly distributed RBCs with no cell overlap). The thin, feathered end should be at least 2 cm long, and the film should occupy the central area of the slide, with free margins on both sides (see Figure 7.20). The presence of long streamers of blood indicates that the slide used as a spreader was dirty or chipped. Streaks in the film are usually caused by dirt, and holes in the film indicate the presence of grease on the slide. After the film has air dried (do not apply heat), it may be stained. The necessity for fixation before staining will depend on the stain selected. Since Giemsa is an aqueous stain, laking of the RBCs occurs during the staining process. However, when Wright's stain is used, a fixing agent is incorporated into the stain, so that laking of the blood films must be done prior to staining. If you are using one of the rapid blood stains, read the package insert to confirm laking and/or fixation requirements for the thin and thick blood films.

The instrument-prepared monolayer method or coverslip methods generally do not provide the best morphology for malarial parasites within the RBCs. However, the selection of a slide preparation method can be dictated by personal preference, since a malarial infection can be diagnosed from either type of slide.

General Comments

P. vivax tends to invade RBCs with the Duffy blood group antigens. The FyFy Duffy blood group genotype is found predominantly in Black individuals in Africa and the United States, who are the only groups completely resistant to infection by *P. vivax*. However, there are confirmed *P. vivax* infections among Duffy-negative populations, indicating possible alternate invasion pathways.

Within an hour, sporozoites from the mosquito are carried via the blood to the liver, where they penetrate parenchymal cells, initiating the preerythrocytic or primary exoerythrocytic cycle. The sporozoites become round or oval and begin dividing, resulting in many liver merozoites. The merozoites leave the liver and invade the RBCs, initiating the erythrocytic cycle. In *P. vivax* and *P. ovale*, a secondary or dormant schizogony occurs from organisms that remain quiescent in the liver until later; they are called hypnozoites. Delayed schizogony does not occur in *P. falciparum*, *P. knowlesi*, or *Plasmodium malariae*.

If the RBC infection is not eliminated by the immune system or by therapy and the numbers in the RBCs begin to increase again with subsequent clinical symptoms, this is called a recrudescence; it can occur with all species. If the erythrocytic infection is eliminated and a relapse due to a new invasion of the RBCs from the liver occurs later, this is called a true relapse. It occurs only with *P. vivax* and *P. ovale*.

The merozoite (young trophozoite) is vacuolated, ring shaped, and uninucleate. Once the nucleus begins to divide, the trophozoite is called a developing schizont; the mature schizont contains merozoites, which are released into the bloodstream. Merozoites invade RBCs, in which a new cycle of erythrocytic schizogony begins. After several erythrocytic generations, some of the merozoites undergo development into the male and female gametocytes.

Excess protein and hematin left over from the metabolism of hemoglobin combine to form malarial pigment; this is found in all five species. True stippling (evidence of RBC membrane damage; also called Schüffner's dots) is seen only in *P. vivax* and *P. ovale* malaria.

Additional Information

P. vivax infects only the reticulocytes, and the parasitemia is usually limited to 2 to 4% of the available RBCs. Splenomegaly occurs during the first few weeks, and the spleen becomes hard during a chronic infection. If therapy is given early, the spleen returns to normal size. Leukopenia is seen; leukocytosis may be present during the febrile episodes. Total plasma proteins are unchanged, although the albumin level may be low and the globulin fraction may be elevated due to antibody development. Serum potassium may also be increased.

When requests for malarial smears are received in the laboratory, the following patient history information should be provided to the laboratorian. (i) Where has the patient been and what was the date of return to the United States? (ii) Has malaria ever been diagnosed in the patient before? If so, what species was identified? (iii) What medication (prophylaxis or other) has the patient received, and how often? When was the last dose taken? (iv) Has the patient ever received a blood transfusion? Is there a possibility of another type of needle transmission (drug user)? (v) When was the blood specimen drawn, and was the patient symptomatic at the time? Is there any evidence of a fever periodicity? The answers may help eliminate the possibility of infection with *P. falciparum*

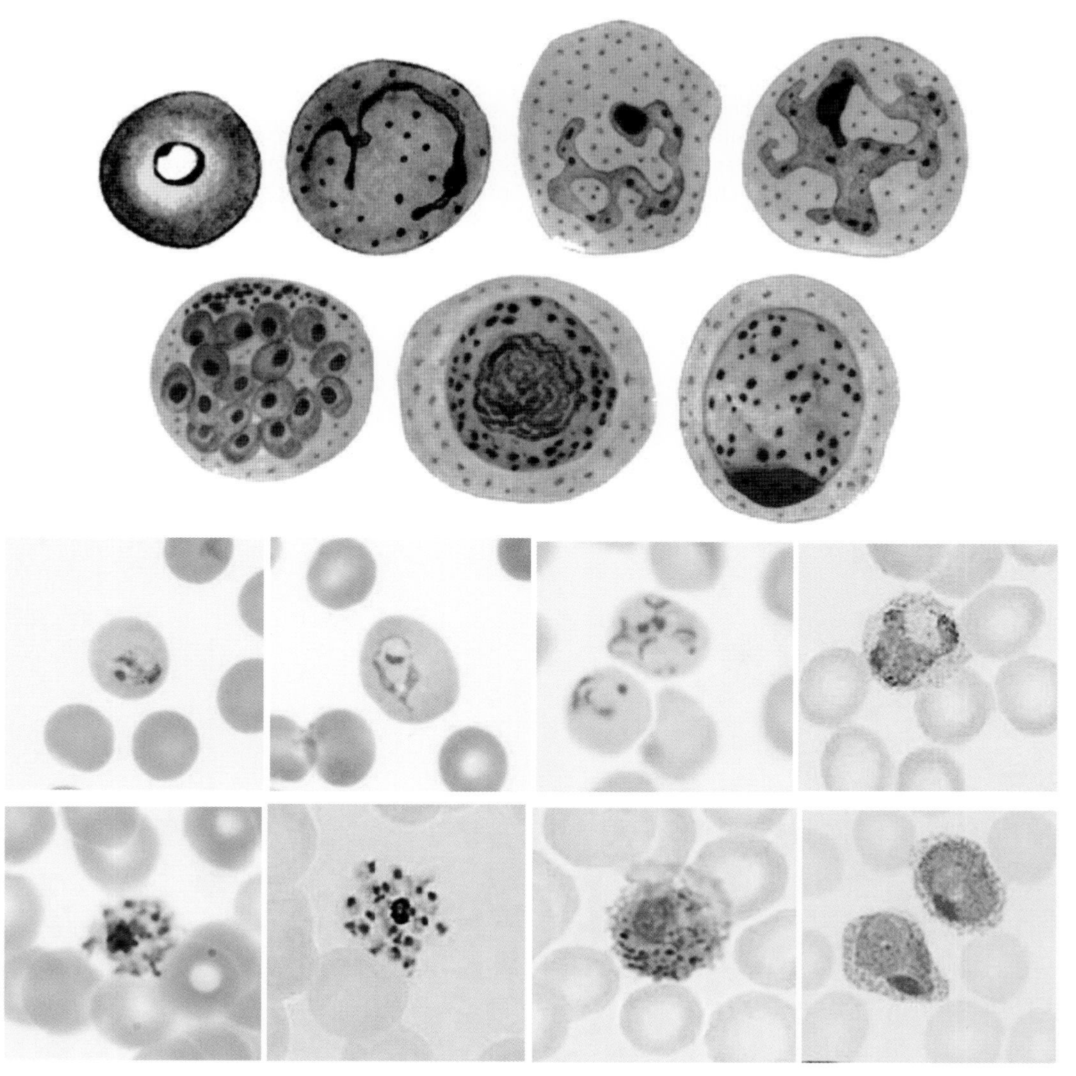

Figure 7.21 ***Plasmodium vivax.*** **(Row 1)** Illustrations of early ring form, developing ring form, developing rings (trophozoites) (ameboid), and older trophozoite (ameboid). Note that the ring forms are very pleomorphic (ameboid). **(Row 2)** Illustrations of developing schizont, mature schizont containing approximately 16 to 18 merozoites (seen as separate structures), male macrogametocyte, and female macrogametocyte (note the presence of Schüffner's dots). Also note the enlarged infected RBCs. **(Row 3)** Developing and older trophozoites (although no Schüffner's dots are visible, this is probably evidence of collection in EDTA and delayed blood film preparation). **(Row 4)** Developing schizont; mature schizont; male gametocyte; two female gametocytes.

or *P. knowlesi*, usually the only species that can rapidly cause death. However, in many areas, *P. vivax* is considered the second lethal cause of malaria after *P. falciparum*.

The primary clinical attack usually occurs 7 to 10 days after infection, although there are strain differences, with a much longer incubation period being possible. Symptoms such as headache, photophobia, muscle aches, anorexia, nausea, and sometimes vomiting may occur before organisms can be detected in the bloodstream. Alternatively, the parasites can be found in the bloodstream several days before symptoms appear.

During the first few days, the patient may not exhibit a typical paroxysm pattern but rather may have a steady low-grade fever or an irregular remittent fever pattern. Once the typical paroxysms begin, after an irregular periodicity, a regular 48-h cycle is established. An untreated primary attack may last from 3 weeks to 2 months or longer. The paroxysms become less severe and more irregular in frequency and then stop altogether. In 50% of patients, relapses may occur after weeks, months, or up to 5 years. Severe complications are rare in *P. vivax* infections, although coma, sudden death, and other signs of cerebral involvement have been reported. Severe sequelae can be seen in cases of primaquine-tolerant or primaquine-resistant cases.

Certain clinical features, such as vomiting, abdominal pain, headache, altered consciousness, cough with breathlessness, and hepatosplenomegaly, as well as laboratory parameters such as severe thrombocytopenia, leukopenia, raised total bilirubin, elevated serum creatinine, and prolonged prothrombin time, may indicate the onset of severe malaria. Cases demonstrating the above warning signs should be closely monitored to prevent adverse outcomes.

Plasmodium falciparum

Pathogenic	Yes
Disease	Malignant tertian malaria, periodicity of approximately 36–48 h
Acquired	Bite of female anopheline mosquito, blood; shared needles; congenital infections
Body site	Liver, RBCs
Symptoms	Few; anemia, splenomegaly, paroxysm (cold stage, fever, and sweats)
Clinical specimen	Blood, multiple draws (EDTA)
Epidemiology	Tropics, mosquito-to-human and human-to-human transmission
Control	Vector control, avoiding shared needles, checking blood supply; various vaccines for *Plasmodium* spp. are undergoing trials (*Plasmodium falciparum* sporozoite malaria vaccine, PfSPZ; transmission-blocking vaccines [TBV] against forms in the mosquito stomach; RH5 blood stage *P. falciparum* antigen; VAR2CSA is the leading antigen for vaccine to protect pregnant women against placental malaria)

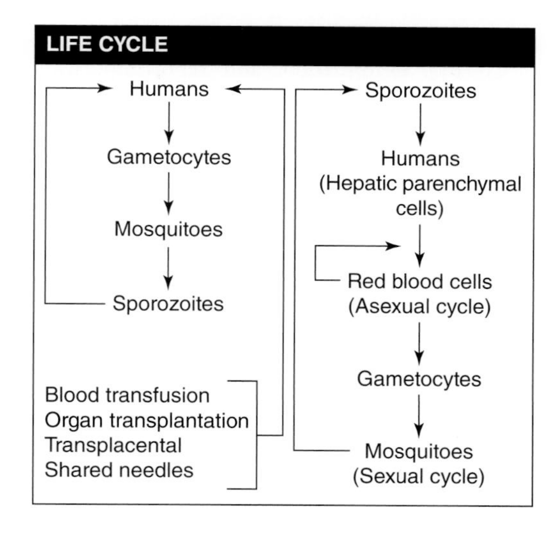

Diagnosis

Malaria has probably had a greater impact on world history than any other infectious disease. It has been responsible for the outcome of wars, population movements, and the growth and development of various nations throughout the world. Before the American Civil War, malaria was found as far north as southern Canada; however, there has been a gradual decline, and by the early 1950s it was no longer an endemic disease within the United States. It is still a very common disease in many parts of the world, particularly in tropical and subtropical areas. Although over 1,000 cases of malaria are reported to the Centers for Disease Control and Prevention each year, the actual number of cases may be much larger. There has been a definite increase in the number of cases of *Plasmodium falciparum* malaria reported, which may be related to increased resistance to chloroquine.

Often, organism recovery and identification may be more difficult than the textbooks imply. It is very important that this fact be recognized, particularly when dealing with a possibly fatal infection with *P. falciparum* or *Plasmodium knowlesi*. There are distinct differences between the clinical presentations of (i) a patient who has had prior exposure to malaria and contains residual antibody (i.e., an individual in an area of endemicity) and (ii) an immunologically naive patient who has no prior exposure to malaria and no antibody (i.e., a traveler). Patients from areas of endemicity with residual antibody tend to become symptomatic with a higher parasitemia and may present with typical periodic fevers. In the case of *P. falciparum*, gametocytes may be present and should be reported (this impacts therapy).

Immunologically naive travelers may present with very nonspecific symptoms and a very low parasitemia. *P. falciparum* gametocytes are often not yet present, since they take approximately 2 weeks to appear in the life cycle. Often, the thin blood films may appear to be negative, while the thick films reveal the presence of *Plasmodium* spp. Handling of specimens for blood parasites should be considered stat (orders, collection, processing, examination, and reporting). Even with a low parasitemia, patients can be quite ill.

Both thick and thin blood films should be prepared immediately *on admission of the patient* (to the clinic or emergency room or in the hospital), and at least 300 oil immersion fields should be examined on the thick and thin films before a negative report is issued. Since one set of negative films does not rule out malaria, additional blood specimens should be examined over 36 h. Although Giemsa stain is used for parasitic blood work, the organisms can also be seen with other blood stains, such as Wright's, Wright-Giemsa, or any of the rapid blood stains. Blood collected with EDTA anticoagulant is acceptable; however, if the blood remains in the tube for any length of time, true stippling may not be visible within the infected RBCs (e.g., for *P. vivax* and/or *P. ovale*); organism morphology changes, and some of the parasites disintegrate (after about 4 to 6 h). Also, the proper ratio between blood and anticoagulant is necessary for good organism morphology. Excess protein and hematin left over from the metabolism of hemoglobin combine to form malarial pigment; this is found in all four species. True stippling (evidence of RBC membrane damage; also called Schüffner's dots) is seen only in *P. vivax* and *P. ovale* malaria.

Diagnostic Tips

In addition to the comments in the *P. vivax* section, you will need to consider the following recommendations.

Patients with no prior exposure to malaria will usually present with very low parasitemias and nonspecific symptoms (low, steady fever with no periodicity and malaise); thus, the possibility of malaria may not be considered. Malaria can mimic many other diseases, such as gastroenteritis, pneumonia, meningitis, encephalitis, and hepatitis. Other possible symptoms include lethargy, anorexia, nausea, vomiting, diarrhea, and headache.

P. falciparum tends to invade RBCs of all ages, and the proportion of infected cells may exceed 50%, particularly in patients in areas of endemicity with constant exposure. Schizogony occurs in the internal organs (spleen, liver, bone marrow, etc.) rather than in the circulating blood. Thus, on the blood films, ring forms, gametocytes, and rarely mature schizonts are seen; no developing stages are seen. Ischemia caused by the plugging of vessels within these organs by masses of parasitized RBCs produces various symptoms, depending on the organ involved. It has been suggested that a decrease in the ability of the RBCs to change shape when passing through capillaries or the splenic filter may lead to plugging of the vessels. If the first set of blood films is positive for ring forms only, the second set of blood films may provide more specific identification information. If the second set of blood films also contains only ring forms, this is a very valuable clue: the species identification is *P. falciparum*. Any of the other species would also contain developing forms on the blood films.

All positive results must be reported with the parasitemia counts (as number of infected RBCs per 100 RBCs). Each subsequent set of blood films must also include parasitemia information to confirm treatment efficacy. Remember to report gametocytes if present; this information will impact therapy.

General Comments

Within an hour, sporozoites from the mosquito are carried via the blood to the liver, where they penetrate parenchymal cells, initiating the preerythrocytic or primary exoerythrocytic cycle. The sporozoites become round or oval and begin dividing, resulting in many liver merozoites. The merozoites leave the liver and invade the RBCs, initiating the erythrocytic cycle. In *P. vivax* and *P. ovale*, a secondary or dormant schizogony occurs from organisms that remain quiescent in the liver until later; they are called hypnozoites. Delayed schizogony does not occur in *P. falciparum*, *P. knowlesi*, or *P. malariae*.

If the RBC infection is not eliminated by the immune system or by therapy and the numbers in the RBCs begin to increase again with subsequent clinical symptoms, this is called a recrudescence; it can occur with all species. If the erythrocytic infection is eliminated and a relapse due to a new invasion of the RBCs from the liver occurs later, this is called a true relapse. It occurs only with *P. vivax* and *P. ovale*.

The merozoite (young trophozoite) is vacuolated, ring shaped, and uninucleate. Once the nucleus begins to divide, the trophozoite is called a developing schizont; the mature schizont contains merozoites which are released into the bloodstream. Merozoites invade RBCs, in which a new cycle of erythrocytic schizogony begins. After several erythrocytic generations, some of the merozoites undergo development into the male and female gametocytes.

Coincident infection with more than one species is more common than previously suspected. Dual *P. falciparum-P. vivax* infections are found, as are *P. falciparum-P. malariae* infections. *P. ovale* is common in areas of Africa with populations refractory to *P. vivax*, preventing coinfections. These species coexist in New Guinea.

Additional Information

Cerebral malaria is most often seen in *P. falciparum* malaria, although it can occur in the other types as well. If the onset is gradual, the patient may become disoriented or violent or may develop severe headaches and fall into a coma. Some patients with no prior symptoms may suddenly become comatose. Physical signs of central nervous system (CNS) involvement are quite variable, and there is no real correlation between the severity of the symptoms and the peripheral-blood parasitemia. It has been shown that patients with cerebral malaria were infected with RBC rosette-forming *P. falciparum* and that plasma from these patients generally had no antirosetting activity. A rosette usually consists of a parasitized RBC surrounded by three or more uninfected RBCs. Interaction with adjacent uninfected RBCs in rosettes appears to be mediated by knobs seen on the parasitized RBC. In contrast, *P. falciparum* parasites from patients with mild malaria lacked the rosetting phenotype or had a much lower rosetting rate. Also, antirosetting activity has been detected in the plasma of these patients. These findings strongly support the idea that RBC rosetting contributes to the pathogenesis of cerebral malaria, while antirosetting antibodies offer protection against these clinical sequelae.

Extreme fevers, 107°F (ca. 41.7°C) or higher, may develop in a relatively uncomplicated attack of malaria or may develop as another manifestation of cerebral malaria. Without vigorous therapy, the patient usually dies. Cerebral malaria is considered the most serious complication and the major cause of death from infection with *P. falciparum*; this complication accounts for up to 10% of all *P. falciparum*-infected patients admitted to the hospital and for 80% of fatal cases.

When requests for malarial smears are received in the laboratory, the following patient history information should be provided to the laboratorian. (i) Where has the patient been and what was the date of return to the United States? (ii) Has malaria ever been diagnosed in the patient before? If so, what species was identified? (iii) What medication (prophylaxis or other) has the patient received, and how often? When was the last dose taken? (iv) Has the patient ever received a blood transfusion? Are there other needle transmission possibilities (drug user)? (v) When was the blood specimen drawn, and was the patient symptomatic at the time? Is there any evidence of a fever periodicity? Answers to such questions may help eliminate the possibility of infection with *P. falciparum* or *P. knowlesi*, usually the only species that can rapidly cause death. However, *P. vivax* is often considered the second most lethal malarial agent after *P. falciparum*.

Accurate species diagnosis is essential for good patient management, since it may determine which drug(s) is indicated. Some patients do not yet have the crescent-shaped gametocytes in the blood. Low parasitemias with the delicate ring forms may be missed; oil immersion examination at ×1,000 is mandatory.

Onset of a *P. falciparum* malaria attack is 8 to 12 days after infection and is preceded by 3 to 4 days of vague symptoms, such as aches, pains, headache, fatigue, anorexia, or nausea. The onset involves fever, severe headache, and nausea and vomiting, with occasional severe epigastric pain. There may be only a feeling of chilliness. **Periodicity of the cycle is not established during the early stages, and the presumptive diagnosis may be totally unrelated to a possible malaria infection.** If the fever does develop a synchronous cycle, it is usually a cycle of somewhat less than 48 h.

An untreated primary attack of *P. falciparum* malaria usually ends within 2 to 3 weeks. True relapses from the liver do not occur, and recrudescences after a year are rare. Severe or fatal complications of *P. falciparum* malaria can occur at any time and are related to the plugging of vessels in the internal organs. *The symptoms depend on the organ(s) involved.* Severe complications may not correlate with the parasitemia seen in the peripheral blood. Disseminated intravascular coagulation is a rare complication of high parasite burden; vascular endothelial damage from endotoxins and bound parasitized blood cells may lead to clot formation in small vessels.

Cerebral malaria involving disorientation, violent episodes, severe headaches, and coma is most often seen in *P. falciparum* malaria. There is no real correlation between the severity of symptoms and the peripheral-blood parasitemia.

Malaria is one of the few parasitic infections considered immediately life-threatening, and a patient with *P. falciparum* malaria is a medical emergency. Any laboratory providing the expertise to identify malarial parasites should do so on a 24-h basis, 7 days/week.

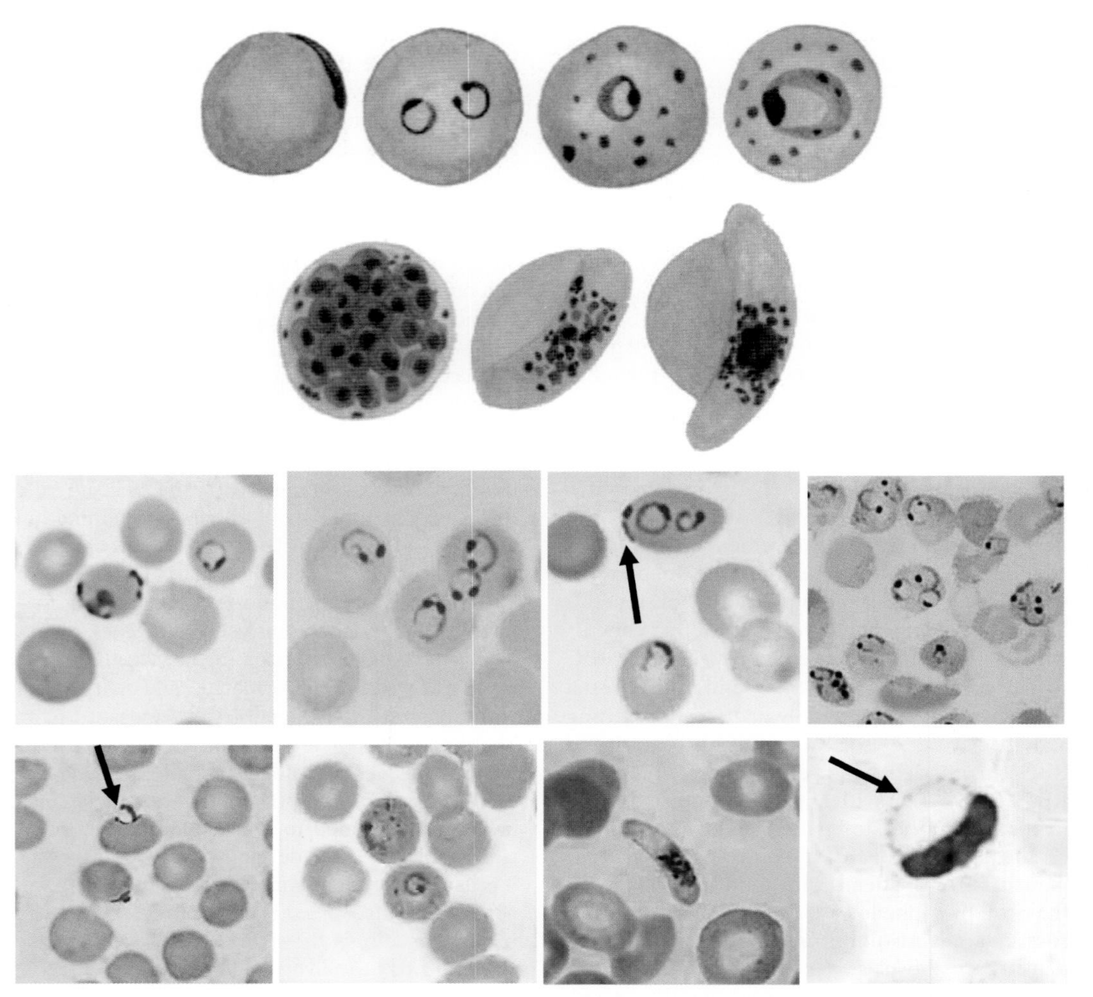

Figure 7.22 ***Plasmodium falciparum.*** **(Row 1)** Illustrations of ring forms. **(Row 2)** Illustrations of mature schizont, male gametocyte (not quite yet in crescent shape), and mature female macrogametocyte (note the crescent shape of gametocyte within the RBC). **(Row 3)** Multiple ring forms; ring forms with two nuclei (headphone appearance); ring in the accolé or appliqué form (arrow); double rings per cell (very high parasitemia). **(Row 4)** Ring form protruding from the RBC (arrow; very characteristic of *P. falciparum*); developing rings (RBCs contain Maurer's clefts [dots]); female macrogametocyte; male microgametocyte (note the crescent shape of gametocytes). In the last image, note the faint outline of the RBC (arrow). Although the RBC outline is not always seen, the gametocytes are still intracellular (within the RBCs).

Plasmodium malariae

Pathogenic	Yes
Disease	Quartan malaria; low parasitemia in asymptomatic patients can persist for very long times, often years; periodicity of approximately 72 h
Acquired	Bite of female anopheline mosquito, blood, shared needles, congenital infections
Body site	Liver, RBCs
Symptoms	Few; anemia, splenomegaly, paroxysm (cold stage, fever, and sweats)
Clinical specimen	Blood, multiple draws (EDTA)
Epidemiology	Sporadic distribution; mosquito-to-human and human-to-human transmission
Control	Vector control, avoiding shared needles, checking blood supply; various vaccine trials are under way for *Plasmodium* spp.

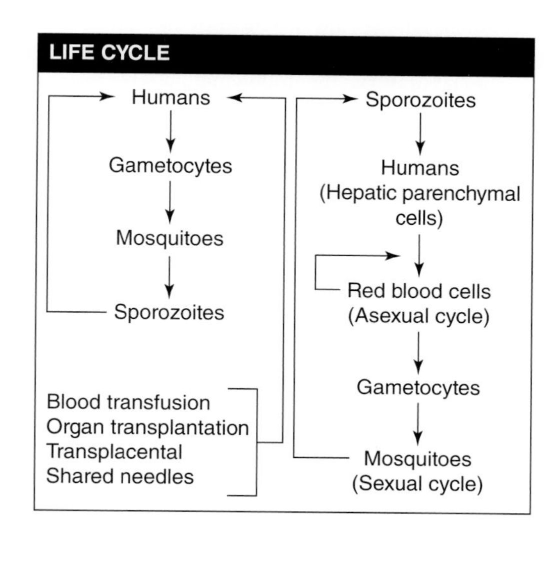

Diagnosis

Plasmodium malariae invades primarily the older, more mature RBCs, so that the number of infected cells is somewhat limited. The incubation period between infection and symptoms may be much longer than that seen with *Plasmodium vivax* or *Plasmodium ovale* malaria, ranging from about 27 to 40 days. Parasites can be found in the bloodstream several days before the initial attack, and the prodromal symptoms may resemble those of *P. vivax* malaria. A regular periodicity is seen from the beginning, with a more severe paroxysm, including a longer cold stage and more severe symptoms during the hot stage. Collapse during the sweating phase is not uncommon.

Both thick and thin blood films should be prepared *on admission of the patient* (to the clinic or emergency room or in house), and at least 300 oil immersion fields should be examined on the thick and thin films before a negative report is issued. Since one set of negative films does not rule out malaria, additional blood specimens should be examined over 36 h. Giemsa stain is recommended for all parasitic blood work; the organisms can also be seen with other blood stains such as Wright's stain. Blood collected with EDTA anticoagulant is acceptable; however, if the blood remains in the tube for any length of time, true stippling may not be visible within the infected RBCs (e.g., for *P. vivax*); organism morphology changes, and some of the parasites disintegrate (after 4 to 6 h). Also, the proper ratio between blood and anticoagulant is necessary for good organism morphology. Excess protein and hematin left over from the metabolism of hemoglobin combine to form malarial pigment; this is found in all five species. True stippling (evidence of RBC membrane damage; also called Schüffner's dots) is seen only in *P. vivax* and *P. ovale* malaria.

Diagnostic Tips

Since *P. malariae* tends to infect older RBCs, the infected RBCs will be small to normal size; this characteristic can be very helpful in arriving at the correct species identification. Characteristic forms will include the "band" trophozoite stages and the "rosette" mature schizont. Some infected patients will remain asymptomatic with very low parasite numbers for many years.

If the test is positive, but you are unable to identify the organisms to the species level, report comments are extremely helpful for the physician. An example would be "*Plasmodium* spp. seen; unable to rule out *Plasmodium falciparum* or *Plasmodium knowlesi* (the two most pathogenic)."

When requests for malarial smears are received in the laboratory, the following patient history information should be provided to the laboratorian. (i) Where has the patient been and what was the date of return to the United States? (ii) Has malaria ever been diagnosed in the patient before? If so, what species? (iii) What medication (prophylaxis or otherwise) has the patient received, and how often? When was the last dose taken? (iv) Has the patient ever received a blood transfusion? Are there other needle transmission possibilities (drug user)? (v) When was the blood specimen drawn, and was the patient symptomatic at the time? Is there any evidence of fever periodicity? The answers may help eliminate the possibility of infection with *P. falciparum* and *P. knowlesi*, usually the only species that can rapidly cause death. However, *P. vivax* in many areas is considered the second lethal malarial infection after *P. falciparum*.

General Comments

P. malariae is relatively common in tropical Africa and the southwest Pacific region. Incidence figures vary from below 1% to as high as 40% in specific areas of West Africa and Indonesia. In patients who do not receive therapy, the infection persists for many years; these individuals may be asymptomatic and serve as potential blood donors. Approximately 25% of blood donor-related cases of malaria are due to *P. malariae*.

Within an hour, sporozoites from the mosquito are carried via the blood to the liver, where they penetrate parenchymal cells, initiating the preerythrocytic or primary exoerythrocytic cycle. The sporozoites become round or oval and begin dividing, resulting in many liver merozoites. The merozoites leave the liver and invade the RBCs, initiating the erythrocytic cycle. In *P. vivax* and *P. ovale*, a secondary or dormant schizogony occurs from organisms that remain quiescent in the liver until later; they are called hypnozoites. Delayed schizogony does not occur with *P. falciparum*, *P. knowlesi*, or *P. malariae*.

If the RBC infection is not eliminated by the immune system or by therapy and the numbers in the RBCs begin to increase again with subsequent clinical symptoms, this is called a recrudescence; it can occur with all species. If the erythrocytic infection is eliminated and a relapse due to a new invasion of the RBCs from the liver occurs later, this is called a true relapse. It occurs only with *P. vivax* and *P. ovale*. In *P. malariae* infections, the low-grade parasitemia can persist for very long times, even years, in asymptomatic patients.

The merozoite (young trophozoite) is vacuolated, ring shaped, and uninucleate. Once the nucleus begins to divide, the trophozoite is called a developing schizont; the mature schizont contains merozoites, which are released into the bloodstream. Merozoites invade RBCs, in which a new cycle of erythrocytic schizogony begins. After several erythrocytic generations, some of the merozoites undergo development into the male and female gametocytes.

Additional Information

P. malariae invades primarily the older RBCs, so the number of infected cells is limited. The incubation period between infection and symptoms may be much longer than that seen with *P. vivax* or *P. ovale* malaria, ranging from about 27 to 40 days. Parasites can be found in the bloodstream several days before the initial attack, and the prodromal symptoms may resemble those of *P. vivax* malaria. A regular periodicity is seen from the beginning, with a more severe paroxysm, including a longer cold stage and more severe symptoms during the hot stage. Collapse during the sweating phase is not uncommon.

Proteinuria is common in *P. malariae* infections; in children it may be associated with clinical signs of the nephrotic syndrome. Kidney problems may result from deposition within the glomeruli of circulating antigen-antibody complexes in an antigen-excess situation in a chronic infection. Apparently, the nephrotic syndrome associated with *P. malariae* infections is unaffected by administration of steroids. A membranoproliferative type of glomerulonephritis with relatively sparse proliferation of endothelial and mesangial cells is the most common type of lesion seen in quartan malaria. When immunofluorescence is used, granular deposits of IgM, IgG, and C3 are seen. Since chronic glomerular disease associated with *P. malariae* infections is usually not reversible with therapy, genetic and environmental factors may play a role in the nephrotic syndrome.

The infection may end with spontaneous recovery, or there may be a recrudescence or series of recrudescences over many years. These patients are left with a latent infection and persisting low-grade parasitemia for many years. Diagnosis at this point in the disease may be difficult due to low parasitemias and negative results using the rapid test formats. Deaths associated with *P. malariae* are primarily from end-stage renal disease.

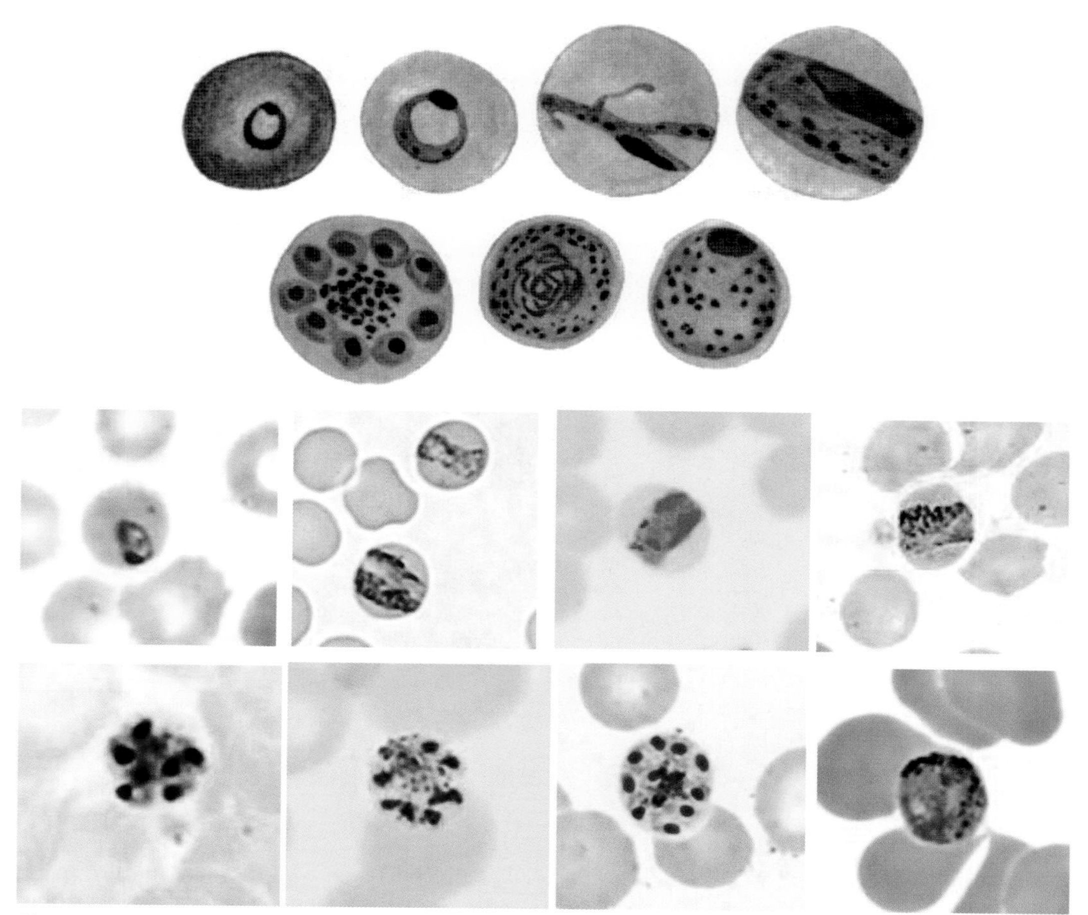

Figure 7.23 ***Plasmodium malariae.*** **(Row 1)** Illustrations of young rings (first two images), typical band form, and band form. **(Row 2)** Illustrations of mature schizont, male gametocyte, and female gametocyte. **(Row 3)** Typical ring; two band forms; band form; band form (note the dark malarial pigment at the top of the band form in the last image). **(Row 4)** Developing schizont; developing schizont (note the excess malarial pigment in the center); mature schizont showing the rosette formation; female macrogametocyte.

Plasmodium ovale wallickeri
Plasmodium ovale curtisi

Pathogenic	Yes, periodicity of approximately 48 h (like *Plasmodium vivax*)
Disease	Ovale malaria
Acquired	Bite of female anopheline mosquito, blood, shared needles, congenital infections
Body site	Liver, RBCs
Symptoms	Few; anemia, splenomegaly, paroxysm (cold stage, fever, and sweats)
Clinical specimen	Blood, multiple draws (EDTA)
Epidemiology	Central West Africa and some South Pacific islands; mosquito-to-human and human-to-human transmission
Control	Vector control, avoiding shared needles, checking blood supply; vaccine is possible in the near future

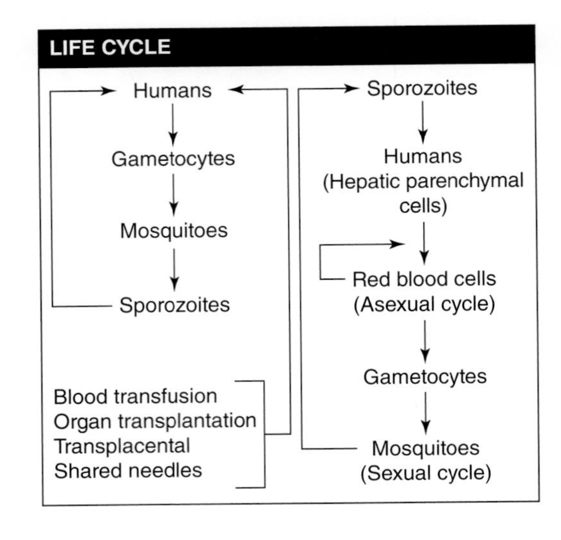

Diagnosis

Although malaria is no longer endemic within the United States, it is life-threatening, and laboratory requests for blood smear examination and organism identification should be treated as stat requests. Frequently, for a number of reasons, organism recovery and identification may be more difficult than the textbooks imply, particularly in immunologically naive patients with no past exposure to malaria (travelers versus patients from areas where the disease is endemic). It is very important that this fact be recognized, particularly when dealing with a possibly fatal infection with *Plasmodium falciparum* or *Plasmodium knowlesi.*

Both thick and thin blood films should be prepared *on admission of the patient* (to the clinic or emergency room or in house), and at least 300 oil immersion fields should be examined on the thick and thin films before a negative report is issued. Since one set of negative films does not rule out malaria, additional blood specimens should be examined over 36 h. Giemsa stain is recommended for all parasitic blood work; the organisms can also be seen with other blood stains such as Wright's stain. Blood collected with EDTA anticoagulant is acceptable; however, if the blood remains in the tube for any length of time, true stippling may not be visible within the infected RBCs (e.g., for *Plasmodium vivax*), organism morphology changes, and some of the parasites will disintegrate (after 4 to 6 h). Also, the proper ratio between blood and anticoagulant is necessary for good organism morphology.

Excess protein and hematin left over from the metabolism of hemoglobin combine to form malarial pigment; this is found in all five species. True stippling (evidence of RBC membrane damage; also called Schüffner's dots) is seen only in *P. vivax* and *Plasmodium ovale* malaria.

Diagnostic Tips

When requests for malarial smears are received in the laboratory, the following patient history information should be provided to the laboratorian. (i) Where has the patient been and what was the date of return to the United States? (ii) Has malaria ever been diagnosed in the patient before? If so, what species? (iii) What medication (prophylaxis or otherwise) has the patient received, and how often? When was the last dose taken? (iv) Has the patient ever received a blood transfusion? Are there other needle transmission possibilities (drug user)? (v) When was the blood specimen drawn, and was the patient symptomatic at the time? Is there any evidence of fever periodicity? Answers to such questions may help eliminate the possibility of infection with *P. falciparum* or *P. knowlesi*, usually the only species that can rapidly lead to death. However, in some areas, *P. vivax* is considered the second most lethal malarial agent after *P. falciparum*.

With all human malarial species (*P. vivax*, *P. falciparum*, *P. malariae*, *P. ovale* [two subspecies], and *P. knowlesi*), if chloroquine prophylaxis is taken, the number of infected RBCs on the blood films may be reduced. It is very important for the microbiologist to know if the patient has taken any medication(s) in the 72 hours prior to blood draws. If so, the amount of time spent on blood film examination will need to be increased (both thick and thin films).

P. ovale resembles *P. vivax*; both species infect young RBCs (enlarged RBCs). Schüffner's dots are somewhat larger and more easily visible in *P. ovale*; however, remember that if the blood has been in EDTA anticoagulant several hours before smear preparation, the dots may not be visible in either species.

Typical key features include RBCs with fimbriated edges, nonameboid trophozoites, and some oval RBCs. Search the blood films for populations of organisms; some outlier RBC parasite morphologies cannot be identified to the species level. Although two subspecies have been identified (*P. ovale wallikeri* and *P. ovale curtisi*), they cannot be differentiated on the basis of morphology during examination of stained blood films.

General Comments

P. ovale is found primarily in tropical Africa and New Guinea and can be found in up to 15% of returning travelers. In areas of Africa where the population is Duffy antigen negative, *P. ovale* tends to be more common than *P. vivax*, which tends to infect those with Duffy-positive blood group genotypes. Severe complications, such as spleen rupture, severe anemia, and acute respiratory distress syndrome, may occur in patients with *P. ovale* malaria. Thus, the global burden of *P. ovale* infection might have been underestimated.

Molecular methods have confirmed the existence of two distinct nonrecombining species of *P. ovale* (classic type *P. ovale curtisi* and variant type *P. ovale wallikeri*). In one study, a significant finding was more severe thrombocytopenia among patients with *P. ovale wallikeri* infection than among those with *P. ovale curtisi* infection.

Within an hour, sporozoites from the mosquito are carried via the blood to the liver, where they penetrate parenchymal cells, initiating the preerythrocytic or primary exoerythrocytic cycle. The sporozoites become round or oval and begin dividing, resulting in many liver merozoites. The merozoites leave the liver and invade the RBCs, initiating the erythrocytic cycle. In *P. vivax* and *P. ovale*, a secondary or dormant schizogony occurs from organisms that remain quiescent in the liver until later; they are called hypnozoites. Delayed schizogony does not occur in *P. falciparum*, *P. knowlesi*, or *P. malariae*.

If the RBC infection is not eliminated by the immune system or by therapy and the numbers in the RBCs begin to increase again with subsequent clinical symptoms, this is called a recrudescence; it can occur with all species. If the erythrocytic infection is eliminated and a relapse due to a new invasion of the RBCs from the liver occurs later, this is called a true relapse. It occurs only with *P. vivax* and *P. ovale*.

The merozoite (young trophozoite) is vacuolated, ring shaped, and uninucleate. Once the nucleus begins to divide, the trophozoite is called a developing schizont; the mature schizont contains merozoites which are released into the bloodstream. Merozoites invade RBCs, in which a new cycle of erythrocytic schizogony begins. After several erythrocytic generations, some of the merozoites undergo development into the male and female gametocytes.

Additional Information

P. ovale infects only the reticulocytes, and the parasitemia is usually limited to 2 to 5% of available RBCs. Splenomegaly occurs during the first few weeks, and the spleen becomes hard during a chronic infection. If therapy is given early, the spleen will return to normal size. Leukopenia is seen; leukocytosis may be present during the febrile episodes. Total plasma proteins are unchanged, although the albumin may be low and the globulin fraction may be elevated due to antibody development. Serum potassium may also be increased.

The primary clinical attack usually occurs 7 to 10 days after infection, although there are strain differences, with a much longer incubation period being possible. Symptoms such as headache, photophobia, muscle aches, anorexia, nausea, and vomiting may occur before organisms can be detected in the bloodstream. Alternatively, the parasites can be found in the bloodstream before symptoms appear.

During the first few days, the patient may not exhibit a typical paroxysm pattern but rather may have a steady low-grade fever or an irregular remittent fever pattern. Once the typical paroxysms begin, after an irregular periodicity, a regular 48-h cycle is established. An untreated primary attack may last from 3 weeks to 2 months or longer. The paroxysms become less severe and more irregular and then stop altogether. In 50% of patients, relapses may occur after weeks, months, or up to 5 years. Severe complications are rare in *P. ovale* infections,

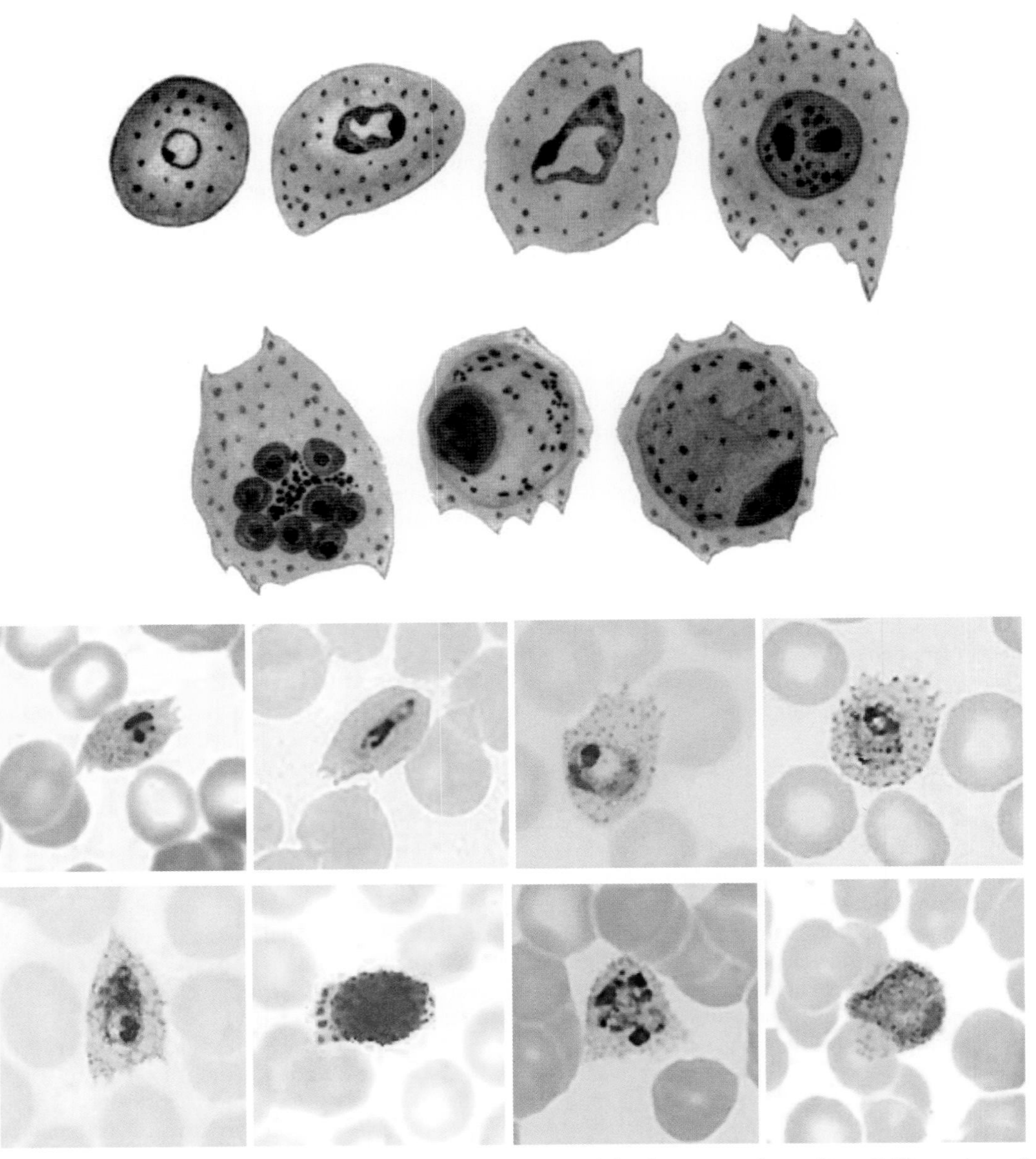

Figure 7.24 ***Plasmodium ovale*****.** **(Row 1)** Illustrations of young and developing ring forms. **(Row 2)** Illustrations of mature schizont, male gametocyte, and female gametocyte (note the fimbriated edges of RBCs and somewhat oval shape of some infected RBCs; also note the appearance of Schüffner's dots [true stippling]). **(Row 3)** Four developing ring/trophozoite forms (note that the last two images contain true stippling [Schüffner's dots]; dots appear later in the cycle in *Plasmodium vivax*). **(Row 4)** Two developing trophozoites; developing schizont; mature female macrogametocyte. Note the enlarged RBCs and the oval shape and fimbriated edges of some of the RBCs.

although coma and sudden death or other symptoms of cerebral involvement can occur. *P. ovale* malaria is usually less severe than *P. vivax*, tends to relapse less frequently, and usually ends with spontaneous recovery, often after only 6 to 10 paroxysms. The incubation period is similar to that of *P. vivax* malaria, and symptoms are less frequent and severe, with a lower fever and a lack of typical rigors.

The geographic range includes tropical Africa, the Middle East, Papua New Guinea, and Irian Jaya in Indonesia. However, infections in Southeast Asia may cause benign and relapsing malaria. In both Southeast Asia and Africa, two different types of *P. ovale* circulate in humans. Human infections with variant-type *P. ovale* are associated with higher parasitemias, with possible clinical relevance.

Plasmodium knowlesi

Pathogenic	Yes, periodicity of approximately 24 h (most rapid of all human malarias)
Disease	Simian malaria
Acquired	Bite of female anopheline mosquito, blood, shared needles, congenital infections
Body site	Liver, RBCs
Symptoms	Similar to those seen in *P. falciparum* malaria (febrile illness); 10% progress to severe disease (respiratory distress, renal failure)
Clinical specimen	Blood, multiple draws (EDTA)
Epidemiology	Southeast Asian countries, especially east Malaysia; mosquito-to-human and human-to-human transmission (suspected in certain geographic areas)
Control	Vector control, avoiding shared needles, checking blood supply; vaccine possibilities under discussion

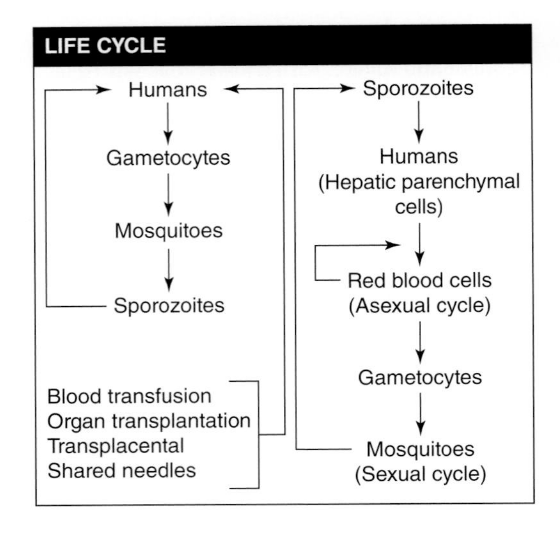

Diagnosis

Although malaria is no longer endemic within the United States, it is life-threatening, and laboratory requests for blood smear examination and organism identification should be treated as STAT requests. Since *Plasmodium knowlesi* is less well known than the other human *Plasmodium* spp., organism recovery and identification may be more difficult than normal. It is very important that this fact be recognized, particularly when dealing with a possibly fatal infection with *Plasmodium falciparum* or *P. knowlesi*. Although in these patients, malaria may be diagnosed, identification to the species level has often been incorrect.

Both thick and thin blood films should be prepared *on admission of the patient* (to the clinic or emergency room or in house), and at least 300 oil immersion fields should be examined on the thick and thin films before a negative report is issued. Since one set of negative films does not rule out malaria, additional blood specimens should be examined over 36 h. Giemsa stain is recommended for all parasitic blood work; the organisms can also be seen with other blood stains, such as Wright's stain. Blood collected with EDTA anticoagulant is acceptable; however, if the blood remains in the tube for any length of time, true stippling may not be visible within the infected RBCs (e.g., for *Plasmodium vivax*); organism morphology changes, and some of the parasites will disintegrate (after 4 to 6 h). Also, the proper ratio between blood and anticoagulant is necessary for good organism morphology.

Excess protein and hematin left over from the metabolism of hemoglobin combine to form malarial pigment; this is found in all five species. True stippling (evidence of RBC membrane damage; also called Schüffner's dots) is seen only in *P. vivax* and *Plasmodium ovale* malaria.

Diagnostic Tips

Rapid diagnostic tests are insensitive and nonspecific for zoonotic species; however, in very rare instances the test may be positive. These cases probably have a relatively high parasitemia. Rapid tests that can detect *P. knowlesi* are under development. In routine microscopy, *P. knowlesi* has morphologic characteristics of both *P. falciparum* and *Plasmodium malariae* and may be misdiagnosed as *P. malariae*, *P. falciparum/P. malariae* (mixed infection), or another *Plasmodium* species.

Accurate detection requires the use of sensitive and specific molecular tools such as PCR. According to results obtained with these methods, *P. knowlesi* is the most common cause of malaria across Malaysia and some regions of western Indonesia and is the only cause of malaria in Brunei and Singapore. Although molecular testing of clinical cases is now routine in Malaysia, molecular surveillance for zoonotic malaria elsewhere in Southeast Asia is more limited. PCR has ability to detect infections with parasitemia as low as 22 parasites/ml of blood. Two commercial malaria PCR assays (RealStar Malaria S&T PCR kit 1.0 [Altona Diagnostics] and FTD malaria differentiation [Fast Track Diagnostics; does not detect *P. knowlesi*] multiplex real-time PCR assays) are suitable for discriminating the five *Plasmodium* species in clinical samples and can provide additional information in cases of microscopically uncertain findings. However, these products are currently not licensed with Health Canada and are not FDA cleared.

Asymptomatic cases of *P. knowlesi* have also been confirmed. Despite rarely causing clinical disease, submicroscopic malaria infections play a role in malaria transmission. Individuals with these extremely low parasitemia infections can infect the mosquitoes.

General Comments

The true prevalence of *P. knowlesi* in the Southeast Asia region is still undetermined, and when or where this organism first caused a natural infection in a human is unknown. The first reported infection in a human was in 1965 in peninsular Malaysia. Confirmed human cases are increasing, especially in Malaysian Borneo, although this may be the result of increased awareness of, and testing for, *P. knowlesi* (simian malaria).

The natural hosts for *P. knowlesi* are the cynomolgus monkeys, the long-tailed and pig-tailed macaques (*Macaca fascicularis* and *Macaca nemestrina*, respectively), found throughout Southeast Asia. These two species coexist throughout much of this region. Zoonotic *P. knowlesi* malaria has now been reported from all countries of Southeast Asia where humans come into contact with *Anopheles leucosphyrus* group mosquitoes and the macaque reservoir host. The potential disease load that can be associated with the transfer of parasites between humans and primate reservoirs is currently unknown.

Counterfeiting of drugs is now considered one of the most critical public health problems for malaria treatment in developing countries. Although the problem of modern-day counterfeit drugs was first addressed in 1985 at a conference in Nairobi, Kenya, the counterfeit market continues to grow and has become worldwide during the last decade. The World Health Organization (WHO) estimates that 50% of medicines available via the Internet are fake.

Counterfeiting of drugs is estimated by WHO at more than US$35 billion and represents more than 15% of the pharmaceutical market worldwide; this percentage increases to more than 60% in developing countries (35 to 90% in Southeast Asia).

The situation is even worse in sub-Saharan Africa, where the burden of malaria is the greatest. In 2001, WHO recommended that African countries where malaria is endemic consider changing to artemisinin derivatives as first-line treatment for malaria. However, adopting this new policy was difficult due to the high cost and relative shortage of the raw plant material source (*Artemisia annua*). Thus, the high cost and shortage of drugs led to the spread of fake artemisinin, which places many thousands of African children at risk.

Although *P. knowlesi* can cause high parasitemia similar to that seen with *P. falciparum*, severe disease most often occurs with a relatively low parasitemia. Patients with a parasitemia of >15,000 parasites/μl are 16 times more likely to develop severe malaria. In areas of Southeast Asia where the infection is endemic, prompt and accurate diagnosis and early treatment are mandatory. The recommended intravenous artesunate therapy for those with a parasitemia of >15,000/μl represents a lower threshold than that recommended for *P. falciparum*. Age distribution and parasitemia differ markedly in *P. knowlesi* malaria compared to that caused by human-only species, with both uncomplicated and severe disease occurring at low parasitemia. Severe *P. knowlesi* malaria tends to occur only in adults; however, anemia is more common in children despite lower parasitemias.

Additional Information

Although human-to-human transmission has been demonstrated experimentally and is suspected in certain geographic areas, current information indicates that natural human transmission of *P. knowlesi* remains zoonotic. Reasons for the rise in *P. knowlesi* malaria incidence include greater human exposure resulting from changes in land use and deforestation, both of which impact vector characteristics and the macaques, as well as possible parasite adaptations.

Severe *P. knowlesi* malaria characteristics are similar to those found in severe *P. falciparum* malaria in adults and include hyperparasitemia, jaundice, acute kidney injury, respiratory distress, shock, and metabolic acidosis.

In *P. falciparum* malaria, severe disease is characterized by cytoadherence of infected red blood cells to activated endothelium, leading to microvascular sequestration and impaired organ perfusion. Decreased deformability of both infected and uninfected RBCs is an additional key contributor to microvascular obstruction. In severe *P. knowlesi* malaria, a single autopsy report revealed widespread microvascular accumulation of infected RBCs. However, endothelial cytoadherence was not clearly evident, and intercellular adhesion molecule-1, which mediates cytoadherence to brain endothelial cells in *P. falciparum* malaria, has not been detected.

The RBC diameter exceeds the capillary diameter by several micrometers. Therefore, the ability of RBCs to deform as they pass through capillaries is mandatory in maintaining normal microvascular flow. It has been demonstrated that in human adults with *P. knowlesi* malaria, deformability of RBCs is reduced in proportion to disease severity. Also, intravascular hemolysis is independently associated with markers of disease severity, including lactate, microvascular dysfunction, and creatinine, suggesting that hemolysis likely contributes to impaired tissue perfusion and organ dysfunction in *P. knowlesi* malaria. Apparently, intravascular hemolysis is increased in severe *P. knowlesi* malaria, and to a greater extent than that seen with *P. falciparum* malaria.

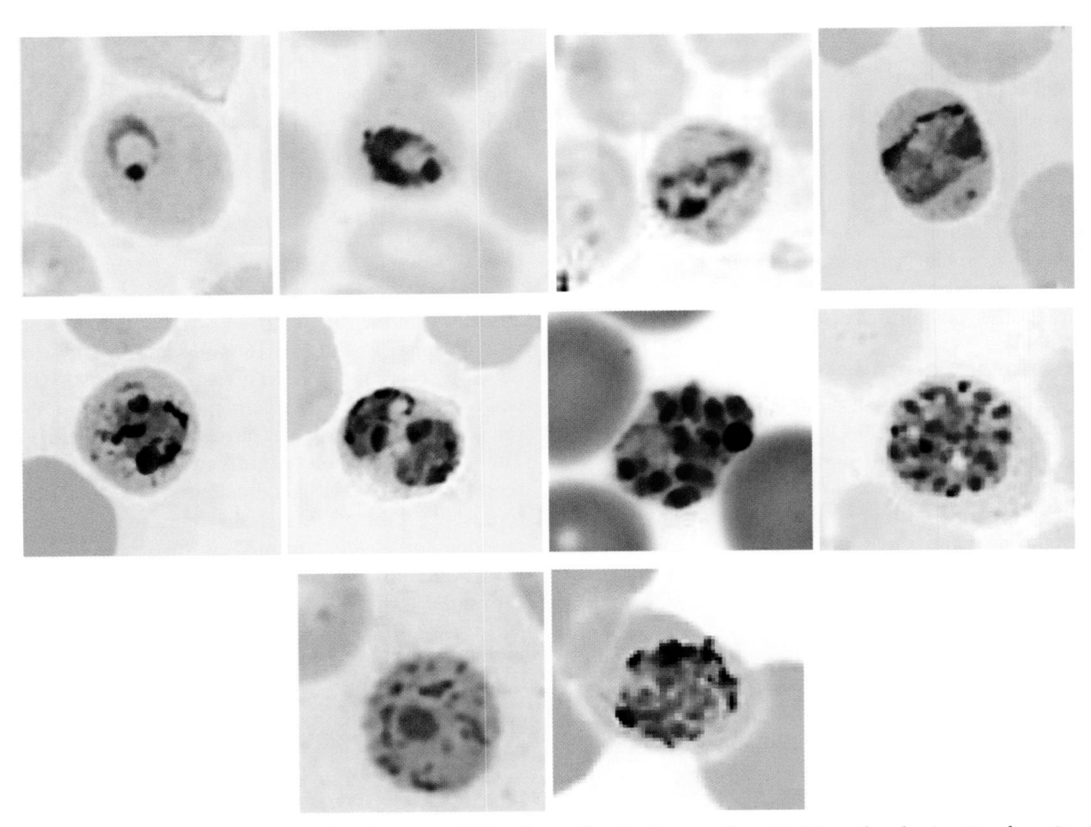

Figure 7.25 ***Plasmodium knowlesi*****.** **(Top row)** Ring form; developing trophozoite/ring; developing trophozoite as band form; more mature band form. **(Middle row)** Developing schizont; developing schizont; older developing schizont (reprinted from Millar SB, Cox-Singh J, *Clin Microbiol Infect* **21**:640-648, 2015, with permission from Elsevier, Ltd.); mature schizont. **(Bottom row)** Mature macrogametocyte; developing microgametocyte (courtesy of Wadsworth Center, New York State Department of Health).

Malaria

Key Diagnostic Points

1. Blood films should be prepared on admission of the patient (ordering, collection, processing, examination, and reporting on a STAT basis). A fever pattern may not be apparent early in the course of the infection (immunologically naive patients [travelers]); symptoms may be completely random and may mimic any other condition with vague complaints.
2. Both thick and thin blood films should be prepared. At least 300 oil immersion fields (×1,000) on both thick and thin films should be examined before the specimen is considered negative.
3. Wright's, Wright-Giemsa, Giemsa, or a rapid stain can be used. The majority of the original organism descriptions were based on Giemsa stain. However, if the white blood cells appear to be well stained, any blood parasites present will also be well stained. The WBCs on the patient smear serve as the QC organism; there is no need to use a *Plasmodium*-positive slide for QC.
4. Malarial parasites may be missed with the use of automated differential instruments. Even with technologist review of the smears, a light parasitemia is very likely to be missed.
5. The number of oil immersion fields examined may have to be increased if the patient has had any prophylactic medication during the past 48 h (the number of infected cells may be decreased on the blood films).
6. *One negative set of blood smears does not rule out malaria. Quantitate organisms from every positive blood specimen.* The same method for calculating parasitemia should be used for each subsequent positive blood specimen.
7. In spite of new technology, serial thick-film parasite counts are a simple, cheap, rapid, and reliable method for identifying patients at high risk of recrudescence due to drug resistance and treatment failure.
8. If you are using any of the alternative methods, make sure you thoroughly understand the pros and cons of each compared with the thick and thin blood film methods.

Report Comments

Report comments can be extremely helpful in conveying information to the physician. Depending on the results of diagnostic testing, the following information can lead to improved patient care and clinical outcomes. The report is provided with the following comments:

1. **No parasites seen:** The submission of a single blood specimen does not rule out malaria; submit additional blood samples every 4 to 6 h for 3 days if malaria remains a consideration.

   Interpretation/discussion: It is important to make sure that the physician knows that examination of a single blood specimen will not rule out malaria.
2. ***Plasmodium* spp. seen:** Unable to rule out *Plasmodium falciparum* or *Plasmodium knowlesi.*

   Interpretation/discussion: Since *P. falciparum* and *P. knowlesi* cause the most serious illness, it is important to let the physician know that these species have NOT been ruled out.
3. ***Plasmodium* spp., possible mixed infection:** Unable to rule out *Plasmodium falciparum* or *Plasmodium knowlesi.*

   Interpretation/discussion: Since *P. falciparum* and *P. knowlesi* cause the most serious illness, it is important to let the physician know these species have NOT been ruled out.
4. *Plasmodium malariae*: Unable to rule out *Plasmodium knowlesi.*

   Interpretation/discussion: If the patient has traveled to the area where *P. knowlesi* is endemic, it may be impossible to differentiate between *P. malariae* (band forms) and *P. knowlesi.*
5. *Plasmodium falciparum*: Unable to rule out *Plasmodium knowlesi.*

   Interpretation/discussion: If the patient has traveled to the area where *P. knowlesi* is endemic, it may be impossible to differentiate between *P. falciparum* (ring forms) and *P. knowlesi.*
6. **Negative for parasites using automated hematology analyzer**: Automated hematology analyzers do not detect low malaria parasitemias seen in immunologically naive patients (travelers).

Interpretation/discussion: Patients who have never been exposed to malaria (immunologically naive) become symptomatic with very low parasitemias that will not be detected using automation (0.001 to 0.0001%).

7. **Negative for malaria using the BinaxNOW rapid test:** This result does not rule out the possibility of a malaria infection. Blood should be submitted for stat thick and thin blood film preparation, examination, and reporting.

Interpretation/discussion: The maximum sensitivity of this rapid test occurs at 0.1% parasitemia. Patients (immunologically naive travelers) may present to the emergency room or clinic with a parasitemia much lower than 0.1%, leading to a false-negative report. Also, this rapid test is not designed to identify *P. malariae*, *P. ovale*, and *P. knowlesi*; the results are most clinically relevant for *P. falciparum* and *P. vivax*. The BinaxNOW test is FDA cleared for use in the United States.

Babesia spp. (*Babesia microti*, *B. duncani*, *B. divergens*, *B. venatorum*)

Pathogenic	Yes
Disease	Babesiosis
Acquired	Bite of various ticks; blood, shared needles, congenital infections
Body site	RBCs
Symptoms	Few; general malaise followed by fever, shaking chills, profuse sweating, arthralgias, myalgias, fatigue, and weakness; some species more pathogenic than others, particularly in compromised patients.
Clinical specimen	Blood, multiple draws (EDTA)
Epidemiology	Sporadic distribution; tick-to-human and human-to-human transmission
Control	Vector control, avoiding shared needles, checking blood supply

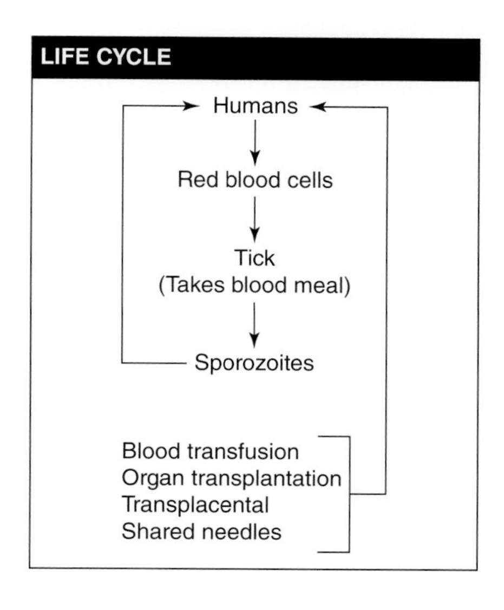

Diagnosis

When blood parasites are discussed, malaria is usually considered the most important and well-known infection. However, another organism also causes human infection. Recognition of the spectrum of disease caused by organisms of the genus *Babesia* has recently resulted in an increase in the number of cases reported. Increased interest has also been generated because the infection can be transmitted via transfusion, can cause serious illness (particularly in compromised patients), and is difficult to treat.

Both thick and thin blood films should be prepared *on admission of the patient* (to the clinic or emergency room or in house), and at least 300 oil immersion fields should be examined on the thick and thin films before a negative report is issued. Since one set of negative films does not rule out babesiosis, additional blood specimens collected over 36 h should be examined. Any of the blood stains, such as Giemsa stain, Wright's stain, or any of the rapid blood stains, can be used for diagnostic work. Blood collected with EDTA anticoagulant is recommended. Also, the proper ratio between blood and anticoagulant is necessary for good organism morphology (fill the tubes to the top).

Potential diagnostic problems with automated differential instruments have been reported. A case of babesiosis and two cases of malaria were missed when blood smears were examined by these methods, resulting in delayed therapy. The number of fields scanned by a technologist on instrument-read smears is quite small; thus, light parasitemia is almost sure to be missed. Although these instruments are not designed to detect intracellular blood parasites, routine use of the automated systems may pose serious diagnostic problems, particularly if the suspected diagnosis is not conveyed to the laboratory. Most documented human cases have a low parasitemia; thus, both thick and thin blood films stained with any of the blood stains must be examined. Although identification to the species level cannot be achieved using microscopy, any patient history related to geographic exposure could suggest various species options.

Preliminary studies are ongoing to develop semiautomated or automated methods for the detection and identification of *Plasmodium* and other blood parasite groups from digital stained blood film images. However, as with the automated hematology analyzers, low parasitemias may yield false-negative results. This may be a potential problem with travelers who present with low parasitemias but who may be seriously ill.

Diagnostic Tips

It is now well established that under blood-banking conditions (4°C for 30 days), *Babesia microti* can remain infective, and transfusion-acquired infection with this parasite could occur from blood stored for the normal storage time.

Often, when the diagnosis of babesiosis is considered, only a single blood specimen is submitted to the laboratory for examination; however, single films or specimens cannot be relied on to exclude the diagnosis, especially with a low parasitemia. There does not have to be any significant history of travel outside the United States. If organisms are seen, they tend to mimic *Plasmodium falciparum* rings. Information regarding travel history and/or possible exposure to tick-infested areas, recent blood transfusion, and splenectomy is very important.

No morphological stages other than ring forms will be seen. The classic arrangement of the four rings (Maltese cross) is not always seen. Numerous Maltese cross configurations have been seen in a case of *Babesia duncani* infection. Both thick and thin blood films should be prepared. At least 200 to 300 oil immersion fields should be examined before the smears are considered negative. *Babesia* parasites may be missed by automated differential instruments. Even with technologist review of the smears, a light parasitemia is very likely to be missed.

Because the ring forms are quite small, it is important that the thick blood film be thoroughly examined; at least 300 fields need to be examined. Also, remember that small forms may be missed using a 60× oil immersion objective, rather than the recommended 100× objective for the stained blood films. Wright's, Wright-Giemsa, Giemsa, or rapid stain can be used; if the WBC morphology is acceptable, any parasites will also exhibit typical morphology. Color variation is not critical.

General Comments

Cases of babesiosis have been documented worldwide, and several outbreaks in humans have occurred in the northeastern United States, particularly in Long Island, Cape Cod, and the islands off the East Coast. Although there are many species of *Babesia*, *B. microti* is the cause of most human infections in the United States. Infections by *B. microti* have also been reported in the Upper Midwest, particularly Minnesota and Wisconsin. A lower incidence of babesiosis is observed in western regions of the United States, where it is usually caused by infection with *B. duncani*. *B. duncani* is harder to treat than *B. microti* and typically requires a longer course of treatment. In some cases, *Babesia* infections can be refractory, and recrudescence of infection may occur.

Infection is transmitted by several species of ticks in which the sexual multiplication cycle occurs. When a tick takes a blood meal, the infective forms are introduced into the human host. The organisms infect the RBCs, in which they appear as pleomorphic, ringlike structures. They resemble the early trophozoite (ring) forms of malarial parasites, particularly *P. falciparum*. The organisms measure 1.0 to 5.0 μm, the RBCs are not enlarged or pale, and the cells do not contain stippling. Malarial pigment is never seen. The early form contains very little cytoplasm and has a very small nucleus. In mature forms, two or more chromatin dots may be seen. Occasionally a tetrad formation, referred to as a Maltese cross, is seen.

Babesia spp. are grouped into two groups. The small *Babesia* spp. (1.0 to 2.5 μm) include *B. gibsoni*, *B. microti*, and *B. rodhaini*. The large *Babesia* spp. (2.5 to 5.0 μm) include *B. bovis*, *B. caballi*, and *B. canis*. These phenotypic classifications are, for the most part, consistent with genetic characterization based on nuclear small-subunit ribosomal DNA sequences. The small babesias are more closely related to *Theileria* spp. than are the larger organisms. The one exception is *Babesia divergens*, which appears small on blood smears but is genetically related to the large babesias. The two primary pathogens of humans are *B. microti* and *B. duncani* (western United States). While significant knowledge about *B. microti* has been gained over the past few years, nothing is known about *B. duncani* biology, pathogenesis, mode of transmission, or sensitivity to recommended drugs. *B. duncani* has also been found in the eastern United States.

Babesiosis is usually self-limiting; there are probably also asymptomatic and undiagnosed carriers, and the mortality rate has been reported at about 5%. However, infections in Europe caused by *B. divergens* are far more serious, with a mortality rate of 42%. Fatal cases of *B. divergens* infection have been reported in France, Britain, Ireland, Spain, Sweden, Switzerland, the former Yugoslavia, and the former Soviet Union.

Additional Information

Babesia infections described in California and in other parts of the world are quite different from those seen in the northeastern United States, where the infection is most often subclinical. Infections in the western United States (*B. duncani*) and Europe (*B. divergens*) tend to present as a fulminating, febrile, hemolytic disease with severe sequelae impacting splenectomized or immunosuppressed individuals. In immunocompetent individuals, the illness is also more serious than that seen with *B. microti* (primarily eastern United States). Early case reports of *B. microti* infection are presented below.

A case of transfusion-induced babesiosis accompanied by disseminated intravascular coagulopathy was detected in 1982 in an elderly patient. Symptoms included fever, chills, nausea, arthralgias, and lethargy, which began after the patient received 2 units of packed RBCs during surgery. One of the donors had a high titer of immunofluorescent antibody (IFA) against *B. microti*, but the infection was confirmed only by hamster inoculation.

The public health notice that was distributed to the Nantucket Board of Health indicated that *Babesia* parasites are transmitted from mice to humans primarily by the bite of the deer tick, *Ixodes scapularis*, and that persons with the infection may be asymptomatic. The notice indicated that prevention still relies on avoiding ticks or removing them promptly once detected, since there are apparently no fully effective tick repellents. If symptoms appear 1 to 2 weeks after a tick bite, a physician should be consulted.

In patients from Nantucket Island, MA, who were not splenectomized, symptoms began 10 to 20 days after a tick bite and continued for several weeks. There was general malaise followed by fever, shaking chills, profuse sweating, arthralgias, myalgias, fatigue, and weakness. Hepatosplenomegaly was present, and five patients had slightly elevated serum bilirubin and transaminase levels due to hemolytic anemia.

During summer 1980, six patients from Shelter Island, NY, and eastern Long Island, NY, were diagnosed as having *B. microti* babesiosis. Symptoms lasted for 19 to 24 days in five patients and included fever, shaking chills, dark urine, and headache, as well as anorexia, malaise, and lethargy. The splenectomized patient also had the most severe illness. Parasitemia could not be detected in one patient, was 5% in four of the six, and was 80 to 90% in the splenectomized patient. Hamster inoculation produced a patent parasitemia in all cases.

By the end of 1991, 13 cases of babesiosis had been reported in Connecticut; this was the largest number of human cases reported in the mainland United States. Information suggests that 12 infections were acquired via tick bites and 1 was acquired from a blood transfusion. Ages ranged from 61 to 95 for those with tick-acquired

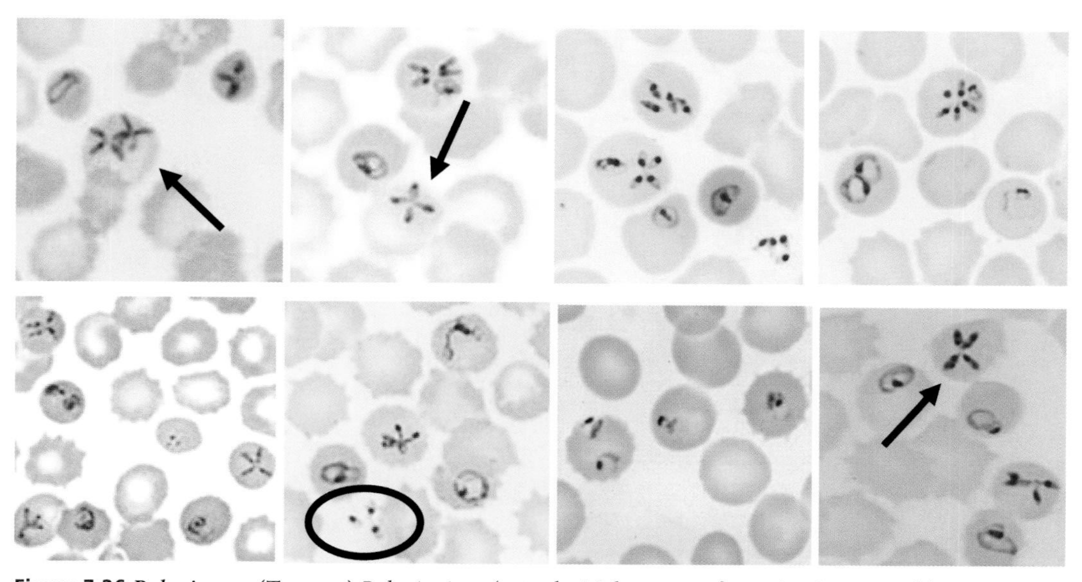

Figure 7.26 ***Babesia*** **spp.** **(Top row)** *Babesia* rings (note the Maltese cross formation in some of the RBCs more commonly seen in *B. duncani* [arrows]. **(Bottom row)** *Babesia* rings (note the three rings outside of the RBC in the second image [oval], a situation that occurs with *Babesia* but very rarely with *Plasmodium* spp. unless the parasitemia is quite high).

infections. Two patients died with active infections, and one died from chronic obstructive pulmonary disease. *B. microti* was isolated in Syrian hamsters given blood from 7 of 12 patients tested. Babesiosis is thus endemic in New Jersey and other areas of the northeast.

Blood donors in the United States have not been screened routinely for human babesiosis; however, there are now FDA-cleared serological and molecular tests to detect *B. microti* in the blood supply. *B. duncani* will most likely be included as well. Because donors of blood, plasma, living organs, and tissues can be asymptomatic and have very low parasitemia, they need to be screened for *B. duncani* and *B. microti*. In states where the organism is endemic, donor seroprevalence can be as high as 2%, with incidence ranging from 1 case per 604 to 1 case per 100,000 RBC units transfused.

In 2018, an Ad Hoc Babesia Policy Working Group concluded that a regional approach to donor screening in states where the disease is endemic was appropriate, since the intervention would be applied where the risk was highest and cost appropriately allocated to the risk (see Ward et al., Suggested Reading). Nucleic acid testing using an rRNA template was the recommended most cost-effective intervention. This approach resulted in no wasted units and captured numbers of infections similar to those detected with antibody assays plus DNA-based PCR.

Toxoplasma gondii

Pathogenic	Yes
Disease	Toxoplasmosis
Acquired	Ingestion of oocysts from cat feces and/or contaminated food and drinking water; ingestion of rare or raw infected meats; congenital; transfusion, transplantation
Body site	Multiple tissues (muscle, brain)
Symptoms	Asymptomatic to severe (immunocompromised, congenital)
Clinical specimen	Serum, serologic testing recommended
Epidemiology	Worldwide
Control	Hand hygiene; proper cooking of meat; keep cats indoors; change litter box every 2 days (wear mask/gloves)

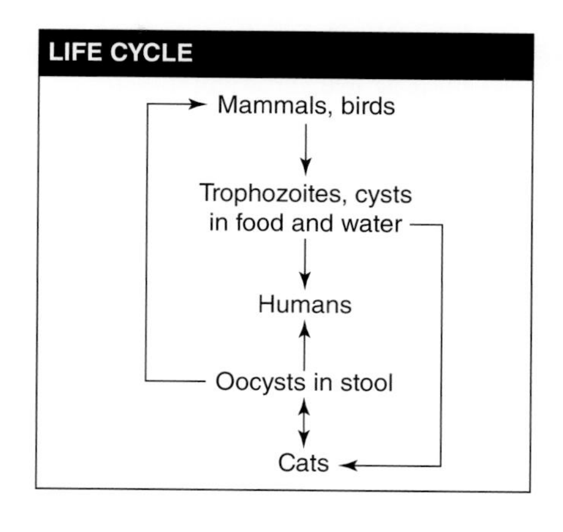

Diagnosis

Toxoplasmosis, caused by infection with the coccidian *Toxoplasma gondii*, is a significant public health problem worldwide. An estimated 8 to 22% of people in the United States are infected; the same prevalence exists in the United Kingdom. In Central America, South America, and continental Europe, estimates of infection range from 30 to 90%.

These infections have significant consequences affecting mortality and quality of life. In the United States, over a million people are infected each year, and approximately 2,839 develop symptomatic ocular disease annually.

Toxoplasmosis can be diagnosed by using various serologic procedures, histologic findings from the examination of biopsy specimens, and isolation of the organism, either in a tissue culture system or by animal inoculation. Individuals with positive serologic tests have been exposed to *Toxoplasma* organisms and may have the resting stages within their tissues. This is why histologic identification of the organism and its recovery either in a tissue culture system or by animal inoculation may be misleading, since the organisms could be isolated yet might not be the etiologic agent of disease. Thus, serologic tests are often recommended as the diagnostic approach of choice; however, serologic diagnosis of toxoplasmosis is very complex. Two situations in which organism detection may be significant are (i) when smears and/or tissue cultures from cerebrospinal fluid (CSF) are tachyzoite positive and (ii) when tachyzoites are seen in Giemsa-stained smears of bronchoalveolar lavage (BAL) fluid from patients with acute pulmonary disease, some tachyzoites being extracellular and some intracellular.

In an immunodeficient or immunosuppressed patient, a presumptive diagnosis of toxoplasmosis can be made by observing an elevated serologic titer and the presence of the clinical syndrome, which would include neurologic symptoms. However, in certain patients with monoclonal gammopathies, titers of antibody to *T. gondii* may be extremely high but may not reflect the cause of the clinical condition.

Cysts are formed in chronic infections, and organisms within the cyst wall are strongly periodic acid-Schiff stain positive. During the acute phase, there may be groups of tachyzoites that appear to be cysts; however, they are not strongly periodic acid-Schiff stain positive, and these have been termed pseudocysts.

Diagnostic Tips

Toxoplasmosis is usually asymptomatic or with mild flu-like symptoms. Fewer than 10% of infected persons develop a pattern of symptoms including fever, headache, and aching limbs, along with lymphadenitis, especially of the cervical and occipital lymph nodes. These lymph nodes become hard and tender to the touch and may remain swollen for several weeks. In a few cases, an uncharacteristic maculopapular rash, reactive arthritis, hepatosplenomegaly, myocarditis, and pneumonia have been seen. Clinically significant expressions include congenital and ocular toxoplasmosis and, in immunosuppressed patients, organ symptoms that can often be life-threatening, such as encephalitis and/or pneumonia.

Diagnostic testing involves primarily serological procedures, such as enzyme-linked immunosorbent assay (ELISA) and immunoblotting. If an acute infection is suspected, the patient's serum should be tested for *T. gondii*-specific IgG and IgM antibodies.

In prenatal testing, if high IgM and low IgG antibody titers are found, this suggests acute toxoplasmosis; the suspicion can be confirmed in a follow-up test 2 to 3 weeks later by a significant rise in IgG antibodies and possibly also IgA antibodies.

In ocular testing, clinical examination often shows characteristic retinal findings; therefore, no further diagnostic workup is necessary. Apparently, ocular toxoplasmosis is the most common etiology of posterior uveitis worldwide. PCR of ocular fluids is used in patients with atypical clinical presentations; however, a negative PCR result for *Toxoplasma gondii* does not rule out the diagnosis, and the clinical response to therapy may be necessary to confirm the diagnosis.

In immunocompromised patients, serological tests are unreliable, especially in those who are HIV positive. It is important to have a very sensitive serological test for IgG to avoid false-negative serology, since immunocompromised patients present with a very low IgG level. The results are often equivocal, regardless of the technique used, when the IgG concentrations are close to the threshold value of the assay. Toxoplasmosis is the most common cause of neurological disease in HIV-positive patients and often results in severe pathology or death. In these cases, the use of PCR is recommended for demonstration of the actual organisms.

General Comments

There are three infectious stages of *T. gondii*: the tachyzoites (in groups or clones), the bradyzoites (in tissue cysts), and the sporozoites (in oocysts). Tachyzoites rapidly multiply in any cell of the intermediate host (many other animals and humans) and in nonintestinal epithelial cells of the definitive host (cats). Bradyzoites are found within the tissue cysts and multiply very slowly; the cyst may contain few to hundreds of organisms, and intramuscular cysts may reach 100 µm in size. Although the tissue cysts may develop in visceral organs such as the lungs, liver, and kidneys, they are more prevalent in neural and muscular tissues, including the brain, eyes, and skeletal and cardiac muscle. Intact tissue cysts can persist for the life of the host and do not cause an inflammatory response.

When the tachyzoites are actively proliferating, they invade adjacent cells from the original infected cell as it ruptures. This process creates continually expanding focal lesions. Once the cysts are formed, the process becomes quiescent, with little or no multiplication and spread. In immunocompromised or immunodeficient patients, a cyst rupture or primary exposure to the organisms often leads to lesions. The organisms can be disseminated via the lymphatics and the bloodstream to other tissues. Disintegration of cysts may give rise to clinical encephalitis in the presence of apparently adequate immunity.

Toxoplasmosis is categorized into four groups: (i) disease acquired in immunocompetent patients; (ii) acquired or reactivated disease in immunosuppressed or immunodeficient patients; (iii) congenital disease; and (iv) ocular disease.

Additional Information

In 90% of immunocompetent individuals, no clinical symptoms are seen during the acute infection. However, 10 to 20% of patients with acute infection may develop painless cervical lymphadenopathy, which is benign and self-limiting, with symptoms resolving within weeks to months. Acute visceral manifestations are rarely seen.

In immunocompromised patients, underlying conditions that may influence disease include malignancies (such as Hodgkin's disease, non-Hodgkin's lymphomas, leukemias, and solid tumors), collagen vascular disease, organ transplantation, and AIDS. In immunocompromised patients, the CNS is primarily involved, with diffuse encephalopathy, meningoencephalitis, or cerebral mass lesions. More than 50% of these patients show altered mental state, motor impairment, seizures, abnormal reflexes, and other neurologic sequelae. Most patients receiving chemotherapy for toxoplasmosis will improve significantly or have complete remission. However, in those with AIDS, therapy must be continued for long periods to maintain a clinical response.

In transplant recipients, disease severity depends on previous exposure to *T. gondii* by donor and recipient, the type of organ transplanted, and the level of immunosuppression of the patient. Disease can be due to reactivation of a latent infection or an acute primary infection acquired directly from the transplanted organ. Stem cell transplant (SCT) recipients are particularly susceptible to severe toxoplasmosis, primarily due to reactivation of a latent infection. If SCT patients have a positive serology prior to transplantation, they are at risk for severe disseminated disease. Seronegative cardiac transplant recipients who receive an organ from a seropositive donor may develop toxoplasmic myocarditis; the symptoms may mimic organ rejection.

AIDS patients may develop disease when their $CD4^+$ T-lymphocyte count falls below 100,000/ml. Fever and malaise usually precede the first neurologic symptoms; headache, confusion, seizures, or other focal signs strongly suggest the diagnosis of toxoplasmosis.

Toxoplasma encephalitis is fatal if untreated. Psychiatric manifestations of *T. gondii* are also seen in immunocompromised individuals with AIDS in whom latent infections have become reactivated. Altered mental status may occur in approximately 60% of patients, with symptoms including delusions, auditory hallucinations, and thought disorders.

Congenital infections result from the transfer of parasites from mother to the fetus when she acquires a primary infection during pregnancy. At birth or soon thereafter, symptoms in these infants may include retinochoroiditis, cerebral calcification, and occasionally hydrocephalus or microcephaly. Symptoms of congenital CNS involvement may not appear until several years later.

Congenital transmission can occur (rarely) even though the mother is immune; reinfection of the mother during pregnancy is possible, particularly if she is exposed to large numbers of infective cysts and/or oocysts.

Hydrocephalus, cerebral calcifications, and chorioretinitis resulting in mental retardation,

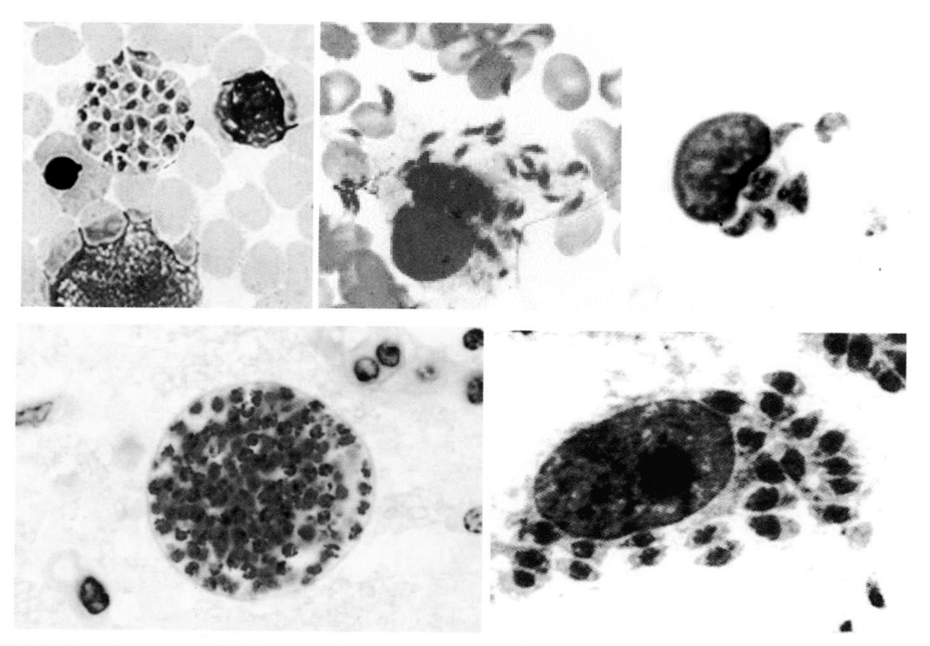

Figure 7.27 ***Toxoplasma gondii.*** **(Top row)** *T. gondii* tachyzoites in bone marrow; tachyzoites in bone marrow (reprinted from Chin CK, Finlayson J, *Ann Hematol Oncol* 5[8]:1223, 2018 © 2018 Finlayson et al, CC-BY 4.0); tachyzoites in blood smear. **(Bottom row)** *T. gondii* tissue cyst containing bradyzoites; intracellular tachyzoites in tissue culture.

epilepsy, and impaired vision represent the most severe form of the disease. Cerebral lesions may calcify, providing retrospective signs of congenital infection.

Chorioretinitis in immunocompetent patients is generally due to an earlier congenital infection. Patients may be asymptomatic until the second or third decade, at which time cysts may rupture, with eye lesions developing. Chorioretinitis is usually bilateral in patients with congenital infections and unilateral in patients with recently acquired infections.

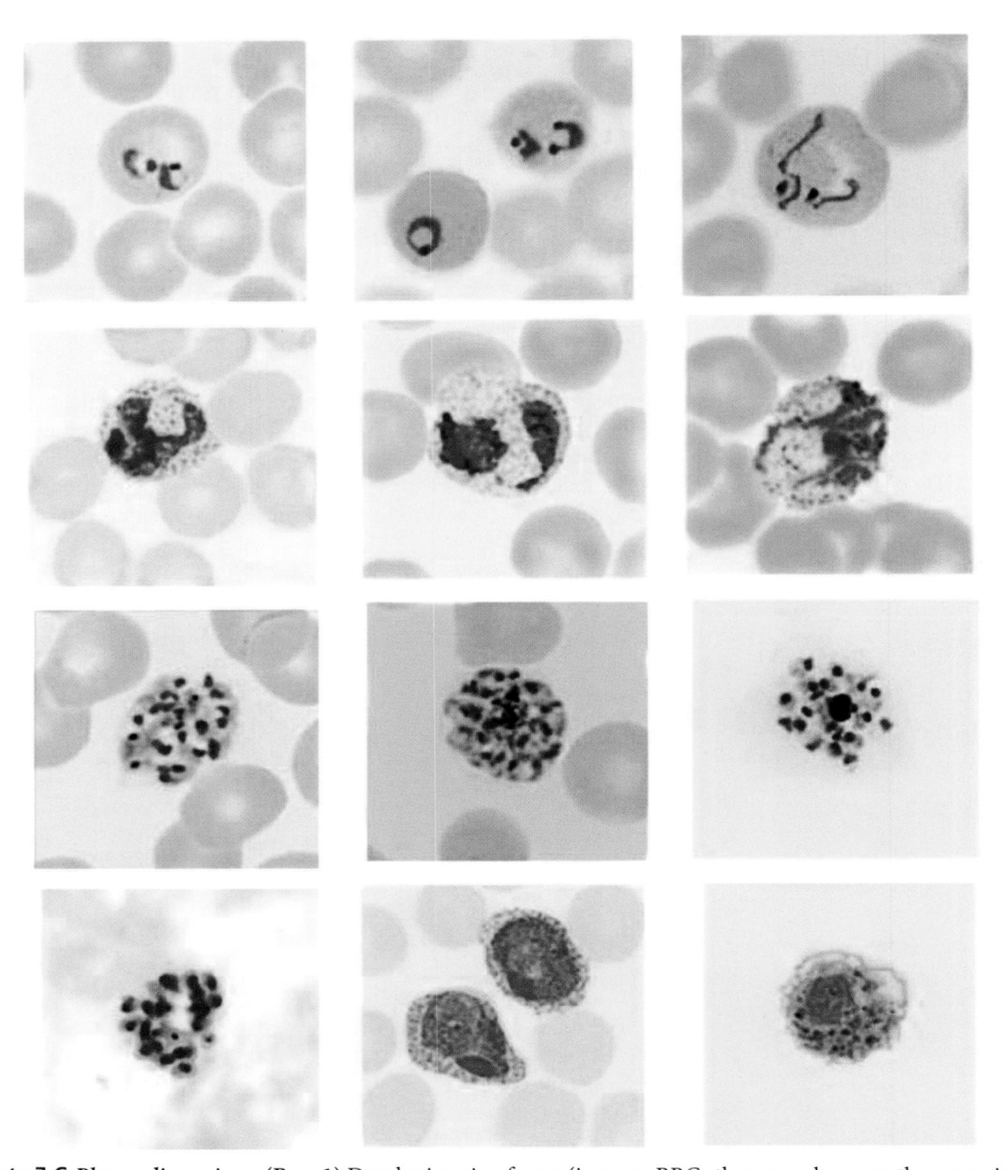

Plate 7.G ***Plasmodium vivax*****.** **(Row 1)** Developing ring forms (in some RBCs there may be more than one ring; this is more commonly seen with *Plasmodium falciparum* but does occur in *P. vivax* infections). **(Row 2)** Developing trophozoites (note ameboid forms and Schüffner's dots). **(Row 3)** Developing schizonts. **(Row 4)** (left to right) Mature schizont; female macrogametocytes; and male microgametocyte.

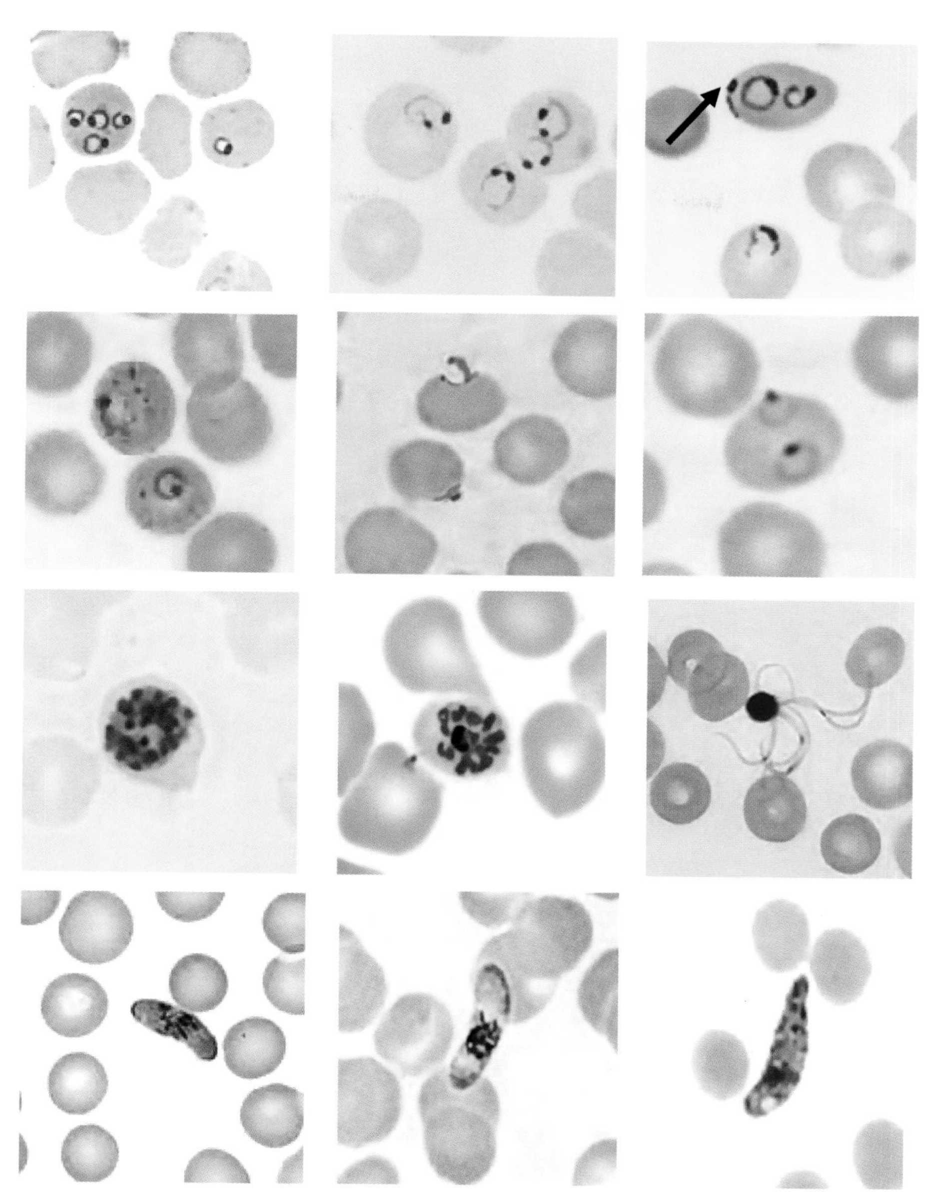

Plate 7.H *Plasmodium falciparum*. **(Row 1)** Developing ring forms of *P. falciparum* (note multiple rings per cell, "headphone" rings, and, in the third image, an appliqué form [arrow]). **(Row 2)** Ring forms (in the first image note the Maurer's clefts [not true stippling]); developing ring forms; developing ring forms. **(Row 3)** Developing schizont; mature schizont; exflagellation of the male microgametocyte. Exflagellation can occur in any of the species of *Plasmodium* in a tube of EDTA blood if the cap is removed and the blood cools to room temperature; parasites react as if they are in the mosquito. These fragments have been misidentified as spirochetes. **(Row 4)** Gametocyte; gametocyte (both crescent-shaped) (although the RBC outline is not visible, these stages are within the RBC); ookinete (occurs in the mosquito cycle and can mimic a gametocyte).

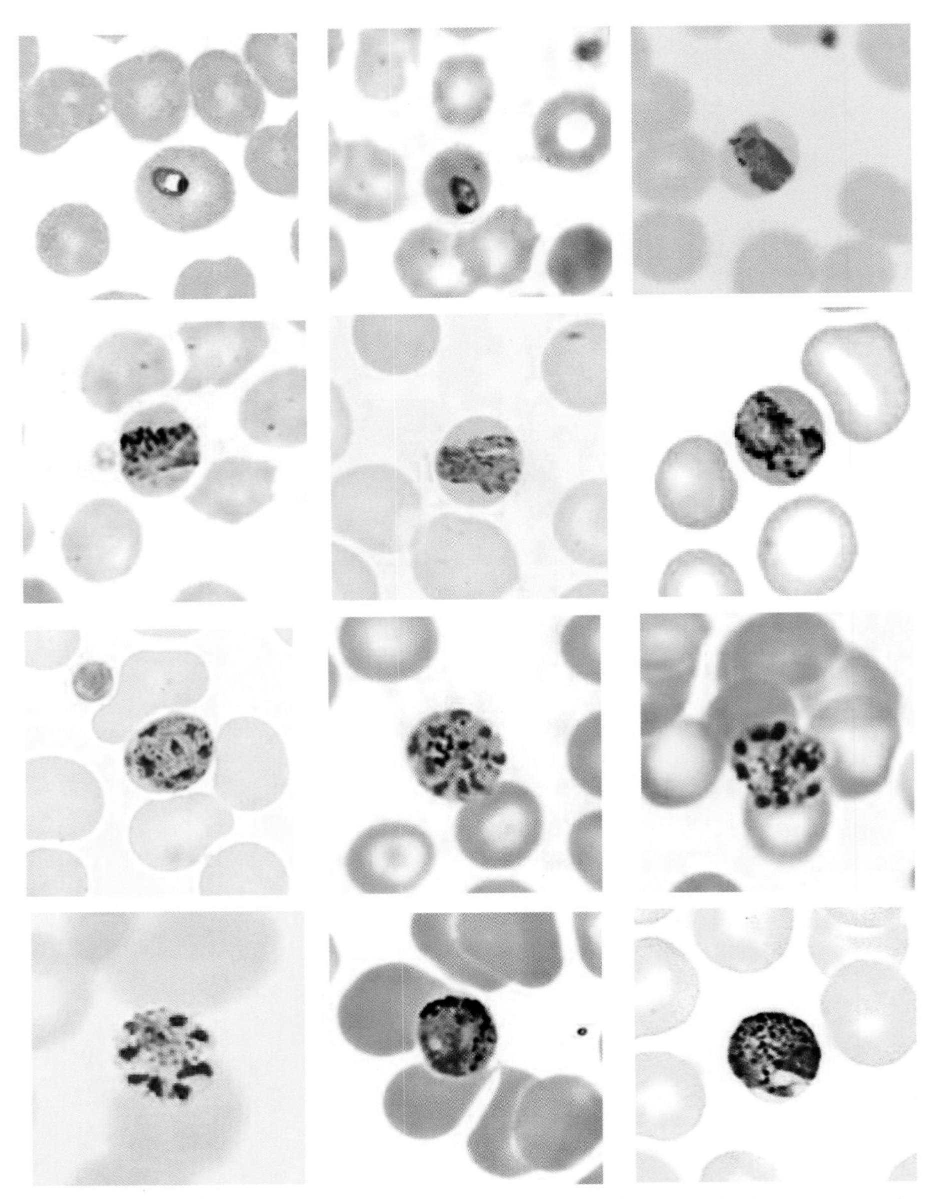

Plate 7.1 *Plasmodium malariae*. **(Rows 1 and 2)** Developing trophozoites (note the band form configuration). **(Row 3)** Developing schizonts. **(Row 4)** Mature rosette schizont and mature male and female gametocytes.

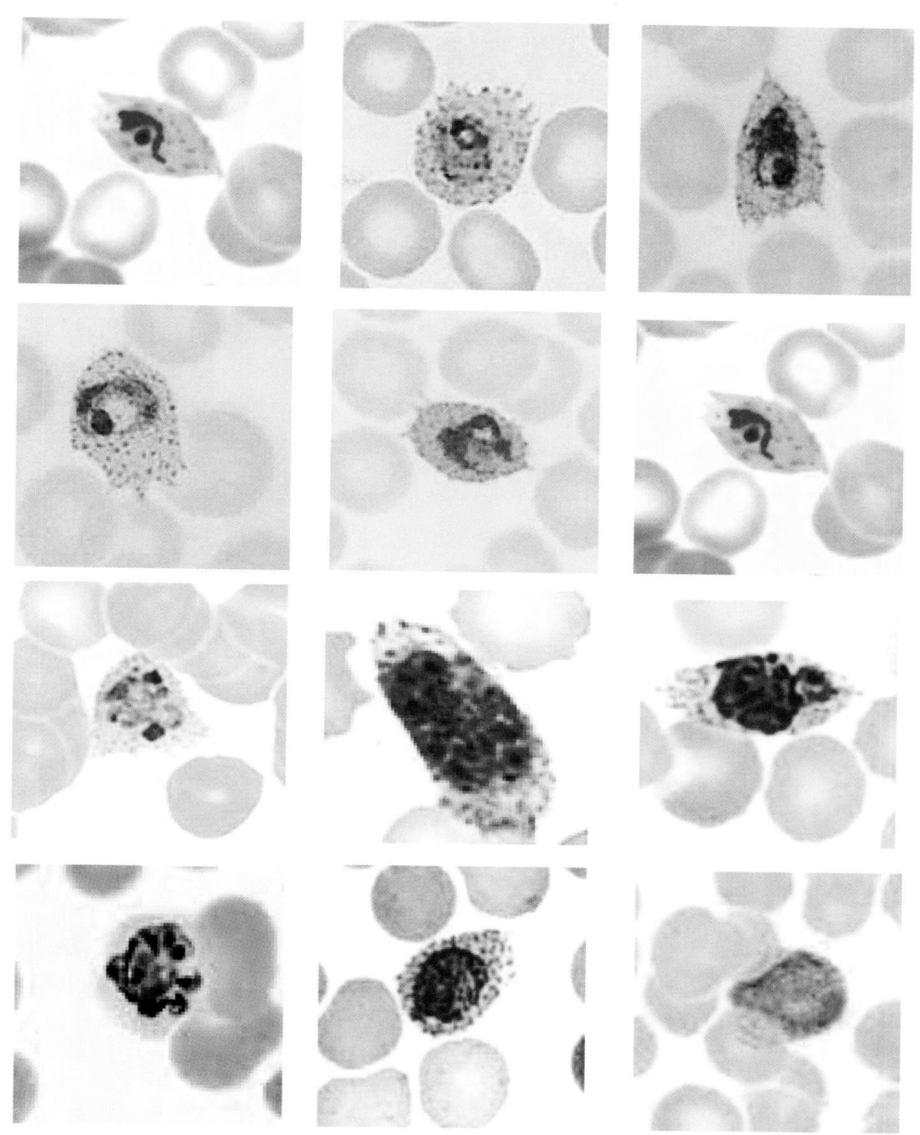

Plate 7.J ***Plasmodium ovale.*** **(Rows 1 and 2)** Developing trophozoites. The parasites are much less ameboid than *Plasmodium vivax*; note the presence of Schüffner's dots and the oval RBCs. **(Row 3)** Developing schizonts. **(Row 4)** Mature schizont and male and female gametocytes.

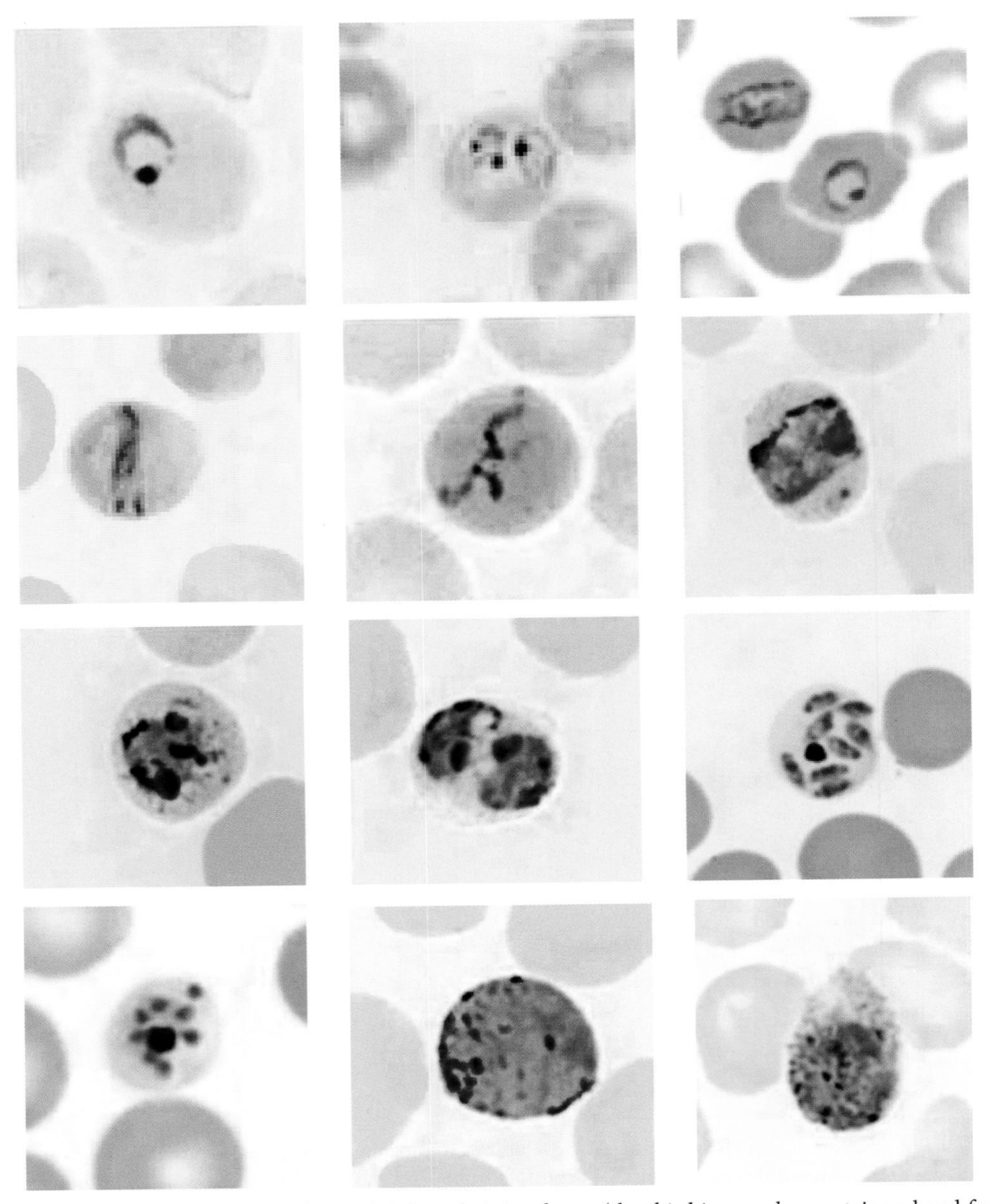

Plate 7.K *Plasmodium knowlesi*. (**Row 1**) *P. knowlesi* ring forms (the third image also contains a band form). (**Row 2**) Developing trophozoites in band form configurations. (**Row 3**) Developing schizonts. (**Row 4**) Immature schizont and male and female gametocytes (last image courtesy of Wadsworth Center, New York State Department of Health).

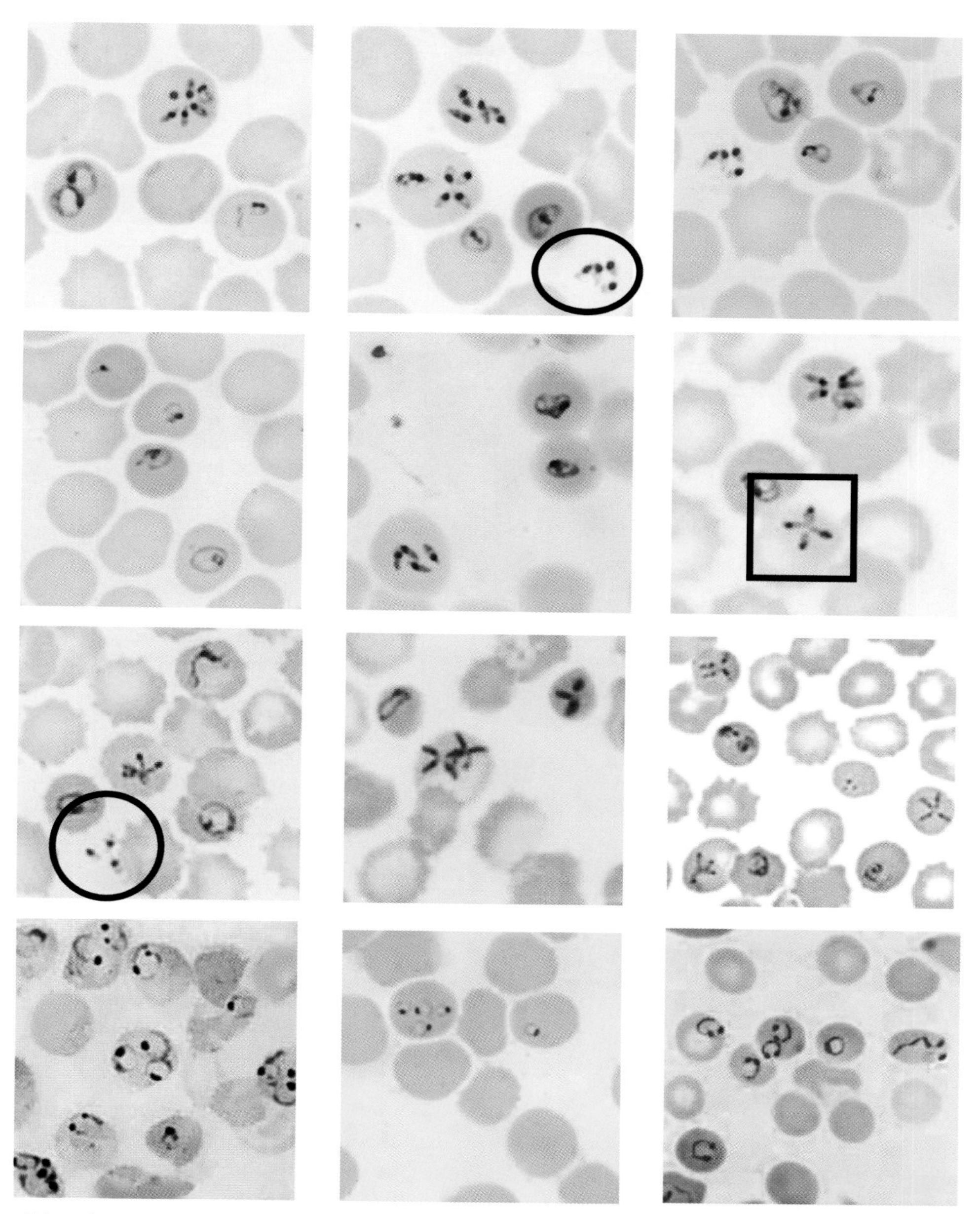

Plate 7.L *Babesia* **spp. (Rows 1, 2, and 3)** Examples of blood films containing the ring-like forms of *Babesia* spp. Various ring forms are shown, with multiple rings per cell and some rings present outside the red blood cells, which is relatively common with *Babesia* spp. (oval). The typical Maltese cross configuration of the four rings (square) is also seen. **(Row 4)** *Plasmodium falciparum* rings for comparison. It is rare to see *Plasmodium* rings outside the RBCs unless the parasitemia is quite high.

Leishmania spp.

Pathogenic	Yes
Disease	Leishmaniasis (cutaneous, mucocutaneous, visceral, American tegumentary) (WHO neglected tropical disease)
Acquired	Bite of sand flies; blood, shared needles, congenital infections
Body site	Reticuloendothelial system
Symptoms	Papule, ulcers (cutaneous); destruction of cartilage (mucocutaneous); splenomegaly and hepatomegaly (visceral)
Clinical specimen	*Cutaneous/mucocutaneous/ American tegumentary*: wall of the lesion (punch biopsy) *Visceral*: blood (EDTA), bone marrow aspirate
Epidemiology	Old and New World; sand fly-to-human and human-to-human transmission
Control	Vector control, avoiding shared needles, checking blood supply

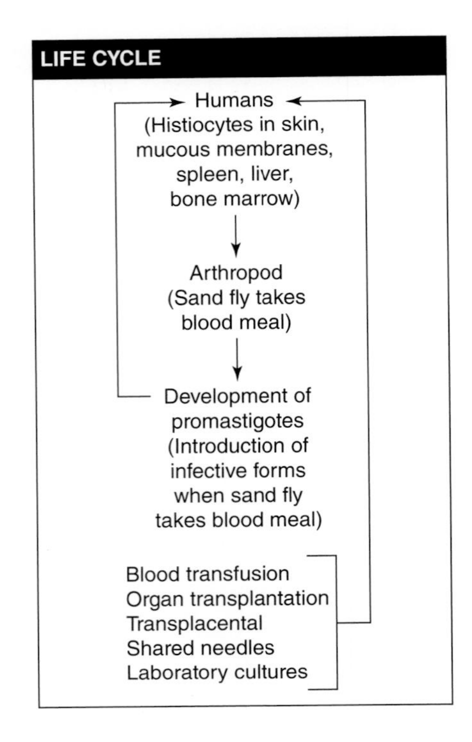

Diagnosis

Leishmaniasis has been identified as high-priority disease by the World Health Organization. It is caused by protozoa belonging to the genus *Leishmania,* and it is transmitted by the bite of the phlebotomine sand fly, a 2- to 3-millimeter-long insect vector found throughout the world's intertropical and temperate regions. The disease occurs in six clinical forms: self-healing or chronic cutaneous leishmaniasis (CL), diffuse cutaneous leishmaniasis (DCL), mutilating mucosal or mucocutaneous leishmaniasis (ML MCL), life-threatening visceral leishmaniasis (VL), post-kala-azar dermal leishmaniasis (PKDL), and American tegumentary leishmaniasis (ATL) (some publications indicate that this is another designation for cutaneous leishmaniasis [ATL/CL]). These forms vary in degree of severity, with VL being by far the most devastating and having the highest mortality.

Leishmaniasis is endemic in large areas of the tropics, subtropics, and Mediterranean basin, covering close to 100 countries. There are at present ~350 million people at risk and ~12 million cases, with an estimated worldwide annual incidence of 0.7 to 1.2 million cases of cutaneous leishmaniasis and 0.2 to 0.4 million cases of visceral leishmaniasis.

In areas where infection is endemic, the diagnosis may be made on clinical grounds. The infection may go unrecognized in other areas. Definitive diagnosis depends on detecting either the amastigotes in clinical specimens or the promastigotes in culture. Cutaneous and mucocutaneous leishmaniasis may have to be differentiated from tropical ulcers, syphilis, yaws, leprosy, South American blastomycosis, sporotrichosis, pyogenic nodules or abscesses, furuncles, insect bites, cutaneous tuberculosis, and atypical mycobacterial infections of the skin.

Detection of parasites in aspirates, scrapings, or biopsy tissues depends on organism distribution, the host immune response, bacterial infections in the lesions, and whether the specimen is from an active or healing lesion. Samples should be collected from the most active lesion. All lesions should be cleaned with 70% alcohol, and extraneous material should be removed to minimize the risk of bacterial or fungal contamination of subsequent cultures. Material collected from the center of a necrotic ulcer may reveal only pyogenic organisms. If specimens are collected from the margins by aspiration, scraping, or biopsy, local anesthesia may be used. Three biopsy specimens should be divided (for cultures, touch preparations, and histopathology). The core of tissue from a biopsy can be used to make imprints or touch preparations on a slide.

Diagnostic Tips

Cutaneous leishmaniasis is the most common leishmanial syndrome worldwide and the one most likely to be encountered in patients in North America. The skin lesions of CL are usually painless and chronic, often occurring at sites of infected sand fly bites. Slow spontaneous healing as cell-mediated immunity develops is the usual natural history, accelerated by antileishmanial therapy.

Cutaneous. Punch biopsy specimens should be collected at the margins or base (one sterile for culture, two nonsterile for touch/squash preparations and histopathology; two sent to microbiology, one sent to pathology) of the most active lesion (papule or ulcer) The lesion must be cleaned of cellular debris and eschar or exudates prior to tissue collection, to reduce the chances of culture contamination with fungi or bacteria. Make sure the tissue is kept moist to prevent drying.

Mucosal. Biopsy specimens should be obtained (including specimens for culture).

Visceral. Tissue aspirates or biopsy specimens should be obtained for smears, histopathology, parasite culture, and molecular testing. Bone marrow is preferred, while liver, enlarged lymph nodes, and whole blood buffy coat are other possible tissue options. Serum should also be collected for antibody testing. In immunocompetent patients, buffy coat examination, culture, and molecular testing are highly recommended. Patients with kala-azar have hypergammaglobulinemia; the Formol-gel procedure may be helpful.

All tissue imprints or smears should be stained with one of the blood stains. If the WBCs look fine, then any parasites present will also look fine and will exhibit the same staining characteristics as the WBCs. Amastigotes should be found within macrophages or close to disrupted cells.

Specimens must be taken aseptically for culture, and control organism cultures should be set up at the same time. Cultures need to be checked weekly for 4 weeks before they are declared negative. If organisms are isolated in culture, species identification can be done using DNA-based assays or isoenzyme analysis. Molecular tests should be performed; they are the most sensitive options. Skin testing is not recommended, since there are no standardized, approved, or commercially available skin test products for North America. A list of reference diagnostic laboratories in North America can be found at the following site: https://academic.oup.com/cid/article/63/12/e202/2645609 (accessed 2 October 2019).

Tests for antileishmanial antibodies are not recommended for CL and should not be performed as the sole diagnostic assay for VL. Antibodies may be undetectable or present at low levels in persons with VL who are immunocompromised because of concurrent HIV/AIDS or for other reasons. Serologic tests cannot be used to assess treatment efficacy. Also, in some patients, antileishmanial antibodies can be detected years after successful therapy.

General Comments

The parasite life cycle has two distinct phases. The organism is engulfed by reticuloendothelial cells (RE cells) of the mammalian host, where it can be found in the amastigote form (Leishman-Donovan body) within the phagocytic cell. Amastigote forms are small, round or oval bodies of 3 to 5 μm; the large nucleus and small kinetoplast can be seen with Giemsa or Wright's stain, and the short intracytoplasmic portion of the flagellum may also be seen. The amastigote multiplies by binary fission in the macrophage parasitophorous vacuole until the cell is destroyed; liberated parasites are phagocytized by other RE cells or ingested by the insect vector.

On ingestion during a blood meal by the phlebotomine sand fly vector, the amastigote transforms into the promastigote, a motile, slender organism (10 to 15 μm) with a single anterior flagellum. Promastigotes multiply by longitudinal fission in the insect gut, attaching to the gut wall by their flagella. Stages found in the vector vary from rounded or stumpy forms to elongated, highly motile metacyclic promastigotes. The metacyclic promastigotes migrate to the sand fly hypostome, where they are inoculated into humans during the next blood meal.

Leishmaniasis is diagnosed each year in the United States in immigrants from countries with endemic infection, military personnel, and travelers. There is potential for more infections in areas of endemic infection in Texas and Arizona.

VL caused by *Leishmania infantum* has been reported in foxhound kennels in various locations within the United States. Organisms can remain latent for years; even when the potential exposure is long past, leishmaniasis should be considered, particularly in immunocompromised patients.

Human cutaneous leishmaniasis is endemic in the United States (as of 2018) and, at least regionally, is acquired endemically more frequently than it is via travel. Data strongly suggest making leishmaniasis a federally reportable disease. There are indications that the number of citizens exposed to leishmaniasis will double by 2080. Despite the World Health Organization classifying the United States as a country where leishmaniasis is endemic in 2015, as of December 2020 leishmaniasis is still not a federally reportable disease. Autochthonous cases in the United States have been identified in Texas; there was no history of travel outside the United States for these patients.

Additional Information

Leishmaniasis is principally a zoonosis, although in certain areas it is endemic, with human-vector-human transmission. It is caused by a diverse group of agents of increasing public health importance (more than 400,000 new cases are reported annually). Approximately 350 million people are at risk of acquiring leishmaniasis, with 12 million currently infected. Depending on the species, *Leishmania* infection can result in cutaneous, mucocutaneous, or visceral disease.

From August 2002 to February 2004, over 500 confirmed cases of CL were found in military personnel serving in Afghanistan, Iraq, and Kuwait. Most were probably acquired in Iraq, with *Leishmania major* being confirmed as the agent by isoenzyme electrophoresis. Based on data from Fort Campbell, KY, ca. 1% of troops returning from Iraq were diagnosed with CL, most by laboratory confirmation including PCR.

VL is very prevalent in AIDS patients in some areas, with many cases being subclinical. Subclinical VL occurs at any stage of HIV-1 infection, but symptomatic cases appear mainly when severe immunosuppression is present. HIV-*Leishmania* coinfection is increasingly common in the Mediterranean basin, especially in Spain, France, and Italy.

Leishmaniasis should be suspected in individuals who resided in or traveled to areas where the disease is endemic. Diagnosis of VL would be supported by findings of remittent fevers, hypergammaglobulinemia with anemia, circulating immune complexes, rheumatoid factors, weight loss, leukemia, and hypersplenism. Differential diagnosis should include African trypanosomiasis, brucellosis, endocarditis, malaria, schistosomiasis, tuberculosis, typhoid, cirrhosis, leukemia, and lymphoma. Lesions caused by post-kala-azar dermal leishmaniasis are confused with those of leprosy.

Reduction of reservoir hosts (rodents and dogs) and treatment of infected individuals have been successful in limiting transmission.

Cutaneous lesions are usually single, self-limiting papules, nodules, or ulcers, found mostly on the face and ears (60% of cases); they are painless and often heal spontaneously within a few months.

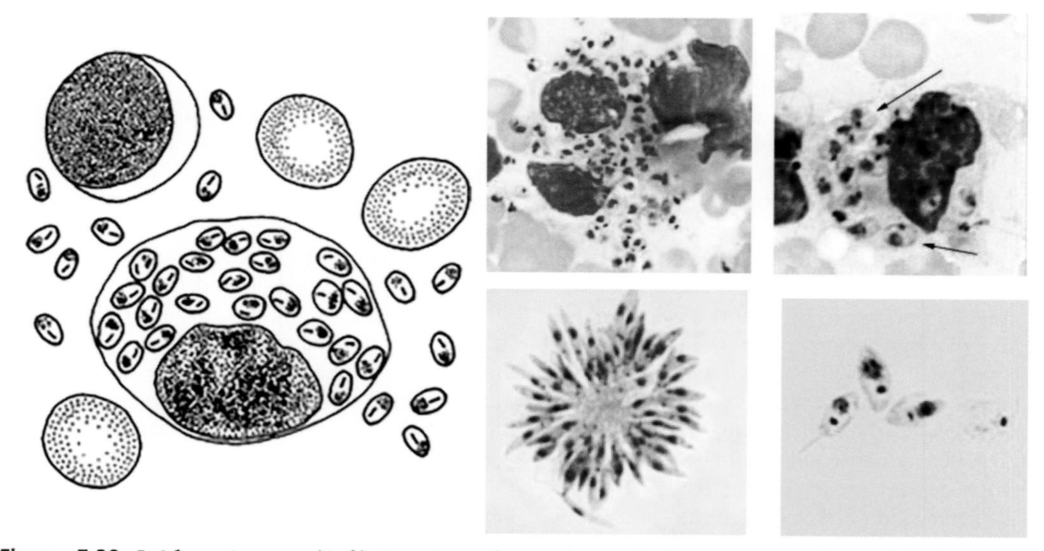

Figure 7.28 *Leishmania* spp. **(Left)** Drawing of amastigotes in bone marrow (visceral leishmaniasis). **(Top row)** Amastigotes in bone marrow (arrows). **(Bottom row)** Promastigote stages from culture.

Early signs of ML are nasal inflammation and stuffiness. Metastatic spread to the nasal or oral mucosa may occur immediately or many years later. Mucosal disease is more prevalent in males than in females; patients are both serologically and skin test positive. There is progressive ulceration and erosion of the soft tissue and cartilage, leading to loss of the lips, soft parts of the nose, and soft palate. Mucosal lesions do not heal spontaneously; secondary bacterial infections are frequent and may be fatal. Aspiration pneumonia is a common complication. Mortality is low, but disfigurement can be substantial.

Symptoms of VL may be vague, while acute onset may resemble typhoid fever, malaria, acute Chagas' disease, amebic liver abscess, or other febrile diseases. Symptoms include fever, anorexia, malaise, weight loss, and often diarrhea. Fever may occur at irregular intervals; once infection is established, a double (dromedary) or triple fever peak may be seen daily. Clinical signs include nontender hepatomegaly and splenomegaly, lymphadenopathy, and occasional acute abdominal pain; darkening of facial, hand, foot, and abdominal skin (kala-azar) is often seen in light-colored persons in India. Patients with kala-azar have anemia (normocytic and normochromic unless there is an underlying iron deficiency), eosinopenia, neutropenia, thrombocytopenia, and hypergammaglobulinemia (polyclonal B-cell activation).

Trypanosoma brucei gambiense (West) *T. brucei rhodesiense* (East)

Pathogenic	Yes
Disease	African trypanosomiasis (WHO neglected tropical disease)
Acquired	Tsetse fly; blood, shared needles, congenital infections
Body site	Blood, CNS
Symptoms	Acute febrile illness, leading to sleeping sickness (chronic CNS invasion)
Clinical specimen	Blood (buffy coat preparations, culture)
Epidemiology	Africa; vector-to-human and human-to-human transmission
Control	Vector control, avoiding shared needles, checking blood supply

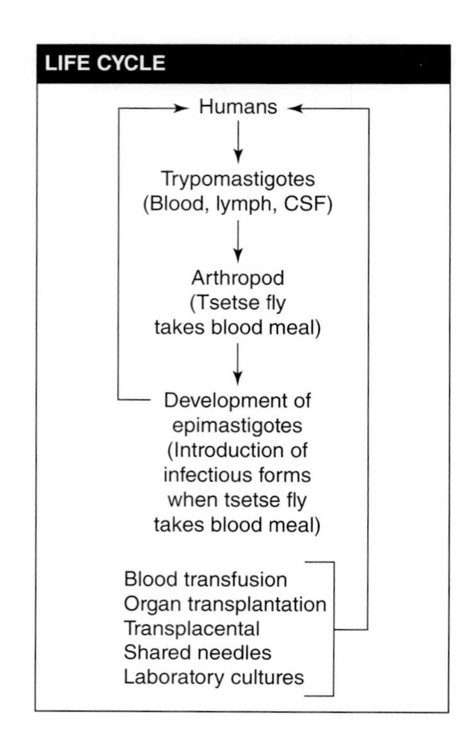

Diagnosis

Human African trypanosomiasis (HAT) caused by *Trypanosoma brucei gambiense* is a more chronic infection than that caused by *Trypanosoma brucei rhodesiense*. The initial stage often presents with a painless eschar at the point of infection; patients generally do not recall such a lesion, primarily due to the lack of symptoms. The first stage includes irregular fever, enlargement of the lymph nodes (particularly those of the posterior triangle of the neck [Winterbottom's sign]), delayed sensation to pain (Kerandel's sign), erythematous skin rashes, headaches, malaise, and arthralgias. In the second stage, patients will develop somnolence, fatigue, neurological deficits, tremors, ataxia, seizures, comas, and eventually death.

HAT caused by *T. brucei gambiense* is a more acute, rapid infection. Patients typically present with a painful eschar and a rapidly progressing illness characterized by fevers, rash, fatigue, and myalgias. Within a few weeks to months, the disease progresses to the second stage, with symptoms identical to that of *T. brucei gambiense* HAT but with a much-accelerated course that quickly leads to death.

Definitive diagnosis depends on demonstration of trypomastigotes in blood, lymph node aspirates, sternum bone marrow, and CSF. There is a better chance of detecting organisms in body fluids in infections caused by *T. brucei rhodesiense* than *T. brucei gambiense*. Because of periodicity, parasite numbers in the blood may vary; therefore, multiple specimens should be collected and a number of techniques used to detect the trypomastigotes.

Health care personnel must use standard precautions when handling specimens from suspected cases of African trypanosomiasis, because the trypomastigote is highly infectious. Blood can be collected from either finger-stick or venipuncture (with EDTA). Multiple blood examinations should be performed before trypanosomiasis is ruled out. Parasites are found in large numbers in blood during the febrile period and in small numbers when the patient is afebrile. If CSF is examined, 5 ml or more should be collected.

In addition to thin and thick blood films, a buffy coat concentration method is recommended. Parasites can be detected on thick blood smears when numbers are greater than 2,000/ml and with hematocrit capillary tube concentration when they are greater than 100/ml. CSF examination must be done with centrifuged sediments.

Diagnostic Tips

Trypomastigotes may be detected in aspirates of the chancre and enlarged lymph node in addition to blood and CSF. Concentration techniques, such as centrifugation of CSF and blood, should be used in addition to thin and thick smears (blood stains). Remember that trypomastigotes are present in largest numbers in the blood during febrile periods; they tend to be more numerous in the blood in Rhodesian trypanosomiasis. If the parasitemia is low, examination of multiple daily blood samples may be necessary to detect the parasites. Blood and CSF specimens should be examined during therapy and 1 to 2 months post-therapy. If CSF is examined, a volume greater than 1 ml, preferably 5 ml or more, should be collected. It is important to remember that CSF examination must be conducted with centrifuged sediments.

Parasites can be detected on thin blood films with a detection limit at approximately 1 parasite/200 microscopic fields (high dry power magnification, ×400); they can be detected on thick blood smears when the numbers are greater than 2,000/ml and with hematocrit capillary tube concentration when they are greater than 100/ml.

T. brucei rhodesiense is more adaptable to cultivation than is *T. brucei gambiense*. A number of media have been described; however, cultivation is not a practical diagnostic approach. Again, the use of precautions specific for blood-borne pathogens is mandatory.

A simple and rapid test, the card indirect agglutination trypanosomiasis test (TrypTect CIATT) (latex agglutination) (Brentec Diagnostics, Nairobi, Kenya), is available, primarily in areas of endemic infection, for the detection of circulating antigens in persons with African trypanosomiasis. The test is normally performed on a drop of freshly collected heparinized blood and is followed by a more specific confirmation test on diluted blood, plasma, or serum. The sensitivity of the test (95.8% for *T. brucei gambiense* and 97.7% for *T. brucei rhodesiense*) is significantly higher than that of lymph node puncture, microhematocrit centrifugation, and CSF examination after single and double centrifugation. Its specificity is excellent, and it has a high positive predictive value. This test can be used for evaluating therapeutic cure, as well as for diagnosis.

The card agglutination test for trypanosomiasis antibody detection (CATT/*T. b. gambiense*) is more sensitive than methods that require visualization of the trypomastigotes with clinical specimens and can be performed with blood, plasma, or serum. The test is produced by the Institute of Tropical Medicine in Antwerp, Belgium, and is widely used for mass population screening. This approach can also be used with blood-impregnated filter paper.

General Comments

In blood, the trypanosomes move rapidly among the RBCs. Trypomastigotes are 14 to 33 μm long and 1.5 to 3.5 μm wide. At the organism's posterior end are the kinetoplast and the remaining intracytoplasmic flagellum (axoneme), which may not be noticeable. The flagellum and undulating membrane arise from the kinetoplast. The flagellum runs along the edge of the undulating membrane until the undulating membrane merges with the trypanosome body at the organism's anterior end. At this point, the flagellum becomes free to extend beyond the body. Trypanosomal forms are ingested by the tsetse fly (*Glossina* spp.) when a blood meal is taken. Once the short, stumpy trypomastigote reaches the midgut of the fly, it transforms into a long, slender procyclic stage. The organisms multiply in the lumen of the midgut and hindgut of the fly. After about 2 weeks, they migrate to the salivary glands through the hypopharynx and salivary ducts, where they attach to the epithelial cells of the salivary ducts and then transform to their epimastigote forms. In the epimastigote forms, the nucleus is posterior to the kinetoplast, in contrast to the trypomastigote. Within the salivary gland, metacyclic (infective) forms develop from the epimastigotes in 2 to 5 days. With development of the metacyclic forms, the tsetse fly is infective and can introduce these forms into the puncture wound when the next blood meal is taken. The entire developmental cycle in the fly takes about 3 weeks. Once infected, the tsetse fly remains infected for life.

Additional Information

African trypanosomiasis is limited to the tsetse fly belt of Central Africa, where it has been responsible for some of the most serious obstacles to economic and social development in Africa. Within this area, the vast majority of tsetse flies prefer animal blood, which limits the raising of livestock. Over 50 million people are at risk, and there are approximately 25,000 new cases per year. Human infections are caused by *T. brucei gambiense* and *T. brucei rhodesiense*.

The trypanosomal (trypomastigote) forms can be found in the blood, CSF, lymph node aspirates, and fluid aspirated from the trypanosomal chancre (if one forms at the site of the tsetse fly bite). The trypomastigote forms multiply by longitudinal binary fission. The forms range from long, slender-bodied organisms with a long flagellum (trypomastigote) (≥30 µm long) to short, fat, stumpy forms without a free flagellum (ca. 15 µm long). The short, stumpy forms do not divide in the bloodstream but are the infective stage for the tsetse fly.

Metacyclic trypomastigote stages are introduced via the tsetse fly bite and set up a local inflammatory reaction. A nodule or chancre may develop after a few days and resolves spontaneously within 1 to 2 weeks; this is seen frequently in white Europeans but rarely in patients indigenous to the area. The trypomastigotes enter the bloodstream, causing a symptom-free low-grade parasitemia that may continue for months. This is stage I disease (systemic trypanosomiasis; no CNS involvement). Parasites may be difficult to detect, even on thick blood film examinations. The infection may self-cure during this period.

Symptoms include remittent, irregular fevers with night sweats. Headaches, malaise, and anorexia frequently accompany the fevers. Febrile periods alternate with periods of no fever. Many trypomastigotes are found in circulating blood during fevers; few are seen when fever is absent. Enlarged lymph nodes are soft, painless, and nontender; enlarged posterior cervical nodes (Winterbottom's sign) are the most common. The spleen and liver are enlarged. With Gambian trypanosomiasis, the blood-lymphatic stage may last for years before sleeping sickness syndrome occurs. The Rhodesian form causes a more rapid fulminant disease; death may occur in months rather than years.

Trypomastigote invasion of the CNS causes sleeping sickness (stage II disease). Trypomastigotes are found primarily in the frontal lobes, pons, and medulla. Behavioral and personality changes occur during CNS invasion. Gambian trypanosomiasis is characterized by steady progressive meningoencephalitis, apathy, confusion,

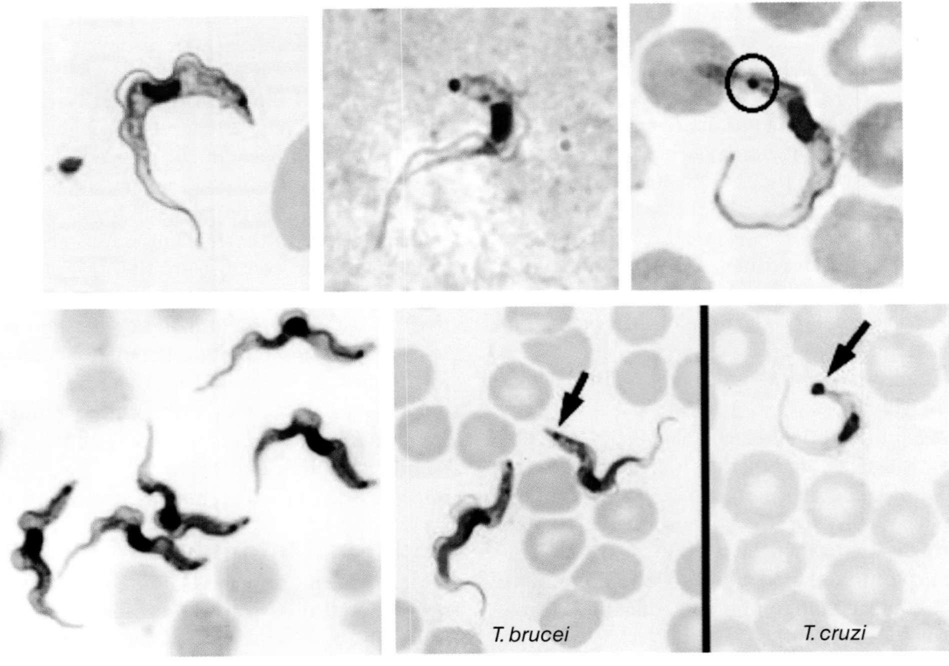

Figure 7.29 African trypanosomes (*Trypanosoma brucei gambiense, T. brucei rhodesiense*). (Top row) *T. brucei gambiense* or *T. brucei rhodesiense* trypomastigotes (stained with any blood stain [Giemsa, Wright-Giemsa combination, or any of the rapid blood stains]). Note the very small kinetoplast; however, this kinetoplast is a bit larger than normal (oval). **(Bottom row)** African trypomastigotes dividing in the blood; combination image of *T. brucei gambiense* or *T. brucei rhodesiense* and *Trypanosoma cruzi* trypomastigotes (note the very small kinetoplast in the African trypomastigote and the very large kinetoplast in the American trypomastigote [arrows]).

fatigue, coordination loss, and somnolence. In the terminal phase, the patient becomes emaciated, progressing to profound coma and death, usually from secondary infection.

Laboratory findings include anemia, granulocytopenia, reduction in platelets, increased sedimentation rate, polyclonal B-cell activation with a marked increase in serum IgM, heterophile and anti-DNA antibodies, rheumatoid factor, and circulating immune complexes.

Due to antigenic variation, a small fraction of the parasite population is able to evade the mammalian host humoral immune response and proliferate until the new surface antigen coat is recognized by a new generation of specific antibodies, mainly of the IgM type. Up to 1,000 different genes encoding the variant surface glycoproteins are present in the *T. brucei* genome. This phenomenon explains the fluctuating number of circulating trypanosomes in the patient's blood, which contributes to the limited sensitivity of parasite detection methods in clinical practice. The sustained high IgM levels are a result of the parasite producing variable antigen types which allow the organism to evade the patient's defense system. **In an immunocompetent host, the absence of elevated serum IgM rules out trypanosomiasis.**

Trypanosoma cruzi

Pathogenic	Yes
Disease	American trypanosomiasis, Chagas' disease (WHO neglected tropical disease)
Acquired	Triatomid bug (feces), blood, shared needles, congenital infections, organ transplants, contaminated food or drinks
Body site	Blood, all tissues
Symptoms	Acute febrile illness, lymphadenopathy and myocarditis, chagoma or Romaña's sign; cardiomegaly, cardiac conduction defects, severe constipation, or dysphagia
Clinical specimen	Blood (buffy coat preparations, culture, xenodiagnosis)
Epidemiology	American continents; vector-to-human and human-to-human transmission
Control	Vector control, avoiding shared needles, checking blood supply and donor organs

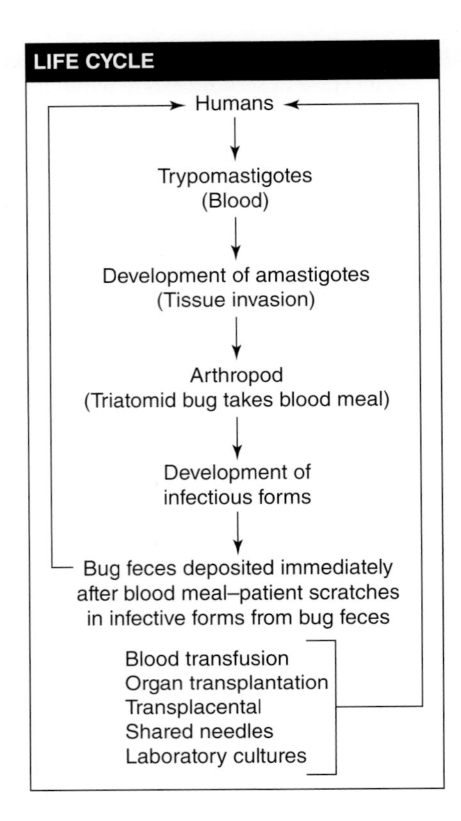

Diagnosis

Chagas' disease is a zoonosis caused by the protozoan hemoflagellate *Trypanosoma cruzi*, which is transmitted to humans mainly by insects of the subfamily Triatominae (Hemiptera: Reduviidae) through contact of skin and mucous membranes with feces and other secretions of these insects. Other means of transmission are oral transmission, congenital transmission, blood transfusion, organ transplant, and laboratory accidents. The differential diagnosis of acute Chagas' infection includes brucellosis, endocarditis, salmonellosis, schistosomiasis, toxoplasmosis, tuberculosis, connective tissue diseases, and leukemia. Chronic Chagas' disease with cardiomyopathy may be confused with endocarditis, ischemic heart disease, and rheumatic heart disease.

Trypomastigotes may be detected in the blood in young children; however, in chronic disease, this stage is rare or absent except during fever episodes. They may be detected by using thin and thick blood films or buffy coat concentration techniques (most sensitive). Any blood stains can be used for trypomastigote and amastigote stages. *Trypanosoma rangeli* trypomastigote infections may have to be differentiated from those with *T. cruzi*. *T. cruzi* trypomastigotes are usually C or U shaped on fixed blood films and have a large oval kinetoplast at the posterior end. *T. rangeli* trypomastigotes have a smaller kinetoplast near the posterior end and do not have an amastigote stage.

Aspirates from chagomas and enlarged lymph nodes can be examined for amastigotes and trypomastigotes. Histologic examination of biopsy specimens may also be done. Aspirates, blood, and tissues can be cultured. The medium of choice is Novy-MacNeal-Nicolle medium. Cultures should be incubated at 25°C and observed for epimastigote stages for up to 30 days before being considered negative.

Diagnostic Tips

Laboratory personnel must use caution when handling potentially infective blood specimens and/or cultures; the trypomastigotes are highly infective. Although trypomastigotes are prevalent in the blood of patients with acute Chagas' disease, organism numbers are much smaller in the indeterminate and chronic stages of the infection. *T. rangeli* (nonpathogenic) cannot be differentiated from *T. cruzi* on the basis of parasite morphology; patient information regarding geographic exposure is required for more appropriate interpretation of laboratory results.

In addition to thin and thick blood smears, concentration methods should be used to concentrate the trypomastigotes in the blood. Immunoassays for antigen detection are now available and are highly sensitive and specific. Alternative methods such as culture and serologic testing can be used; however, these approaches may not be feasible without the use of a reference laboratory. Molecular methods are also being more widely used.

The chronic phase normally has an asymptomatic clinical course lasting 2 to 3 decades, after which approximately 10% and 20% of infected individuals develop digestive and heart complications, respectively. Due to low parasitemia and high levels of specific anti-*T. cruzi* antibodies, diagnosis in this chronic phase is traditionally performed by serological methods, including ELISAs, indirect immunofluorescence assays, and indirect hemagglutination inhibition assays. Agreement in at least two tests must be obtained due to wide parasite antigenic variability. Three ELISAs (Hemagen, Ortho, and Wiener) and one rapid test (InBios) are FDA cleared, but comparative data in U.S. populations are sparse. Within the United States, the best current testing algorithm for blood donor testing would employ a high-sensitivity screening test followed by a high-specificity confirmatory test.

The risk of *T. cruzi* transmission with liver or kidney transplantation is much less than that seen with heart transplantation. For recipients with Chagas' cardiomyopathy undergoing heart transplant, clinical and laboratory protocols are mandatory to assess *T. cruzi* reactivation. It is important that both organ donors and recipients undergo surveillance testing for Chagas' disease. Laboratory monitoring of peripheral blood with PCR can identify reactivation prior to the occurrence of symptoms and allograft damage.

General Comments

In humans, *T. cruzi* can be found in two forms, amastigotes and trypomastigotes. The trypomastigote does not divide in the blood but carries the infection to all parts of the body. The amastigote multiplies within virtually any cell, preferring cells of the reticuloendothelial system, cardiac muscle, skeletal muscle, smooth muscle, and neuroglia. The amastigote is indistinguishable from those found in leishmanial infections. It is 2 to 6 μm in diameter and contains a large nucleus and rod-shaped kinetoplast that stains red or violet with Giemsa stain.

The disease is transmitted to humans through the bite wound caused by reduviid bugs. Metacyclic trypomastigotes are released with feces during a blood meal, and the feces are rubbed or scratched into the bite wound or onto mucosal surfaces, an action stimulated by the allergic reaction to the insect's saliva. The metacyclic forms invade local tissues, transform to the amastigote, and multiply within the cells. Once inside a cell, the trypomastigote loses its flagellum and undulating membrane and divides by binary fission to form amastigotes. The amastigotes continue to divide, and they eventually fill and destroy the infected cell. Both amastigote and trypomastigote forms are released from the cell.

Trypomastigotes are ingested by the reduviid bug during a blood meal. They transform into epimastigotes that multiply in the posterior portion of the midgut. Metacyclic trypomastigotes develop from the epimastigotes (10 days) and are passed in the feces to infect humans when rubbed into the insect's puncture wound or rubbed onto exposed mucous membranes.

In the Brazilian Amazon, the suspected source of infection in an outbreak of acute Chagas' disease was açai berry juice. Patient blood and juice samples contained *Trypanosoma cruzi* TcIV, indicating oral transmission of the infective forms. Opportunities for contamination of the juice include the attraction of contaminated triatomines by the light used during nighttime açai pulp extraction. Another possibility is that contamination occurs during collection and processing of açai berries without use of proper hygiene before mashing. Triatomine infestation of Amazonian palm trees also supports the potential for oral *T. cruzi* contamination of humans.

Additional Information

Chagas' disease is a zoonosis in the American continents and involves triatomid (reduviid) bugs living in close association with human reservoirs (dogs, cats, armadillos, opossums, raccoons, and rodents). There are ca. 100 million persons at risk of infection, of whom 16 million to 18 million are actually infected. There are ca. 50,000 deaths per year due to Chagas' disease. In some areas, ca. 10% of all adult deaths are due to Chagas' disease. Human infections occur mainly in rural areas where poor sanitary and socioeconomic conditions and poor housing provide excellent breeding places for reduviid bugs and allow maximum contact between the vector and humans. Reduviids in the United States have not adapted to household habitation.

Although it is less common than cardiac involvement, patients may have dilation of the digestive tract with or without cardiomyopathy. These symptoms are most often seen in the esophagus and colon as a result of neuronal destruction. Megaesophagus, characterized by dysphagia, chest pain, regurgitation, and malnutrition, is related to loss of contractility of the lower esophagus. Hypersalivation may occur, leading to aspiration with repeated bouts of aspiration pneumonia. Megacolon results in constipation, abdominal pain, and inability to discharge feces. In some individuals, there may be acute obstruction leading to perforation, septicemia, and death.

In children younger than 5, the disease is most severe; in older children and adults, it is more chronic. An erythematous painful nodule (chagoma) may form on the face; it may take 2 to 3 months to subside. Amastigotes or trypomastigotes may be aspirated from the chagoma; if the route of inoculation is the ocular mucosa, edema of the eyelids and conjunctivitis may occur (Romaña's sign). The infective stages spread to the regional lymph nodes, which become enlarged, hard, and tender. Trypomastigotes appear in the blood about 10 days after infection and persist through the acute phase. These stages are rare or absent during the chronic phase.

Acute infections are characterized by high fevers, which may be intermittent, remitting, or continuous; hepatosplenomegaly; myalgia; erythematous rash; acute myocarditis; lymphadenopathy; and subcutaneous edema of the face, legs, and feet. There may be meningoencephalitis, which has a very poor prognosis. Myocarditis is manifested by electrocardiographic changes, tachycardia, chest pain, and weakness. Amastigotes proliferate within and destroy the cardiac muscle cells, causing conduction defects and loss of heart contractility. Death may occur due to myocardial insufficiency.

Death may occur within a few weeks or months, or the patient may enter the chronic phase. This phase may be asymptomatic; organisms are seldom seen in peripheral blood, but transmission by blood transfusion is a serious problem in areas where the disease is endemic. Also, recrudescence of *T. cruzi* infections in immunosuppressed patients, particularly transplantation patients, is a grave concern.

Chronic cardiomyopathy appears to fulfill many of the criteria for autoimmune diseases. However, validation of the target antigens must involve induction of cardiac lesions after immunization or passive transfer of antigen-specific T cells. Aberrant T-cell activation may also result in cardiac injury. Not all researchers agree that autoantibodies play a role in pathogenesis. Also, some propose a "combined" theory to explain the sequence of events leading to chronic myocarditis. The actual clinical course may vary from heart failure to a slow but progressive loss of cardiac function, with possible ventricular rupture and thromboemboli.

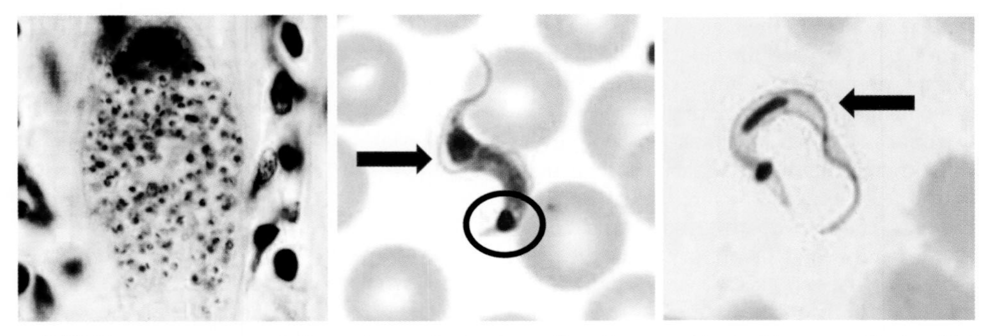

Figure 7.30 *Trypanosoma cruzi. T. cruzi* amastigotes in cardiac tissue and typical trypomastigotes. Note the large kinetoplast, much larger than that seen in the African sleeping sickness trypomastigotes (circle); also note the undulating membrane, very faint in the first trypomastigote (arrows).

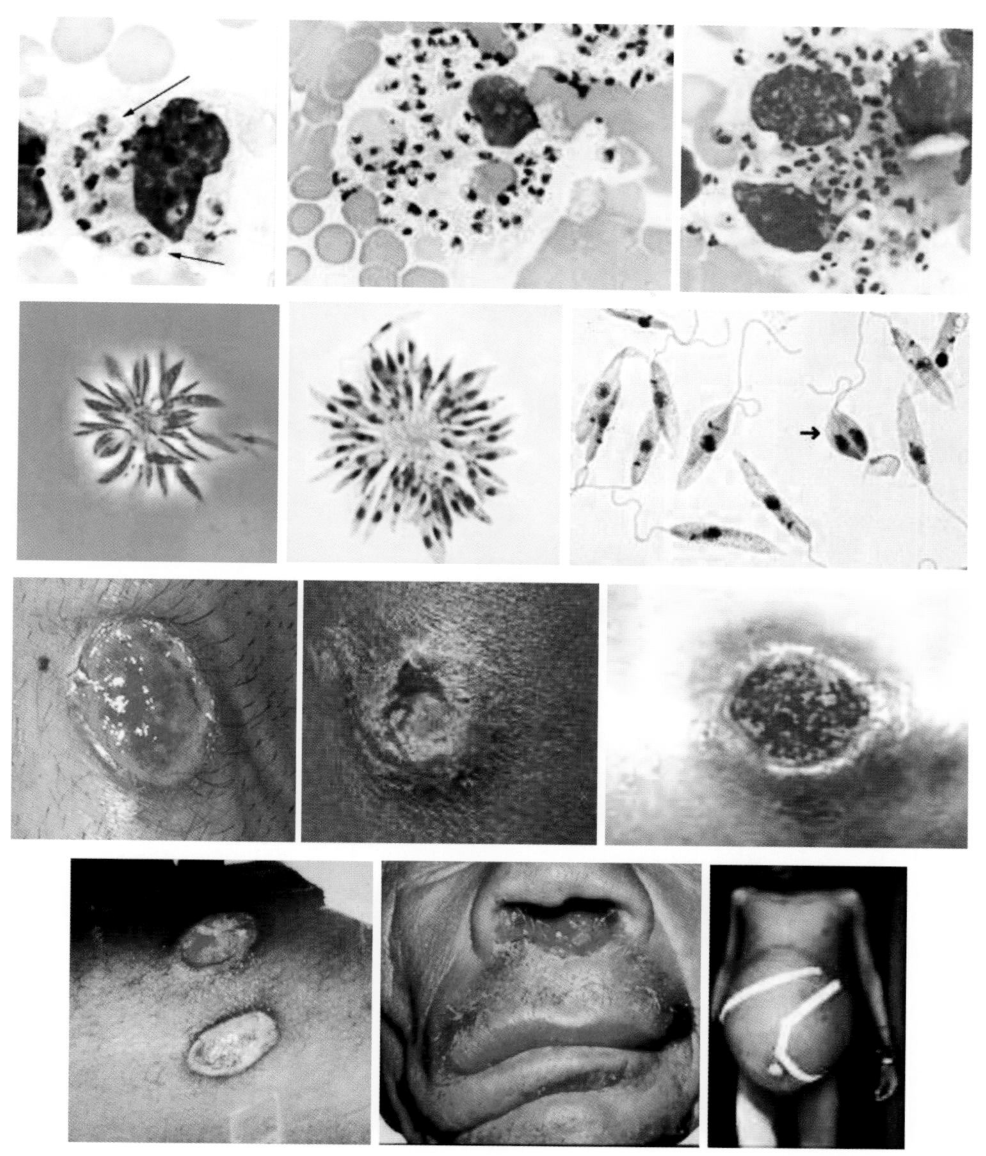

Plate 7.M (**Row 1**) *Leishmania* amastigotes within macrophage (note the nucleus and bar within each amastigote) (courtesy of CDC DpDx, https://www.cdc.gov/dpdx/leishmaniasis/index.html); *Leishmania donovani* in bone marrow; amastigotes in bone marrow. (**Row 2**) *Leishmania donovani* promastigotes in culture (wet mount); promastigotes stained with a blood stain; individual promastigotes (higher power) (note the dividing organisms [arrow]). (**Row 3**) *Leishmania* cutaneous wet lesion (active); *Leishmania* cutaneous dry lesion; *Leishmania* cutaneous lesion. (**Row 4**) *Leishmania* cutaneous lesions (the bandage on the original lesion slipped, inoculating the skin; hence, there are two lesions); *Leishmania* mucocutaneous lesions (note the lack of nasal septum); *L. donovani* infection (visceral leishmaniasis [kala-azar], hepatomegaly, and splenomegaly).

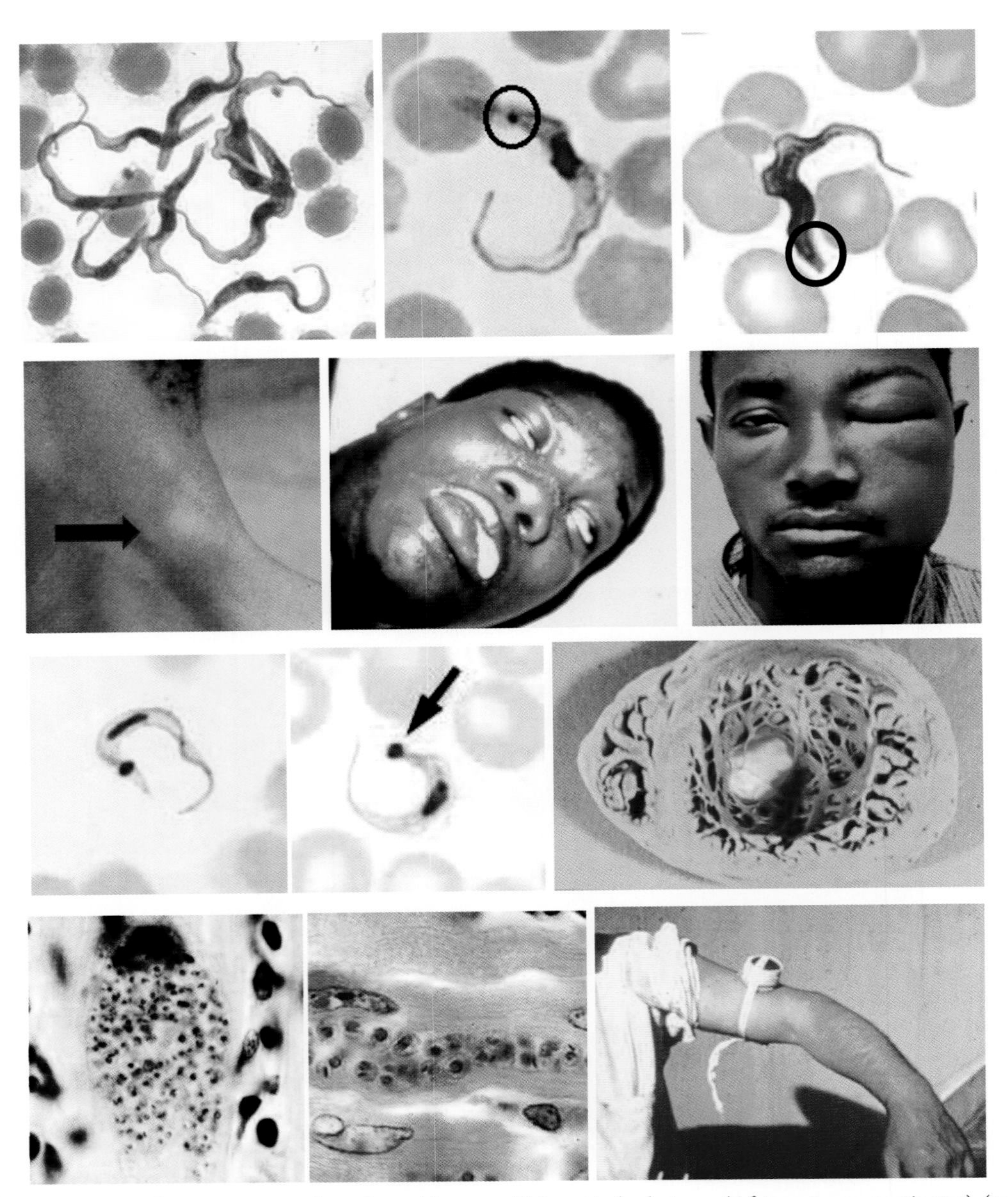

Plate 7.N **(Row 1)** *Trypanosoma brucei gambiense* or *T. brucei rhodesiense* (African trypomastigotes) (note the small kinetoplasts [circles]). **(Row 2)** (left to right) Winterbottom's sign (African trypanosomiasis; swollen lymph node at posterior cervical region); patient with sleeping sickness; Romaña's sign (Chagas' disease; edema of the eyelid). **(Row 3)** *Trypanosoma cruzi* trypomastigotes (left and middle) (note the large kinetoplast [arrow]); heart showing cardiomyopathy (dilation and thinning of the apical myocardium and marked concentric muscular hypertrophy) (right) (from the collection of Herman Zaiman, "A Presentation of Pictorial Parasites"). **(Row 4)** *T. cruzi* amastigotes in tissue; *T. cruzi* amastigotes in tissue (higher magnification); xenodiagnosis (trypanosome-free bugs are allowed to feed on individuals suspected of having Chagas' disease; if organisms are present in the blood meal, the parasites multiply and can be detected in the bug's intestinal contents, which should be examined for 3 months).

Naegleria fowleri

Pathogenic	Yes; rare but almost always fatal
Disease	Primary amebic meningoencephalitis (PAM)
Acquired	Contaminated water, olfactory epithelium, possible airborne exposure
Body site	CNS; hemorrhagic necrosis
Symptoms	Headache, nausea, vomiting, confusion, fever, stiff neck, seizures, coma
Clinical specimen	Brain biopsy specimen, CSF wet preparation, culture, IFA of tissue, PCR
Epidemiology	Worldwide
Control	Avoid potentially contaminated water exposure (including water used in neti pots)

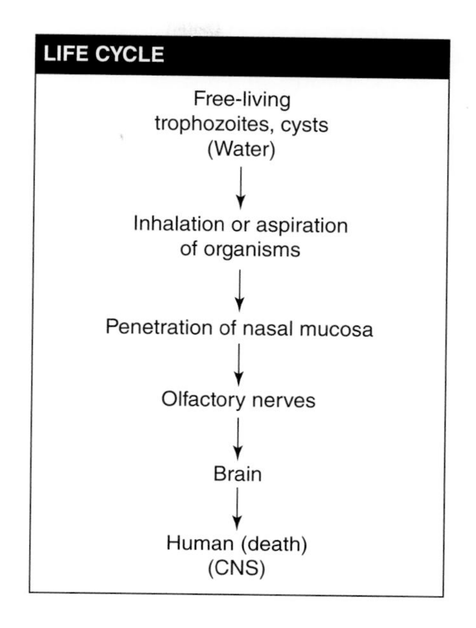

Diagnosis

Infections caused by small, free-living amebae belonging to the genera *Naegleria*, *Acanthamoeba*, *Balamuthia*, and *Sappinia* are generally not very well known or recognized clinically. Also, methods for laboratory diagnosis are unfamiliar and not routinely offered by most laboratories. However, infections caused by these organisms include primary amebic meningoencephalitis (PAM) caused by *Naegleria fowleri*; granulomatous amebic encephalitis (GAE) caused by *Acanthamoeba* spp. and *Balamuthia mandrillaris*; and amebic keratitis, cutaneous lesions, and sinusitis caused by *Acanthamoeba* spp. Another organism, recently identified, has been linked to encephalitis; *Sappinia diploidea* has been confirmed as a newly recognized human pathogen.

Diagnostic testing for the free-living amebae should always be considered a stat request (specimen collection, processing, testing, and reporting). Clinical and laboratory data usually cannot be used to differentiate pyogenic meningitis from PAM, and so the diagnosis may have to be reached by a process of elimination. A high index of suspicion is mandatory for early diagnosis. Most cases are associated with exposure to contaminated water through swimming or bathing. The rapidly fatal course of 3 to 6 days after the beginning of symptoms (with an incubation period of 1 day to 2 weeks) requires early diagnosis and immediate chemotherapy if the patient is to survive.

Analysis of CSF shows decreased glucose and increased protein levels. Leukocyte levels may range from several hundred to >20,000 cells/mm^3. Gram stains and bacterial cultures of CSF are negative; however, the Gram stain background can incorrectly be identified as positive with bacteria, leading to a misdiagnosis and incorrect therapy.

In cases of presumptive pyogenic meningitis in which no bacteria are identified in the CSF, the appearance of basal arachnoiditis on a computed tomography (CT) scan should alert the staff to the possibility of acute PAM.

A definite diagnosis can be made by demonstration of the amebae in the CSF or in biopsy specimens. Either CSF or sedimented CSF should be placed on a slide, under a coverslip, and observed for motile trophozoites; smears can also be stained with any of the blood stains. CSF, exudate, or tissue fragments can be examined by light microscopy or phase-contrast microscopy. It is very easy to confuse leukocytes and amebae, particularly when CSF is examined in a counting chamber; hence the recommendation to use just a regular slide and coverslip. Motility may vary, so the main differential characteristic is the spherical nucleus with a large karyosome. Specimens should never be refrigerated prior to examination. When the CSF is centrifuged, low speeds ($250 \times g$) should be used so that the trophozoites are not damaged.

If *Acanthamoeba* is involved, cysts may also be seen in specimens from CNS infection. Unfortunately, most cases of PAM are diagnosed at autopsy; confirmation of these tissue findings must include culture and/or special staining with monoclonal reagents in IFA procedures. Organisms can also be cultured on nonnutrient agar plated with *Escherichia coli*.

In general, serologic tests have not been helpful in diagnosis. The disease progresses so rapidly that the patient is unable to mount an immune response.

The risk of transmission of *N. fowleri* by donor organs has not been clarified, and no practical test is available to ensure that donor organs are organism free. Also, no prophylactic drug regimen to treat transplant recipients has been established.

Diagnostic Tips

Clinical specimens should never be refrigerated prior to examination. If *N. fowleri* is the causative agent, trophozoites only are normally seen. If the infecting organism is *Acanthamoeba*, cysts may also be seen in specimens from patients with CNS infection.

Beware of the false-positive Gram stain, especially since PAM usually mimics bacterial meningitis (CSF contains increased numbers of leukocytes, increased protein concentrations, and decreased sugar concentrations). The *Naegleria* trophozoites mimic leukocytes in a counting chamber. Motility is more likely to be seen if a drop of CSF is placed directly on a slide and a coverslip is added. Phase-contrast optics are recommended; however, if regular bright-field microscopy is used, the light level should be low. The slide can also be warmed to 35°C to stimulate trophozoite motility. Low-speed centrifugation is recommended for CSF ($150 \times g$ for 5 min).

Organisms can be cultured on nonnutrient agar plated with *Escherichia coli*. Incubate the plate in room air at 35 to 37°C for tissues from the CNS and at 30°C for tissues from other sites. Using the low-power objective (10×), observe the plates daily for 7 days. Remember that *B. mandrillaris* does not grow on agar plates seeded with bacteria. Standard precautions should be used when handling these specimens and cultures. Procedures should be performed in a biological safety cabinet.

In cases of presumptive pyogenic meningitis in which no bacteria are identified in the CSF, the CT appearance of basal arachnoiditis (obliteration of basal cisterns in the precontrast scan with marked enhancement after the administration of intravenous contrast medium) should alert the staff to the possibility of acute PAM.

General Comments

N. fowleri is ubiquitous and found mostly in freshwater lakes, hot water springs, poorly chlorinated pools, and thermally polluted water bodies worldwide. It has not been found in seawater. There have been 34 cases reported in the United States from 2008 through 2017, and a total of 143 infections were reported to the CDC from 1962 through 2017. Most cases of infection caused by *N. fowleri* have occurred through recreational freshwater exposure when swimming or diving. Unfortunately, fewer than 30% of cases from 1937 to 2013 were diagnosed premortem in the United States, probably due to late consideration of the diagnosis, difficulty detecting the organism in CSF, or rapid death.

The amebae may enter the nasal cavity by inhalation or aspiration of water, dust, or aerosols containing the trophozoites or cysts. They then penetrate the nasal mucosa and migrate via the olfactory nerves to the brain. The nasopharyngeal mucosa shows ulceration, and the olfactory nerves are inflamed and necrotic. Hemorrhagic necrosis is seen primarily in the olfactory bulbs and the base of the brain. Trophozoites can be found in the meninges, perivascular spaces, and sanguinopurulent exudates. PAM is an acute, suppurative infection of the brain and meninges. With extremely rare exceptions, the disease is rapidly fatal in humans.

Early symptoms include vague upper respiratory distress, headache, lethargy, and occasionally olfactory problems. The acute phase includes sore throat, stuffy blocked or discharging nose, and severe headache. Progressive symptoms include pyrexia, vomiting, and stiffness of the neck. Mental confusion and coma usually occur approximately 3 to 5 days prior to death. The cause of death is usually cardiorespiratory arrest and pulmonary edema.

Description

The trophozoites can occur in two forms, ameboid and flagellate. Motility can be observed in hanging-drop preparations from cultures of CSF; the ameboid form (the only form recognized in humans) is elongate with a broad anterior end and tapered posterior end. The size ranges from 7 to 35 μm; the diameter of the rounded forms is usually 15 μm. There is a large central karyosome and no peripheral nuclear chromatin. The cytoplasm is somewhat granular and contains vacuoles. The ameboid-form organisms change to the transient, pear-shaped flagellate form when transferred from culture or teased from tissue into water and maintained at 27 to 37°C. The change may occur very quickly (within a few hours) or may take as long as 20 h. The flagellate form has two flagella at the broad end. Motility is typical, with either spinning or jerky movements. These flagellate forms do not divide, but when the flagella are lost, the ameboid forms resume reproduction.

Cysts from nature and from agar cultures look the same and have a single nucleus almost identical to that seen in the trophozoite. They are generally round, measuring 7 to 15 μm, and there is a thick double wall.

Additional Information

In cases where nasal irrigation is required, such as use of a neti pot, or in religious practices, such as ablution, water that is sterile, distilled, filtered

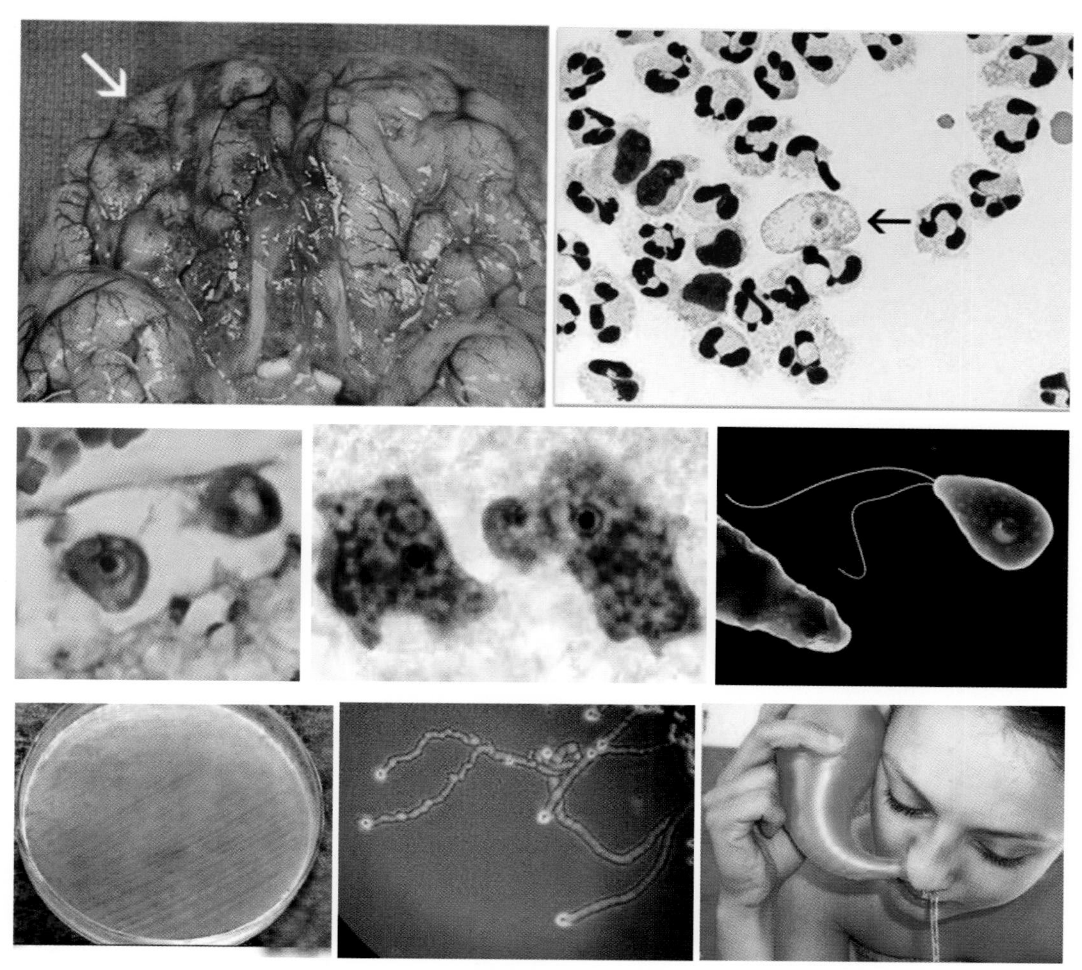

Figure 7.31 *Naegleria fowleri.* **(Top row)** Primary amebic meningoencephalitis (PAM) (note the extensive exudate and hemorrhage of the frontal cerebral cortex [arrow]); *N. fowleri* trophozoite in CSF cytospin preparation stained with Wright-Giemsa (courtesy of CDC DpDx, https://www.cdc.gov/dpdx/enterobiasis/index.html). **(Middle row)** Trophozoites in brain tissue; trophozoites (note large karyosome within the nucleus); flagellated form (courtesy of CDC, National Center for Emerging and Zoonotic Infectious Diseases [NCEZID], Division of Foodborne, Waterborne, and Environmental Diseases [DFWED]). **(Bottom row)** Nonnutrient agar plate seeded with *Escherichia coli*; trophozoites on the surface of an agar plate (growth on nonnutrient agar with bacterial overlay as a food source) (note the movement tracks on the agar); use of the neti pot for nasal irrigation.

(filters labeled for cyst removal, having an absolute pore size of 1 µm or smaller), or boiled (at least 1 min at sea level [3 min above 6,500 feet] and allowed to cool] should be used. Adequately disinfecting water bodies like pools and public water system with chlorine can also prevent infection.

Virulence factors include rapid motility of trophozoites, strong chemotaxis toward acetylcholine and other olfactory nerve products, surface proteins homologous to epithelial cell receptors that promote adhesion, resistance to complement-mediated immunity, a variety of enzymatic breakdown products, and intense stimulus of proinflammatory cytokines. Most cases of PAM have been reported in individuals with normal immunity. Those who have been tested demonstrated normal levels of mucosal IgA and IgG, including *Naegleria*-specific antibody; however, this antibody does not appear to be protective.

Swabs from the nose, mouth, and pharynx of healthy individuals have yielded *N. fowleri*. Also, antibody surveys suggest that *N. fowleri* exposure without apparent sequelae may be much more common than previously thought.

Acanthamoeba spp.
Balamuthia mandrillaris
Sappinia diploidea

Pathogenic	Yes
Disease	Amebic keratitis, amebic encephalitis (granulomatous amebic encephalitis [GAE])
Acquired	Contaminated water for keratitis (not linked to *Balamuthia;* unknown for *Sappinia*)
Body site	Eyes, skin, lungs, CNS
Symptoms	*Eyes*: Keratitis, corneal ulceration, retinitis
	Skin, lungs: Nonhealing lesions, sinusitis
	CNS: Chronic, slow meningoencephalitis
Clinical specimen	*Eyes*: Corneal scrapings, opened lens solutions
	CNS: CSF
Epidemiology	Worldwide, primarily environment-to-human transmission
Control	Refrain from using homemade or outdated lens solutions; avoid water exposure (immunocompromised patients)

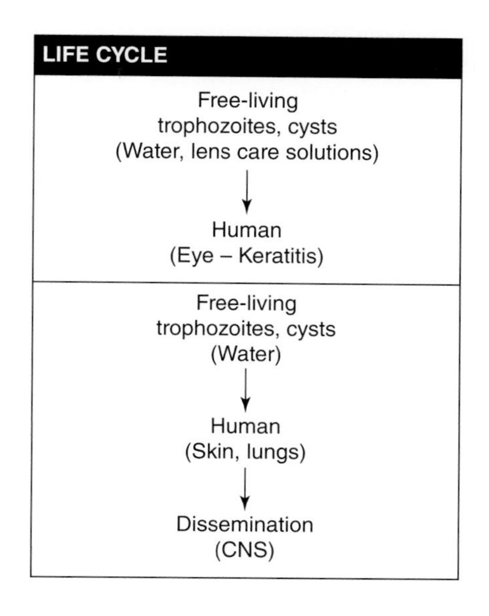

Diagnosis: *Acanthamoeba* spp.

Keratitis, uveitis, and corneal ulceration have been associated with *Acanthamoeba* spp. Infections have been seen in both hard- and soft-lens wearers. Particular attention has been paid to soft-lens disinfection systems, including homemade saline solutions. Cocontamination of lens systems with bacteria may be a prime factor in the development of amebic keratitis. The onset of *Acanthamoeba* corneal infection can vary tremendously; however, two factors often appear to be involved: trauma and contaminated water.

Meningoencephalitis caused by *Acanthamoeba* spp. may present as an acute suppurative inflammation of the brain and meninges like that seen with *Naegleria fowleri* infection. However, *Acanthamoeba* spp. generally cause a more chronic form of meningoencephalitis. Granulomatous amebic encephalitis (GAE) is characterized by confusion, dizziness, drowsiness, nausea, vomiting, headache, lethargy, stiff neck, seizures, and sometimes hemiparesis. GAE or disseminated disease can occur at any time of the year. The trophozoite is the infective stage for all clinical presentations; however, unlike *Naegleria*, both trophozoites and cysts have been recovered in biopsied tissue. While GAE has been documented in patients with normal immunity, *Acanthamoeba* GAE most often occurs in individuals with a variety of immunocompromising conditions, including advanced HIV disease, malnutrition, diabetes, renal failure, liver cirrhosis, receipt of steroids or chemotherapy, lymphoproliferative disorders, solid organ or bone marrow transplants, and agammaglobulinemia.

The differential diagnosis should include other space-occupying lesions of the CNS (tumor, abscess, fungal infection, etc.). Predisposing factors include Hodgkin's disease, diabetes, alcoholism, pregnancy, and steroid therapy. Organisms have also been found in the adrenal gland, brain, eyes, kidneys, liver, pancreas, skin, spleen, thyroid gland, and uterus.

A definite diagnosis can be made by demonstration of the amebae in the CSF or biopsy specimens. CSF or sedimented CSF should be placed on a slide, under a coverslip, and observed for motile trophozoites; smears can also be stained with Wright's or Giemsa stain. CSF, exudate, or tissue fragments can be examined by light microscopy or phase-contrast microscopy. It is very easy to confuse leukocytes and amebae, particularly when CSF is examined by using a counting chamber; hence the recommendation to use just a regular slide and coverslip. Motility may vary, so the main differential characteristic is the spherical nucleus with a large karyosome.

Diagnostic Tips

There are a number of published cases that emphasize the need for clinicians to consider *Acanthamoeba* sp. infection in the differential diagnosis of eye infections that fail to respond to antibacterial, antifungal, or antiviral therapy. These infections are often due to direct exposure of the eyes to contaminated materials or solutions. It is now recognized that the wearing of contact lenses is the leading risk factor for keratitis. Conditions which are linked with disease include the use of homemade saline solutions, poor contact lens hygiene, and corneal abrasions.

When corneal abrasions occur, the disease process is usually more rapid, with ulceration, corneal infiltration, iritis, scleritis, severe pain, hypopyon (pus in the anterior chamber), and loss of vision. When this process occurs in an individual who wears contact lenses, the onset is more gradual but the results are often the same.

Decreased corneal sensation has contributed to the misdiagnosis of *Acanthamoeba* keratitis as herpes simplex keratitis; this error can also be attributed to the presence of irregular epithelial lesions, stromal infiltrative keratitis, and edema, which are commonly seen in herpes simplex keratitis. *Acanthamoeba* keratitis may be present as a secondary or opportunistic infection in patients with herpes simplex keratitis. Unfortunately, as a result, treatment can be delayed for 2 weeks to 3 months. The presence of nonhealing corneal ulcers and the presence of ring infiltrates are also clinical signs that alert the ophthalmologist to the possibility of amebic infection.

Clinical manifestations of patients with AIDS and *Acanthamoeba* infection may include general complaints such as fever and chills, nasal congestion or lesions, neurologic symptoms, and musculoskeletal and cutaneous lesions. Some patients, especially those with AIDS, can develop erythematous nodules, chronic ulcerative skin lesions, or abscesses. Although CSF lymphocytosis is seen in non-AIDS patients, the CSF may reveal no cells in HIV-positive patients. These patients also tend to exhibit chronic sinusitis, otitis, and cutaneous lesions. Patients with CNS symptoms and AIDS may be misdiagnosed with toxoplasmosis, although serologic tests for *Toxoplasma* may be negative. Other patients have been misdiagnosed with CNS vasculitis, squamous cell carcinoma, or bacterial meningitis.

As with *N. fowleri*, beware of a false-positive Gram stain (the leukocyte count in CSF may be elevated). Examine clinical material on a slide with a coverslip; do not use a counting chamber (organisms in CSF look like leukocytes).

Nonnutrient agar with Page's saline and an *Escherichia coli* overlay can be tried for culture recovery. Various other media containing different agar bases, some of which contain horse, sheep, or rabbit blood, are also available. Tissue stains are also effective, and the use of calcofluor white to visualize the double-walled cyst is recommended.

Balamuthia mandrillaris has also been shown not to grow well on *E. coli*-seeded nonnutrient agar plates. In the diagnostic laboratory, these organisms can be cultured in mammalian cell cultures; some success has been obtained with monkey kidney cells and with MRC, HEp-2, and diploid macrophage cell lines. Generally, electron microscopy and histochemical methods are required for definitive identification of *B. mandrillaris*. An immunofluorescence test using species-specific sera is the most reliable means of distinguishing between *Acanthamoeba* and *Balamuthia* spp.

General Comments

Acanthamoebae are found worldwide, and contact with these organisms is unavoidable. Trophozoites and cysts may be recovered from vegetation (including vegetables), soil, water (salt or freshwater), warm and cold locations (including Antarctica), the surface interface between water and air, and even dust and air. These free-living amebae have been isolated from cooling towers for electrical and nuclear power plants. Other sources include bottled mineral and distilled water, humidifiers, air-conditioning units, and residential plumbing. *Acanthamoeba* spp. have been found in recalled contact lens solutions, as well as contact lenses and storage cases belonging to those with keratitis. Organisms can also be found in the nose, throat, sinus, urine, and sputum of hospitalized individuals.

A contact lens can act as a mechanical vector for transport of amebae from the storage case to the cornea. Another consideration involves the potential infection of the nasal cavity via lacrimal drainage, a condition that probably causes no problems in healthy individuals. It remains unknown whether these organisms in the nasopharynx of an immunocompromised individual increase the risk of GAE.

The disease caused by *B. mandrillaris* is very similar to GAE caused by *Acanthamoeba* spp. and has an unknown incubation period. The clinical course tends to be subacute or chronic and is usually not associated with swimming in freshwater. No characteristic clinical symptoms, laboratory findings, or radiologic indicators have been found to be diagnostic for GAE.

In general, neuroimaging findings show heterogeneous, hyperdense, nonenhancing, space-occupying lesions. Whether single or multiple, they involve mainly the cerebral cortex and subcortical white matter. These findings suggest a CNS neoplasm, tuberculoma, or septic infarcts. Patients complain of headaches, nausea, vomiting, fever, visual disturbances, dysphagias, seizures, and hemiparesis. There may also be a wide range in terms of the clinical course, from a few days to several months. In immunocompetent hosts, an inflammatory response is mounted and amebae are surrounded by macrophages, lymphocytes, and neutrophils. However, with rare exceptions, these patients also tend to die of severe CNS disease.

Like *Acanthamoeba*, *B. mandrillaris* is not likely to be seen on CSF wet mount. *B. mandrillaris* feeds not on bacteria but on mammalian cells; thus, culture on agar seeded with *Escherichia coli* will not detect it. Growth in axenic culture (sterile, cell-free) media is possible but slow and thus a suboptimal diagnostic tool.

Sappinia diploidea is a newly recognized human pathogen causing amebic encephalitis. The infection has been well documented in a single patient. Trophozoites measure from 40 to 70 μm and have a distinctive double nucleus. Transmission electron microscopy has confirmed that the trophozoites contain two nuclei attached to each other by connecting perpendicular filaments. The trophozoites had ingested host blood cells and stained brightly with Giemsa and periodic acid-Schiff stains. Cysts were not seen, but their presence in the human host was not excluded. These and other morphologic characteristics led to the diagnosis of infection with *S. diploidea*, demonstrated to be a human pathogen for the first time.

Description (Trophozoites)

Motile *Acanthamoeba* organisms have spine-like pseudopods; however, progressive movement is usually not very evident. There is a wide range (25 to 40 μm), with the average diameter of the trophozoites being 30 μm. The nucleus has the typical large karyosome, like that in *N. fowleri*. This morphology can be seen on a wet preparation.

B. mandrillaris trophozoites are usually irregular in shape, and actively feeding amebae may measure 12 to 60 μm long (normal, 30 μm). In tissue culture, broad pseudopodia are usually seen; however, as the monolayer cells are destroyed, the trophozoites develop fingerlike pseudopodia. These organisms do not grow using the seeded-agar-plate method.

Sappinia trophozoites measure 40 to 70 μm and have a distinctive double nucleus. Transmission electron microscopy confirms that they contain two nuclei attached to each other by connecting perpendicular filaments. The trophozoites ingest host blood cells and stain brightly with Giemsa and periodic acid-Schiff stains.

Warning: It is important that all patients with unresponsive microbial keratitis, even those who do not wear contact lenses, be evaluated for possible *Acanthamoeba* infection.

Description (Cysts)

Acanthamoeba cysts are usually round with a single nucleus; they also have the large karyosome seen in the trophozoite nucleus. The double wall is usually visible, with the slightly wrinkled outer cyst wall and a polyhedral inner cyst wall. This cyst morphology can be seen in organisms cultured on agar plates. Cyst formation occurs under adverse environmental conditions; cysts are resistant to biocides, chlorination, and antibiotics and can survive low temperatures (0 to 2°C).

Balamuthia cysts are usually spherical (6 to 30 μm in diameter). Under electron microscopy, they are characterized by having three layers in the cyst wall: an outer wrinkled ectocyst, a middle structureless mesocyst, and an inner thin endocyst. Under light microscopy, they appear to have two walls, an outer irregular wall and an inner round wall.

Sappinia cysts were not identified in the first human infection, but they cannot be excluded as being present in the human host. Based on recent data, it appears that *S. diploidea* may have a sessile, bicellular cyst.

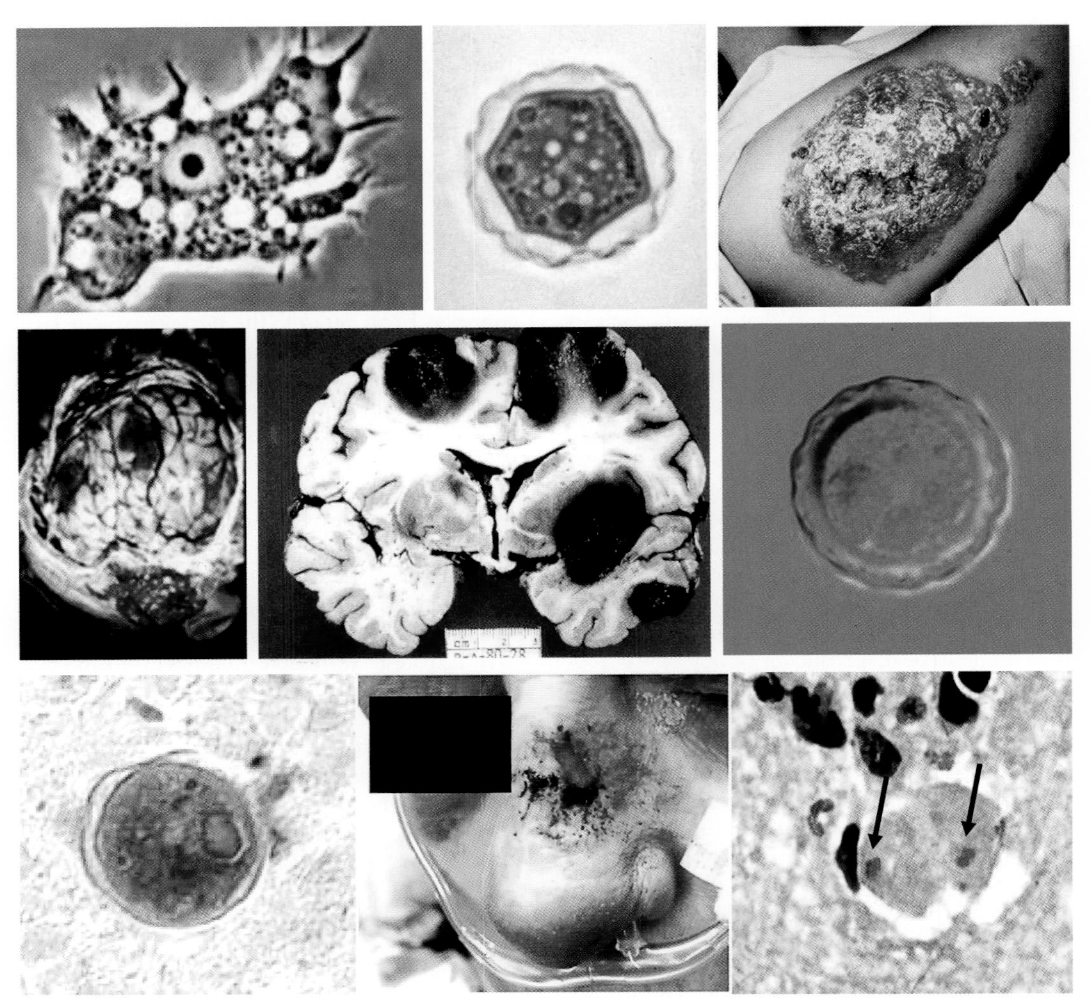

Figure 7.32 Free-living amebae. (Top row) *Acanthamoeba* trophozoite (note the spiky pseudopods) (this trophozoite looks very much like that of *Balamuthia*); cyst (note the double wall); *Acanthamoeba* skin lesion (courtesy of Dr. George H. Healy, CDC). **(Middle row)** *Acanthamoeba* brain lesion; *Acanthamoeba* brain lesion; *Balamuthia* cyst (courtesy of CDC DpDx, https://www.cdc.gov/dpdx/freelivingamebic/index.html). **(Bottom row)** *Balamuthia* cyst (courtesy of CDC, https://www.cdc.gov/parasites/balamuthia/diagnosis-hcp.html); *Balamuthia* nasal lesion (reprinted with permission from Piper KJ, Foster H, et al, *Int J Infect Dis* 77:18–22, ©2018 the authors, published by Elsevier Ltd.); *Sappinia* trophozoite in CNS section (note the double nuclei (arrows) (courtesy of G. Visvesvara, CDC).

PROTOZOA • Flagellates (Other Body Sites)

Trichomonas vaginalis

Pathogenic	Yes
Disease	Trichomoniasis
Acquired	Direct sexual contact; contaminated towels and underclothes (rare)
Body site	Urogenital tract
Symptoms	Vaginal pruritus, discharge, urethritis
Clinical specimen	Vaginal and urethral discharges and prostatic secretions
Epidemiology	Worldwide, primarily human-to-human transmission
Control	Treat asymptomatic males, promote awareness of sexual transmission

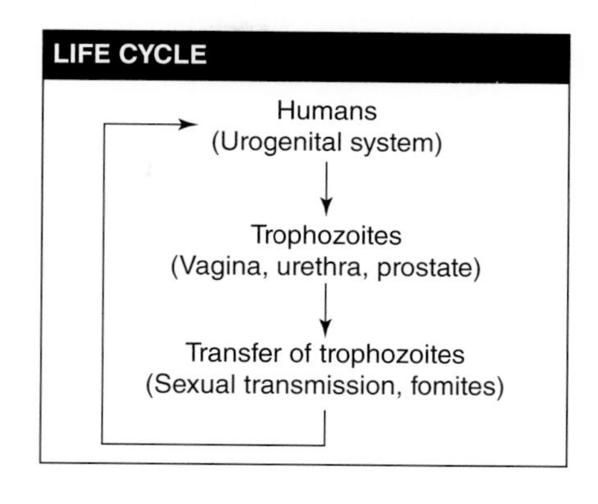

Diagnosis

Trichomonas vaginalis is the most common curable sexually transmitted infection worldwide. Identification is usually based on examination of wet preparations of vaginal and urethral discharges and prostatic secretions. Several specimens may have to be examined to detect the organisms. The specimen should be diluted with a drop of saline and examined under low power with reduced illumination for the presence of actively moving organisms; urine sediment can be examined in the same way. As the jerky motility of the trophozoite diminishes, it may be possible to see the movement of the undulating membrane, particularly under high dry power. Specimens should never be refrigerated.

Since the morphology of nonpathogenic *Pentatrichomonas hominis* from stool is very similar to that of pathogenic *T. vaginalis*, it is important to ensure that the specimen is not contaminated with fecal material.

Diagnostic tests other than wet preparations, such as permanent stains, fluorescent stains, culture, probes, and rapid tests (dipsticks), can also be used. Organisms may be difficult to recognize in permanent stains; however, if a dry smear is submitted to the laboratory, Giemsa or Papanicolaou stain can be used. Chronic *Trichomonas* infections may cause atypical cellular changes that can be misinterpreted, particularly on the Papanicolaou smear. Organisms are routinely missed on Gram stains. Diagnosis should be confirmed by observation of motile organisms either from the direct wet mount or from appropriate culture media. Results obtained with rapid dipstick tests are also excellent.

Diagnostic Tips

When performing a wet mount, it is critical that the specimen be examined as quickly as possible; organism deterioration occurs within hours. If examination is delayed, the specimen should be fixed for subsequent permanent staining.

If culture techniques are used, it is mandatory that the specimen be collected correctly, immediately inoculated into the proper medium, and properly incubated. If these requirements are not met, a false-negative result may be obtained. Although culture is more sensitive than wet mounts, this approach is not always used because of cost. It is important to use a positive control when culturing patient specimens.

The BD MAX vaginal panel is highly sensitive and specific and simplifies the identification of infectious vaginitis. In one study, 15 of 1,000 patients were positive for *T. vaginalis* with a sensitivity/specificity of 100%. Another option is the BD Affirm VPIII microbial identification test (Affirm), which demonstrates sensitivity and specificity equal to those of the BD MAX vaginal panel.

General Comments

T. vaginalis infection is classified as a neglected parasitic infection because it disproportionately affects underserved communities and contributes to reproductive health disparities. Trichomoniasis is linked to pelvic inflammatory disorder, premature birth, low birth weight, infertility, and endometriosis.

T. vaginalis is an anaerobic unicellular, flagellated eukaryote, which has only the trophozoite stage in its life cycle, and it is very similar in morphology to other trichomonads. The trophozoite is 7 to 23 µm long and 5 to 15 µm wide. The axostyle is clearly visible, and the undulating membrane stops halfway down the side of the trophozoite. The nuclear chromatin is uniformly distributed, and there are abundant siderophil granules that are particularly evident around the axostyle. Normal body sites for these organisms include the vagina and prostate. Apparently, the organisms feed on the mucosal surface of the vagina, where bacteria and leukocytes are found. The preferred pH for good growth is slightly alkaline or acid, not the normal pH of the healthy vagina. Although the organisms can be recovered in urine, in urethral discharge, or after prostatic massage, the pH preference of the organisms in males has not been determined. Often the organisms are recovered in the spun urine sediment from both male and female patients. This organism, like the other trichomonads, divides by binary fission. There are no known cyst forms for this organism.

Additional Information Related to Diagnosis

Several techniques used concurrently may increase organism recovery. Both monoclonal antibodies and DNA probes for the detection of *T. vaginalis* are very effective. An EIA has been developed to detect *T. vaginalis* antigen from vaginal swabs. The predictive value of a positive test was >80% and that of a negative test was almost 100% (482 women in the study). A rapid latex agglutination test for diagnosing *T. vaginalis* is also available, and results for 395 women attending a genitourinary medicine clinic indicated a sensitivity of 95% for the latex agglutination test and the EIA compared with 74% for microscopy and 76% for culture in Oxoid media. Some of the new culture pouch techniques are more sensitive than the older culture methods.

This envelope (pouch) approach allows both immediate examination and culture in one self-contained system. In a group of 62 positive patients, wet mounts (direct or from the envelope) were equal in sensitivity (66%); however, values were 89% with *Trichomonas* medium no. 2 (Oxoid) and 97% with PEM-TV. This system is commercially available as the InPouch TV (Biomed Diagnostics, San Jose, CA), which serves as the specimen transport container, the growth chamber during incubation, and the "slide" during microscopy. Once inoculated, it requires no opening for examination, and positive growth occurs within 5 days.

Various dipsticks are also available (OSOM [Sekisui Diagnostics, Burlington, MA] and XenoStrip-Tv [Xenotope Diagnostics, Inc., San Antonio, TX]). However, apparently the OSOM *Trichomonas* test is unable to accurately diagnose *T. vaginalis* from urine in men.

The GeneXpert TV test for women and men is a moderately complex test, requires a small platform, and can be performed in <1 hour. The sensitivity compared with wet preparation or culture was 96.4% for self-collected vaginal swabs, 98.9% for endocervical specimens, and 98.4% for female urine. For men, sensitivity for urine samples was excellent (97.2%). The specificity for all assays was excellent (see Gaydos et al., Suggested Reading).

For clinical settings in which vaginal specimens are not available and culture is not an option, urine-based PCR-EIA may be another option.

Additional Information

T. vaginalis is site specific and usually cannot survive outside the urogenital system. After introduction, proliferation begins, with resulting inflammation and many trophozoites in tissues and secretions. Vaginal secretions have been described as being liquid, greenish or yellowish, sometimes frothy, and foul smelling. As the infection becomes more chronic, the purulent discharge diminishes, with a decrease in the number of organisms. The normal incubation period is 4 to 28 days. The onset of symptoms such as vaginal or vulval pruritus and discharge is often sudden and occurs during or after menstruation as a result of the increased vaginal acidity. About 20% of women with vaginal trichomoniasis have dysuria, which may occur before any other symptoms. Infection in males may be latent, with essentially no symptoms, or may be present as self-limiting, persistent, or recurring urethritis. *T. vaginalis* has been detected in 10 to 20% of subjects with nonspecific urethritis and in 20 to 30% of those whose sexual partners had vaginitis. There is also an association with increased HIV transmission and cervical dysplasia.

T. vaginalis caused respiratory distress in a full-term, normal male infant after delivery. A wet preparation of thick, white sputum showed few leukocytes and motile flagellates, which were identified as *T. vaginalis*. This study supports previous data indicating that this organism may cause neonatal pneumonia.

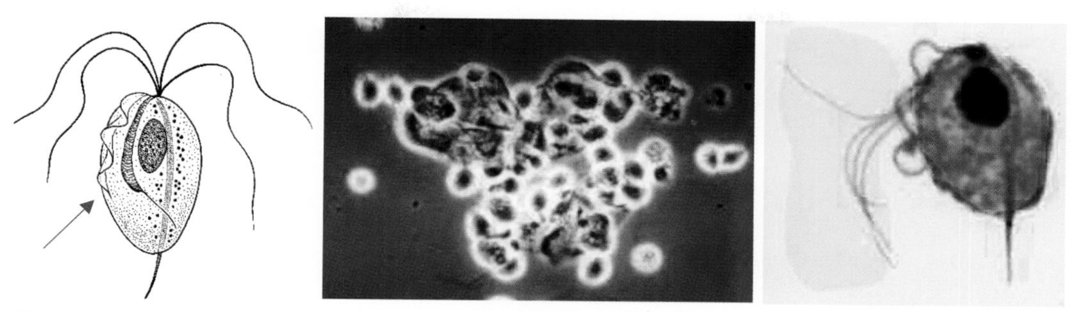

Figure 7.33 ***Trichomonas vaginalis.*** Drawing of *T. vaginalis* (note that the undulating membrane stops about halfway down the organism (arrow); wet mount of organisms; stained *T. vaginalis* trophozoite.

Ascaris lumbricoides

Pathogenic	Yes
Disease	Ascariasis (WHO neglected tropical disease)
Acquired	Fecal-oral transmission of infective eggs in contaminated food or water
Body site	Intestine, larvae in lungs
Symptoms	Pneumonitis, vague intestinal complaints or asymptomatic
Clinical specimen	Stool
Epidemiology	Worldwide, primarily human-to-human transmission
Control	Improved hygiene, adequate disposal of fecal waste, adequate washing of contaminated fruits and vegetables

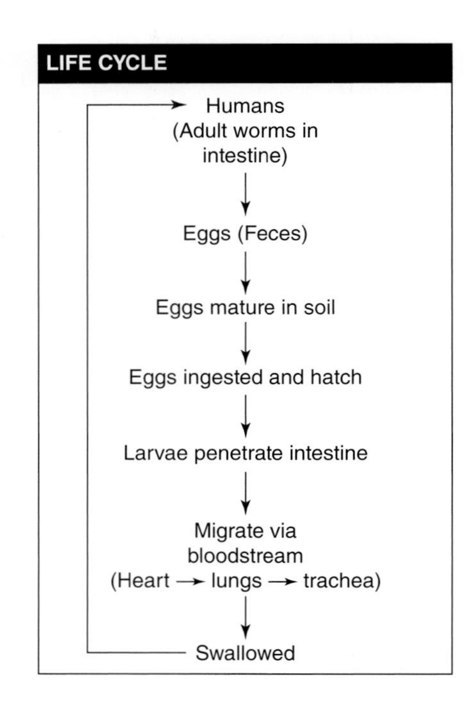

Diagnosis

The standard O&P exam is recommended for recovery and identification of *Ascaris lumbricoides* eggs in stool specimens, primarily from the wet-preparation examination of the concentration sediment. Diagnosis in the larval migration phase of the infection is based on finding larvae in sputum or gastric washings. The typical Loeffler's syndrome is more likely in areas where transmission is highly seasonal. Diagnosis in the intestinal phase is based on finding eggs (unfertilized or fertilized) or adult worms in the stool. The eggs are most easily seen on a direct wet smear or a wet preparation of the concentration sediment.

Diagnostic Tips

Unfertilized *Ascaris* eggs do not float in the zinc sulfate flotation concentration method (they are too heavy). Also, if too much iodine is added to the wet preparations, the eggs may look like very dark debris. Eggs may be very difficult to identify on a permanent stained smear because of stain retention and asymmetric shape.

Intestinal disease can often be diagnosed from radiographs of the gastrointestinal tract, where the worm intestinal tract may be visualized. This may be particularly obvious when two worms are lying parallel, like "trolley car lines." Involvement of body sites may cause specific symptoms indicative of bowel obstruction, biliary or pancreatic duct blockage, appendicitis, or peritonitis. Therapy targets specific symptoms and involved areas.

General Comments

Human infection is acquired through ingestion of embryonated eggs from contaminated soil. Approximately one billion people worldwide are infected with this nematode. On ingestion, the eggs hatch in the stomach and duodenum, where the larvae actively penetrate the intestinal wall; they are carried to the right heart via the hepatic portal circulation and then into the pulmonary circulation, where they are filtered out by the capillaries. After ca. 10 days in the lungs, they break into the alveoli, migrate via the bronchi until they reach the trachea and pharynx, and then are swallowed. They mature and mate in the intestine and produce eggs, which are passed in the stool.

The entire developmental process from egg ingestion to egg passage from the adult female takes 8 to 12 weeks. During her life span, she may deposit a total of 27,000,000 eggs. Both unfertilized and fertilized eggs are passed. Often only female worms are recovered from the intestine. Fertilized eggs become infective within 2 weeks if they are in moist, warm soil, where they may remain viable for months or even years. Often both types of eggs are found in the same stool specimen. The total absence of fertilized eggs means that only female worms are present in the intestine.

In children, particularly those under 5, there may be severe nutritional impairment related to the worm burden. Direct effects include increased fecal nitrogen and fecal fat and impaired carbohydrate absorption, all of which return to normal with elimination of the adult worms. Worms can also be spontaneously passed without any therapy.

Developmental disabilities are a significant and frequently undetected health problem in developing countries; malnutrition associated with intestinal helminth infections may be an important contributory factor in these disabilities.

Description (Eggs)

The fertilized egg is broadly oval, with a thick, mammillated (bumpy, tuberculated) coat, usually bile stained a golden brown; it measures up to 75 µm long and 50 µm wide.

Fertilized eggs may be heavy and may not float in the zinc sulfate flotation concentration procedure. When seen on a wet preparation, the eggs may resemble debris if too much iodine is used and the appearance is quite dark. Also, when these eggs are seen in a permanent stained smear, they may appear to be debris.

The mammillated coat in fertilized eggs is less pronounced than that in the unfertilized eggs. In proficiency testing specimens, some fertilized eggs continue to mature, and you may actually see motile larvae within the eggshell.

Unfertilized eggs are usually more oval, measure up to 90 µm long, and may have a pronounced mammillated coat or an extremely minimal mammillated layer.

Unfertilized eggs may be heavy and do not float in the zinc sulfate flotation concentration procedure. On a wet preparation, they may resemble debris if too much iodine is used and the appearance is quite dark. In a permanent stained smear, they may appear to be debris; normally they stain very dark and no longer resemble helminth eggs. The mammillated shell may be quite pronounced and irregular, more so than that of the fertilized eggs. In some cases, the mammillated shell is somewhat minimal.

Description (Adult Worms)

Adult *A. lumbricoides* organisms are nematodes (roundworms) that parasitize the human intestine. The adult worms are the largest of the common nematode parasites of humans (adult females, 20 to 35 cm; adult males, 15 to 30 cm) and have a curved posterior end. They are cylindrical, with a tapering anterior end, and three well-developed lips at the anterior end are characteristic. The fresh worms tend to be cream in color. *A. lumbricoides* is the primary species involved in human infections globally. Adult roundworms can be stimulated to migrate to any orifice by stressful conditions, such as gastrointestinal disease, fever, anesthesia, and anthelmintic drugs. Based on worm migrations, complications such as acute cholecystitis, acute cholangitis, and acute pancreatitis caused by ascariasis of bile or pancreatic ducts have been reported. More serious complications of *Ascaris* infection may occur when a large worm burden is present in the lumen of the intestine, such as intestinal obstruction, intussusception, volvulus, or even gangrene.

Additional Information

The minimum prepatent period after ingestion of the infective eggs is 60 days. Infection may be aborted by spontaneous passage of adult worms within about a year.

Eggs are difficult to kill while they are in the soil, especially clay soil under favorable environmental conditions. In some countries where infections are common, mass population treatment plans have been used with great success, even in areas with high reinfection rates. The use of human feces, or "night soil," for fertilization of crops should be recognized as a potential hazard. Any vegetables or fruits from such fields should not be eaten raw or unprocessed. Even with proper pretreatment of night soil, *Ascaris* eggs remain viable and infective more often than eggs of any other helminth species.

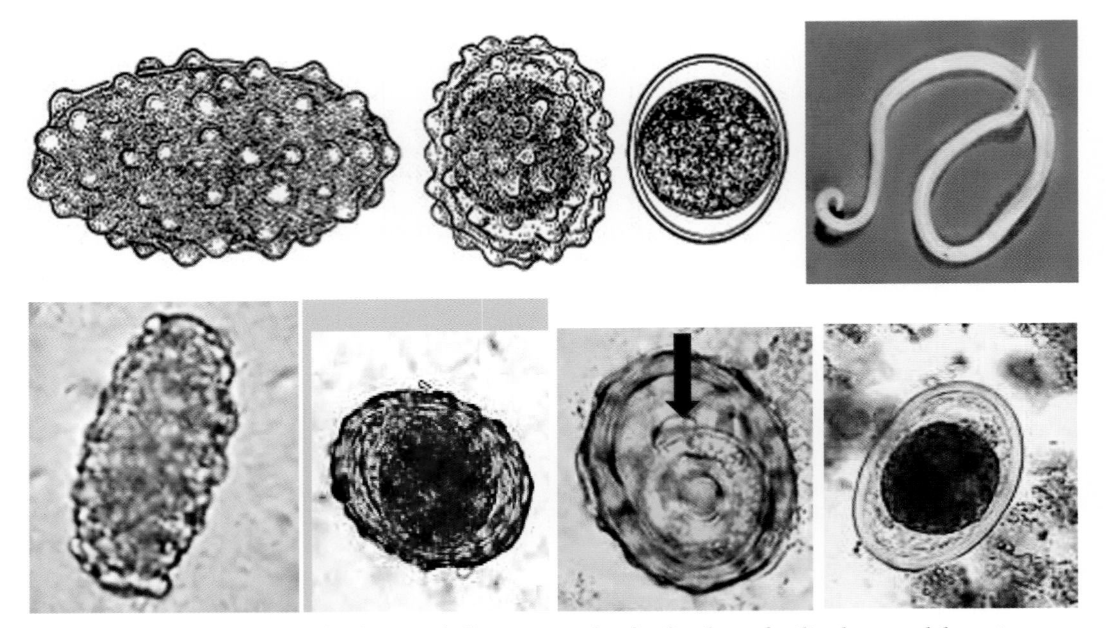

Figure 7.34 *Ascaris lumbricoides*. **(Top row)** Illustrations of unfertilized egg, fertilized egg, and decorticate egg (which has lost the bumpy coat); image of the adult male worm. **(Bottom row)** Unfertilized egg (note the very bumpy shell and somewhat elongated shape); fertilized egg; fertilized egg containing a larva (arrow); decorticate (no bumpy shell) fertilized egg.

NEMATODES • Intestinal

Trichuris trichiura
Capillaria philippinensis

Pathogenic	Yes
Disease	Trichuriasis (WHO neglected tropical disease)
Acquired	Ingestion of infective eggs from contaminated soil, food, or water
Body site	Intestine
Symptoms	Abdominal cramps, rectal tenesmus or rectal prolapse (children), or asymptomatic
Clinical specimen	Stool
Epidemiology	Worldwide, primarily human-to-human transmission
Control	Improved hygiene, adequate disposal of fecal waste, adequate washing of contaminated fruits and vegetables

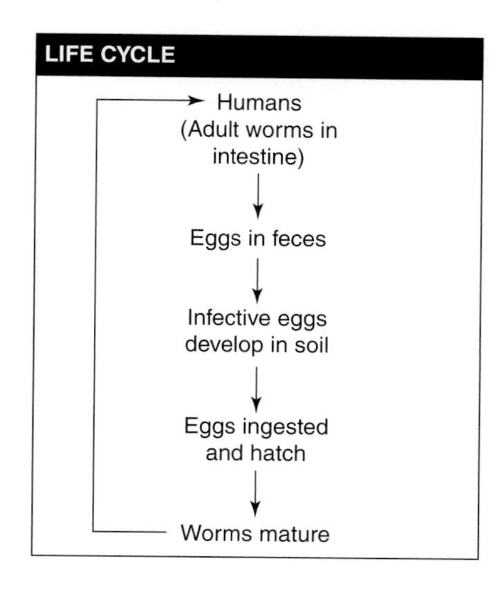

Diagnosis

The parasite *Trichuris trichiura* (whipworm) lives primarily in the cecum and appendix but can also be found in large numbers in the colon and rectum. Light infections are asymptomatic, but heavy infections may cause diarrhea, at times containing mucus and blood.

The standard O&P exam is recommended for recovery and identification of *T. trichiura* eggs in stool specimens, primarily from the wet-preparation examination of the concentration sediment. Most whipworm infections can be easily diagnosed by finding these characteristic eggs in the stool. The eggs may be quantitated (rare, few, moderate, or many), since light infections usually cause no problems and do not require therapy. However, quantitation information is not commonly included, and patients are generally treated regardless of the number of eggs seen. Rarely, the adult worms may be seen during colonoscopy.

Dysentery caused by *T. trichiura* and dysentery caused by *Entamoeba histolytica* are very similar; however, whipworm dysentery is usually more chronic, associated with malnutrition, and likely to cause rectal prolapse. Recovery and identification of the eggs or protozoan trophozoites differentiate the two infections. In severe infections, the adult worms are usually visible on the rectal mucosa.

Petechial hemorrhage, edema, inflammation, and mucosal bleeding develop, and heavy infections can cause rectal prolapse. Small amounts of blood (0.005 ml per worm) are lost each day by seepage at the attachment site. Colitis/proctitis, anemia, clubbing of fingers, and growth retardation are also reported to be associated with heavy whipworm infections.

Diagnostic Tips

T. trichiura eggs submitted in stool preserved with PVA do not concentrate as well as those preserved in formalin. However, very few laboratories perform this type of concentration, so the problem is not widespread.

The eggs can usually be identified from the permanent stained smear, but morphology is more easily seen in the wet-smear preparations. Typical morphology may be seen on a stained smear; *Trichuris* eggs are smaller than those of *Ascaris* and do not tend to overstain, thus they look more like typical eggs and less like debris.

General Comments

T. trichiura (whipworm) infection is more common in warm, moist areas of the world and is often seen in conjunction with *Ascaris* infections. Worm burdens vary considerably; individuals with few worms are unaffected by the presence of these parasites. Prevalence rates of 20 to 25% have been reported from the southern United States.

Whipworms are much larger than pinworms, being 35 to 50 mm long (female) and 30 to 45 mm long (male); the male has a 360° coil at the caudal extremity. Adult worms are rarely recovered from the stool, since they are attached to the wall of the intestine.

Human infection is acquired through ingestion of fully embryonated eggs from the soil. The eggs hatch in the small intestine and eventually attach to the mucosa in the large intestine. The adults mature in about 3 months and begin egg production.

Trichuris dysentery syndrome develops in some children because they have a defect in antiparasitic cell-mediated immunity. Probably both cellular and humoral responses are required to eliminate worms from the colon.

Differential diagnosis of chronic diarrhea includes celiac disease, inflammatory bowel disease, and irritable bowel syndrome. Heavily infected patients may present with a chronic dysentery-like syndrome leading to anemia and growth retardation. Diarrhea without blood and mucus may last for years. However, once blood is evident, medical intervention may be required; in some cases, the diagnosis requires colonoscopy, and prolonged therapy may be necessary to eliminate the parasites.

Description (Eggs)

The eggs are barrel shaped with clear, mucoid-appearing polar plugs. They are 50 to 54 μm long and 22 to 23 μm wide. They are passed in the unsegmented stage and require 10 to 14 days in moist soil for embryonation to occur.

Distorted eggs that are much larger than normal have been seen following therapy with mebendazole and with other drugs. This is not common but should be considered if distorted eggs are seen. There are also some reports in the literature that *Trichuris vulpis* (dog whipworm) eggs have been recovered in human stools. These eggs tend to be larger (70 to 80 μm long by 30 to 42 μm wide) and have prominent but small polar plugs compared with those of *T. trichiura*.

These nematode eggs are probably the easiest to identify; the shape is very consistent in wet preparations, and the eggs maintain their shape in permanent stained smears. However, on the permanent stained smear, they tend to stain dark and may be mistaken for debris.

Description (Adults)

T. trichiura is known as whipworm because the long, narrow anterior end and the shorter, more robust posterior end give the worm the look of a whip. The pinkish-white worms are threaded through the mucosa and attach by their anterior end. Females (approximately 45 mm long) are larger than males; they are bluntly rounded posteriorly, whereas the males have a coiled posterior. It is unusual to find adult worms in the stool unless the infection is quite severe with a very heavy worm burden. The majority of cases are diagnosed based on the presence of the typical eggs.

Additional Information

Although eosinophils and Charcot-Leyden crystals are present in the stool in patients with dysentery, a peripheral eosinophilia on the differential smear is not always seen, and the degree of eosinophilia may not correlate with the severity of infection (it rarely exceeds 15%). Heavy infections are rare in developed countries, as are complications requiring surgical intervention.

Inflammatory bowel disease, including Crohn's disease, probably occurs from a failure to downregulate a chronic Th1 intestinal inflammatory process. Induction of a Th2 immune response by intestinal helminths reduces the Th1 inflammatory process. Ulcerative colitis is more common in Western industrialized countries than in underdeveloped countries, particularly those where helminth infections are common. People with helminth infections exhibit altered immunologic antigen responses. Helminths prevent or alleviate colitis through the induction of regulatory T cells and modulatory cytokines. The use of *Trichuris suis* in the therapy of ulcerative colitis has been controversial for years. However, studies have shown improvement in 43.3% of patients receiving ova treatment, compared with 16.7% who received placebo.

Complications can occur and include rectal bleeding, bowel obstruction, appendicitis, hepatobiliary disease, and pancreatic pseudocyst.

General Comments on *Capillaria philippinensis*

Capillaria philippinensis worms live in the mucosa of the small bowel, most commonly the jejunum. Human infection is initiated by the ingestion of raw fish; the infective larvae are located in the mucosa of the fish intestine. Specific details of the life cycle are not clearly understood; however, there may be an internal autoinfection capability. The female worms produce eggs that resemble those of *T. trichiura*. They are thick shelled, with less prominent polar plugs, and the shell is striated. The eggs are somewhat smaller than those of *T. trichiura*, measure 36 to 45 by 20 μm, and require 10 to 14 days in the soil to embryonate and 3 weeks to develop into the infective form in fish.

Symptoms are related to the worm burden; with large numbers of worms, there may be intestinal malabsorption and fluid loss along with electrolyte and plasma protein imbalance. The wall of the small intestine is thickened and indurated and contains many larval and adult worms. Watery stools are passed, with fluid loss of several liters. Patients lose weight rapidly and develop muscle wasting, abdominal distention, and edema. Death from pneumonia, heart failure, hypokalemia, or cerebral edema may occur within several weeks to a few months. In some cases, patients reported chronic abdominal pain and diarrhea over a period of many months prior to diagnosis. On gastroduodenoscopy and subsequent histology, the jejunal mucosa revealed flattened villi, crypt proliferation, acute inflammation, and eosinophilic granulomata.

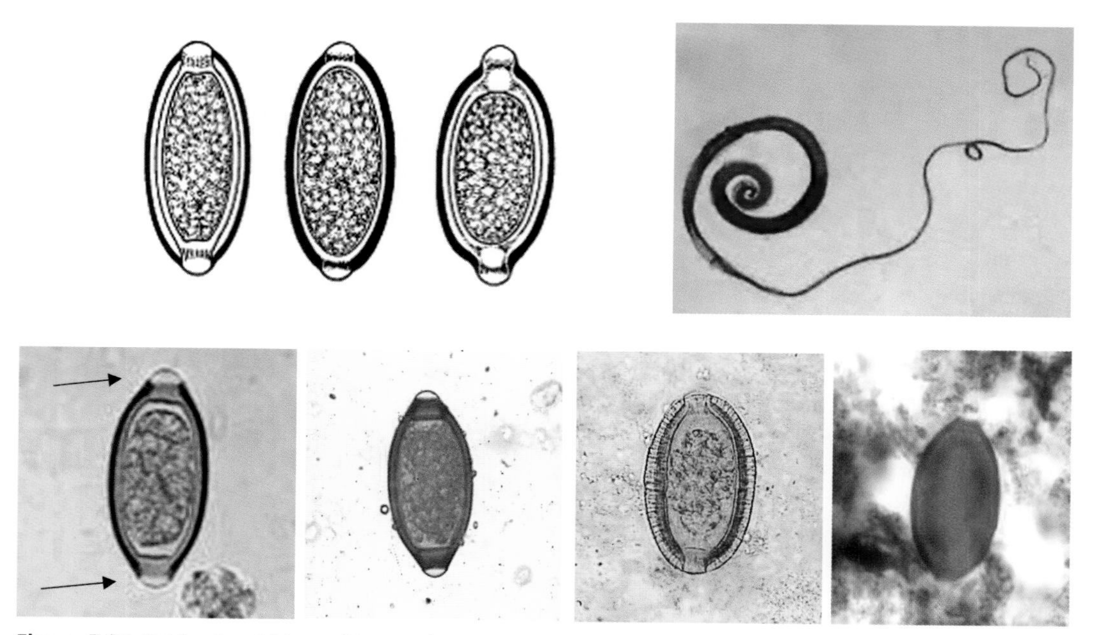

Figure 7.35 *Trichuris trichiura.* **(Top row)** Drawings of *T. trichiura* eggs; adult worm (the small end is the head end) (from the collection of Herman Zaiman, "A Presentation of Pictorial Parasites"). **(Bottom row)** Typical eggs seen in wet mounts stained with iodine (first two images), showing the barrel shape and polar plugs (arrows); *Capillaria* egg, which has striations on the eggshell that are not found on the *Trichuris* eggshell; egg from a permanent stained smear (Wheatley's trichrome; note that the internal morphology is not clear, as in the wet-mount images).

Necator americanus *Ancylostoma duodenale* *Ancylostoma ceylanicum* (Hookworms) *Trichostrongylus* spp.

Pathogenic	Yes
Disease	Hookworm disease (WHO neglected tropical disease)
Acquired	Skin penetration of filariform infective larvae from contaminated soil
Body site	Intestine, larvae in lungs
Symptoms	Pneumonitis, vague intestinal complaints; can be asymptomatic; blood loss anemia in heavy infections
Clinical specimen	Stool
Epidemiology	Worldwide, primarily human-to-human transmission
Control	Improved hygiene, adequate disposal of fecal waste, adequate washing of contaminated fruits and vegetables

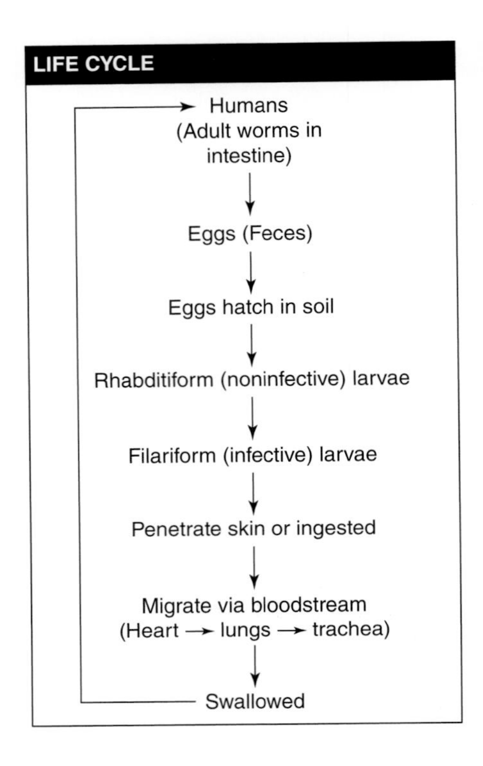

Diagnosis

Hookworm infection is difficult to differentiate clinically from other parasitic infections and certain other diseases. Diagnosis is made by demonstrating eggs in stool specimens. The two genera cannot be distinguished on the basis of their eggs. Direct microscopic examination of the stools may provide a diagnosis in heavy infections, but a concentration method should be used in most cases.

The standard O&P exam is recommended for recovery and identification of hookworm eggs in stool specimens, primarily from the wet-preparation examination of the concentration sediment. In the larval migration phase of the infection, diagnosis can rarely be made by finding the larvae in sputum or in gastric washings.

Diagnosis in the intestinal phase is based on finding the eggs (55 to 79 μm by 35 to 47 μm, 8- to 16-cell stage of development) in the stool. Adult worms are rarely seen. The eggs are most easily seen on a direct wet smear or a wet preparation of the concentration sediment. They appear shrunken and darkly colored in most permanent stained preparations.

Egg counts of 5/mg of stool are rarely clinically significant, counts of more than 20/mg are usually associated with symptoms, and counts of 50/mg or more represent very heavy worm burdens. For the most part, egg counts are rarely performed in routine clinical laboratories.

Diagnostic Tips

Hookworm eggs have a thin shell; when iodine is added to the wet preparation, there is a clear space between the eggshell and developing embryo. Eggs may be very difficult to identify on a permanent stained smear because of stain retention and potential collapse.

If the stool specimen is stored at room temperature (no preservative) for more than 24 h, the larvae continue to mature and hatch. These larvae must be differentiated from *Strongyloides* larvae, since therapy may be different for the two infections.

General Comments

Infection is acquired through skin penetration by filariform larvae from the soil. During skin penetration, infective hookworm larvae encounter hyaluronic acid; hookworm hyaluronidase activity has now been confirmed and can facilitate passage of the infective larvae through the epidermis and dermis during larval migration. After skin penetration, the larvae are carried first by the venules to the right heart and then to the lungs. Larvae invade the alveoli, migrate via the bronchi until they reach the trachea and pharynx, and are swallowed, bringing them to the small intestine, where they reside. They attach to the mucosa via a temporary mouth structure, mature sexually, and finally develop the permanent characteristic mouth structure that they use to attach to the mucosa. Any pneumonitis due to migrating larvae depends on the burden. These larvae do not cause the level of sensitization seen with *Ascaris* or *Strongyloides* infection.

Symptoms during the intestinal phase of infection are caused by (i) necrosis of the intestinal tissue within the adult worm mouth and (ii) blood loss by direct ingestion of blood by the worms and continued blood loss from the original attachment site possibly as a result of anticoagulant secreted by the worm. Patients with acute infections may experience fatigue, nausea, vomiting, abdominal pain, diarrhea with black to red stools, weakness, and pallor. Heavy worm burdens in young children may have serious sequelae, including death.

In chronic infections, the main clinical finding is iron deficiency anemia (microcytic, hypochromic) with pallor, edema of the face and feet, listlessness, and hemoglobin levels of 5 g/dl or less. There may be cardiomegaly and both mental and physical retardation.

Description (Eggs)

The eggs are usually in the early cleavage stage when passed in the stool. They are oval (ca. 60 μm long by 40 μm wide) with broadly rounded ends. They have a clear space between the developing embryo and the thin eggshell; this feature can be easily seen in the wet preparation stained with D'Antoni's iodine.

Egg survival and larval development are maximum in moist, shady, warm soil (sandy loam), where larvae hatch from the eggs within 1 to 2 days. The infective filariform larvae develop within 5 to 8 days and may remain viable in the soil for several weeks.

Description (Adults)

Adult males are 7 to 11 mm long by 0.4 to 0.5 mm wide. *Ancylostoma* worms are larger than *Necator* worms. Adults are rarely seen, since they remain firmly attached to the intestinal mucosa via well-developed mouth parts (teeth in *Ancylostoma duodenale* and cutting plates in *Necator americanus*).

Females begin to deposit eggs 5 months or more after initial infection. If mature filariform larvae of *A. duodenale* are swallowed, they can develop into mature worms in the intestine without migrating through the lungs.

Additional Information

Eosinophilia is common, usually develops 25 to 35 days after exposure, and peaks about 1 (*N. americanus*) to 2 (*A. duodenale*) months later. Eosinophilia may reach approximately 18%.

Hookworms cause gastrointestinal blood loss. Significant loss can occur with *A. duodenale*, but its impact on the iron status of populations is no greater than that of *N. americanus*. Patients with a heavy worm burden can lose up to 250 ml of blood per day.

Hookworm-mediated immunosuppression affects vaccine development. Unless preexisting infections are cured, vaccine-induced immunity may be affected. Also, if the vaccine is only partially effective, vaccine recipients may be susceptible to reinfection.

Helminth infection may protect against asthma and malaria; however, it may increase susceptibility to HIV/AIDS or tuberculosis. People with hookworm infection are also more likely to be infected with *Ascaris lumbricoides* and *Trichuris trichiura*, findings that can only partially be explained by overlapping areas of endemicity and potential exposure.

A. duodenale is found primarily in southern Europe, the north coast of Africa, northern India, northern China, and Japan. *N. americanus* is found throughout the southern United States, the Caribbean, Central America, northern South America, central and southern Africa, southern Asia, Melanesia, and Polynesia. In some areas, such as northern Ghana, both hookworms are present; mixed infections have been confirmed by PCR. Overlapping geographic areas continue to become more common.

Molecular-type surveys in Asia have shown that *Ancylostoma ceylanicum* is the second most common hookworm species infecting humans, accounting for between 6 and 23% of total hookworm infections. *A. ceylanicum* mimics the clinical picture produced by the anthroponotic hookworms of "ground itch" and moderate to severe abdominal pain in the acute phase. Natural human infections with *A. ceylanicum* have been reported in almost all areas in which the parasite is endemic in dogs and cats.

General Comments on *Trichostrongylus* spp.

Trichostrongylus spp. are small worms, similar to hookworms, and live embedded in the mucosa of the small intestine. Unlike the adult hookworms, the adult worms have no distinct buccal capsule with special mouth parts (teeth or cutting plates).

Infection in humans is acquired through ingestion of infective larvae contaminating plant material. After reaching the small intestine, the larvae mature in 3 to 4 weeks without any migratory pathway through the lungs. The eggs are very similar to those of hookworms, being oval and somewhat longer, with the ends being more pointed than those of hookworm eggs. The eggs may hatch within 24 h under favorable conditions (warm, moist soil) and develop into infective larvae after about 60 h.

Hemorrhage and desquamation may occur (similar to findings in hookworm infection); however, symptoms are usually not clinically significant unless several hundred worms are present. Symptoms include epigastric pain, diarrhea, anorexia, nausea, dizziness, and generalized fatigue or malaise; eosinophilia is usually present. Heavy worm loads may lead to the development of anemia and cholecystitis, as the worms enter the biliary tract. The definitive diagnosis can be made by identification of eggs in the stool.

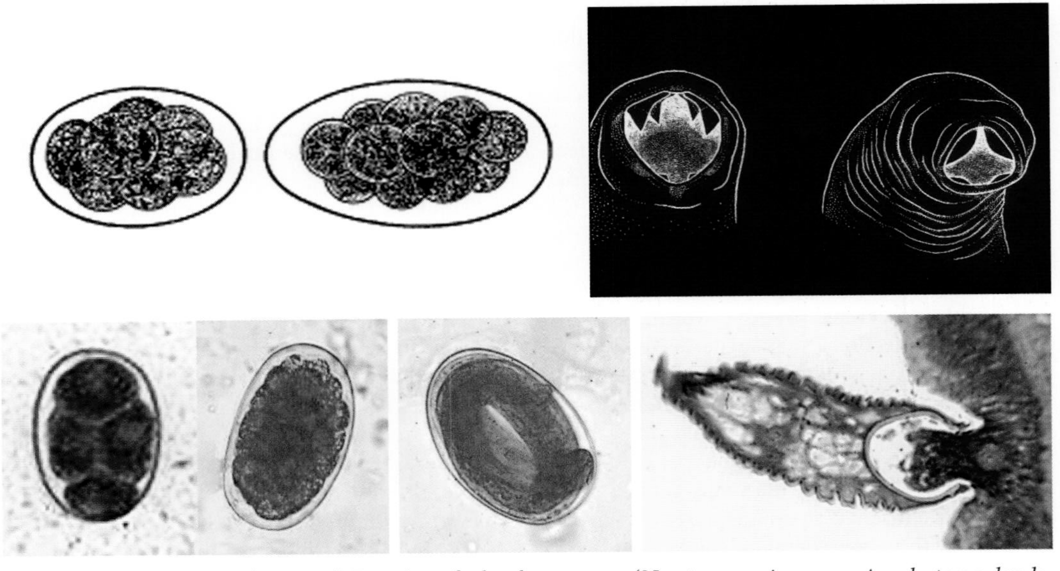

Figure 7.36 Hookworm. (Top row) Drawing of a hookworm egg (*Necator americanus* or *Ancylostoma duodenale*); drawing of a *Trichostrongylus* egg (it appears longer, with one end more pointed than the hookworm egg); mouth parts of adult *Ancylostoma* (left; note the teeth) and *Necator* (right; note the cutting plates). **(Bottom row)** Three typical hookworm eggs, the third of which contains a larval worm (this suggests that fresh, unpreserved stool was left at room temperature for some time before being examined or being placed in fixative; if this egg hatched, the rhabditiform larva would have to be differentiated from that of *Strongyloides stercoralis*); adult hookworm with a section of the worm attached to the mucosa (from the collection of Herman Zaiman, "A Presentation of Pictorial Parasites").

Strongyloides stercoralis

Pathogenic	Yes
Disease	Strongyloidiasis (hyperinfection syndrome, internal autoinfection)
Acquired	Skin penetration of infective filariform larvae from contaminated soil
Body site	Intestine, larvae in lungs
Symptoms	Pneumonitis, vague intestinal complaints, disseminated; may also be asymptomatic
Clinical specimen	*Intestinal*: Stool
	Disseminated: Sputum, various tissues
Epidemiology	Worldwide, primarily human-to-human transmission
Control	Improved hygiene, adequate disposal of fecal waste, adequate washing of contaminated fruits and vegetables

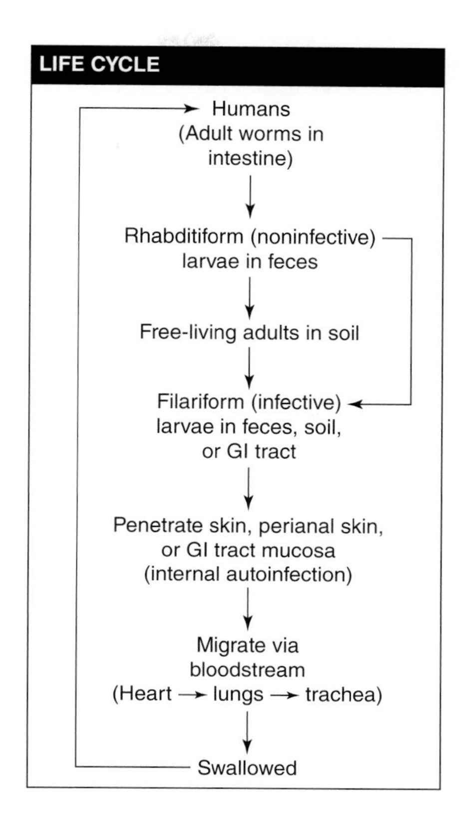

Diagnosis

Strongyloides stercoralis may coexist with hookworms; both require similar soil and climatic conditions for development. Based on the autoinfection aspect of the life cycle, persons who have contracted this infection in areas of endemicity may remain infected for years after leaving such areas (the current record appears to be approximately 55 years). In disseminated strongyloidiasis, the parasite may be found in any part of the body. In pulmonary infections, there may be pneumonia and hemorrhage. Meningitis and/or sepsis (primarily with Gram-negative bacteria from the gut) are also reported. The hyperinfection syndrome (disseminated disease) may be fatal.

The standard O&P exam is the most common method used for recovery and identification of *Strongyloides* larvae in stool specimens, primarily from the wet-preparation examination of the concentration sediment. In the larval migration phase of the infection, diagnosis is occasionally made by finding the larvae in sputum or in gastric washings. During the intestinal phase, the diagnosis is based on finding rhabditiform larvae in the stool. Adult worms and eggs are rarely seen, except in a very heavy infection. The larvae are most easily seen on a direct wet smear or a wet preparation of the concentration sediment. Transplant patients (kidney, bone marrow) are also subject to complications with strongyloidiasis. The American Society for Transplantation's guidelines recommend screening solid organ transplant recipients, but not donors, to assess the risk for reactivation of chronic infection in those from areas in which *Strongyloides* is endemic. However, donor-derived infection might be more common than previously believed, and screening of donors from areas of endemicity might help to protect organ recipients. Rapid communication among transplant centers with patients who received organs from a single donor also is essential. *Strongyloides* hyperinfection can happen any time after transplantation; however, this condition usually occurs within the first 3 months during times of increased immunosuppression.

It is very important to consider this infection in military personnel and travelers who, many years earlier, were in an area where the infection is endemic. More than 30 to 40 years after acquisition of the original infection, persistent, undiagnosed disease can be found in these individuals. If for any reason these individuals become immunocompromised, the result can be disseminated disease leading to the hyperinfection syndrome and death.

Diagnostic Tips

The noninfective rhabditiform larvae can transform to the infective filariform larvae, both of which can be found in the stool, and must be differentiated from hookworm larvae, since therapy may be different.

The efficacy of sampling of duodenal contents is controversial. Other techniques include the Entero-Test capsule, special concentration techniques (Baermann), and larva culture (Harada-Mori, petri dish); the sensitivities of these methods vary.

Agar plate cultures are recommended and are more sensitive than some other available diagnostic methods; stool is placed on agar plates, which are sealed to prevent accidental infections and held for 2 days at room temperature. As the larvae crawl over the agar, they carry bacteria with them, creating visible tracks over the agar. The plates are examined under the microscope for confirmation of larvae, the surface of the agar is then washed with 10% formalin, and larval identification is confirmed via wet examination of the sediment from the formalin washings.

Larvae and/or eggs seen in a permanent stained fecal smear are very difficult to identify; the organisms retain the stain and can resemble debris.

General Comments

Human infection is acquired by skin penetration of the filariform larvae (infective larvae) from the soil. After penetration of the skin, the larvae are carried via the cutaneous blood vessels to the lungs, where they break out of the pulmonary capillaries into the alveoli. They migrate via the respiratory tree to the trachea and pharynx, are swallowed, and enter the mucosa in the duodenum and upper jejunum. Development usually takes about 2 weeks; the females then begin egg production. The eggs usually hatch, and the rhabditiform larvae (noninfective larvae) pass out of the intestinal tract in the feces onto the soil, where they develop into free-living male and female worms, eventually producing infective filariform larvae (egg, noninfective larvae, and infective larvae). In temperate climates, the free-living male and female worms do not develop; however, the rhabditiform larvae develop into the filariform (infective) larvae, which are ready to infect the next host through skin penetration.

The pulmonary route is one of several possible pathways to the duodenum, regardless of whether the larvae penetrated the skin or the intestine, and may not be as universally applicable as was once thought.

Initial skin penetration usually causes very little reaction, although there may be some pruritus and erythema. Pulmonary symptoms, intestinal pain, sepsis, and meningitis with intestinal bacterial flora can also be seen, particularly in immunocompromised patients with disseminated disease (hyperinfection).

Description (Eggs and Larvae)

The eggs are oval and thin shelled and are 50 to 58 μm long by 30 to 34 μm wide (generally a bit smaller than hookworm eggs).

The rhabditiform larvae that are passed out in the stool are up to 380 μm long by 20 μm wide, with a muscular esophagus (club-shaped anterior, then a restriction, and a posterior bulb). There is a genital primordium packet of cells, which is fairly obvious and can be seen about two-thirds of the way back from the anterior end. A key morphologic difference between these larvae and those of hookworm is the length of the mouth opening (buccal capsule). The opening in the rhabditiform larvae of *S. stercoralis* is very short (only a few micrometers), while the mouth opening in hookworm rhabditiform larvae is approximately three times as long. These differences can be seen by examining the larvae under the microscope using the low-power (10×) or high dry power (40×) objectives. The filariform larvae are long and slender (up to 630 μm long by 16 μm wide) and may remain viable in soil or water for several days.

Additional Information

When autoinfection occurs, some rhabditiform larvae in the intestine develop into filariform larvae while passing through the bowel. These larvae can reinfect the host by (i) invading the intestinal mucosa, traveling via the portal system to the lungs, and returning to the intestine or (ii) being passed out in the feces and penetrating the host on reaching the perianal or perineal skin. However, the pulmonary route is just one of several possible pathways to the duodenum, regardless of whether the larvae penetrated the skin or the intestine.

Filarial larval migration through the lungs may stimulate symptoms; some patients are asymptomatic, while others may have pneumonia. With a heavy infective dose or in the hyperinfection syndrome, individuals often develop cough, shortness of breath, wheezing, fever, and transient pulmonary infiltrates (Loeffler's syndrome). Larvae may be found in the sputum.

Immunocompromised patients with disseminated strongyloidiasis may have sepsis and meningitis, and larvae can be found in almost every body site (reported at autopsy). In these cases, eggs and both types of larvae (rhabditiform, filariform) can be found in the stool specimen.

Debilitated or immunocompromised patients should always be suspected of having strongyloidiasis, particularly if there are unexplained bouts of diarrhea and abdominal pain, repeated episodes of sepsis or meningitis with intestinal bacteria, or unexplained eosinophilia. However, similar infections have been seen in AIDS and non-AIDS patients. Diagnosing intestinal parasites in HIV-AIDS patients is necessary, especially in chronic alcoholics and those not on antiretroviral treatment. It is also important to check prehematopoietic stem cell transplantation patients in areas of endemicity.

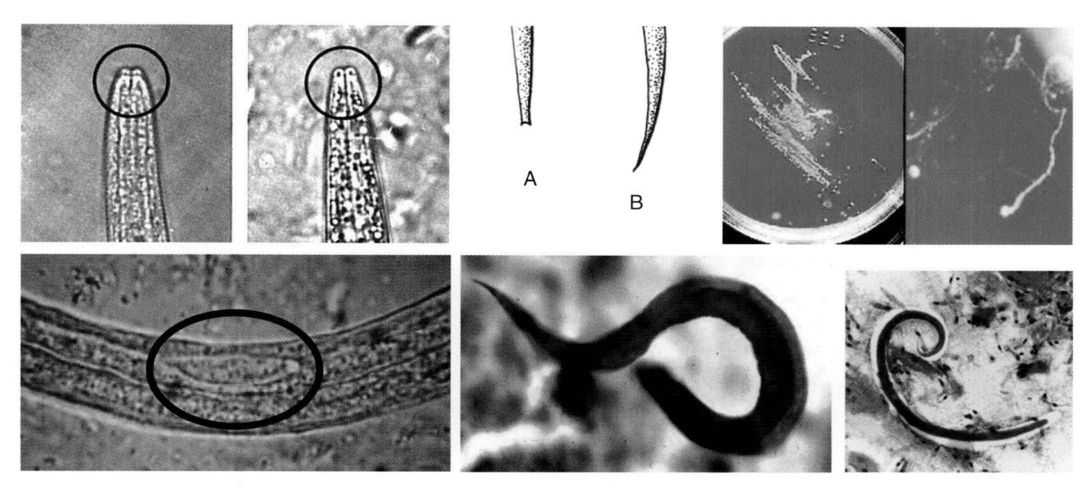

Figure 7.37 ***Strongyloides stercoralis.*** **(Top row)** Short mouth opening of *Strongyloides* rhabditiform larva; longer mouth opening of hookworm rhabditiform larva; drawing of the slit in the tail of a *Strongyloides* filariform larva (A) and the pointed tail of a hookworm filariform larva (B); agar plate culture (note the random tracks on the agar). **(Bottom row)** Packet of genital primordial cells of the *Strongyloides* rhabditiform larva (oval); rhabditiform larva in permanent stained smear (internal morphology not visible; wet mounts are recommended); H&E stain of BAL fluid showing a *Strongyloides stercoralis* filariform larva (reproduction authorized by the Editorial Committee of *Biomédica*; published in Rodríguez-Pérez EG, Arce-Mendoza A, Saldívar-Palacios R, Escandón-Vargas K, *Biomédica* **40**[Suppl 1]:32–36, 2020, https://doi.org/10.7705/biomedica.5071).

Enterobius vermicularis

Pathogenic	Yes
Disease	Enterobiasis
Acquired	Fecal-oral transmission of infective eggs via contaminated food or water or on fingers and hands; airborne eggs can also be inhaled and ingested
Body site	Intestine
Symptoms	Anal itching may be the only symptom; nervousness, insomnia, nightmares; occasional vaginal discharge
Clinical specimen	Adhesive tape anal swab
Epidemiology	Worldwide, primarily human-to-human transmission
Control	Improved hygiene, adequate disposal of fecal waste, adequate washing of hands

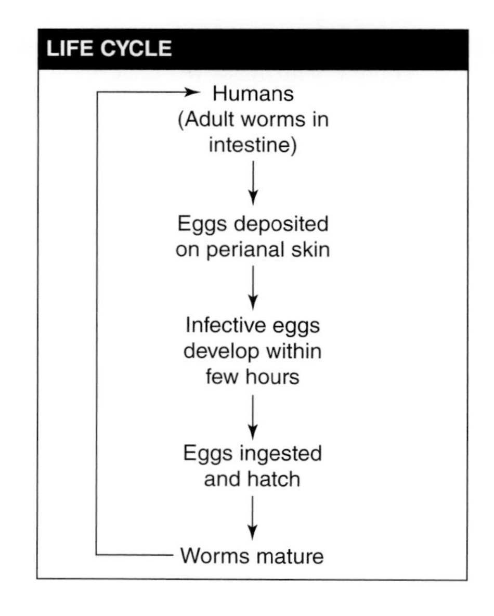

Diagnosis

Enterobius vermicularis is thought to cause the world's most common human parasitic infection. It has been said, "You had this infection as a child; you have it now; or you will get it again when you have children!"

Diagnosis depends on finding eggs or adult worms. This is normally done by sampling the perianal and perineal skin with cellulose tape (Scotch tape), which is applied sticky side down to the skin. The tape is transferred to a glass slide and examined under the microscope for eggs or adult worms.

Since the female worms migrate on a sporadic basis, a series of four to six consecutive tapes may be necessary to demonstrate the infection. The samples are taken late in the evening, when the patient has been sleeping for several hours, or first thing in the morning before the patient takes a shower or goes to the bathroom.

Adult worms may be found on or under the surface of the stool specimen, particularly in children. Eggs are occasionally recovered in stool, but this is an incidental finding, and stool is not the specimen of choice.

Treatment often includes counseling for the parents, who may be very upset at learning that their children have "worms." They may not realize how prevalent the infection is, particularly in children, or that many children will never have any symptoms or sequelae from the infection.

Since this infection is so common and transmission is so easy (anus-to-mouth contamination, soiled nightclothes, airborne eggs, and contaminated furniture, toys, and other objects), prevention is marginal.

Diagnostic Tips

The cellulose tape (Scotch tape) preparation is recommended as the diagnostic test of choice (a minimum of four to six consecutive negative tapes is required to rule out the infection). Prior to microscopic examination, lift one side of the tape, apply 1 drop of toluene or xylene, and press the tape down on the glass slide. The preparation will then be clear, and the eggs will be visible.

If Magic Mend tape is accidentally submitted, microscopic clarity can be restored by adding some immersion oil to the top of the tape. Commercial paddles or other collection devices are also acceptable. Although the paraffin swab is also an option, this method is seldom used. Because it is unlikely to routinely receive four to six consecutive tapes, therapy can be initiated on the basis of symptoms alone.

General Comments

E. vermicularis is thought to cause the world's most common human parasitic infection, regardless of geographic area and/or socioeconomic class. The infection is more prevalent in cool and temperate zones where people tend to bathe less often and change their underclothes less frequently. Prevalence in children can be high.

Infection is initiated by the ingestion of infective eggs, which hatch in the intestine (cecal region), where they develop into adult worms. It probably takes about 1 month for the female to mature and begin egg production. After fertilizing the female worms, the males usually die and may be passed out in the stool. Almost the entire body of gravid females is filled with eggs. At this point, the female migrates down the colon and out the anus, where the eggs are deposited on the perianal and perineal skin.

Occasionally the female worm migrates into the vagina. After egg deposition, the female worm probably returns to the intestine. Occasionally when the bolus of stool passes out of the anus, adult worms are found on the surface. Adult worms may also be picked up on the Scotch tape preparations used to diagnose this infection. Although egg deposition usually does not occur in the intestine, some eggs may be recovered in the stool. The eggs are fully embryonated and infective within a few hours. Transmission is often attributed to ingestion of infective eggs by nail biting and inadequate hand washing, but airborne eggs can also be inhaled and ingested.

The most striking symptom is pruritus, which is caused by migration of female worms from the anus onto the perianal skin to deposit eggs. The occasional intense itching results in scratching and occasional scarification. In most infected people this is the only symptom, and many individuals remain asymptomatic. Eosinophilia may or may not be present.

The degree of infection varies tremendously. As many as 5,000 worms have been removed from a single patient, but most cases average less than 1 migrating worm per evening. Women are symptomatic three times as often as men and young people more frequently than older people.

Description (Eggs)

The eggs have been described as footballs with one side flattened. They are oval, compressed laterally, and flattened on one side and measure 50 to 60 μm long by 20 to 30 μm wide. Occasionally, eggs are seen to contain a fully developed embryo; these eggs are infective within a few hours after being deposited.

Description (Adults)

The female worm is 8 to 13 mm long by 0.3 to 0.5 mm wide and has a pointed tail (hence the name pinworm). The male is much smaller, measuring 2 to 5 mm long by 0.1 to 0.2 mm wide, and has a curved caudal end. The males are rarely seen because they die shortly after copulation and are expelled.

The adult female is often white (or light in color), and the pointed tail is very obvious. There may be multiple worms on or below the surface of the stool specimen (fresh specimen); however, the adult worms are not seen frequently, even on the Scotch tape preparations.

The adult female worms are full of eggs and often "explode" when they migrate out of the body and begin to dry out.

Additional Information

Although tissue invasion has been attributed to the pinworm, it is uncommon. Enterobiasis is an uncommon cause of acute appendicitis in U.S. children; however, it may be associated with acute appendicitis, chronic appendicitis, ruptured appendix, or no significant symptoms. Confirmation as the actual cause of appendicitis remains somewhat controversial in the literature.

Other uncommon ectopic sites include the peritoneal cavity, lung, liver, urinary tract, and natal cleft. Pathologic examination usually shows chronic granulomatous inflammation with or without central necrosis, which is surrounded by polymorphonuclear neutrophilic leukocytes, eosinophils, and fibroblasts. Macrophages, giant cells, epithelioid cells, and Charcot-Leyden crystals may also be present. During a case of suspected transverse colon carcinoma, histologic examination of the mass revealed eggs of *E. vermicularis* embedded in granulomatous tissue in the submucosa of the colon; no malignancy was found. Apparently, this is the first report of enterobiasis presenting as colon carcinoma.

The eggs are infectious within a few hours and *under most circumstances, total prevention is neither realistic nor possible.* If a patient tends to be symptomatic with this infection, therapy may have to be repeated due to possible reinfection.

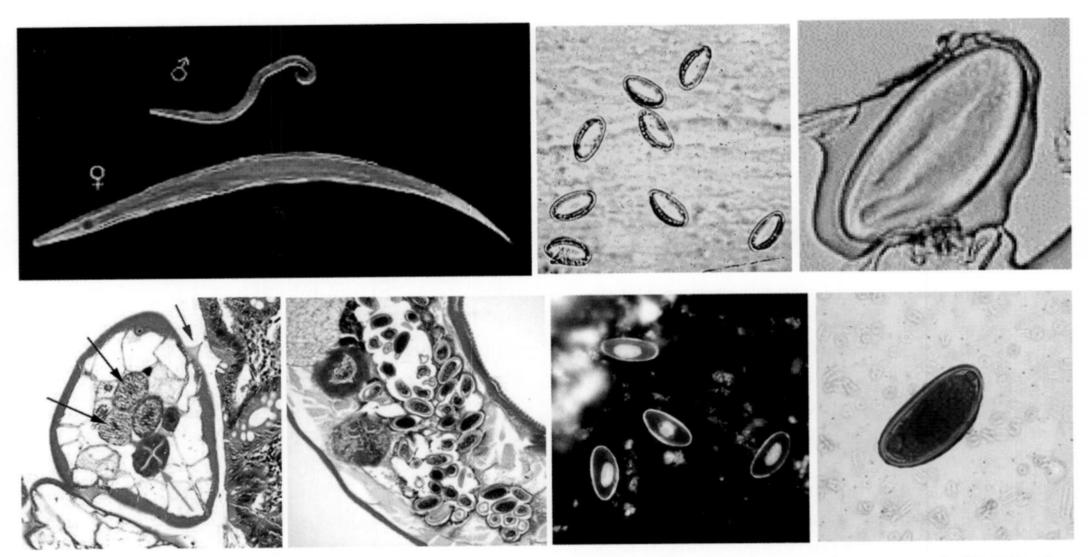

Figure 7.38 *Enterobius vermicularis*. **(Top row)** Adult male and female pinworms (courtesy of CDC DpDx, https://www.cdc.gov/dpdx/enterobiasis/index.html); pinworm eggs on cellulose tape; pinworm egg showing the larval form. **(Bottom row)** Cross section of an adult female *E. vermicularis* (note the presence of the alae [blue arrow], intestine [green arrow], and ovaries [black arrows]); longitudinal section of an adult female *E. vermicularis* (note the presence of many eggs); *E. vermicularis* eggs viewed under UV microscopy (images in this row courtesy of CDC DpDx, https://www.cdc.gov/dpdx/enterobiasis/index.html).

Ancylostoma braziliense
Ancylostoma caninum
Uncinaria stenocephala
(Dog and Cat Hookworms)

Pathogenic	Yes
Disease	Cutaneous larva migrans (CLM)
Acquired	Skin penetration by filariform (infective) larvae from contaminated soil, food, or water
Body site	Skin (serpiginous tracks)
Symptoms	Inflammatory response or migration of larvae in deeper tissues; intense itching
Clinical specimen	None; visual inspection of characteristic linear tunnels or tracks in the skin
Epidemiology	Worldwide, primarily dog/cat-to-human transmission
Control	Improved hygiene, adequate disposal of fecal waste, covering of all sandboxes where pets may defecate and children play

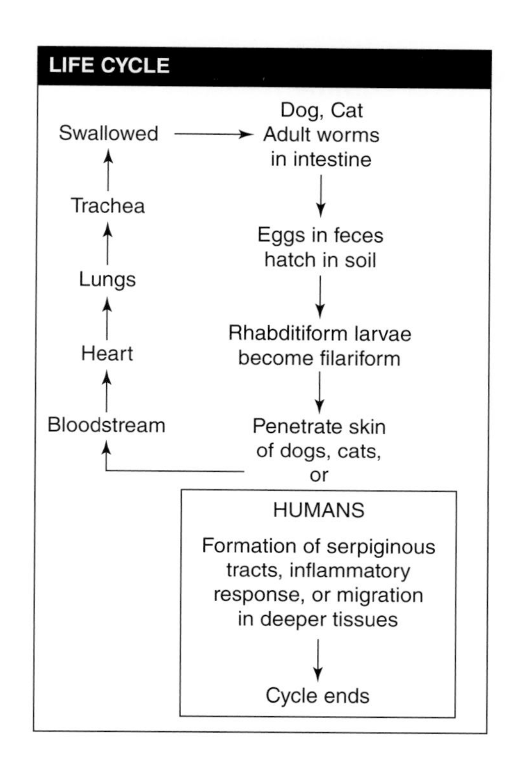

Diagnosis

The most common etiologic agent of cutaneous larva migrans (CLM) in the southern United States is *Ancylostoma braziliense*, a very common hookworm of dogs and cats. *Ancylostoma caninum*, the common hookworm of dogs, has been implicated in cases of CLM. Other species are also capable of producing CLM, although they are less common than *A. braziliense*.

Diagnosis is usually based on the characteristic linear tunnels or tracks and a history of possible exposure. Biopsy is not recommended. However, newer PCR methods for detection and identification of larvae in human tissues may improve the test results. There may be elevated eosinophilia (peripheral or sputum).

Patients with a relevant travel history may have travel-related skin diseases such as CLM as well as more common entities. To prevent and manage skin-related morbidity during travel, international travelers should avoid direct contact with sand, soil, and animals.

In 2006, the director of a children's aquatic sports day camp notified the Miami-Dade County Health Department of three campers who had received a diagnosis of CLM, or creeping eruption, a skin condition typically caused by *Ancylostoma* larvae. The investigation identified exposure to cat feces in a playground sandbox as the likely source of infection. Although CLM outbreaks are rarely reported to the Florida Department of Health, CLM is a potential health hazard in Florida. This disease cluster highlights the importance of appropriate environmental hygiene practices and education in preventing CLM.

Diagnostic Tips

Blood tests are not necessary for diagnosis. Eosinophilia is found in fewer than 40% of patients with

CLM and is nonspecific. A biopsy is not recommended (poor sensitivity), and while secondary changes and infiltrates may be helpful, it is not necessary to confirm this clinical diagnosis.

Scabies, loiasis, myiasis, schistosomiasis, tinea corporis, and contact dermatitis may have some overlapping features. However, these are all easily differentiated by the lack of serpentine migration. The most similar disease is the migrating lesion of *Strongyloides stercoralis*, termed larva currens.

General Comments

Human infection is acquired through skin penetration by infective larvae from the soil. These larvae can also cause infection when ingested. When larvae penetrate the skin, they produce pruritic papules, which after several days become linear tracks that are elevated and vesicular. The larva continues to migrate several centimeters each day, and the older portion of the track dries and becomes scarred. This process is associated with severe pruritus, and scratching can lead to secondary infection. Secondary bacterial infections often occur as a result of intense scratching of the tracks.

The larvae migrate in the epidermis just above the basal layer and rarely penetrate the dermis. Larval secretions containing proteolytic enzymes may cause inflammatory reactions associated with intense itching as the lesion progresses. Although the larvae cannot reach the intestine to complete their life cycle in the human host, they do occasionally migrate to the lungs, where they produce pulmonary infiltrates. Both larvae and eosinophils have been demonstrated in the sputum of patients with pulmonary involvement. The larvae die without reproducing, and the disease is self-limiting.

In hookworm folliculitis, the histologic picture is characterized by an eosinophilic folliculitis due to an inflammatory reaction to the presence of larvae trapped within the follicular canal. Only a few cases of hookworm folliculitis have been reported in the literature; however, this presentation should be recognized as one of the less typical presentations of CLM.

Additional Information

Larvae that first enter the skin and cause creeping eruption may later migrate to the deeper tissues (lungs), leading to pneumonitis with larval recovery in the sputum. Peripheral eosinophilia, as well as many eosinophils and Charcot-Leyden crystals in the sputum, may also be present.

Primary eosinophilic gastrointestinal disorders (eosinophilic esophagitis, eosinophilic gastritis, eosinophilic gastroenteritis, eosinophilic enteritis, and eosinophilic colitis) selectively affect the gastrointestinal tract with eosinophil-rich inflammation in the absence of known causes of eosinophilia, including parasitic infections, drug reactions, and malignancy. Eosinophils are important components of the gastrointestinal mucosal immune system, and these disorders involve IgE-mediated and delayed Th2-type responses.

Segmental eosinophilic inflammation of the gastrointestinal tract may be isolated or part of a multisystem problem. An increasing number of cases have been reported in northern Queensland, Australia. All of the patients were Caucasians with a wide age range and no previous illness who had severe abdominal pain, occasional diarrhea, weight loss, and dark stools; all cases were associated with eosinophilia and elevated serum IgE levels.

In one patient, a single adult *A. caninum* worm was found in a segment of inflamed ileum. Human hookworms do not occur in urban Australia, and no hookworm eggs were being passed in the stool. All the patients were closely associated with dogs, most of which had hookworms. Also, all patients treated with anthelmintic agents showed a return to normal peripheral-blood eosinophil counts.

The similarities among the reported cases implicate *A. caninum* as the cause of eosinophilic enteritis (EE). It has been speculated that *A. caninum* causes human EE by inducing allergic responses to its secretions, including cysteine proteinases, which are involved in pathogenesis in other parasites. Immunologic studies also suggest that this parasite is a major cause of EE and peripheral-blood eosinophilia. Although there are other causes, this disease entity may become more commonly recognized in other areas of the world, confirming the causative agent as the common dog hookworm.

In addition to pharmacologic therapy, banning dogs from beaches may decrease deposition of larvae into the soil. Although towels do not provide adequate protection against transmission, wearing protective footwear is generally effective.

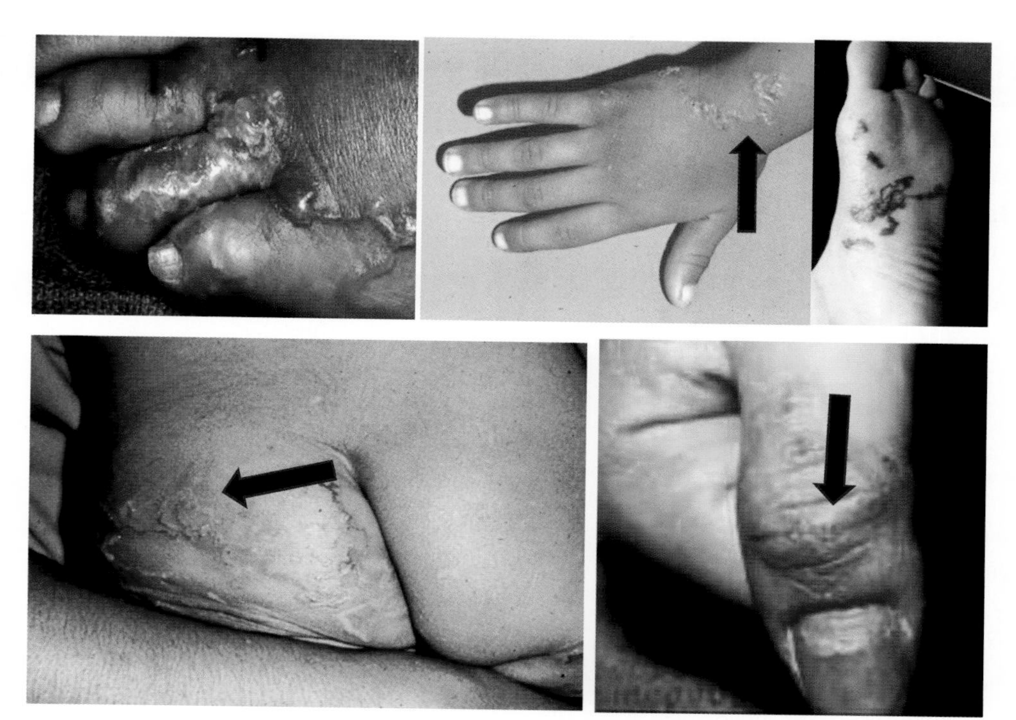

Figure 7.39 Cutaneous larva migrans. (Top row) Linear tracks on the top of the foot; linear tracks on the hand (arrow) (from the collection of Herman Zaiman, "A Presentation of Pictorial Parasites"); linear tracks on the bottom of the foot (courtesy of the Florida Department of Health, Duval County Epidemiology). **(Bottom row)** Linear tracks on the buttocks of a child (who had sat down in a sandbox containing sand contaminated with dog/cat hookworm larvae); linear tracks on thumb (both images from the collection of Herman Zaiman, "A Presentation of Pictorial Parasites").

Toxocara canis
Toxocara cati
(Dog and Cat Ascarid Worms)

Pathogenic	Yes
Disease	Visceral larva migrans (VLM), ocular larva migrans (OLM), neural larva migrans (NLM) (all considered neglected diseases)
Acquired	Egg ingestion from the soil via contaminated food or water
Body site	Tissues, including the eyes
Symptoms	Fever, hepatomegaly, hyperglobulinemia, pulmonary infiltrates, cough, neurologic disturbances, endophthalmitis
Clinical specimen	Serum (EIA)
Epidemiology	Worldwide; primarily dog/cat-to-human transmission
Control	Improved hygiene, adequate disposal of fecal waste, covering all sandboxes where pets may defecate and children play, deworming pets

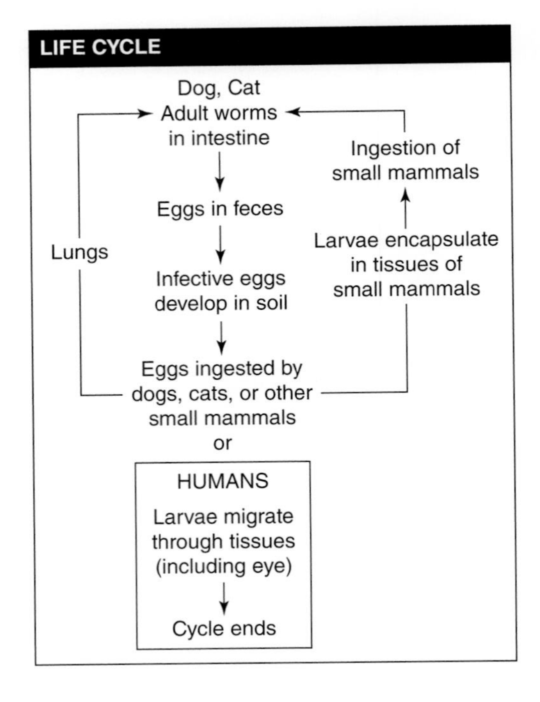

Diagnosis

Toxocariasis is a prevalent zoonosis with a significant socioeconomic impact, particularly on impoverished communities around the world. Humans are considered an accidental or aberrant host; *Toxocara* larvae cannot develop into adult worms inside the human body. VLM symptoms caused by *Toxocara* spp. must be differentiated from those caused by other tissue-migrating helminths (ascarids, hookworms, filariae, *Strongyloides* spp., and *Trichinella* spp.) and other hypereosinophilic syndromes. OLM can be confused with retinoblastoma, ocular tumors, developmental anomalies, exudative retinitis, trauma, and other childhood eye problems. It should be considered in any child with unilateral vision loss and strabismus who has raised, unilateral, whitish or gray lesions in the fundus. Peripheral eosinophilia may be absent. VLM should be suspected in pediatric patients with unexplained febrile illness and eosinophilia, especially if there is a history of pica and if hepatosplenomegaly and multisystem disease occur.

The diagnosis can be confirmed only by identification of larvae in autopsy or biopsy specimens. However, if children have *Ascaris* or *Trichuris* infections, toxocariasis might be considered, since all three infections are transmitted via ingestion of contaminated soil. Since biopsy specimens are usually not recommended, serologic testing is the most appropriate approach. EIA is recommended; it is highly specific and shows no cross-reactions with sera from patients infected with other common human parasites. The diagnostic titers vary between VLM throughout the body and OLM. Titers of 1:32 and 1:8 are diagnostic for VLM and OLM, respectively.

Diagnostic Tips

Biopsy specimens are usually not recommended; serologic testing is the most common approach and is recommended. Serum samples can be sent to the appropriate state Department of Public Health (check the applicable state submission requirements). These specimens are often sent to the CDC. History information is required, and each sample must be identified as serum or eye fluid, so that a correct interpretation of the results can be made.

Tissue specimens containing larvae can be referred to a reference center (university or Armed Forces Institute of Pathology, Washington, D.C.).

Life Cycle

Humans acquire infection by ingesting infective eggs of *Toxocara canis* or *Toxocara cati*. Pups are often infected by vertical transfer of larvae from their dams transplacentally or via nursing and can begin shedding eggs by 2 weeks of age. In cats, lactogenic but not transplacental transmission occurs. Occasionally, in some developed countries, urban and rural foxes are the primary source of eggs and human infection. Kittens and pups recover at 3 to 6 months of age. Infections in older animals are acquired by ingestion of infective eggs from soil or ingestion of larvae in infected rodents, birds, or other paratenic hosts. Eggs are shed in the feces and take about 2 to 3 weeks to mature and become infective.

Eating raw cows' liver is the main route for acquiring toxocariasis in Japan and Korea, whereas stray dogs and cats spreading eggs in the environment are the main source of infection for people in India and Southeast Asian countries. In developed countries, including the United States, France, and Austria, patients are infected via contact with soil contaminated with *Toxocara* eggs, for example in playgrounds, sandpits, and gardens. Information also implicates ingestion of uncooked meats as a potential cause of human toxocariasis.

After the eggs are accidentally ingested by a human, larvae hatch in the small intestine, penetrating the intestinal mucosa and migrating to the liver. During migration, the larvae do not mature, even if they make their way back into the intestine. The larvae are usually <0.5 mm long and 20 μm wide.

Most infections are probably asymptomatic. VLM occurs mostly in younger children (ca. 3 years), while OLM is more likely in older children (ca. 8 years). This does not tend to be the case with *Baylisascaris procyonis* (raccoon roundworm), where most cases of VLM, NLM, and OLM occur in very young children.

Additional Information

Toxocariasis has an increasing adverse impact on human health, particularly in underprivileged, tropical, and subtropical communities worldwide. Although tens of millions of people, especially children, will be exposed to and/or infected with *Toxocara* species, there is limited information on the relationship between seropositivity and disease (toxocariasis) on a global scale.

Toxocara spp. cause zoonotic infections worldwide, which may be much more common than previously thought. Infection rates in dogs are 2 to 90%, with the highest rates being in pups via transmission from their dams. The overall incidence of infected dogs older than 6 months is probably less than 10%.

Clinical symptoms depend on the number of migrating larvae and the tissue(s) involved. Infections range from asymptomatic to severe disease. Larvae often remain in the liver and/or lungs, where they become encapsulated in dense fibrous tissue. Other larvae continue to migrate, causing inflammation and granuloma formation. The most outstanding feature is a high peripheral eosinophilia, which may reach 90%. The overall severity of the clinical picture depends on the initial dose of infective eggs. As few as 200 *T. canis* larvae in small children may produce a peripheral eosinophilia of 20 to 40% for more than a year, with no other symptoms. Patients with 50% eosinophilia usually have symptoms, which might include fever, hepatomegaly, hyperglobulinemia, pulmonary infiltrates, cough, neurologic disturbances, and endophthalmitis. Although rare, CNS involvement can cause seizures, neuropsychiatric symptoms, or encephalopathy.

The relationship between asthma and covert toxocariasis remains unclear; however, data indicate a seroprevalence of anti-*T. canis* antigen (E/S antigen) of 26.3% in asthmatic patients and 4.5% in the controls. This suggests that asthmatic patients with anti-*Toxocara* IgE and IgG may have suffered a covert infection with *Toxocara* spp.

Evidence suggests that ocular disease can occur in the absence of systemic involvement and vice versa for VLM. Although these facts may be explained by possible *Toxocara* strain differences, VLM may reflect the consequences of the host inflammatory response to waves of migrating larvae, while OLM may occur in individuals who have not become sensitized.

Although many cases of VLM are diagnosed by serologic testing, toxocariasis has generally been defined as an infection with *Toxocara* spp., with no attempt to identify the species involved. Using preabsorbed sera, the ability to distinguish between the two species should be helpful in further biological, epidemiologic, and clinical studies. Although the currently recommended serologic test is EIA, a measurable titer does not always represent current infection. A small percentage of the U.S. population (2.8%) exhibits a positive titer that reflects the prevalence of asymptomatic toxocariasis.

The probability of hepatic toxocariasis can be evaluated using imaging techniques and ultrasonography. Findings include focal ill-defined hepatic lesions, hepatosplenomegaly, biliary dilatation, sludge, and periportal lymph node enlargement.

Several other animal parasites have been associated with visceral larva migrans-like syndromes. These include *Ascaris suum*, *Capillaria hepatica*, *Angiostrongylus cantonensis*, *Angiostrongylus costaricensis*, and *Baylisascaris procyonis*. The tissue phase of such human helminths as *Strongyloides stercoralis* and *Ascaris lumbricoides* can also produce similar clinical syndromes. Larvae of species of *Anisakis* and closely related nematodes of marine mammals have been reported to invade the stomach and other areas of the gastrointestinal tract of humans.

The following preventive measures are recommended: regularly deworming dogs and cats, beginning at 2 weeks of age; removing cat and dog feces from around homes and children's playgrounds; covering children's sandboxes when not in use; washing hands regularly after handling soil and before eating; and teaching children not to put dirty objects into their mouths.

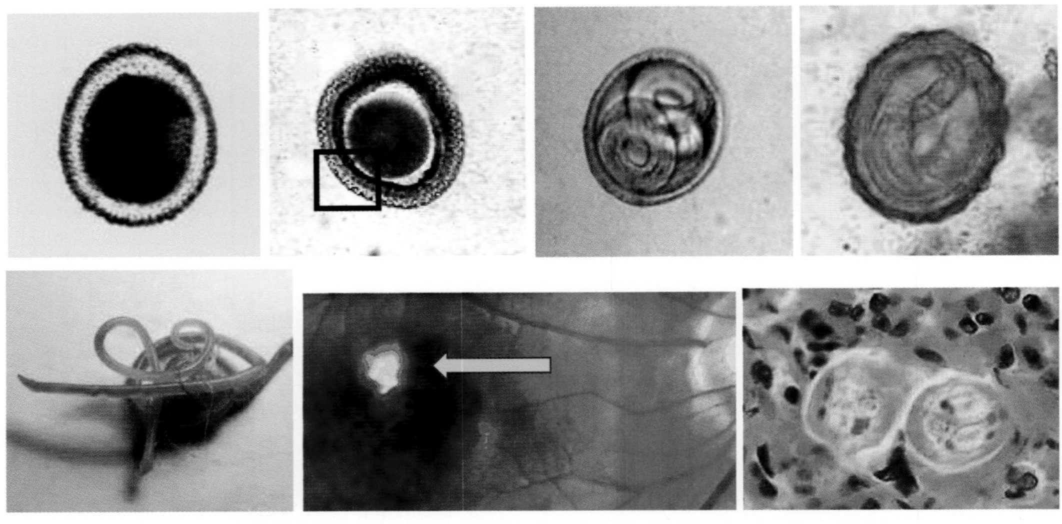

Figure 7.40 ***Toxocara*** **spp. (Top row)** *Toxocara* eggs (note the dimpled shells [square]). The last two eggs contain larvae. **(Bottom row)** Adult *Toxocara* worms; ocular larva migrans (toxocariasis) (note the white, elevated granuloma on the retina (arrow) (courtesy of Despommier D, *Clin Microbiol Rev* **16**:265–272, 2003, doi: 10.1128/CMR.16.2.265-272.2003); cross section of *Toxocara* sp. in liver, H&E (courtesy of the CDC Public Health Image Library).

Dracunculus medinensis

Pathogenic	Yes
Disease	Dracunculiasis (Guinea worm disease) (WHO neglected tropical disease)
Acquired	Ingestion of infective copepods
Body site	Deep connective tissue, subcutaneous tissues
Symptoms	Erythema, tenderness in area of blister; urticarial rash, intense pruritus, nausea, vomiting, diarrhea, or asthmatic attacks
Clinical specimen	*Blister:* Adult worm; calcified worms/radiography
	Muscle: Biopsy, routine histology
	Serum: EIA, bentonite flocculation
Epidemiology	Angola, Cameroon, Chad, Ethiopia, Mali, South Sudan (eradication within reach)
Control	Anthelmintic treatment of dogs, thorough cooking of fish, burying fish entrails; there may also be additional paratenic hosts (other than dogs or fish)

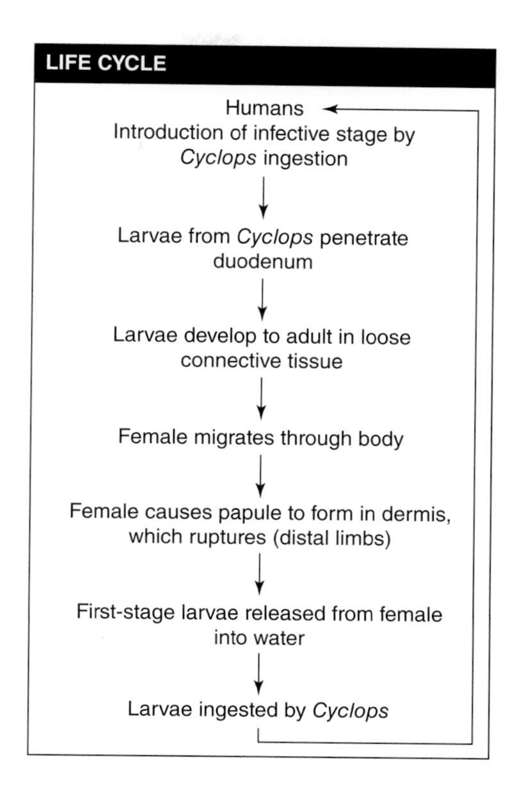

Diagnosis

Human infection is acquired from ingestion of infected copepods (*Cyclops* water fleas). The released larvae penetrate the duodenal mucosa and develop in the loose connective tissue. The possibility also exists that paratenic hosts, such as fish, tadpoles and frogs, are important means of transporting infective larvae of *Dracunculus* species up the food chain, thus facilitating transmission to the definitive hosts.

Diagnosis can be confirmed at the time the cutaneous lesion forms, with subsequent appearance of the adult worm. Infected lesions must be distinguished from carbuncles, deep cellulitis, focal myositis or periostitis, and even rheumatism. Calcified worms may also be found in subcutaneous tissues by radiography. They may appear as linear densities (up to 25 cm), tightly coiled structures, or sometimes nodules.

For centuries, the worms have been removed by slowly being wound around a stick. This approach works well unless the worm is accidentally broken and secondary infection occurs. Allergic manifestations can be decreased by using epinephrine.

Diagnostic Tips

Diagnosis is mainly by clinical presentation. Peripheral eosinophilia can be present in the blood work. Immunoglobulin G4 levels might be elevated. If the worms die before they emerge from the skin, they may calcify and can be visible on X rays. The clinical picture can mimic filariasis and/or cellulitis.

General Comments

The worms are very long, with the females measuring up to 1 m in length by 2 mm in width. The male is much smaller and inconspicuous (2 cm long). The worms mature in the deep connective tissue, and the females migrate to the subcutaneous tissues when they are gravid and contain coiled uteri filled with rhabditiform larvae. Maturation takes approximately 1 year. At this stage in the life cycle, the female migrates to the skin and a papule is formed in the dermis, usually by the ankles or feet (although papules can be anywhere on the body). The papule changes into a blister within 24 h to several days. Eventually, the blister ulcerates, and on contact with freshwater, a portion of the uterus prolapses through the worm's body wall, bursts open, and discharges thousands of larvae into the water. This may happen several times until all of the larvae are discharged. The larvae are then ingested by an appropriate species of *Cyclops*. Development takes about 8 days before the larvae are infective for humans. Although the adult worms are often described as creamy white, there are reports of red worms that appear to be female *Dracunculus medinensis*.

Additional Information

After ingestion of an infected copepod, no specific pathologic changes are seen with larval penetration into the deep connective tissues and maturation of the worms. Once the gravid female begins to migrate to the skin, some erythema and tenderness in the area where the blister will form may occur. Several hours before blister formation, some systemic reactions may occur (urticarial rash, intense pruritus, nausea, vomiting, diarrhea, or asthmatic attacks). The lesion develops as a reddish papule, measuring 2 to 7 cm in diameter. Symptoms usually subside when the lesion ruptures, discharging both the larvae and worm metabolites.

If the worms are removed at this time, healing usually occurs with no problems. If the worm is damaged or broken during removal, there may be an intense inflammatory reaction with possible cellulitis along the worm's migratory track. If secondary infection occurs, there may be more serious sequelae, including arthritis, synovitis, and other symptoms, depending on the site of the lesion. Abscesses can occur when the worms migrate to other tissue sites, like the lung, pericardium, and spinal cord. Occasionally sepsis can occur due to systemic infections and cellulitis.

Comments on Potential for Total Eradication

The number of cases of dracunculiasis (Guinea worm disease) decreased from an estimated 3.5 million in 1986 to 28 in 2018. Emergence of Guinea worm infections in dogs has complicated eradication efforts.

During January to June 2019, the number of human dracunculiasis cases reported increased to 25 cases in three countries (Angola, Cameroon, and Chad) and 1,346 infected domestic dogs were reported; Ethiopia, Mali, and South Sudan reported no human cases.

Villages where endemic transmission of dracunculiasis has ended (i.e., zero human cases or animal infections reported for ≥12 consecutive months) are kept under active surveillance for 2 additional years. Active surveillance involves daily searches of households by village volunteers (supported by their supervisors) for persons or animals with signs of dracunculiasis. An imported human case or animal infection is one resulting from ingestion of contaminated water in a place other than the community where the case or infection is detected and reported. Since 2012, no internationally imported cases or infections have been reported.

WHO certifies that a country is dracunculiasis-free after adequate nationwide surveillance for ≥3 consecutive years with no indigenous human cases or animal infections. An indigenous dracunculiasis human case or animal infection is defined as an infection consisting of a skin lesion or lesions with emergence of one or more Guinea worms in a person or animal who had no history of travel outside their residential locality during the preceding year. Although eradication is within reach, it is complicated by civil unrest, insecurity, and continuing epidemiologic and zoologic questions.

Apparently dracunculiasis is widespread throughout the eastern United States and Canada, and *Dracunculus* infections in dogs are more common than is revealed in published literature. However, more research is needed to understand the epidemiology, including transmission route(s), prevalence, and distribution, of this parasite.

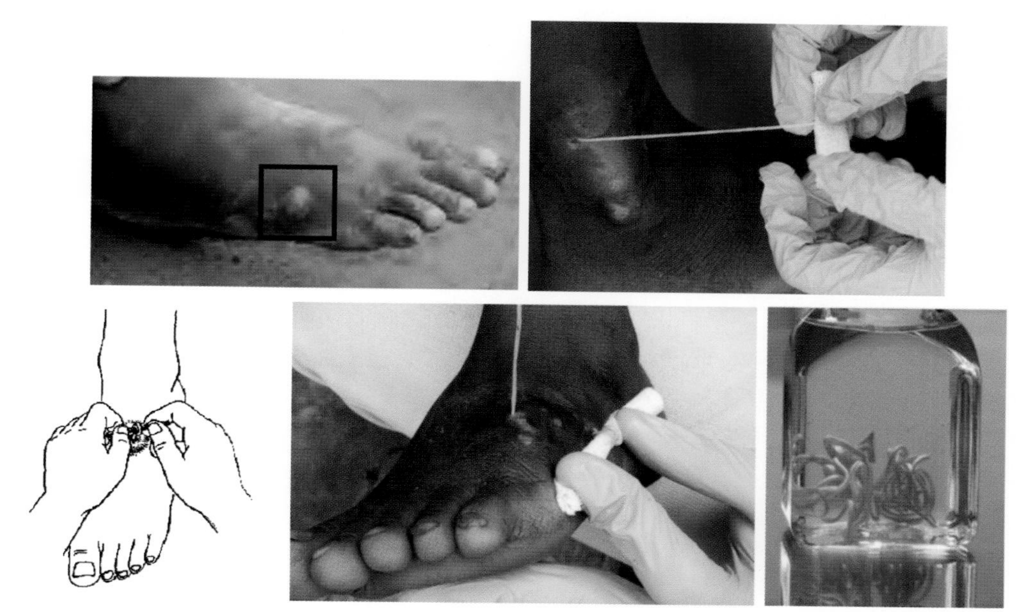

Figure 7.41 ***Dracunculus medinensis.*** **(Top)** *Dracunculus medinensis* (Guinea worm) blister (courtesy of Global 2000/The Carter Center, Atlanta, GA); removal of the larvae (worm) from ankle (The Carter Center/L. Gubb). **(Bottom)** Drawing (Sharon Belkin) and image (The Carter Center/L. Gubb) of blister showing Guinea worm removal; worms (The Carter Center).

Trichinella spiralis

Pathogenic	Yes
Disease	Trichinosis
Acquired	Ingestion of infective raw or poorly cooked meat (pork, bear, walrus, horse, or any other carnivorous or omnivorous mammal)
Body site	Intestine, striated muscle
Symptoms	Diarrhea (first 24 h), muscle pain, fever; depends on tissues infected
Clinical specimen	*Intestinal*: Stool (rarely requested)
	Muscle: Biopsy, routine histology
	Serum: EIA, bentonite flocculation
Epidemiology	Worldwide, primarily animal-to-human transmission
Control	Adequate cooking of meat, use of cooked rather than raw garbage for pigs

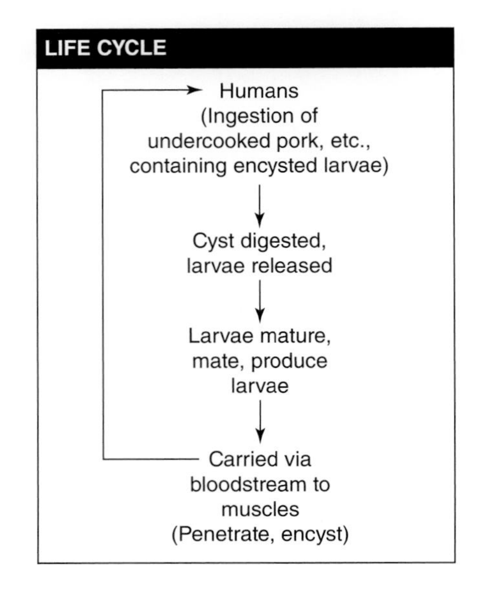

Diagnosis

The severity of the disease is proportionate to the number of larvae ingested. In heavy infections, the clinical symptoms correlate with the biologic stages of *Trichinella* as it completes its life cycle. The first clue may be a history of possible ingestion of raw or rare pork or other infected meat. There may also be other individuals from the same group with similar symptoms. Trichinosis should always be included in the differential diagnosis of any patient with periorbital edema, fever, myositis, and eosinophilia, regardless of whether a complete history of ingestion of raw or poorly cooked possibly infected meat consumption is available. If present, subconjunctival and subungual splinter hemorrhages also add support to such a presumptive diagnosis.

Muscle biopsy (gastrocnemius, deltoid, and biceps) specimens may be examined by compressing the tissue between two slides and checking the preparation under low power (10× objective). This method does not show positive results until 2 to 3 weeks after the onset of the illness.

Serologic tests are also very helpful, the standard two being EIA and bentonite flocculation (BF), which are recommended for trichinosis. EIA is used for routine screening, and all EIA-positive specimens are tested by BF for confirmation. A positive reaction with both tests indicates infection with *Trichinella spiralis* within the last few years. Titers tend to peak in the second or third months postinfection and then decline over a few years.

Diagnostic Tips

The history and clinical findings may suggest possible trichinosis (consumption of rare or raw infected meat). Remember to check hematology results for a possible eosinophilia (which can reach 50% or higher).

Examination of suspect meat using compression slides may reveal larvae (artificial digestion procedure). Treatment of a portion of the muscle for several hours with pepsin and hydrochloric acid to liberate the encysted larvae, followed by microscopic examination of the concentrated sediment for larvae, may improve the diagnostic yield. Note that not all species of larvae form the capsule; however, the unencapsulated larvae can still be seen in a squash preparation of biopsy material.

Larvae or adult worms are rarely recovered in fecal specimens during the intestinal phase (diarrhea). However, this could occur as an incidental finding. Serologic tests for antibody detection may be very helpful; coproantigen detection tests are being developed.

General Comments

Although recommendations have been made to use several species designations, some publications still use the single species designation *T. spiralis*. Genetic relationships among many *Trichinella* isolates are being assessed by dot blot hybridization, restriction endonuclease, and gel electrophoresis techniques. Taxonomic changes will continue to occur.

Human infection is initiated by ingestion of raw or poorly cooked pork, bear, walrus, or horse meat, or meat from other mammals (carnivores and omnivores) containing viable, infective larvae. After digestion, the excysted larvae invade the intestinal mucosa, develop through four larval stages, mature, and mate by day 2. By day 6 of infection, the female worms begin to deposit motile larvae, which are carried by the intestinal lymphatic system or mesenteric venules to the body tissues, primarily striated muscle. Deposition of larvae continues for 4 to 6 weeks, with each female producing up to 1,500 larvae in the nonimmune host. Newborn larvae can penetrate almost any tissue but can continue their development only in striated muscle cells. With the exception of *Trichinella pseudospiralis*, *T. papuae*, and *T. zimbabwensis*, invasion of striated muscle cells stimulates the development of nurse cells. As the larvae begin to coil, the nurse cell completes the formation of the cyst within ca. 2 to 3 weeks.

The most active muscles, which have the greatest blood supply, including the diaphragm, muscles of the larynx, tongue, jaws, neck, and ribs, the biceps, the gastrocnemius, and others, are invaded.

The cyst wall develops from the host's immune response to the presence of the larvae, and the encysted larvae may remain viable for many years, although calcification can occur within 6 to 9 months. Just five larvae per gram of body muscle can cause death, although 1,000 larvae per g have been recovered from individuals who died from causes other than trichinosis.

Description (Adults, Larvae, Cyst)

The adult viviparous female (4 mm by 60 μm) is larger than the male (1.5 mm by 40 μm) and may produce 1,000 to 10,000 larvae during her 6-week life span. The infective larvae (about 1 mm long) become encysted in striated muscle, where they may retain their viability and infectivity for years. In the human host, the cyst measures about 400 by 260 μm, and within the cyst, the coiled larva is 800 to 1,000 μm long. At this point, the larvae are fully infective.

Additional Information

Preventive measures for pork containing temperate-zone strains include refrigeration at 5°F (−15°C) for not less than 20 days, at −10°F for 10 days, or at −20°F for 6 days or deep freezing (−37°C). Smoking, salting, and drying are not effective. In 1981, the USDA issued a news release suggesting that microwave cooking might not kill the larvae. The current recommendation states that "*all parts* of pork muscle tissue must be heated to a temperature not lower than 137°F (58.3°C)." An internal meat thermometer should be used when cooking pork. Reduction in the number of cases is due primarily to regulations requiring heat treatment of garbage and low-temperature storage of the meat. Occasional outbreaks are frequently due to problems with feeding, processing, and cooking of pigs raised for home use. Although freezing meat at 5°F (−15°C) for 20 days is recommended, freezing may not be adequate to eliminate cold-resistant strains of *Trichinella*.

Recent information also confirms the need to review the intentional feeding of animal products and kitchen waste to horses, a high-risk practice requiring implementation of regulations to ensure that such feeds are safe for horses, as is currently required for feeding to swine.

Symptoms of trichinosis are generally separated into three phases. Phase 1 is related to the presence of the parasite in the host prior to muscle invasion. Phase 2 is related to the inflammatory and allergic reactions due to muscle invasion; there may also be an incubation period of up to 50 days. Phase 3 is the convalescent phase or chronic period.

Damage can be classified as (i) intestinal and (ii) muscle penetration and larvae encapsulation. Damage caused in any phase of the infection is usually based on the original number of ingested cysts. Early symptoms include diarrhea, nausea, abdominal cramps, and general malaise, all of which may suggest food poisoning, particularly if several people are involved. Diarrhea can be prolonged, lasting up to 14 weeks (average, 5.8 weeks), with little or no muscle symptoms. It is unknown whether this new clinical presentation is related to variant biological behavior of Arctic *Trichinella* organisms, to previous exposure to the parasite, or to other factors. During muscle invasion, there may be fever, facial (periorbital) edema, and muscle pain, swelling, and weakness. The extraocular muscles are usually the first to be involved, followed by the muscles of the jaw and neck, limb flexors, and back. Problems in chewing, swallowing, or breathing can be seen. The most severe symptom is myocarditis (after week 3); death may occur between weeks 4 and 8. Peripheral eosinophilia of at least 20%, often over 50%, and possibly up to 90% is present during the muscle invasion phase of the infection.

Other severe symptoms may involve the CNS. Although *Trichinella* encephalitis is rare, it is life-threatening. CT, angiogram, and electroencephalogram are of no diagnostic assistance. Up to 20% of patients with trichinosis have CNS involvement; the mortality rate may reach 50% if they are not treated. As a reminder, although most infections are self-limiting, serious complications or death may result from invasion of the heart, lungs, or CNS.

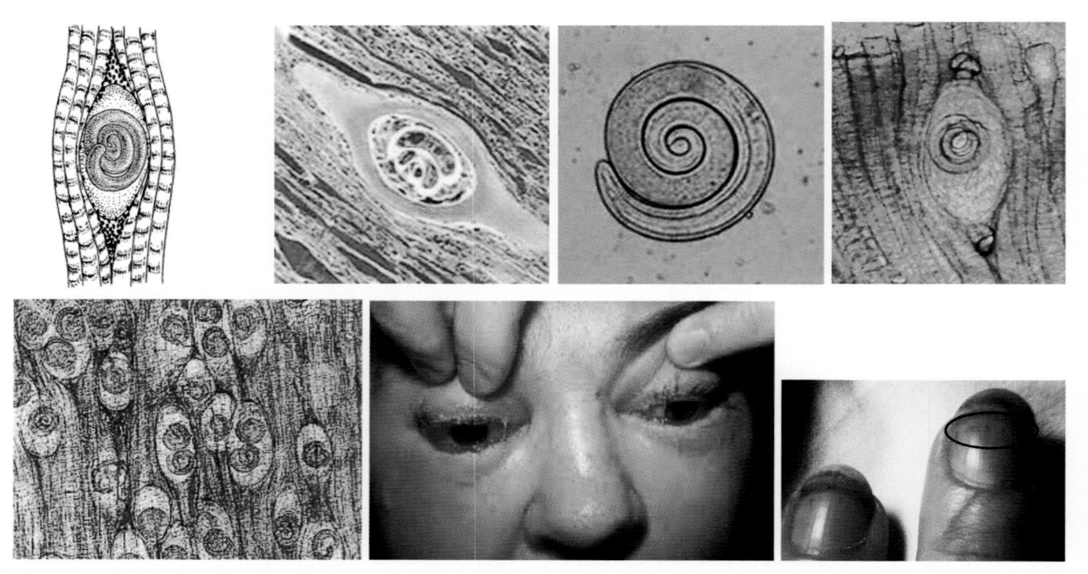

Figure 7.42 ***Trichinella*** **spp. (Top row)** Drawing of an encysted larva (illustration by Sharon Belkin); three photographs of larvae, the one on the left being sectioned (histopathology). **(Bottom row)** Encysted larvae; periorbital swelling and eye irritation (courtesy of CDC PHIL, #342, CDC/Emory Univ., Dr. Thomas F. Sellers); example of splinter hemorrhages in nail (courtesy of CDC PHIL, https://phil.cdc.gov/Details.aspx?pid=343).

Filarial Worms

Pathogenic	Yes, some more than others
Disease	Filariasis (WHO neglected tropical disease)
Acquired	Bite of various arthropods
Body site	Blood, lymph, tissues
Symptoms	Few to elephantiasis; depends on body site (lymphatics, tissues, etc.)
Clinical specimen	Blood, multiple draws (in EDTA); skin snips
Epidemiology	Sporadic distribution; arthropod-to-human transmission
Control	Vector control

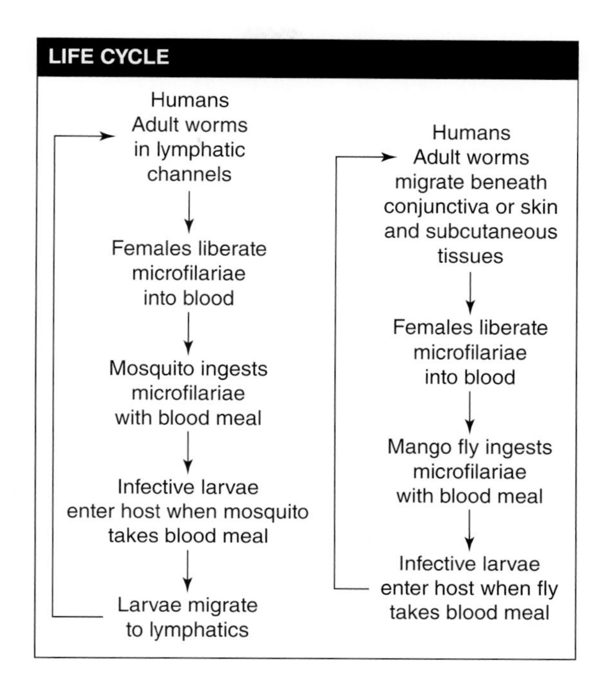

Diagnosis

The filarial nematodes are a group of arthropod-borne worms that reside in the subcutaneous tissues, deep connective tissues, lymphatic system, or body cavities of humans. Some adult filarial worms can survive in the human host for many years, causing a number of chronic and debilitating symptoms, including inflammatory reactions.

Presumptive diagnosis of filariasis must include lymphangitis and lymphedema. Confirmation of filarial infections is based on detecting microfilariae in blood or tissues. Microfilariae can be identified to the species level by the presence or absence of a sheath and the position of body nuclei in stained specimens. Definitive diagnosis is based on detecting microfilariae of *Wuchereria bancrofti*, *Brugia malayi*, *Loa loa*, *Mansonella ozzardi*, and *Mansonella perstans* in the circulating blood. Microfilariae of *Onchocerca volvulus* and *Mansonella streptocerca* are detected primarily in the skin, although they are occasionally detected in the blood. Microfilariae may also be found in hydrocele fluid and urine, particularly in patients who have high microfilaremias or have been treated recently with diethylcarbamazine.

The optimal time for drawing blood to detect periodic infections of *W. bancrofti*, *B. malayi*, and *Brugia timori* is between 10 p.m. and 4 a.m. Blood to detect subperiodic species of *W. bancrofti* and *B. malayi* may be drawn any time. Blood for *L. loa* should be drawn between 10 a.m. and 4 p.m., and blood to detect *Mansonella* infections can be drawn at any time. Finger-stick or earlobe blood may be taken for direct wet, thin, and thick blood smears. Blood films may be stained with Giemsa or Delafield's hematoxylin stain. Giemsa stain does not stain the microfilarial sheath adequately, although hematoxylin stains do. Examination of a blood film for microfilariae should include low-power review of the *entire* film. Sheathed microfilariae often lose their sheaths when drying on thick films.

Serologic tests are more meaningful in patients who have not resided in the areas of endemicity for extended periods.

Diagnostic Tips

A travel and geographic history should be obtained to determine the best type of specimen and optimal collection time for the filarial infection suspected. In addition to multiple thin and thick blood films, Knott or membrane concentration techniques should be used to detect microfilariae normally found in the peripheral blood. It is important to examine every portion of the thin and thick blood films; microfilariae are often found at the outside edges or in the original drop from which the thin film was pulled. All thin or thick blood films should first be examined using the 10× objective (low power). If immediate examination is undertaken at a higher magnification, the microfilariae may be missed, particularly if the parasite numbers are low.

Giemsa stain does not stain the *W. bancrofti* sheath as well as a hematoxylin-based stain (Delafield's hematoxylin). Also, Giemsa stains the sheath of *Brugia* sp. pink but does not stain the sheath of *B. timori,* a species found in the islands near Indonesia.

Antigen detection tests are commercially available and may be very helpful in the detection of circulating filarial antigens. PCR may prove to be valuable in the diagnosis of lymphatic filariasis; however, these procedures are often limited to research facilities. Some testing is now available commercially. Ultrasonography has proven to be very valuable in assessing lymphatic filariasis in both adults and children; this approach can be much more sensitive than a physical examination alone.

Any patient from the area of endemicity for *M. streptocerca* who presents with hypopigmented macules and pruritus should be suspected of having the infection. However, this appearance can be confused with leprosy, and some patients are misdiagnosed and inappropriately treated for leprosy for long periods. The diagnosis is made by the detection of microfilariae from skin snips taken over the scapula and examined as wet mounts.

In areas of endemic onchocerciasis, the clinical diagnosis is not difficult when individuals present with typical features such as hanging groin, leopard skin, skin atrophy, or subcutaneous nodules. However, skin lesions must be differentiated from those caused by scabies, insect bites, streptocerciasis, contact dermatitis, hypersensitivity reactions, traumatic or inflammatory depigmentation, tuberculoid leprosy, dermatomycoses, and treponematoses.

Microfilariae of *Dirofilaria* cannot be found in the blood or tissues, and serologic results lack sensitivity and specificity. Some patients have a moderate eosinophilia. Diagnosis can be confirmed by the identification of worms in surgical or autopsy specimens. Because the immature larvae may be detected in only a few microscopic sections, careful histologic examination is necessary. The cuticle of nematodes contains chitin, which can be stained with nonspecific whiteners such as calcofluor white. Dirofilaria larvae stained with calcofluor white can be easily recognized in tissue sections, whereas the parasite may be difficult to identify using routine histologic stains. Microfilariae may not be present in the blood during the early and late stages of the disease. However, a history of recurrent episodes of lymphangitis and lymphadenitis may form the basis for a presumptive diagnosis.

General Comments

Infections are transmitted to humans by the bites of obligate blood-sucking arthropods that had become infected through ingesting larvae (microfilariae) in a blood meal from a mammalian host. Each parasite has a complex life cycle, and human infections are not readily established unless there is intense and prolonged exposure to infective larvae. After exposure, it may take years before significant pathologic changes in the human host are evident. Adult stages inhabit the lymphatic system, subcutaneous tissues, or deep connective tissues. Adult females produce microfilariae, i.e., prelarvae that may retain the egg membrane (sheathed microfilariae) or may lose it (unsheathed microfilariae). Once released by the female worm, the highly motile and threadlike microfilariae can be detected in the peripheral blood or cutaneous tissues, depending on the species. They may survive for 1 to 2 years, are not infective for other vertebrate hosts, and do not develop further in the host.

Adult *O. volvulus* worms (which cause a WHO neglected tropical disease) lie within fibrous tissue capsules in the dermis and subcutaneous tissues, and the microfilariae are usually nearby to be ingested by a species of *Simulium* (blackfly or buffalo gnat). Humans are infected when bitten by the infected fly, and larvae are deposited into the bite site. Microfilariae are normally found in the dermis.

When *Dirofilaria* worms lodge in the pulmonary artery branches, they cause an infarct. These lesions are usually on the periphery of the lungs and are sharply defined (coin lesion). There is a central necrotic area surrounded by a granulomatous inflammation and a fibrous wall. Dead or dying worms may be found in the lesion. Reported cases highlight the morphologic variation seen in human pulmonary dirofilariasis and emphasize the need to consider this diagnosis in all cases of necrotizing granulomas of the lungs.

Additional Information

Three species, *W. bancrofti*, *B. malayi*, and *O. volvulus*, account for most infections. There are 90 million people currently infected (two-thirds live in China, India, and Indonesia) with *W. bancrofti*, *B. malayi*, and *B. timori*.

In lymphatic filariasis with *W. bancrofti* and *B. malayi*, disease is caused by the presence of worms in the regional lymphatic vessels and, particularly, by the host response to the worms and worm products. The microfilariae are released into the blood. Infections involving small numbers of worms are often asymptomatic. Early symptoms consist of intermittent fever and enlarged, tender lymph nodes. The inguinal lymph nodes are very often involved. The lymphatic vessels that drain into the lymph nodes and that harbor the developing and adult worms also become inflamed and painful. In more chronic infections, there may be pain also in the epididymis and testes. Swollen lymphatics may burst and drain into the genitourinary system; the resulting chyluria may be noticed and may serve as a stimulus to see a physician.

In a small number of chronic cases, permanent lymphatic dysfunction caused by repeated exposure to infection over a number of years results in permanent lymphatic blockage and massive lymphedema. The exact cause of elephantiasis is not fully understood; however, repeated exposure appears to result in abnormally large amounts of collagenous material and fibrosis of the tissue around the affected lymphatics.

O. volvulus has infected 45 million to 50 million people in Africa and Central and South America, of whom approximately 1 million are blind. Symptoms of heavy *O. volvulus* infections include dermatitis, onchocercomas (subcutaneous nodules containing adult worms), lymphadenitis, and blindness. Individuals with onchocerciasis may have clinically normal skin, whereas others may have pruritus and disfiguring skin lesions. The pruritus may cause sleeplessness, fatigue, and weakness. The skin may be painful, hot, and edematous, eventually resulting in permanent thickening. Acute attacks of onchodermatitis have resulted in a purplish skin discoloration known as *mal morado* in Central America. Chronically infected skin loses its elasticity and becomes thickened. In Central America, these skin changes are frequently seen as noticeable thickening of the earlobes and thickening of facial skin to mimic leonine facies (seen in lepromatous leprosy). In Africa, the same skin changes occurring in the hip region produce a condition known as hanging groin, which may predispose infected individuals to inguinal and femoral hernias. Lymphadenopathy in the inguinal and femoral areas is common among Africans.

Note: Patients who have been treated for lymphatic filariasis should have blood specimens reexamined for microfilariae 2 to 6 weeks posttherapy. Onchocerciasis patients should have skin snips examined 3 to 6 months posttherapy.

The clinical manifestations of filariasis vary and may depend on host factors and parasite strains. Some patients harbor adult worms without a peripheral microfilaremia, or the microfilaremia may be too low to be detected by the usual laboratory procedures. Other patients have a heavy microfilaremia but are clinically asymptomatic. Many patients who were thought to be microfilaremic but clinically asymptomatic have symptoms of elephantiasis.

Early manifestations include high fevers (filarial or elephantoid fever), lymphangitis, and lymphadenitis. Filarial fever usually begins with a high fever and chills that last 1 to 5 days before spontaneously subsiding; patients often do not have a microfilaremia. The lymphangitis extends distally from the affected nodes where the filarial worms reside. Lymphadenitis and lymphangitis develop in the lower extremities more commonly than in the upper. There can also be genital (almost exclusively a feature of *W. bancrofti* infection) and breast involvement. The lymph nodes most often affected are the epitrochlear and femoral nodes. They are firm, discrete, and tender and tend to remain enlarged. The lymph vessel is indurated and inflamed. The overlying skin is tense, erythematous, and hot, and the surrounding area is edematous. Occasionally abscesses form at the lymph node or along the lymphatic system and may take 2 to 3 months to heal.

In *Dirofilaria* infections, adult worms reside in the right heart of dogs and microfilariae are found in the blood, where they are ingested by mosquitoes or *Simulium* black flies. After biological development in the vector, dogs and humans may be infected with the infective larvae during a blood meal. Development to a mature adult takes approximately 180 days in the dog. In humans, the worms do not reach maturity and no microfilariae can be detected. The cuticle of nematodes contains chitin, which can be stained with nonspecific whiteners such as calcofluor white. *Dirofilaria* larvae stained with calcofluor white can be easily recognized in tissue sections, whereas the parasite may be difficult to identify using routine histologic stains.

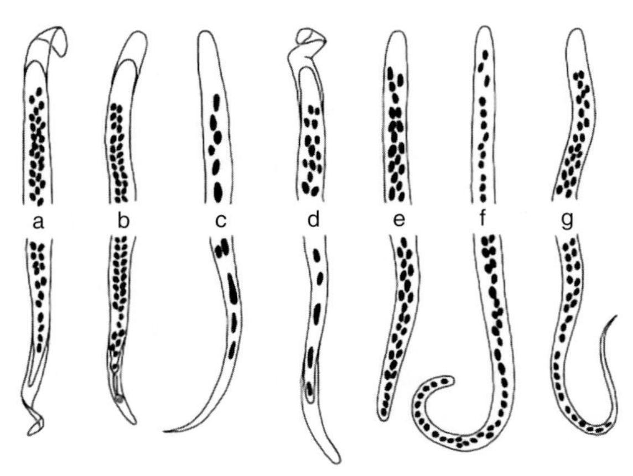

Figure 7.43 Filarial infections. Diagrams of human microfilariae—*Wuchereria bancrofti*, *Brugia malayi*, *Onchocerca volvulus*, *Loa loa*, *Mansonella perstans*, *Mansonella streptocerca*, and *Mansonella ozzardi*. (Illustration by Nobuko Kitamura.)

Plate 7.0 **(Row 1)** *Brugia malayi* microfilaria (note the pink sheath and two terminal nuclei); *B. malayi* tail (again, note the sheath and two terminal nuclei [circle]); *B. malayi* head (sheath not visible). **(Row 2)** *Loa loa* microfilaria (note the sheath); *L. loa* tail (note that nuclei run all the way to the end of the tail [circle]; the sheath is not visible); *L. loa* head (the sheath is visible [arrow]). **(Row 3)** *L. loa* in eye (left and middle); *L. loa* Calabar swelling (note the swollen left knee) (right) (all three images from the collection of Herman Zaiman, "A Presentation of Pictorial Parasites"). **(Row 4)** Head of *Wuchereria bancrofti* microfilaria (note the nuclei and sheath); *W. bancrofti* tail (note that the nuclei end before the end of the tail [oval]; the sheath is visible); elephantiasis (occurs with certain types of filariasis) (right image from the collection of Herman Zaiman, "A Presentation of Pictorial Parasites").

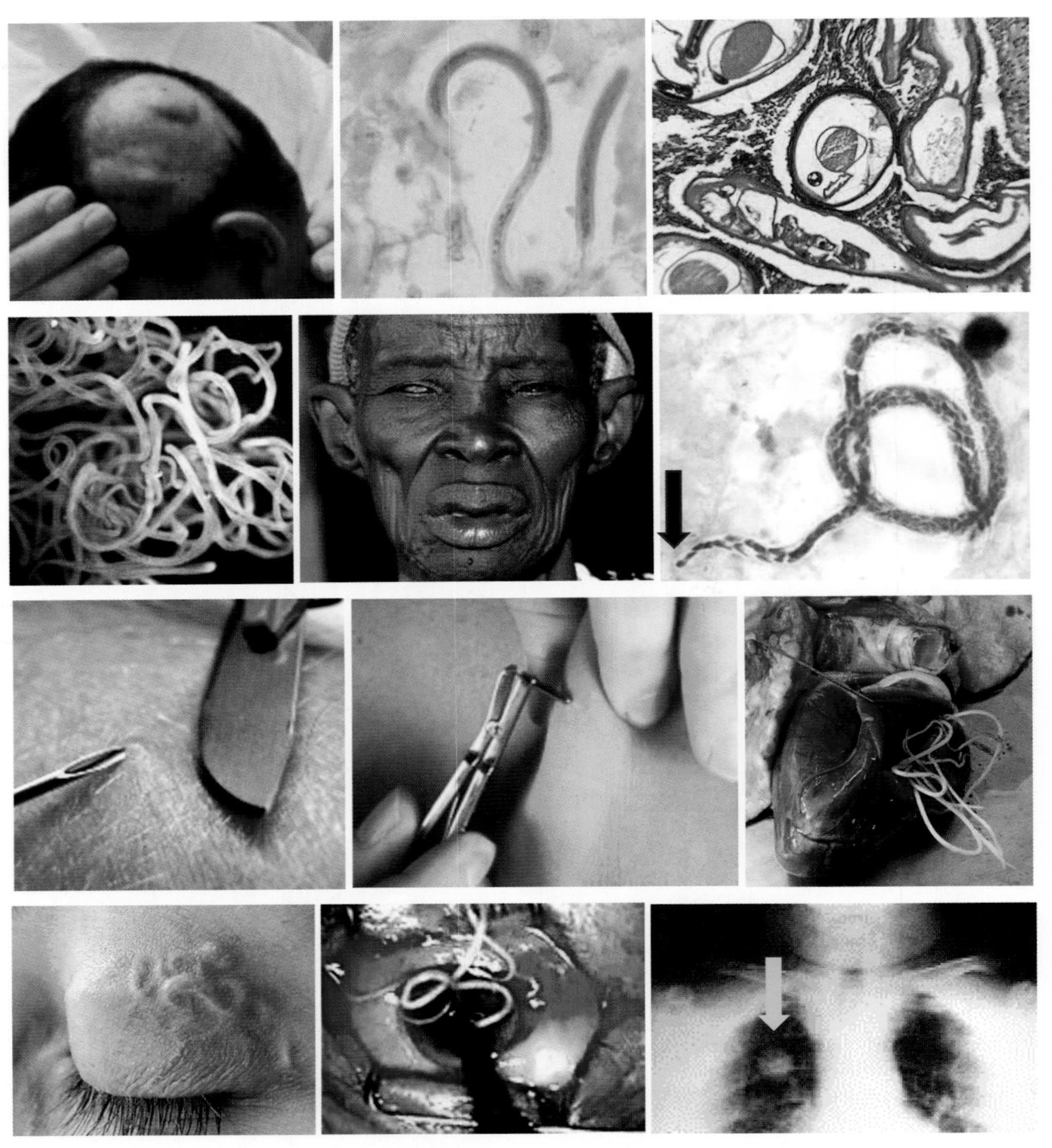

Plate 7.P **(Row 1)** *Onchocerca volvulus* (nodule on scalp); *O. volvulus* (organisms seen in skin biopsy) (courtesy of CDC DpDx, https://www.cdc.gov/dpdx/onchocerciasis/index.html); *O. volvulus* (cross section of worms within a nodule). **(Row 2)** *O. volvulus* (worms removed from nodule); patient with *O. volvulus* infection (example of "river blindness"); *Mansonella ozzardi* (this microfilaria does not have a sheath; note the terminal nucleus at the tip of the tail [arrow]). **(Row 3)** A physician takes a skin sample from a patient for a skin snip biopsy by elevating a piece of skin with a needle and shaving it off with a scalpel (there should be little or no blood); a physician takes a skin sample from a patient for a skin snip biopsy using a sclerocorneal biopsy punch (images courtesy of Thomas B. Nutman, MD, Laboratory of Parasitic Diseases, NIAID, NIH); *Dirofilaria* sp. adult worm from dog's heart. **(Row 4)** *Dirofilaria* in eyelid; removal of *Dirofilaria* adult worm from eye; radiograph showing a *Dirofilaria* "coin lesion" in the lung (arrow) (all three images from the collection of Herman Zaiman, "A Presentation of Pictorial Parasites").

Taenia saginata

Pathogenic	Yes
Disease	Beef tapeworm disease, taeniasis
Acquired	Ingestion of infective cysts (immature tapeworm larvae, cysticerci) from infected (raw or poorly cooked) beef
Body site	Intestine
Symptoms	Vague to none, discomfort or embarrassment caused by proglottids crawling from the anus
Clinical specimen	Stool or individual proglottids
Epidemiology	Worldwide, primarily human-to-human transmission
Control	Cattle should not be allowed to graze on ground contaminated by human sewage

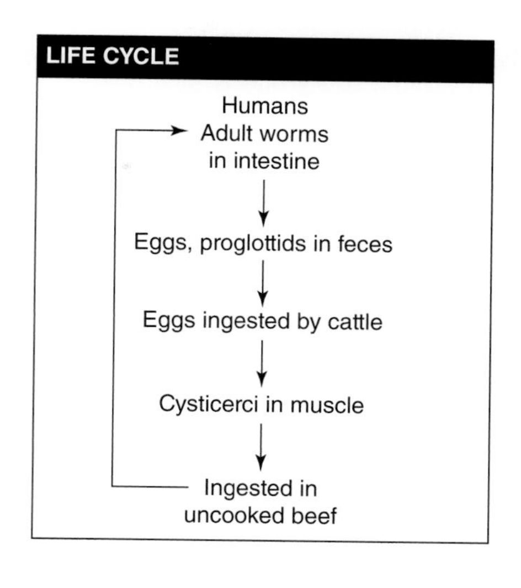

Diagnosis

Few symptoms are associated with the presence of the adult worm in the intestine. Although rare symptoms (obstruction, diarrhea, hunger pains, weight loss, or appendicitis) have been reported, the most common complaint is the discomfort and embarrassment caused by the proglottids crawling from the anus. This may be the first clue that the patient has a tapeworm infection. Occasionally, the proglottids are also seen on the surface of the stool after it is passed.

The standard O&P exam is recommended for recovery and identification of *Taenia* sp. eggs in stool specimens, primarily from the wet-preparation examination of the concentration sediment. The eggs are most easily seen on a direct wet smear or a wet preparation of the concentration sediment.

Proglottids may be received by the laboratory for identification. If they are gravid, the uterine structure (branch counts) can be used to identify the proglottid to the species level.

Diagnostic Tips

Since the eggs of *Taenia saginata* and *Taenia solium* look identical, identification to the species level is normally based on recovery and examination of gravid proglottids, in which the main lateral branches are counted (count on one side only; 15 to 20 for *T. saginata* and 7 to 13 for *T. solium*). Often the gravid proglottids of *T. saginata* are somewhat larger than those of *T. solium*, but this difference may be minimal or impossible to detect. Preliminary examination of the gravid proglottid may not allow identification without clearing or injection of the uterine branches with India ink. The scolex has four suckers and no hooks.

Any patient with *Taenia* eggs recovered in the stool should be cautioned to use good hygiene. These patients should be treated as soon as possible to avoid the potential danger of accidental infection with the eggs, which may lead to cysticercosis (*T. solium*).

General Comments

T. saginata was apparently differentiated from *T. solium* in the late 1700s; however, cattle were not identified as the intermediate host until 1863. This infection occurs worldwide and is generally much more common than *T. solium* infection, particularly in the United States. The overall impact on human health is much lower than that of *T. solium*, since *T. saginata* cysticercosis is quite rare.

The life cycle is very similar to that of *T. solium*. Infection with the adult worm is initiated by ingestion of raw or poorly cooked beef containing encysted *T. saginata* larvae. The larva is digested out of the meat in the stomach, and the tapeworm evaginates in the upper small intestine and attaches to the intestinal mucosa, where the adult worm matures in 5 to 12 weeks. The adult worm can grow up to 25 ft but often measures only about half this length. The scolex is "unarmed" and has four suckers with no hooks. The proglottids usually number 1,000 to 2,000, with the mature proglottids being broader than long and the gravid proglottids being narrower and longer. Although a single worm is usually found, multiple worms can be present (personal observation).

The eggs can remain viable in the soil for days to weeks. On ingestion by cattle, the oncospheres hatch in the duodenum, penetrate the intestinal wall, and are carried via the lymphatics or bloodstream, where they are filtered out in the striated muscle. They develop into the bladder worm, or cysticercus, within approximately 70 days. The mature cysticercus is 7.5 to 10 mm wide by 4 to 6 mm long and contains the immature scolex, which has no hooks (unarmed). Other animals that harbor cysticerci include buffalo, giraffe, llama, and possibly reindeer. Actual cases of human cysticercosis with *T. saginata* are rare, and some reported cases may have been inaccurately diagnosed.

Description (Eggs)

The eggs are usually spheroidal and yellow-brown. They are thick-shelled eggs, measuring 31 to 43 μm; they contain a six-hooked oncosphere (embryo). They are routinely found in the stool, even if gravid proglottids are not found in the specimen.

T. saginata eggs look like those of *T. solium* (pork tapeworm). Since *T. solium* eggs are infectious for humans, all clinical specimens containing proglottids and/or eggs must be handled very carefully to avoid accidental ingestion of infective *T. solium* eggs and resulting cysticercosis. Prior to therapy, the patient should be advised to use excellent hygiene to avoid possible ingestion of infective eggs.

In proficiency testing specimens, the eggs may be harvested from gravid or mature proglottids. Some eggs have a gelatinous coating that may be confusing. This coating is generally found on eggs that have been recovered from the mature proglottids and not from the gravid proglottids at the end of the strobila. This extra coating should not cause confusion; the six-hooked embryo can still be seen within the eggshell. The true tapeworm egg has a radially striated shell that is relatively thick.

Description (Adults)

The adult tapeworm consists of the attachment organ (scolex), to which is attached a chain of segments or proglottids called the strobila. Each proglottid contains a male and female reproductive system. The proglottids are classified as immature, mature, or gravid; gravid proglottids are found at the end of the strobila and contain the fully developed uterus full of eggs. The branched uterine structure in the gravid proglottids is often used as the main criterion for identification to the species level. The scolex and eggs can also be used to identify a cestode to the species level. The adult worms can reach about 15 to 20 ft long and may survive for up to 25 years.

The scolex is quadrate; it has four suckers and no rostellum or hooklets.

The gravid proglottids can be found in feces and are longer than wide (19 by 17 mm) with 15 to 20 lateral branches on each side of the central uterine stem. They usually appear singly and can actively crawl like an inchworm; they may actually migrate under a fresh stool specimen. The proglottids are usually shed singly, rather than several linked together in a chain (occasionally *T. solium* proglottids are passed with a couple linked together).

Additional Information

Beef inspection for the presence of cysticerci is the best preventive measure. Beef must be thoroughly cooked in areas of endemicity—to at least 56°C throughout the meat. Freezing at −10°C for 10 days usually is lethal to *Taenia* cysticerci, but they can withstand 70 days at 0°C.

Recent epidemiologic studies of taeniasis in Southeast Asia indicate the presence of a form of human *Taenia* sp. which can be distinguished from *T. saginata* and *T. solium*. This newly recognized cestode was originally called the Taiwan *Taenia* sp. and was first found in Taiwanese aboriginals. However, it is now referred to as the Asian *Taenia* sp. or *Taenia asiatica*, since it has also been found in a number of other Asian countries.

Cysticerci of *T. asiatica* develop in the liver of pigs, cattle, and goats and are much smaller than those of *T. saginata* or *T. solium*. The adult worm matures in 10 to 12 weeks, measuring 4 to 8 m long, and contains 300 to 1,000 proglottids. In some indigenous populations in Taiwan, 10 to 20% of the people are infected, primarily from the ingestion of pork meat and viscera. Multiple worms are often present, and in one area of Taiwan during the late 1990s, the infection rate with *T. asiatica* was 11%. A history of eating raw pig liver is relevant for *T. asiatica*.

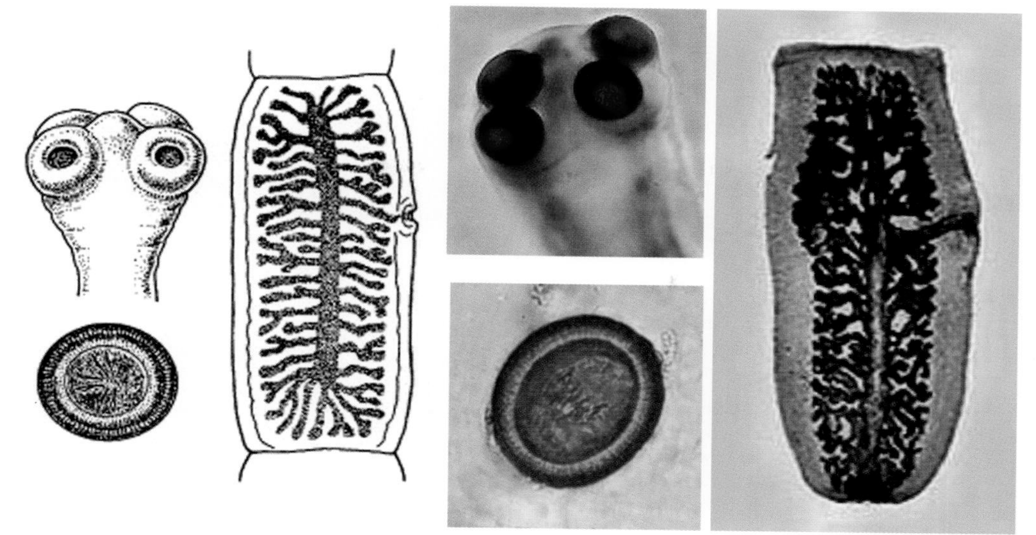

Figure 7.44 *Taenia saginata.* Drawings depicting the scolex, egg, and gravid proglottid of *T. saginata*; photographs of the same structures. Note the number of uterine branches (many more than in *Taenia solium*). Also, note that the scolex does not contain hooklets, like that of *T. solium*. The hooklets present on the oncosphere (developing within the egg shell) have no relationship to the hooklets on the scolex. Those on the oncosphere are present in all species of *Taenia* and are used to help the oncosphere break out of the egg. **NOTE:** Although Ziehl-Neelsen staining has been used to try to differentiate *T. saginata* from *T. solium* eggs, this approach is neither very sensitive nor completely specific. In some cases, *T. solium* eggs stained entirely magenta, while eggs of *T. saginata* were blue/purple. This staining approach is interesting, but not that clinically relevant or accurate.

Taenia solium

Pathogenic	Yes
Disease	Pork tapeworm disease, taeniasis, cysticercosis
Acquired	*Adult worm*: Ingestion of infective cysts (immature tapeworm larvae, cysticerci) from infected (raw or poorly cooked) pork
	Cysticercosis: Ingestion of infective eggs from adult tapeworm; development of cysticerci in human tissues, similar to that seen in the pig (WHO neglected tropical disease)
Body site	*Adult worm*: Intestine
	Cysticerci: Primarily CNS and muscle
Symptoms	*Adult worm*: Vague to none, discomfort or embarrassment caused by proglottids crawling from the anus
	Cysticercosis: Dependent on the body site
Clinical specimen	Stool or individual proglottids
Epidemiology	Worldwide, primarily human-to-human transmission
Control	Pigs should not be allowed to graze on ground contaminated by human sewage

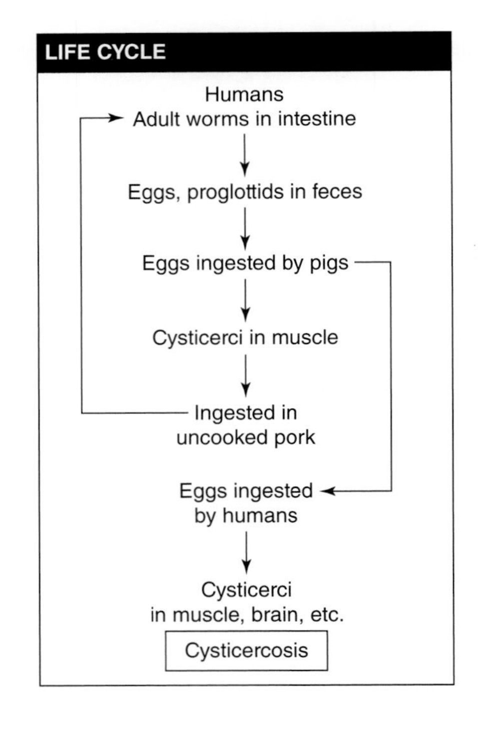

Diagnosis

The standard O&P exam is recommended for recovery and identification of *Taenia solium* eggs in stool specimens, primarily from the wet-preparation examination of the concentration sediment. The eggs are most easily seen on a direct wet smear or a wet preparation of the concentration sediment.

Computed tomography (CT) is superior to magnetic resonance imaging (MRI) for demonstrating small calcifications. However, MRI shows cysts in some locations (cerebral convexity, ventricular ependyma) better than CT, is more sensitive than CT to demonstrate surrounding edema, and may show internal changes indicating the death of cysticerci.

There are two available serologic tests to detect cysticercosis, the enzyme-linked immunoelectrotransfer blot (EITB) and commercial enzyme-linked immunoassays. The immunoblot is the test preferred by CDC, because its sensitivity and specificity have been well characterized in published analyses (test available at CDC).

Diagnostic Tips

Since *T. solium* and *T. saginata* eggs look identical, identification to species level is normally based on the recovery and examination of gravid proglottids, in which the main lateral branches are counted (7 to 13 for *T. solium* and 15 to 20 for *T. saginata*). Often the gravid proglottids of *T. saginata* are larger than those of *T. solium*, but this difference may be minimal or impossible to detect. The scolex has four suckers and an armed rostellum (hooklets present). Preliminary examination of the gravid proglottid may not allow identification without clearing or injection of the uterine branches with India ink.

Because of the possible danger of egg ingestion, specimens and proglottids should be handled with extreme care. The proglottid uterine structure will be packed with viable, infectious eggs (*T. solium*), even if the proglottids are received in fixative. Any patient with *Taenia* eggs recovered in the stool should be cautioned to use good hygiene. These patients should be treated as soon as possible to avoid the potential danger of accidental infection with the eggs, which may lead to cysticercosis (*T. solium*).

Guidelines published in *Clinical Infectious Diseases* by the Infectious Diseases Society of America (IDSA) and the American Society of Tropical Medicine & Hygiene (ASTMH) (https://www.ncbi.nlm.nih.gov/pmc/articles/PMC6248812/) address a range of different forms the disease can take. To confirm an infection, an EITB is preferred over an enzyme-linked immunosorbent assay (ELISA), according to these guidelines, because the EITB is more sensitive. Although serologic reagents for antibody detection and antigen detection tests have been developed, not all reagents produce sensitive and specific results. Consultation with the CDC and/or your state public health laboratories may be helpful. (See White et al., Suggested Reading.)

General Comments

T. solium infection may have been recognized since biblical times, with the life cycle being delineated in the mid-1850s. The tapeworm is found in many parts of the world and is considered an important human parasite where raw or poorly cooked pork is eaten.

Cysticercosis infections with *T. solium* (larvae) are relatively common in certain parts of the world but rare in the United States. This extraintestinal infection is far more serious than that caused by the adult worm in the intestine.

Infection with the adult worm is initiated by ingestion of raw or poorly cooked pork containing encysted *T. solium* larvae. The larva is digested out of the meat in the stomach, and the tapeworm head evaginates in the upper small intestine, attaches to the intestinal mucosa, and grows to the adult worm within 5 to 12 weeks. Although usually a single worm is present, there may be multiple worms in the intestine. The adult worm reaches 2 to 7 m long and may survive 25 years or more.

On ingestion by hogs or humans, the eggs hatch in the duodenum or jejunum after exposure to gastric juice in the stomach. The released oncospheres penetrate the intestinal wall, are carried via the mesenteric venules throughout the body, and are filtered out in the subcutaneous and intramuscular tissues, the eye, the brain, and other body sites.

Description (Eggs and Larvae)

Eggs usually pass from the uterus through the ruptured wall, where the proglottids break off from the strobila. The eggs are round or slightly oval (31 to 43 μm), have a thick striated shell, and contain a six-hooked embryo or oncosphere. They may remain viable in the soil for weeks.

T. solium eggs look like those of *T. saginata* (beef tapeworm). Since *T. solium* eggs are infectious for humans, all clinical specimens containing proglottids and/or eggs must be handled very carefully to avoid accidental ingestion of infective *T. solium* eggs and resulting cysticercosis. In proficiency testing specimens, the eggs may be harvested from gravid or mature proglottids. In some cases, the eggs have a gelatinous coating that may be confusing.

The cysticercus is an ovoid, milky white bladder with the head invaginated into the bladder. The host generally produces a fibrous capsule around the bladders, unless they are located in the brain, particularly the ventricles. The laboratory infrequently sees material from CNS tissues.

Description (Adult)

The adult tapeworm is composed of the attachment organ (scolex), to which is attached a chain of segments or proglottids called the strobila. Each proglottid contains a male and female reproductive system. The proglottids are classified as immature, mature, or gravid; gravid proglottids are found at the end of the strobila and contain the fully developed uterus full of eggs. The branched uterine structure in the gravid proglottids is often used as the main criterion for identification to the species level. The scolex and eggs can also be used to identify a cestode to the species level. The adult worms can reach about 15 to 20 ft long and may survive for up to 25 years.

The scolex of *T. solium* is quadrate; it has four suckers and a rounded rostellum with a double set of hooklets.

The gravid proglottids are found in feces and are approximately square with 7 to 13 lateral branches on each side of the central uterine stem. They usually appear singly and can actively crawl like an inchworm; they may migrate under a fresh stool specimen, although this type of motility is more often seen in *T. saginata*. Occasionally two or three proglottids are passed together (attached).

Adult-Onset Epilepsy

The presence of cysticerci in the brain is the most common parasitic infection of the human nervous system and the most common cause of adult-onset epilepsy throughout the world. In Latin America, it is unusual for both brain and muscle cysticercosis to occur in the same patient; fewer than 6% of patients have cysticerci in both sites. However, elsewhere, subcutaneous involvement by *T. solium* cysticerci has been found in up to 78.5% of patients with cerebral cysticercosis. Possible reasons for such differences include (i) the immune status of the patient, (ii) the human leukocyte antigen type, (iii) the nutritional status of the patient, (iv) the burden of eggs infecting the patient, and (v) a difference in the strains of *T. solium*.

Racemose Cysticercosis

The racemose, or proliferating, form of the cysticercus is composed of several bladders that are connected and are often found in the brain, particularly the fourth ventricle and subarachnoid space (see images below). Within these bladders, no scolices are found and the growth may resemble that of a metastatic tumor. Other larval cestodes may also have a racemose growth pattern, and it may be very difficult to differentiate a *T. solium* racemose cysticercus from other types histologically. The prognosis for such infections is very poor. Racemose cysticerci are usually not found in infected children and may be either a degenerative form or formed in response to infection at a different anatomic site. Cervical spinal cysticercosis is usually associated with the racemose type within the posterior fossa.

Diagnosis of racemose cysticercosis is based on the combination of clinical, epidemiologic, radiographic, and immunologic information. Compared with the more normal presentation of cysticercosis, which most commonly presents as seizures, due to its extra-axial location, racemose neurocysticercosis presents with raised intracranial pressure and meningitis and frequently requires neurosurgical intervention. Racemose cysticercosis is rare in the cerebellar hemisphere, but neurocysticercosis should be taken into consideration as a differential diagnosis of multiple cystic lesions in the cerebellum.

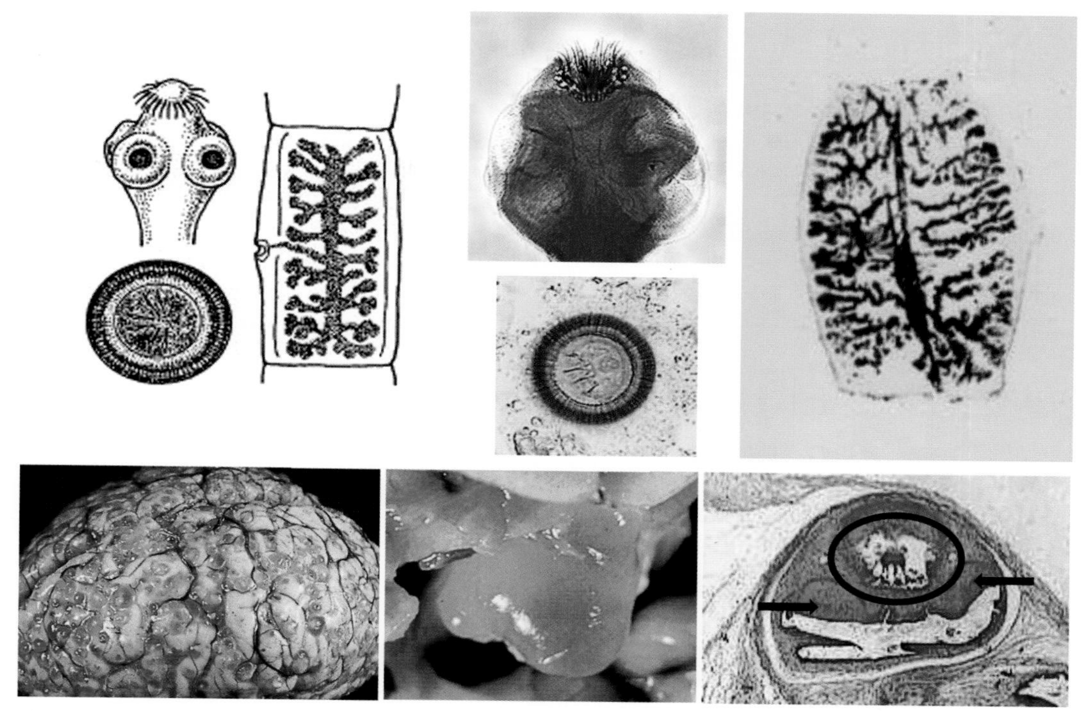

Figure 7.45 ***Taenia solium.*** **(Top row)** Drawings depicting the scolex, egg, and gravid proglottid of *T. solium*; photographs showing the same structures. Note the number of uterine branches (many fewer than that in *Taenia saginata)*. Also note that the scolex has hooklets, unlike that of *T. saginata*. The hooklets present on the oncosphere (developing within the egg shell) have no relationship to the hooklets on the scolex. Those on the oncosphere are present in all species of *Taenia* and are used to help the oncosphere break out of the egg. **(Bottom row)** Gross specimen of brain containing many cysticerci; single cysticercus in the brain (from the collection of Herman Zaiman, "A Presentation of Pictorial Parasites"); single cysticercus from brain (note the suckers [arrows] and hooklets [circle]) (courtesy of Dr. Luciano S. Queiroz, Department of Pathology, School of Medical Sciences, State University of Campinas [UNICAMP], Campinas, Sao Paulo, Brazil, http://anatpat.unicamp.br/).

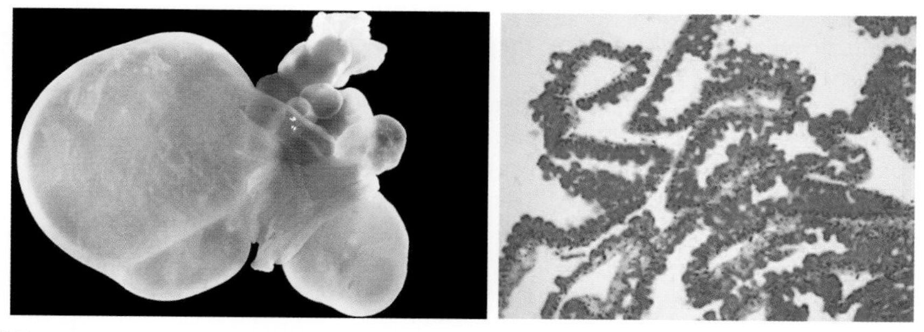

Figure 7.46 **Racemose cysticercosis.** Racemose form of *Taenia solium* cysticercus with several interconnected bladders of various sizes (gross specimen) (courtesy of Armed Forces Institute of Pathology, Joint Pathology Center / Mary KlassenFischer and Ronald C. Neafie); section of proliferating bladder wall of racemose cysticercus, demonstrating multiple layers (courtesy of Dr. Luciano S. Queiroz, Department of Pathology, School of Medical Sciences, State University of Campinas [UNICAMP], Campinas, Sao Paulo, Brazil, http://anatpat.unicamp.br/).

Diphyllobothrium latum

Pathogenic	Yes
Disease	Broad fish tapeworm disease (diphyllobothriasis)
Acquired	*Adult worm*: Ingestion of infective larvae from infected freshwater fish (raw or poorly cooked fish)
Body site	*Adult worm*: Intestine
Symptoms	*Adult worm*: Vague to none
Clinical specimen	*Intestinal*: Stool or proglottids
Epidemiology	Worldwide, primarily human-to-human transmission
Control	Dogs may serve as carriers of the adult worms, and infected dogs should be periodically treated

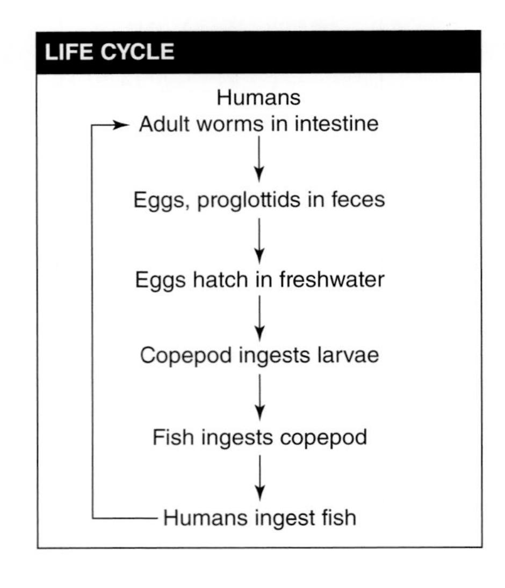

Diagnosis

Diphyllobothrium latum belongs to the pseudophyllidean tapeworm group, which is characterized by having a scolex with two bothria (sucking organs) rather than the typical four suckers seen in the *Taenia* tapeworms. The distribution of this worm is worldwide, with various increased outbreaks reported from time to time.

Diagnosis is usually based on recovery and identification of the characteristic eggs or proglottids. The eggs are unembryonated at the time they are passed in the stool.

Proglottids are often passed in chains (a few inches to several feet), and this is a clue to identifying *D. latum*. The overall proglottid morphology with the rosette uterine structure also facilitates identification.

Diagnostic Tips

If the egg operculum, or cap, is difficult to see, the coverslip of the wet preparation can be tapped and the pressure may cause the operculum to pop open, making it more visible. The light should be somewhat reduced to allow the operculum to be seen more easily. *D. latum* eggs can be confused with those of *Paragonimus westermani*; however, *P. westermani* eggs have opercular shoulders into which the operculum fits (like a teapot lid), while eggs of *D. latum* have a very smooth opercular outline with no shoulders.

Often, the proglottids are passed in chains (a few inches to a few feet), unlike those seen with the *Taenia* group. Remember, the gravid proglottids are much wider than long, and the uterine structure is typically in a rosette shape.

For patients with vague abdominal complaints or unexplained low vitamin B_{12} levels, it might be worth inquiring about the types of fish consumed, particularly including sushi and sashimi, and it might be worth checking the stool for *Diphyllobothrium* ova.

Handling of the specimens presents no danger in terms of cysticercosis; however, all fecal specimens should be considered potentially infectious (due to other organisms).

General Comments

Symptoms depend on the number of worms present, the amounts and types of by-products produced by the worm, the patient's reaction to these by-products, and the absorption of various metabolites by the worms. There may be occasional intestinal obstruction, diarrhea, abdominal pain, or anemia. If the worm is attached at the jejunal level, there may very rarely be vitamin B_{12} deficiency resembling pernicious anemia. Heavy or long-term infections may cause megaloblastic anemia due to parasite-mediated dissociation of the vitamin B_{12}-intrinsic factor complex within the gut lumen, making B_{12} unavailable to the host. This is more common in Finland, where some individuals have a genetic predisposition to pernicious anemia. In patients without this genetic predisposition, symptoms of *D. latum* infection may be absent or minimal, consisting of a slight leukocytosis with eosinophilia. The infection has been associated with a condition similar to pernicious anemia. However, this condition is rarely seen, as a result of improved diet, prenatal care, and available treatment.

Infection with the adult worm is acquired by the ingestion of raw, poorly cooked, or pickled freshwater fish (pike, perch, lawyer, salmon, trout, white fish, grayling, ruff, turbot, etc.) containing the encysted plerocercoid larvae. Plerocercoids in fish are quickly killed by thorough cooking, freezing at −10°C for 15 minutes, or thorough pickling. After ingestion, the worm matures, with egg production beginning in week 5 or 6. The adult worm is ca. 10 m long and contains up to 3,000 proglottids.

After developing for 2 weeks in freshwater, the eggs hatch, and the ciliated coracidium larvae are ingested by the first intermediate host, the copepod. The copepods, containing the second larval stage (procercoid), are then ingested by fish, which may be ingested by larger fish. The final fish intermediate host may contain many plerocercoid larvae, which initiate the infection with the adult worm when ingested by humans.

Description (Eggs)

The eggs are broadly oval and operculated. They are yellow-brown and measure 58 to 75 μm by 40 to 50 μm in feces. They usually have a small knob at the abopercular end. The eggs are unembryonated when passed in the feces.

Description (Adults)

The scolex of *D. latum* is elongate and spoon shaped and has two long sucking grooves, one on the dorsal surface and one on the ventral surface. The mature and gravid proglottids are wider than long, with the main reproductive structures (mainly the uterus) in the center of the gravid proglottid. This configuration of the uterine structure is called a rosette. Identification to species level is usually based on this typical morphology of the gravid proglottids. Both eggs and proglottids may be found in the stool. Often a partial chain of proglottids may be passed (a few inches to several feet).

Additional Information

In areas where human infection is rare, *D. latum* infections have been found in other mammals. However, the natural transmission cycle from mammals other than humans does not seem to be sustained. Infection can result from ingestion of infected raw freshwater fish that has been shipped under refrigeration to areas where the infection is not endemic. Preventive measures include thorough cooking of all freshwater fish or freezing for 24 to 48 h at −18°C. This infection has been called the Jewish housewives' disease, since the individual preparing the food may sample the dish (e.g., gefilte fish) prior to cooking and acquire the infection. Other groups who tend to eat raw or insufficiently cooked fish include Russians, Finns, and Scandinavians. Raw fish marinated in lime juice (ceviche) is also a source of infection (*D. pacificum*) in Latin America. Since domestic dogs can serve as reservoir hosts, infected dogs should be periodically treated. Other factors include the continued dumping of wastewater into lakes and the possibility of animal reservoirs.

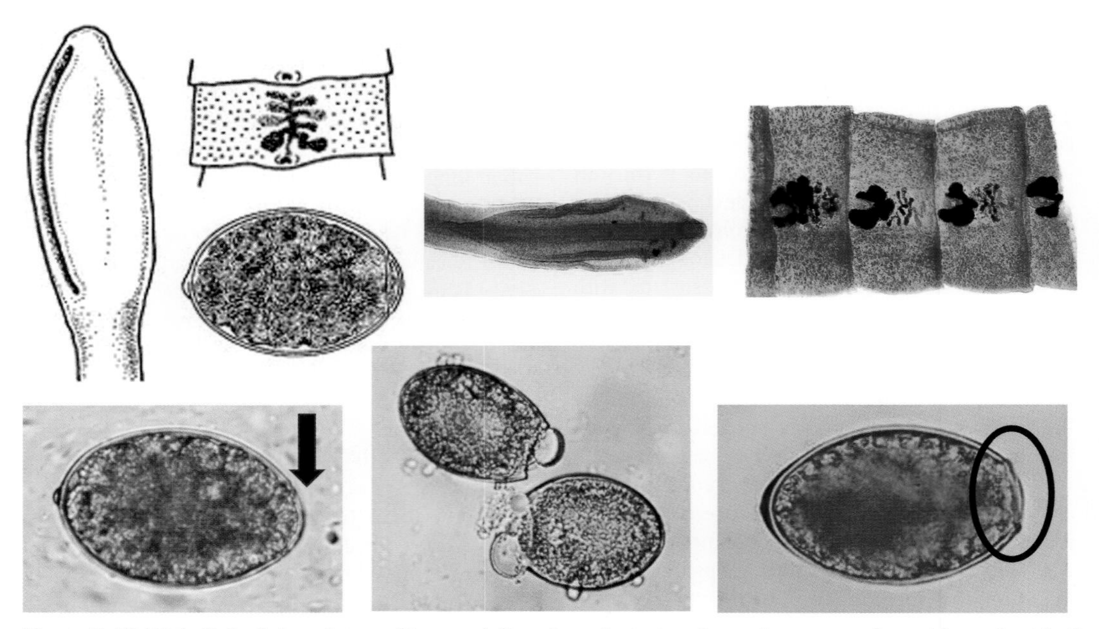

Figure 7.47 ***Diphyllobothrium latum*****.** **(Top row)** Drawings depicting the scolex, egg, and gravid proglottid of *D. latum*; photographs showing the scolex and proglottids. The gravid proglottids are often passed as a chain (which can be several feet long). Some of the eggs are seen with popped-open opercula (trapdoors). **(Bottom row)** *D. latum* egg (note the smooth egg outline [arrow] and bump on the shell at the abopercular end); two eggs whose opercula have popped open; *Paragonimus* sp. egg (lung fluke) (note the opercular shoulders into which the operculum fits, like a teapot lid [oval]).

Hymenolepis (Rodentolepis) nana

Pathogenic	Yes
Disease	Dwarf tapeworm disease
Acquired	Ingestion of infective eggs or (less commonly) cysts (immature tapeworm larvae, cysticercoid) from infected grain beetles
Body site	Intestine
Symptoms	Vague to none; some patients have low-grade eosinophilia (5% or more)
Clinical specimen	Stool
Epidemiology	Worldwide, primarily human-to-human transmission
Control	Improved hygiene, adequate disposal of fecal waste, adequate washing of fruits and vegetables

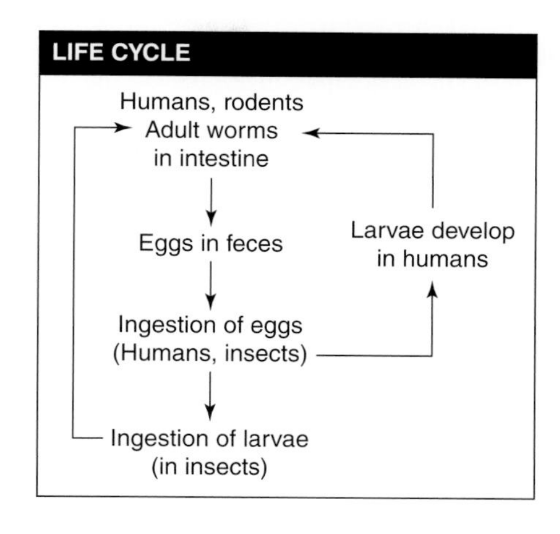

Diagnosis

Hymenolepis nana is called the dwarf tapeworm and has a worldwide distribution. The fact that an intermediate host is not required in the life cycle was determined in the late 1800s. For this reason, *H. nana* is considered the most common tapeworm throughout the world.

The adult worm or proglottids are rarely seen in the stool; therefore, diagnosis is based on recovery and identification of the characteristic eggs. They are most easily identified in fresh specimens or those preserved in formalin-based fixatives. Morphologic characteristics of eggs in specimens in preservative containing PVA are not as well delineated as those in formalin-fixed specimens. These thin-shelled eggs also tend to collapse on the permanent stained smear and may be difficult to identify; the direct wet film or concentration wet mount is recommended. The standard O&P exam is recommended for recovery and identification of *H. nana* eggs in stool specimens, primarily from the wet-preparation examination of the concentration sediment. The eggs are most easily seen on a direct wet smear or a wet preparation of the concentration sediment.

H. nana is the only human tapeworm in which the intermediate host is not necessary and transmission is from person to person via the eggs. Children are usually infected more often than adults. Good personal hygiene is an important preventive measure. Infection from rats and mice is always a possibility, as is the accidental ingestion of infective insect intermediate hosts.

Diagnostic Tips

Adult worms or proglottids are rarely seen in the stool. Eggs with the characteristic thin shell, six-hooked oncosphere, and polar filaments are diagnostic. The shape varies from round to oval. Egg morphology is more easily seen in fresh specimens or those preserved in non-PVA-based fixatives. The eggs are infectious for humans, and unpreserved stools should be handled with caution. Eggs in a permanent stained fecal smear tend to retain a lot of stain (i.e., they may appear like debris) and are often distorted. If a suspicious structure is seen on the permanent stain, an additional wet mount can be examined for egg confirmation.

General Comments

The infection is most common in children, although adults are also infected. Worldwide, 50 million to 75 million people are infected. There has been some discussion regarding placing *H. nana* in the genus *Rodentolepis*.

Infection is usually acquired by ingestion of *H. nana* eggs, primarily from human stool. The eggs hatch in the stomach or small intestine, and the liberated larvae, or oncospheres, penetrate the villi in the upper small intestine. The larvae develop into the cysticercoid stage in the tissue and migrate back into the lumen of the small intestine, where they attach to the mucosa. The adult worms mature within several weeks.

Although accidental ingestion of the insect intermediate host can result in development of the adult worms, this mode of infection is probably not common.

Heavy human infection can be attributed to internal autoinfection in which the eggs hatch in the intestine and follow the normal life cycle to the adult worm. This autoinfection feature of the life cycle can lead to complications in the immunocompromised patient.

Description (Eggs)

Since *H. nana* and *Hymenolepis diminuta* eggs look very much alike, identification depends on seeing the polar filaments in *H. nana* eggs. The eggs are round or oval with a thin shell and are 30 to 47 μm in diameter. The oncosphere has two polar thickenings from which arise polar filaments that lie between the oncosphere and the shell.

In proficiency testing specimens, the eggs may not always have hooklets visible within the oncosphere. You may have to look very carefully for the polar filaments. These filaments and hooklets are more easily seen in stool specimens that have not been stored for a long time in formalin.

Description (Adults)

The scolex has four suckers and a short rostellum with hooks. The adult worm is rarely seen in the stool. The eggs are released by disintegration of the gravid proglottids, are passed out in the stool, and are immediately infectious.

The worms are very small compared with *Taenia* worms and measure up to 40 mm long. The more worms present, the shorter the total length of each worm.

Additional Information

An infection with *H. nana* may cause no symptoms even with a heavy worm burden. Some patients complain of headache, dizziness, anorexia, abdominal pain, diarrhea, or possibly irritability. Some patients have low-grade eosinophilia (5% or more).

Heavy human infection can be attributed to internal autoinfection in which the eggs hatch in the intestine and follow the normal life cycle to the adult worm. This autoinfection feature of the life cycle can lead to complications in the compromised patient. Since infection is also from person to person via the eggs, good personal hygiene is an important preventive measure. Infection from rats and mice is always a possibility, as is the accidental ingestion of infective insect intermediate hosts.

Direct human-to-human transmission is the most common route of infection with *H. nana*, particularly in environments with numerous infections due to poor hygiene and inadequate sanitation. However, it is still considered a zoonosis because infected rodents, such as mice and rats, and arthropod intermediate hosts represent a reservoir of infection.

Although molecular evidence requires additional clarification, data suggest that there may be two strains of *H. nana* that are maintained in zoonotic and non-zoonotic cycles. It is also suggested that isolates of *H. nana* in Australia exist as two morphologically identical but genetically different species. *H. nana* can also infect lower primates, and in such cases, humans appear to be the source of infection in a reverse zoonosis situation.

In areas where transmission is high, even with effective therapy, reinfection is very common. Without improved hygiene and sanitation, control is almost impossible.

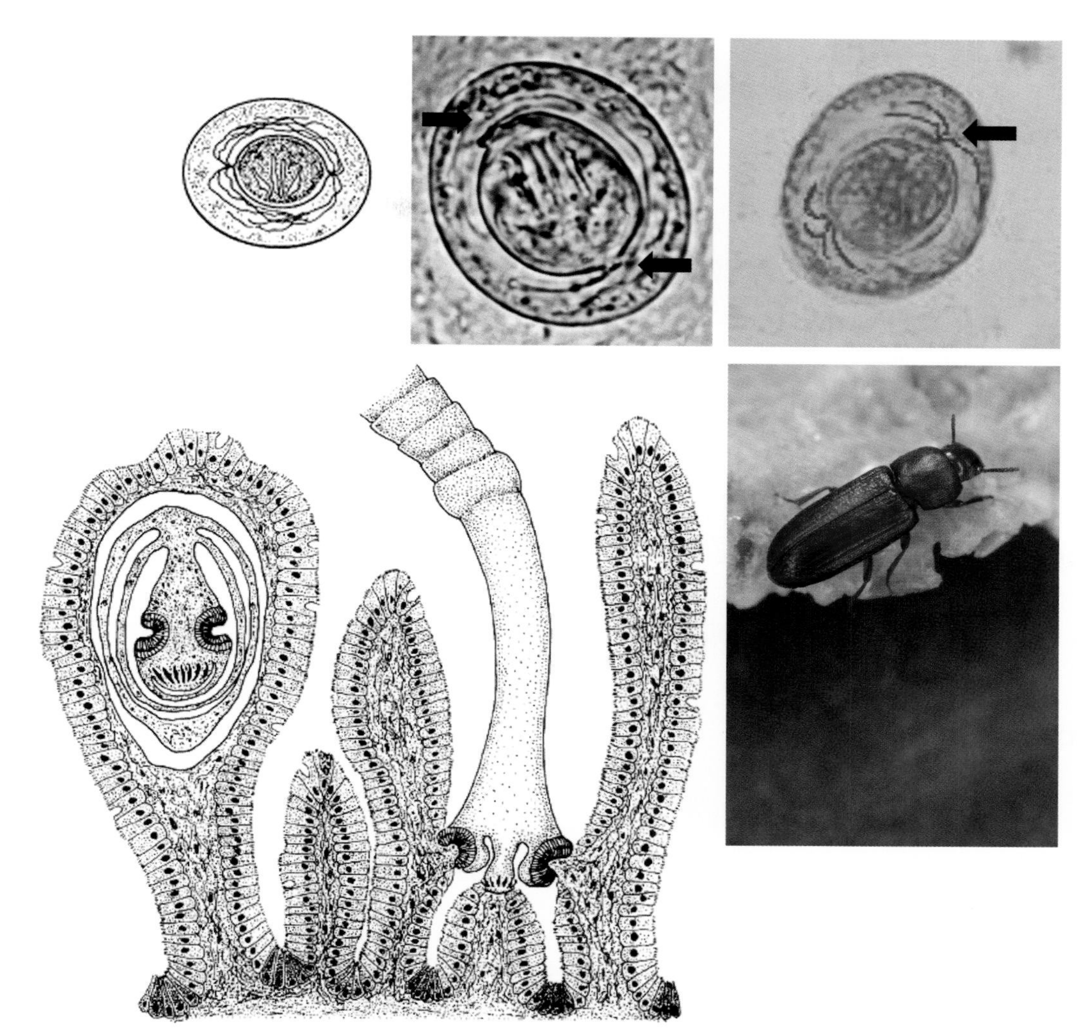

Figure 7.48 ***Hymenolepis nana.*** **(Top row)** Drawing depicting the *H. nana* egg; photographs showing the same structures. Note the six-hooked embryo (oncosphere) and the polar filaments that lie between the oncosphere and the thin eggshell (arrows). The key difference between this egg and that of the rat tapeworm (*Hymenolepis diminuta*) is the lack of polar filaments in the latter. **(Bottom row)** Diagram of both a cysticercoid and adult worm in the small intestine (illustration by Sharon Belkin); *Tribolium* sp. flour beetle, which can be found in stored grains, cereals, etc. (courtesy Peggy Greb, USDA Agricultural Research Service, Bugwood.org).

Hymenolepis diminuta

Pathogenic	Yes
Disease	Rat tapeworm disease
Acquired	Ingestion of immature tapeworm larvae (cysticercoid) from infected arthropods
Body site	Intestine
Symptoms	Vague to none
Clinical specimen	Stool
Epidemiology	Worldwide
Control	Improved hygiene, adequate disposal of fecal waste, adequate washing of contaminated fruits and vegetables

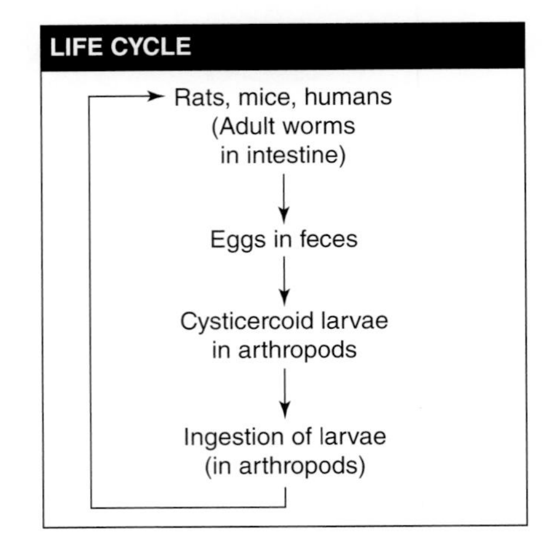

Diagnosis

Although *Hymenolepis diminuta* is commonly found in rats and mice, it is infrequently found in humans. It has a worldwide distribution in normal hosts, and fewer than 500 human cases, primarily from India, the former Soviet Union, Japan, Italy, and certain areas of the southern United States (Tennessee, Georgia, and Texas), have been reported.

The adult worms or proglottids usually disintegrate in the gut; therefore, diagnosis is based on recovery and identification of the characteristic eggs. They are most easily identified in fresh specimens or those preserved in formalin-based fixatives. Morphologic characteristics of eggs in specimens in preservatives containing PVA are not as well delineated as those in formalin-fixed specimens.

The standard O&P exam is recommended for recovery and identification of *H. diminuta* eggs in stool specimens, primarily from the wet-preparation examination of the concentration sediment. The eggs are most easily seen on a direct wet smear or a wet preparation of the concentration sediment.

Diagnostic Tips

In proficiency testing specimens, the eggs may not always have hooklets visible within the oncosphere. The hooklets are more easily seen in stool specimens that have not been stored for a long time in formalin. *H. diminuta* thin-shelled eggs also tend to collapse on the permanent stained smear and may be difficult to identify, also due to intense stain retention; the direct wet film or concentration wet mount is recommended. Some pollen grains closely mimic *Hymenolepis* spp. eggs.

General Comments

The life cycle is very similar to that of *H. nana*; however, the arthropod intermediate host is obligatory. A number of different arthropods (lepidopterans, earwigs, myriapods, larval fleas, and beetles) can serve as intermediate hosts. After egg ingestion by the arthropods, the cysticercoid stage forms (similar in morphology to the cysticercoid stage of *H. nana*). Infection by accidental ingestion of the infected arthropod containing cysticercoids results in development of the adult worm in the intestine. Unlike *H. nana* eggs, *H. diminuta* eggs are not infectious from person to person. As with most tapeworms, the infection is usually tolerated very well by the host, with few if any symptoms.

This infection is a true zoonosis, since infected rats infect insects, which in turn are consumed by humans.

Description (Eggs)

Since *H. nana* and *H. diminuta* eggs look very much alike, identification depends on seeing the polar filaments in *H. nana* eggs. *H. diminuta* eggs are round to oval with a thin shell and are 60 to 79 μm in diameter. The oncosphere has no polar thickenings and no polar filaments.

Description (Adults)

The scolex has four suckers and no hooks. The adult worm is rarely seen in the stool. Compared with *Taenia* worms (4.5 to 6 m), the worms are very small. *H. diminuta* is larger than *H. nana*, reaching about 1 m in single infections. The more worms present, the shorter the total length of each worm. The eggs are released by disintegration of the gravid proglottids and are passed out in the stool.

Additional Information

As with most tapeworms, the infection is usually tolerated very well by the host, with few if any symptoms. Most infections occur in children younger than 3 years of age; however, infected adults have also been reported. Symptoms include diarrhea, anorexia, nausea, headache, and dizziness; they are most common in children with a heavy infection.

Since the infection is acquired from the accidental ingestion of infected intermediate arthropod hosts, avoidance of this type of exposure is recommended. Possible situations include swallowing ectoparasites from the rodent host or accidentally ingesting beetles in precooked cereals. Rat control programs might also decrease the possibility of human exposure.

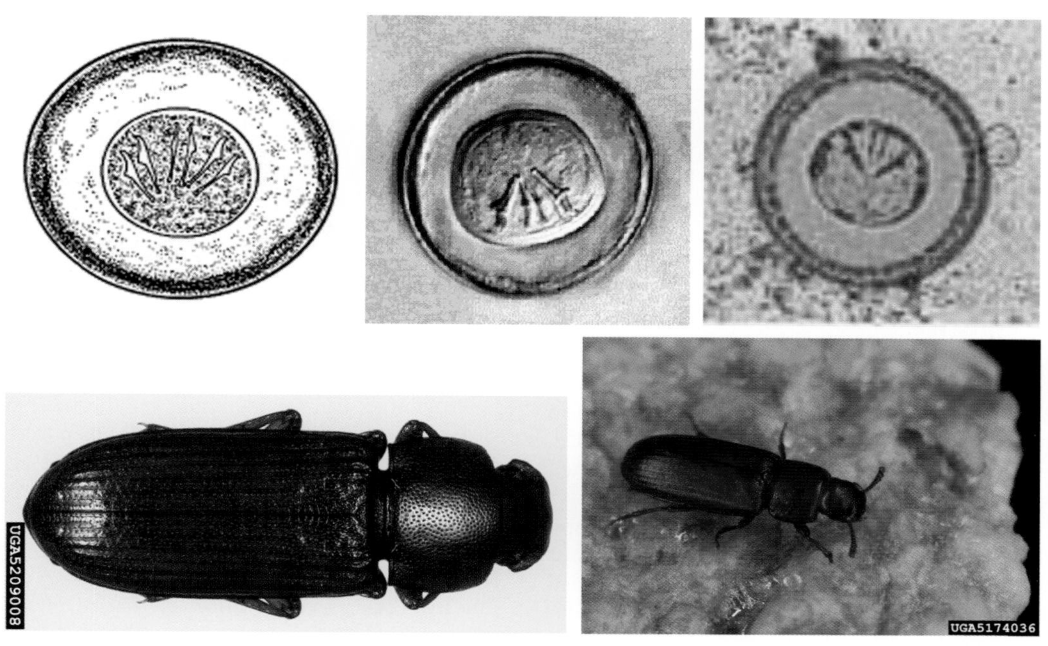

Figure 7.49 ***Hymenolepis diminuta.*** **(Top row)** Drawing depicting the *H. diminuta* egg; photographs showing the same structures. Note the six-hooked embryo (oncosphere) and the lack of polar filaments. The key difference between this egg and that of *Hymenolepis nana* is the presence of polar filaments between the oncosphere and the thin eggshell in *H. nana* eggs. **(Bottom row)** *Tribolium confusum* (common pest found in stored grain products and flour), intermediate host of *Hymenolepis* spp. (courtesy of Natasha Wright, Braman Termite & Pest Elimination, Bugwood.org, CC-BY 3.0); red flour beetle, *Tribolium castaneum*, on cereal flake (courtesy Peggy Greb, USDA Agricultural Research Service, Bugwood.org).

Dipylidium caninum

Pathogenic	Yes
Disease	Dog and cat tapeworm disease
Acquired	Ingestion of immature tapeworm larvae (cysticercoid) from infected arthropods
Body site	Intestine
Symptoms	Vague to none
Clinical specimen	Stool
Epidemiology	Worldwide
Control	Improved hygiene, adequate disposal of fecal waste, adequate washing of contaminated fruits and vegetables

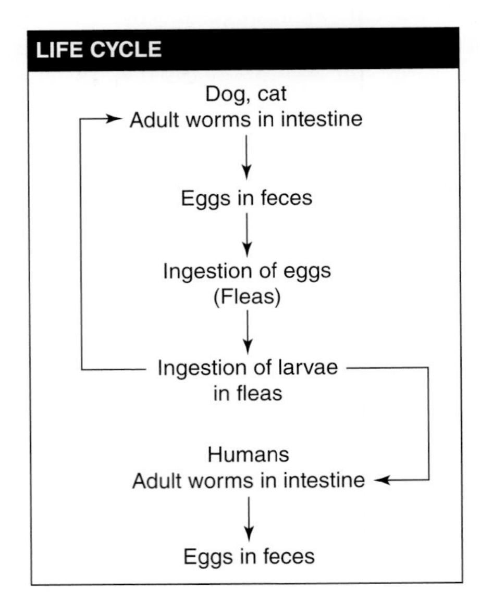

Diagnosis

Dipylidium caninum is commonly found throughout the world in dogs and cats, both domestic and wild. Human infections have also been reported from many areas of the world, including the United States, and most reported infections have been in children.

Diagnosis depends on the recovery and identification of the characteristic egg packets or the proglottids. They are most easily identified in fresh specimens or those preserved in formalin-based fixatives. These egg packets also tend to collapse on the permanent stained smear and may be difficult to identify; the direct wet film or concentration wet mount is recommended.

The standard O&P exam is recommended for recovery and identification of *D. caninum* eggs in stool specimens, primarily from the wet-preparation examination of the concentration sediment. The eggs are most easily seen on a direct wet smear or a wet preparation of the concentration sediment.

Diagnostic Tips

The egg packets are very characteristic and are frequently used to make the diagnosis. The individual eggs within the packet tend to have a thin shell and contain a six-hooked oncosphere. Morphologically, the individual eggs are similar to *Taenia* spp. eggs; normally, there may be 8 to 10 (or more) eggs per packet.

Proglottids can be identified on the basis of shape (they resemble cucumber seeds when fresh, rice grains when dry) and the presence of egg packets that can be recovered from a ruptured proglottid.

General Comments

The life cycle is very similar to that of *Hymenolepis diminuta*, in which the arthropod is an obligatory intermediate host. The adult worms are found in the dog or cat intestine, and gravid proglottids separate from the strobila and may migrate singly or in short chains out of the anus. The eggs are ingested by the larval stages of the dog, cat, or human flea, where they develop into cysticercoid larvae. When these fleas are ingested by the definitive host (dogs, cats, or humans), the adult worm develops within 3 to 4 weeks.

This infection is a true zoonosis, since infected dogs and cats infect fleas (*Ctenocephalides canis* or *Ctenocephalides cati*), which in turn are accidentally consumed by humans. Children may be more likely to accidentally swallow the infected fleas or may be more susceptible to infection.

Periodic administration of anthelmintic agents to dogs and cats and the use of flea powders help reduce the risk of infection.

Based on genetic differences and biological observations, there may be two distinct species within *D. caninum*, one genotype found in fleas and proglottids shed by domestic dogs and the other genotype found in fleas and proglottids shed by domestic cats. Additional studies will be required to clarify these findings.

Description (Eggs)

Groups of eggs (egg packets) may be found in the stool. Each egg measures 25 to 40 μm and contains the six-hooked oncosphere. The individual eggs may closely resemble those of *Taenia* spp., particularly if they are released from the egg packet.

Description (Adults)

The adult worms measure 10 to 70 cm long and have a scolex with four suckers and an armed rostellum. The single proglottids have been described as looking like cucumber seeds (shiny and moist) when moist and like rice grains when dry.

Additional Information

The symptoms are related to the worm burden; however, in most patients (usually children), they consist of indigestion and appetite loss. Awareness of the infection may be due to the migration of proglottids from the anus. This infection is relatively benign but should be considered in children with gastrointestinal symptoms and eosinophilia.

The prevalence of parasites in shelter cats suggests that the prevalence of infection is much higher than what is generally reported for client-owned animals, highlighting the importance of using appropriate fecal diagnostic techniques to detect gastrointestinal and respiratory parasites on newly adopted cats.

Although praziquantel is a highly effective therapy, from 2016 to 2018 a population of dogs was identified in which cestode infections could not be eliminated even with increased doses, dose frequency, and length of treatment. Isolates resistant to therapy were identified and characterized. Based on these findings, it is important for physicians to be aware of these possible cestode therapy limitations in human and veterinary practices.

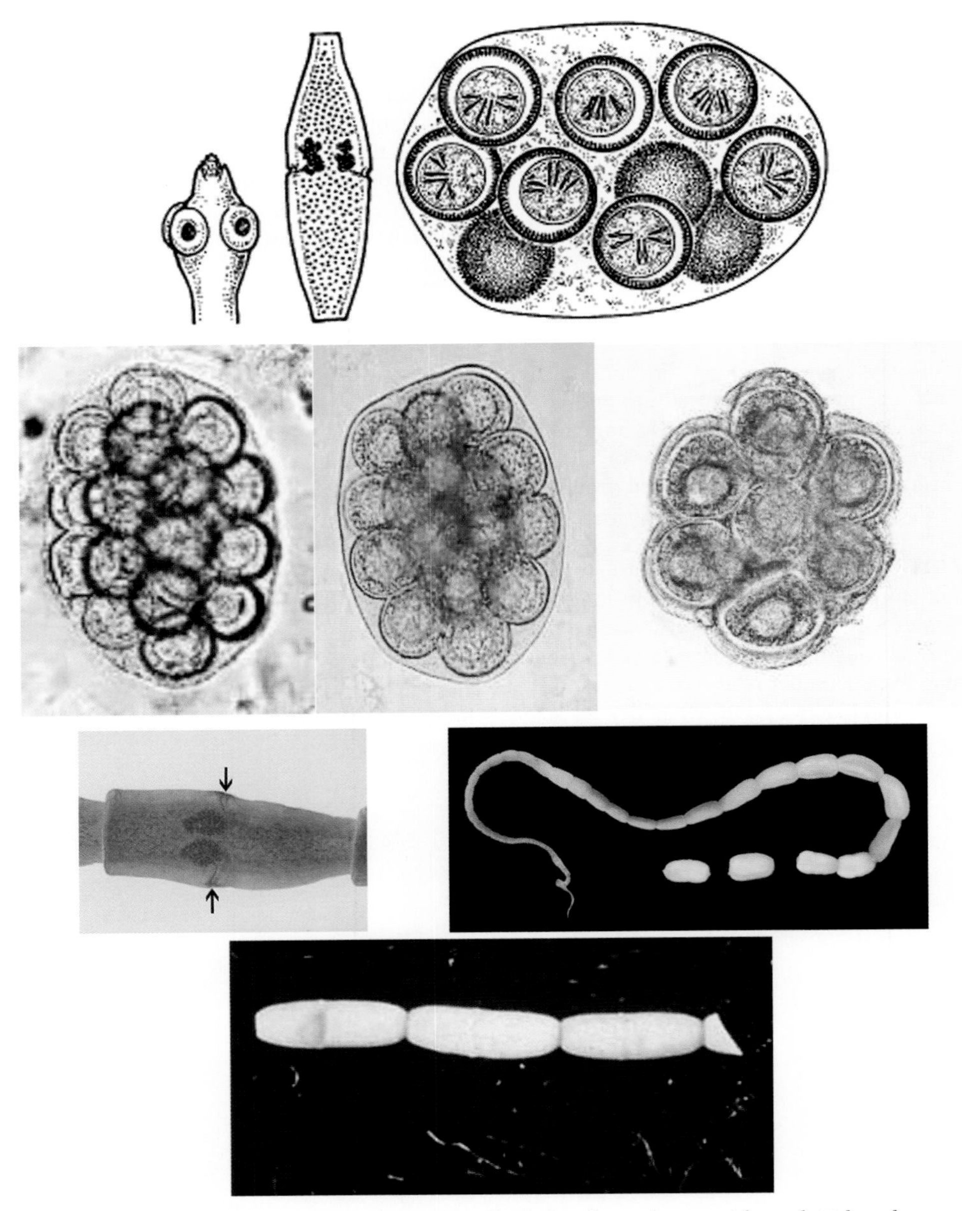

Figure 7.50 ***Dipylidium caninum*****.** **(Row 1)** Drawings depicting the scolex, gravid proglottid, and egg packet of *D. caninum*. **(Row 2)** Typical egg packets. Note the six-hooked embryo (oncosphere) and striated shells of the individual eggs within the egg packet. Individually, the eggs normally resemble *Taenia* eggs. **(Row 3)** Stained proglottid of *D. caninum* (the genital pores are visible in the carmine-stained proglottid [arrows]); string of proglottids (both images courtesy of CDC DpDx, https://www.cdc.gov/dpdx/dipylidium/). **(Row 4)** Individual proglottids (courtesy of Dr. Andrew Peregrine and Ontario Veterinary College).

Echinococcus granulosus
Echinococcus multilocularis
Echinococcus vogeli
Echinococcus oligarthrus

Pathogenic	Yes
Disease	Echinococcosis (WHO neglected tropical disease), hydatid disease (cystic disease); “sheep-sheepdog disease”
Acquired	Ingestion of eggs from the tapeworm in the dog’s intestine
Body site	Hydatid cysts in liver, lungs, and other tissues
Symptoms	Vague to none
Clinical specimen	Aspirated fluid from hydatid cyst; routine histology
Epidemiology	Worldwide
Control	Improved adult dog tapeworm prophylaxis, no feeding of sheep entrails to dogs

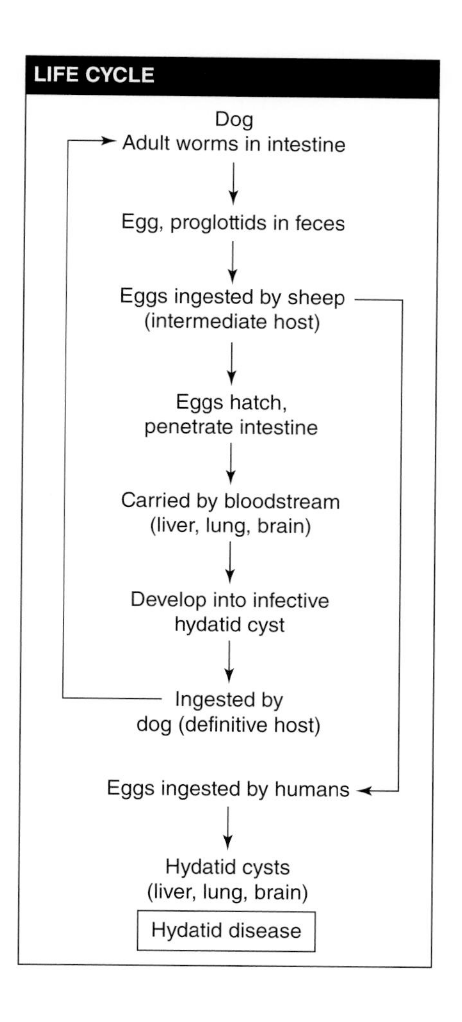

Diagnosis

Currently, four species are recognized within the genus *Echinococcus*: *Echinococcus granulosus* (cystic disease), *E. multilocularis* (alveolar disease), *E. vogeli* (polycystic disease), and *E. oligarthrus* (unicystic/neotropical disease). Hydatid cysts should be considered in patients with abdominal masses with no clearly defined diagnosis. Eosinophilia is present in 20 to 25% of patients but is merely suggestive. Many asymptomatic cysts are first discovered after radiologic studies. The cyst usually has a well-defined margin with occasional fluid level markings. These studies can also be helpful in diagnosing osseous involvement. Scans may also demonstrate a space-occupying lesion, particularly in the liver.

Serologic tests are available, including an enzyme-linked immunoelectrotransfer blot (EITB) test, which apparently offers greater sensitivity and specificity than do the EIA and arc-5 double-diffusion assay (DD5); when the tests were run simultaneously, the greatest number of cases was detected by using a combination of the EITB and DD5 tests. Newer EIA procedures appear to provide sensitivity and specificity greater than 90% compared to the EIBT.

Diagnostic Tips

Presumptive diagnosis may be based on history, radiographic studies, or scans. Additional supportive data may be acquired from immunologic tests.

Microscopic examination of hydatid cyst fluid may reveal the hydatid sand (protoscolices) or, under certain circumstances, just the hooklets. If the cyst fluid is thick or purulent, it may have to be subjected to a digestion procedure prior to examination; the hooklets survive this treatment and can then be seen and identified. Light, fluorescence, and epifluorescence microscopy can be used to visualize the hooklets; some approaches require staining, and some do not. Although not common, isolated hooks in the sputum suggest rupture of a lung cyst.

Once the cyst is discovered and surgical removal is selected, some of the cyst fluid can be aspirated and submitted for microscopic examination to detect the presence of hydatid sand, confirming the diagnosis. This procedure is risky because of possible fluid and/or tissue leakage or dissemination. Cyst aspiration is usually performed at the time of surgery. Hydatid sand is not always present. Also, if the cyst is old, the daughter cysts and/or scolices may have disintegrated, so only the hooklets are left. These may be difficult to find and identify if the cyst contains debris.

If a drop of centrifuged fluid is placed on a slide, another slide is placed on top, and the two slides are rubbed back and forth over the fluid, the grating of the hooks on the glass may be felt and heard (this method used on hydatid sand sounds like glass grating on sand grains). If the individual scolices are intact, routine microscopic examination of the centrifuged fluid as a wet mount will confirm the diagnosis. If the cyst is sterile (no daughter cysts or scolices), the diagnosis can be confirmed histologically from the cyst wall.

General Comments

E. granulosus adult worms are very small (3 to 6 mm long) and consist of a scolex, neck, and only a single proglottid at each stage of development (immature, mature, and gravid). There may be several hundred worms in the intestine of the canine host (usually the dog). The worms may survive in the host for up to 20 months, and each gravid proglottid contains few eggs compared with some of the other, larger tapeworms. After the gravid proglottids and eggs are passed in the feces, they may be swallowed by an intermediate host, including humans, where they hatch in the duodenum. The released oncospheres penetrate the intestine and are carried via the bloodstream, where they are filtered out in the various organs. The most common site in humans is the liver (60 to 70% of cases). Usually by the fifth month, the wall of the hydatid cyst has become differentiated into an outer friable, laminated, nonnucleated layer and an inner nucleated germinal layer. Various daughter cysts (brood capsules) bud off from the inner germinal layer and may remain attached or float free in the interior of the fluid-filled cyst. The individual scolices bud off from the inner wall of the daughter cysts; these scolices and free daughter cysts are called hydatid sand. Each scolex normally invaginates to protect the hooklets. Although not every cyst produces daughter cysts and/or scolices, this general tissue organization is called a unilocular cyst, in which the cyst contents are held within a single limiting cyst wall.

Additional Information

Hydatid disease in humans is potentially dangerous; however, size and organ location greatly influence the outcome. Most hydatid cysts reside in the liver, causing symptoms including chronic abdominal discomfort, occasionally with a palpable or visible abdominal mass. If cysts are in a vital area or bone (osseous cysts), even relatively small cysts can cause severe damage. Some unilocular cysts remain undetected for many years until they become large enough to crowd other organs. Cysts in the lungs are usually asymptomatic until there is cough, shortness of breath, or chest pain.

During the life of the cyst, there may be small fluid leaks into the systemic circulation that sensitize the patient. Later, if the cyst bursts or there is a large fluid leak, serious allergic sequelae, including anaphylactic reactions, may occur. Release of cyst tissue may lead to abscess formation, emboli, and/or the development of additional young cysts at secondary sites.

The percentage of infected hosts varies throughout the world, but human infection is still much less common than infection of any of the reservoir hosts. The risk of infection depends on the association between humans and dogs. Those at high risk include populations where dogs are used to herd sheep and are also intimate members of the family, often having unrestricted access to the house and family members. Cystic echinococcosis has been recorded in 21 of China's 31 provinces, autonomous regions, and municipalities (approximately 87% of the territory). Hydatid disease caused by *E. granulosus* is a zoonosis of major public health concern throughout Latin America, particularly in the Andean and South Cone regions. It is also widely found throughout Arab North Africa and the Middle East. In areas of endemic infection around the world, the practice of giving raw viscera of slaughtered livestock to the dogs enhances transmission; however, in areas where this practice has been curtailed, prevalence figures have decreased.

Although the alveolar form of hydatid disease with *E. multilocularis* has been found in other tissues, the liver is the most common site. The disease may resemble a slowly growing carcinoma and may cause symptoms of intrahepatic portal hypertension. The cyst itself is composed of many irregular cavities with little or no fluid, rare or no free scolices, and often central necrosis and cavitation of the lesion. Because the growth of the alveolar cyst is very slow, some patients may be infected for 20 or 30 years before exhibiting any symptoms. In patients who remain untreated, the mortality rate can exceed 90%.

E. vogeli was observed as the etiologic agent of polycystic disease in humans in areas within South America. The liver was the most common site, followed by the lungs, mesentery, spleen, and pancreas. The main symptoms included abdominal pain, hepatomegaly, jaundice, weight loss, anemia, fever, hemoptysis, palpable abdominal masses, and signs of portal hypertension. The most common clinical presentation involved the abdomen, with hard, round masses in or connected to the liver. In 25% of cases, there were also signs of portal hypertension; all of these patients died, either from the disease or from surgical complications. Calcified, round structures were observed in the liver, suggestive of calcified polycystic hydatids. This disease is chronic and can last for over 20 years; in advanced cases the mortality is high.

The first three known cases of *E. oligarthrus* have been documented, with the first two involving infection in the eye and the third involving infection in the heart. During surgery for a suspected tumor in the eye, fluid and internal tissues were taken from the cyst, which was also later removed. A few protoscolices and hooklets were found, and *E. oligarthrus* was identified.

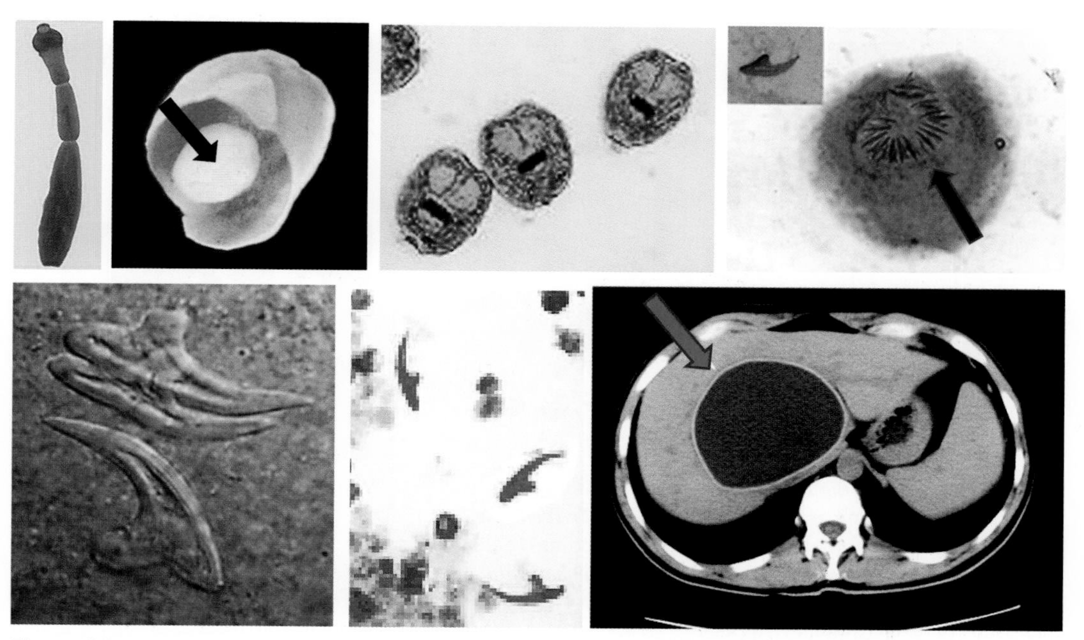

Figure 7.51 *Echinococcus granulosus.* **(Top row)** Adult worm (three proglottids); hydatid cyst containing a daughter cyst (arrow); immature scolices (hydatid sand; the dark line indicates hooklets); hooklets (arrow); enlarged stained hooklet (inset). **(Bottom row)** Enlarged hooklets (unstained); hooklets stained using the Ryan modified blue trichrome stain; hydatid cyst in the liver of an 11-year-old girl (red arrow) (reprinted with permission from Yuksel M, Demirpolat G, Sever A, et al, *Korean J Radiol* 8[6]:531–540, © 2007 The Korean Radiological Society).

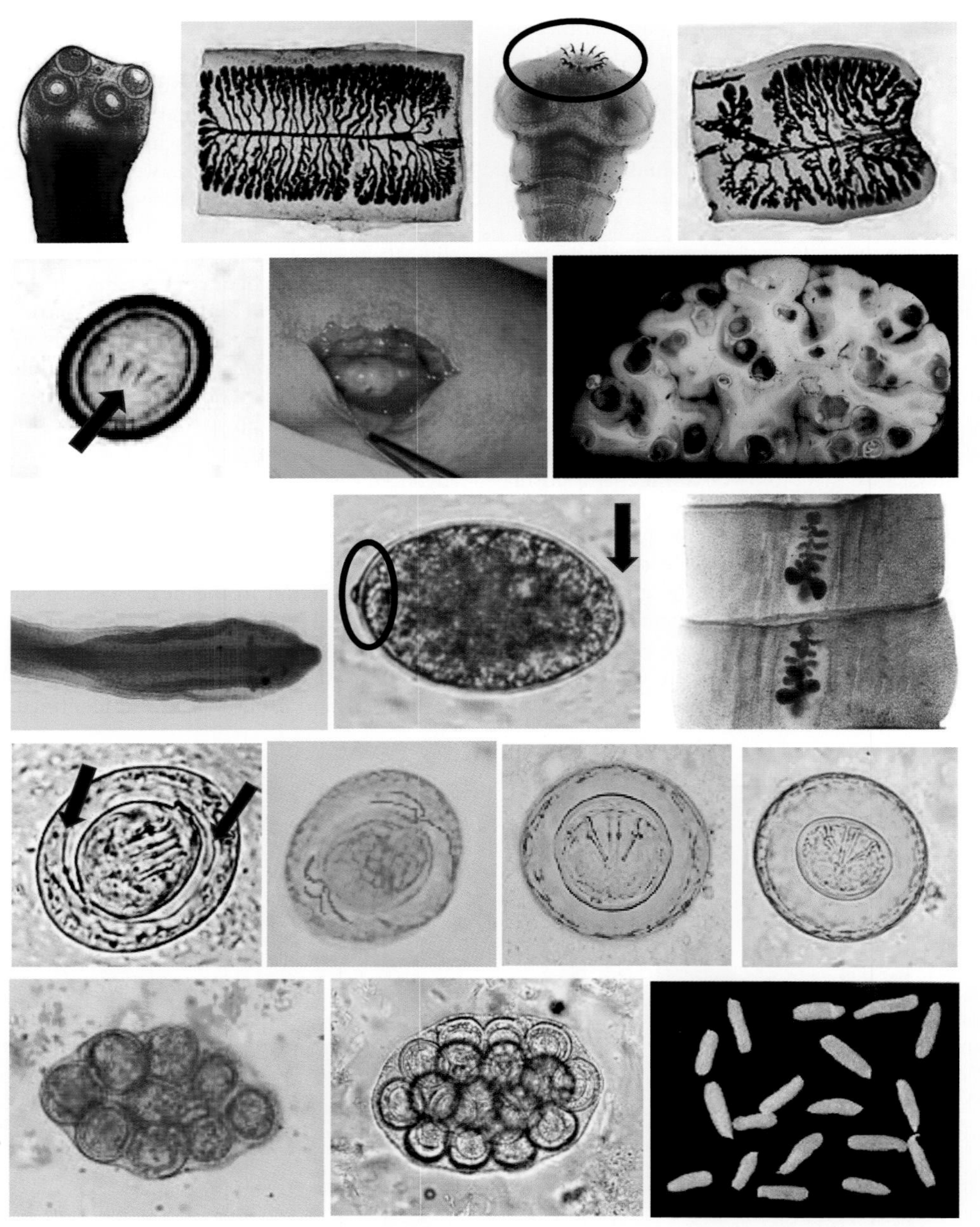

Plate 7.Q **(Row 1)** *Taenia saginata* scolex; *T. saginata* gravid proglottid; *Taenia solium* scolex (note the rostellar hooks [oval]); *T. solium* gravid proglottid. **(Row 2)** *Taenia* sp. egg (*Taenia* cannot be identified to the species level based on egg morphology without special staining) (note the oncosphere hooklets [arrow]); cysticercus in arm; numerous cysticerci in brain. **(Row 3)** *Diphyllobothrium latum* scolex (note two elongate, shallow sucking grooves [bothroid type scolex] rather than *Taenia*-type suckers); *D. latum* operculated egg (note the smooth shell [no opercular shoulders in shell] [arrow] and bump at the abopercular end [oval]); *D. latum* gravid proglottid showing rosette-shaped ovaries (three images courtesy of the CDC Public Health Image Library and CDC DpDx). **(Row 4)** *Hymenolepis nana* egg (note polar filaments [arrows]); *H. nana* egg; *Hymenolepis diminuta* egg (note the absence of polar filaments); *H. diminuta* egg. **(Row 5)** *Dipylidium caninum*/*D. cati* egg packet (iodine stain); *D. caninum*/*D. cati* egg packet (unstained); *D. caninum*/*D. cati* proglottids.

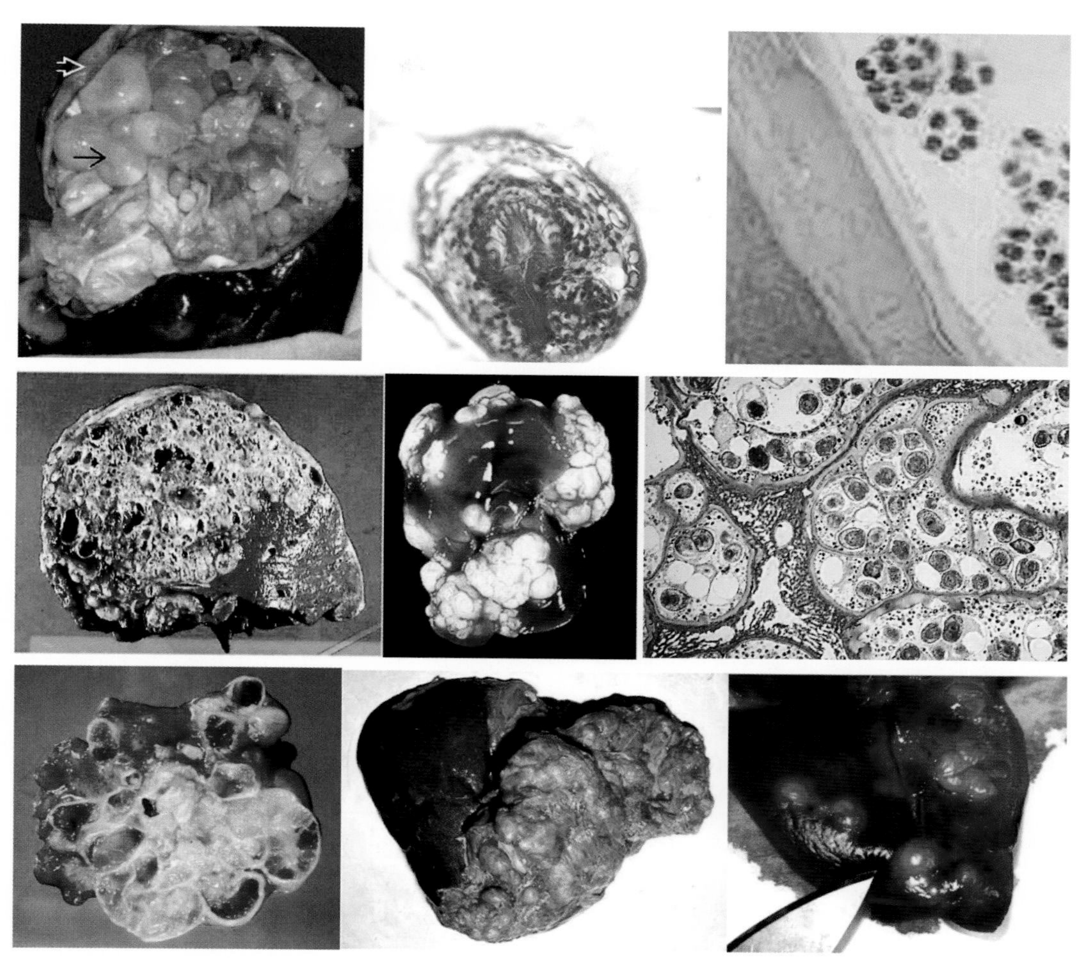

Plate 7.R (Top row) *Echinococcus granulosus* hydatid cyst (cystic disease) containing daughter cysts in liver (courtesy Kari D. Caradine, MD, https://basicmedicalkey.com/echinococcosis/); magnified view of an *E. granulosus* protoscolex, H&E stain (note the row of hooklets; courtesy of the CDC Public Health Image Library); cross section of hydatid cyst (structures, left to right: host tissue, acellular laminated layer, nucleate germinal layer, and numerous protoscolices). **(Middle row)** *Echinococcus multilocularis* hepatic (alveolar) echinococcosis (note that there is no single limiting membrane); alveolar hydatid disease (rodent tissue); alveolar hydatid cyst (routine histology) (this mimics metastatic cancer—no limiting membrane and multichambered appearance). **(Bottom row)** *Echinococcus vogeli* hydatid cysts: polycystic lesion of the pericardium (reprinted from D'Alessandro A, Rausch RL, Cuello C, Aristibal N, *Am J Trop Med Hyg* **28**:303–317, 1979); polycystic echinococcosis in the human liver (courtesy of Octavio Sousa, Centro de Investigación y Diagnóstico de Enfermedades Parasitarias, Facultad de Medicina, República de Panamá); *Echinococcus oligarthrus* unicystic/neotropical hydatid cyst in agouti (note liver with multiple cysts in the liver parenchyma) (reprinted from Arrabal JP, Avila HG, Rivero MR, et al, *Vet Parasitol* **240**:60–67, 2017, with permission from Elsevier, Ltd.).

Fasciolopsis buski

Pathogenic	Yes
Disease	Fasciolopsiasis, giant intestinal fluke disease
Acquired	Ingestion of infective metacercariae encysted on plant material
Body site	Intestine
Symptoms	None, or abdominal pain and diarrhea
Clinical specimen	Stool
Epidemiology	Bangladesh, Cambodia, central and southern China, India, Indonesia, Laos, Malaysia, Pakistan, Taiwan, Thailand, and Vietnam; human-to-human and animal-to human transmission (dogs, pigs, and rabbits)
Control	Improved hygiene, adequate disposal of fecal waste, adequate washing of contaminated plant material

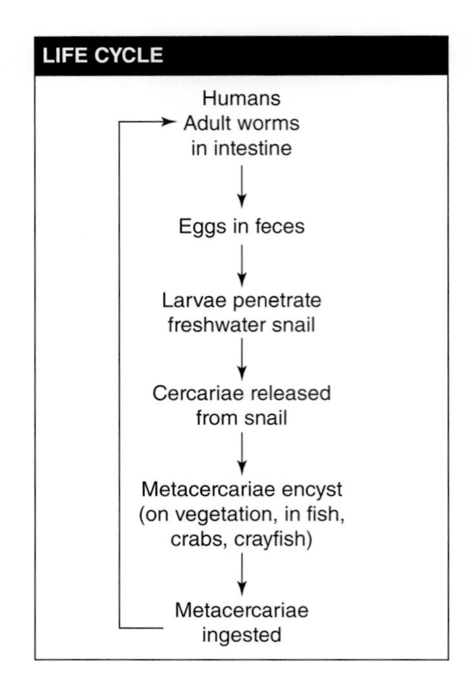

Diagnosis

Trematodes are parasitic flatworms with unique life cycles involving sexual reproduction in mammalian and other vertebrate definitive hosts and asexual reproduction in snail intermediate hosts. The trematodes are divided into four groups in humans: (i) the hermaphroditic liver flukes, which reside in the bile ducts and infect humans on ingestion of watercress (*Fasciola*) or raw fish (*Clonorchis* and *Opisthorchis*); (ii) the hermaphroditic intestinal fluke (*Fasciolopsis*), which infects humans on ingestion of water chestnuts; (iii) the hermaphroditic lung fluke (*Paragonimus*), which infects humans on ingestion of raw crabs or crayfish; and (iv) the bisexual blood flukes (*Schistosoma*), which live in the intestinal or vesical (urinary bladder) venules and infect humans by direct penetration through the skin.

The standard O&P exam is recommended for recovery and identification of *Fasciolopsis* eggs in stool specimens, primarily from the wet-preparation examination of the concentration sediment. The eggs are detected in feces by direct microscopy or by concentration techniques. Infrequently, adult worms are detected in the stool in very heavy infections or during therapy.

Diagnostic Tips

While eggs can be found in the stool, adult worms are rarely seen other than in heavy infections. The zinc sulfate flotation method is not recommended because of the operculum and the fragility of the eggshell. In addition to the surface material, the sediment would have to be examined, because the eggs sink to the bottom. The routine formalin-ether sedimentation technique is recommended instead. The sedimented material can be examined with or without iodine. The eggs are brownish yellow and may contain fully formed miracidia, depending on the species. The less mature the egg, the more difficult it may be to see the actual operculum—it blends into the shell outline, and the "breaks" in the shell may be difficult to identify. This egg has no opercular shoulders; therefore, it is difficult to see where the operculum breaks in the shell actually occur.

Humans become infected by ingesting the raw or undercooked plants containing the metacercariae. The metacercariae excyst, attach to the duodenal or jejunal mucosa, and develop into adult worms within 3 months. The adult life span seldom exceeds 1 year.

The eggs of *Echinostoma ilocanum*, *Fasciola hepatica*, *Fasciola buski*, and *Gastrodiscoides hominis* are similar in size and shape; therefore, an exact diagnosis cannot be made from examining the eggs. It is possible to detect adult worms in the stool in heavy infections when they lose their ability to remain attached to the intestinal mucosa.

General Comments

Eggs deposited by the adult worms are passed in the feces. They contain immature miracidia when passed in the stool and require a period for embryonation in the outside environment. All eggs of intestinal trematodes have an operculum from which the miracidium can escape. *Fasciolopsis* eggs hatch in freshwater to release a free-swimming miracidium larva that must find the snail host to which it is adapted or die. Once the miracidium has penetrated the snail's soft tissues, it begins to develop into a first-generation sporocyst (an elongated sac without distinct internal structures in which germ balls proliferate). These germ balls develop into rediae that contain a mouth, pharynx, blind cecum, and birth pore. Within the rediae, the germ balls again proliferate, developing into cercariae. On reaching maturity, the cercariae escape from the snail host into the water. *Fasciolopsis* rediae develop a second generation of rediae before forming cercariae. The cercariae are free-living in the water, and they must find their next intermediate host or die. When the appropriate host or vegetation is found, the cercariae lose their tails and encyst (becoming metacercariae). *Fasciolopsis* cercariae encyst on aquatic vegetation. On ingestion of uncooked vegetation or of raw or inadequately cooked mollusks or fish, the metacercariae excyst in the small intestine and develop into mature hermaphroditic adults in the intestinal tract.

Description (Eggs)

The eggs of *E. ilocanum*, *F. hepatica*, *F. buski*, and *G. hominis* are similar in size and shape, so exact identification cannot be made from examining the eggs in wet mounts of fecal concentration sediment. Since the eggs of *Fasciola* and *Fasciolopsis* (the more common of the four trematodes) are so similar in size and shape, if either is found, the report can indicate that the two trematodes cannot be separated into appropriate genera on the basis of egg morphology. The eggs are ellipsoidal, operculate, and yellow brown. They measure 130 to 140 µm by 80 to 85 µm, with the operculum found at the more pointed end of the transparent eggshell.

Description (Adults)

The adult worms live in the small intestines of pigs and humans, where the worms lay unembryonated eggs that are then passed from the intestinal lumen with the feces. *Fasciolopsis buski* is also known as the giant intestinal fluke and is one of the largest parasites to infect humans, measuring 20 to 75 mm in length, 8 to 20 mm in width, and 0.5 to 3 mm in thickness. The adult worms are fleshy, somewhat pale, and elongate-ovoid and have no cephalic cone structures like that seen in the liver fluke, *F. hepatica*.

Additional Information

F. buski reservoir hosts include dogs, pigs, and rabbits. The infection is common in Bangladesh, Cambodia, central and southern China, India, Indonesia, Laos, Malaysia, Pakistan, Taiwan, Thailand, and Vietnam; it has also been found in Japan. Drainage of farm waste, use of manure for cultivation, and defecation in or near ponds or lakes that contain snails from the family Planorbidae, with water plants acting as vectors, support the life cycle. Metacercariae encyst on freshwater vegetation such as water chestnuts, bamboo shoots, or water caltrops, and the infection is acquired when these infested plants are consumed raw or the outer coat is peeled off the nut with the teeth, resulting in accidental ingestion. To prevent infection, plants should be cooked or immersed in boiling water for a few seconds before being eaten or peeled. In areas of endemicity, the use of unsterilized night soil for fertilizer should be prohibited.

The disease occurs focally and is most prevalent in school-age children. In areas of endemic infection, the prevalence in children ranges from 57% in mainland China to 25% in Taiwan and from 50% in Bangladesh and 60% in India to 10% in Thailand. Control programs are not fully successful because of long-standing traditions of eating raw aquatic plants and using untreated water.

In light infections, the adult parasites inhabit the duodenum and jejunum; in heavy infections, they reside in the stomach and most of the intestinal tract. The attachment of worms to the mucosal wall produces local inflammation with hypersecretion of mucus, hemorrhage, ulceration, and possible abscess formation. In heavy infections, the worms may cause bowel obstruction, acute ileus, and absorption of toxic or allergic worm metabolites, producing general edema and ascites. A marked eosinophilia and leukocytosis are common. Few symptoms are associated with light infections, but in heavier infections the patient may experience abdominal pain and diarrhea. In heavy infections, the stools are profuse and yellow-green and contain increased amounts of undigested food, suggesting a malabsorption process. The disease can be fatal, depending on the worm burden.

Sequence analysis comparisons indicate that *F. buski* isolates from China and India may represent different taxonomic groups, while *F. buski* isolates from Vietnam and China represent the same species. Additional studies from more geographical areas are required to better understand the *F. buski* species complex.

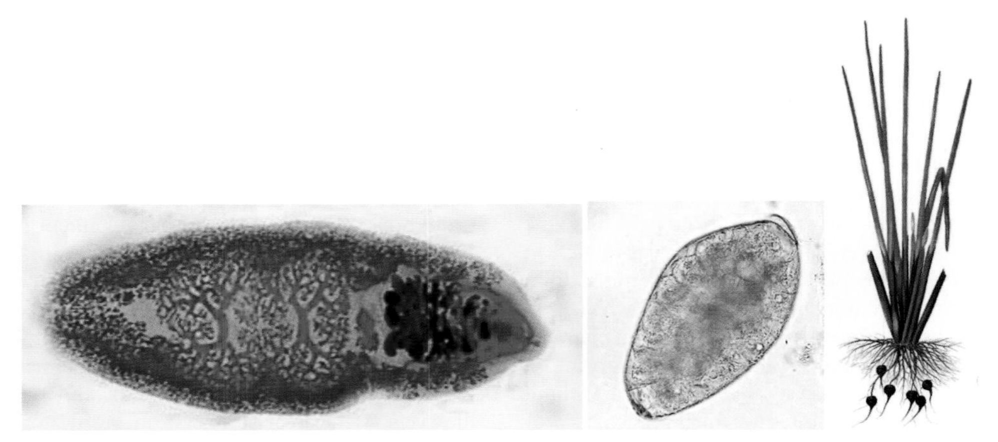

Figure 7.52 ***Fasciolopsis buski.*** *F. buski* adult worm; *F. buski* egg (the operculum blends into the shell; there are no opercular shoulders into which the operculum fits; note that the operculum has popped open; this egg and that of several other trematodes, including *Fasciola hepatica*, look almost identical and cannot be differentiated visually); water chestnut plant on which the infective metacercariae are encysted. Note the small brown chestnuts on the plant; the outer coating of the chestnut is often peeled with the teeth, thus providing entry via ingestion of the encysted metacercariae.

Paragonimus westermani
Paragonimus mexicanus
Paragonimus kellicotti
Depending on area of endemicity, a number of other species have been confirmed

Pathogenic	Yes
Disease	Paragonimiasis
Acquired	Ingestion of encysted metacercariae in fish, crabs, and crayfish
Body site	Lungs, extrapulmonary sites
Symptoms	None, or chronic cough, vague chest pains, hemoptysis
Clinical specimen	*Intestinal*: Stool
	Lung: Sputum
Epidemiology	Far East; North, Central, and South America; human-to-human and animal-to-human transmission (dogs, cats)
Control	Improved hygiene, adequate disposal of fecal waste, adequate cooking of fish, crabs, crayfish

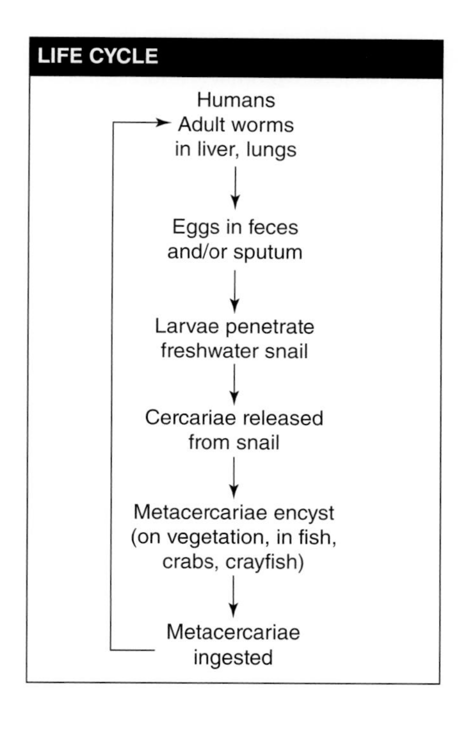

Diagnosis

Paragonimus mexicanus is an important human pathogen in Central and South America, while *Paragonimus kellicotti* infections are found in North and South America. Most paragonimiasis infections are caused by *Paragonimus westermani*, which is the focus of this section. Individuals with symptoms of chronic cough, vague chest pains, and hemoptysis who have resided in an area where infections are endemic and have a history of eating raw crayfish or crabs should be suspected of having paragonimiasis.

Paragonimus eggs can be detected in sputum and in stool; both specimens should be tested to improve the overall detection rate. Small numbers of eggs are present intermittently in the sputum and feces. For light infections, up to seven sputum examinations have been recommended. Frequently, pulmonary paragonimiasis is misdiagnosed as pulmonary tuberculosis. The Ziehl-Neelsen method for detecting mycobacteria destroys the eggs of *Paragonimus*. The typical findings of cough, hemoptysis, and eggs in the feces or sputum may be absent in patients with ectopic or pleural infection.

Skull films generally reveal round or oval cystic calcifications, described as looking like bubbles. CT scans reveal multilocular cysts with edema, migration tracks, peripheral density, bronchial wall thickening, and centrilobular nodules; ring enhancement may be visible after the use of contrast. MRI demonstrates areas of granulomatous inflammation surrounding the lesions; MRI of the brain may reveal multiple conglomerated iso-signal-intensity or low-signal-intensity round nodules with peripheral rim enhancement. CT and MRI of the liver may reveal a cluster of small cysts with rim enhancement in the subcapsular area. Solitary nodular lesions often mimic lung cancer, tuberculosis, or fungal diseases. When a pulmonary mass lesion or empyema is detected in patients who live in areas of endemic infection, paragonimiasis should always be included in the differential diagnosis.

Immunodiagnostic tests have been used to diagnose pulmonary and extrapulmonary infections. These serologic assays are available in areas of endemicity or in specialized diagnostic centers.

Diagnostic Tips

Paragonimus eggs may be confused with *Diphyllobothrium latum* eggs because of similarities in their size and shape. However, most *Paragonimus* eggs have opercular shoulders and a marked thickening at the abopercular end, unlike *D. latum* eggs.

The sedimentation concentration method should be used; because the eggs are operculated, they do not float in the zinc sulfate flotation concentration method. It is important not to add too much iodine to the wet preparation, or the eggs will stain very darkly and resemble debris. The wet preparation can be examined using the 10× (low-power) objective; these eggs are large enough that they can usually be seen under this magnification. The wet preparation should not be so thick that the eggs are obscured by normal stool debris.

When looking at sputum specimens for eggs, it is necessary to carefully examine any blood-tinged flecks for eggs; the egg clusters have been described as looking like iron filings. Remember, the Ziehl-Neelsen method for detecting mycobacteria destroys the eggs of *Paragonimus*. In light infections, multiple stool and sputum specimens may be needed before the eggs are detected. If available, antigen or antibody detection may be helpful in confirming the infection.

General Comments

The eggs hatch in water in 2 to 3 weeks, releasing a miracidium to infect a susceptible snail host. Cercariae are released after sporocyst and redia generations. Crabs and crayfish are infected by cercariae via the gill chamber or on ingestion of an infected snail. Cercariae encyst in the gill vessels and muscles. Humans are infected by ingesting uncooked crabs or crayfish containing metacercariae. The metacercariae excyst in the duodenum and migrate through the intestinal wall into the abdominal cavity. The larvae migrate around or through the diaphragm into the pleural cavity and the lungs. The larvae mature to adults in the vicinity of the bronchioles, where they discharge their eggs into the bronchial secretions.

The most serious consequence of paragonimiasis is cerebral complications, which are commonly found in younger age groups. Most patients with extrapulmonary lesions have an associated lung lesion or a history of lung disease. Symptoms include fever, headache, nausea, vomiting, visual disturbances, motor weakness, localized or generalized paralysis, and possibly death.

Pulmonary paragonimiasis is rarely fatal; however, cerebral disease is characterized by chronic morbidity and symptoms including epilepsy, dementia, and other neurologic sequelae. About 5% of patients with cerebral disease die due to hemorrhage in the first 2 years of the disease.

Other body sites that have been infected include the breast, lymph nodes, heart, pericardium, mediastinum, kidney, adrenal gland, omentum, bone marrow, stomach wall, bladder, spleen, pancreas, and reproductive organs. While ectopic lesions are usually thought to be caused by worm migration, dissemination of eggs to other body sites may also be responsible.

Description (Eggs)

Paragonimus and *D. latum* eggs are similar in size and shape. Remember to look for the opercular shoulders and thickening at the abopercular end of the *Paragonimus* eggs. *D. latum* eggs do have an operculum (the outline of the egg is smooth with no "teapot lid" opercular shoulders). Also, there is a knob at the abopercular end in *D. latum* eggs, although it is often difficult to see. Eggs deposited by the worms are ovoid, brownish yellow, unembryonated, and thick shelled, with an operculum at one end and opercular shoulders. Eggs measure 80 to 120 μm by 45 to 65 μm. Since the eggs of the two trematodes mentioned above are so similar in size and shape, it is important to measure the eggs and review several in order to assign the eggs to the correct genus.

Description (Adults)

The adult worm is a plump, ovoid, reddish brown fluke found encapsulated in the lung. Eggs escape from the encapsulated tissue through the bronchioles, are coughed up and voided in the sputum, or are swallowed and passed out in the feces. Although these worms are hermaphroditic, two worms are usually required for fertilization to occur. The worms can live as long at 20 years, but most die after about 6 years.

Additional Information

Migration of larval forms through the intestinal wall into the abdominal cavity is generally not associated with symptoms. Once the larvae have reached the peritoneal cavity, they migrate through organs and tissues, producing localized hemorrhage and leukocytic infiltrates. When they reach the lungs and mature, a pronounced tissue reaction occurs with infiltration of eosinophils and neutrophils. A fibrotic capsule forms around the worm. The cysts contain purulent fluid with flecks, or "iron filings," composed of brownish yellow eggs. Many cysts perforate the bronchioles, releasing their contents of eggs, necrotic debris, metabolic by-products, and blood into the respiratory tract. The eggs may also enter the pulmonary tissue or be carried by the circulatory system to other body sites, where they cause a granulomatous reaction. Larval forms may end up in many ectopic sites other than the lungs. Cysts have been detected in the liver, intestinal wall, muscles, brain, and peritoneum.

Symptoms of paragonimiasis depend largely on the worm burden of the host and are usually insidious in onset and mild in chronic cases. Light infections may be asymptomatic, although peripheral-blood eosinophilia and lung lesions may be noted on X rays. As the cyst ruptures, a cough develops with increased production of viscous blood-tinged sputum (rusty sputum which may have a foul fish odor) and increasing chest pain. The patient may experience increasing dyspnea with chronic bronchitis and be misdiagnosed as having tuberculosis. There will generally be a moderately high peripheral-blood eosinophilia and leukocytosis with elevated serum IgG and IgE.

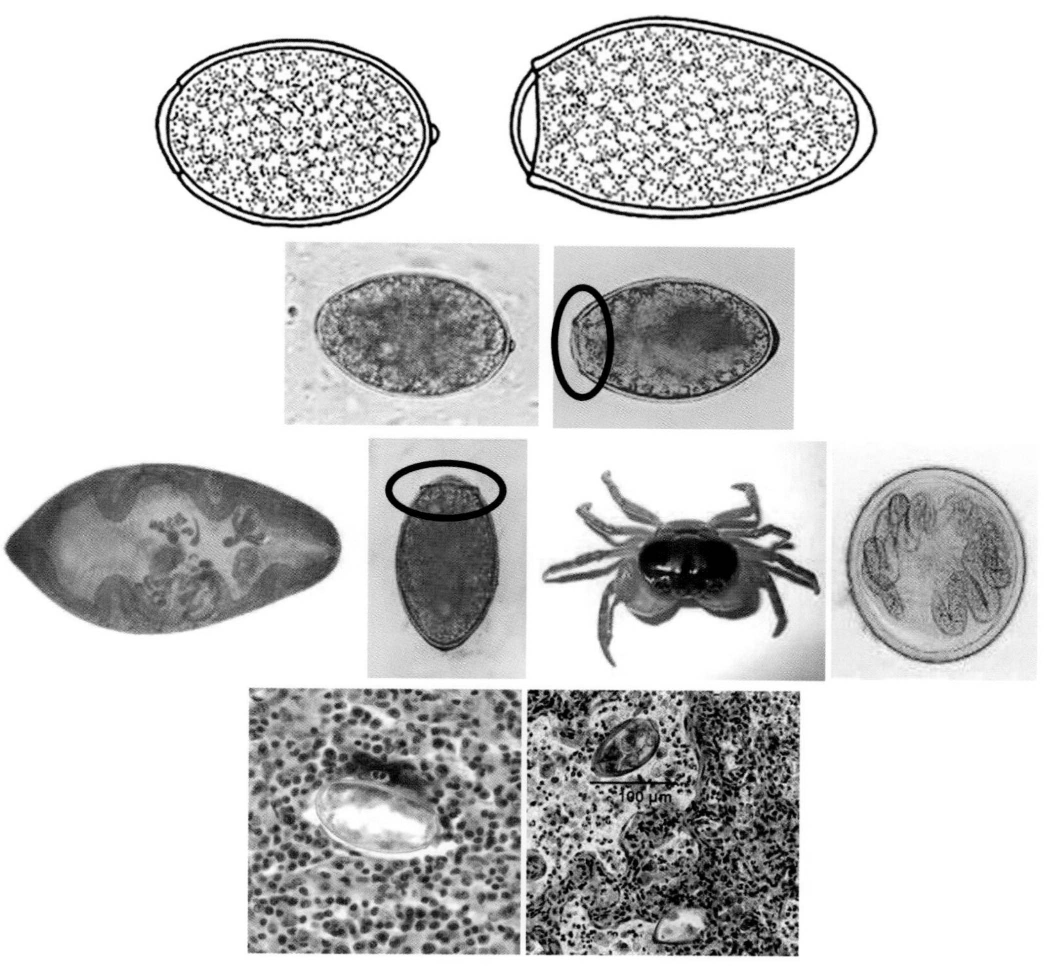

Figure 7.53 ***Paragonimus*** **spp. (Top row)** Drawings showing *Diphyllobothrium latum* (A) and *Paragonimus westermani* (B). Although the eggs are different sizes, there is some similarity between the two. The photograph shows the same two eggs, *D. latum* (left) and *P. westermani*. Note the opercular shoulders visible with *P. westermani* (oval). It is critical that these eggs be measured carefully, including specimens used for proficiency testing. **(Middle row)** *Paragonimus* adult fluke; *P. westermani* egg (note the opercular shoulders [oval]); intermediate host (crab); metacercariae that can be found in crab and crayfish. **(Bottom row)** *Paragonimus* eggs in lung tissue (courtesy of CDC DpDx, https://www.cdc.gov/dpdx/paragonimiasis/index.html).

Fasciola hepatica

Pathogenic	Yes
Disease	Fascioliasis (WHO neglected tropical disease)
Acquired	Ingestion of infective metacercariae encysted on plant material
Body site	Bile ducts
Symptoms	None to abdominal pain and diarrhea
Clinical specimen	Stool
Epidemiology	Bolivia, Ecuador, Egypt, France, Iran, Peru, and Portugal; human-to-human and animal-to-human transmission (cattle, goats, sheep)
Control	Improved hygiene, adequate disposal of fecal waste, adequate washing of contaminated plant material

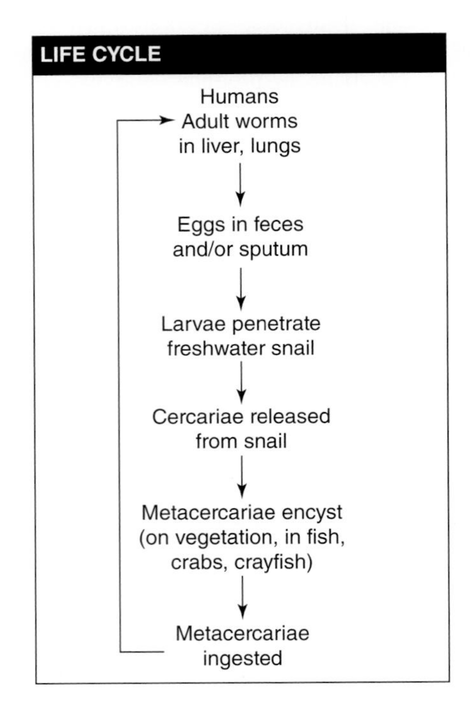

Diagnosis

Fascioliasis (sheep liver fluke disease) is primarily a zoonotic disease that causes liver infections with adult flukes. Over 2.4 million people are infected with this parasite worldwide, including Europe, the Americas, northern Asia, Oceania, Africa, New Zealand, Tasmania, Great Britain, Iceland, Cyprus, Corsica, Sardinia, Sicily, Japan, Papua New Guinea, the Philippines, and the Caribbean. A few cases have been documented in Hawaii, California, and Florida. This infection has been recognized by the World Health Organization as an important health problem, particularly in the Andean countries of Peru, Bolivia, and Chile.

The standard O&P exam is recommended for recovery and identification of *Fasciola hepatica* eggs in stool specimens, primarily from the wet-preparation examination of the concentration sediment. The eggs are detected in feces by direct microscopy or by concentration techniques. Infrequently, adult worms are detected in the stool in very heavy infections or when the patient is undergoing therapy.

The zinc sulfate flotation method is not recommended because of the operculum and the fragility of the eggshell. In addition to the surface material, one would have to examine the sediment because the eggs sink to the bottom.

Imaging methods include ultrasound, computer tomography, magnetic resonance imaging, and tissue harmonic imaging. These imaging methods are able to reveal the pathology of the bile ducts. Ultrasound can detect biliary stones, dilatation, and fibrosis due to liver fluke infections, while computer tomography and magnetic resonance imaging can determine the lumen diameter of bile ducts, fibrosis, calcification, and epithelial hyperplasia. Tissue harmonic imaging is useful for the observation of bile duct wall trauma and stones.

Diagnostic Tips

Multiple stool specimens may be required to find the trematode eggs. In a very light infection, eggs may not be recovered; however, the patient might be asymptomatic. The sedimentation concentration method should be used; because the eggs are operculated, they do not float in the zinc sulfate flotation concentration method. The routine formalin-ether sedimentation technique is recommended instead. The sedimented material can be examined with or without iodine. The eggs appear brownish yellow and may contain fully formed miracidia, depending on the species. It is important not to add too much iodine to the wet preparation, or the eggs will stain very darkly and resemble debris. The wet preparation can be examined using the 10× (low-power) objective; these eggs are large enough that they can usually be seen using this magnification. Eggs of *F. hepatica* can resemble those of *Fasciola gigantica* (liver fluke) and *Fasciolopsis buski* (intestinal fluke). The wet preparation should not be so thick that the eggs are obscured by normal stool debris. If available, antigen or antibody detection may be helpful in confirming the infection.

General Comments

Adult worms, which may live for 9 years in the bile ducts, produce eggs that are carried by the bile fluid into the intestinal lumen and passed into the environment with the feces. The miracidium develops within 1 to 2 weeks and escapes from the egg to infect the snail intermediate host, *Lymnaea* sp. Cercariae are liberated from the snail after a sporocyst generation and two or three redia generations. Cercariae encyst on water vegetation, e.g., watercress. Humans are infected by ingesting this vegetation raw. Metacercariae excyst in the duodenum and migrate through the intestinal wall into the peritoneal cavity. Larvae enter the liver by penetrating the capsule (Glisson's capsule) and inhabit the liver parenchyma for up to 9 weeks; they finally enter the bile ducts, where they mature and produce eggs, which are passed out in feces.

The incubation period ranges from a few days to a few months. Symptoms reflect the phase of the infection, as well as the number of parasites in the host. In the acute phase, symptoms may occur for weeks to months. In the more chronic phases of the disease, the patient generally has few or no symptoms once the flukes have lodged in the biliary passages. Other body sites include the intestinal wall, lungs, heart, brain, and skin. Symptoms mimic those seen with visceral larva migrans and include vague abdominal pain.

Description (Eggs)

E. ilocanum, *F. hepatica*, *F. buski*, and *G. hominis* eggs are similar in size and shape, so exact identification cannot be made from examining the eggs. Since *Fasciola* and *Fasciolopsis* eggs are so similar, if either is found, the report can indicate that the two trematodes cannot be separated into different genera on the basis of egg morphology. Multiple stool examinations may be needed to detect light infections.

The eggs are unembryonated, operculated, large, ovoid, and brownish yellow and measure 130 to 150 μm by 63 to 90 μm. Humans become infected by ingesting raw or undercooked plants containing the metacercariae. The metacercariae excyst, attach to the duodenal or jejunal mucosa, and develop into adult worms within 3 months. The adult worms can grow to 30 mm long and 13 mm wide and can live for more than 10 years.

Description (Adults)

Adult *F. hepatica* organisms can be found in the stool in heavy infections, when they can no longer remain attached to the intestinal mucosa. Humans are infected by ingestion of uncooked aquatic vegetation on which metacercariae are encysted. Metacercariae excyst in the duodenum and migrate through the intestinal wall into the peritoneal cavity. The larvae enter the liver by penetrating the capsule (Glisson's capsule) and wander through the liver parenchyma for up to 9 weeks. The larvae finally enter the bile ducts, where they mature and produce eggs, which are passed out in the feces. The adult worms can attain a length of 20 to 30 mm and a width of about 8 to 13 mm and can live for more than 10 years.

Additional Information

Fascioliasis is primarily a zoonotic disease involving liver infections with adult flukes. In areas of endemicity where uncooked goat and sheep livers may be eaten, such as Lebanon, adult worms may attach to the pharyngeal mucosa, causing suffocation (halzoun syndrome). This condition is temporary, although distressing. The adult worms may lodge on the pharyngeal mucosa, causing edema and congestion of the soft palate, pharynx, larynx, nasal fossae, and eustachian tube. Symptoms include dyspnea, dysphagia, deafness, and occasionally suffocation. Some cases may be caused by infection with larval linguatulids, rather than adult worms of *F. hepatica*.

Eggs may be detected in the stools of individuals who have eaten *F. hepatica*-infected liver, yielding an erroneous laboratory result (a "spurious" infection). True and spurious infections can be distinguished by giving the patient a liver-free diet for at least 3 days. If the patient continues to pass eggs in the stool, the infection is probably genuine.

The incubation period for fascioliasis can range from a few days to a few months. The degree of damage depends on the worm burden. In the acute phase, symptoms may be present over a period of weeks to months and include dyspepsia, fever, right upper quadrant pain, anorexia, hepatomegaly, splenomegaly, ascites, urticaria, respiratory symptoms, and jaundice. Metacercarial larvae do not cause significant pathologic damage until they begin to migrate through the liver parenchyma. Linear lesions of 1 cm or greater can be found. Hyperplasia of the bile ducts occurs, possibly as a result of toxic products produced by the larvae. Symptoms associated with this migratory phase include fever, epigastric and right upper quadrant pain, and urticaria, while some patients remain asymptomatic. Leukocytosis, eosinophilia, and mild to moderate anemia occur in many patients. Levels of IgG, IgM, and IgE in serum are usually elevated. In sheep, the migratory phase produces such extensive liver parenchyma damage that the disease is known as liver rot.

Larvae may be found in ectopic foci after penetrating the peritoneal cavity. Worms in human infections have been discovered in many areas of the body other than the liver.

Once the worms are established in the bile ducts and have matured, they produce considerable damage from mechanical irritation and metabolic by-products as well as obstruction. The degree of pathology depends on the number of flukes penetrating the liver.

The infection produces hyperplasia of the biliary epithelium and duct fibrosis with portal or total biliary obstruction. The gallbladder undergoes similar damage and may even harbor adult worms. Adult worms may reinvade the liver parenchyma, producing abscesses.

Triclabendazole (TCBZ) is the only chemical that kills early immature and adult *F. hepatica*; however, widespread resistance to the drug greatly complicates fluke control in livestock and humans. The mode of action of TCBZ and the mechanism(s) underlying parasite resistance to the drug are not known. Human infections with TCBZ-resistant *F. hepatica* have been reported in The Netherlands, Chile, Turkey, and Peru. Currently, research is ongoing for a liver fluke vaccine for livestock.

Figure 7.54 ***Fasciola hepatica.*** **(Top row)** Drawing of *F. hepatica* egg (the operculum blends into the shell; there are no opercular shoulders into which the operculum fits); egg in which the operculum has popped open (this egg and that of *Fasciolopsis buski* and several other trematodes look almost identical and cannot be differentiated visually); adult fluke (note the cone-shaped anterior end) (courtesy of CDC, https://www.cdc.gov/parasites/fasciola/index.html). **(Bottom row)** Watercress; *Fossaria bulamoides*, a snail host for *F. hepatica* in the western United States (© Guido & Philippe Poppe, www.conchology.be); adult *F. hepatica* (fresh specimen) (courtesy of I. Flukeman, CC-BY-SA Wikimedia Commons).

Clonorchis (*Opisthorchis*) *sinensis* (Chinese liver fluke)

Pathogenic	Yes
Disease	Clonorchiasis; sequelae may include cholangiocarcinoma
Acquired	Ingestion of infective metacercariae encysted in raw or poorly cooked freshwater fish
Body site	Bile duct and liver
Symptoms	None, or acute pancreatitis, cholecystitis, and cholelithiasis (may be the result of worm invasion); biliary tract obstruction
Clinical specimen	Stool
Epidemiology	China, Japan, Korea, Malaysia, Singapore, Taiwan, and Vietnam; human-to-human and animal-to-human transmission (dogs, cats, fish-eating mammals)
Control	Improved hygiene, adequate disposal of fecal waste, adequate cooking of freshwater fish

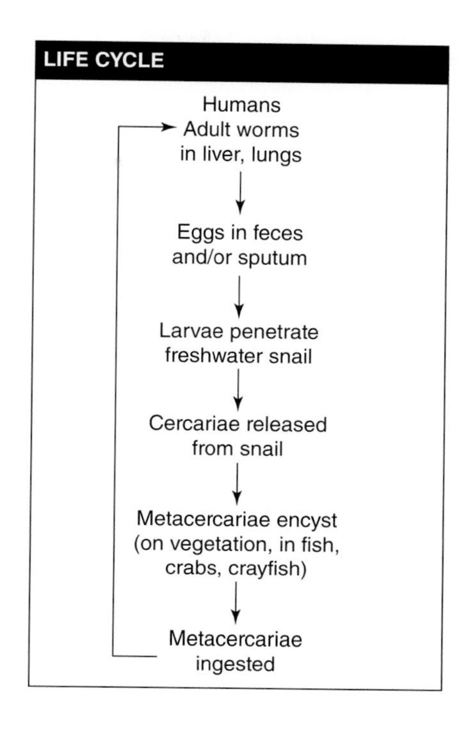

Diagnosis

Clonorchis, *Opisthorchis*, and *Fasciola* spp. are trematodes that parasitize the biliary ducts of humans. *Clonorchis* and *Opisthorchis* spp. are narrow, elongate worms that localize in the more distal, smaller ducts of the biliary tree. *Fasciola hepatica*, because of its much larger size, resides in the larger bile ducts and gallbladder.

The standard O&P exam is recommended for recovery and identification of *Clonorchis sinensis* eggs in stool specimens, primarily from the wet-preparation examination of the concentration sediment. The eggs are detected in feces by direct microscopy or concentration techniques. Infrequently, adult worms are detected in the stool in very heavy infections or during therapy.

Cholangiography, ultrasonography, and liver scans may reveal lesions consistent with liver fluke infection. Clonorchiasis is often diagnosed incidentally during abdominal ultrasonography, since symptoms tend to be nonspecific. However, none of these methods are any more sensitive than fecal examination because of their limited sensitivity and specificity.

Diagnostic Tips

Multiple stool specimens may be required to find the eggs. The zinc sulfate flotation method is not recommended because of the operculum and the fragility of the eggshell. In addition to the surface material, one would have to examine the sediment because the eggs sink to the bottom. The routine formalin-ether sedimentation technique is recommended instead. The sedimented material can be examined with or without iodine. The eggs appear brownish yellow and may contain fully formed miracidia, depending on the species. It is important not to add too much iodine to the wet preparation, or the small eggs will stain very darkly and resemble debris. Although the wet preparation can be examined using the 10× (low-power) objective, these eggs measure approximately 30 µm and can easily be missed. It is recommended that the 40× (high dry power) objective be used for the wet-preparation examination before providing the final report.

The eggs are similar in size and shape to those of *Heterophyes heterophyes* and *Metagonimus yokogawai* and cannot be readily differentiated. If a patient has not resided in or recently visited areas where infections are endemic, the infection is probably due to *C. sinensis* or *Opisthorchis viverrini*. The infection may be confirmed by detecting eggs in the bile fluid (duodenal aspirate), by recovering adult worms, or from the clinical history. Some strains produce eggs with a comma-shaped appendage at the abopercular end. Multiple egg measurements are usually required to determine size differences, but absolute identification of the small trematode eggs can be very difficult.

In patients with biliary obstruction, eggs are not found in the stool specimens; needle aspiration, surgery, or autopsy specimens may be required to confirm their presence. In these patients, biliary obstruction must be differentiated from enlarged gallbladder, cholangitis with jaundice, liver carcinoma, and cholangiocarcinoma. Cholangiography, ultrasonography, and liver scans may reveal lesions consistent with infection.

If duodenal sampling is performed (via drainage or the Entero-Test), eggs may be found in this material. If available, antigen or antibody detection may be helpful in confirming the infection.

General Comments

The infections caused by the liver and lung trematodes are food borne and have considerable economic and public health impact. More than 50 million people have acquired food-borne trematode infections. Approximately 601, 293, 91, and 80 million people are at risk of infection with *C. sinensis*, *Paragonimus* spp., *Fasciola* spp., and *Opisthorchis* spp., respectively. Of great public health concern is cholangiocarcinoma associated with *Clonorchis* and *Opisthorchis* infections, severe liver disease associated with *Fasciola* infections, and the misdiagnosis of tuberculosis in those infected with *Paragonimus* spp.

Adult *Clonorchis* worms deposit eggs in the bile ducts, and the eggs are discharged with the bile fluid into the feces and passed out into the environment. Eggs are ingested by the snail host, and the miracidium hatches to infect the snail. Sporocyst and redia generations are produced before cercariae are released to encyst in the skin or flesh of freshwater fish. Humans are infected by ingesting the metacercariae in uncooked fish.

Metacercariae excyst in the duodenum, enter the common bile duct, and travel to the distal bile capillaries, where the worms mature. The life cycle takes approximately 3 months in humans.

Description (Eggs)

The eggs of *C. sinensis*, *H. heterophyes*, and *M. yokogawai* are similar in size and shape, so exact identification may not be possible from examining the eggs. Adult worms can be found in the stool in heavy infections when they can no longer remain attached to the intestinal mucosa.

C. sinensis and *O. viverrini* eggs are fully embryonated when laid and measure 28 to 35 μm by 12 to 19 μm. They are ovoid, with a thick, light brownish yellow shell and an operculum. There are distinct opercular shoulders surrounding the operculum. Since the eggs of the three trematodes mentioned above are so similar, if any of the three are found, the report can indicate that the genus cannot be determined on the basis of egg morphology.

Description (Adults)

An adult *C. sinensis* is a flattened (dorsoventrally) and leaf-shaped fluke. The body is somewhat elongated and slender, measuring 15 to 20 mm in length and 3 to 4 mm in width. It narrows at the anterior region into a small opening called the oral sucker (mouth). From the mouth, two tubes called ceca run the length of body; these are the digestive and excretory tracts. The posterior end is broad and blunt. A poorly developed ventral sucker lies behind the oral sucker. As a hermaphrodite, the adult fluke contains both male and female reproductive organs. A single rounded ovary is at the center of the body, and two testes are towards the posterior end. The uterus from the ovary and seminal ducts from the testes meet and open at the genital pore. The testes are highly branched. Other highly branched organs called vitellaria (or vitelline glands) are distributed on either side of the body.

Additional Information

Light infections generally cause no symptoms. In heavier infections, the patient may experience dull pain and abdominal discomfort that may last 1 to 2 h, often in the afternoon. As the disease progresses, the pain persists and may become so severe that the patient is unable to work. Patients who have had the disease for a long time have liver enlargement with some degree of functional impairment secondary to biliary obstruction. Acute infections caused by ingestion of large numbers of metacercariae will, within a month, cause fever, chills, diarrhea, epigastric pain, enlarged tender liver, and possibly jaundice. The acute symptoms last for about 1 month and subside at about the time eggs are detected in the stool.

As the worms mature, an inflammatory response is seen in the biliary epithelium, related to the intensity and duration of infection. Lesions are due to mechanical irritation and toxic products. In light infections, there is little or no change in liver parenchyma, while in heavy infections, there is thickening and localized dilations of the bile ducts with hyperplasia of the mucinous glands. Biliary tract obstruction causes bile retention, infiltration of lymphocytes and eosinophils, and fibrosis. Many patients have recurrent pyogenic cholangitis. Acute pancreatitis, cholecystitis, and cholelithiasis may be due to worm invasion. Cirrhosis is probably related to malnutrition. Serum IgE and *C. sinensis*-specific IgE levels are elevated in infected individuals. In acute infections, there is an increase in the levels of IgM, followed by IgA and IgG. In chronic infections, the IgA level returns to normal while the IgG and IgM levels remain elevated.

C. sinensis has been linked to neoplasms of the bile duct or cholangiocarcinoma, which are usually seen in areas where clonorchiasis is endemic. There is apparently no direct link between infection and carcinoma; however, one of the first steps in malignant transformation may be induced by the biliary tract hyperplasia caused by the worms.

Identification of patients with deteriorating liver function before cholangiocarcinoma develops remains an important goal in the management of primary sclerosing cholangitis, particularly in areas where *C. sinensis* infection is endemic. Cofactors may play a role; liver flukes are promoters and not initiators of cholangiocarcinoma.

Infection can be prevented by thorough cooking of all freshwater fish. Night soil used without disinfection for fertilizer should not be applied in lakes or ponds containing susceptible snails; night soil should be stored prior to use, since *C. sinensis* eggs die within 2 days when stored at 26°C.

Aquaculture

There has been a dramatic increase in aquaculture of fish during the past 30 to 50 years. Freshwater fish production now accounts for 45.1% of the total aquaculture production. The increased production of grass carp, an important species that serves as an intermediate host for food-borne trematodes, increased from >10,000 tons in 1950 to >3 million tons in 2002. Freshwater fish aquaculture increased from an annual production of 136,000 tons in 1952 to 16.6 million tons 50 years later in China, the Republic of Korea, and Vietnam. It is estimated that by 2030, at least half of the globally consumed fish will come from aquaculture farming. Growth of this industry must be monitored for potential problems related to increased disease in which parasitic infection is transmitted through the ingestion of raw or poorly cooked fish.

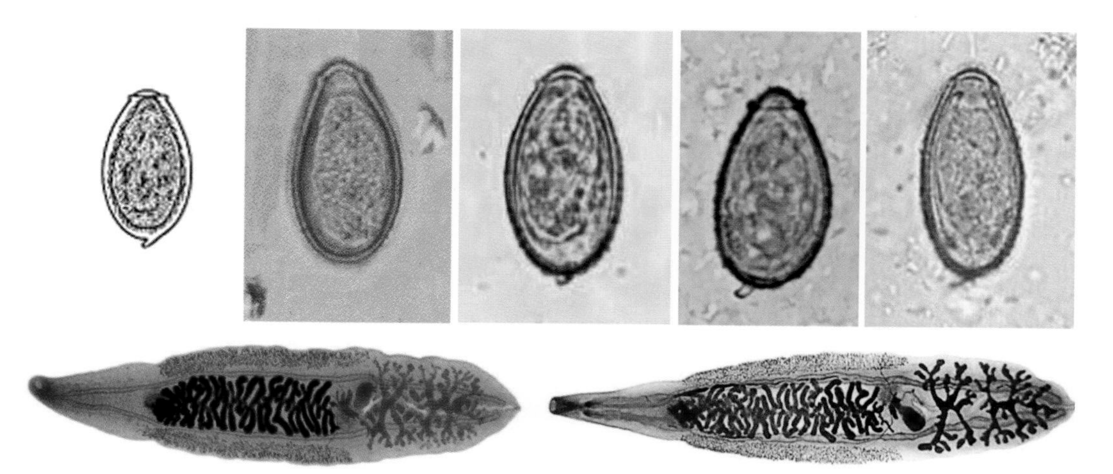

Figure 7.55 ***Clonorchis sinensis.*** **(Top row)** Drawing and photographs of eggs of *C. sinensis.* **(Bottom row)** *Clonorchis sinensis* adults (10 to 25 mm long by 3 to 5 mm wide) (courtesy of CDC DpDx, https://www.cdc.gov/dpdx/clonorchiasis/index.html [left] and Armed Forces Institute of Pathology [right]).

Schistosoma spp. (*Schistosoma mansoni, S. haematobium, S. japonicum, S. mekongi, S. malayensis, S. intercalatum*)

Pathogenic	Yes
Disease	Schistosomiasis (WHO neglected tropical disease)
Acquired	Skin penetration by cercariae released from freshwater snails
Body site	Veins overlying intestine/bladder
Symptoms	Cercarial dermatitis, high fever, hepatosplenomegaly, lymphadenopathy, eosinophilia, and dysentery
Clinical specimen	*Intestinal*: Stool
	Bladder: Urine
Epidemiology	*S. mansoni*: western and central Africa, Egypt, Madagascar, Arabian Peninsula, Brazil, Suriname, Venezuela, West Indies
	S. japonicum: Far East, China, Indonesia, Japan, Philippines
	S. haematobium: Africa, Asia Minor, Cyprus, islands off Africa's east coast, southern Portugal; focus in India
	S. mekongi: Mekong River basin in Kampuchea, Laos, and Thailand
Control	Improved hygiene, adequate disposal of fecal and urine waste, snail control

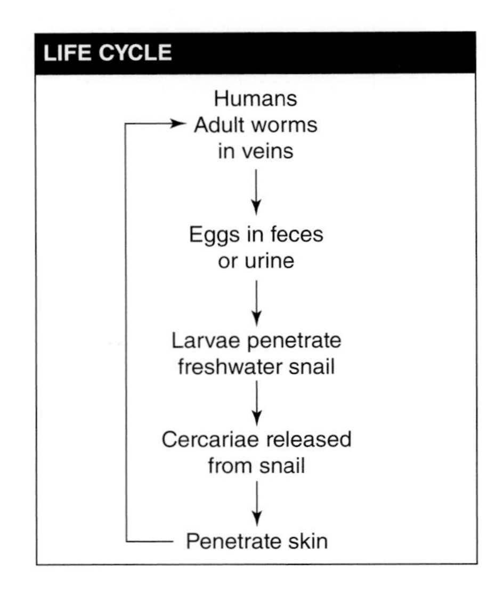

Diagnosis

Schistosomes belong to the phylum *Platyhelminthes*, family *Schistosomatidae*, and are a group of digenetic, dioecious trematodes requiring definitive and intermediate hosts to complete their life cycles. Four species are important agents of human disease: *Schistosoma mansoni*, *S. japonicum*, *S. mekongi*, and *S. haematobium*. *S. intercalatum* is of less epidemiologic importance. *S. malayensis* is a member of the *S. japonicum* complex and is found in Peninsular Malaysia; however, the number of cases is small. Schistosomiasis affects between 200 million and 300 million people in 77 countries throughout the world.

The standard O&P exam is recommended for recovery and identification of schistosome eggs in stool and urine specimens, primarily from the wet-preparation examination of the concentration sediment. Spot urine samples and 24-h urine samples should be examined, as well as several stools. All specimens should be collected without preservatives and concentrated without formalin or water; if eggs are found, it is important to determine egg viability. Viable or nonviable eggs will help determine the therapeutic approach for the patient.

Diagnostic Tips (Stool)

Eggs cannot be detected in stool until the worms mature (this may take 4 to 7 weeks after initial infection). In very light or chronic infections, eggs may be very difficult to detect in stool; therefore, multiple stool examinations may be required. Biopsy and/or immunologic tests for antigen or antibody may be helpful in diagnosing infection in these patients. All specimens must be collected without preservatives. It is important to be able to distinguish living from dead eggs.

In active infections, eggs should contain live or mature miracidia. Examination to confirm flame cell activity must be done on fresh specimens using the microscopic wet mount or the hatching test; no preservatives can be used prior to the wet-mount test or the hatching test. Any rinses should be done using saline; this will prevent the eggs from prematurely hatching. Do not use the zinc sulfate flotation concentration; eggs do not float. The sedimentation method is recommended. It may be necessary to tap the coverslip to move the eggs; the lateral spine may not be visible if the egg is turned on its side. Occasionally, eggs of *S. mansoni* are detected in the urine (crossover phenomenon).

Patients who have been treated should have follow-up O&P examinations for up to 1 year to evaluate treatment. In active infections, the eggs should contain live or mature miracidia. Examination for the confirmation of flame cell activity must be performed with fresh specimens, using the microscopic wet-mount test or the hatching test; no preservatives can be used prior to these two tests. During the hatching test, the light must not come too close to the surface of the water, since excess heat can kill the liberated larvae. Also, the water should be examined about every 30 min for up to 4 h.

Diagnostic Tips (Urine)

Eggs cannot be detected in urine until the worms mature (this may take up to 3 months after initial infection). In very light or chronic infections, eggs may be very difficult to detect in urine; therefore, multiple urine examinations may be required. Biopsy and/or immunologic tests may be helpful in diagnosing infection in these patients. Both 24-h and spot urine samples should be examined as wet mounts (after concentration using no preservatives [use saline so hatching will not occur]); the urine specimens should be collected with no preservatives. It is important to be able to distinguish living from dead eggs. Do not use the zinc sulfate flotation concentration; eggs do not float. These eggs are also occasionally recovered in stool, and so both urine and stool specimens should be examined. The membrane filtration technique using Nuclepore filters can be very helpful in diagnosing infection with *S. haematobium*.

Patients who have been treated should have follow-up O&P examinations for up to 1 year to evaluate treatment. In active infections, the eggs should contain live or mature miracidia. Examination for the confirmation of flame cell activity must be performed on fresh specimens using the wet-mount or hatching test; no preservatives can be used prior to the wet-mount or hatching test. If the specimens are concentrated, use saline rather than water to prevent premature hatching of the eggs. During the hatching test, the light must not come too close to the surface of the water, since excess heat can kill the liberated larvae. Also, the water should be examined about every 30 min for up to 4 h. It is important to remember that the small and less commonly seen miracidium larvae of *S. haematobium* may be present in the urine; motility in unpreserved specimens or stained morphology could confirm this diagnosis. It is also possible to see *S. haematobium* eggs in semen specimens, even when repeated urinary and fecal examinations and serologic tests are negative.

General Comments

Humans are the definitive host for *S. mansoni*, *S. japonicum*, *S. haematobium*, *S. mekongi*, *S. malayi*, and *S. intercalatum*. *S. mattheei*, which causes infections in sheep, cattle, and horses, also infects humans and can cause disease.

Humans are infected by penetration of cercariae through intact skin. Cercariae contain glands whose material is used to penetrate skin and a bifurcated tail that is lost when the cercariae penetrate the skin. A pruritic rash usually appears after exposure to cercariae. After entry, the organism is termed a schistosomulum; it migrates through the tissues and finally invades a blood vessel, where it is carried to the lungs and then the liver. Once within the liver sinusoids, the worms mature into adults. *S. haematobium* adults are found primarily in the blood vessels of the vesical, prostate, and uterine plexuses. *S. mansoni* and *S. japonicum* adults are found in the inferior and superior mesenteric veins.

Egg deposition occurs in the small venules of the intestine and rectum (*S. japonicum* and *S. mansoni*) or the venules of the bladder (*S. haematobium*). A mature female worm produces 300 to 3,000 eggs per day depending on species. The eggs are immature when first laid and take ca. 10 days to develop a mature miracidium. Egg deposition takes place intravascularly, and the eggs work their way through the tissues either into the lumen of the bladder and urethra (*S. haematobium*) or the intestine (*S. mansoni* and *S. japonicum*) to be released from the body in the urine or feces.

Schistosome eggs are nonoperculate. Egg deposition takes place intravascularly. Many of the eggs laid are swept away by the bloodstream and become lodged in the microvasculature of the liver and other organs. About half of the eggs swept away become embedded in the venule walls. The presence of eggs in the tissues stimulates granuloma formation; the eggs die, calcify, and are eventually absorbed by the host.

Description (Eggs)

S. mansoni eggs are yellow-brown (114 to 180 μm long by 45 to 73 μm wide), elongate, and ovoid and have a large lateral spine projecting from the egg near one end. *S. haematobium* eggs are yellow-brown and have a distinct terminal spine (112 to 170 μm by 40 to 70 μm). *S. japonicum* eggs are more spherical (55 to 85 μm by 40 to 60 μm). A minute lateral spine occurs at one end in some strains. *S. mekongi* eggs are similar in shape to *S. japonicum* eggs but smaller. They are subspherical (30 to 55 μm by 50 to 65 μm) and have a small lateral spine near one end.

Description (Adults)

The adult male and female can reach lengths of 1.2 and 1.6 cm, respectively. They have oral and ventral (acetabulum) suckers at the anterior end. The worm surface is a tegument containing a syncytium of cells. The male worm's body, which is flattened behind the ventral sucker, appears cylindrical as it curves to form the gynecophoral canal to clasp the female worm. The female worm is long, slender, and cylindrical in cross section. While held in the gynecophoral canal of the male, the female ingests 10 times more red blood cells than does the male. The tegument of the male contains many prominent tuberculations, while the tegument of the female is devoid of the tuberculations. The uterus of the female is short and usually contains only one egg at a time. After mating, the female leaves the male and migrates against the flow of blood to the small venules of the intestine. Initial egg production begins 4 to 7 weeks after infection.

Additional Information

In Egypt, approximately 20% of the population is infected; prevalence rates in some villages are ca. 85%. In China, there are 1.52 million infected individuals. About 10% of infected people have serious disease; this represents 20 million to 30 million individuals worldwide. About half of the 180 million to 270 million infected individuals have symptoms.

Schistosomiasis symptoms are related to the stage of infection, previous host exposure, worm burden, and host response. Syndromes include cercarial dermatitis, acute schistosomiasis (Katayama fever), and tissue changes resulting from egg deposition.

Cercarial dermatitis follows skin penetration by cercariae and may partly be due to previous host sensitization. Few symptoms are associated with primary exposure, but both humoral and cellular immune responses are elicited later. After cercarial skin penetration, petechial hemorrhages with edema and pruritus occur. The subsequent maculopapular rash, which may become vesicular, may last 36 h or more. Cercarial dermatitis is more common with *S. haematobium* and *S. mansoni* infections. Dermatitis is a constant feature of human infection with avian schistosomes, with cercarial death occurring in the subcutaneous tissues and immediate hypersensitivity reactions at the invasion sites. Previous contact with cercariae leads to a more immediate intense immune response.

Eggs become trapped in the fine venules and are able to pass through the tissues, escaping into the intestine or bladder. They liberate soluble antigens, evoking minute abscesses which facilitate their passage into the lumen. When they pass through the wall of the intestine or bladder, symptoms include fever, abdominal pain, liver tenderness, urticaria, and general malaise. In *S. haematobium* infections, there may be hematuria at the end of micturition and possibly dysuria; in *S. japonicum* and *S. mansoni* infections, blood and mucus occur in the stools, and the patient may have diarrhea or dysentery.

Although many eggs remain where deposited, others are swept into the circulation and filtered out in the liver, leading to hepatosplenic schistosomiasis. Hepatosplenomegaly is common in chronic *S. japonicum* and *S. mansoni* infections but less pronounced in *S. haematobium* infections. In some areas, *S. mansoni*, *S. japonicum*, and viral hepatitis are the most common causes of chronic liver disease.

Disease syndromes associated with schistosomiasis are related to the stage of infection, previous host exposure, worm burden, and host response. Syndromes include cercarial dermatitis, acute schistosomiasis (Katayama fever), and related tissue changes resulting from egg deposition. It has been noted that schistosomiasis exerts disruptive influences on the nutritional reserves and growth of humans from middle childhood through adolescence. Clinical features of acute infection are high fever, hepatosplenomegaly, lymphadenopathy, eosinophilia, and dysentery. In chronic disease, symptoms include fever, abdominal pain, liver tenderness, urticaria, and general malaise.

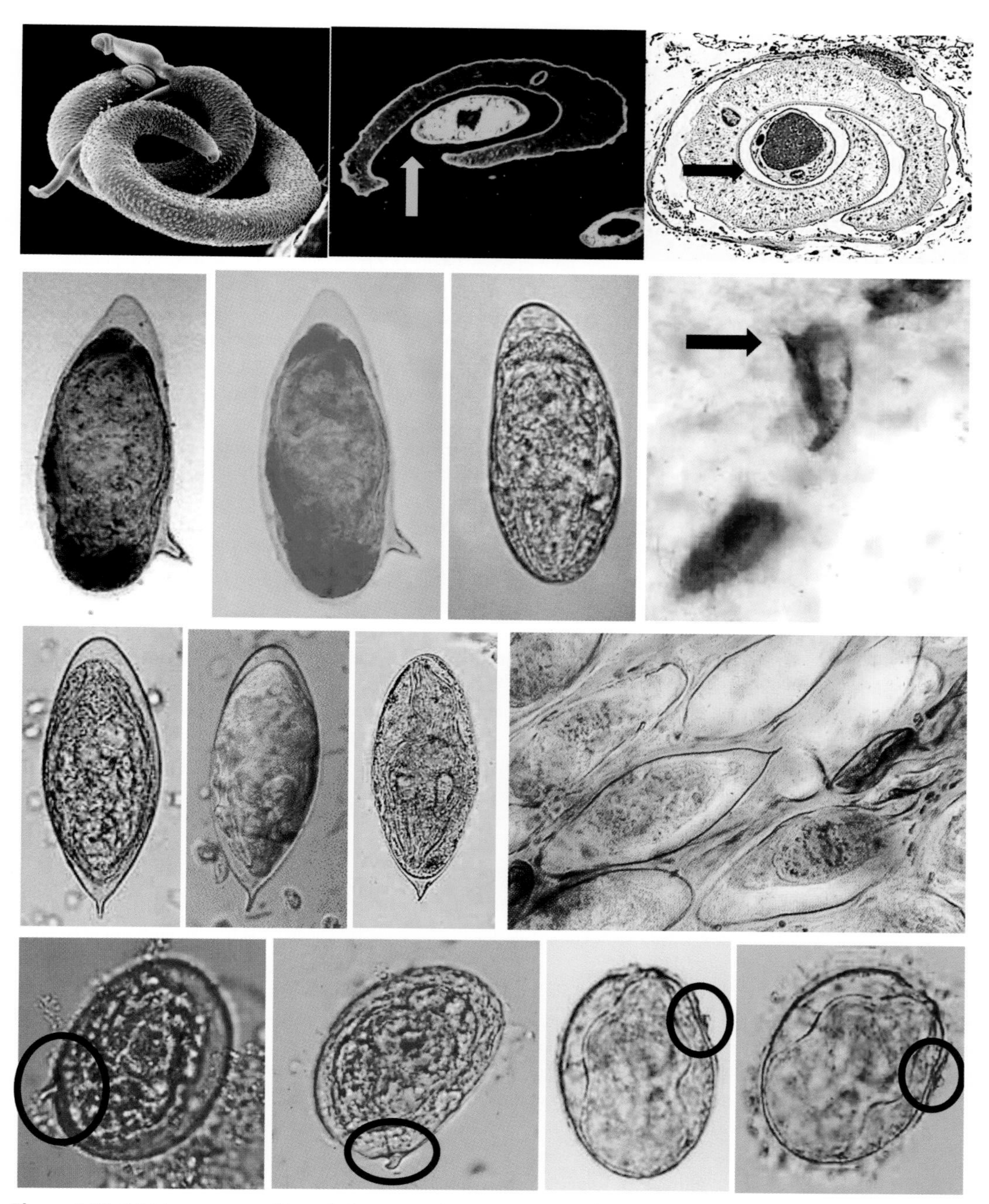

Figure 7.56 ***Schistosoma*** **spp.** **(Row 1)** Scanning electron micrograph of male and female schistosomes *in copula* (courtesy of Marietta Voge); cross section (indirect fluorescent antibody stain) (courtesy of CDC PHIL, #644, CDC/Marianna Wilson, 1972); cross section of adult worm pair in the mesenteric venule of a mouse, H&E stain (reprinted from Colley DG, Secor WE, PLoS Negl Trop Dis 1[3]:e32, 2007, https://doi.org/10.1371/journal.pntd.0000032, CC-BY 4.0). The female is seen within the male worm in both sections (arrows). **(Row 2)** *Schistosoma mansoni* eggs (note the large lateral spines; in the third egg, the lateral spine is not visible—this may happen if the egg is turned the wrong way in the wet preparation; tap on the coverslip to move the eggs into a different position); *S. mansoni* egg in tissue (note the lateral spine [arrow]). **(Row 3)** Three images of *Schistosoma haematobium* eggs (note the large terminal spine); eggs in a bladder biopsy specimen (higher magnification) (from the collection of Herman Zaiman, "A Presentation of Pictorial Parasites"). **(Row 4)** Two images of *Schistosoma japonicum* eggs (note the small lateral spines [ovals]); two images of *Schistosoma mekongi* eggs (note the small lateral spines [ovals]).

Suggested Reading

Almeria S, Cinar HN, Dubey JP. 2019. *Cyclospora cayetanensis* and cyclosporiasis: an update. *Microorganisms* **7**:317.

Cakir F, Cicek M, Yildirim IH. 2019. Determination of the subtypes of *Blastocystis* spp. and evaluate the effect of these subtypes on pathogenicity. *Acta Parasitol* **64**:7–12.

Garcia LS. 2016. *Diagnostic Medical Parasitology*, 6th ed. ASM Press, Washington, DC.

Garcia LS, Arrowood M, Kokoskin E, Paltridge GP, Pillai DR, Procop GW, Ryan N, Shimizu RY, Visvesvara G. 2017. Practical guidance for clinical microbiology laboratories: laboratory diagnosis of parasites from the gastrointestinal tract. *Clin Microbiol Rev* **31**:e00025–e17.

Garcia LS, Shimizu RY. 1997. Evaluation of nine immunoassay kits (enzyme immunoassay and direct fluorescence) for detection of *Giardia lamblia* and *Cryptosporidium parvum* in human fecal specimens. *J Clin Microbiol* **35**:1526–1529.

Garcia LS, Shimizu RY. 2000. Detection of *Giardia lamblia* and *Cryptosporidium parvum* antigens in human fecal specimens using the ColorPAC combination rapid solid-phase qualitative immunochromatographic assay. *J Clin Microbiol* **38**:1267–1268.

Garcia LS, Shimizu RY, Bernard CN. 2000. Detection of *Giardia lamblia, Entamoeba histolytica/Entamoeba dispar*, and *Cryptosporidium parvum* antigens in human fecal specimens using the triage parasite panel enzyme immunoassay. *J Clin Microbiol* **38**:3337–3340.

Garcia LS, Shimizu RY, Brewer TC, Bruckner DA. 1983. Evaluation of intestinal parasite morphology in polyvinyl alcohol preservative: comparison of copper sulfate and mercuric chloride bases for use in Schaudinn fixative. *J Clin Microbiol* **17**:1092–1095.

Garcia LS, Shimizu RY, Bruckner DA. 1986. Blood parasites: problems in diagnosis using automated differential instrumentation. *Diagn Microbiol Infect Dis* **4**:173–176.

Garcia LS, Shimizu RY, Shum A, Bruckner DA. 1993. Evaluation of intestinal protozoan morphology in polyvinyl alcohol preservative: comparison of zinc sulfate- and mercuric chloride-based compounds for use in Schaudinn's fixative. *J Clin Microbiol* **31**:307–310.

Gaydos CA, Klausner JD, Pai NP, Kelly H, Coltart C, Peeling RW. 2017. Rapid and point-of-care tests for the diagnosis of *Trichomonas vaginalis* in women and men. *Sex Transm Infect* **93**(S4):S31–S35.

Hanscheid T, Grobusch MP. 2017. Modern hematology analyzers are very useful for diagnosis of malaria and, crucially, may help avoid misdiagnosis. *J Clin Microbiol* **55**:3303–3304.

Hanson KL, Cartwright CP. 2001. Use of an enzyme immunoassay does not eliminate the need to analyze multiple stool specimens for sensitive detection of *Giardia lamblia. J Clin Microbiol* **39**:474–477.

Hiatt RA, Markell EK, Ng E. 1995. How many stool examinations are necessary to detect pathogenic intestinal protozoa? *Am J Trop Med Hyg* **53**:36–39.

Leber AL (ed). 2016. *Clinical Microbiology Procedures Handbook*, 4th ed. ASM Press, Washington, D.C.

Markell EK, Udkow MP. 1986. *Blastocystis hominis*: pathogen or fellow traveler? *Am J Trop Med Hyg* **35**:1023–1026.

Mathis A, Deplazes P. 1995. PCR and in vitro cultivation for detection of *Leishmania* spp. in diagnostic samples from humans and dogs. *J Clin Microbiol* **33**:1145–1149.

Mathison BA, Pritt BS. 2017. Update on malaria diagnostics and test utilization. *J Clin Microbiol* **55**:2009–2017.

National Committee for Clinical Laboratory Standards. 2000. Use of blood film examination for parasites. Approved standard M15-A. National Committee for Clinical Laboratory Standards, Wayne, PA.

National Committee for Clinical Laboratory Standards/Clinical and Laboratory Standards Institute. 2005. Procedures for the recovery and identification of parasites from the intestinal tract. Approved standard M28-A2. Clinical and Laboratory Standards Institute, Wayne, PA.

Stanley SL Jr. 2003. Amoebiasis. *Lancet* **361**:1025–1034.

Stark D, Garcia LS, Barratt JL, Phillips O, Roberts T, Marriott D, Harkness J, Ellis JT. 2014. Description of *Dientamoeba fragilis* cyst and precystic forms from human samples. *J Clin Microbiol* **52**:2680–2683.

Ward SJ, Stramer SL, Szczepiorkowski ZM. 2018. Assessing the risk of Babesia to the United States blood supply using a risk-based decision-making approach: report of AABB's Ad Hoc Babesia Policy Working Group (original report). *Transfusion* **58**:1916–1923.

Weber R, Schwartz DA, Deplazes P. 1999. Laboratory diagnosis of microsporidiosis, p 315–362. *In* Wittner ML, Weiss LM (ed), *The Microsporidia and Microsporidiosis*. ASM Press, Washington, DC.

White AC Jr, Coyle CM, Rajshekhar V, Singh G, Hauser WA, Mohanty A, Garcia HH, Nash TE. 2018. Diagnosis and treatment of neurocysticercosis: 2017 clinical practice guidelines by the Infectious Diseases Society of America (IDSA) and the American Society of Tropical Medicine and Hygiene (ASTMH). *Clin Infect Dis* **66**:e49–e75.

World Health Organization. 2017. WHO-FIND malaria RDT evaluation programme. http://www.who.int/malaria/areas/diagnosis/rapid-diagnostic-tests/rdt-evaluation-programme/en/

SECTION 8

Common Problems in Parasite Identification

Practical Guide to Diagnostic Parasitology, Third Edition. Lynne S. Garcia
 DOI: 10.1128/9781683673637.ch08

The drawings are reprinted from Garcia LS, *Diagnostic Medical Parasitology*, 6th ed., ASM Press, Washington, DC, 2016. Images are from the author and/or PARA-SITE Online (Medical Chemical Corporation, http://www.med-chem.com/para-site.php?url=home).

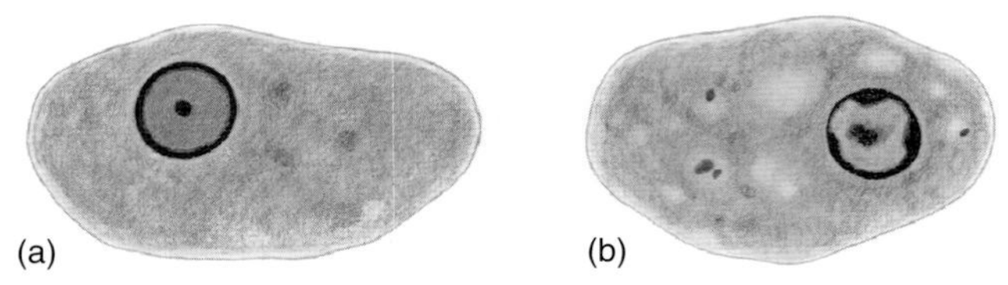

Figure 8.1 (a) *Entamoeba histolytica*/*E. dispar* trophozoite. Note the evenly arranged nuclear chromatin, central compact karyosome, and relatively "clean" cytoplasm. (b) *Entamoeba coli* trophozoite. Note the unevenly arranged nuclear chromatin, eccentric karyosome, and "messy" cytoplasm. These characteristics are very representative of the two organisms. (Illustration by Sharon Belkin.)

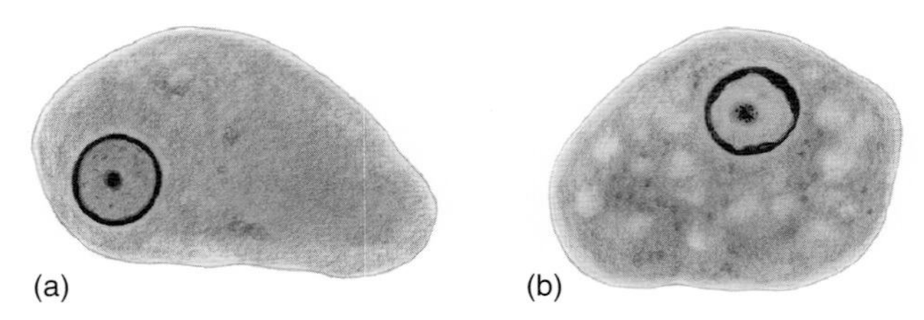

Figure 8.2 (a) *Entamoeba histolytica*/*E. dispar* trophozoite. Note the evenly arranged nuclear chromatin, central compact karyosome, and "clean" cytoplasm. (b) *Entamoeba coli* trophozoite. Note that the nuclear chromatin appears to be evenly arranged, the karyosome is central (but more diffuse), and the cytoplasm is "messy," with numerous vacuoles and ingested debris. The nuclei of these two organisms tend to resemble one another (very common finding in routine clinical specimens). However, the karyosome in *E. coli* tends to be larger and more blot-like. (Illustration by Sharon Belkin.)

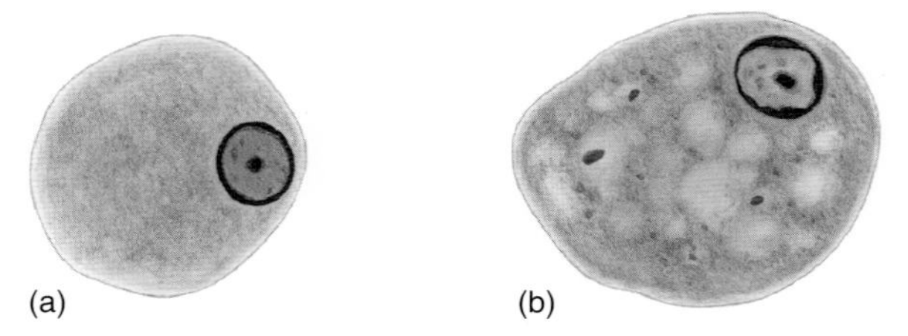

Figure 8.3 (a) *Entamoeba histolytica*/*E. dispar* trophozoite. Again, note the typical morphology (evenly arranged nuclear chromatin, central compact karyosome, and relatively "clean" cytoplasm). (b) *Entamoeba coli* trophozoite. Although the nuclear chromatin is eccentric, note that the karyosome seems to be compact and central. However, note the various vacuoles containing ingested debris. These organisms show some characteristics that are very similar (very typical in clinical specimens). (Illustration by Sharon Belkin.)

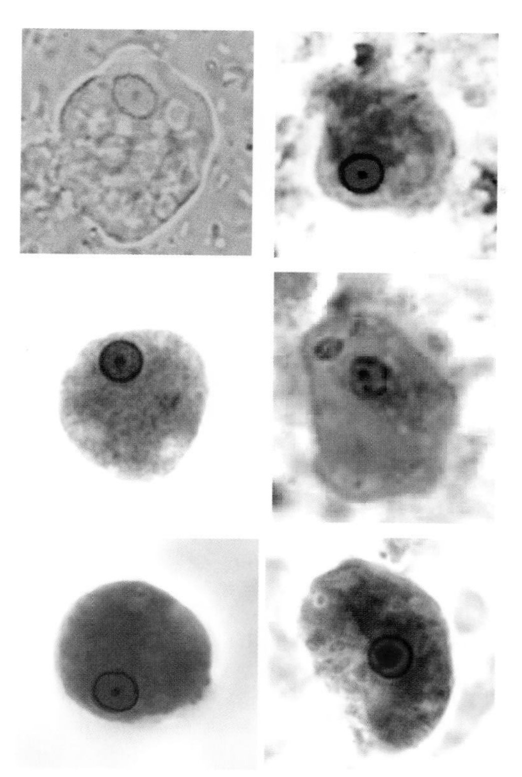

Figure 8.4 **(Top row)** *Entamoeba histolytica/E. dispar* trophozoite (wet mount; note the evenly arranged nuclear chromatin and central compact karyosome); *E. histolytica/E. dispar* trophozoite permanent stain. **(Middle row)** *E. histolytica/E. dispar* trophozoite; *Entamoeba coli* trophozoite (note the uneven nuclear chromatin and eccentric karyosome. **(Bottom row)** *E. histolytica/E. dispar* trophozoite; *E. coli* trophozoite (note the large blot-like karyosome).

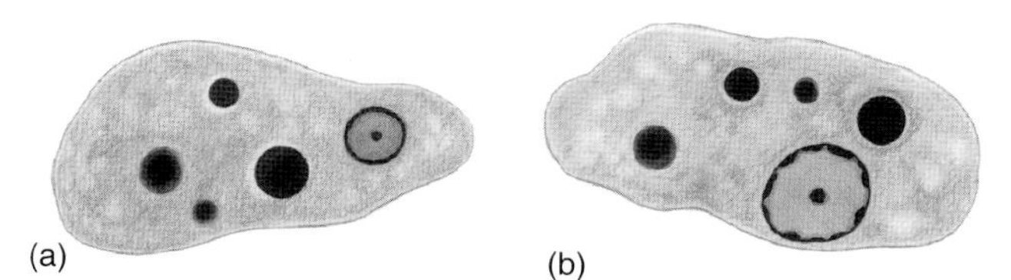

Figure 8.5 (a) *Entamoeba histolytica* trophozoite. Note the evenly arranged nuclear chromatin, central compact karyosome, and red blood cells (RBCs) in the cytoplasm. (b) Human macrophage. The key difference between the macrophage nucleus and that of *E. histolytica* is the size. Usually, the ratio of nucleus to cytoplasm in a macrophage is approximately 1:6 or 1:8, while the true organism has a nucleus/cytoplasm ratio of approximately 1:10 or 1:12. The macrophage also contains ingested RBCs. In cases of diarrhea or dysentery, trophozoites of *E. histolytica* and macrophages can often be confused, occasionally leading to a false-positive diagnosis of amebiasis when no parasites are present. Both the actual trophozoite and the macrophage may also be seen without ingested RBCs, and they can mimic one another. (Illustration by Sharon Belkin.)

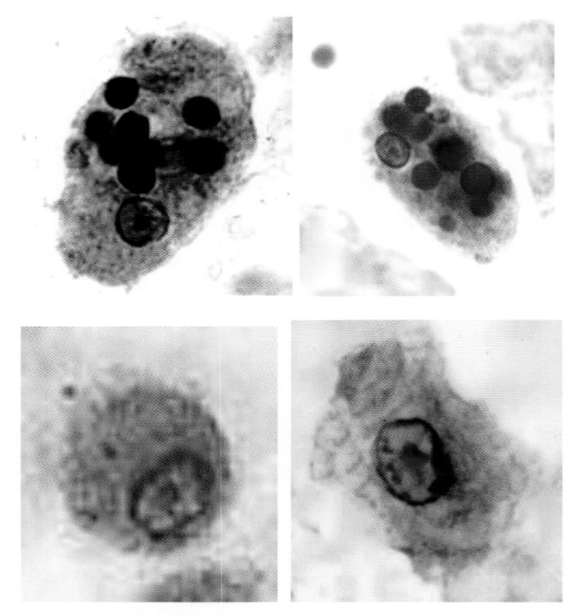

Figure 8.6 (Top row) *Entamoeba histolytica* trophozoites on permanent stained smears; note the evenly arranged nuclear chromatin, central compact karyosome, and ingested RBCs. The presence of the ingested RBCs indicates the presence of the true pathogen, *E. histolytica*. Without the presence of ingested RBCs, the organism would have been identified as *E. histolytica/E. dispar* (it is impossible to determine pathogenicity from the morphology on the permanent stained smear). **(Bottom row)** Two macrophages from a human fecal specimen. The main difference between the macrophage nucleus and that of *E. histolytica* is the size. Usually the ratio of nucleus to cytoplasm in a macrophage is approximately 1:6 or 1:8, while the true organism has a nucleus-to-cytoplasm ratio of approximately 1:10 or 1:12. The top left image is stained with iron-hematoxylin; the remainder are stained with Wheatley's trichrome stain (routine trichrome for fecal specimens).

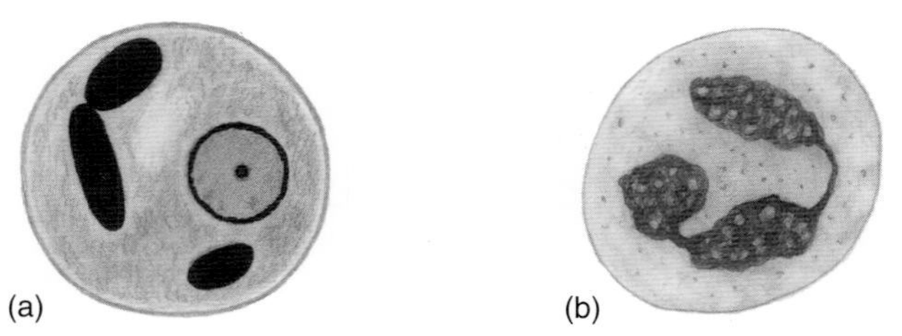

Figure 8.7 (a) *Entamoeba histolytica/E. dispar* precyst. Note the enlarged nucleus (prior to division) with evenly arranged nuclear chromatin and central compact karyosome. Chromatoidal bars (rounded ends with smooth edges) are also present in the cytoplasm. (b) Polymorphonuclear leukocyte (PMN). The nucleus is somewhat lobed (normal morphology) and represents a PMN that has not been in the gut very long. Occasionally, the positioning of the chromatoidal bars of the parasite and the lobed nucleus of the PMN will mimic one another. The chromatoidal bars stain more intensely, but the shapes can overlap, as seen here. (Illustration by Sharon Belkin.)

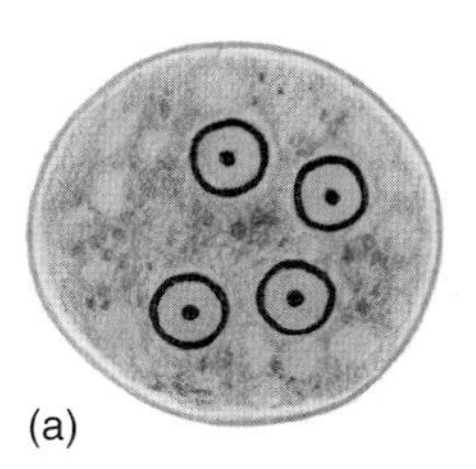

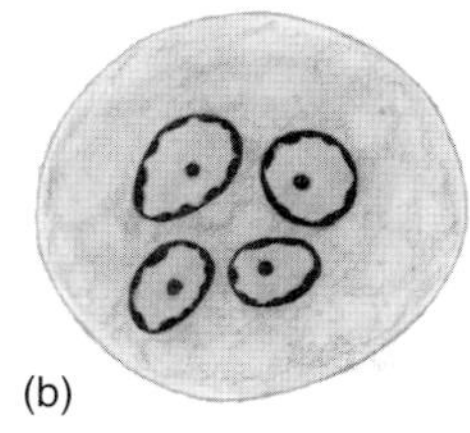

Figure 8.8 (a) *Entamoeba histolytica/E. dispar* mature cyst. Note that the four nuclei are very consistent in size and shape. (b) Polymorphonuclear leukocyte (PMN). Note that the normal lobed nucleus has now broken into four fragments, which mimic four nuclei with peripheral chromatin and central karyosomes. When PMNs have been in the gut for some time and have begun to disintegrate, the nuclear morphology can mimic that seen in an *E. histolytica/E. dispar* cyst. However, human cells are often seen in the stool of patients with diarrhea; with rapid passage of the gastrointestinal tract contents, there is not time for amebic cysts to form. Therefore, for patients with diarrhea and/or dysentery, if "organisms" that resemble the cell in panel b are seen, think first of PMNs, not *E. histolytica/E. dispar* cysts. Also, the four fragments seen in the PMN occupy more space within the cell than the space occupied by the actual *E. histolytica/E. dispar* four nuclei. (Illustration by Sharon Belkin.)

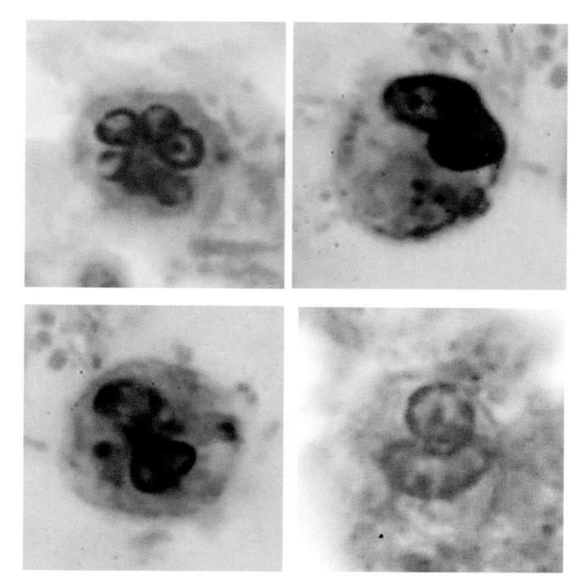

Figure 8.9 **(Top row)** PMNs. Note the appearance of "multiple nuclei" in the cell on the left, while the cell on the right contains the lobed nucleus, which has not yet fragmented to mimic an *Entamoeba* cyst. **(Bottom row)** PMNs in which the lobed nuclei have not yet fragmented; these cells should not be confused with *Entamoeba* cysts. When PMNs have been in the gut for some time and have begun to disintegrate, the nuclear morphology can mimic that seen in an *Entamoeba* cyst. However, remember that in patients with diarrhea, the gut contents move too rapidly for cysts to form; thus, cells with what appear to be multiple nuclei are actually PMNs.

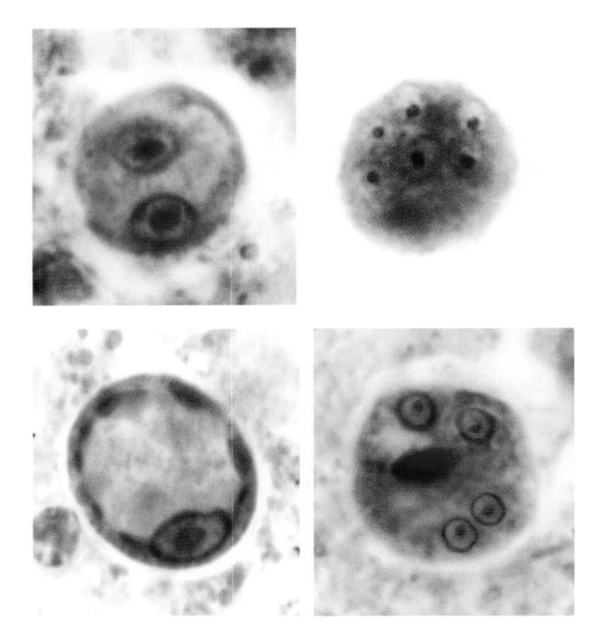

Figure 8.10 **(Top row)** *Entamoeba coli* precyst with two enlarged nuclei, one on each side of the precyst; *E. coli* cyst (five or more nuclei). These cells could be confused with PMNs. **(Bottom row)** *Entamoeba histolytica/E. dispar* precyst with a single enlarged nucleus; *E. histolytica/E. dispar* mature cyst (containing four nuclei and one chromatoidal bar). All cells were stained with Wheatley's trichrome stain (routine trichrome for fecal specimens).

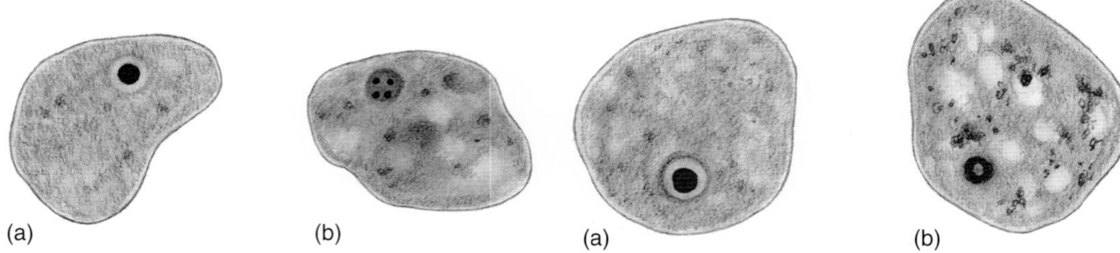

Figure 8.11 (a) *Endolimax nana* trophozoite. This organism is characterized by a large karyosome with no peripheral chromatin, although many nuclear variations are normally seen in any positive specimen. (b) *Dientamoeba fragilis* trophozoite. Normally, the nuclear chromatin is fragmented into several dots (often a tetrad arrangement). The cytoplasm is normally more "junky" than that seen in *E. nana*. If the morphology is typical, as in these two illustrations, differentiating between these two organisms is not very difficult. However, the morphologies of the two are often very similar. (Illustration by Sharon Belkin.)

Figure 8.12 (a) *Endolimax nana* trophozoite. Note that the karyosome is large and surrounded by a halo, with very little chromatin on the nuclear membrane. (b) *Dientamoeba fragilis* trophozoite. In this organism, the karyosome is beginning to fragment and there is a slight clearing in the center of the nuclear chromatin. If the nuclear chromatin has not become fragmented, *D. fragilis* trophozoites can very easily mimic *E. nana* trophozoites. This could lead to a report indicating that no pathogens were present when, in fact, *D. fragilis* is now considered a definite cause of symptoms. (Illustration by Sharon Belkin.)

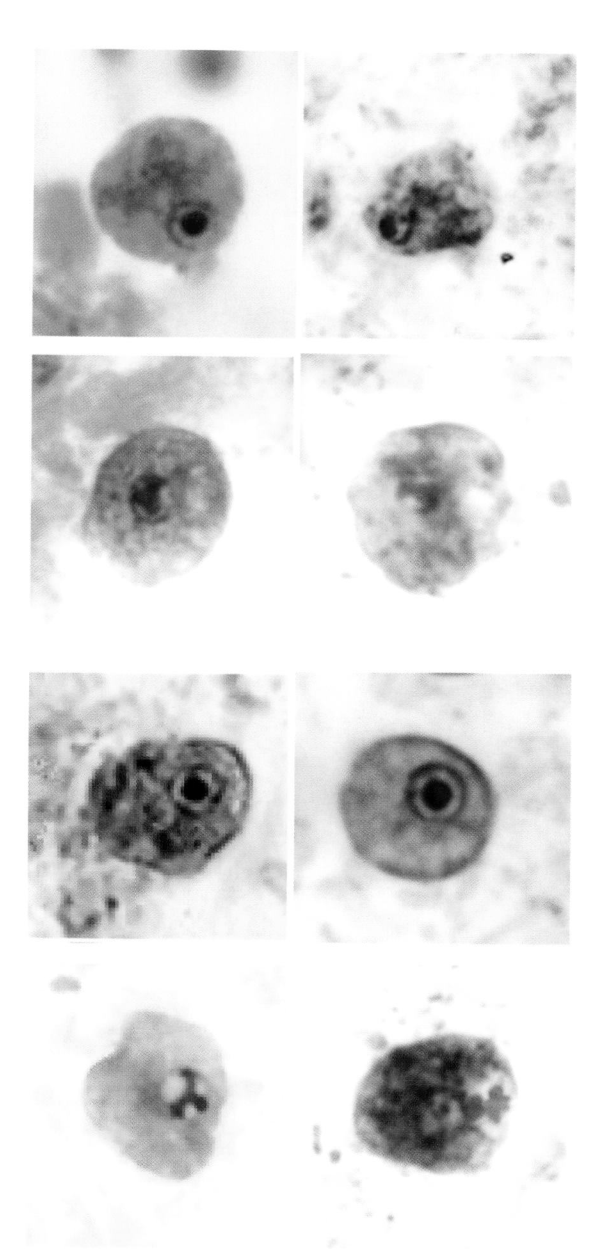

Figure 8.13 *Endolimax nana* trophozoites. Note the tremendous nuclear variation; there is more nuclear variation in this organism than any of the other intestinal protozoa. The karyosome appears quite different from organism to organism; trophozoites can mimic *Dientamoeba fragilis* and/or *Entamoeba hartmanni* (particularly if the nucleus appears to have peripheral chromatin like the organisms in row 3).

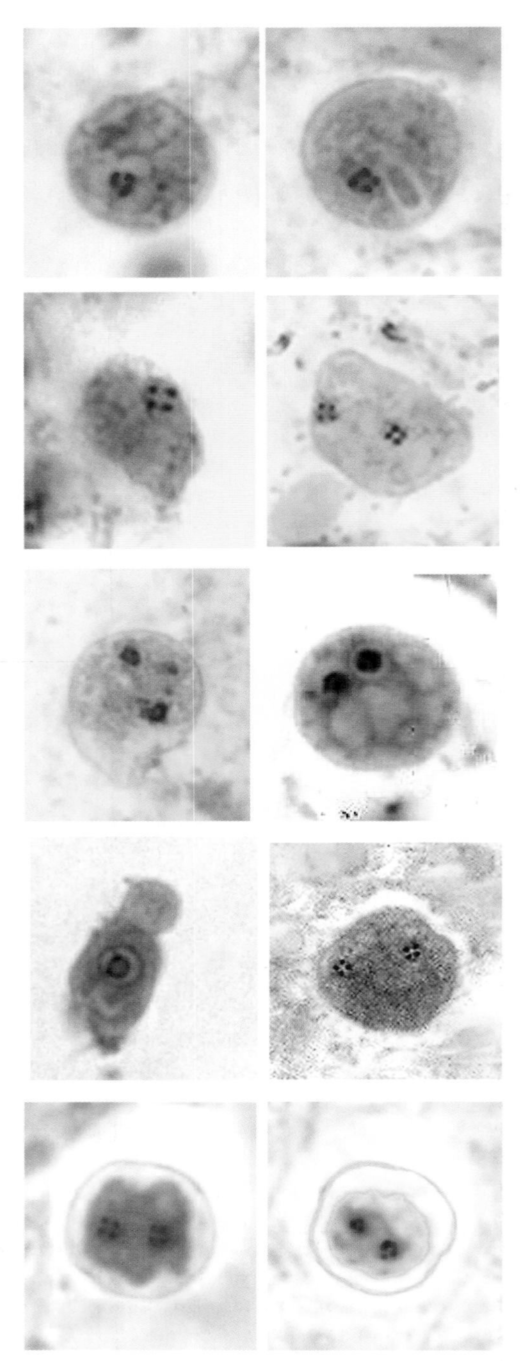

Figure 8.14 *Dientamoeba fragilis* trophozoites. (**Rows 1 to 3**) Some organisms have a single nucleus, while others have two nuclei; the nuclei tend to fragment into several chromatin dots. (**Row 4**) The image on the left shows the clearing within the karyosome prior to fragmentation into chromatin dots. The image on the right is a trophozoite with two fragmented nuclei. Both specimens in this row were stained with iron-hematoxylin. All others were stained with Wheatley's routine trichrome stain. (**Row 5**) These two images show the newly confirmed cyst stage, with the double cyst wall (Stark D, et al, *J Clin Microbiol* **52**:2680–2683, 2014). However, the cystic forms tend to be rare in human specimens, a probable factor in the original failure to report and confirm these forms.

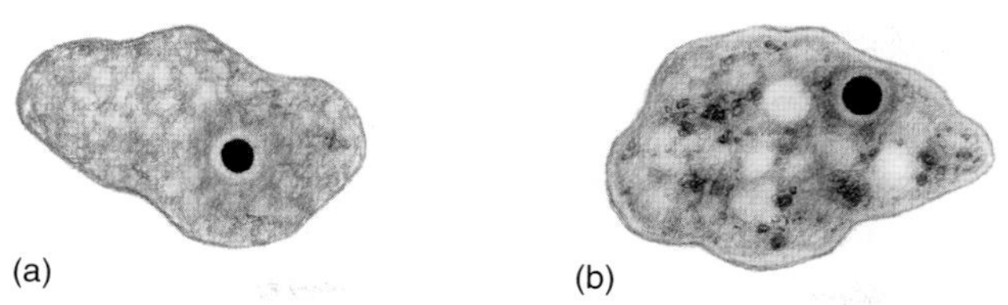

Figure 8.15 (a) *Endolimax nana* trophozoite. Note the large karyosome surrounded by a clear space. The cytoplasm is relatively clean. (b) *Iodamoeba bütschlii* trophozoite. Although the karyosome is similar to that of *E. nana*, note that the cytoplasm in *I. bütschlii* is much more heavily vacuolated and contains ingested debris. Often, these two trophozoites cannot be differentiated; proficiency testing specimens will contain cysts if the two organisms need to be differentiated one from the other. However, the differences in the cytoplasm are often helpful. There is a definite size overlap between the two genera. (Illustration by Sharon Belkin.)

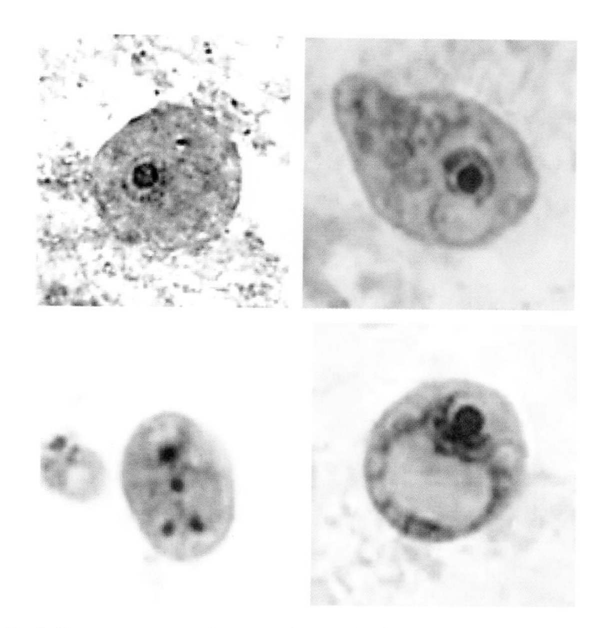

Figure 8.16 (Top row) *Endolimax nana* trophozoite (note the large single karyosome without any peripheral chromatin); *Iodamoeba bütschlii* trophozoite (note the large karyosome, some light peripheral nuclear chromatin, and the messy/dirty cytoplasm containing many vacuoles). **(Bottom row)** *E. nana* cyst (note the four karyosomes with no peripheral chromatin; the cyst shape tends to be somewhat round to oval); *I. bütschlii* cyst (note the large vacuole, single nucleus with large karyosome, and chromatin granules at the lower edge of the nucleus [basket nucleus]).

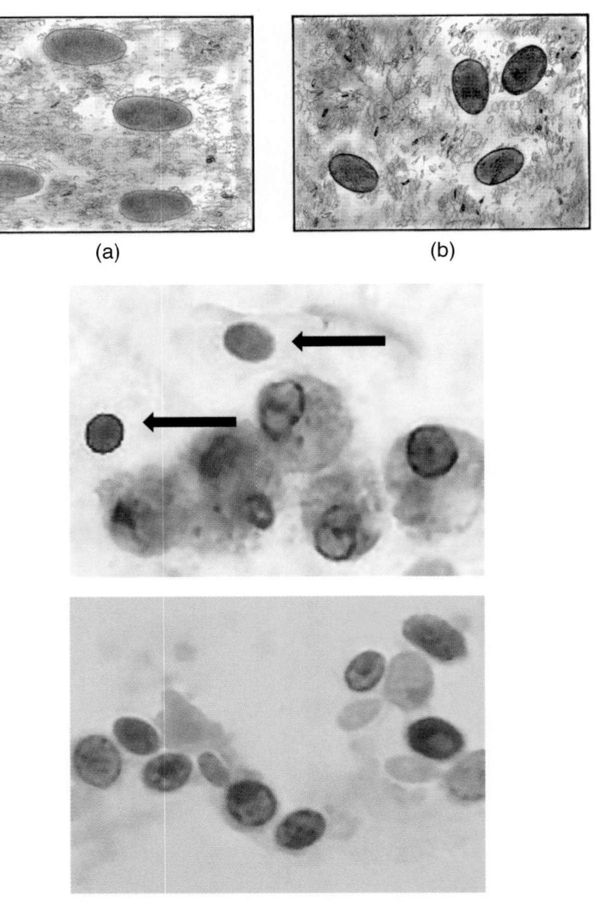

Figure 8.17 **(Top row)** (a) RBCs on a stained fecal smear. Note that the cells are very pleomorphic but tend to be positioned in the direction in which the stool was spread onto the slide. (b) Yeast cells on a stained fecal smear. These cells tend to remain oval and are not aligned in any particular way on the smear. These differences are important when the differential identification is between *Entamoeba histolytica* containing RBCs and *Entamoeba coli* containing ingested yeast cells. If RBCs or yeast cells are identified in the cytoplasm of an organism, they must also be visible in the background of the stained fecal smear. (Illustration by Sharon Belkin.) **(Middle row)** PMNs and RBCs (arrows). The RBC cell shape can vary tremendously. **(Bottom row)** Yeast cells in fecal material. Note that the shape tends to be consistent and there appears to be no particular direction of placement, as is seen frequently with RBCs.

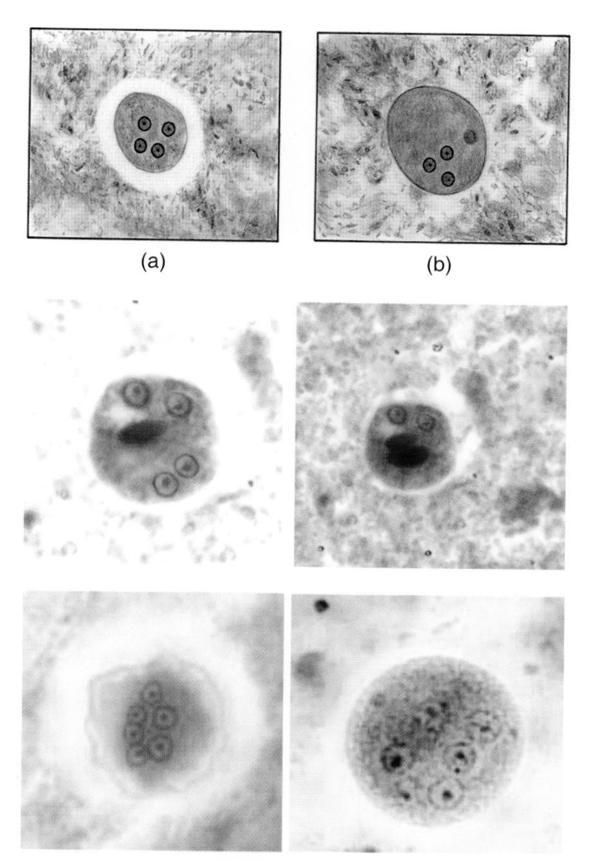

Figure 8.18 **(Top row)** (a) *Entamoeba histolytica/E. dispar* cyst. Note the shrinkage due to dehydrating agents in the staining process. (b) *E. histolytica/E. dispar cyst.* In this case, the cyst exhibits no shrinkage. Only three of the four nuclei are in focus. Normally, this type of shrinkage is seen with protozoan cysts and is particularly important when a species is measured and identified as either *E. histolytica/E. dispar* or *Entamoeba hartmanni.* The whole area, including the halo, must be measured prior to species identification. If just the cyst is measured, the organism would be identified as *E. hartmanni* (nonpathogenic) rather than *E. histolytica/E. dispar* (possibly pathogenic). (Illustration by Sharon Belkin.) **(Middle row)** *E. histolytica/E. dispar* cyst with four nuclei and one chromatoidal bar (note the shrinkage around the cyst); *E. hartmanni* cyst with two of the four nuclei visible and two chromatoidal bars (note also the shrinkage around the cyst wall; this entire area would need to be measured for accurate species identification. **(Bottom row)** *Entamoeba coli* cysts. Note the shrinkage around the cyst wall (left); however, the identification to species level could be made on the basis of five or more visible nuclei (both images).

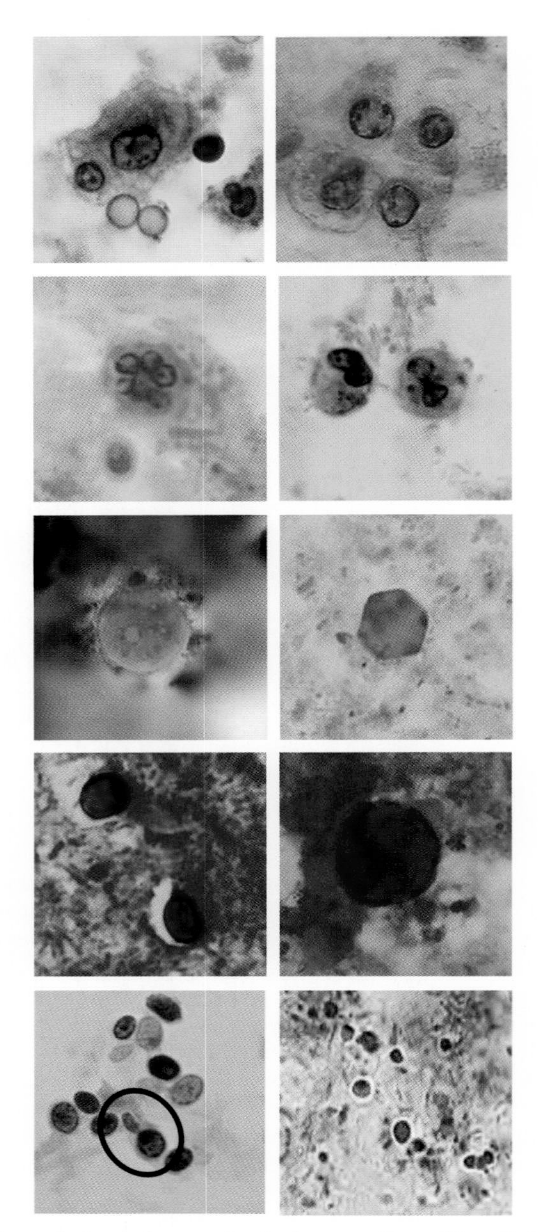

Figure 8.19 Various structures that may be seen in stool preparations. **(Row 1)** Macrophage; epithelial cells. Both can be confused with *Entamoeba histolytica*/*E. dispar* trophozoites. **(Row 2)** Polymorphonuclear leukocyte with a fragmented nucleus (left) and artifact (right) that can be confused with *Entamoeba* sp. cysts. **(Row 3)** Two artifacts that can resemble protozoan cysts. **(Row 4)** Yeast cells (left) and an artifact (right) that can be confused with *Cryptosporidium* spp. and *Cyclospora cayetanensis*, respectively, on positive acid-fast stains; it is important to measure the structures/organisms carefully before confirming organism identification. **(Row 5)** Yeast cells, which can be confused with microsporidial spores (however, notice the budding cell within the circle); artifacts that can also be confused with microsporidial spores (these were thought to be small yeast cells).

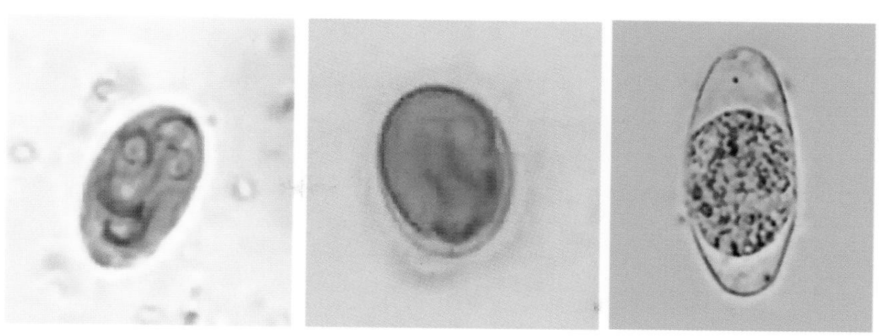

Figure 8.20 *Giardia lamblia* and *Cystoisospora belli*. Two images of *Giardia lamblia* (also called *G. duodenalis* or *G. intestinalis*) with iodine stain (normally measures approximately 14 µm in length); *Cystoisospora belli*, iodine wet mount (normally measures 20 to 33 µm in length). Note the size differences of the two organisms. In wet mounts particularly, these two organisms can be confused; it is mandatory that measurements be taken to eliminate possible confusion.

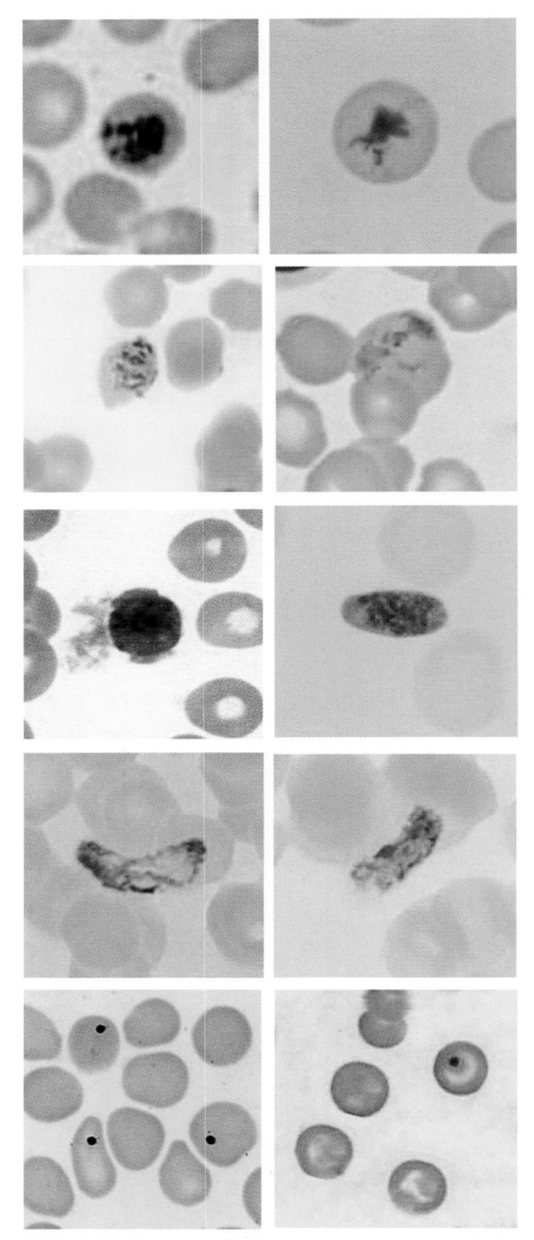

Figure 8.21 (**Rows 1 and 2**) Stain deposition on the surface of uninfected RBCs that could easily be confused with developing *Plasmodium* sp. stages. (**Row 3**) *Plasmodium falciparum* gametocytes that have rounded up and no longer appear as the typical crescent-shaped gametocytes that are normally seen (this could be due to low temperatures and/or storage for too many hours in EDTA-blood). (**Row 4**) Developing *Plasmodium vivax* trophozoites that appear to resemble *P. falciparum* gametocytes (found on blood smears prepared from EDTA-blood that had been collected more than 8 h previously). (**Row 5**) RBCs containing Howell-Jolly bodies that could be confused with very small, young ring forms of *Plasmodium* spp.

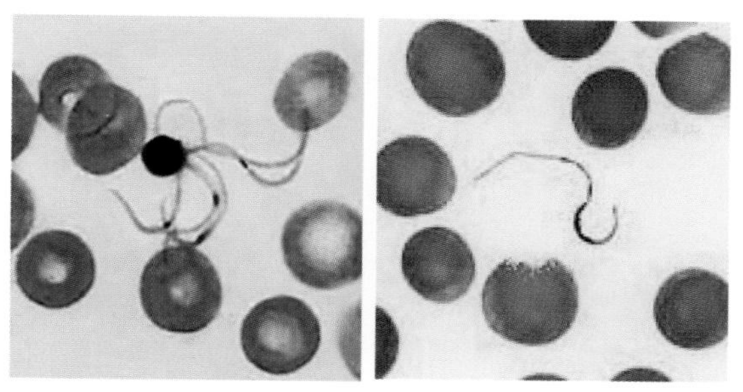

Figure 8.22 Male *Plasmodium* gametocyte (microgametocyte) undergoing exflagellation; single strand. These microgametes could easily be confused with some type of spirochete. These forms were seen in blood films prepared from blood stored for longer than 12 h in EDTA prior to additional smear preparation.

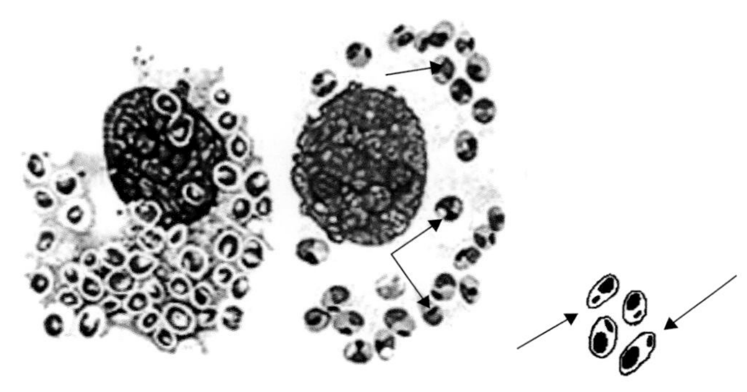

Figure 8.23 **(Left)** *Histoplasma capsulatum*. **(Middle)** *Leishmania donovani*. Note that the *Leishmania* amastigotes have the bar (arrows), while the *Histoplasma* amastigotes do not; *Histoplasma* also has the halo around the organisms (left image). **(Right)** Illustration of amastigotes; note the bars (arrows).

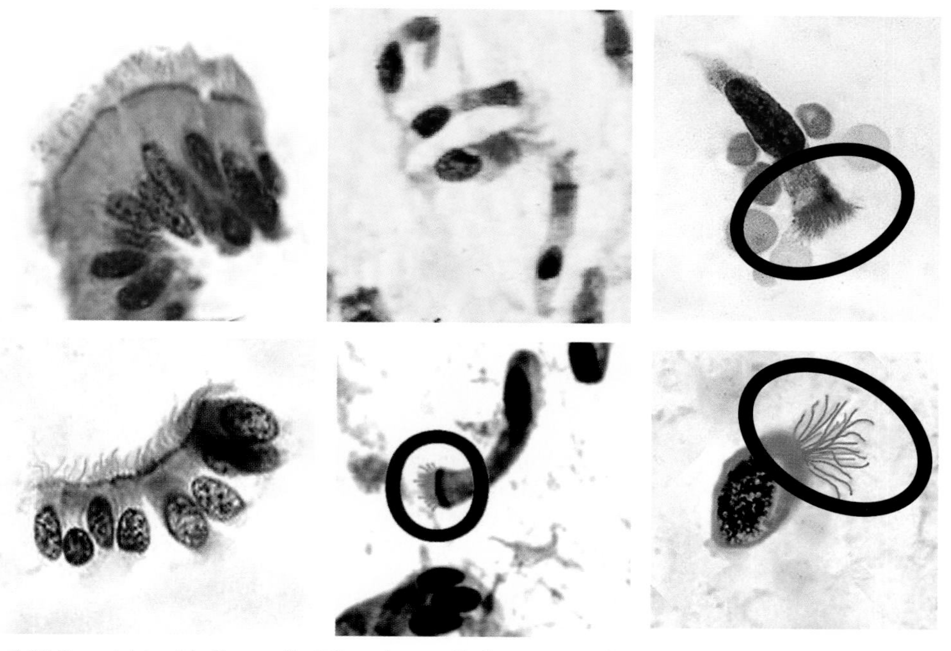

Figure 8.24 Bronchial epithelium cells. When these cells disintegrate, the ciliary tufts (ovals) may be visible and may be confused with protozoan flagellates or ciliates (detached ciliary tufts are also known as ciliocytophthoria).

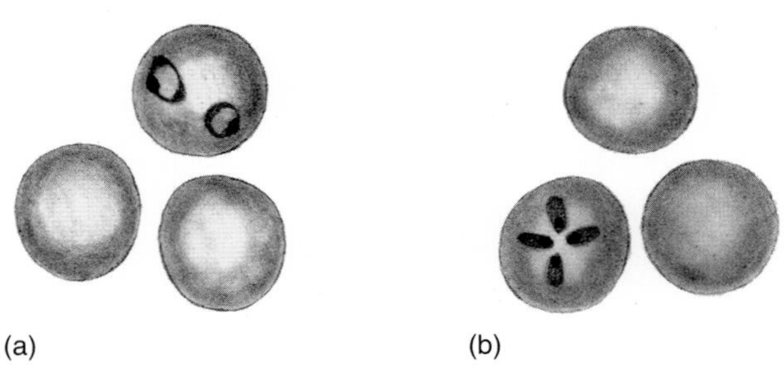

Figure 8.25 (a) *Plasmodium falciparum* rings. Note the two rings in the RBC. Multiple rings per cell are more typical of *P. falciparum* than of the other species causing human malaria. (b) *Babesia* rings. One of the RBCs contains four small *Babesia* rings. This particular arrangement is called the Maltese cross and is diagnostic for *Babesia* spp. (although this configuration is not always seen). *Babesia* infections can be confused with cases of *P. falciparum* malaria, primarily because multiple rings can be seen in the RBCs. Another difference involves ring morphology. *Babesia* rings are often of various sizes and tend to be very pleomorphic, while those of *P. falciparum* tend to be more consistent in size and shape. (Illustration by Sharon Belkin.)

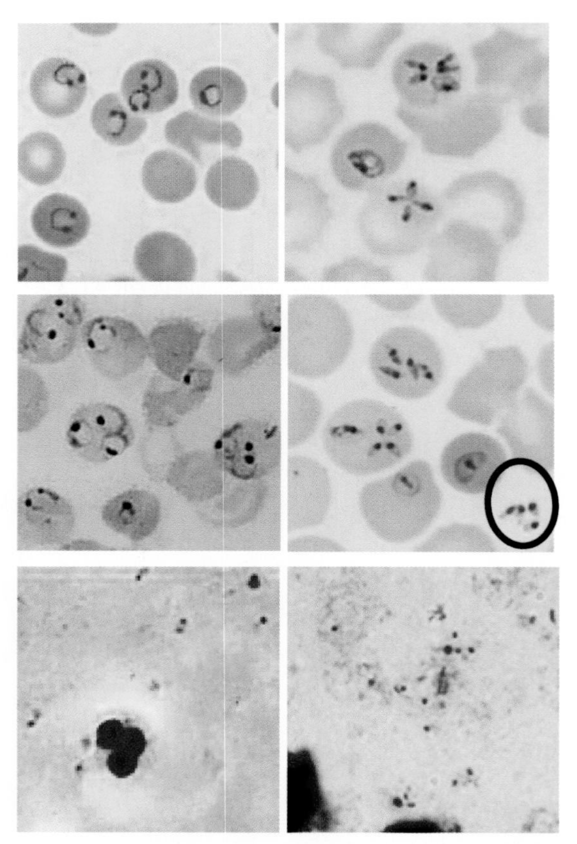

Figure 8.26 **(Top row)** *Plasmodium falciparum* rings (note the "clean" morphology and headphone appearance of some of the rings); *Babesia* spp. (note the "messy/pleomorphic" rings and the presence of the Maltese cross configuration of four rings within a single RBC; this does not occur in every species of *Babesia*). **(Middle row)** *Plasmodium falciparum* ring forms in a heavy infection (note the clearly defined rings); *Babesia* spp. (note the smaller rings; also note the rings outside the RBCs [oval], which do not occur in malarial infections [with the exception of a very heavy parasitemia]). **(Bottom row)** *Plasmodium falciparum* rings in a thick film (note the nuclei and the cytoplasm portion of the ring forms); *Babesia* sp. thick film (note that the nuclei are seen and that they appear smaller than those seen in the malaria thick film; however, differentiation of organisms based on thick-film morphology can be quite difficult; identification is best achieved from the thin blood film).

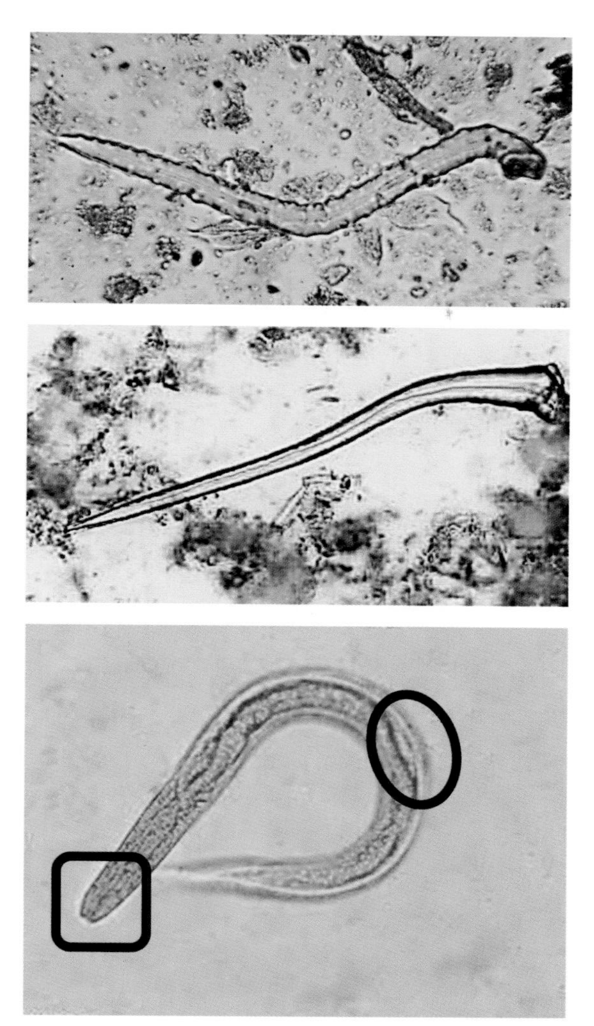

Figure 8.27 **(Top)** Root hair. **(Middle)** Root hair. Note that there is no internal structure visible within the root hairs. **(Bottom)** *Strongyloides stercoralis* rhabditiform larva. Note the short buccal cavity at the head end of the larva (square) and the genital primordial packet of cells within the curved portion of the body (oval).

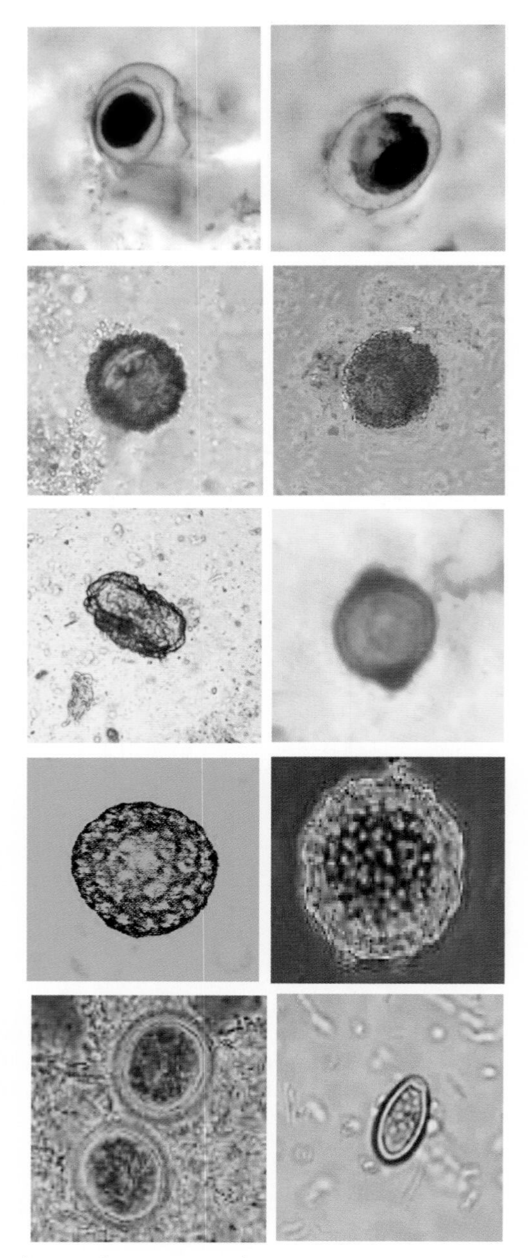

Figure 8.28 Various artifacts that may be seen in stool preparations (wet mounts or permanent stained smears). Many of these structures are pollen grains or egglike objects. Visually, they can be confused with eggs of helminths, such as *Hymenolepis nana, Ascaris lumbricoides,* hookworm, *Enterobius vermicularis,* and some small trematodes.

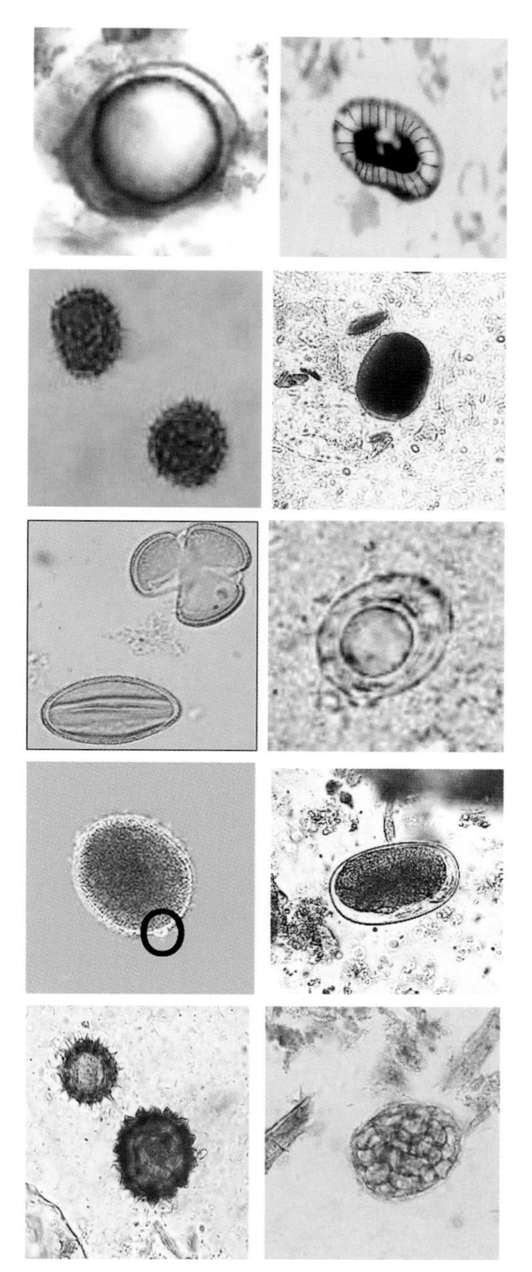

Figure 8.29 Various artifacts that may be seen in stool preparations (wet mounts). Note the egg-like structure in the fourth row (left). There is a small bubble (circle) that mimics the small knob found at the abopercular end of a *Diphyllobothrium latum* egg.

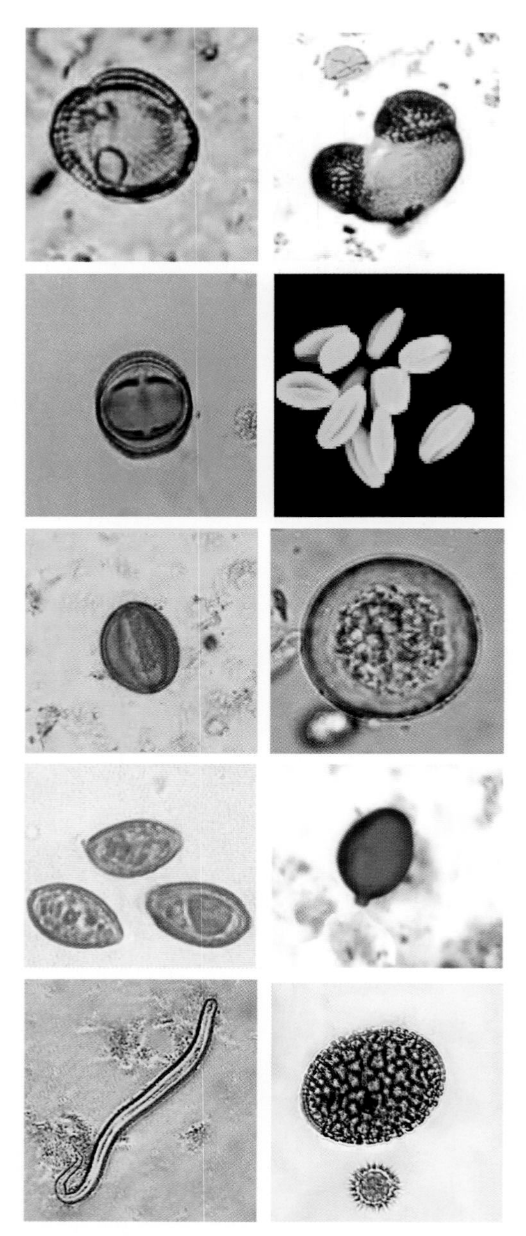

Figure 8.30 Various types of pollen grains and a root hair (fifth row, left). These structures can mimic various helminth eggs (*Ascaris lumbricoides* and *Trichuris trichiura*), as well as nematode larvae. (Left image in the third row courtesy of the CDC Public Health Image Library.)

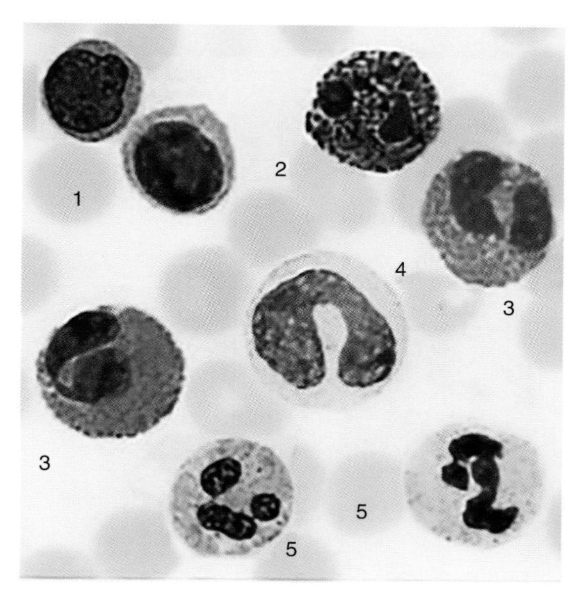

Figure 8.31 White blood cells in a stained blood film. (1) Lymphocytes. (2) Basophil. (3) Eosinophils. (4) Monocyte. (5) Polymorphonuclear leukocytes.

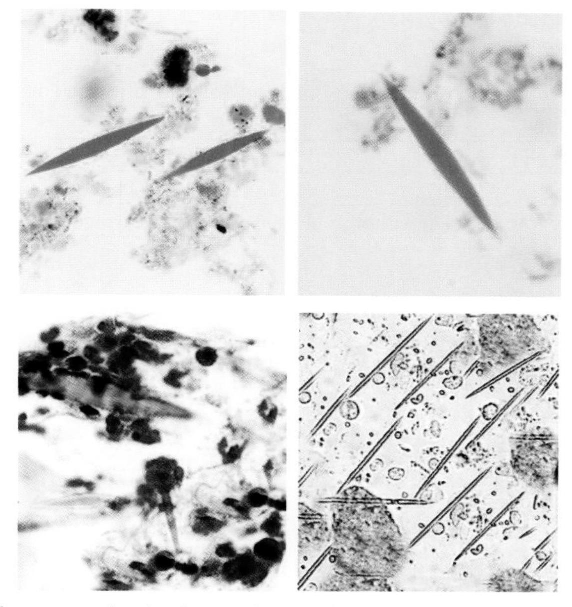

Figure 8.32 **(Top row)** Charcot-Leyden (CL) crystals in trichrome-stained fecal smear. **(Bottom row)** CL crystals in a sputum specimen; pineapple crystals in stool (these may be confused with CL crystals; however, the pineapple crystals are much more slender). CL crystals are formed from the breakdown products of eosinophils and basophils and may be present in the stool, sputum, or other specimens with or without intact eosinophils. The CL crystals tend to stain red to red-purple on the permanent stained fecal smears, often darker than nuclear material; although the shape is consistent, there is a large size range in a single fecal smear or sputum mount.

Table 8.1 ***Entamoeba*** **sp. trophozoites versus macrophages**

Characteristic	*Entamoeba histolytica*	*E. dispar* *E. moshkovskii* *E. bangladeshi*	Macrophages
Size	12–25 μm	12–25 μm	Comparable to protozoa
RBCs	Round within vacuoles or may not be present	No RBCs	No RBCs
Nuclear:cytoplasmic ratio	1:10–1:12	1:10–1:12	1:6–1:8
Permanent stain	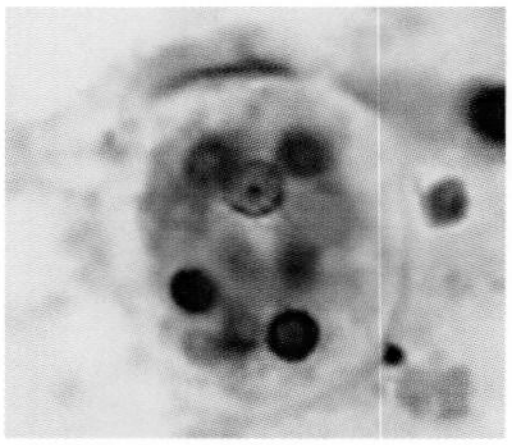	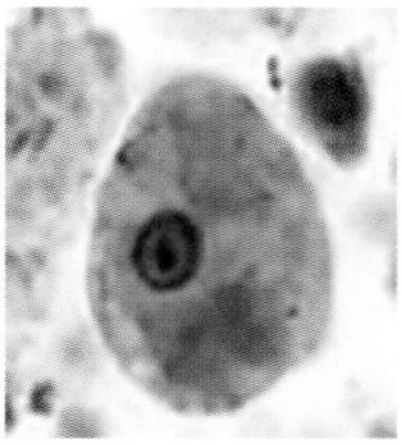	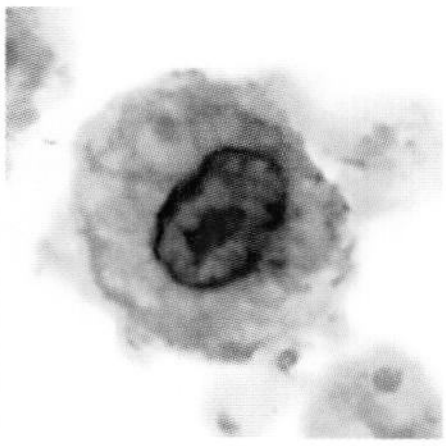

Table 8.2 ***Entamoeba*** **sp. cysts versus PMNs**

Characteristic	*Entamoeba histolytica* *E. dispar* *E. moshkovskii* *E. bangladeshi*	Polymorphonuclear leukocytes
Size	10–15 μm	10–15 μm
RBCs	Not present	Not present
Number of nuclei	4	4 or 5; may be lobed (2–5 nuclear lobes)
Shape of nuclei	Usually round and consistently shaped	Shapes are irregular and may vary tremendously
Permanent stain	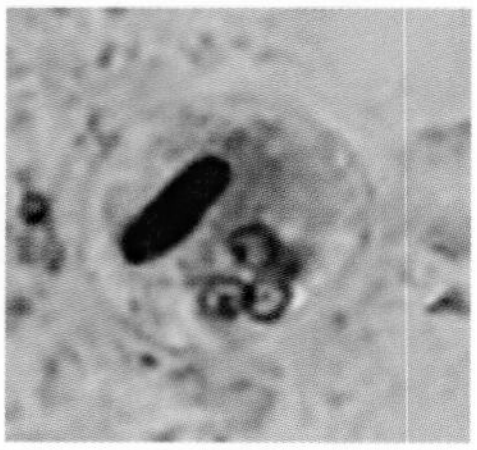	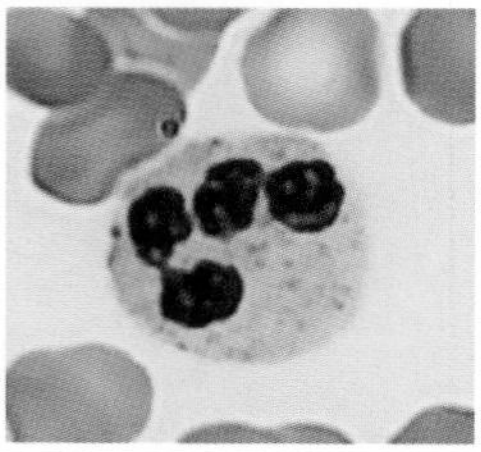
Permanent stain	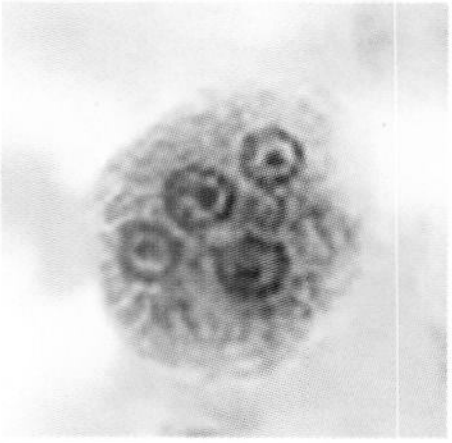	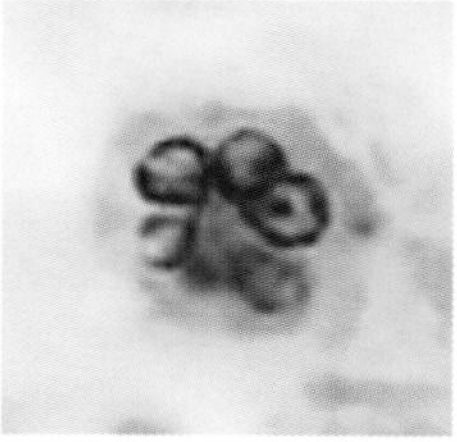

Table 8.3 ***Entamoeba histolytica*** **versus** ***Entamoeba coli*** **precysts and cysts**

Characteristic	*Entamoeba histolytica* *E. dispar* *E. moshkovskii* *E. bangladeshi*	*Entamoeba coli*
Size	10–15 μm	10–15 μm
RBCs	Not present	Not present
Precyst: no. of nuclei	1	2, one at each side
Precyst: shape of nuclei	Usually round	Often elongated
Permanent stain	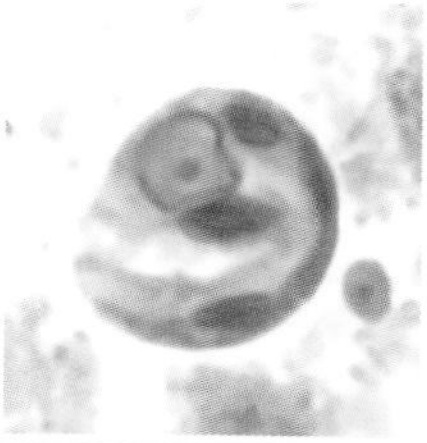	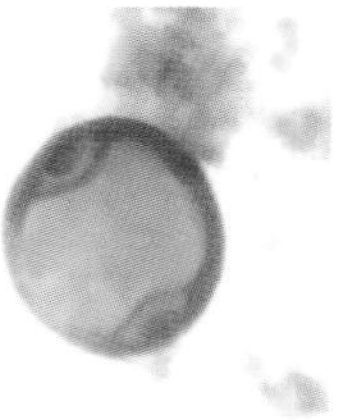
Permanent stain	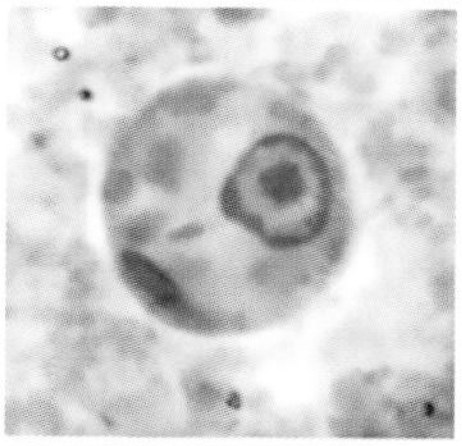	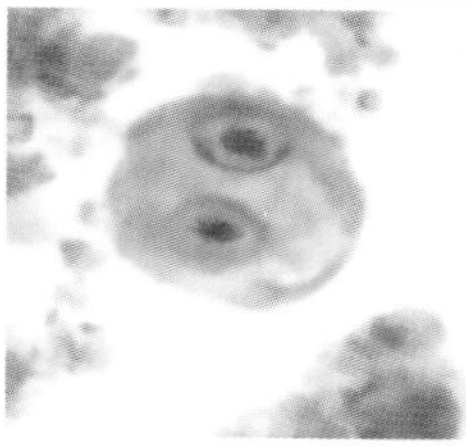
Cysts: no. of nuclei	4; some may not be in plane of focus	5 or more; some may not be in plane of focus
Cysts: chromatoidal bars	Chromatoidal bars have smooth, rounded ends	Chromatoidal bars have sharp, spiked ends
Permanent stain	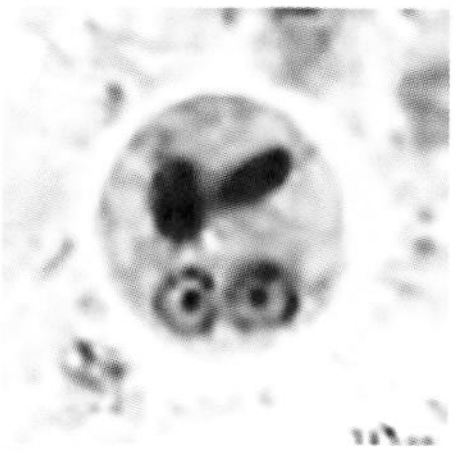	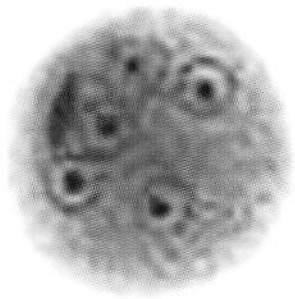
Cysts: chromatoidal bars	Chromatoidal bars may not always be visible	Chromatoidal bars visible
Permanent stain	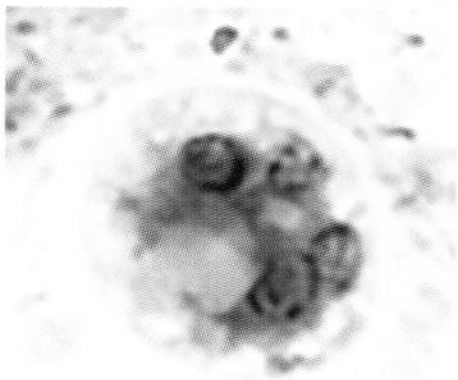	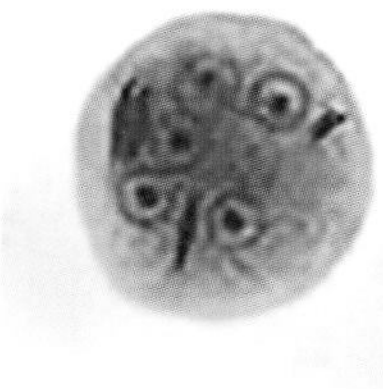

Table 8.4 *Endolimax nana* versus *Dientamoeba fragilis*			
Characteristic	***Endolimax nana***	***Dientamoeba fragilis***	**Comments**
Trophozoite: size	6–12 μm	5–15 μm	Both have wide size and shape range
No. of nuclei	1; tremendous nuclear variation in trophozoites	1 or 2; chromatin often fragmented, 1 nucleus	Large overlap in morphology; 2 nuclei
Permanent stain	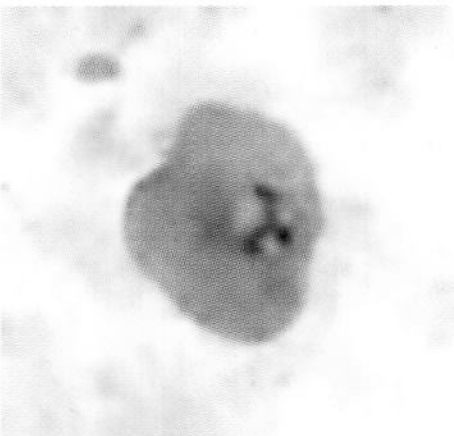	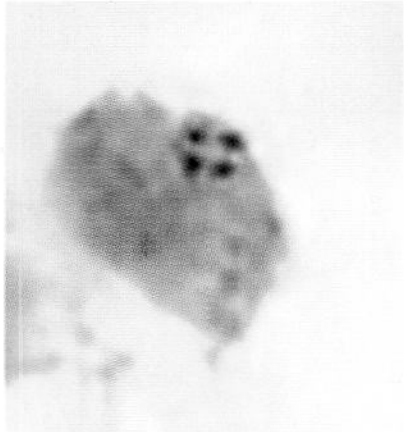	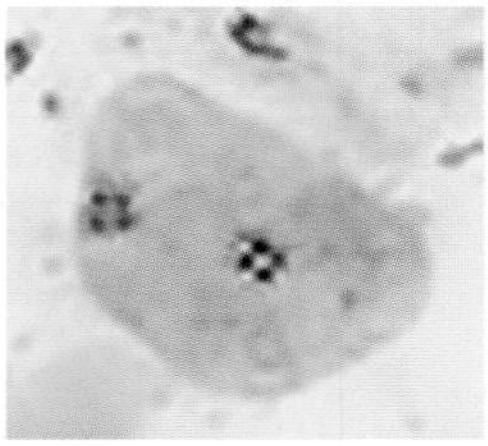
Cyst: size	5–10 μm	5–10 μm	
Permanent stain	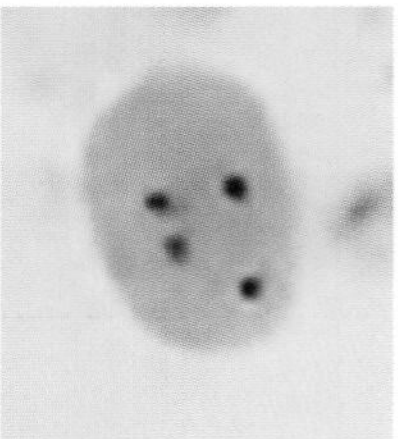	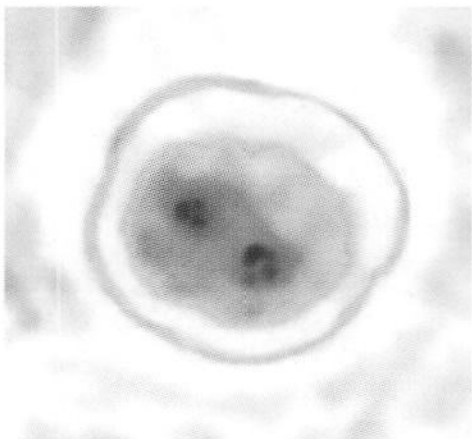	If only a single nucleus is seen, it may be very difficult to differentiate the two organisms. *D. fragilis* is as common as *Giardia*; however, the number of *D. fragilis* cysts in a clinical specimen is quite small, and they may be missed.

SECTION 9

Identification Aids

DIAGNOSTIC CONSIDERATIONS
PROTOZOA
HELMINTHS
BLOOD PARASITES

Abbreviations

BAL, bronchoalveolar lavage fluid; CL, cutaneous leishmaniasis; CLM, cutaneous larva migrans; CNS, central nervous system; CSF, cerebrospinal fluid; DCL, diffuse cutaneous leishmaniasis; DFA, direct fluorescent-antibody assay; DL, dermal leishmaniasis; EIA, enzyme immunoassay; ELISA, enzyme-linked immunosorbent assay; EM, electron microscopy; FA, fluorescent-antibody assay; GI, gastrointestinal; IF, immunofluorescence detection; MCL, mucocutaneous leishmaniasis; NAATs, nucleic acid amplification tests; NNN, Novy-MacNeal-Nicolle; O&P exam, ova and parasite examination; OLM, ocular larva migrans; PAS, periodic acid-Schiff; PMNs, polymorphonuclear leukocytes; POC, point of care; RBCs, red blood cells; RUO, research use only; SAF, sodium acetate-acetic acid-formalin; sp gr, specific gravity; ST, sequence type; VL, visceral leishmaniasis; VLM, visceral larva migrans; WBCs, white blood cells.

Practical Guide to Diagnostic Parasitology, Third Edition. Lynne S. Garcia
 DOI: 10.1128/9781683673637.ch09

DIAGNOSTIC CONSIDERATIONS

Table 9.1 Rapid diagnostic procedures

Traditional procedures	Additional options	Comments
Fecal specimens (permanent stained smear)		
Microscopic examination of stained fecal smear using oil immersion objectives Note: Remember that the total O&P exam is recommended when the fecal immunoassays (limited to specific organisms) are negative and the patient remains symptomatic.	Fecal immunoassays: fluorescence/microscopy, ELISA plates, lateral membrane flow immunochromatographic assay/cartridge Fecal immunoassays are generally simple to perform and allow a large number of tests to be performed at one time, thereby reducing overall costs. **A major disadvantage of antigen detection (fecal immunoassays) in stool specimens is that the method can detect only one or two pathogens at one time. A routine O&P exam must be performed to detect other parasitic pathogens.** The current commercially available antigen tests (DFA, EIA, and lateral-flow cartridges) are more sensitive and more specific than routine microscopy. Current testing is available for *Entamoeba histolytica*, the *Entamoeba histolytica*/*E. dispar* group, *Giardia lamblia*, and *Cryptosporidium* spp. Diagnostic reagents are also in development for some of the other intestinal protozoa. There are also several molecular tests that are in clinical trials for the detection of select gastrointestinal parasites. These tests are molecular gastrointestinal parasite panels and target the most commonly occurring parasitic pathogens found in stool.	Batch testing is possible, the methods use very few steps, and the visual assessment of positive or negative is quite easy. These tests can be very beneficial in the absence of trained microscopists. However, in patients who remain symptomatic after a negative result, the O&P exam should always remain as an option. **BOTH THE O&P EXAM AND FECAL IMMUNOASSAYS SHOULD BE A PART OF THE LABORATORY TEST MENU; SPECIFIC PATIENT HISTORIES WILL DICTATE CORRECT ORDERING OF THE MOST RELEVANT TEST OPTION.** Although tests for *E. histolytica* or the *E. histolytica*/*E. dispar* group are limited to the use of fresh or fresh frozen fecal specimens, they can be helpful in the absence of trained microscopists. Tests for *Giardia* and/or *Cryptosporidium* can be performed on fresh, frozen, unfixed, or formalin-fixed (5%, 10% SAF) fecal specimens. Cary-Blair and some of the single-vial fecal parasite fixatives can also be used. Check with the manufacturer. One of the parasite panels will include *Blastocystis*, *Cryptosporidium*, *Cyclospora*, *Dientamoeba*, *E. histolytica*, *Giardia*, microsporidia, and *Strongyloides*. FDA clearance for this panel is pending and is anticipated in 2021 (VERIGENE II GI panel).
Blood specimens (stained blood films)		
Microscopic examination of stained blood films using oil immersion objectives It is important to remember that a minimum of 300 oil immersion fields (using the 100× oil immersion objective) should be examined on both thick and thin blood films.	Rapid malaria testing: The BinaxNOW malaria test (Abbott) is a rapid immunodiagnostic assay for differentiation and detection of circulating *Plasmodium falciparum* antigen and the antigen common to all malarial species: *Plasmodium vivax*, *Plasmodium ovale*, and *Plasmodium malariae* in whole blood. It is FDA cleared (June 2007). Test line 1 positive = *P. falciparum* Test line 2 positive = *P. vivax*, *P. malariae*, or *P. ovale* Test lines 1 and 2 positive = *P. falciparum* and possible mixed infection	BinaxNOW is positive at 0.1% (5,000/µl) Sensitivity 99.7% (*P. falciparum*) 93.5% (*P. vivax*) Specificity 94.2% (*P. falciparum*) 99.8% (*P. vivax*) It is important to remember that many patients may have a parasitemia much lower than the level of detection of the malaria rapid test. Their parasitemia may be as low as 0.01%. Other tests are in use throughout the world but are not yet FDA cleared.

Table 9.2 Diagnostic characteristics for organisms in wet mounts (direct or concentration sediment)

Specimen	Protozoa	Helminths
Stool, other specimens from GI tract, urogenital system	Size, shape, stage (trophozoite, precyst, cyst, oocyst); motility (fresh specimens only), refractility, cytoplasm inclusions (chromatoidal bars, glycogen vacuoles, axonemes, axostyles, median bodies, sporozoites)	Egg, larva, or adult; size, internal structure Egg: shell characteristics (striations, bumpy or smooth, thick or thin), embryonation, opercular shoulders, abopercular thickenings or projections, hooklets, polar filaments, polar plugs, spines Larva: head/tail morphology, digestive tract Adult: nematode, cestode, or trematode

Table 9.3 Diagnostic characteristics for organisms in permanent stained smears (e.g., Wheatley's trichrome, iron-hematoxylin)

Specimen	Protozoa	Helminths
Stool, other specimens from GI tract and urogenital system	Size, shape, stage (trophozoite, precyst, cyst, oocyst, spore) Nuclear arrangement, cytoplasm inclusions (chromatoidal bars, vacuoles, axonemes, axostyles, median bodies, sporozoites, polar tubules)	Eggs, larvae, and/or adults may not be identified because of excess stain retention or distortion

Identification Key 9.1 Identification of intestinal amebae (permanent stained smear)

1. Trophozoites present	2
Cysts present	7
2. Trophozoites measure >12 μm	3
Trophozoites measure <12 μm	4
3. Karyosome central, compact; peripheral nuclear chromatin evenly arranged; "clean" cytoplasm (size of debris consistent) (RBCs present or absent)	*Entamoeba histolytica*[a]
Karyosome eccentric, spread out (blot-like); peripheral nuclear chromatin unevenly arranged (may be more even with trichrome stain); "dirty" cytoplasm (size and amount of debris more extensive than *E. histolytica*)	*Entamoeba coli*
4. Peripheral nuclear chromatin	5
Other than above	6
5. Karyosome central, compact (nucleus looks like "bull's-eye" target); peripheral nuclear chromatin evenly arranged; "clean" cytoplasm	*Entamoeba hartmanni*
Karyosome large, blot-like; extensive nuclear variation (can mimic *Dientamoeba fragilis* and *Iodamoeba bütschlii*)	*Endolimax nana*
6. No peripheral chromatin, large karyosome, junky cytoplasm; if peripheral chromatin is present, looks like "basket" nucleus	*Iodamoeba bütschlii*
No peripheral chromatin, variable karyosome, clean cytoplasm	*Endolimax nana*

Table continues on next page

Identification Key 9.1 Identification of intestinal amebae (permanent stained smear) (*continued*)	
7. Cysts measure >10 μm (including any shrinkage halo)	8
Cysts measure <10 μm (including any shrinkage halo)	10
8. Single *Entamoeba*-like nucleus with large inclusion mass	*Entamoeba polecki*[b]
Multiple nuclei	9
9. 4 *Entamoeba*-like nuclei; chromatoidal bars have smooth, rounded ends	*Entamoeba histolytica*[a]
≥5 *Entamoeba*-like nuclei, chromatoidal bars have sharp, pointed ends	*Entamoeba coli*
10. Single nucleus (may be "basket" nucleus) (can also be seen in the trophozoite), large glycogen vacuole	*Iodamoeba bütschlii*
Multiple nuclei	11
11. 4 *Entamoeba*-like nuclei, chromatoidal bars have smooth, rounded ends (nuclei may also number only 2)	*Entamoeba hartmanni*
4 karyosomes, no peripheral chromatin (karyosome exhibits tremendous nuclear variability), round to oval shape	*Endolimax nana*

[a]"*Entamoeba histolytica*" here refers to the *Entamoeba histolytica*/*Entamoeba dispar* complex or group. *E. histolytica* (pathogen) can be determined by finding RBCs in the cytoplasm of the trophozoites or by using a molecular method to determine the true pathogen (*Entamoeba histolytica*) rather than the complex. Otherwise, on the basis of morphology, *E. histolytica* (pathogen) and *E. dispar* (nonpathogen) cannot be differentiated and should be reported as *Entamoeba histolytica*/*E. dispar* complex or group. Although *Entamoeba moshkovskii* and *Entamoeba bangladeshi* look very similar to the rest of the *Entamoeba* complex, they are not as relevant for patient care and are not included with the report of *E. histolytica*/*E. dispar* complex.

[b]It is very difficult to differentiate *Entamoeba polecki* trophozoites from *E. histolytica* or *E. coli*.

Identification Key 9.2 Identification of intestinal flagellates	
1. Trophozoites present	2
Cysts present	7
2. Pear or lemon shaped	3
Other shape	6
3. 2 nuclei, sucking disk present	*Giardia lamblia*
1 nucleus present	4
4. Costa (supporting structure for undulating membrane) and undulating membrane entire length of body	*Pentatrichomonas hominis*
No costa	5
5. Cytostome (primitive feeding groove) present, >10 μm	*Chilomastix mesnili*
Cytostome present, <10 μm	*Retortamonas intestinalis* or *Enteromonas hominis*
6. Ameba shaped, 1 or 2 fragmented nuclei	*Dientamoeba fragilis*[a]
Oval shaped, 1 nucleus	*Enteromonas hominis*
7. Oval or round cyst	8
Pear- or lemon-shaped cyst	9

Identification Key 9.2 (*continued*)	
8. 4 nuclei, median bodies, axoneme, >10 μm	*Giardia lamblia*
2 nuclei, no fibrils, <10 μm	*Enteromonas hominis*
9. 1 nucleus, "shepherd's crook" fibril	*Chilomastix mesnili*
1 nucleus, "bird's beak" fibril	*Retortamonas intestinalis*

[a] Although the *Dientamoeba fragilis* cyst has been confirmed, it looks very similar to the trophozoite form and will be difficult to differentiate from the trophozoites. The cysts tend to be present in the fecal specimen in very low numbers (~1%), so they will also be difficult to find.

Identification Key 9.3 Identification of helminth eggs[a]	
1. Eggs nonoperculate (no "trap door"), spherical or subspherical, containing a 6-hooked embryo (oncosphere); thick or thin shell	2
Eggs other than described above	5
2. Eggs passed separately	3
Eggs passed in packets of 12 or more	*Dipylidium caninum* (dog tapeworm)
3. Thick, radially striated shell (6-hooked oncosphere may not be visible in every egg from formalinized fecal specimens) (eggs cannot be identified to species level without special stains)	*Taenia* spp. (*T. saginata*, beef tapeworm; *T. solium*, pork tapeworm)
Thin shell, clear space between shell and developing embryo	4
4. Polar filaments (filamentous strands) present between thin shell and 6-hooked embryo	*Hymenolepis nana* (dwarf tapeworm)
No polar filaments (filamentous strands) present between shell and embryo; somewhat larger	*Hymenolepis diminuta* (rat tapeworm)
5. Egg operculate, generally oval ("trap door" at one end of egg)	6
Egg nonoperculate, generally oval	10
6. Egg <35 μm long	*Clonorchis* (*Opisthorchis*) spp. (Chinese liver fluke), *Heterophyes heterophyes*, *Metagonimus yokogawai*
Egg ≥38 μm long	7
7. Egg 38–45 μm long	*Dicrocoelium dendriticum*
Egg >60 μm long	8
8. Egg with opercular shoulders into which the operculum fits (looks like teapot lid and flange into which lid fits), abopercular end thickened (sometimes hard to see)	*Paragonimus* spp. (lung fluke)
Egg without opercular shoulders	9

Table continues on next page

Identification Key 9.3 Identification of helminth eggs[a] *(continued)*	
9. Egg >85 μm long, operculum break in shell sometimes hard to see—smooth transition from shell to operculum (tap on coverslip to get operculum to pop open)	*Fasciolopsis buski* (giant intestinal fluke), *Fasciola hepatica* (sheep liver fluke), *Echinostoma* spp.
Egg <75 μm long, operculum break in shell sometimes hard to see—smooth transition from shell to operculum (tap on coverslip to get operculum to pop open)	*Diphyllobothrium latum* (broad fish tapeworm)
10. Egg ≥75 μm long, spined, ciliated miracidium larva may be seen	11
Egg <75 μm long, not spined	13
11. Spine terminal (check for egg viability)	*Schistosoma haematobium* (blood fluke, from urine)
Spine lateral	12
12. Lateral spine very short (hard to see) (check for egg viability)	*Schistosoma japonicum* (blood fluke, from stool)
Lateral spine prominent and easily seen (check for egg viability)	*Schistosoma mansoni* (blood fluke, from stool)
13. Egg with thick, tuberculated (mammillated/bumpy) outer layer (in decorticate eggs, this outer layer will be absent; can occur in both fertilized and unfertilized eggs)	*Ascaris lumbricoides* (large roundworm)
Egg without thick, tuberculated outer layer	14
14. Egg barrel shaped, with clear polar plugs	15
Egg not barrel shaped, no polar plugs	16
15. Shell nonstriated	*Trichuris trichiura* (whipworm)
Shell striated	*Capillaria hepatica*
16. Egg flattened on one side, may contain larva	*Enterobius vermicularis* (pinworm)
Egg symmetrical	17
17. Egg bluntly rounded at ends, 56–76 μm long, thin shell (contains developing embryo at 8- to 16-cell stage of development)	Hookworm
Egg pointed at one or both ends, 73–95 μm long	*Trichostrongylus* spp.

[a]It is important to remember that occasionally schistosome eggs can be found in clinical specimens that do not represent their normal site; e.g., eggs that are normally found in urine (*Schistosoma haematobium*) can be found in stool. The same situation occurs for those eggs normally found in stool (*S. mansoni, S. japonicum*), which can occasionally be found in urine. Any patient suspected of schistosomiasis, regardless of species suspected, should submit both stool and urine for examination.

Identification Key 9.4 Identification of microfilariae	
1. Larvae sheathed	2
Larvae unsheathed	*Mansonella ozzardi*, *Mansonella perstans*
2. Tail nuclei do not extend to the tip of the tail	*Wuchereria bancrofti*
Tail nuclei extend to the tip of the tail	3
3. Nuclei not continuous (two nuclei at the tip of the tail)	*Brugia malayi*
Nuclei form continuous row in the tail	*Loa loa*

Table 9.4 Intestinal protozoa: trophozoites of common amebae

Characteristic	*Entamoeba histolytica* *Entamoeba dispar*	*Entamoeba hartmanni*	*Entamoeba coli*	*Endolimax nana*	*Iodamoeba bütschlii*
Size[a] (diameter or length)	12–60 µm; usual range, 15–20 µm; invasive forms (*E. histolytica*) may be over 20 µm	5–12 µm; usual range, 8–10 µm	15–50 µm; usual range, 20–25 µm	6–12 µm; usual range, 8–10 µm	8–20 µm; usual range, 12–15 µm
Motility	Progressive, with hyaline, fingerlike pseudopodia; motility may be rapid	Usually nonprogressive	Sluggish nondirectional, with blunt, granular pseudopodia	Sluggish, usually nonprogressive	Fairly active and progressive in freshly passed stool; sluggish in older stools
Nucleus: visibility and no.	Difficult to see in unstained preparations; 1 nucleus	Usually not seen in unstained preparations; 1 nucleus	Often visible in unstained preparation; 1 nucleus	Occasionally visible in unstained preparations; 1 nucleus	Usually not visible in unstained preparations; 1 nucleus
Peripheral nuclear chromatin (stained)	Fine granules, uniform in size and usually evenly distributed; may have beaded appearance	Nucleus may stain more darkly than *E. histolytica* although morphology is similar; chromatin may appear as solid ring rather than beaded (trichrome)	May be clumped and unevenly arranged on the membrane; may also appear as solid ring with no beads/clumps; mimics *Entamoeba histolytica*/*E. dispar* group	Usually no peripheral chromatin; nuclear chromatin may be quite variable	Usually no peripheral chromatin; if present, is usually clumped on one side (basket nucleus); also see "Karyosome" below
Karyosome (stained)	Small, usually compact; centrally located but may also be eccentric	Usually small and compact; may be centrally located or eccentric; often described as looking like a bull's-eye or target nucleus	Large, not compact; may or may not be eccentric; may be diffuse and darkly stained	Large, irregularly shaped; may appear blot-like; many nuclear variations are common; may mimic *E hartmanni* or *Dientamoeba fragilis*	Large, may be surrounded by refractile granules that are difficult to see (basket nucleus)

Table continues on next page

Table 9.4 Intestinal protozoa: trophozoites of common amebae (*continued*)

Characteristic	*Entamoeba histolytica* *Entamoeba dispar*	*Entamoeba hartmanni*	*Entamoeba coli*	*Endolimax nana*	*Iodamoeba bütschlii*
Cytoplasm appearance (stained)	Finely granular, ground-glass appearance; clear differentiation of ectoplasm and endoplasm; if present, vacuoles are usually small	Finely granular	Granular, with little differentiation into ectoplasm and endoplasm; usually vacuolated	Granular, vacuolated	Granular, may be heavily vacuolated
Inclusions (stained)	Both organisms may contain bacteria; presence of RBCs diagnostic for the true pathogen, *E. histolytica* *E. dispar* does not contain RBCs[b]	May contain bacteria; no RBCs	Bacteria, yeast, other debris	Bacteria	Bacteria and other debris; usually more obvious than inclusions seen in *E. nana*

[a] These sizes refer to wet-preparation measurements. Organisms on a permanent stained smear may be 1 to 1.5 µm smaller due to artificial shrinkage.

[b] *E. histolytica* and *E. dispar* look identical; unless the trophozoite contains ingested RBCs, the organism identification should read "*Entamoeba histolytica*/*E. dispar* group. *E. moshkovskii* and *E. bangladeshi* also resemble the *E. histolytica*/*E. dispar* group."

Table 9.5 Intestinal protozoa: cysts of common amebae

Characteristic	*Entamoeba histolytica* *Entamoeba dispar*[a]	*Entamoeba hartmanni*	*Entamoeba coli*	*Endolimax nana*	*Iodamoeba bütschlii*
Size[b] (diameter or length)	10–20 µm; usual range, 12–15 µm	5–10 µm; usual range, 6–8 µm	10–35 µm; usual range, 15–25 µm	5–10 µm; usual range, 6–8 µm	5–20 µm; usual range, 10–12 µm
Shape	Usually spherical	Usually spherical	Usually spherical; may be oval, triangular, or other shapes; may be distorted on permanent stained slide due to inadequate fixative penetration	Usually oval, may be round	May vary from oval to round; cyst may collapse due to large glycogen vacuole space
Nucleus: no. and visibility	Mature cyst, 4; immature cyst, 1 or 2 nuclei; nuclear characteristics difficult to see on wet preparation	Mature cyst, 4; immature cyst, 1 or 2 nuclei; two nucleated cysts very common	Mature cyst, 8; occasionally 16 or more nuclei may be seen; immature cysts with 2 or more nuclei are occasionally seen	Mature cyst, 4; immature cysts, 2; very rarely seen and may resemble cysts of *Enteromonas hominis*	Mature cyst, 1
Peripheral chromatin (stained)	Peripheral chromatin present; fine, uniform granules, evenly distributed; nuclear characteristics may not be as clearly visible as in trophozoite; usually paler than that seen in trophozoites	Fine granules evenly distributed on the membrane; nuclear characteristics may be difficult to see	Coarsely granular; may be clumped and unevenly arranged on membrane; nuclear characteristics not as clearly defined as in trophozoite; may resemble *E. histolytica*	No peripheral chromatin	No peripheral chromatin
Karyosome (stained)	Small, compact, usually centrally located but occasionally eccentric	Small, compact, usually centrally located	Large, may be compact and/or eccentric; occasionally centrally located	Smaller than karyosome seen in trophozoites, but generally larger than those of the genus *Entamoeba*	Larger, usually eccentric refractile granules may be on one side of karyosome (basket nucleus)

Table continues on next page

Table 9.5 Intestinal protozoa: cysts of common amebae (*continued*)

Characteristic	*Entamoeba histolytica* *Entamoeba dispar*[a]	*Entamoeba hartmanni*	*Entamoeba coli*	*Endolimax nana*	*Iodamoeba bütschlii*
Cytoplasm, chromatoidal bodies (stained)	May be present; bodies usually elongate, with blunt, rounded, smooth edges; may be round or oval	Usually present; bodies usually elongate with blunt, rounded, smooth edges; may be round or oval	May be present (less frequently than *E. histolytica*); splinter shaped with rough, pointed ends	Rare chromatoidal bodies present; occasionally small granules or inclusions seen; fine linear chromatoidal bodies may be faintly visible on well-stained smears	No chromatoidal bodies present; occasionally small granules may be present
Glycogen (stained with iodine)	May be diffuse or absent in mature cyst; clumped chromatin mass may be present in early cysts (stains reddish brown with iodine)	May or may not be present as in *E. histolytica*	May be diffuse or absent in mature cysts; clumped mass occasionally seen in mature cysts (stains reddish brown with iodine)	Usually diffuse if present (will stain reddish brown with iodine)	Large, compact, well-defined mass (will stain reddish brown with iodine)

[a]*Entamoeba polecki* cysts contain only a single nucleus. Cysts of *E. moshkovskii* and *E. bangladeshi* resemble those of the *Entamoeba histolytica*/*E. dispar* group.

[b]Wet-preparation measurements (in permanent stains, organisms usually measure 1 to 2 μm less). When a halo (clear space) is seen around a cyst, remember to measure the halo as a part of the total cyst size; the appearance of a halo is due to artificial shrinkage during processing.

Table 9.6 Intestinal protozoa: trophozoites of less common amebae

Characteristic	*Entamoeba polecki*	*Entamoeba gingivalis*
Source	Intestinal tract, stool	Mouth (poor oral hygiene)
Size (diameter or length)	10–12 μm	5–20 μm (average, 10–15 μm)
Motility	Usually nonprogressive, sluggish (like *E. coli*)	Multiple pseudopodia, varying from long and lobose to short and blunt
Nucleus: visibility and no.	Can occasionally be seen on a wet preparation; intermediate between *E. histolytica* and *E. coli*; 1 nucleus	Similar to *E. histolytica*; 1 nucleus
Peripheral chromatin (stained)	Fine granules (may be interspersed with large granules) evenly arranged on membrane; chromatin may also be clumped at one or both edges of membrane	Fine granules, closely packed
Karyosome (stained)	Small, usually centrally located	Small, well defined, usually centrally located
Cytoplasm appearance (stained)	Finely granular	Finely granular
Inclusion mass (10–12 μm)	Large inclusion; may only be one large mass	None
Inclusions (stained)	May contain other ingested debris such as bacteria, but inclusion mass is most noticeable	Ingested epithelial cells and host leukocytes; other debris may be present; very rare to see ingested RBCs

Table 9.7 Intestinal protozoa: cysts of less common amebae

Characteristic	*Entamoeba polecki*	*Entamoeba gingivalis*
Size (diameter or length)	5–11 μm	No defined cyst stage[a]
Shape	Usually spherical	
Nucleus: no. and visibility	Mature cyst, 1; may be visible in wet preparations (rarely 2 or 4 nuclei)	
Peripheral chromatin (stained)	Similar to that seen in the trophozoite	
Karyosome	Similar to that seen in the trophozoite	
Cytoplasm: chromatoidal bodies, glycogen, inclusions (stained)	Abundant chromatoidal bodies with angular pointed ends; also, threadlike chromatoidal bodies may be present; half of cysts contain a large spherical or ovoidal inclusion mass	

[a]Although cysts have been described in nature for this organism, generally it is assumed that no cyst is formed. However, the complete life cycle has not been completely defined (as of 2020).

Table 9.8 Morphologic criteria used to identify *Blastocystis* spp.		
Species	**Shape and size**	**Other features**
Blastocystis spp.	Organisms are generally round, measure approximately 6 to 40 μm, and are usually characterized by a large, central body (looks like a large vacuole) surrounded by small, multiple nuclei; central-body area can stain various colors or remain clear[a]	The more amebic form can be seen in diarrheal fluid, but will be difficult to identify; due to variation in size, may be confused with various yeast cells
Subtypes	Genetic diversity revisions have led to the identification of 17 subtypes within the *Blastocystis* genus, and 9 (ST1 to ST8, ST12) have been reported in humans with varying prevalence. Approximately half of these human subtypes are pathogenic and half are nonpathogenic.	All 9 subtypes found in humans are potentially zoonotic. Pathogenic subtypes have been described, and their presence is frequently associated with symptoms in humans. The subtypes may eventually be named as species; classification changes are under discussion. Positive correlation between high bacterial abundance and *Blastocystis* has been reported.

[a] In permanent stained fecal smears made with Wheatley's trichrome, the central-body area can be clear or shades of red, green, and/or blue.

Table 9.9 Intestinal protozoa: trophozoites of flagellates

Species	Shape and size	Motility	No. of nuclei and visibility	No. of flagella (usually difficult to see)	Other features
Dientamoeba fragilis	Shaped like amebae; 5–15 μm; usual range, 9–12 μm	Usually nonprogressive; pseudopodia are angular, serrated, or broad lobed and almost transparent	Percentage may vary, but 40% of organisms have 1 nucleus and 60% have 2 nuclei; not visible in unstained preparations; no peripheral chromatin; karyosome is composed of a cluster of 4–8 granules	Internal flagella; however, they are not visible.	Cytoplasm finely granular and may be vacuolated with ingested bacteria, yeasts, and other debris; may be great variation in size and shape on a single smear
Giardia lamblia (*G. duodenalis*, *G. intestinalis*)	Pear shaped; 10–20 μm; width, 5–15 μm	"Falling leaf" motility may be difficult to see if organism is in mucus; slight flutter of flagella may be visible using low light (duodenal aspirate or mucus from Entero-Test capsule)	2; not visible in unstained mounts	4 lateral, 2 ventral, 2 caudal	Sucking disc occupying 1/2–3/4 of ventral surface; pear-shaped front view; spoon-shaped side view
Chilomastix mesnili	Pear shaped; 6–24 μm; usual range, 10–15 μm; width, 4–8 μm	Stiff, rotary	1; not visible in unstained mounts	3 anterior, 1 in cytostome	Prominent cytostome extending 1/3–1/2 length of body; spiral groove across ventral surface
Pentatrichomonas hominis	Pear shaped; 5–15 μm; usual range, 7–9 μm; width, 7–10 μm	Jerky, rapid	1; not visible in unstained mounts	3–5 anterior, 1 posterior	Undulating membrane extends length of the body; posterior flagellum extends free beyond end of body.
Trichomonas tenax	Pear shaped; 5–12 μm; average, 6.5–7.5 μm; width, 7–9 μm	Jerky, rapid	1; not visible in unstained mounts	4 anterior, 1 posterior	Seen only in preparations from mouth; axostyle (slender rod) protrudes beyond the posterior end and may be visible; posterior flagellum extends only halfway down body and there is no free end
Enteromonas hominis	Oval; 4–10 μm; usual range, 8–9 μm; width, 5–6 μm	Jerky	1; not visible in unstained mounts	3 anterior, 1 posterior	One side of body flattened; posterior flagellum extends free posteriorly or laterally
Retortamonas intestinalis	Pear shaped or oval; 4–9 μm; usual range, 6–7 μm; width, 3–4 μm	Jerky	1; not visible in unstained mount	1 anterior, 1 posterior	Prominent cytostome extending approximately 1/2 length of body

Table 9.10 Intestinal protozoa: cysts of flagellates

Species	Size	Shape	No. of nuclei	Other features
Pentatrichomonas hominis, Trichomonas tenax	No cyst stage			
Dientamoeba fragilis	Precyst: 4–5 μm (usual range, 4 μm) Cyst: 4–6 μm (usual range, 5 μm)	The precystic stage is characterized by a compact spherical shape with a reduction in size of up to 50% from "normal" trophozoites. These forms range in size from 4 to 5 μm. The cytoplasm is darkly staining, indicating a denser structure than what is found in normal trophozoites. The cytoplasm is homogeneous and rarely contains any inclusions.	2; each nucleus contains a large central karyosome with a delicate nuclear membrane.	Distinct cyst wall (~5 μm in diameter) with a clear zone around the cyst. A space is present between the cyst wall and the organism enclosed within the cyst wall. The nucleus is often fragmented into distinct granules of chromatin. Cysts are rarely seen in human clinical specimens; also, they can be very difficult to differentiate from the trophozoite stages.
Giardia lamblia (*G. duodenalis, G. intestinalis*)	8–19 μm; usual range, 11–14 μm; width, 7–10 μm	Oval or ellipsoidal, or may appear round	4; not distinct in unstained preparations; usually located at one end	Longitudinal fibers in cysts may be visible in unstained preparations; deeply staining median bodies usually lie across the longitudinal fibers; there is often shrinkage and the cytoplasm pulls away from the cyst wall; there may also be a halo effect around the outside of the cyst wall due to shrinkage caused by dehydrating reagents
Chilomastix mesnili	6–10 μm; usual range, 7–9 μm; width, 4–6 μm	Lemon or pear shaped with anterior hyaline knob	1; not distinct in unstained preparations	Cytostome with supporting fibrils, usually visible in stained preparation; curved fibril along side of cytostome usually referred to as a shepherd's crook
Enteromonas hominis	4–10 μm; usual range, 6–8 μm; width, 4–6 μm	Elongate or oval	1–4; usually 2 lying at opposite ends of cyst; not visible in unstained mounts	Resembles *E. nana* cyst; fibrils or flagella usually not seen
Retortamonas intestinalis	4–9 μm; usual range, 4–7 μm; width, 5 μm	Pear shaped or slightly lemon shaped	1; not visible in unstained mounts	Resembles *Chilomastix* cyst; shadow outline of cytostome with supporting fibrils extends above nucleus; bird's beak fibril arrangement

Table 9.11 Intestinal protozoa: ciliate (*Balantidium coli*)

Form	Shape and size	Motility	No. of nuclei	Other features
Trophozoite	Ovoid with tapering anterior end; 50–100 μm in length; 40–70 μm in width; usual range, 40–50 μm	Ciliates: rotary, boring; may be rapid	1 large kidney-shaped macronucleus; 1 small round micronucleus, which is difficult to see even in the stained smear; macronucleus may be visible in unstained preparation	Body covered with cilia, which tend to be longer near cytostome; cytoplasm may be vacuolated
Cyst	Spherical or oval; 50–70 μm; usual range. 50–55 μm		1 large macronucleus visible in unstained preparation; micronucleus difficult to see	Macronucleus and contractile vacuole are visible in young cysts; in older cysts, internal structure appears granular; cilia difficult to see within the cyst wall

Table 9.12 Apicomplexa

Species	Shape and size	Other features
Cryptosporidium spp. (Apicomplexa)[a]	Oocyst generally round; 4–6 μm; each mature oocyst contains sporozoites	Oocyst usual diagnostic stage in stool. Various other stages in life cycle can be seen in biopsy specimens taken from Gl tract (brush border of epithelial cells) and possibly other tissues (respiratory tract, biliary tract).
Cyclospora cayetanensis (Apicomplexa, coccidia)	Organisms are generally round; 8–10 μm; they mimic *Cryptosporidium* (acid fast) but are larger	In wet smears, they look like nonrefractile spheres; they also autofluoresce with epifluorescence; they are acid-fast variable from no color to light pink to deep red; those that do not stain may appear wrinkled; in a trichrome-stained stool smear they will appear as clear, round, somewhat wrinkled objects; they cause diarrhea in both immunocompetent and immunosuppressed patients.
Cystoisospora belli (Apicomplexa, coccidia)	Ellipsoidal oocyst; usual range, 20–30 μm long; 10–19 μm wide; sporocysts are rarely seen broken out of oocysts but measure 9–11 μm	Mature oocyst contains 2 sporocysts with 4 sporozoites each; usual diagnostic stage in feces is immature oocyst containing spherical mass of protoplasm (diarrhea) (intestinal tract).
Sarcocystis hominis (beef) *S. suihominis* (pork) (Apicomplexa, coccidia)	Oocyst is thin-walled and contains two mature sporocysts, each containing four sporozoites; frequently thin oocyst wall ruptures; ovoidal sporocysts each measure 9–16 μm long and 7.5–12 μm wide	Thin-walled oocyst or ovoidal sporocysts occur in stool (intestinal tract). Another species has been identified from the ingestion of raw horse meat (*S. fayeri*).

Table continues on next page

Table 9.12 Apicomplexa (*continued*)

Species	Shape and size	Other features
Toxoplasma gondii (Apicomplexa, coccidia)	The tachyzoite is the fast-replicating haploid form that disseminates throughout the host and at which the immune response is generally directed (crescent shaped; 4–6 μm by 2–3 μm). In specific tissues and cell types, tachyzoites convert to bradyzoites, a slowly replicating form that makes cysts. Cysts contain many bradyzoites (resemble *Leishmania* amastigotes, but without the linear bar/kinetoplast). The cysts containing many bradyzoites may reach 100 μm, particularly in muscle; usually maximum of 70 μm in brain.	Cysts appear to evade the immune response, thus allowing *T. gondii* to establish a persistent infection for the life of the host. In humans, *T. gondii* encysts and persists in the brain and in cardiac and skeletal muscle.

[a]*Cryptosporidium* sp. is no longer considered a coccidian, but remains in the Apicomplexa.

Table 9.13 Microsporidia (related to the Fungi): general information

Microsporidian (human)	Normal site	General comments	Other information
Anncaliia algerae	Eye, muscle	Routine histology (fair); acid-fast, PAS stains recommended (spores); for other specimens (stool, urine, etc.), modified trichrome stains, optical brightening agents, polyclonal or monoclonal[a] antibody-based kits (FA, Western blot), antigen detection kits (FA, EIA); animal inoculation not recommended—lab animals may carry occult infections; electron microscopy may be necessary	These organisms have been found as insect or other animal parasites; route of infection may be ingestion, inhalation, or direct inoculation (eye); well documented as emerging opportunistic infection in AIDS patients. Many organisms have a wide host range.
Anncaliia connori	Systemic		
Anncaliia vesicularum (former genera, *Brachiola* and *Nosema*)	Systemic		
Encephalitozoon cuniculi (former genus, *Nosema*)	Eyes		
Encephalitozoon hellem	Small intestine		
Encephalitozoon intestinalis (former genus, *Septata*)	Small intestine, biliary tract		
Enterocytozoon bieneusi	Eyes		
Microsporidium africanum	Eyes		
Microsporidium ceylonensis	Muscle		
Microsporidium CU (*Endoreticulatus*-like)	Eyes		
Nosema ocularum	Muscle		
Pleistophora ronneafiei	Eyes, systemic		
Trachipleistophora anthropopthera	Eyes, muscle		
Trachipleistophora hominis	Muscle, systemic		
Tubulinosema acridophagus	Eyes, urinary tract		
Vittaforma corneae (formerly *Nosema corneum*)			

[a]Polyclonal antibody, heterogeneous mix of antibodies derived from the immune response of multiple B cells; each one recognizes a different epitope on the same antigen. Monoclonal antibody, antibodies that come from a single B-cell parent clone; they only recognize a single epitope per antigen.

Table 9.14 Microsporidia: recommended diagnostic techniques		
Technique	**Use[a]**	**Comments**
Light microscopy		
Stool specimens		
Modified trichrome	++	Reliable, available; light infections difficult to confirm
Giemsa	–	Not recommended for routine use, hard to read
Optical brightening agents	++	Calcofluor white, Fungi-Fluor, Uvitex 2B; sensitive but nonspecific
Antigen detection, automated		Commercial availability limits use; products in development not yet FDA cleared; be sure to check parasite listing for *Encephalitozoon intestinalis* and *Enterocytozoon bieneusi* (Verigene now RUO, 2020)
Other bodily fluids		
Modified trichrome	++	Reliable, available; light infections difficult to identify
Giemsa	+	Urine, conjunctival swab, BAL, CSF, duodenal aspirate
Optical brightening agents	++	Calcofluor white, Fungi-Fluor, Uvitex 2B; sensitive but nonspecific
IF technique	++	Commercial availability limits use; products in development
Routine histology		
Hematoxylin and eosin	+	Sensitivity uncertain with low parasite numbers
PAS	+	Controversy over effectiveness
Modified Gram stains (Brown-Brenn, Brown-Hopps)	++	Sensitive, generally recommended
Giemsa	+	Sensitivity uncertain with low parasite numbers
Warthin-Starry	+	Not standardized, may not be necessary
Modified trichrome	++	Reliable, sensitive
IF technique	++	Commercial availability limits use; products in development
Electron microscopy		
Bodily fluid	+	Specific, sensitivity low, used for identification to species level
Tissue sections	++	Gold standard for confirmation, but sensitivity lower than detection of spores in stool or urine; used for identification to species level
Serologic antibody detection	–	Reagents not commercially available; preliminary results controversial
Cultures	–	Generally used in the research setting; continued advances in culture options and organism survival/growth
PCR	–	Availability limited to research laboratories; studies ongoing, including automated molecular panels

[a]–, not available or recommended for routine use; +, reported; ++, in general use (probably most widely used).

Table 9.15 Comparison of *Naegleria fowleri*, *Acanthamoeba* spp., *Balamuthia mandrillaris*, and *Sappinia diploidea*

Characteristic	*Naegleria fowleri*	*Acanthamoeba* spp.	*Balamuthia mandrillaris*	*Sappinia diploidea*
Trophozoite	Biphasic (amebic) and flagellate forms; 8 to 15 µm; lobate, rounded pseudopodia (amebic form)	Large (15 to 25 µm); no flagella; filiform, spiky pseudopodia	Large (15 to 60 µm); no flagella; extensive branching of pseudopodia	Large (40 to 70 µm); no flagella; distinctive double nucleus
Cysts	Not present in tissue; small, smooth, rounded	Present in tissue; large with wrinkled double wall	Present in tissue; large (15 to 30 µm), may be binucleate, irregular double wall	Bicellular cysts; not seen in first reported case of amebic encephalitis with this organism
Growth on media	Requires living cells (bacteria or cell culture); does not grow with >0.4% NaCl	May grow without bacteria; not affected by 0.85% NaCl	Will not grow well on bacterium-seeded nonnutrient agar plates; tissue culture recommended	Limited growth at 37°C, *Escherichia coli* nonnutrient agar, 72 h
Appearance in tissue[a]	Smaller than *Acanthamoeba;* dense endoplasm; less distinct nuclear staining	Large; rounded; less endoplasm; nucleus more distinct	Large; rounded; difficult to differentiate from *Acanthamoeba* spp.	Trophozoites ingested RBCs and stained brightly with Giemsa and PAS

[a]*Entamoeba histolytica* has a delicate nuclear membrane and a small, pale-staining nucleolus. Freshwater amebae have a distinct nuclear membrane and a large, deep-staining nucleolus.

Table 9.16 Characteristics of *Trichomonas vaginalis*

Characteristic	Feature in *T. vaginalis*
Shape and size	Pear shaped, 7–23 μm long (avg, 13 μm); width, 5–15 μm
Motility	Jerky, rapid
No. of nuclei and visibility	1; not visible in unstained mounts
No. of flagella (usually difficult to see)	3–5 anterior, 1 posterior
Other features	Seen in urine, urethral discharges, and vaginal smears; undulating membrane extends 1/2 length of body; no free posterior flagellum; axostyle easily seen
Infective stage	Trophozoite
Usual location	Vagina (male, urethra)
Striking clinical findings	Leukorrhea, pruritus vulva (thin white urethral discharge in male)
Other sites of infection	Urethra (prostate in male)
Stage usually recovered during clinical phase	Trophozoite only—no cyst
Other complications	Preterm birth, pelvic inflammatory disease, coinfection with other sexually transmitted infections, increased risk of acquisition or transmission of HIV
Diagnosis	Direct microscopy wet mounts poor; culture more sensitive, but many infections missed; NAATs now available and much more sensitive (some are automated, FDA cleared, and commercially available); other nonautomated POC options are also available, including high-sensitivity testing for both male and female urine specimens.

Table 9.17 Key characteristics of intestinal tract and urogenital system protozoa

Organism	Trophozoites	Cysts	Comments
Amebae			
Entamoeba histolytica (pathogenic)	Cytoplasm clean; presence of RBCs is diagnostic, but may also contain some ingested bacteria; peripheral nuclear chromatin is evenly distributed with central, compact karyosome.	Mature cyst contains 4 nuclei; chromatoidal bars have smooth, rounded ends. *E. histolytica* and *E. dispar* cannot be differentiated on the basis of cyst morphology.	Considered pathogenic; should be reported as *Entamoeba histolytica* if trophozoites contain ingested RBCs and should be reported to the public health department; trophozoites can be confused with macrophages and cysts with WBCs in the stool. Cysts and trophozoites containing no ingested RBCs should be reported as *Entamoeba histolytica/E. dispar* complex or group.
Entamoeba dispar (nonpathogenic) *Entamoeba moshkovskii* and *Entamoeba bangladeshi* resemble organisms within the *Entamoeba histolytica/ E. dispar* complex.	Morphology identical to that of *E. histolytica* (confirmed by presence of RBCs in cytoplasm). If no ingested RBCs present, specific fecal immunoassays are available for confirmation of the species designation.	Mature cyst identical morphology to that of *E. histolytica*. *E. histolytica* and *E. dispar* cannot be differentiated on the basis of cyst morphology.	Nonpathogenic; morphology resembles that of *E. histolytica*; these organisms (no ingested RBCs) should be reported as *Entamoeba histolytica/E. dispar*. Immunoassay reagents are now available to differentiate pathogenic *E. histolytica* and nonpathogenic *E. dispar*; some laboratories may decide to use these reagents on a routine basis, depending on positivity rate and cost.
Entamoeba hartmanni (nonpathogenic)	Looks identical to *E. histolytica*, but smaller (<12 µm); RBCs are not ingested.	Mature cyst contains 4 nuclei, but often stops at 2; chromatoidal bars often present and look like those found in *E. histolytica*; size, <10 µm.	Shrinkage occurs on the permanent stain (especially in the cyst form), *E. histolytica* may be below the 12- and 10-µm cutoffs; can be as much as 1.5 µm below the limits quoted for wet-preparation measurements.
Entamoeba coli (nonpathogenic)	Cytoplasm dirty, may contain ingested bacteria/debris; peripheral nuclear chromatin is unevenly distributed with a large, eccentric karyosome.	Mature cyst contains 8 nuclei, may see more; chromatoidal bars (if present) tend to have sharp, pointed ends.	If a smear is too thick or thin and if stain is too dark or light, then *E. histolytica* and *E. coli* can often be confused. There is much overlap in morphology.
Endolimax nana (nonpathogenic)	Cytoplasm clean, not diagnostic, with a great deal of nuclear variation; there may even be some peripheral nuclear chromatin. Normally only karyosomes are visible.	Cyst is round or oval with the 4 nuclear karyosomes being visible.	There is more nuclear variation in this ameba than any others. It can be confused with *Dientamoeba fragilis* and/or *E. hartmanni.*

Table 9.17 (*continued*)

Organism	Trophozoites	Cysts	Comments
Iodamoeba bütschlii (nonpathogenic)	Cytoplasm contains much debris; organisms usually larger than *E. nana* but may look similar; large karyosome. The basket nucleus may appear in the trophozoites.	Cyst contains single nucleus (may be a basket nucleus) with bits of nuclear chromatin arranged on the nuclear membrane (karyosome is the basket, bits of chromatin are the handle); large glycogen vacuole	Glycogen vacuole will stain brown with the addition of iodine in the wet preparation; basket nucleus more common in cyst but can be seen in trophozoite; vacuole may be so large, the cyst collapses on itself.
Flagellates	**Trophozoites**	**Cysts**	
Giardia lamblia (pathogenic)	Trophozoites are teardrop shaped from the front, like a curved spoon from the side; contain nuclei, linear axonemes, and curved median bodies.	Cysts are round or oval, containing multiple nuclei, axonemes, and median bodies.	Organisms live in the duodenum, and multiple stools may be negative; may have to use additional sampling techniques (aspiration, Entero-Test) or fecal immunoassays (much more sensitive than routine O&P exam). If fecal immunoassays used, two stool specimens should be tested before indicating that the patient is negative for *G. lamblia* antigen.
Chilomastix mesnili (nonpathogenic)	Trophozoites are teardrop shaped; cytostome must be visible for ID.	Cyst is lemon shaped with 1 nucleus and a curved fibril called a shepherd's crook.	Cyst can be identified much more easily than the trophozoite form. Trophozoites will look like some of the other small flagellates.
Dientamoeba fragilis (pathogenic)	Cytoplasm contains debris; may contain 1 or 2 nuclei (chromatin often fragmented into 4 dots). Flagella are internal and not seen in normal laboratory stains (requires EM).	Distinct cyst wall (~5 μm in diameter) with a clear zone around the cyst. A space is present between the cyst wall and the organism enclosed within the cyst wall. The nucleus is often fragmented into distinct granules of chromatin.	Tremendous size and shape range on a single smear; trophozoites with 1 nucleus can resemble *E. nana*.
Trichomonas vaginalis (pathogenic)	Supporting rod (axostyle) is present; undulating membrane comes halfway down the organism; small dots may be seen in the cytoplasm along the axostyle.	NO KNOWN CYST FORM	Recovered from genitourinary system; often diagnosed at bedside with wet preparation (motility).

Table continues on next page

Table 9.17 Key characteristics of intestinal tract and urogenital system protozoa (*continued*)

Organism	Trophozoites	Cysts	Comments
Pentatrichomonas hominis (nonpathogenic)	Supporting rod (axostyle) is present; undulating membrane comes all the way down the organism; small dots may be seen in the cytoplasm along the axostyle.	NO KNOWN CYST FORM	Recovered in stool; trophozoites may resemble other small flagellate trophozoites.
Ciliates	**Trophozoites**	**Cysts**	
Balantidium coli (pathogenic)	Very large trophozoites (50–100 μm long) covered with cilia; large bean-shaped nucleus present. Micronucleus is very difficult to see.	Morphology not significant with exception of large, bean-shaped nucleus	Rarely seen in the U.S.; causes severe diarrhea with large fluid loss; will be seen in proficiency testing specimens
Apicomplexa	**Trophozoites or tissue stages**	**Cysts or other stages in specimen**	
Cryptosporidium spp. (pathogenic)	Seen in intestinal mucosa (edge of brush border), gallbladder, and lung; seen in biopsy specimens	Oocysts seen in stool and/or sputum; organisms acid fast, measure 4–6 μm; hard to find if present in small numbers. Organism morphology does not differentiate between *C. parvum* (humans and animals) and *C. hominis* (humans).	Chronic infection in the compromised host (internal autoinfective cycle), self-cure in the immunocompetent host; numbers of oocysts correlate with stool consistency; can cause severe, watery diarrhea; oocysts are immediately infective when passed. Fecal immunoassays very sensitive (more so than special modified acid-fast stains)—one specimen is sufficient to rule in or out *Cryptosporidium* antigen in stool.
Cyclospora cayetanensis (pathogenic)	Biopsy specimens can be differentiated from those infected with other coccidia; based on the fact that patients are immunocompetent, biopsy specimens will probably rarely be required or requested.	Oocysts are seen in stool; approximately 8 to 10 μm in size; unsporulated, thus difficult to recognize as coccidia; mimic *Cryptosporidium* on modified acid-fast stained smears. Not all oocysts will stain red to purple, and some will not retain stain; thus, they are considered modified acid-fast variable (1% acid destain recommended; stronger destain will remove too much stain). Often oocysts will resemble wrinkled cellophane with no internal structure (do not contain any sporozoites when passed).	To date, most of these infections are associated with the immunocompetent individual but may also be seen in the immunosuppressed patient; may be associated with travelers' diarrhea. Safranin stains are also recommended. Many foodborne outbreaks documented, primarily with raspberries, strawberries, mesclun, basil, and snow peas (all of which were imported into the U.S.). Oocysts not infective when passed.

Table 9.17 (*continued*)

Organism	Trophozoites	Cysts	Comments
Cystoisospora belli (pathogenic)	Seen in intestinal mucosal cells; seen in biopsy specimens; does not seem to be as common as *Cryptosporidium* spp.	Oocysts seen in stool; organisms are acid fast; best technique is concentration, not permanent stained smear.	Thought to be the only *Cystoisospora* sp. that infects humans; oocysts are not immediately infective when passed.
Microsporidia	**Trophozoite or tissue stages**	**Cyst or other stage in specimen**	
Anncaliia *Encephalitozoon* *Enterocytozoon* *Microsporidium* *Nosema* *Pleistophora* *Trachipleistophora* *Tubulinosema* *Vittaforma* (pathogenic)	Developing stages sometimes difficult to identify; spores can be identified by size, shape, and presence of polar tubules.	Depending on the genus involved, spores could be identified in stool or urine using the modified trichrome stain, brightening reagent stains, or immunoassay reagents (not yet commercially available). Tissue stains (tissue Gram stains) are also used for biopsy specimens.	Spores are generally quite small (1 to 1.5 μm for *Enterocytozoon*) and can easily be confused with other organisms or artifacts (particularly in stool). These infections have most often been diagnosed in immunosuppressed patients; however, due to the lack of commercial fecal immunoassays, data on infections in the immunocompetent host are relatively sparse.

HELMINTHS

Table 9.18 Normal life spans of the most common intestinal nematodes

Nematode	Life span	Comments
Ascaris lumbricoides	1 year	Infection may be aborted by spontaneous passage of adult worms.
Enterobius vermicularis	Several months to years	Reinfection due to both self-infection and outside sources is extremely common.
Trichuris trichiura	Several years	Often accompanies *Ascaris* infection (both are acquired by egg ingestion from contaminated soil).
Hookworms		Symptoms are directly related to worm burden; many infections are asymptomatic.
Necator americanus	4–20 years	
Ancylostoma duodenale	5–7 years	
Strongyloides stercoralis	30+ years (current record is 55 years)	Autoinfection capability can lead to dissemination and the hyperinfection syndrome in the compromised host.

Table 9.19 Characteristics of the most common intestinal nematodes

Characteristic	*Ascaris lumbricoides*	*Enterobius vermicularis*	*Trichuris trichiura*	Hookworms: *Necator americanus*, *Ancylostoma duodenale*	*Trichostrongylus* spp.	*Strongyloides stercoralis*
Usual time to infective stage	2–3 weeks in soil; second-stage larva in egg	4–6 h; first-stage larva in egg	2–3 weeks in soil; first-stage larva in egg	5–7 days in soil; free, third-stage larva	3–5 days in soil; free, third-stage larva	5–7 days in soil; free, third-stage larva
Mode of infection	Ingestion of infective egg	Ingestion of infective egg	Ingestion of infective egg	Skin penetration by *Necator*; ingestion and skin penetration by *Ancylostoma*	Ingestion of third-stage larva	Skin penetration from larvae in the soil
Development and location in human host	Obligatory larval migration through heart, lungs, trachea (larvae swallowed), and intestine; adults in small intestine	Direct development to adult in intestinal tract; adults in cecum, appendix, colon, and rectum	Direct development to adult in intestinal tract; adults in cecum, appendix, and colon	Larval migration same as *Ascaris*; adults attached to mucosa of small intestine	Direct development to adult in intestinal tract; adults in small intestine	Larval same as *Ascaris*, adult females in mucosal epithelium of small intestine; autoinfection may occur
Prepatent period	2 months	3–4 weeks	3 months	6–8 weeks	2–3 weeks	2–4 weeks
Normal life span	Up to 1 year or slightly longer	1–2 months	15 or more years; usually 5–10 years	15 or more years; usually 5–10 years	Up to a year or slightly longer	Up to many years (30+)
Diagnosis by usual means	Bile-stained, mammillated, thick-shelled eggs (45–75 μm by 35–50 μm) in 1-cell stage in feces; infertile eggs (85–95 μm by 43–47 μm) have thinner shells/distorted mammillations; mature or immature adults may be found in feces or may spontaneously migrate out of anus, mouth, or nares	Smooth, thick-shelled eggs (50–60 μm by 20–32 μm) in cellulose tape preparations; may be seen in feces; adult or immature worms may be found in feces	Unembryonated, bile-stained, thick shelled eggs (50–54 μm by 20–23 μm) have mucoid plugs at each end; in feces	Thin-shelled eggs (56–75 μm by 36–40 μm) in 4- to 8-cell stage in feces; clear space seen between developing embryo and egg shell	Large, thin-shelled eggs (73–95 μm by 40–50 μm), tapered at one end, in feces; inner membrane of egg frequently wrinkled; egg already in advanced cleavage when passed	First-stage larvae (108–380 μm by 14–20 μm) in feces; rhabditiform larvae have a short buccal chamber and a prominent, conspicuous genital primordium

Table continues on next page

Table 9.19 Characteristics of the most common intestinal nematodes (*continued*)

Characteristic	*Ascaris lumbricoides*	*Enterobius vermicularis*	*Trichuris trichiura*	Hookworms: *Necator americanus*, *Ancylostoma duodenale*	*Trichostrongylus* spp.	*Strongyloides stercoralis*
Diagnostic problems	Fertile eggs may lose outer mammillated layer (decorticate eggs); infertile eggs may be difficult to recognize; also, will not float in usual solution of $ZnSO_4$ (sp gr, 1.18) used for concentration	Eggs not usually seen in feces; cellulose tape method should be used to demonstrate eggs from perianal region; six consecutive negative tapes required to rule out infection; often patients treated symptomatically	Rarely presents a problem; routine stool examination	Eggs of the two species are indistinguishable from one another; if eggs hatch in feces due to delay in examination, these first-stage rhabditiform larvae must be differentiated from *Strongyloides* larvae.	May be confused with hookworm eggs; however, measurements are different (hookworm eggs tend to be smaller and more rounded at both ends)	Larvae may be passed sporadically and may be found only by concentration procedures or use of Entero-Test or duodenal intubation; culture recommended; serology also useful but has about the same sensitivity as culture
Clinical notes	Due to potential migration of adult worms (fever, drugs, anesthetics), all infections should be treated; pulmonary symptoms may be present during larval migration (prior to egg recovery in the stool); eosinophils present, but not impressive	Generally, only symptomatic patients treated due to high reinfection rate. There may or may not be eosinophilia.	Light infections usually not treated; patients may be asymptomatic, with eggs an incidental finding; *Ascaris* and *Trichuris* infections often found together; moderate eosinophilia in heavy infections (usually does not exceed 15%)	Skin penetration by larvae produces allergic reaction (ground itch, CLM); pulmonary symptoms usually present only in heavy infection; iron deficiency, anemia, and eosinophilia up to 70% may be present	Rarely seen in U.S., common in Asia, Europe, Middle East, and Africa; light infections usually not treated	Hyperinfections may lead to death in the compromised or immunosuppressed host; patient may become symptomatic many years after original infection (without additional exposure) eosinophilia, 10–40% or higher

Table 9.20 Tissue nematodes

Name	Means of acquisition	Location in body	Symptoms	Diagnosis
Trichinella spiralis	Ingestion of raw or rare meats (pork, bear, walrus, horse, other carnivores and/or omnivores)	Active muscles contain encysted larvae (diaphragm, tongue, larynx, neck, ribs, biceps, gastrocnemius).	Diarrhea (larval migration through intestinal mucosa); nausea, abdominal cramps, general malaise. Muscle invasion: periorbital edema, pain, swelling, weakness, difficulties in swallowing, breathing, etc. Most severe symptom: myocarditis. High eosinophilia (20–90%)	Biopsy or autopsy specimen (muscle) compression smear or routine histology. Artificial digestion of muscle to release larvae. (Larvae are very infective and precautions should be taken.) Serologies can be very helpful.
Baylisascaris procyonis	Ingestion of viable eggs in the soil (most likely from raccoon feces)	Central nervous system and eye contain larvae.	Eosinophilic meningitis, unilateral neuroretinitis	Biopsy or autopsy specimen, routine histology. Eggs from raccoon measure 80 μm long by 65 μm wide, have a thick shell with a finely granulated surface; resemble *Ascaris lumbricoides* eggs
Lagochilascaris minor	Life cycle and route of human infection unknown; ingestion of viable eggs in the soil suspected	Adult worms, larvae, and eggs occur in life cycle within human lesions (neck, throat, nasal sinuses, tonsillar tissue, mastoids, brain, lungs).	Pustule, swelling, pus in lesions; chronic granulomatous inflammation	Identification of adult worms, larvae, or eggs from lesions, sinus tracts, biopsy or autopsy specimens
Toxocara canis and *T. cati* (visceral larva migrans)	Ingestion of infective eggs (dog/cat ascarids) from fecal material in the soil	Usually the liver; migratory pathway may include the lungs and even back to the intestine.	Migration of larvae may cause inflammation and granuloma formation. May be fever, hepatomegaly, pulmonary infiltrates, cough, neurological pathology. High eosinophilia (up to 90%; 20–50% common)	Confirmation at autopsy. Serologic test (ocular fluids as well as serum if eye involved)
Ancylostoma braziliense or *A. caninum* (CLM)	Skin penetration of filariform/infective larvae of dog/cat hookworms. Infection can also occur via ingestion of infective larvae.	Larval migration in the skin produces linear/raised/vesicular tracts. Can be on any area of the body	Intense itching, pneumonitis (if larvae migrate to deeper tissues)	Picture of linear tracts. Possible removal of larva from tunnel

Table continues on next page

Table 9.20 Tissue nematodes (*continued*)

Name	Means of acquisition	Location in body	Symptoms	Diagnosis
Dracunculus medinensis[a]	Ingestion of infected copepod/water flea (*Cyclops*)	Adult worms develop in deep connective tissue. Gravid female migrates to feet and ankles (can occur anywhere), where blister forms for larval deposition into the water through the ruptured blister on the skin.	Before blister formation: erythema, tenderness, urticarial rash, intense itching, nausea, vomiting, diarrhea, or asthmatic attacks. If secondary infection occurs there may be cellulitis, arthritis, myositis, etc.	Formation of cutaneous lesion with appearance of adult female worm depositing larvae into the water. Calcified worms can also be found on X ray.
Angiostrongylus cantonensis (eosinophilic meningitis)	Accidental ingestion of infective larvae in slugs, snails, or land planarians	Brain tissue, eye (rare), lung tissue (rare)	Severe headache, convulsions, limb weakness, paresthesia, vomiting, fever, eosinophilia up to 90%	Presumptive: severe headache, meningitis or meningoencephalitis, fever, ocular involvement. Definitive: examination of tissues (surgical specimens)
Angiostrongylus costaricensis	Accidental ingestion of slugs, often on contaminated salad vegetables	Bowel wall	Pain, tenderness, palpable tumor-like mass in right lower quadrant, fever, diarrhea, vomiting, eosinophilia (60%), and leukocytosis	Worm recovery and clinical history
Gnathostoma spinigerum	Ingestion of raw, poorly cooked, pickled freshwater fish or chicken (and other birds), frogs, or snakes	Migration of larvae in deep cutaneous or subcutaneous tissues (may appear anywhere), eye, or cerebrospinal fluid (less common)	Migratory swellings (hard, nonpitting) with inflammation, redness, pain	Worm recovery and clinical history
Anisakis spp. *Phocanema* spp. *Contracaecum* spp.	Ingestion of raw, pickled, salted, or smoked saltwater fish	Wall of gastrointestinal tract	Nausea, vomiting; may mimic gastric/duodenal ulcer, carcinoma, appendicitis; positive stool occult blood	Worm recovery and clinical history
Capillaria hepatica	Accidental ingestion of eggs from soil	Liver	May mimic hepatitis, amebic abscess, other infections involving the liver	Histologic identification
Thelazia spp.	Larval deposition by flies	Conjunctival sacs, migrating over cornea	Excessive lacrimation, itching, pain (feeling of foreign object in eye)	Worm recovery and identification (from eye)
Dirofilaria spp.	Larval deposition by mosquitoes in human subcutaneous tissues	Pulmonary tissue; dead worms produce coin lesions in lung; some in subcutaneous nodules (orbit, scrotum, breast, arm, leg); subconjunctival	Itching, burning sensation in eye; painless mass in lower eyelid; *Dirofilaria* should be considered in all cases of subcutaneous inflammatory or tumor-like lesion of unknown etiology.	Worm recovery and identification (surgical or autopsy specimens); positive filarial serology can be helpful but does not exclude carcinoma.

[a]*Dracunculus medinensis* is an ancient disease first mentioned over 4,500 years ago. Called the “fiery serpent” in the Bible, its modern name translates as “little dragon from Medina.” Some speculate that the single snake entwined around the staff of knowledge and wisdom in the Aesculapius symbol represents *Dracunculus*.

Table 9.21 *Trichinella spiralis*: life cycle stages and clinical conditions

Stage in the life cycle	Beginning of symptoms	Clinical condition[a]
Excysted larvae enter intestinal mucosa	2–4 h 24 h	Gastrointestinal symptoms
Worms mature, mate	30 h	
Females deposit larvae (muscle); invasion begins	Day 6 Day 7	Facial edema/fever
Heaviest muscle invasion	Day 10 Day 11	Maximum fever (40–41°C) Muscle inflammation, pain
Decrease in larval deposition	Day 14	Eosinophilia, antibody
Larvae differentiated	Day 17 Day 20	Maximum eosinophilia
Encapsulation of larvae	Day 21	Myocarditis, neurologic symptoms
Intestine free of adult worms	Day 23 Day 26	Respiratory symptoms
Encapsulation almost complete	Month 1 Month 2	Fever subsides
Adult worms die	Month 3	Death from myocarditis or encephalitis
Cyst calcification begins	Month 6 Month 8	Slow convalescence Myocarditis and neurologic symptoms subside
Cyst calcification usually complete	Year 1	
Most larvae still viable within calcified cyst	Year 6	

[a]Symptoms depend on worm burden; some patients with a very low load may exhibit no symptoms during the infection.

Table 9.22 Characteristics of human microfilariae

Species	Geographic area(s)	Vector	Features of microfilariae				
			Location	Periodicity	Sheath	Mean length (range) (μm)	Tail nuclei
Wuchereria bancrofti	Tropics and subtropics worldwide	Mosquito	Blood, hydrocele fluid	Nocturnal, subperiodic	+ Not stained using Giemsa	260 (244–296)	Nuclei do not extend to tail tip
Brugia malayi	Southeast Asia	Mosquito	Blood	Nocturnal, subperiodic	+ Stains pink using Giemsa	220 (177–230)	Subterminal and terminal nuclei
Brugia timori	Islands of Timor and Lesser Sunda (southwestern Pacific)	Mosquito	Blood	Nocturnal	+	310 (290–325)	Subterminal and terminal nuclei
Loa loa	Africa	Mango fly	Blood	Diurnal	+	275 (250–300)	Nuclei continuous to tail tip
Mansonella perstans	Africa and South America	Midge	Blood	None	–	195 (190–200)	Nuclei continuous to tail tip
Mansonella ozzardi	Central and South America	Midge	Blood	None	–	200 (173–240)	Nuclei do not extend to tail tip
Mansonella streptocerca	Africa	Midge	Skin	None	–	210 (180–240)	Nuclei in single row to tail tip, tail curved
Onchocerca volvulus	Africa and Central and South America	Blackfly	Skin	None	–	254 (221–287)	Nuclei do not extend to tail tip

Table 9.23 Characteristics of cestode parasites (intestinal)

Characteristic	*Diphyllobothrium latum*	*Taenia saginata*	*Taenia solium*	*Hymenolepis nana*	*Hymenolepis diminuta*	*Dipylidium caninum*
Intermediate hosts	Two: copepods and fish	One: cattle	One: hog	One: various arthropods (beetles, fleas); or none	One: various arthropods (beetles, fleas)	One: various arthropods (fleas, dog lice)
Mode of infection	Ingestion of plerocercoid (sparganum) in flesh of infected fish	Ingestion of cysticercus in infected beef	Ingestion of cysticercus in infected pork	Ingestion of cysticercoid in infected arthropod or by direct ingestion of egg; autoinfection may also occur	Ingestion of cysticercoid in infected arthropod	Ingestion of cysticercoid in fleas, lice
Prepatent period	3–5 weeks	3–5 months	3–5 months	2–3 weeks	~3 weeks	3–4 weeks
Normal life span	Up to 25 years	Up to 25 years	Up to 25 years	Perhaps many years due to autoinfection	Usually <1 year	Usually <1 year
Length	4–10 m	4–8 m	3–5 m	2.5–4.0 cm	20–60 cm	10–70 cm
Scolex	Spatulate, 3 by 1 mm; no rostellum or hooklets; has 2 shallow grooves (bothria)	Quadrate, 1- to 2-mm diameter; no rostellum or hooklets; 4 suckers	Quadrate, 1-mm diameter; has rostellum and hooklets, 4 suckers	Knoblike but not usually seen; has rostellum and hooklets; 4 suckers	Knoblike but not usually seen; has rostellum but no hooklets; 4 suckers	0.2–0.5 mm in diameter; has conical/retractile rostellum armed with 4–7 rows of small hooklets; 4 suckers
Usual means of diagnosis	Ovoid, operculate yellow-brown eggs (58–75 µm by 40–50 µm) in feces; egg usually has small knob at abopercular end; proglottids may be passed, usually in chain of segments (few cm to 0.5 m long); proglottids wider than long (3 by 11 mm) and have rosette-shaped central uterus	Gravid proglottids in feces; they are longer than wide (19 by 17 mm) and have 15–20 lateral branches on each side of central uterine stem; they usually appear singly; spheroidal yellow-brown, thick-shelled eggs (31–43 µm in diameter) containing an oncosphere may be found in feces	Gravid proglottids in feces; they are longer than wide (11 by 5 mm) and have 7–13 lateral branches on each side of central uterine stem; usually appear in chain of 5–6 segments; spheroidal, yellow-brown, thick-shelled eggs (31–43 µm) containing an oncosphere may be found in feces	Nearly spheroidal pale, thin-shelled eggs (30–47 µm in diameter) in feces; oncosphere surrounded by rigid membrane, which has two polar thickenings from which 4–8 filaments (polar filaments) extend into the space between the oncosphere and thin outer shell	Large, ovoid, yellowish, moderately thick-shelled eggs (70–85 µm by 60–80 µm) in feces; egg contains oncosphere, but no polar filaments	Gravid proglottids (8–23 mm long) containing compartmented cluster of eggs in feces; proglottids have genital pores at both lateral margins; occasionally may see individual oncospheres (20–33 µm in diameter) in feces

Table continues on next page

Table 9.23 Characteristics of cestode parasites (intestinal) (*continued*)

Characteristic	*Diphyllobothrium latum*	*Taenia saginata*	*Taenia solium*	*Hymenolepis nana*	*Hymenolepis diminuta*	*Dipylidium caninum*
Diagnostic problems or notes	Eggs are sometimes confused with eggs of *Paragonimus*; eggs are unembryonated when passed in feces	Eggs are identical to those of *Taenia solium;* ordinarily one can distinguish between species only by examination of gravid proglottids; eggs often confused with pollen grains (handle all proglottids with extreme care)	Eggs are identical to those of *T. saginata;* one is less likely to find eggs in feces than with *T. saginata* (handle all proglottids with extreme care, as *T. solium* eggs are infective to humans)	Sometimes confused with eggs of *Hymenolepis diminuta;* rodents serve as reservoir hosts	Should not be confused with *H. nana*, as eggs lack polar filaments; rodents serve as reservoir hosts	Gravid proglottids resemble rice grains (dry) or cucumber seeds (moist); dogs and cats serve as reservoir hosts

Table 9.24 Tissue cestodes

Characteristic	*Echinococcus granulosus*	*Echinococcus multilocularis*	*Echinococcus vogeli* *E. oligarthrus*	*Multiceps* spp.	*Spirometra* spp. *Diphyllobothrium* spp.
Disease	Hydatid disease (cystic)	Hydatid disease (alveolar)	Hydatid disease (polycystic)	Coenurosis	Sparganosis
Geographic location of the parasite	Worldwide	North America, northern and central Eurasia	Central and South America (85% in Brazil, Colombia, Ecuador, Argentina)	Worldwide	Worldwide, more common in China, Japan, Southeast Asia
Definitive host(s)	Domestic dog, wild canids (coyote, dingo, red fox, etc.)	Red fox, arctic fox, raccoon dog, coyote, domestic dog, cat	Bush dog, domestic dog	Dogs, other canids	Dogs, cats
Means of acquisition	Egg ingestion (dogs)	Egg ingestion (foxes, cats)	Egg ingestion (dogs [*E. vogeli*], felids [*E. oligarthrus*])	Egg ingestion (dogs, other canids)	(i) Ingestion of infected *Cyclops* spp. (procercoid); (ii) ingestion of raw infected flesh of amphibian, reptiles, birds, mammals (spargana); (iii) local application of raw infected flesh as a poultice (spargana)
Intermediate host(s)	Primarily ungulates (sheep, cattle, swine, horses), also marsupials	Rodents (voles, lemmings, shrews, mice), other small mammals	Paca, agouti, spiny rat	Sheep, goats, cattle, horses, lagomorphs, rodents	Frogs, mammals
Stage of organism found in tissue	Larval form (fluid-filled unilocular hydatid cyst), contains protoscolices and daughter cysts, limiting membrane; 1–>15 cm; visceral, primarily liver and lungs	Larval form (alveolar hydatid cyst), no limiting membrane, usually sterile with no protoscolices; visceral, primarily liver	Larval form (fluid-filled polycystic hydatid cyst), scolex visible in wet mounts, large hooklets (38–46 µm long) and small hooklets (30–37 µm long); visceral, liver, abdomen, lungs (*E. vogeli*); eye sockets, heart (*E. oligarthrus*), vesicles partitioned by septa, protoscolices present	Larval form (intermediate between cysticercus and hydatid cyst)	(i) Procercoid; (ii) spargana; (iii) spargana (see above)

Table continues on next page

Table 9.24 Tissue cestodes (*continued*)

Characteristic	*Echinococcus granulosus*	*Echinococcus multilocularis*	*Echinococcus vogeli* *E. oligarthrus*	*Multiceps* spp.	*Spirometra* spp. *Diphyllobothrium* spp.
Type of growth in humans	Concentric expansion	Exogenous proliferation, tumorlike, similar to metastatic growth	Exogenous and endogenous proliferation (*E. vogeli*), expansive, no indication of exogenous proliferation (*E. oligarthrus*).	Multiple scolices, but no daughter cysts	More elongate, wormlike structure, no suckers or hooks; resemble narrow tapeworm proglottids, motile
Location in body	Liver (60%), lung (20%), kidney (4%), muscle (4%), spleen (3%), soft tissues (3%), brain (3%), bones (2%), other (1%)	All sites as for *E. granulosus*; most common site is liver, metastases in lungs, brain, bones, etc.	Liver, lungs (15%) (*E. vogeli*); eye and heart (*E. oligarthrus*)	Most often in central nervous system	Most tissues have been involved; depends on the site of the poultice application
Symptoms	Depends on cyst location, usually mechanical from enlarging cyst; may also be allergic reactions from cyst fluid leakage	Hepatic disease resembles slowly growing mucoid carcinoma (no fever); hepato- and splenomegaly, jaundice, ascites	Hepatomegaly, palpable peritoneal masses, jaundice (*E. vogeli*); of 3 known human cases, 2 involved retro-ocular locations with exophthalmos (*E. oligarthrus*)	Like space-occupying lesion in the central nervous system, similar to tumor, rarely muscles or subcutaneous tissues	Edema, pain, irritation, inflammation, toxemia, eye damage, elephantiasis if lymphatics involved, slowly growing, tender, subcutaneous nodules (may be migratory), ocular sparganosis
Treatment	Surgical removal; albendazole, praziquantel	Albendazole, praziquantel; surgery not recommended	Surgery plus albendazole; difficult to treat	Rarely diagnosed preoperatively	Surgical removal and drainage

Table 9.25 Characteristics of intestinal trematodes

Species	Distribution	Agent of infection	Size (µm)	Egg morphology	Reservoir hosts	Comments
Fasciolopsis buski	Far East	Water chestnut, bamboo shoots, water caltrop	130–140 by 80–85	Unembryonated operculated	Dogs, pigs, rabbits	The less mature the egg, the more difficult it may be to see the operculum—it blends into the shell outline, and the breaks in the shell may be hard to identify. The egg has no opercular shoulders; thus, it is difficult to see where the operculum breaks occur.
Echinostoma ilocanum	Far East	Mollusks	86–116 by 59–69	Unembryonated operculated	Rats, dogs	Eggs are usually smaller than those of *F. buski*; some strains produce eggs that overlap in size; operculum may be hard to see.
Heterophyes heterophyes	Far East, Middle East	Freshwater fish	27–30 by 15–17	Embryonated operculated with opercular shoulders	Fish-eating mammals	Eggs have very inconspicuous opercular shoulders and, unlike *C. sinensis*, lack the "seated" operculum and knob at the abopercular end. Due to their small size, these eggs may be missed using the low power of the microscope; high dry power is recommended.
Metagonimus yokogawai	Far East, former Soviet Union, Israel, Spain	Freshwater fish	26–28 by 15–17	Embryonated operculated with opercular shoulders	Fish-eating mammals	Eggs have inconspicuous opercular shoulders, but a more obvious operculum than *H. heterophyes*. Due to their small size, eggs may be missed using low power; high dry power is recommended.
Gastrodiscoides hominis	Far East, Middle East, former Soviet Union	Freshwater fish	60–70 by 150	Unembryonated operculated	Pigs, deer mice, rats	Eggs are more slender than those of *F. buski*, but they are much alike; eggs have no opercular shoulders; difficult to see where the operculum breaks in the shell occur.

Table 9.26 Characteristics of liver and lung trematodes

Species	Geographic area(s)	Reservoir hosts	Agent of transmission (by ingestion)	Egg size (μm)	Characteristics
Clonorchis sinensis	Far East	Dogs, cats, other fish-eating mammals	Uncooked fish	28–35 by 12–19	Embryonated, operculated; very prominent opercular shoulders; comma appendage at abopercular end
Opisthorchis viverrini	Northern Thailand, Laos	Dogs, cats, other fish-eating mammals	Uncooked fish	19–29 by 12–17	Embryonated, operculated; prominent opercular shoulders; has a seated operculum and may or may not have a knob at the abopercular end; eggs tend to be broader with less prominent shoulders than *C. sinensis* eggs.
Opisthorchis felineus	Poland, Germany, Russian Federation, Kazakhstan, western Siberia	Dogs, cats, other fish-eating mammals	Uncooked fish	28–30 by 11–16	Embryonated, operculated; prominent opercular shoulders; has a seated operculum and may or may not have a knob at the abopercular end; eggs tend to be broader with less prominent shoulders than *C. sinensis* eggs.
Fasciola hepatica	Worldwide, mixed *F. hepatica* and *F. gigantica* infections have been reported from Pakistan.	Herbivores	Uncooked water plants	130–150 by 63–90	Unembryonated, operculated; the less mature the egg, the more difficult it is to see the operculum—it blends into the shell outline, and the breaks in the shell may be hard to identify. No opercular shoulders, so it is difficult to see where the operculum breaks in the shell occur. Eggs resemble those of *F. buski*, *E. ilocanum*, *F. gigantica*, and *G. hominis*; eggs may have thickening at abopercular end of the shell (unlike eggs of *F. buski*).

Table 9.26 (continued)

Species	Geographic area(s)	Reservoir hosts	Agent of transmission (by ingestion)	Egg size (µm)	Characteristics
Dicrocoelium dendriticum, Dicrocoelium hospes, Eurytrema pancreaticum	Europe, Turkey, northern Africa, Far East, China, Japan, North and South America	Cattle, sheep, deer, water buffalo	Ants, grasshoppers, crickets	38–45 by 22–30; dark brown	Embryonated, operculated, thick shell; eggs have a thick, dark brown shell and essentially no opercular shoulders; they cannot be differentiated from each other.
Paragonimus westermani	Far East, Africa	Dogs, cats, tigers, lions	Crabs, crayfish	80–120 by 45–65	Unembryonated, operculated; opercular shoulders; eggs have a moderately thick, dark golden-brown shell, a prominent operculum, opercular shoulders, and a thickened abopercular end; may be confused with *D. latum* eggs (smaller, abopercular knob, no opercular shoulders)
Paragonimus mexicanus[a]	Central and South America	Opossum, cats, dogs	Crabs	Avg, 79 by 48	Unembryonated, operculated; thin, irregular undulations on outer shell; eggs have prominent operculum, opercular shoulders, and a thickened abopercular end; may be confused with *D. latum* eggs (smaller abopercular knob, no opercular shoulders); eggs smaller than those of *P. westermani*; also golden-brown shell
Paragonimus kellicotti	North and South America	Mink, cat, dog, pig	Crabs, crayfish	75–118 by 48–68	Unembryonated, operculated; slight thickening at abopercular end; tapers more sharply than in *P. westermani*

Table 9.27 Human paragonimiasis

Species	Geographic distribution	Disease (source)	Avg egg size and characteristics
Paragonimus westermani[a]	Asia	Pulmonary (crabs, crayfish, freshwater shrimp)	85–100 by 47 μm; abopercular thickening
Paragonimus heterotremus[a]	China, Laos, Thailand	Pulmonary (crabs, raw shrimp salad in Thailand)	86 by 48 μm; uniform thickness
Paragonimus mexicanus[a] (*P. peruvianus*, *P. ecuadoriensis*[b])	Central and South America	Pulmonary (crabs)	79 by 48 μm; undulated shell
Paragonimus africanus	Nigeria, Cameroon	Pulmonary (crabs)	90 by 50 μm; abopercular thickening
Paragonimus kellicotti	North America	Pulmonary (crabs, crayfish)	75–118 by 48–68 μm; abopercular thickening; tapers more than *P. westermani*
Paragonimus miyazakii	Japan	Pleural (crabs, raw juice of crabs, crayfish)	75 by 43 μm; uniform thickness; eggs not normally seen
Paragonimus philippinensis	Philippines	Pulmonary (crabs)	79 by 50 μm; abopercular thickening
Paragonimus skrjabini	China	Pleural, subcutaneous nodules (crabs)	75 by 48 μm; uniform thickness
Paragonimus hueitungensis	China	Migratory, subcutaneous nodules (crabs)	75 by 46 μm; thin shell; visible opercular shoulders, small knob at abopercular end
Paragonimus uterobilateralis	Cameroon, Guinea, Liberia, Nigeria	Pulmonary (crabs)	70 by 45 μm; abopercular thickening of shell

[a]Pathogenic organism most frequently isolated from humans.
[b]Synonyms for *P. mexicanus*.

Table 9.28 Characteristics of blood trematodes

Species	Geographic areas	Reservoir hosts	Intermediate snail host genus	Diagnostic specimen	Egg size (μm)	Egg morphology	Comments
Schistosoma mansoni	Africa, Malagasy, West Indies, Suriname, Brazil, Venezuela	Humans, non-human primates	*Biomphalaria*	Stool, rectal biopsy, serology	114–180 by 45–73	Elongate, prominent lateral spine; acid-fast positive	In wet preparation, egg may be turned to lateral spine and not visible; evidence of flame cell activity is proof of viability; can use hatching test (unpreserved specimens) to confirm; occasionally found in urine
S. japonicum	China, Indonesia, Japan, Philippines	Dogs, cats, cattle, water buffalo, pigs	*Oncomelania*	Stool, rectal biopsy, serology	55–85 by 40–60	Oval, minute lateral spine; acid-fast positive	Eggs can mimic debris in the wet preparation; small spine difficult to see; debris often clings to surface of egg shell; less likely to be found in urine, but possible
S. mekongi	Mekong River basin	Humans, dogs, rodents	*Lithoglyphopsis*	Stool, rectal biopsy, serology	30–55 by 50–65	Oval, minute lateral spine	Eggs look very much like those of *S. japonicum*; lateral spine may be hard to see.
S. haematobium	Africa, Middle East, India, Portugal	Humans	*Bulinus*	Urine, stool (some cases), serology	112–170 by 40–70	Elongate, terminal spine; acid-fast negative	If present, easy to identify; membrane filter approach can be used (more effective than urine sedimentation); eggs occasionally found in stool; hatching test on urine (unpreserved) sediment is relevant.
S. intercalatum	Central and western Africa	Humans	*Bulinus*	Stool, rectal biopsy, serology	140–240 by 50–85	Elongate, terminal spine; acid-fast positive	Egg resembles *S. haematobium* but is found in stool rather than urine.

Table 9.29 Key characteristics of helminths[a]		
Helminths	**Diagnostic stage**	**Comments**
Nematodes (roundworms)		
Ascaris lumbricoides (pathogenic)	Eggs: Both fertilized (oval to round with thick, mammillated/tuberculated shell) and unfertilized (tend to be more oval/elongate with bumpy shell exaggerated) eggs can be found in stool. Adult worms: 10–12 in., found in stool. Rarely (in severe infections), migrating larvae can be found in sputum.	Unfertilized eggs do not float in flotation concentration method; adult worms have tendency to migrate when irritated (anesthesia, high fever); thus, patients from areas of endemicity should be checked for infection prior to elective surgery. Often dual infections with *Trichuris trichiura* may be seen.
Trichuris trichiura (whipworm) (pathogenic)	Eggs: Barrel shaped with two clear, polar plugs. Eggs should be quantitated (rare, few, etc.), since light infections may not be treated. Adult worms: Rarely seen	Dual infections with *Ascaris* sp. may be seen (both infections acquired from egg ingestion in contaminated soil); in severe infections, rectal prolapse may occur in children or bloody diarrhea can be mistaken for amebiasis (severe cases usually not seen in the U.S.).
Enterobius vermicularis (pinworm) (pathogenic)	Egg: Football shaped with one flattened side Adult worm: About 3/8 in. long, white with pointed tail; female migrates from the anus and deposits eggs on the perianal skin.	May cause symptoms in some patients (itching). Test of choice is adhesive tape preparation; 6 consecutive tapes necessary to rule out infection; symptomatic patients often treated without actual confirmation of infection; eggs become infective within a few hours; reinfection very common, particularly among children.
Ancylostoma duodenale (Old World hookworm) *Necator americanus* (New World hookworm) (pathogenic)	Eggs: Eggs of these species are identical, oval with broadly rounded ends, thin shell, clear space between shell and developing embryo (8- to 16-cell stage). Adult worms: Rarely seen in clinical specimens	May cause symptoms in some patients (blood loss anemia on the differential smear in heavy infections). If stool remains unpreserved for several hours or days, the eggs may continue to develop and hatch; rhabditiform larvae may resemble those of *Strongyloides stercoralis.*
Strongyloides stercoralis (pathogenic)	Rhabditiform larvae (noninfective) usually found in the stool (short buccal cavity or capsule with large, genital primordial packet of cells ("short and sexy"); in very heavy infections, larvae can occasionally be found in sputum and/or filariform (infective) larvae can be found in stool (slit in the tail).	May see unexplained eosinophilia, abdominal pain, unexplained episodes of sepsis and/or meningitis, pneumonia (migrating larvae) in the compromised patient. Potential for internal autoinfection can maintain low-level infections for many years after the patient has left the area of endemicity (patient will be asymptomatic with an elevated eosinophilia); hyperinfection can occur in the compromised patient (leading to disseminated strongyloidiasis and death).
Ancylostoma braziliensis (dog/cat hookworm) (pathogenic)	Humans are accidental hosts; larvae wander through the outer layer of skin, creating tracks (severe itching, eosinophilia); no practical microbiological diagnostic tests	Cause of CLM. Typical setup for infection: dogs and cats defecate in sandboxes, hookworm eggs hatch and penetrate human skin when in contact with infected sand/soil

Table 9.29 *(continued)*		
Helminths	**Diagnostic stage**	**Comments**
Toxocara cati or *T. caninum* (dog/cat ascarid) (pathogenic)	Humans are accidental hosts; ingestion of dog/cat ascarid eggs in contaminated soil; larvae wander through deep tissues (including the eye); can be mistaken for cancer of the eye; serologies helpful for confirmation; eosinophilia	Cause of VLM and OLM. Requests for laboratory services often originate in the ophthalmology clinic.
Cestodes (tapeworms)		
Taenia saginata (beef tapeworm) (pathogenic)	Scolex (4 suckers, no hooklets), gravid proglottid (>12 branches on a single side) are diagnostic; eggs indicate *Taenia* spp. only (thick, striated shell, containing a six-hooked embryo or oncosphere); worm usually around 12 ft in length	Adult worm can cause symptoms in some individuals. Ingestion of raw/poorly cooked beef; usually only a single worm per patient; individual proglottids may crawl from the anus; proglottids can be injected with India ink in order to see the uterine branches for identification.
Taenia solium (pork tapeworm) (pathogenic)	Scolex (4 suckers with hooklets), gravid proglottid (<12 branches on a single side) are diagnostic; eggs indicate *Taenia* spp. only (thick, striated shell, containing a six-hooked embryo or oncosphere), worm usually around 12 ft in length	Adult worm causes gastrointestinal complaints in some individuals; cysticercosis (accidental ingestion of eggs) can cause severe symptoms in the CNS. Ingestion of raw/poorly cooked pork; usually only a single worm per patient; occasionally 2 or 3 proglottids may be passed (hooked together); proglottids can be injected with India ink in order to see the uterine branches for identification; cysticerci are normally small and contained within an enclosing membrane; occasionally they may develop as the "racemose" type where the worm tissue grows in the body like a metastatic cancer.
Diphyllobothrium latum (broad fish tapeworm) (pathogenic)	Scolex (lateral sucking grooves), gravid proglottid (wider than long, reproductive structures in the center rosette); eggs are operculated but have no opercular shoulders.	Causes gastrointestinal complaints in some individuals. Ingestion of raw/poorly cooked freshwater fish; life cycle has 2 intermediate hosts (copepod, fish); worm may reach 30 ft in length; associated with vitamin B_{12} deficiency in genetically susceptible groups (Scandinavians)
Hymenolepis nana (dwarf tapeworm) (pathogenic)	Adult worm not normally seen; egg round to oval, thin shell, containing a six-hooked embryo or oncosphere with polar filaments lying between the embryo and eggshell	Causes gastrointestinal complaints in some individuals. Ingestion of eggs (only life cycle where intermediate host [grain beetle] can be bypassed); life cycle of egg to larval form to adult can be completed in the human; most common tapeworm in the world
Hymenolepis diminuta (rat tapeworm) (pathogenic)	Adult worm not normally seen; egg round to oval, thin shell, containing a six-hooked embryo or oncosphere with NO polar filaments lying between the embryo and eggshell	Uncommon; egg can be confused with *H. nana*. Eggs will be submitted in proficiency testing specimens and must be differentiated from those of *H. nana*.

Table continues on next page

Table 9.29 Key characteristics of helminths[a] (*continued*)		
Helminths	**Diagnostic stage**	**Comments**
Echinococcus granulosus (pathogenic)	Adult worm found only in the carnivore (dog); hydatid cysts develop (primarily in the liver) when humans accidentally ingest eggs from the dog tapeworms; cyst contains daughter cysts and many scolices; laboratory should examine fluid aspirated from cyst at surgery.	Humans are accidental intermediate hosts; normal life cycle is sheep/dog, with the hydatid cysts developing in the liver, lung, etc., of the sheep. Humans may be unaware of the infection unless fluid leaks from the cyst (can trigger an anaphylactic reaction) or pain is felt at the cyst location.
Echinococcus multilocularis (pathogenic)	Adult worm found only in carnivores (fox, wolf); hydatid cysts develop (primarily in the liver) when humans accidentally ingest eggs from the carnivore tapeworms; cyst grows like a metastatic cancer with no limiting membrane.	Humans are accidental intermediate hosts; prognosis in this infection is poor; surgical removal of the tapeworm tissue very difficult. Found in Canada, Alaska, and less frequently in the northern U.S. However, this infection is being seen more often and the geographic range is moving further south within the U.S.
Trematodes (flukes)		
Fasciolopsis buski (giant intestinal fluke) (pathogenic)	Eggs are found in stool; very large and operculated (morphology like that of *Fasciola hepatica* eggs)	Symptoms depend on worm burden; acquired from ingestion of plant material on which metacercariae have encysted (water chestnuts); worms hermaphroditic
Fasciola hepatica (sheep liver fluke) (pathogenic)	Eggs are found in stool; cannot be differentiated from those of *F. buski*	Symptoms depend on worm burden; acquired from ingestion of plant material on which metacercariae have encysted (watercress); worms hermaphroditic
Clonorchis sinensis (Chinese liver fluke) (pathogenic)	Eggs are found in stool; very small (<35 μm); operculated with shoulders into which the operculum fits	Symptoms depend on worm burden; acquired from ingestion of raw fish; eggs can be missed unless ×400 power is used for examination; eggs can resemble those of *Metagonimus yokogawai* and *Heterophyes heterophyes* (small intestinal flukes); worms hermaphroditic; linked to cholangiocarcinoma
Paragonimus westermani (lung fluke) (pathogenic)	Eggs are coughed up in sputum (brownish "iron filings" are the egg packets); can be recovered in sputum or stool (if swallowed); operculated with shoulders into which operculum fits with a thickened abopercular end	Symptoms depend on worm burden and egg deposition; acquired from ingestion of raw crabs or crayfish; eggs can be confused with those of *D. latum*; infections are seen in Asia (infections with *P. mexicanus* found in Central and South America, while infections with *P. kellicotti* are found in North America); worms hermaphroditic but often cross-fertilize with another worm if present.

Table 9.29 (continued)

Helminths	Diagnostic stage	Comments
Schistosoma mansoni (blood fluke) (pathogenic)	Eggs recovered in stool (large lateral spine); specimens should be collected with no preservatives (to detect egg viability); worms in veins of large intestine	Acquired from skin penetration of single cercariae from the freshwater snail; pathology caused by body's immune response to the presence of eggs in tissues; adult worms in veins cause no problems; adult worms are separate sexes. Occasionally, eggs may be found in urine; both urine and stool should be examined in any patient suspected of having schistosomiasis.
Schistosoma haematobium (blood fluke) (pathogenic)	Eggs recovered in urine (large terminal spine); specimens should be collected with no preservatives (to detect egg viability); in veins of bladder	Acquired from skin penetration of single cercariae from the freshwater snail; pathology caused by worms is like that with *S. mansoni*; 24-h and spot urine samples should be collected; chronic infection has association with bladder cancer; adult worms are separate sexes. Occasionally, eggs may be found in stool; both urine and stool should be examined in any patient suspected of having schistosomiasis.
Schistosoma japonicum (blood fluke) (pathogenic)	Eggs recovered in stool (very small lateral spine); specimens should be collected with no preservatives (to detect egg viability); worms in veins of small intestine	Acquired from skin penetration of multiple cercariae from the freshwater snail; pathology as with *S. mansoni*; infection usually most severe of the three due to original loading infective dose of cercariae from the freshwater snail (multiple cercariae stick together). Pathology is associated with egg production, which is greatest in *S. japonicum* infections.

[a]1 in. = 2.54 cm; 1 ft = 30.48 cm.

BLOOD PARASITES

Table 9.30 Malaria characteristics with fresh blood or EDTA-blood[a]
Plasmodium vivax (benign tertian malaria) 1. 48-h cycle, exoerythrocytic cycle persists 2. Tends to infect young cells 3. Enlarged RBCs 4. Schüffner's dots (true stippling) after 8–10 h 5. Delicate ring 6. Very ameboid trophozoite 7. Mature schizont contains 12–24 merozoites
Plasmodium malariae (quartan malaria) 1. 72-h cycle (long incubation period) 2. Tends to infect old cells 3. Normal size RBCs 4. No stippling 5. Thick ring, large nucleus 6. Trophozoite tends to form bands across the cell 7. Mature schizont contains 6–12 merozoites
Plasmodium ovale 1. 48-h cycle, exoerythrocytic cycle persists 2. Tends to infect young cells 3. Enlarged RBCs with fimbriated edges (oval); usually one end of RBC only (different from crenated RBCs) 4. Schüffner's dots appear in the beginning (in RBCs with very young ring forms, in contrast to *P. vivax*) 5. Smaller ring than *P. vivax* 6. Trophozoite less ameboid than that of *P. vivax* 7. Mature schizont contains average 8 merozoites
Plasmodium falciparum (malignant tertian malaria) 1. 36- to 48-h cycle 2. Tends to infect any cell regardless of age; thus, very heavy infection may result 3. All sizes of RBCs 4. No Schüffner's dots (Maurer's dots: may be larger, single dots, bluish) 5. Multiple rings/cell (only young rings, gametocytes, and occasional mature schizonts are seen in peripheral blood) 6. Delicate rings, may have two dots of chromatin/ring, appliqué or accolé forms 7. Crescent-shaped gametocytes
Plasmodium knowlesi (simian malaria)[b] 1. 24-h cycle 2. Tends to infect any cell regardless of age; thus, very heavy infection may result 3. All sizes of RBCs, but most tend to be normal size 4. No Schüffner's dots (faint, clumpy dots later in cycle) 5. Multiple rings/cell (may have 2 or 3) 6. Delicate rings, may have two or three dots of chromatin/ring, appliqué forms 7. Band form trophozoites commonly seen 8. Mature schizont contains 16 merozoites, no rosettes 9. Gametocytes round, tend to fill the cell

[a]Fresh blood or blood collected using EDTA with no extended lag time (preparation of thick and thin blood films within 60 min of collection).
[b]Early stages mimic *P. falciparum*; later stages mimic *P. malariae*.

Table 9.31 Potential problems with using EDTA anticoagulant for the preparation of thin and thick blood films[a]

Potential problem	Comments
Adhesion to the slide; blood falls off slide during staining	Incorrect ratio of anticoagulant to blood; fill tube completely with blood (7 ml or pediatric draw tube).
Distortion of parasites; same type of distortion can also be seen after blood is refrigerated (not recommended).	Prolonged storage of blood in EDTA may lead to distortion: trophozoites (*P. vivax*) and gametocytes (*P. falciparum*) tend to round up, thus mimicking *P. malariae*.
Change in ring form size	Ring forms of *P. falciparum* will continue to enlarge, thus resembling rings of the other species. Typical "small" rings will appear larger than usual.
Use of EDTA anticoagulant (used primarily by the hematology laboratory because the cellular components and morphology of the blood cells are preserved). 1. Blood smears for differentials from acceptable specimens should be prepared within 2 h of collection. 2. Blood counts from acceptable venipuncture specimens should be performed within 6 h of collection. 3. **FOR NORMAL PARASITE MORPHOLOGY, THICK AND THIN BLOOD FILMS FOR BLOOD PARASITES (PARTICULARLY AGENTS OF MALARIA) SHOULD BE PREPARED WITHIN 60 MIN OR LESS FROM THE TIME OF BLOOD COLLECTION. PARASITES WILL BEGIN TO DISAPPEAR WITHIN 4–6 HOURS FROM BLOOD HELD IN EDTA.** Underfilling the EDTA blood collection tube can lead to erroneously low blood cell counts and hematocrits, morphologic changes to RBCs, and staining alteration. Excess EDTA can shrink RBCs. Conversely, overfilling the blood collection tube will not allow the tube to be properly mixed and may lead to platelet clumping and clotting.	EDTA prevents coagulation of blood by chelating calcium. Calcium is necessary in the coagulation cascade, and its removal inhibits and stops a series of events, both intrinsic and extrinsic, which cause clotting. In some individuals, EDTA may cause inaccurate platelet results. These anomalies, platelet clumping and platelet satellitism, may be the result of changes in the membrane structure occurring when the calcium ion is removed by the chelating agent, allowing the binding of preformed antibodies. Proper mixing of the whole-blood specimen ensures that EDTA is dispersed throughout the sample. Evacuated blood collection tubes with EDTA should be mixed by 8–10 end-over-end inversions immediately following venipuncture collection. Microcollection tubes with EDTA should be mixed by 10 complete end-over-end inversions immediately following collection. They should then be inverted an additional 20 times prior to analysis.
Loss of Schüffner's dots (stippling) in *P. vivax* and *P. ovale*	Schüffner's dots (true stippling) occur in both *P. vivax* and *P. ovale*; in the absence of stippling, identification to the species level may not be possible; problems may be related to buffer pH and/or storage of blood in EDTA for more than 1 h.
Prolonged storage of EDTA blood (room temperature with stopper removed)	The pH, CO_2 level, and temperature changes may mimic conditions within the mosquito. Thus, exflagellation of the male gametocyte may occur while still in the tube of blood prior to thin and thick blood film preparation. Microgametes may be confused with *Borrelia* or may be ignored as debris.
Release of merozoites from the schizonts into the blood. Normally, merozoites are not found outside the RBCs, in contrast to *Babesia* spp., where rings may be seen outside the RBCs. In severe malaria infections, rings outside the RBCs may be seen.	Small rings may be seen outside the RBCs or appear to be appliqué forms, thus suggesting *P. falciparum*. It is important to differentiate these true rings (both cytoplasmic and nuclear colors) from platelets (uniform color).
Incorrect submission of blood in heparin	EDTA has less impact on parasite morphology than heparin.

[a]If blood films are prepared within recommended time guidelines (60 min or less from time of collection in EDTA), both thick and thin films should provide typical parasite morphology; however, if blood film preparation is delayed, many artifacts will be seen and parasites will begin to disappear from the blood after ~4 to 6 h.

Table 9.32 Plasmodia in Giemsa-stained thin blood smears[a]

Characteristic	*Plasmodium vivax*	*Plasmodium malariae*	*Plasmodium falciparum*	*Plasmodium ovale*	*Plasmodium knowlesi*
Persistence of exo-erythrocytic cycle	Yes	No	No	Yes	No
Relapses	Yes	No, but long-term recrudescences are recognized	No long-term relapses	Possible, but usually spontaneous recovery	Recrudescence
Time of cycle	44–48 h	72 h	36–48 h	48 h	24 h
Appearance of parasitized RBCs; size and shape	1.5–2 times larger than normal; oval to normal; may be normal size until ring fills 1/2 of cell; young RBCs	Normal shape; size may be normal or slightly smaller; old RBCs	Both normal; all RBCs	60% of cells larger than normal and oval; 20% have irregular, frayed edges; young RBCs	Stages mimic those seen with *P. malariae*; all RBCs; normal-size RBCs
Schüffner's dots (eosinophilic stippling)	Usually present in all cells except early ring forms	None	None; occasionally comma-like red dots are present (Maurer's dots)	Present in all stages, including early ring forms; dots may be larger and darker than in *P. vivax*	No true stippling; may be clumpy dots later on in cycle (golden-brown with Giemsa stain)
Color of cytoplasm	Decolorized, pale	Normal	Normal, bluish tinge at times	Decolorized, pale	Similar to *P. malariae*
Multiple rings/cell	Occasional	Rare	Common	Occasional	Common
Developmental stages present in peripheral blood	All stages present	Ring forms few, as ring stage is brief; mostly growing and mature trophozoites and schizonts	Young ring forms and no older stages; few gametocytes	All stages present	All stages present
Appearance of parasite; young trophozoite (early ring form)	Ring is 1/3 diameter of cell, cytoplasmic circle around vacuole; heavy chromatin dot	Ring often smaller than in *P. vivax*, occupying 1/8 of cell; heavy chromatin dot; vacuole at times filled in; pigment forms early	Delicate, small ring with small chromatin dot (frequently 2); scanty cytoplasm around small vacuoles; sometimes at edge of RBC (appliqué form) or filamentous slender form; may have multiple rings per cell	Ring is larger and more ameboid than in *P. vivax*, otherwise similar to *P. vivax*	Early rings mimic *P. falciparum*; later stages mimic *P. malariae;* delicate rings, may have two dots of chromatin/ring, appliqué or accolé forms, two rings/RBC

Table 9.32 (*continued*)

Characteristic	*Plasmodium vivax*	*Plasmodium malariae*	*Plasmodium falciparum*	*Plasmodium ovale*	*Plasmodium knowlesi*
Growing trophozoite	Multishaped irregular ameboid parasite; streamers of cytoplasm close to large chromatin dot; vacuole retained until close to maturity; increasing amounts of brown pigment	Nonameboid rounded or band-shaped solid forms; chromatin may be hidden by coarse dark brown pigment.	Heavy ring forms; fine pigment grains	Ring shape maintained until late in development; nonameboid compared to *P. vivax*	Mimics *P. malariae*; band-form trophozoites common; may be highly ameboid
Mature trophozoite	Irregular ameboid mass; 1 or more small vacuoles retained until schizont stage; fills almost entire cell; fine brown pigment	Vacuoles disappear early; cytoplasm compact, oval, band shaped, or nearly round almost filling cell; chromatin may be hidden by peripheral coarse dark brown pigment	Not seen in peripheral blood (except in severe infections); development of all phases following ring form occurs in capillaries of viscera	Compact; vacuoles disappear; pigment dark brown, less than in *P. malariae*	Band forms common; mimics *P. malariae*
Schizont (presegmenter)	Progressive chromatin division; cytoplastic bands containing clumps of brown pigment	Similar to *P. vivax* except smaller; darker, larger pigment granules peripheral or central	Not seen in peripheral blood (see above)	Smaller and more compact than *P. vivax*	Similar to *P. malariae*
Mature schizont	Merozoites, 16 (12 to 24), each with chromatin and cytoplasm, filling entire RBC, which can hardly be seen	8 (6 to 12) merozoites in rosettes or irregular clusters filling normal-sized cells, which can hardly be seen; central arrangement of brown-green pigment	Not seen in peripheral blood	3/4 of cells occupied by 8 (8 to 12) merozoites in rosettes or irregular clusters	Merozoites, 16; may not be seen in peripheral blood
Macrogametocyte	Rounded or oval homogeneous cytoplasm; diffuse delicate light brown pigment throughout parasite; eccentric compact chromatin	Similar to *P. vivax*, but fewer; pigment darker and coarser	Sex differentiation difficult; crescent or sausage shapes characteristic; may appear in showers, black pigment near chromatin dot, which is often central	Smaller than *P. vivax*	Round; may mimic *P. vivax*

Table continues on next page

Table 9.32 Plasmodia in Giemsa-stained thin blood smears[a] (*continued*)

Characteristic	*Plasmodium vivax*	*Plasmodium malariae*	*Plasmodium falciparum*	*Plasmodium ovale*	*Plasmodium knowlesi*
Microgametocyte	Large pink to purple chromatin mass surrounded by pale or colorless halo; evenly distributed pigment	Similar to *P. vivax*, but fewer; pigment darker and coarser	Same as macrogametocyte (described above)	Smaller than *P. vivax*	Round; may mimic *P. vivax*
Main criteria	Large pale RBC; trophozoite irregular; pigment usually present; Schüffner's dots not always present; several phases of growth seen in one smear; gametocytes appear as early as third day; relapse	RBC normal in size and color; trophozoites compact, stain usually intense, band forms not always seen; coarse pigment; no stippling of RBCs; gametocytes appear after a few weeks; recrudescence	Development following ring stage takes place in blood vessels of internal organs; delicate ring forms and crescent-shaped gametocytes are only forms normally seen in peripheral blood; gametocytes appear after 7–10 days; recrudescence	RBC enlarged, oval, with fimbriated edges; Schüffner's dots seen in all stages; gametocytes appear after 4 days or as late as 18 days; relapse (controversial)	RBC normal size; early trophozoites and mature schizonts not always seen; most infections misidentified as *P. malariae*; early rings resemble *P. falciparum*; no. of infected RBCs not limited; clumpy golden-brown granules; recrudescence; Southeast Asia, Malaysia

[a]Other blood stains are perfectly acceptable; if the PMNs look acceptable/normal on the stained blood films, any parasites present will also exhibit typical morphology. Other acceptable stains include Wright's stain, Wright-Giemsa combination stain, Field's stain, and rapid stains (Diff-Quik [various manufacturers] and Wright's Dip Stat stain set [Medical Chemical Corp., Torrance, CA]). Color variation is normal, even with Giemsa stain.

Table 9.33 Relevant issues for handling requests for identification of infectious blood parasites	
Process	**Comments**
Test ordering	Should always be handled as a stat procedure; the most relevant orders involve blood collection for thick and thin blood film preparation; since suspected organisms are frequently not known, this stat approach should apply to all orders. If *Plasmodium* spp. are suspected, remember that collection should not depend on the patient's fever pattern; a true pattern is often not seen in patients with an early infection. The rule is to draw the blood immediately; do not wait for a fever spike that may never come.
Issues related to stat requests	Patients with malaria may appear for diagnostic blood work when least expected. Laboratory personnel should be aware of the stat nature of such requests and the importance of obtaining some specific patient history information. The typical textbook presentation of the blood smears may not be seen by the technologist. It is very important that the smears be examined at length and under oil immersion. The most important thing to remember is that even though a low parasitemia may be present on the blood smears, the patient may still be faced with a serious or even life-threatening disease.
Relevant patient information	When requests for malarial smears are received in the laboratory, some patient history information should be made available to the laboratorian. This information should include the following. Often, the physician may have to be contacted. 1. Where has the patient been, and what was the date of return to the United States? ("Where do you live?" —this has relevance to "airport" malaria.) 2. Has malaria ever been diagnosed in the patient before? If so, what species was identified? 3. What medication (prophylaxis or otherwise) has the patient received, and how often? When was the last dose taken? 4. Has the patient ever received a blood transfusion? Is there a possibility of other needle transmission (drug user)? 5. When was the blood specimen drawn, and was the patient symptomatic at the time? Is there any evidence of a fever periodicity? Answers to such questions may help eliminate the possibility of infection with *P. falciparum* or *P. knowlesi*, usually the only species that can rapidly lead to death.
Specimen collection	Collect finger-stick blood (no longer performed in most U.S. laboratories) or draw blood into an EDTA (lavender-top) tube. Make sure the tube is filled, thoroughly mixed, and immediately transported to the laboratory (all in a stat manner).
Specimen receipt by the laboratory	Check information for name, birthdate, etc.; remember that the suspected infection is rarely included. The order may specify blood for possible parasites (generally referring to thick and thin blood films). If a tube with a green top (heparin) is accidentally submitted, do not reject the specimen. Since the order is stat, process the heparin tube and immediately submit a request for a lavender-top EDTA tube (stat collection).
Specimen processing	Immediately prepare several thick and thin blood films (number will depend on laboratory-recommended numbers). DO NOT remove the EDTA tube stopper and let the blood stand for an hour or more before processing; dramatic morphology changes and/or parasite loss will occur if *Plasmodium* organisms are present. Morphology changes occur within 1–2 h, and parasite loss occurs at 4–6 h. Automation is not recommended; low-parasitemia infections will be routinely missed.
Life cycle within a tube of blood (EDTA)	Unfortunately, when blood tubes are held at room temperature, especially with the cap removed, the process of exflagellation can occur within the blood. Subsequently, when the blood films are prepared, these microgametocytes can be confused with *Borrelia* spp. While the organisms are within the tube of blood, the life cycle continues, unlike the situation when finger-stick blood films are prepared. Thus, any delay in thick and thin blood film preparation can result in a loss of organisms, as well as multiple artifact problems.

Table continues on next page

Table 9.33 Relevant issues for handling requests for identification of infectious blood parasites (*continued*)

Process	Comments
Stain options	Although for many years, Giemsa stain has been the stain of choice, the parasites can also be seen on blood films stained with Wright's stain, a Wright-Giemsa combination stain, or one of the more rapid stains, such as Diff-Quik (various manufacturers), Wright's Dip Stat stain (Medical Chemical Corp., Torrance, CA), or Field's stain.
Quality control	Wright's, Wright-Giemsa, Giemsa, or a rapid stain can be used. The majority of the original organism descriptions were based on Giemsa stain. However, if the WBCs appear to be well stained, any blood parasites present will also be well stained. The WBCs on the patient smear serve as the QC organism; there is no need to use a *Plasmodium*-positive slide for QC.
Stained blood film examination	It is recommended that both thick and thin blood films be prepared on admission of the patient, and at least 200 to 300 oil immersion fields should be examined on both films before a negative report is issued. The number of oil immersion fields examined may have to be increased if the patient has had any prophylactic medication during the past 48 h (the number of infected cells may be decreased on the blood films). Since one set of negative films will not rule out malaria, additional blood specimens should be examined over a 36-h time frame. Although a 60× oil immersion objective may be used for screening the blood films, 200–300 oil immersion fields should be examined using the 100× oil immersion objective before reporting the specimen as negative.
Parasitemia	*Quantitate organisms from every positive blood specimen.* The same method for calculating parasitemia should be used for each subsequent positive blood specimen.
Report comments	**Report comments can be extremely helpful in conveying information to the physician. Depending on the results of diagnostic testing, the following information can lead to improved patient care and clinical outcomes. Several representative comments are seen below.** 1. **No parasites seen:** The submission of a single blood specimen will not rule out malaria; submit additional blood samples every 4–6 h for 3 days if malaria remains a consideration. *Interpretation/Discussion*: It is important to make sure the physician knows that examination of a single blood specimen will not rule out malaria. 2. ***Plasmodium* spp. seen:** Unable to rule out *Plasmodium falciparum* or *Plasmodium knowlesi.* *Interpretation/Discussion:* Since *P. falciparum* and *P. knowlesi* cause the most serious illness, it is important to let the physician know these species have NOT been ruled out. 3. ***Plasmodium* spp., possible mixed infection:** Unable to rule out *Plasmodium falciparum* or *Plasmodium knowlesi.* *Interpretation/Discussion:* Since *P. falciparum* and *P. knowlesi* cause the most serious illness, it is important to let the physician know these species have NOT been ruled out. 4. ***Plasmodium malariae*:** Unable to rule out *Plasmodium knowlesi.* *Interpretation/Discussion:* If the patient has traveled to the area where *P. knowlesi* is endemic, it may be impossible to differentiate between *P. malariae* (band forms) and *P. knowlesi.* 5. ***Plasmodium falciparum*:** Unable to rule out *Plasmodium knowlesi.* *Interpretation/Discussion:* If the patient has traveled to the area where *P. knowlesi* is endemic, it may be impossible to differentiate between *P. falciparum* (ring forms) and *P. knowlesi.* 6. **Negative for parasites using automated hematology analyzer:** Automated hematology analyzers will not detect low malaria parasitemias seen in immunologically naive patients (travelers). *Interpretation/Discussion:* Patients who have never been exposed to malaria (immunologically naive) will become symptomatic with very low parasitemias that will not be detected using automation (0.001 to 0.0001%).

Table 9.33 (*continued*)

Process	Comments
	7. **Negative for malaria using the BinaxNOW rapid test:** This result does not rule out the possibility of a malaria infection. Blood should be submitted for stat thick and thin blood film preparation, examination, and reporting. *Interpretation/Discussion:* The maximum sensitivity of this rapid test occurs at 0.1% parasitemia. Patients (immunologically naive travelers) may present to the emergency room or clinic with a parasitemia much lower than 0.1%, leading to a false-negative report. Also, this rapid test is not designed to identify *P. malariae*, *P. ovale*, and *P. knowlesi*; the results are most clinically relevant for *P. falciparum* and *P. vivax*. The BinaxNOW test is FDA cleared for use within the United States.

Table 9.34 Features of human leishmanial infections[a]

Species	Disease type	Humoral antibodies	Delayed hypersensitivity	Parasite quantity	Self cure	Recommended specimen
Leishmania donovani	VL	Abundant	Absent	Absent	Rare	Bone marrow, spleen
	CL	Variable	Present	Present	Yes	Skin macrophages
	DL	Variable	Variable	Variable	Variable	Skin macrophages
Leishmania tropica	CL	Variable	Present	Present	Yes	Skin macrophages
Leishmania major	CL	Present	Present	Present	Rapid	Skin macrophages
Leishmania aethiopica	CL	Variable	Weak	Present	Slow	Skin macrophages
	DCL	Variable	Absent	Abundant	No	Skin macrophages
Leishmania mexicanus	CL	Variable	Present	Present	Yes	Skin macrophages
	DCL	Variable	Absent	Abundant	No	Skin macrophages
Leishmania braziliensis	CL	Present	Present	Present	Yes	Skin macrophages
	MCL	Present	Present	Scant	No	Skin macrophages
	ATL, TL	Present	Present	Variable	Variable	Skin macrophages

[a]For culture, specimens must be collected aseptically; in older lesions, the parasites may be scant and difficult to recover. For isolation, hamsters or culture media (Novy-MacNeal-Nicolle medium and Schneider's *Drosophila* medium with 30% fetal bovine serum) may be used. Serology is most suitable for visceral leishmaniasis but has little value for cutaneous leishmaniasis and limited value for mucocutaneous leishmaniasis. The Montenegro test assesses a delayed hypersensitivity reaction to intradermal injection of cultured parasites. ATL, American tegumentary leishmaniasis; CL, cutaneous leishmaniasis; DCL, diffuse cutaneous leishmaniasis; DL, dermal leishmaniasis; MCL, mucocutaneous leishmaniasis; TL, tegumentary leishmaniasis; VL, visceral leishmaniasis.

Table 9.35 Characteristics of American trypanosomiasis		
Characteristic	**Causative organism**	
	Trypanosoma cruzi	***Trypanosoma rangeli***
Vector	Reduviid bug	Reduviid bug
Primary reservoirs	Opossums, dogs, cats, wild rodents	Wild rodents
Illness	Symptomatic (acute, chronic)	Asymptomatic
Diagnostic stage		
Blood	Trypomastigote	Trypomastigote
Tissue	Amastigote	None
Recommended specimen(s)	Blood, lymph node aspirate, chagoma	Blood

Table 9.36 Characteristics of East and West African trypanosomiasis

Characteristic	East African	West African
Organism	*Trypanosoma brucei rhodesiense*	*Trypanosoma brucei gambiense*
Vector	Tsetse fly, *Glossina morsitans* group	Tsetse fly, *Glossina palpalis* group
Primary reservoirs	Animals	Humans
Illness	Acute (early CNS invasion), <9 months	Chronic (late CNS invasions), months to years
Lymphadenopathy	Minimal	Prominent
Parasitemia	High	Low
Epidemiology	Anthropozoonosis, game parks	Anthroponosis, rural populations
Diagnostic stage	Trypomastigote	Trypomastigote
Recommended specimens	Chancre aspirate, lymph node aspirate, blood, CSF	Chancre aspirate, lymph node aspirate, blood, CSF

Table 9.37 Key characteristics of blood parasites[a]

Protozoan	Diagnostic stage	Comments
Malaria agents		
Plasmodium vivax (benign tertian malaria)	Ameboid rings; presence of Schüffner's dots, beginning in older rings (appear later than those in *Plasmodium ovale*); all stages seen in peripheral blood; enlarged RBCs; mature schizont contains 16–18 merozoites.	Infects young cells; 48-h cycle; large geographic range; tends to result in true relapse from residual liver stages (hypnozoites)
Plasmodium ovale (ovale malaria)	Nonameboid rings; presence of Schüffner's dots, beginning in young rings (appear earlier than those in *Plasmodium vivax*); all stages seen in peripheral blood; enlarged RBCs; mature schizont contains 8–10 merozoites; RBCs may be oval and have fimbriated edges.	Infects young cells; 48-h cycle; narrow geographic range; tends to result in true relapse from residual liver stages (hypnozoites)
Plasmodium malariae (quartan malaria)	Rings are thick; no stippling; all stages seen in peripheral blood; presence of band forms and rosette-shaped mature schizont containing approximately 8 merozoites; normal to small RBCs; lots of malarial pigment	Infects old cells; 72-h cycle; narrow geographic range; associated with recrudescence and nephrotic syndrome; no true relapse

Table continues on next page

Table 9.37 Key characteristics of blood parasites[a] (*continued*)		
Protozoan	**Diagnostic stage**	**Comments**
Plasmodium falciparum (malignant tertian malaria)	Multiple rings; appliqué/accolé forms; no stippling (rare Maurer's clefts); rings and crescent-shaped gametocytes seen in peripheral blood (no other developing stages; rare exception, mature schizont); no limit on number of RBCs that can be infected	Infects all cells, 36- to 48-h cycle; large geographic range; no true relapse; most pathogenic of five species; plugged capillaries can cause severe symptoms/sequelae (cerebral malaria, lysis of RBCs, etc.).
Plasmodium knowlesi (simian malaria)	Multiple rings, appliqué/accolé forms; no stippling, all stages in peripheral smear; early stages mimic *P. falciparum*, while older stages mimic *P. malariae* (including band form trophozoites); no limit on number of RBCs that can be infected	Early trophozoites and mature schizonts not always seen; most infections misidentified as *P. malariae*; early rings resemble *P. falciparum*; number of infected RBCs not limited; clumpy golden-brown granules; recrudescence; Southeast Asia, Malaysia; symptoms can mimic those seen with *P. falciparum*
Babesia spp.	Ring forms only (resemble *P. falciparum* rings); seen in splenectomized patients; endemic in the U.S. and Europe (no travel history necessary); Maltese cross configuration is diagnostic although it is not always present. Parasitemia may be higher than that seen with cases of malaria.	Tick-borne infection; associated with Nantucket Island; infection mimics malaria; ring forms more pleomorphic than in malaria; more rings/cell (usually) than in malaria; endemic in several areas within the U.S.; occasionally organisms may be seen outside the RBCs (unlike malaria merozoites). More severe infections seen with European *B. divergens* and some species in the western U.S.; less severe infections seen with *B. microti*, which occurs in the eastern U.S. (exceptions with compromised patients)
Trypanosomes		
Trypanosoma brucei gambiense (West African sleeping sickness)	Trypomastigotes long, slender, with typical undulating membrane; lymph nodes/blood can be sampled; microhematocrit tube concentration helpful; examine spinal fluid in later stages of the infection.	Tsetse fly vector; tends to be chronic infection, exhibiting the real symptoms of sleeping sickness, often over a period of months
Trypanosoma brucei rhodesiense (East African sleeping sickness)	Trypomastigotes long, slender, with typical undulating membrane; lymph nodes/blood can be sampled; microhematocrit tube concentration helpful; examine spinal fluid in later stages of the infection.	Tsetse fly vector; tends to be more severe, short-lived infection (particularly in children); patient may die before progressive symptoms of sleeping sickness appear over weeks rather than months.
Trypanosoma cruzi (Chagas' disease) (American trypanosomiasis); well-established zoonotic cycle of *T. cruzi* in the southern half of the United States	Trypomastigotes short, stumpy, often curved in C shape; blood sampled early in infection; trypomastigotes enter striated muscle (heart, gastrointestinal tract) and transform into the amastigote form.	Reduviid bug vector (kissing bug); chronic in adults; severe in young children; great morbidity associated with cardiac failure and loss of muscle contractility in heart and gastrointestinal tract. Organisms and disease now endemic in Texas

Table 9.37 (*continued*)

Protozoan	Diagnostic stage	Comments
Leishmania spp. (cutaneous) (not actually a blood parasite, but presented for comparison with *L. donovani*)	Amastigotes found in macrophages of skin; presence of intracellular forms containing nucleus and kinetoplast is diagnostic.	Sand fly vector; organisms recovered from site of lesion only; specimens can be stained or cultured in NNN and/or Schneider's medium; animal inoculation rarely used (hamster)
Leishmania braziliensis (mucocutaneous) (not actually a blood parasite, but presented for comparison with *L. donovani*)	Amastigotes found in macrophages of skin and mucous membranes; presence of intracellular forms containing nucleus and kinetoplast is diagnostic.	Sand fly vector; organisms recovered from site of lesion only; specimens can be stained or cultured in NNN and/or Schneider's medium; animal inoculation rarely used (hamster)
Leishmania donovani (visceral)	Amastigotes found throughout the reticuloendothelial system, spleen, liver, bone marrow, etc.; presence of intracellular forms containing nucleus and kinetoplast is diagnostic.	Sand fly vector; organisms recovered from buffy coat (rarely found), bone marrow aspirate, spleen or liver puncture (rarely performed); specimens can be stained or cultured in NNN and/or Schneider's medium; animal inoculation rarely used (hamster); cause of kala-azar
Helminths		
Wuchereria bancrofti (pathogenicity due to presence of adult worms)	Microfilaria sheathed, clear space at end of tail; nocturnal periodicity seen; elephantiasis is seen in chronic infections	Mosquito vector; microfilariae recovered in blood (membrane filtration, Knott concentrate, thick films); hematoxylin will stain sheath (Giemsa will not stain sheath).
Brugia malayi (pathogenicity due to presence of adult worms)	Microfilaria sheathed, subterminal and terminal nuclei at end of tail; nocturnal periodicity seen; elephantiasis is seen in chronic infections.	Mosquito vector; microfilariae recovered in blood (membrane filtration, Knott concentrate, thick films); hematoxylin will stain sheath (Giemsa will stain sheath pink).
Loa loa (African eye worm) (pathogenicity due to presence of adult worms)	Microfilaria sheathed, nuclei continuous to tip of tail; diurnal periodicity; adult worm may cross the conjunctiva of the eye.	Mango fly vector; history of Calabar swellings, worms in the eye; microfilariae difficult to recover from blood; hematoxylin will stain sheath.
Mansonella spp. (mild pathogenicity due to presence of adult worms)	Microfilaria unsheathed, nuclei may or may not extend to tip of tail (depending on species); nonperiodic; symptoms usually absent or mild	Midge or black fly vector; microfilariae recovered in blood (membrane filtration, Knott concentrate, thick films)
Mansonella streptocerca (mild pathogenicity due to presence of adult worms and/or microfilariae)	Microfilaria unsheathed, nuclei extend to tip of tail; when immobile, curved like shepherd's crook; adults in dermal tissues	Midge vector; microfilariae found in skin snips; microfilarial tails are split rather than blunt.
Onchocerca volvulus (pathogenicity due to presence of microfilariae)	Microfilaria unsheathed, nuclei do not extend to tip of tail; adults in nodules	Black fly vector; microfilariae found in skin snips; microfilariae migrate to optic nerve, cause of river blindness

[a]Modified from Leber AL (ed), Section 9: Parasitology. *In Clinical Microbiology Procedures Handbook*, 4th ed, ASM Press, Washington, DC, 2016.

Entamoeba histolytica	*Entamoeba hartmanni*	*Entamoeba coli*	*Entamoeba polecki*	*Endolimax nana*	*Iodamoeba bütschlii*	*Blastocystis* spp.
						0 20 µm

Figure 9.1 Intestinal amebae of humans. (**Top row**) Trophozoites. *Entamoeba histolytica* is shown with ingested RBCs. This is the only microscopic finding allowing differentiation of pathogenic *E. histolytica* from nonpathogenic *E. dispar*; all cysts and trophozoites containing no RBCs must be reported as *Entamoeba histolytica/E. dispar*. An ameboid form of *Blastocystis* spp. is rarely seen and difficult to identify. (**Middle row**) Cysts. For *Blastocystis* spp., the central-body form is depicted. (**Bottom row**) Trophozoite nuclei, shown in proportion. (From Novak-Weekley S, Leber AL, *in* Carroll KC, Pfaller MA, Landry ML, McAdam AJ, Patel R, Richter SS, Warnock DW [ed], *Manual of Clinical Microbiology*, 12th ed, ASM Press, Washington, DC, 2019.)

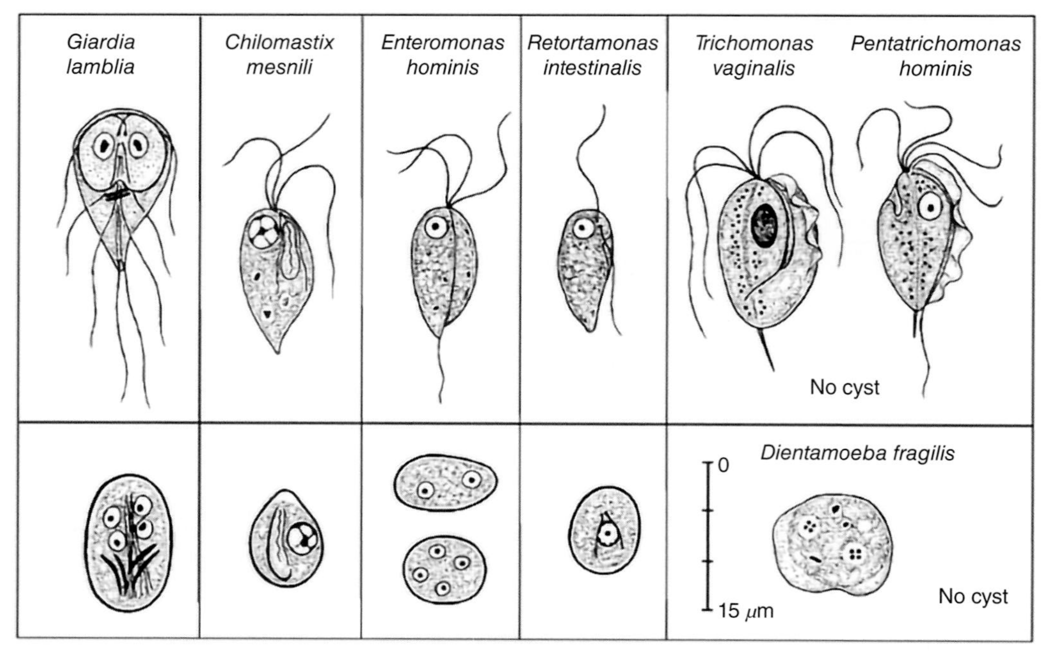

Figure 9.2 Intestinal and urogenital flagellates of humans. (**Top row**) Trophozoites. *Trichomonas vaginalis* is found in urogenital sites; all other flagellates are intestinal. (**Bottom row**) Cysts. A *Dientamoeba fragilis* trophozoite is shown; there is no cyst stage. (Illustration by Sharon Belkin.) (Modified from Novak-Weekley S, Leber AL, *in* Carroll KC, Pfaller MA, Landry ML, McAdam AJ, Patel R, Richter SS, Warnock DW [ed], *Manual of Clinical Microbiology*, 12th ed, ASM Press, Washington, DC, 2019.)

Clonorchis sinensis
27–35 μm long
12–19 μm wide

Hookworm
56–75 μm long
36–40 μm wide

Taenia spp.
31–43 μm diameter

Diphyllobothrium latum
58–75 μm long
40–50 μm wide

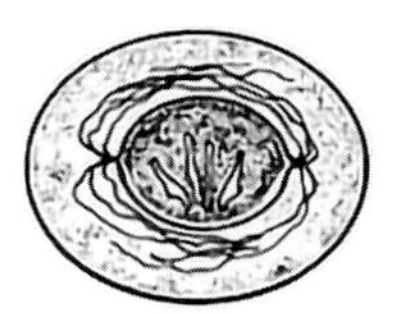

Hymenolepis nana
30–47 μm diameter

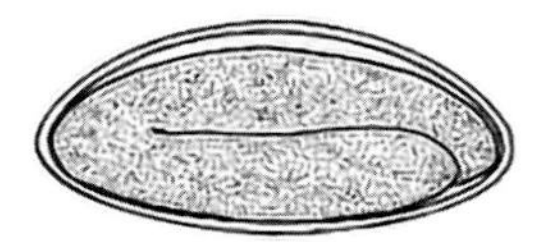

Enterobius vermicularis
70–85 μm long
60–80 μm wide

Trichuris trichiura
50–54 μm long
20–23 μm wide

Hymenolepis diminuta
70–85 μm long
60–80 μm wide

Ascaris lumbricoides (fertile egg)
45–75 μm long
35–50 μm wide

Trichostrongylus
73–95 μm long
40–50 μm wide

Figure 9.3 Helminth eggs depicted in the order of size (smallest to largest). (*Continued next page*)

Ascaris lumbricoides
(unfertilized egg)
85–95 μm long
43–47 μm wide

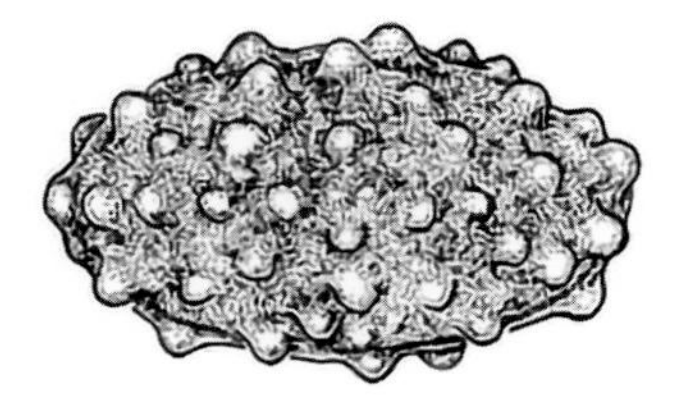

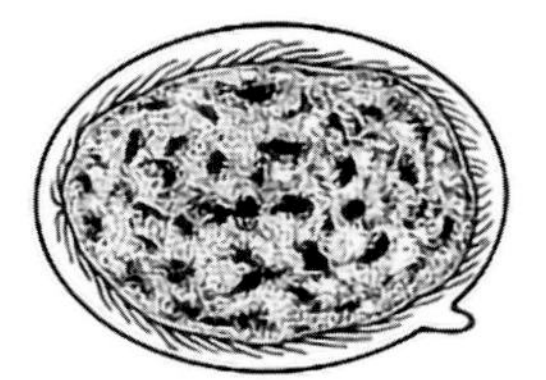

Schistosoma japonicum
70–100 μm long
55–65 μm wide

Paragonimus westermani
80–120 μm long
48–60 μm wide

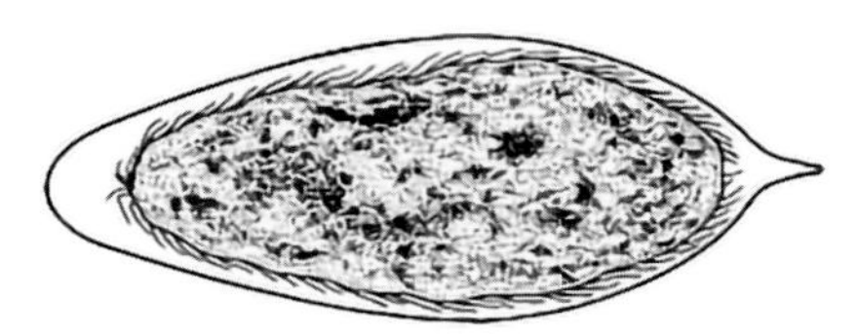

Schistosoma haematobium
112–170 μm long
40–70 μm wide

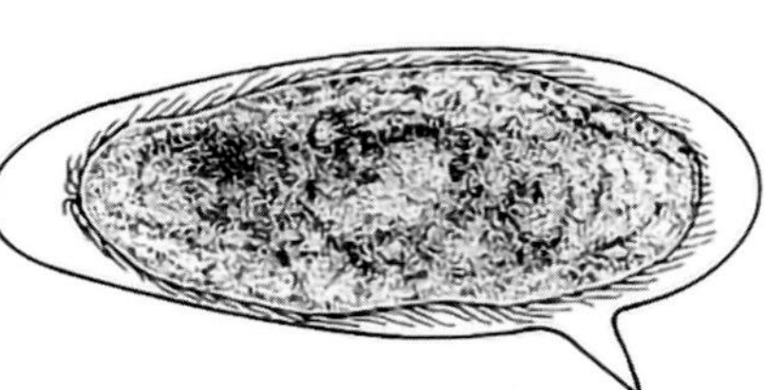

Schistosoma mansoni
114–180 μm long
45–70 μm wide

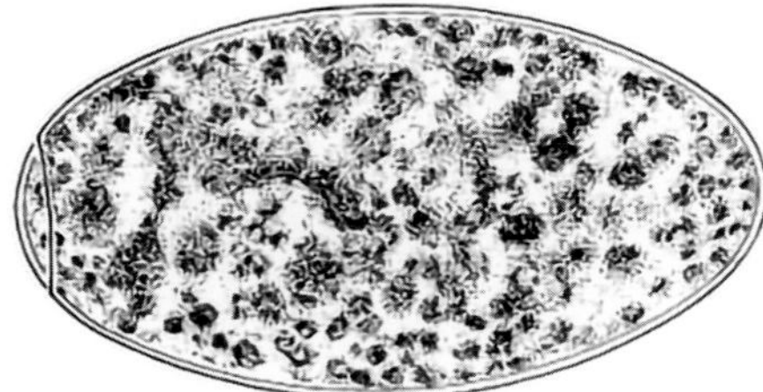

Fasciola hepatica or
Fasciolopsis buski
130–140 μm long
80–80 μm wide

Figure 9.3 *(continued)*

Index

Practical Guide to Diagnostic Parasitology, Third Edition. Lynne S. Garcia

DOI: 10.1128/9781683673637.index

B

C

E

F

G

H

I

K

L

M

N

O

P

Q

R

S

T

U

V

W

X

Z